INTRODUCTION T

CHEMICAL ENGINEERING TH

INTRODUCTION TO
CHEMICAL ENGINEERING
THERMODYNAMICS

SECOND EDITION

GOPINATH HALDER

Associate Professor
Department of Chemical Engineering
National Institute of Technology Durgapur
West Bengal

PHI Learning Private Limited

Delhi-110092
2014

₹ 495.00

INTRODUCTION TO CHEMICAL ENGINEERING THERMODYNAMICS, Second Edition
Gopinath Halder

ISBN-978-81-203-4897-4

The export rights of this book are vested solely with the publisher.

Second Printing (Second Edition)   ...   .   **August, 2014**

Published by Asoke K. Ghosh, PHI Learning Private Limited, Rimjhim House, 111, Patparganj Industrial Estate, Delhi-110092 and Printed by Raj Press, New Delhi-110012.

To
My late grandfather *Bankubihari Goswami*
My daughter *Swagata*
and
My son *Souris*

Contents

Preface

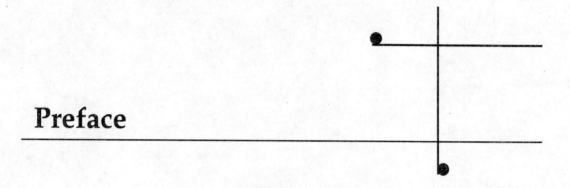

I am very much pleased to present the 'Second edition' of the book entitled '*Introduction to Chemical Engineering Thermodynamics*' which covers the syllabus of thermodynamics course offered precisely to B.E./B.Tech students of chemical, mechanical, pharmaceutical engineering leather technology, food technology and biotechnology disciplines of various Indian universities. The intact book has been painstakingly revised and some of the chapters have been augmented with the addition of new topics of interests and worked-out examples. In addition to six model question papers presented in the first edition, one more '*Solved Model Question Paper*' has been incorporated at the end of the book. Besides, an appreciable quantities of objective type questions have been added to the existing section of '*Multiple Choice Questions*'. Typographical errors and printing mistakes have been corrected as far as possible.

I take this opportunity to express my gratefulness to Ms Pushpita Ghosh, Managing Director and Mr Darshan Kumar, former Executive Editor and Mr Ajai Kumar Lal Das, Assistant Production Manager of PHI Learning for their inspiration and valuable guidance in printing the book.

I am extremely thankful to the editorial staff of PHI Learning, especially Ms Babita Mishra, for their help in converting the book into a good shape and dimension and those who are directly or indirectly associated with the said task.

I express my indebtedness to my dear colleagues of Chemical Engineering Department of National Institute of Technology Durgapur for extending their generous cooperation and moral support for releasing the second edition of the book.

Word of appreciation also goes to my dear students of National Institute of Technology Durgapur and Durgapur Institute of Advanced Technology and Management for their prompt response and positive feedback in regard to the usefulness of the book and to make it errorless.

I hope the book in its present updated form will have an immense value to the readers and will serve its purpose.

Suggestions and constructive criticism from teachers and students towards the further improvement of this book are most welcome.

Gopinath Halder

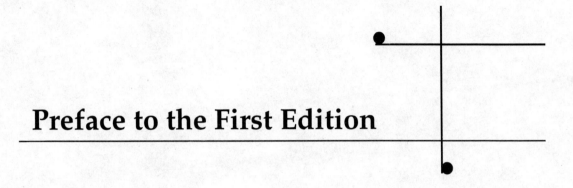

Preface to the First Edition

Chemical Engineering Thermodynamics is one of the vital or core subjects of the Chemical Engineering discipline, specifically at the B.E./B.Tech. level. This subject deals with different forms of energy and their transformation in particular directions and to particular extents.

The book is designed to serve as a textbook for undergraduate students of Chemical Engineering and other related engineering disciplines such as Polymer, Petroleum and Pharmaceutical Engineering.

The book covers the syllabi of several well-recognized institutions such as West Bengal University of Technology, Jadavpur University, the National Institutes of Technology, the Indian Institutes of Technology, Jawaharlal Nehru Technological University, Punjab University, and Anna University very suitably. It emphasizes the relevant concepts with the help of lucid language and clear illustrations, which help the students to develop better understanding of the subject matter.

Problem solving techniques form an important part of the methodology adopted in this book. It will enable the students to develop the ability to solve problems systematically. Numerous solved examples are provided in each chapter. An abundant number of chapter-end exercises and solved model question papers at the end of the book will greatly help in the assessment of the students' learning and understanding ability.

Criticism and suggestions from teachers and students for the improvement of the quality of the book are most welcome. I would sincerely appreciate it if the readers bring to my attention any error that they might come across while going through this book.

Gopinath Halder

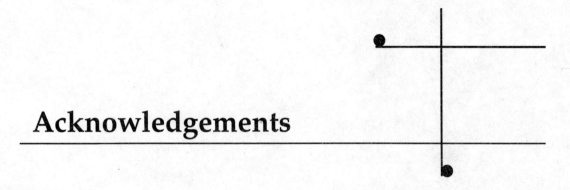

Acknowledgements

I express my sincere gratitude to Prof. Tarkeshwar Kumar, Director, National Institute of Technology Durgapur for providing the infrastructural facility for writing this book. I am deeply indebted to my previous organization, Indian School of Mines Dhanbad, for the inspiration provided to me for this purpose. In fact, the various academic problems faced by my students prompted me to write the book.

I am extremely thankful to Prof. Sukumar Laik, Dean, Students Welfare, Indian School of Mines Dhanbad, for his moral support and encouragement.

I wish to express my gratefulness to Prof. Sujoy Basu, Jadavpur University for his valuable guidance in writing the manuscript as well as his advice for the improvement of the same.

I am indebted to Dr. S.C. Sarkar, my own teacher, for his valuable suggestions for the preparation of the manuscript and his strong support, expressed through positive interactions and discussions.

Words of appreciation must also go to my dear students, especially to Angan Sengupta, Kumar Anupam, Anoar Ali Khan, Seema Halder and Miss Madhubanti Bhattacharya.

I will do injustice if I fail to mention the name of Soumik Deb, Sk. Riazuddin, Madhushree and Kartik Paswan for their outstanding help and cooperation in typesetting various parts of the book using the tool MATHTYPE, and their all-out efforts to assist me in finishing the job within a tight schedule.

I wish to convey my thankfulness to my wife, Mrs Swapna Halder, as without her support and encouragement this long and time-consuming work could not have been completed.

I express my deep sense of gratitude to my late father, Jiban Krishna Halder, and my mother, Bhagabati Halder, for their full support and encouragement for this work.

Gopinath Halder

Introduction and Basic Concepts

1

LEARNING OBJECTIVES

After reading this chapter, you will be able to:

- Know the definition and necessity of the science of thermodynamics
- Appreciate the microscopic and macroscopic approaches to thermodynamics
- Understand the importance of dimensions and units of force, pressure, temperature, work, energy and heat
- Discuss the terminology and fundamental concepts of thermodynamics
- Represent the different thermodynamic processes
- Describe the thermodynamic state, equilibrium, and path functions
- State and explain the zeroth law of thermodynamics
- Develop the ideal gas temperature scale

Before going into a detailed discussion on this subject, it is necessary to know the proper definition and significance of thermodynamics. Thermodynamics is the branch of science which embodies the principle of transformation of energy in the thermodynamic system. This science has a broad application area, starting from microscopic species to industrial and household appliances. The essence of the microscopic and macroscopic approaches to the study of thermodynamics has been discussed. Units and dimensions play an important role in the analysis of thermodynamic processes. Some commonly encountered physical quantities, such as temperature, pressure, energy, heat, work, etc., have been highlighted with the most useful British and SI units for better understanding of unit operations. To learn about thermodynamics, one must know the terminology of the subject properly. The methodology includes the discussion on different thermodynamic systems, processes, system boundaries, properties of substance, equilibrium, state, etc. The importance of establishing the zeroth law of thermodynamics and the development of the ideal gas temperature scale have been discussed.

1.1 THERMODYNAMICS AND ITS APPLICATIONS

Thermodynamics is the science which deals with the energy and its transformation. Nowadays this definition has been broadened to the following familiar version: "Thermodynamics is a science dealing with the quantitative relationship between heat and work". The term 'thermodynamics' is derived from the Greek word *therme* meaning heat and *dynamis* meaning power. This science was born to explain how heat is converted into power. Thermodynamics is based on three simple laws. The first law is represented by the law of conservation of energy, which states that 'Energy can neither be created nor destroyed, it can only be transformed from one form to another'. The first law of thermodynamics is extensively applied to different flow processes and their relevant devices such as venturimeter, nozzle, ejectors, compressors, throttling valves, etc. The science of thermodynamics enables us to calculate the maximum amount of work that can be obtained from heat in a heat engine. It can help us to determine the equilibrium conditions for the physical and chemical changes. This science of energy helps us to select the optimum conditions such as temperature, pressure, and concentration of the reactants for the chemical reactions. Thermodynamics can tell us about the possibility of a chemical reaction, but it cannot give us specific information about the reaction—whether it will take place or not. It also cannot tell us about the rapidity or speed of the reaction and at what rate the system attains equilibrium. It is well known to us that the rate is the ratio of the driving force to the resistance, and the driving force is a vital factor for the attainment of thermodynamic equilibrium in a process. Thermodynamic principles do not permit the estimation of resistance. That is why the rate cannot be determined by thermodynamics. The mechanism by which a chemical reaction goes to completion cannot also be substantiated by thermodynamics. Refrigeration and air-conditioning systems play an important role in providing us comfortable living. The systems work on the principle of the second law of thermodynamics. The science of thermodynamics finds wide application in numerous chemical process industries such as petroleum, food and beverage, textile, pigments, dye, and rubber. The most significant applications of thermodynamics include the establishment of the relationship between the useful properties and variables such as temperature, pressure, and volume.

Hence, the science of thermodynamics can serve the following purposes:

- The estimation of heat and work requirement for any physical and chemical process
- Determination of equilibrium conditions for chemical reaction
- The transfer of chemical species between phases.

1.2 MACROSCOPIC AND MICROSCOPIC APPROACHES

The science of thermodynamics can be better represented and studied by two approaches. These are:

(i) Macroscopic approach—Classical thermodynamics
(ii) Microscopic approach—Statistical thermodynamics.

In the macroscopic approach to the study of thermodynamics, the behaviour of individual molecules is not required to be taken into consideration. This approach is adopted in classical thermodynamics, while in the microscopic approach it is necessary to discuss the average behaviour of a large group of individual molecules with significance, because the molecular kinetic theory has been incorporated as a supplement to thermodynamics for the modern development of this subject. This approach is adopted in statistical thermodynamics. In most of the engineering applications of thermodynamics, the macroscopic approach plays a dominant role to provide effective and valid results.

For example, the certain amount of gas is contained in a vessel. The state of the system can be described by the variables such as temperature, pressure, volume and chemical composition. All are measurable quantities. If the experimenter wants to measure the temperature and pressure, then it is very much possible with the help of a thermometer and pressure gauge respectively. But one does not require to know the behaviour of gas molecules to determine any of the above quantities.

In the macroscopic approach to the study of thermodynamics

   (i)   A small number of variables are needed to explain the thermodynamic state of the system;
   (ii) Measurable variables are required;
   (iii) No consideration of behaviour or structure of the individual molecules is required.

On the other hand, in the microscopic approach

   (i)   A large number of variables are needed to explain the thermodynamic state of the system;
   (ii) Variables cannot be measured;
   (iii) It is necessary to consider the behaviour or structure of the individual molecules.

1.3 IMPORTANCE OF UNITS AND DIMENSIONS

Units and dimensions play an important role in understanding the several physical quantities involved in thermodynamic processes. Physical quantities serve to describe precisely the various natural processes, laws, or phenomena. These physical quantities can be characterized by dimensions such as length, time, temperature, mass, etc., and the magnitude assigned to them are known as units.

The dimension of any physical quantity can be divided into two categories, namely the primary dimension and the secondary or derived dimension.

Primary dimensions are length, time, and mass, and the units are metre, second and kilogram respectively.

A secondary or derived dimension, such as force, is derived from primary dimensions.

$$\text{Force} = \text{Mass} \times \text{Acceleration} = MLT^{-2}$$

The unit of force is kgm/s^2 or Newton (N in short form). This unit is a derived unit.

The units and dimensions of some commonly encountered physical quantities have been discussed in the following sections.

1.3.1 Force

Force is a dimension. Its unit is derived from Newton's second law of motion, which can be mathematically expressed as

$$\text{Force} = \text{Mass} \times \text{Acceleration}$$

or

$$F = ma$$

or

$$F = Cma = \frac{1}{g_c} ma \tag{1.1}$$

where

$$C = \frac{1}{g_c} = \text{constant of proportionality}$$

- The SI (Système International or International System of Units) unit of force is Newton. It is denoted by N. One N is defined as the force required to accelerate a mass of 1 kg at a rate of 1 m/s^2, i.e.

$$1 \text{ N} = 1 \text{ kg-m/s}^2$$

Hence

$$C = \frac{1}{g_c} = 1$$

then Eq. (1.1) yields

$$F = ma$$

- The English unit of force is pound-force (lbf). It is defined as the force required to accelerate a mass of 32.174 lb at a rate of 1ft/s^2. i.e.

$$1 \text{ lbf} = \frac{1}{g_c} \times 1 \text{ lbm} \times 32.174 \text{ ft/s}^2 = \frac{1}{g_c} \times 32.174 \text{ lbm-ft/s}^2$$

where

$$g_c = 32.174 \text{ lbm-ft/s}^2$$

EXAMPLE 1.1   A man weighs 800 N on the earth's surface where the acceleration of gravity is $g = 9.83$ m/s^2. Calculate the weight of the man on the moon where the acceleration due to gravity is $g = 3.2$ m/s^2.

Solution: From the expression of force, the force on the man on the earth's surface is given by

$$F = ma$$

Putting the value of F and a in this equation, we get

$$800 = m \times 9.83 \text{ m/s}^2$$

$$m = \frac{800}{9.83} = 81.38 \text{ kg}$$

On the moon, the weight of the mass is equal to the force acting on the mass on the moon and is given by

$$F = ma = 81.38 \text{ kg} \times 3.2 \text{ m/s}^2 = 260.41 \frac{\text{kg-m}}{\text{s}^2} = 260.41 \text{ N}$$

EXAMPLE 1.2 Estimate the gravitational force on a body of 1.5 kg mass on the earth's surface, given that the radius and mass of the earth are 6000 km and 6×10^{24} kg respectively, and the universal gravitational constant, $G = 6.672 \times 10^{-11}$ Nm^{2}kg^{-2}.

Solution: According to Newton's universal law of gravity

$$F = \frac{Gm_1m_2}{r^2}$$

where

F = Gravitational force

G = Gravitational constant

m_1, m_2 = Masses of the two bodies

r = Distance between the two bodies.

Substituting the values in the preceding expression, we have

$$F = \frac{Gm_1m_2}{r^2} = \frac{6.672 \times 10^{-11} \times 1.5 \times 6 \times 10^{24}}{(6000 \times 10^3)^2} \left[\frac{\text{Nm}^2\text{kg}^{-2} \times \text{kg} \times \text{kg}}{\text{m}^2} \right] = \boxed{16.68 \text{ N}}$$

EXAMPLE 1.3 Assuming that the radius of the moon and the earth are 0.3 km and 1 km respectively, calculate the weight of 1 kg mass on the moon. The mass of the moon is 0.013 relative to the earth's mass.

Solution: According to Newton's universal law of gravitation

$$F = \frac{Gm_1m_2}{r^2}$$

For the earth, with the subscript e being used

$$F_e = \frac{Gm_em_2}{r_e^2}$$

while for the moon, with the subscript m being used,

$$F_m = \frac{Gm_mm_2}{r_m^2}$$

On the moon, the weight of the mass is equal to the force acting on the mass on the moon. Comparing with the earth, we get

$$\frac{F_e}{F_m} = \frac{m_e}{m_m}\left(\frac{r_m^2}{r_e^2}\right) = \frac{m_e}{m_m}\left(\frac{r_m}{r_e}\right)^2 = \frac{1}{0.013}(0.3)^2 = 6.92 \approx \boxed{7}$$

Hence, the mass of 1 kg will weigh approximately 1/7 kg on the moon.

1.3.2 Pressure

Pressure is defined as the normal force exerted by the fluid per unit area. It is a scalar quantity. Mathematically

$$P = \frac{F}{A} \tag{1.2}$$

where

F = Force normal to the area
A = Area of the surface
P = Pressure.

- In SI system, the unit of pressure is Newton per square metre (N/m^2), i.e., one Newton force acting on one square metre area. This is also called Pascal (Pa). The relation between Pascal (N/m^2) and bar, another common SI unit of pressure, can be expressed as

$$1 \text{ Pa} = 1 \text{ N/m}^2$$
$$1 \text{ bar} = 10^5 \text{ Pa} = 10^5 \text{ N/m}^2$$

Again

$$1 \text{ bar} = 10^5 \text{ Pa} = 0.1 \text{ MPa} = 100 \text{ kPa}$$

- In English engineering system of units, the unit of pressure is pound force per square inch ($lbf/inch^2$), or

$$1 \text{ bf/inch}^2 = 144 \text{ lbf/ft}^2$$

- The metric unit of pressure is 1 kgf/cm^2, or

$$1 \text{ kgf/cm}^2 = 10^4 \text{ kgf/m}^2$$

- The standard atmospheric pressure is defined as the pressure produced by a column of mercury 760 mm high (0.76 m or 760 torr) atmosphere. It is expressed as 1 atm. This is another widely used unit in different areas. It can be expressed as

$$1 \text{ standard atmosphere (atm)} = 1.01325 \text{ bar}$$
$$= 1.01325 \times 10^5 \text{ Pa}$$
$$= 760 \text{ mm Hg} = 29.92 \text{ inch Hg}$$

- Most of the pressure gauges indicate pressure relative to the atmospheric pressure, i.e., the difference between the actual (or absolute) pressure and the local atmospheric pressure. The pressure measured by gauge is called the *gauge pressure*. The relation is

$$\text{Absolute pressure} = \text{Gauge pressure} + \text{Atmospheric pressure}$$
$$P_{absolute} = P_{gauge} + P_{atmospheric}$$

EXAMPLE 1.4 A mercury manometer reads 40 cm at 30°C and 1 atm. Determine the absolute pressure, given that $\rho_{Hg} = 14.02 \text{ g/cm}^3$ and $g = 9.792 \text{ m/s}^2$.

Solution: From the definition, we get

$$P = \frac{F}{A}$$

Again, it is well known to us that

$$F = ma$$

Now, the mass of the manometer fluid

$$m = h\rho A$$

where

h = Height of the manometer fluid
ρ = Density of the fluid
A = Cross-sectional area of the column of the fluid.

Then

$$F = ma = h\rho A g \qquad (g = \text{acceleration due to gravity})$$

or

$$\frac{F}{A} = h\rho g$$

Hence, after equating for F/A, we have

$$P = h\rho g$$

Putting the value of h, ρ, and g into the preceding equation, we have

$$P = 40 \text{ cm} \times 14.02 \text{ g/cm}^3 \times 9.792 \text{ m/s}^2 = 5491.35 \ \frac{\text{g-m}}{\text{cm}^2\text{-s}^2}$$

$$= 5.4913 \ \frac{\text{kg-m}}{\text{cm}^2\text{-s}^2} = 5.4913 \text{ N/cm}^{-2} = 54.913 \text{ kPa}$$

EXAMPLE 1.5   The pressure of a gas in a container is measured with a U-tube manometer. The density of the liquid used is 1200 kg/m³. The difference in the height of the fluid between the two limbs is 62 cm and the local atmospheric pressure is 112 kPa. Calculate the absolute pressure within the container.

Solution: The absolute pressure within the container can be determined as

$$P = P_{\text{atm}} + \rho g h$$
$$= 112 \text{ kPa} + (1200 \text{ kg/m}^3 \times 9.81 \text{ m/s}^2 \times 0.62 \text{ m})$$
$$= 112 \text{ kPa} + 7298.64 \text{ Pa}$$
$$= 112 \text{ kPa} + 7.298 \text{ kPa}$$
$$= 119.298 \text{ kPa}$$

1.3.3   Work

Work is a form of energy. It is defined as the product of the force and the displacement in the direction of the applied force. In other way, work is performed whenever a force acts through a distance.

The work done can be mathematically expressed as

$$dW = Fdl \tag{1.3}$$

where F is the applied force and dl the displacement.

That is why the unit of work is Newton-metre (N-m). It is also called Joule. 1 N-m is equivalent to 1 Joule, i.e., 1 N-m = 1 J.

- In SI unit, the unit of work done is kilojoule. 1 kJ = 10^3 J.
- The British unit of work is ft-lbf.

As per the sign convention of work, work done by the system is treated as a positive quantity because of the displacement in the same direction, while the work done on the system by an external agent is regarded as a negative quantity. When the system performs positive work, its surrounding will perform an equal amount of negative work. Therefore, in any process

$$W_{system} + W_{surroundings} = 0$$

Consider a common example—compression or expansion of a gas in a cylinder–piston assembly. The gas is enclosed by a frictionless piston in a cylinder. The piston moves along the length of the cylinder.

The total force acted upon the piston is

$$F = PA \tag{1.4}$$

where

P = Pressure of the gas
A = Area of the piston.

The displacement of the piston in the direction of force is

$$dl = \frac{dV}{A} \tag{1.5}$$

where dV is change in volume of the gas.

Putting the value of dl and F into Eq. (1.3), we get

$$dW = Fdl = PA \cdot \frac{dV}{A} = PdV \tag{1.6}$$

Integrating Eq. (1.6) with integral limits $V = V_1$ and $V = V_2$ as the volume of the gas changes, we get

$$W = \int_{V_1}^{V_2} PdV \tag{1.7}$$

Here, the positive work done implies the amount of work done by the system and the process belongs to the expansion of the gas, because the applied force and the displacement of the piston are in the same direction. The reverse situation occurs for compression of the gas.

The operation and the piston movement in a cylinder–piston arrangement is illustrated in Fig. 1.1(a). The entire process is graphically represented by the P–V diagram shown in Fig. 1.1(b). The shape of the P–V curve and the bounded area can indicate clearly the process, whether it is compression or expansion of the gas.

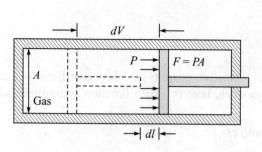

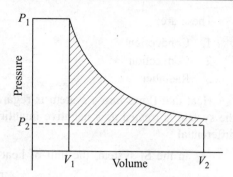

(a) Work done by a system.   (b) Representation of work done on *P–V* diagram.

Fig. 1.1

EXAMPLE 1.6   A force of 150 N is applied to a block which rests on a rough surface, to push back the block by 10 m from the original position. Determine the work done by the system.

Solution: The work done by the system is given by

$$\text{Work done} = \text{Force} \times \text{Displacement}$$
$$= 150 \text{ N} \times 10 \text{ m}$$
$$= 1500 \text{ N-m} = 1500 \text{ J}$$

EXAMPLE 1.7   A gas is enclosed by a movable piston in a cylinder. The gas expands from an initial volume of 3 m³ to a final volume of 5 m³ as a result of 210 kJ of work done on the system by an external source. The pressure of the system remains constant at 560 kPa. Estimate the net work done by the system.

Solution: The work done by the system is given by

$$W = \int_{V_1}^{V_2} P dV = \int_{3}^{5} P dV = 560 \times 10^3 (5 - 3)$$
$$= 11.2 \times 10^5 \text{ N-m} = 11.2 \times 10^5 \text{ J}$$

Again, the system receives 210 kJ of work from the external agent, i.e., the work is done on the system. Hence, the actual work done by the system can be expressed as

$$W = 11.2 \times 10^5 \text{ J} - 2.1 \times 10^5 \text{ J} = 9.1 \times 10^5 \text{ J}$$

1.3.4   Heat

Heat is defined as the form of energy. Whenever temperature difference exists between two regions, heat is transferred from the higher temperature region to the lower temperature region. The flow of heat continues until the difference in temperature is equalized. Heat is the energy in transit. Like other forms of energy, heat cannot be completely converted into work. There are three modes of heat transmission.

These are:

1. Conduction
2. Convection
3. Radiation.

Heat that flows into a system is regarded as a positive quantity and heat that flows out of the system is treated as a negative quantity. Like work, heat is a path function and an inexact differential.

- In the SI system, the unit of heat is Joule (J).

$$1 \text{ J} = 1 \text{ N-m}$$
$$1 \text{ cal} = 4.2 \text{ J}$$

- In English engineering system of unit, the unit of heat is ft-lb.

$$1 \text{ ft-lb} = 1.3558 \text{ J}$$
$$1 \text{ British Thermal Unit (BTU)} = 1055.04 \text{ J}$$

1.3.5 Energy

Energy can be defined as a property which can be transformed into or produced from work. It has the capacity for producing effect. Energy in the form of heat or work can be exchanged between the system and the surroundings. It is available in numerous forms, such as chemical, electrical, mechanical, electromagnetic, radiant, and internal energy.

The total energy E of a system comprises the following two forms. These are:

(a) Macroscopic form, and
(b) Microscopic form.

Macroscopic Form of Energy

It includes the energies that a system possesses with respect to the external reference plane, such as kinetic energy and potential energy. This mode of energy is concerned with motion and influence of some external effects, viz. electricity, magnetism, surface tension, and gravity.

Kinetic Energy: The energy possessed by a system by virtue of its motion is called *kinetic energy*. Consider a system of mass m moving at the velocity V; then the kinetic energy of the system is given by

$$\text{K.E.} = E_K = \frac{1}{2}mV^2 \tag{1.8}$$

Potential Energy: The energy possessed by a system by virtue of its elevation with reference to an arbitrary reference plane is called *potential energy*. Consider a system of mass m is at elevation Z from the earth's surface, then the potential energy of the system is given by

$$\text{P.E.} = E_P = mgZ \tag{1.9}$$

where g is the acceleration due to gravity.

Microscopic Form of Energy

It includes the energies that are related to the molecular structure of a system, such as internal energy.

Internal Energy: Every system has a certain amount of energy with itself. This is known as internal energy. Internal energy is nothing but the sum of all the microscopic forms of energy. It is denoted by U.

Other forms of energy can also be stored within the system. These are electrical, magnetic, surface energy, etc. These energies can be neglected except in some special cases. Hence, the total energy E of a system can be represented as

$$E = \underbrace{E_K + E_P}_{\text{Macro}} + \underbrace{U}_{\text{Micro}} \qquad (1.10)$$

where

E_K = Kinetic energy
E_P = Potential energy
U = Internal energy.

EXAMPLE 1.8 A pump delivers water from a well which is 100 m deep. The local gravitational acceleration is 9.81 m/s². Determine the change in potential energy per kg of water.

Solution: The basis is 1 kg of water. The change in potential energy is given by

$$E_P = mgZ$$

where
$m = 1$ kg
$g = 9.81$ m/s²
$Z = 100$ m.

Putting the values of m, g and Z, we have

$$E_P = mgZ = 1 \text{ kg} \times 9.81 \text{ m/s}^2 \times 100 \text{ m}$$
$$= 981 \text{ N-m}$$
$$= 981 \text{ J}$$

EXAMPLE 1.9 A metal block of mass 15 kg falls freely from rest. What will be its kinetic energy after falling 12 m? Find the velocity of the metal block just after impact.

Solution: The initial potential energy and kinetic energy of the metal block will be exactly equal to the final potential energy and kinetic energy.

At the initial condition, $V_1 = 0$, so the kinetic energy is zero; at the final condition, $Z_1 = 0$, so the potential energy is zero. This can be mathematically expressed as

$$E_{K1} + E_{P1} = E_{K2} + E_{P2}$$

or

$$0 + E_{P1} = E_{K2} + 0$$

or

$$E_{P1} = E_{K2}$$

$$mgZ_1 = \frac{1}{2}mV_2^2$$

Put the values of m, g, Z_1, and V_2, we get

$$15 \text{ kg} \times 9.81 \text{ m/s}^2 \times 12 \text{ m} = \frac{1}{2} \times 15 \text{ kg} \times V_2^2$$

$$V_2 = 15.34 \text{ m/s}$$

The velocity of the metal block is 15.34 m/s.

$$\text{Kinetic energy after falling } 12 \text{ m} = \frac{1}{2} mV^2$$

$$= \frac{1}{2} \times 15 \text{ kg} \times (15.34 \text{ m/s})^2$$

$$= 1764.867 \text{ N-m}$$

$$= 1764.867 \text{ J}$$

1.3.6 Power

Power is defined as the rate at which work is done or the ratio of work done to unit time. It can be expressed as

$$\text{Power} = \frac{\text{Work done}}{\text{Time}} = \frac{\text{Joule}}{\text{second}} = \text{Watt (W)} \tag{1.11}$$

- The SI unit of power is kJ/s or kW.
- The most familiar unit of power is horse power (hp). 1 hp = 745.7 W.

EXAMPLE 1.10 Determine the power required to accelerate a car weighing 1200 kg from a velocity of 10 km/h to 100 km/h in just 1 min on a level road.

Solution: Here, we are required to change the kinetic energy of the body in order to accelerate, and this is the extent of work needed.

Therefore, the work required can be expressed as

$$W = \frac{1}{2}m(V_2^2 - V_1^2)$$

where

$$m = 1200 \text{ kg}$$

$$V_1 = 10,000 \text{ m/h} = \frac{10,000 \text{ m}}{3600 \text{ s}} = 2.777 \text{ m/s}$$

$$V_2 = 100,000 \text{ m/h} = \frac{100,000 \text{ m}}{3600 \text{ s}} = 27.777 \text{ m/s}$$

Substituting the values of m, V_1 and V_2 in the preceding equation, we get

$$W = \frac{1}{2} \times 1200 \text{ kg} \left[\left(27.777 \frac{m}{s} \right)^2 - \left(2.777 \frac{m}{s} \right)^2 \right]$$

$$= 458310 \text{ N-m} = 458310 \text{ J} = 458.31 \text{ kJ}$$

$$\text{Power required} = \frac{\text{Work done}}{\text{Time}}$$

$$= \frac{458.31}{60} \frac{\text{kJ}}{\text{s}}$$

$$= 7.63 \text{ kW}$$

EXAMPLE 1.11   A gas is enclosed by a frictionless piston in a 0.3 m diameter cylinder, and a metal block is placed on the piston. The weight of the piston and the block rested on it together equal 100 kg. The atmospheric pressure and the acceleration due to gravity g are 1.013 bar and 9.792 m/s^2 respectively.

(a) Calculate:

(i)   The force exerted by the atmosphere, the piston and the weight on the gas

(ii)   The pressure of the gas in kPa

(b)   If the gas expands on application of heat, the piston moves upward along with the weight by 0.5 m. Then what is the work done by the gas in kJ?

(c)   What is the change in potential energy of the piston and the weight after the expansion of the gas?

Solution: (a) (i) From the definition of force, we have

$$F = PA = \text{Pressure} \times \text{Area}$$

Here, the force exerted by the atmosphere can be calculated as

$$F = 1.013 \times 10^5 \text{ N/m}^2 \times \frac{\pi}{4} \times (0.3)^2 \text{ m}^2$$

$$= 7160.47 \text{ N}$$

Force exerted by the piston and the metal block

$$= mg = 100 \text{ kg} \times 9.792 \text{ m/s}^2$$

$$= 979.2 \text{ kg-m/s}^2 = 979.2 \text{ N}$$

Hence, the total force acting upon the gas

$$= 7160.47 \text{ N} + 979.2 \text{ N} = 8139.67 \text{ N}$$

(ii)   The pressure of the gas can be estimated, as we know that

$$\text{Pressure} = \frac{\text{Force}}{\text{Area}} = \frac{\text{Force}}{\frac{\pi}{4} d^2} = \frac{8139.67}{\frac{\pi}{4}(0.3)^2} = 1.152 \times 10^5 \text{ N/m}^2$$

(b) The gas expands on application of heat, the volume of the gas goes on increasing, and the piston moves upward.

Work done due to expansion of gas

$$W = \text{Force} \times \text{Displacement}$$
$$= F \, dl = 8139.67 \text{ N} \times 0.5 \text{ m}$$
$$= 4069.835 \text{ N-m} = 4069.835 \text{ J}$$

(c) Change in potential energy of piston and weight after expansion process

$$E_P = mgZ$$
$$= 100 \text{ kg} \times 9.792 \text{ m/s}^2 \times 0.5 \text{ m}$$
$$= 489.6 \text{ kg-m}^2/\text{s}^2$$
$$= 489.6 \text{ N-m} = 489.6 \text{ J}$$

1.4 TERMINOLOGY AND FUNDAMENTAL CONCEPTS

It is essential to have knowledge of some special terms for better understanding of thermodynamics. The terms and their thermodynamic meanings are discussed here with simple schematic representations.

1.4.1 System and Surroundings

System

It is defined as the quantity of matter or a region in a space upon which the attention is concentrated in the analysis of a problem. In other words, a system is defined as *any portion of the universe separated from the surroundings by a real or imaginary boundary*. Examples are a piston, a solution in a test tube, a living organism, a planet, etc.

A typical thermodynamic system is illustrated in Fig. 1.2.

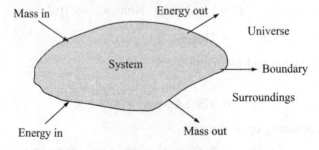

Fig. 1.2 Illustration of a thermodynamic system.

Surroundings

The area or region external to the system is known as the surroundings.

Universe

The system and surroundings together constitute an universe.

Boundary

The system is separated from the surrounding by a real or imaginary surface, which is known as boundary. A boundary has no thickness and it can neither contain any mass nor occupy any volume in space. A boundary is not necessarily rigid.

The boundary may be broadly categorized into two forms:

1. Fixed boundary
2. Moving boundary.

Consider a piston–cylinder assembly as shown in Fig. 1.3. The gas is confined in the cylinder having a closely fitted piston. Now, if the gas is heated, the gas will start expanding and the piston will move upward. Here, the cylinder is the fixed boundary and the piston is the moving boundary.

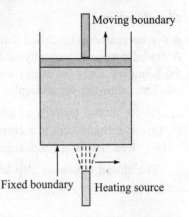

Fig. 1.3 Demonstration of fixed and moving boundaries.

The thermodynamic system can be better classified on the basis of the nature of the boundary and the quantities flowing through it, such as matter, energy, work, heat and entropy.

Open System

A system is said to be open if it can exchange both mass and energy with the surroundings.

An open system is one in which both mass and energy can be transferred from the system to the surroundings and vice versa. It is illustrated in Fig. 1.4.

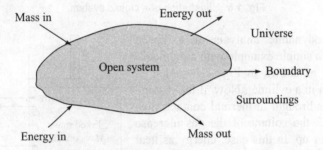

Fig. 1.4 Illustration of an open system.

Most of the engineering devices such as air compressor, turbine, and nozzle belong to the open system. In case of air compressor, air enters at low pressure and leaves at high pressure; here both mass and energy cross the system boundary.

An open system is often called the *control volume*. Actually the region in the space selected for the analysis of an open system is known as the control volume.

Consider a compressor as shown in Fig. 1.5. Here, mass flows into and out of the device, and this can be analyzed as the control volume. In this device, an arbitrary region in space can be identified as the control volume.

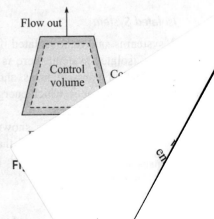

Closed System

A system is said to be closed if it can exchange energy only with its surroundings, not the mass. A closed system is basically of fixed mass, and one in which there is no mass transfer across the boundary. Only the energy can be transferred into or out of the system as shown in Fig. 1.6.

The following are examples of closed system:

(i) A certain quantity of gas confined in a cylinder bounded by a frictionless piston.

(ii) Hot liquid kept in a closed metallic flask

(iii) Reaction in a batch reactor.

The closed system is also known as *control mass*.

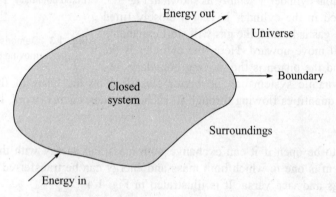

Fig. 1.6 Illustration of a closed system.

For the thermodynamic analysis of a closed system, we can cite a simple example with a cylinder–piston arrangement as shown in Fig. 1.7. A gas is enclosed by a piston in a cylinder. Now, if the piston–cylinder assembly is brought in thermal contact, then the temperature and, the volume of the gas increase and the piston moves up. In this case, energy as heat is transferred to the system from the surroundings, but no mass crosses the system boundary.

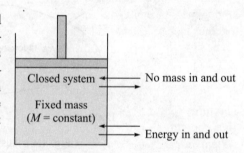

Fig. 1.7 Cylinder–piston assembly—a closed system.

Isolated System

A system is said to be isolated if it can exchange neither energy nor mass with its surroundings. In any isolated system, there is no interaction between the system and the surroundings. The system is of constant mass and energy. In other words, it is a system enclosed by a rigid boundary and one in which energy as well as work is not allowed to cross the boundary. A major example is thermoflask.

Consider a vessel, as shown in Fig. 1.8, containing a liquid which is in contact with its ᵃapour. Now, if the wall and the mouth of the vessel are thermally insulated, then no mass or ᵉrgy can be exchanged with the surroundings.

Isolated systems may be of three kinds:

(i) Thermally isolated system—when the system walls are impermeable to heat flow.
(ii) Mechanically isolated system—when the enclosing walls are rigid.
(iii) Completely isolated system—neither mass nor energy can be exchanged with the system.

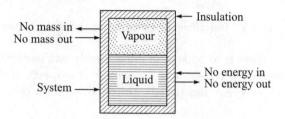

No mass in ←
No mass out → Vapour
← Insulation

System → Liquid
← No energy in
→ No energy out

Fig. 1.8 Schematic representation of an isolated system.

Homogeneous System

A system is considered to be homogeneous when the properties are uniform throughout. A homogeneous system contains necessarily only one phase. A good instance of this type of system is liquid water in a container.

Heterogeneous System

A system is referred to as heterogeneous when the properties are not uniform throughout the system. It consists of more than one phase. It requires the presence of at least two distinct homogeneous phases in order to proceed. For example, water in liquid and vapour phases together in a container, and a mixture of water and toluene, are interesting examples of heterogeneous system. Here, toluene and water are two immiscible liquids.

1.4.2   State of a System

The term *state* refers to the condition in which the system exists. The state of a system in any given condition can be better characterized by certain observable properties of the system. A system in order to be in a state must have definite values assigned to its properties such as temperature, pressure, volume, and composition. Suppose a system is not undergoing any change. At this stage, all the above-mentioned properties can be measured or calculated throughout the system.

Consider a horizontal piston–cylinder arrangement containing a certain amount of gas. The piston is held in place by latches. This stage is designated to be State I. Now, if the restraining latches are removed, the gas inside the cylinder will start expanding to an appreciable extent. This is State II. It is clearly understood from Fig. 1.9 that the system has different sets of properties in two different states. Thus the same gas can exist in a number of different states.

Steady State of a System

The state is the entity that determines the condition of a system. The state of a system interacting with the surroundings is said to be *steady* if it does not vary with time. Simply, the change of state is zero, but the interactions are non-zero. The term 'steady' implies no change with time.

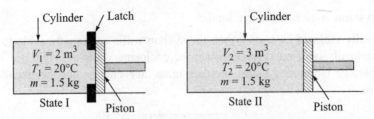

Fig. 1.9 Two different thermodynamic states of a system.

In this context, one thing is to be kept in mind: steady and uniform conditions are not the same thing. *Uniform* means *no change with location over a specified region*. These meanings are consistent with then everyday use (steady girlfriend, uniform properties, etc.).

1.4.3 Thermodynamic Equilibrium

Thermodynamics is very much concerned with the equilibrium state of the system. The term *equilibrium* is used to imply a state of balance of a system. Consider that a system is undergoing a change in a given state. If there is no change in the state of the system, then the system is considered to be in a state of thermodynamic equilibrium. Hence, a system is said to be in equilibrium state if the system has no tendency to undergo any further change, i.e., the properties like pressure, temperature, and composition are uniform in magnitude throughout the system.

It is needless to mention that in the equilibrium state, there are no unbalanced driving forces within the system and so it cannot exchange energy in the form of heat as well as work with the surroundings.

Equilibrium may be of many kinds. A system, in order to be in complete thermodynamic equilibrium, must satisfy the conditions of the following three relevant types of equilibrium:

 (i) Thermal equilibrium
 (ii) Mechanical equilibrium
 (iii) Chemical equilibrium.

Thermal Equilibrium

A system is said to be in thermal equilibrium if the temperature is uniform throughout the system.

Consider two systems A and B. If the thermal equilibrium is attained, then there is no difference in temperature between the interacting systems A and B. In other words, no temperature gradient exists between the two systems. Here, the driving force is temperature.

At this condition, $T_A = T_B$, where T_A and T_B are the temperatures of systems A and B respectively.

Mechanical Equilibrium

A system is said to be in mechanical equilibrium if the pressure is uniform throughout the system. Alternatively, in a state of mechanical equilibrium, there is no difference in pressure at any point with time.

Consider two systems A and B. If there is no imbalance of forces between the different parts of the interacting systems A and B, then the system is in mechanical equilibrium. Here, the driving force is pressure. Thus, mechanical equilibrium implies the uniformity of pressure.

At this condition, $P_A = P_B$, where P_A and P_B are the pressures of systems A and B respectively.

Chemical Equilibrium

This sort of equilibrium is related to chemical composition. A system is said to be in chemical equilibrium if the composition of the system does not change with time.

Consider that two systems A and B are in chemical equilibrium. Then there is no tendency for any chemical reaction to occur between the interacting systems A and B. Here, the driving force is chemical potential. Thus, chemical equilibrium implies the uniformity of chemical potential.

At this condition, $\mu_i^A = \mu_i^B$ for $i = 1, 2, 3, 4, \ldots, n$, where μ_i^A and μ_i^B are the chemical potentials of component i in systems A and B respectively. Hence, a system will be in thermodynamic equilibrium if it fulfils the preceding criteria of three equilibriums simultaneously.

The types and conditions of thermal, mechanical and chemical equilibrium are highlighted in Table 1.1.

Table 1.1 Different types of equilibrium with corresponding driving forces

Types of equilibrium	Driving force
Thermal equilibrium	Temperature
Mechanical equilibrium	Pressure
Chemical equilibrium	Chemical potential

1.4.4 Boundary Wall of a System

The wall or boundary of a thermodynamic system may be categorized in the following ways:

- *Permeable Wall:* It is the wall that allows the passage of both matter and energy.
- *Impermeable Wall:* It is the wall that does not allow but rather prevents the passage of matter.
- *Rigid Wall:* It is the wall whose shape and position are fixed.
- *Adiabatic Wall:* It is the wall that prevents the passage of matter or energy. It does not permit the flow of heat into and out of the system.
- *Diathermal Wall:* It is the wall that prevents the passage of matter but allows the transmission of energy.

1.4.5 Thermodynamic Processes

Any change that a system undergoes from one equilibrium state to another is brought about through a process. Alternatively, the operation which brings about the changes in the state of the system is called the *process*. The major examples are vaporization, freezing, sublimation, expansion, and compression.

There are several thermodynamic processes are such as isothermal process, isochoric process, isobaric process, adiabatic process, cyclic process, and quasi-static process. They have been discussed elaborately in Chapter 2. Here, these processes are outlined in the following way:

- *Isothermal Process*:   It is the process in which the temperature remains constant throughout the system. Here, T = constant or $dT = 0$
- *Isobaric Process*:   It is the process during which the pressure remains constant throughout the system. Here, P = constant or $dP = 0$
- *Isochoric Process:*   It is the process in which the volume remains constant throughout the system. Here, V = constant or $dV = 0$.
- *Adiabatic Process:*   It is the process in which there is no exchange of heat between the system and the surroundings. Here, Q = constant or $dQ = 0$
- *Cyclic Process:*   It is the process in which a system, having undergone a change, returns to its initial state at the end of the process. Here, the path of the process is called a *cycle*. For a cycle, the initial and final states are identical.
- *Quasi-static Process:*   It is the process which takes place very slowly and with infinitesimal driving forces, or, when a process proceeds in such a way that the system remains infinitesimally close to an equilibrium state at all times.

1.4.6   Properties of a System

The characteristics which are experimentally measurable and which enable us to define a system are called its *properties*. The properties can describe properly the state of a system. Hence, the term *property* of a system or substance stands for any of its identifiable or observable characteristics. Some common system properties are pressure, temperature, volume, composition, viscosity, thermal conductivity, refractive index, and dielectric constant.

The properties may broadly be categorized into the following two groups:

1.   Extensive property
2.   Intensive property.

Extensive Property

The properties dependent upon the extent or quantity of the mass of a system are known as extensive properties. Mass, volume, length, surface area, internal energy, enthalpy, etc. are extensive properties. The total value of any extensive property is the sum of the values of the properties of the individual components.

Intensive Property

The properties independent of the extent or quantity of the mass of a system are known as intensive properties. Temperature, pressure, density, specific volume, specific heat, etc. are intensive properties.

Now, it is necessary to remember that the ratio of two extensive properties of a homogeneous system is an intensive property. For example, specific volume, the ratio of volume to mass or volume per unit mass, is an intensive property, while volume and mass are extensive properties. The easy way to define this is given by

$$\text{Specific property} = \frac{\text{Extensive property}}{\text{Unit mass}}$$

1.5 ZEROTH LAW OF THERMODYNAMICS

This law was coined and formulated by R.H. Fowler in 1931. The law states: *If the two bodies are in thermal equilibrium with a third body, they are also in thermal equilibrium with each other*.

Suppose two bodies A and B are brought into thermal contact with each other as shown in Fig. 1.10. It is our common experience that heat will flow from the body at higher temperature to the one at lower temperature. The process continues until thermal equilibrium is established between A and B. Now if the another body C is in thermal equilibrium with A, then B and C in thermal equilibrium. Hence, it can be concluded that A, B and C are in thermal equilibrium.

Fig. 1.10 Demonstration of zeroth law.

Mathematically, if $T_A = T_B$ and $T_C = T_A$, then

$$T_A = T_B = T_C \tag{1.12}$$

1.6 PHASE RULE

A phase is defined as a *physically distinct but homogeneous part of a system separated from other parts by a boundary surface*. The phase rule is a qualitative treatment of systems in equilibrium. It was formulated by J.W. Gibbs in 1876. The phase diagram is fundamentally based on the phase rule. The number of independent variables in a multicomponent and multiphase system is clearly indicated by the phase rule and given by

$$F = C - P + 2 \tag{1.13}$$

where

F = Degrees of freedom
C = Number of components
P = Number of phases.

In the above formulation, 2 is taken into consideration due to the fact that two variables T and P must be specified to describe the state of equilibrium. This rule is widely used for the study of different types of systems in equilibrium. For instance, in a water system

Ice $\leftrightarrow$ Liquid water $\leftrightarrow$ Water vapour

Here, P = number of phases = 3, and C = number of components = 1.

Substituting the value of P and C in Eq. (1.13), we have $F = 1 - 3 + 2 = 0$, i.e., ice, liquid water and water vapour are in equilibrium. Hence, degrees of freedom are determined for the system in equilibrium.

1.7 STATE FUNCTIONS AND PATH FUNCTIONS

The measurable properties of a system which describe the present state of the system are known as *state functions*. The changes in such state functions are independent of how the change is brought about. These depend only upon the initial and final states of the system. Hence, the state functions are fixed for a particular state of a system.

For example, suppose one wants to climb a mountain peak 2 km above his guesthouse. In this case the guesthouse is the initial state and the mountain peak is the final state. He can go up to the mountain by different paths, but the distance is 2 km. The height cannot be changed by choosing another path, although the amount of work done will be different. Here, the vertical distance is a state function. In addition to this, internal energy, enthalpy, entropy, temperature, pressure, etc. are state functions.

The path function is defined as one whose magnitude depends on the path followed during a process as well as on the end-states. Major examples are heat and work.

The state functions have exact differentials and are represented by dH, dT, dP, etc., and the path functions have inexact differentials, denoted by ∂W, ∂q, etc.

For a cyclic process in which the initial and final states are identical,
state function

$$\oint dU = 0$$

and
path function

$$\oint \delta q \neq 0 \qquad \oint \delta W \neq 0$$

Suppose a gas in a cylinder–piston arrangement is allowed to undergo a change from an initial state T_1, P_1 to a final state T_2, P_2. The process is accompanied by change in internal energy (ΔU) and enthalpy (ΔH). These ΔU and ΔH are fixed and so do not depend on the path followed by the process. Here, ΔU and ΔH are state functions, whereas the heat and work interactions are path functions as they depend on the path followed by the process to change from the initial state T_1, P_1 to the final state T_2, P_2.

1.8 TEMPERATURE AND IDEAL GAS TEMPERATURE SCALE

The temperature of a substance is such a thermal property that indicates the direction of flow of heat. There are several temperature measuring devices used in laboratories and industries, such as mercury-in-glass thermometer, optical pyrometer, resistance thermometer, radiation pyrometer, and thermocouple. But the mercury-in-glass thermometer is of common use in measuring temperature. This thermometer consists of an uniform glass tube filled with mercury. On application of heat, the liquid inside the thermometer expands along the length of the tube and indicates the temperature. The length between the upper and lower ends is equally divided into 100 units known as degrees.

The ideal gas temperature scale is developed on the basis of an ideal fluid. This scale is identical with the Kelvin scale. The constant-volume gas thermometer is used to measure the temperature on this scale. This temperature scale could be developed only due to the fact that the thermometer is independent of the fluid used. It will show the same reading at ice and steam points. But at the points somewhere in between, the fluids may vary in their expansion characteristics, as the thermometric property of the fluid may change. This variation necessitates the development of the ideal gas temperature scale, which does not depend upon the nature of the thermometric fluid. An ideal gas obeying the relation $PV = RT$ would be the right choice, as its pressure and volume vary linearly with temperature. We know that all gases behave like ideal gases at sufficiently low pressure. The constant volume gas thermometer, used in measuring temperature on the ideal gas temperature scale, is based on the principle that at low pressure the temperature of the gas is directly proportional to its pressure at constant volume.

In the ***Celsius scale*** (also called the *centigrade scale*), the ice point is 0 and the steam point is 100, and temperature is denoted by °C (degrees Celsius). The calibration of a thermometer is done by first immersing it into an ice bath and adjusting it with the ice point, and then again immersing it into boiling water and adjusting it with the boiling point.

In the ***Kelvin scale***, an ideal gas is used as the thermometric fluid, and the symbol of temperature is K. The word *degree* is not used with temperature in the Kelvin scale, because the word *Kelvin* as a unit is not capitalized.

The relation between the Kelvin temperature scale and the Celsius temperature scale can be expressed as

$$T \text{ (K)} = T \text{ (°C)} + 273.15 \tag{1.14}$$

Apart from the Celsius and Kelvin scales, there are two other thermometer scales used chiefly in the United States. These are the Rankine scale and the Fahrenheit scale.

The Rankine scale is an absolute scale in which the temperature is denoted by R, and the empirical relationship between the Rankine and Kelvin scales is given by

$$T \text{ (R)} = 1.8 \ T \text{ (K)} \tag{1.15}$$

The Fahrenheit scale has a relation with Rankine scale that can be expressed as

$$T \text{ (°F)} = T \text{ (R)} - 459.67 \tag{1.16}$$

The following relation exists between the Fahrenheit and Celsius scales:

$$T \text{ (°F)} = 1.8T \text{ (°C)} + 32 \tag{1.17}$$

A comparative study between the Celsius, Kelvin, Fahrenheit and Rankine scales has been presented in Fig. 1.11 and Table 1.2 for better understanding of the differences in the ice point, steam point and absolute zero among the three scales.

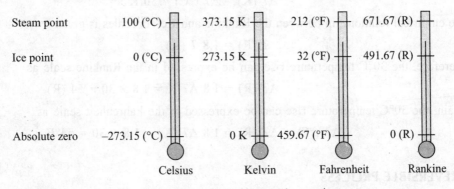

Fig. 1.11 Comparison of temperature scales.

Table 1.2 Different temperature scales

	Celsius	Kelvin	Fahrenheit	Rankine
Steam point	100°C	373.15 K	212°F	671.67 R
Ice point	0°C	273.15 K	32°F	491.67 R
Absolute zero	−273.15°C	0 K	459.67°F	0 R

EXAMPLE 1.12 Find the temperature which has the same value on both the centigrade and Fahrenheit scales.

Solution: We have the relation

$$\frac{C}{5} = \frac{F - 32}{9}$$

Putting
$$C = F, \text{ we get}$$
$$9\,C - 5\,C = -160$$
$$4\,C = -160$$
$$C = -40$$

Hence
$$-40°C = -40°F$$

EXAMPLE 1.13 During the adiabatic compression of a gas in a cylinder, the temperature of the gas rises by 30°C. Express the rise in temperature in terms of the Fahrenheit, Kelvin and Rankine scales.

Solution: We know that the relation between the Kelvin temperature scale and the Celsius temperature scale can be expressed as

$$T\,(K) = T\,(°C) + 273.15$$

Here, the temperature rise is to be expressed in terms of K, but the difference in temperature will be the same in the Kelvin and Celsius scales of temperature.
Therefore

$$\Delta T\,(K) = \Delta T\,(°C) = 30 \text{ K}$$

The empirical relationship between the Rankine and Kelvin scales is given by

$$T\,(R) = 1.8\,T\,(K)$$

Therefore, the 30°C temperature rise can be expressed in the Rankine scale as

$$\Delta T\,(R) = 1.8\,\Delta T \text{ K} = 1.8 \times 30 = 54 \text{ (R)}$$

Again, the 30°C temperature rise can be expressed in the Fahrenheit scale as

$$\Delta T\,(°F) = \Delta T\,(R) = 1.8\,\Delta T\,(K) = 1.8 \times 30 = 54°F$$

1.9 REVERSIBLE PROCESS

A process is said to be *reversible* if both the system and the surroundings can be restored to their respective original states by reversing the direction of the process. A reversible process is one in which the properties of the system at every instant remain uniform when the process takes place. Alternatively, it can be said that a reversible process proceeds with no driving force and can be reversed without causing any change to the surroundings. It is needless to mention that the reversal of the direction of the process by an infinitesimal change causes the system and the surroundings to return to its initial state following the same path. The thermodynamic properties at different stages will be the same in magnitude as in the forward process, but opposite in direction. A reversible process must be carried out at a slow rate, so that sufficient time is allowed for the system to attain an equilibrium.

Consider the following examples:

(i)   Vaporization of a liquid in a closed vessel at a constant temperature
(ii) Chemical reaction in a galvanic cell.

Suppose a gas is confined in a cylinder–piston arrangement as shown in Fig. 1.12 and initially the system is in equilibrium. The system is at a pressure greater than the atmospheric pressure. The piston is loaded with certain amount of sand particles. Now the sand particles are removed and shifted on to an adjacent platform which is at the same elevation as sand. Due to gradual unloading, the piston moves up slowly and the gas expands. During the expansion process, the gas does some work on the surroundings in raising the particles to different elevations. Now, if the sand particles are kept on the piston from the adjacent platform, the piston moves down slowly and the gas gets compressed. This process continues until the system comes back to the initial state. Here, the work done on the system during compression is exactly equal to the work done by the system during expansion, i.e., $W_{compression} = W_{expansion}$. Thus the process can be reversed and the system returned to its initial state. Now the system and the surroundings are in equilibrium with each other.

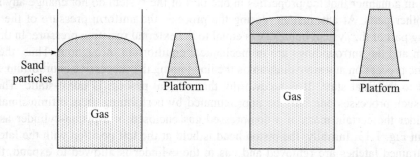

Fig. 1.12 Reversible expansion of a gas.

The characteristics of a reversible process can be summarized as follows:

- A reversible process proceeds with no driving force through a succession of equilibrium steps.
- A reversible process can be restored to its original state at the end of the reverse process.
- The process undergoes an infinitesimal change towards completion.
- During a reversible process the properties of the system remain uniform at every instant.
- In the reverse process, the magnitude of the thermodynamic quantities at different stages will be the same as in the forward process but opposite in direction.

1.10   IRREVERSIBLE PROCESS

Our commonsense tells us that a process which does not satisfy the criterion of the reversible process is known as an *irreversible process*. In this process, change occurs very rapidly and there is no chance of attainment of equilibrium of the system. All spontaneous processes occurring in nature are irreversible. They can not be reversed by the help of any external agency.

For example, the flow of heat from one body to another, free expansion of a gas, and rusting of iron in presence of atmospheric oxygen are irreversible processes.

The following factors are responsible for the irreversibility of a process:

- Friction
- Mixing of two fluids
- Unrestrained expansion
- Heat transfer across a finite temperature difference
- Electrical resistance
- Inelastic deformation of solids
- Chemical reaction.

1.11   QUASI-STATIC PROCESS

When a process occurs in such a way that the system remains nearly in equilibrium (infinitesimally close to an equilibrium state) at all times, it is called a *quasi-static process* or *quasi-equilibrium process*. The term *'quasi'* means almost, i.e. almost equilibrium. Alternatively, a quasi-static process is one which takes place so slowly that allows the system to adjust itself internally in a manner that the properties in one part of the system do not change any faster than those at other parts. At any instant, during the process, the uniform pressure of the system on the moving part of the system boundary is equal to the external resisting pressure. In this process, the system and the surroundings are in mechanical equilibrium throughout. Thus, the path of a quasi-static process on any state diagram is the line joining the successive equilibrium states from the initial to the final state. It is noteworthy that no real process is quasi-static. Therefore, in practice, such processes can only be approximated by performing them infinitesimally slowly.

Consider the certain mass of a compressed gas enclosed in a piston-cylinder assembly as depicted in Fig. 1.13. Initially, the piston head is held at the left position with the latches. Now, if the restrained latches are removed and gas in the cylinder is allowed to expand, then piston moves slowly from left to right.

Let the pressure of the system be P at any moment during the movement of the piston between the two extreme positions, the total volume be V and the cross-sectional area of the piston be A.

Then the total force acting upon the piston is given by

$$F = PA \tag{1.18}$$

The differential moving boundary work done during this process is

$$\partial W_b = F \times dS = (PA)dS = P(A)dS = P(A)\frac{dV}{A} = PdV \tag{1.19}$$

where

    dS = The distance travelled by the piston

    dV = Volume swept by the system boundary.

It is necessary to remember that the pressure P is the absolute pressure which is always positive, but the differential change in volume dV is positive during expansion process only. Therefore, the boundary work is positive during an expansion process and negative during compression process.

The total moving boundary work during the process 1–2 is given by

$$W_b = \int_1^2 PdV \qquad (1.20)$$

In order to evaluate the integral, a functional relation is required between P and V, i.e. $P = f(V)$. Here, $P = f(V)$ represents the process path on PV diagram.

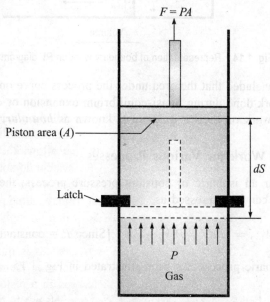

Fig. 1.13 Demonstration of work done during the process in cylinder piston device.

The above quasi-equilibrium expansion process has been described in a PV diagram. In Fig. 1.14, the differential area

$$dA = PdV$$

which is the differential amount of work. The total area A under the process curve 1–2 is obtained by adding these differential areas.

$$\text{Area} = A = \int_1^2 dA = \int_1^2 PdV \qquad (1.21)$$

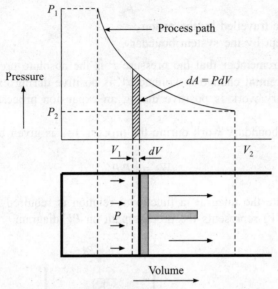

Fig. 1.14 Representation of boundary work on *PV* diagram.

Hence, it can be concluded that the area under the process curve on a *PV* diagram is equal in magnitude to the work done during quasi-equilibrium expansion or compression process of a closed system. This work on the *PV* diagram is known as ***boundary work***.

1.11.1 Quasi-static Work for Various Processes

Isobaric process: For an isobaric or constant pressure process, the quasi-static or quasi-equilibrium work done can be expressed as

$$W_{1-2} = \int_1^2 PdV = P(V_2 - V_1) \quad [\text{Since } P = \text{constant}] \tag{1.22}$$

The work done for isobaric process has been illustrated in Fig. 1.15.

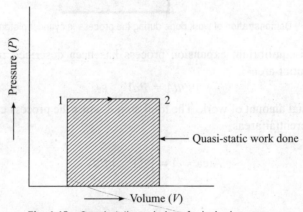

Fig. 1.15 Quasi-static work done for isobaric process.

Isochoric process: For an isochoric or constant volume process, the quasi-static or quasi-equilibrium work done can be represented as

$$W_{1-2} = \int_1^2 PdV = 0 \quad [\text{Since, } V = \text{constant and } dV = 0] \tag{1.23}$$

The work done for isochoric process has been illustrated in Fig. 1.16.

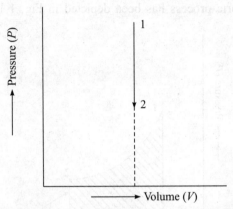

Fig. 1.16 Quasi-static work done for isochoric process.

Isothermal process: For an isothermal or constant temperature process, the quasi-static or quasi-equilibrium work done is given by

$$W_{1-2} = \int_1^2 PdV \tag{1.24}$$

Since an isothermal process is governed by the law

$$PV = C = \text{Constant}$$

Therefore, on substitution of $P = \dfrac{C}{V}$ into Eq. (1.24), it yields

$$W_{1-2} = C\int_1^2 \frac{dV}{V} \tag{1.25}$$

Integrating Eq. (1.25), we get

$$W_{1-2} = C \ln\left(\frac{V_2}{V_1}\right) \tag{1.26}$$

Considering the state 1 of the process, we can write P_1V_1 and Eq. (1.26) can be obtained as

$$W_{1-2} = P_1V_1 \ln\left(\frac{V_2}{V_1}\right) \tag{1.27}$$

or

$$W_{1-2} = P_1V_1 \ln\left(\frac{P_1}{P_2}\right) \tag{1.28}$$

Similarly, for state 2 of the process, work done can be expressed as

$$W_{1-2} = P_2 V_2 \ln\left(\frac{V_2}{V_1}\right) \tag{1.29}$$

or
$$W_{1-2} = P_2 V_2 \ln\left(\frac{P_1}{P_2}\right) \tag{1.30}$$

The work done for isochoric process has been depicted in Fig. 1.17.

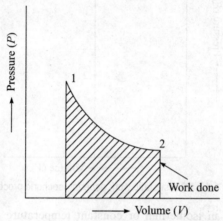

Fig. 1.17   Quasi-static work done for isothermal process.

Polytropic process: The polytropic process is a general process and governed by the following equation:

$$PV^n = C$$

where, the value of n varies from 0 to ∞ and C = constant.

On substitution of the different 'n' values into the above expression, the process becomes isothermal, isochoric or isobaric. These processes are represented in Fig. 1.28.

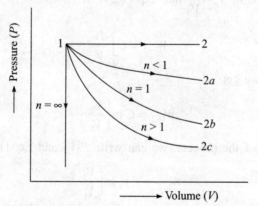

Fig. 1.18   Demonstration of a polytropic process.

For instance,

when $n = 0$, then PV^0 or $P = C$ or Pressure = Constant, i.e. isobaric process

when $n = 1$, then $PV^1 = C$ or $PV = C$ or Pressure $\times$ Volume = Constant, i.e. isothermal process

when $n = \infty$, then PV^∞ or $V = C$ or Volume = Constant, i.e. isochoric process

Considering the general expression $PV^n = C$, the work done for polytropic process can be expressed as

$$W_{1-2} = \int_1^2 PdV = \int_1^2 \frac{C}{V^n} dV \qquad \left[\text{Since, } PV^n = C \text{ or, } P = \frac{C}{V^n} \right]$$

or

$$W_{1-2} = \int_1^2 \frac{C}{V^n} dV = \frac{C}{(1-n)} (V_2^{-n+1} - V_1^{-n+1})$$

or

$$W_{1-2} = \frac{1}{(1-n)} (CV_2^{-n+1} - CV_1^{-n+1})$$

Putting $C = P_2V_2^n$ for the first term and $C = P_1V_1^n$ for the second term in the parenthesis to obtain

$$W_{1-2} = \frac{1}{(1-n)} (P_2V_2^nV_2^{-n+1} - P_1V_1^nV_1^{-n+1})$$

or

$$W_{1-2} = \frac{1}{(1-n)} (P_2V_2 - P_1V_1) \tag{1.31}$$

$$= \frac{(P_1V_1 - P_2V_2)}{(n-1)} \tag{1.32}$$

$$= \frac{nR(T_1 - T_2)}{(n-1)} \tag{1.33}$$

Depending upon the availability of the nature of the data given in the problem, sometimes it is required to convert Eqs. (1.32) and (1.33) towards obtaining the results easily. For example, if the initial conditions along with the pressure ratios are given, then we need to substitute the values in the expression having the form comprising of pressure ratio term.

Hence, in order to convert Eq. (1.32) to other forms, we may proceed as

$$W_{1-2} = \frac{(P_1V_1 - P_2V_2)}{(n-1)}$$

or

$$W_{1-2} = \frac{P_1V_1}{(n-1)} \left(1 - \frac{P_2V_2}{P_1V_1} \right) \tag{1.34}$$

Again, we know that for initial and final states of a polytropic process, it can be expressed as

$$P_1V_1^n = P_2V_2^n$$

or

$$\left(\frac{V_2}{V_1} \right) = \left(\frac{P_1}{P_2} \right)^{\frac{1}{n}} = \left(\frac{P_2}{P_1} \right)^{-\frac{1}{n}} \tag{1.35}$$

On substitution of Eq. (1.35) into Eq. (1.34), it yields

$$W_{1-2} = \frac{P_1 V_1}{(n-1)} \left[1 - \left(\frac{P_2}{P_1} \right) \left(\frac{P_2}{P_1} \right)^{-\frac{1}{n}} \right]$$

or

$$W_{1-2} = \frac{P_1 V_1}{(n-1)} \left[1 - \left(\frac{P_2}{P_1} \right)^{\frac{(n-1)}{n}} \right] \tag{1.36}$$

This is a useful relation to compute the quasi-static or quasi-equilibrium work done of a polytropic process subject to the known value of pressure ratio.

SUMMARY

Thermodynamics, as a science, deals with the principle of energy and its transformation, the feasibility of a chemical process, and to what extent such a process can proceed. It has certain limitations such as the incapability of finding out the mechanism and rate of a chemical process. It is essential to have complete knowledge of the units and dimensions of some commonly encountered physical quantities associated with the processes for the study of thermodynamics. Thermodynamics can be studied in sufficient detail by using microscopic as well as macroscopic approaches. The microscopic approach requires knowledge of the average behaviour and structure of a large group of individual molecules to a high level of significance, while the macroscopic approach does not require such knowledge of the behaviour of individual molecules. The essence of thermodynamics has been discussed with the help of terms such as system, surroundings, boundary, walls, universe, equilibrium, state, and processes. A process is said to be in thermodynamic equilibrium if it satisfies the conditions of thermal, mechanical, and chemical equilibrium. State and path functions are vital factors to be taken into consideration for the study of several properties and processes. Phase rule plays a dominant role in constructing a phase diagram. The necessity of the zeroth law of thermodynamics and the development of the ideal gas temperature scale have been substantiated with examples. Thermodynamics finds extensive application in engineering, chemical, and physical processes. The factors responsible for the reversibility and irreversibility of a chemical process have been enumerated for clear understanding of the process.

KEY TERMS

Adiabatic Process   The process in which there is no exchange of heat between the system and the surroundings.

Boundary   A real or imaginary surface which separates a system from its surroundings.

Chemical Equilibrium   A form of equilibrium in which the composition of the system does not change with time.

Closed System   A system that can exchange only energy with its surroundings, not the mass.

Cyclic Process   The process in which a system, having undergone a change, returns to its initial state at the end of the process.

Dimensions   The entities by which physical quantities can be characterized.

Energy   A property which can be transformed into or produced from work. It has the capacity for producing effect.

Extensive Property   The properties dependent upon the extent or quantity of the mass of a system.

Force   A physical dimension derived from Newton's second law of motion.

Heat   The form of energy that can be transferred from a higher-temperature region to a lower-temperature region whenever a temperature difference exists between them.

Heterogeneous System   A system whose properties are not uniform throughout the system.

Homogeneous System   A system whose properties are uniform throughout.

Ideal Gas Temperature Scale   A temperature scale developed on the basis of an ideal fluid. This scale is identical with the Kelvin scale.

Intensive Property   The properties independent of the extent or quantity of the mass of a system.

Internal Energy   A certain amount of energy that every system has within itself, and is the sum of all the microscopic forms of energy; it is denoted by U.

Irreversible Process   A process which does not satisfy the criterion of a reversible process.

Isobaric Process   The process in which pressure remains constant throughout the system.

Isochoric Process   The process in which volume remains constant throughout the system.

Isolated System   A system that can exchange neither energy nor mass with its surroundings.

Isothermal Process   The process in which temperature remains constant throughout the system.

Kinetic Energy   The energy possessed by a system by virtue of its motion.

Macroscopic Approach   The approach to the study of thermodynamics which does not require knowledge of the behaviour of individual molecules.

Macroscopic Form of Energy   A form which includes the energies that a system possesses with respect to an external reference plane; examples are kinetic energy and potential energy.

Mechanical Equilibrium   A form of equilibrium in which pressure is uniform throughout the system.

Microscopic Approach   The approach to the study of thermodynamics which requires knowledge of the average behaviour of large groups of individual molecules to a high level of significance.

Microscopic Form of Energy A form which includes the energies that are related to the molecular structure of a system, such as internal energy.

Open System A system that can exchange both mass and energy with its surroundings.

Phase Rule A rule involving qualitative treatment of systems in equilibrium, formulated by J.W. Gibbs in 1876; the phase diagram is fundamentally based on the phase rule.

Potential Energy The energy possessed by a system by virtue of its elevation with reference to an arbitrary reference plane.

Power The rate at which work is done, or the ratio of work done to unit time.

Pressure The normal force exerted by a fluid per unit area; it is a scalar quantity.

Quasi-static Process The process which takes place very slowly and with infinitesimal number of driving forces.

Reversible Process A process in which both the system and the surroundings can be restored to their respective original states by reversing the direction of the process.

State Functions The measurable properties of a system which describe the present state of the system.

State of a System The condition in which the system exists.

Steady State of a System The state of a system interacting with the surroundings if it does not vary with time; simply put, in the steady state the change of state is zero but the interactions are non-zero.

Surroundings The area or region external to a system.

System The quantity of matter or a region in space upon which attention is concentrated during the analysis of a problem.

Temperature A physical property which indicates the direction of flow of heat.

Thermal Equilibrium A form of equilibrium in which temperature is uniform throughout the system.

Thermodynamic Equilibrium The state of balance of a system.

Thermodynamics The science which deals with energy and its transformation.

Units The physical quantities assigned to the dimensions for expressing their magnitudes precisely.

Universe The entity that is constituted by the system and its surroundings.

Work A form of energy defined as the product of an applied force and the displacement in the direction of the applied force; in other words, it is the work performed whenever a force acts through a distance.

Zeroth Law of Thermodynamics If the two bodies are in thermal equilibrium with a third body, they are also in thermal equilibrium with each other.

IMPORTANT EQUATIONS

1. Force is given by

$$F = Cma = \frac{1}{g_c} ma \qquad (1.1)$$

where

$$C = \frac{1}{g_c} = \text{Constant of proportionality}$$

2. Pressure is given by

$$P = \frac{F}{A} \qquad (1.2)$$

where

F = Force normal to the area
A = Area of the surface
P = Pressure.

3. Work done is given by

$$dW = Fdl \qquad (1.3)$$

where

F = Applied force
dl = Displacement.

4. Work done by the system within the integral limits V_1 and V_2 is given by

$$W = \int_{V_1}^{V_2} PdV \qquad (1.7)$$

5. Kinetic energy is given by

$$\text{K.E.} = E_K = \frac{1}{2}mV^2 \qquad (1.8)$$

6. Potential energy is given by

$$\text{P.E.} = E_P = mgZ \qquad (1.9)$$

7. Total energy of a system can be expressed as

$$E = \underbrace{E_K + E_P}_{\text{Macro}} + \underbrace{U}_{\text{Micro}} \qquad (1.10)$$

where

E_K = Kinetic energy
E_P = Potential energy
U = Internal energy.

8. The phase rule is given by

$$F = C - P + 2 \qquad (1.13)$$

where

F = Degrees of freedom
C = Number of components
P = Number of phases.

9. The relation between the Kelvin and Celsius temperature scales can be expressed as

$$T\,(\text{K}) = T\,(^{\circ}\text{C}) + 273.15 \qquad (1.14)$$

10. The empirical relationship between the Rankine and Kelvin scales is given by

$$T\,(\text{R}) = 1.8\,T\,(\text{K}) \qquad (1.15)$$

11. The relation between the Fahrenheit and Rankine scales can be expressed as

$$T\,(^{\circ}\text{F}) = T\,(\text{R}) - 459.67 \qquad (1.16)$$

12. The relation between the Fahrenheit and Celsius scales is given by

$$T\,(^{\circ}\text{F}) = 1.8\,T\,(^{\circ}\text{C}) + 32 \qquad (1.17)$$

EXERCISES

A. Review Questions

1. What is thermodynamics?
2. Explain the scope and limitations of thermodynamics.
3. Explain the significance of the microscopic and macroscopic approaches to the study of thermodynamics.
4. How do units and dimensions play an important role in characterizing the physical quantities of thermodynamic systems?
5. Mention the SI and British units of the following:
   (a) Pressure (b) Force (c) Temperature (d) Work done (e) Energy (f) Heat (g) Power.
6. Distinguish between absolute pressure and gauge pressure. How is it related to the other in case of vacuum?
7. Differentiate between open, closed and isolated systems with the help of neat sketches and examples.
8. Write short notes on the following:
   (a) Homogeneous system (b) Heterogeneous system.
9. Write an informatory note on extensive and intensive properties with suitable examples.
10. Point out the extensive and intensive properties from the following list:
    (a) Temperature (b) Pressure (c) Volume (d) Specific heat (e) Internal energy (f) Refractive index (g) Potential energy (h) Enthalpy.
11. Define isothermal, isochoric and isobaric processes with suitable examples.

12. Why is the thermodynamic process also called cyclic process? Explain with suitable example.
13. What conditions must be satisfied for a process to be in thermodynamic equilibrium?
14. What is the criterion for a thermodynamic system to be at steady state?
15. What is the difference between steady and uniform states?
16. State the zeroth law of thermodynamics. How does it play an important role in measuring temperature?
17. State the phase rule and show how it can be mathematically expressed. What is the importance of degree of freedom? Find the value of degree of freedom for the water system.
18. Differentiate between state function and path function. Prove that internal energy is a state function and work is a path function.
19. Mention the importance of the ideal gas temperature scale in expressing temperatures through other scales.
20. Justify the following statement with an example: 'A reversible process proceeds without any driving force'.
21. 'All spontaneous processes are irreversible'. Explain with a common example.
22. What are the factors responsible for the irreversibility of a process?

B. Problems

1. An astronaut weighs 800 N on the earth's surface where the acceleration of gravity is $g = 9.83$ m/s^2. The mass and the radius of the moon are 7.4×10^{22} kg and 3500 km respectively. Find out the value of g on the moon. Also calculate the weight of the astronaut on the moon.
2. Estimate the gravitational force and the acceleration due to gravity on a body of 1.25 kg mass on the earth's surface. The radius and mass of the earth are 6370 km and 6.02×10^{24} kg respectively.
3. A reactor contains a gas mixture of 25 kg NH_3, 15 kg CO and 10 kg C_2H_2. Calculate the total number of moles of the gas mixture present in the reactor and the average molar mass of the mixture.
4. An elevator with a mass of 2000 kg rests at a level of 6 m above the base of the elevator shaft. It is at a height of 60 m when the cable holding it breaks. It falls freely to the base, where it is brought to rest by a strong spring. Assume $g = 9.81$ m/s^2. Estimate:
   (a) The potential energy of the elevator in its initial position
   (b) The potential energy of the elevator in its highest position
   (c) The work done in raising the elevator
   (d) The kinetic energy and the velocity of the elevator just before it strikes the spring
   (e) The potential energy of the spring when the elevator rests on it.
5. At what absolute temperature do the Celsius and Fahrenheit scales show the same numerical value?
6. How can temperature in the Fahrenheit scale be expressed in the Rankine scale? What will the temperature of 160°C be in the Rankine scale of temperature?

7. A man is driving a two-wheeler of 150 kg mass at the speed of 70 km/h. What is its kinetic energy? What amount of work should be done to stop the engine of the two-wheeler, if the driver so wants?

8. N_2 gas is confined in a cylinder of 0.5 m diameter having a closely fitted frictionless piston. The weights of the piston and the block resting on it together equal 185 kg. The atmospheric pressure and the acceleration due to gravity g are 1.013 bar and 9.792 m/s² respectively. Calculate the force exerted by the piston and the weight on the gas respectively.

9. A manometer shows the reading as 60 cm at 30°C and 1 atm. Determine the absolute pressure, given that ρ_{Hg} = 14.02 g/cm³ and g = 0.792 m/s².

10. The reading on a mercury-filled manometer is 70.2 cm at 25°C. Determine the pressure that corresponds to the height of the mercury column, given that g = 9.832 m/s².

11. During the adiabatic compression of a gas in a cylinder, the temperature of the gas rises by 30°C. Express the rise in temperature in terms of the Fahrenheit, Kelvin and Rankine scales.

12. The potential energy of a body of 20 kg mass is 2 kJ. What should be the height of the body from the ground?

13. A spherical ball of mass 10 kg is dropped from a height of 15 m. What is its potential energy at the time of release? Consider the potential energy of the ball as transformed into kinetic energy when it strikes the ground. Determine the velocity at which the ball strikes the ground.

14. Determine the work done while a body of mass 47 kg is lifted through a distance of 20 m. Also calculate the power. The entire process takes time to the tune of 3 min 45 s.

15. An arbitrary temperature scale is proposed, in which 20° is assigned to the ice point and 75° is assigned to the steam point. Derive an equation relating this scale to the Celsius scale.

16. If a Celsius temperature is two-thirds the corresponding Fahrenheit temperature, determine both the temperatures.

17. A skin diver descends to a depth of 30 m in a salt lake where the density is 1030 kg/m³. What is the pressure on the diver's body at this depth?

18. A tank contains 400 kg of a fluid. If the volume of the tank is 2.5 m³, then what is the density of the fluid and what is the specific gravity?

19. A mass of 110 kg is hung from a spring in a local gravitational field where g = 9.806 m/s², and the spring is found to deflect by 30 mm. If the same mass is taken to a planet where g = 5.412 m/s². By how much will the spring deflect if its deflection is directly proportional to the applied force?

20. A pump delivers water from a well that is 50 m deep. Determine the change in potential energy per kg of water. Take g = 9.81 m/s².

21. A U-tube mercury manometer is connected to a venturimeter to measure the pressure drop of water flowing through a pipeline. If the difference in the mercury levels in the tube is 20 cm, then determine the pressure drop, given that the density of mercury = 1.4 × 10⁴ kg/m³ and that of water = 1 × 10³ kg/m³.

22. A vacuum gauge mounted on a condenser reads 0.73 mm Hg. Determine the absolute pressure in the condenser in kPa when the atmospheric pressure is 101.35 kPa.

First Law of Thermodynamics

2

The discussion in this chapter has been started with the demonstration of Joule's classical experiment called the *paddle-wheel experiment* to illustrate the relationship between heat and work in a thermodynamic cycle. This experiment takes a historical step towards the establishment of the first law of thermodynamics. The significant statements and mathematical expression of the first law of thermodynamics have been presented in a simple way for a cyclic process. The sign convention for heat and work interaction adopted in this books is meant in a lucid language to avoid the confusion to the readers. Joule's experiment is again highlighted for finding out the change in internal energy for an ideal gas. No thermal communication takes place between the system and the surroundings in adiabatic process. The work done under adiabatic change in a thermodynamic system is clearly discussed along with reversible compression and expansion process. Heat capacity at constant pressure and volume play a dominant role in deriving several useful thermodynamic relations. These are considered in thermodynamic process calculation. Finally, the energy and mass balance for open system have been represented considering different thermodynamic state functions.

2.1 JOULE'S EXPERIMENT: MECHANICAL EQUIVALENCE OF HEAT

From 1843 to 1847, James Prescott Joule (an English physicist) conducted a series of experiments which showed, the relationship between heat and work in a thermodynamic cycle for a system which in turn led to the development of the first law of thermodynamics. His classical paddle-wheel experiment demonstrated that the mechanical work required for rotating the wheel is transferred to heat, i.e., work is converted into heat. Based on this experiment, he postulated the equivalence of heat.

In his experimental set-up, shown in Fig. 2.1, Joule used a paddle-wheel to agitate an insulated vessel filled with a known amount of fluid. He observed the process cycle in 2 ways. In the first, the amount of work done on the system by means of a paddle-wheel was noted. The work was done by lowering a weight mg through a height z, so that the work done = mgz. This process caused temperature rise of the fluid. Later, this vessel was placed in contact with the water bath and cooled by transferring heat from the fluid to the water as shown in Fig. 2.2. The energy involved in increasing the temperature of the bath was shown to be equal to that supplied by lowering the weight. That is, an exact proportionality was established between the amount of work supplied to the system (W), the rise in temperature, as well as the amount of heat transfer (Q) from the system.

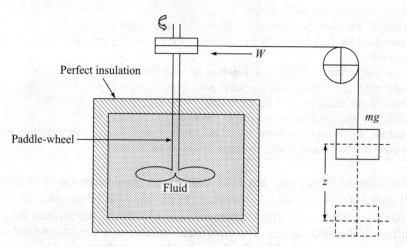

Fig. 2.1 Work supplied to the system.

Joule also performed experiments where electric work was converted to heat, by passing current through a coil immersed in water, and obtained the same result. In this way, Joule illustrated the quantitative relationship between heat and work.

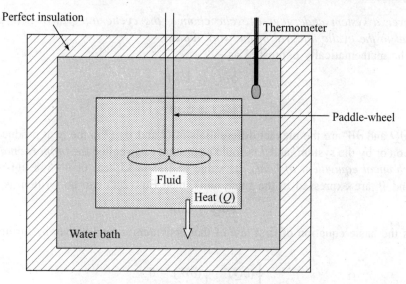

Fig. 2.2 Heat transferred to the water bath.

2.2 STATEMENT OF FIRST LAW OF THERMODYNAMICS

Joule's experiment resulted in the formulation of the first law of thermodynamics. The first law basically represents the law of conservation of energy, which states: *Energy can neither be created nor be destroyed; it may only be converted from one form to another.* On the basis of the relationship between heat and work, the first law of thermodynamics has been established. It can be stated in the following ways:

- The total energy is constant, and when a quantity of one form of energy disappears an exactly equivalent quantity of another form of energy appears.
- The total energy of an isolated system remains constant.
- In a system of constant mass, energy cannot be created or destroyed, by any physical or chemical change, but may be transferred from one form of energy to another form.
- The sum of all the forms of energies in an isolated system is always constant.
- The total sum of mass and energy in the universe is constant.
- The total sum of the energy in the system and the surroundings is equal to a constant value, i.e.

$$E_{\text{System}} + E_{\text{Surroundings}} = \text{Constant}$$

or

$$dE_{\text{System}} + dE_{\text{Surroundings}} = 0$$

Here, E represents the total energy of the system.

2.3 FIRST LAW OF THERMODYNAMICS FOR CYCLIC PROCESS

Joule's experiment led to the following implication of the first law of thermodynamics:

Whenever a system undergoes a cyclic change, the cyclic integral of heat produced is proportional to the cyclic integral of the work done.

It can be mathematically expressed as

$$\oint \partial Q \propto \oint \partial W$$

$$J \oint \partial Q = \oint \partial W \tag{2.1}$$

Here, ∂Q and ∂W are the inexact differentials of Q and W. Q is the heat produced, W is the work done on or by the system, and J is Joule's constant (also called the *proportionality constant* or the *mechanical equivalent of heat*). Its value is 4.1868 kJ/kcal or 4.1868 kgf-m/kcal.

If Q and W are expressed in the same units, then Eq. (2.1) can be written as

$$\oint \partial Q = \oint \partial W \tag{2.2}$$

This is the basic equation of first law of thermodynamics, and it may be further expressed as

$$\oint (\partial Q) - \oint (\partial W) = 0 \tag{2.3}$$

This implies that the total energy of the system in the cycle is constant. One consequence of the first law is that the total energy of the system is a property of the system. This leads to the concept of internal energy.

2.4 CONCEPT OF INTERNAL ENERGY

Internal energy is a state function of a system and an extensive property. *Every system has a certain amount of energy within itself; this is known as internal energy.* Again, in thermodynamics, the internal energy of a system is due to its temperature.

The internal energy of a system does not include macroscopic energies such as kinetic energy, potential energy, and relativistic mass–energy equivalent ($E = mc^2$). The microscopic energy refers to the energy stored in the molecular and atomic structure of the system, which is called the *molecular internal energy* or simply *internal energy*. It is denoted by U.

The internal energy can better be mathematically expressed as

$$U = N\xi$$

where

  N = Total number of molecules in the system
  ξ = Energy of one molecule
   = $\xi_{translational} + \xi_{rotational} + \xi_{vibrational} + \xi_{chemical} + \xi_{electronic} + \xi_{nuclear}$

Hence, the total energy E of a system is given by

$$E = E_K + E_P + U$$

where

  E_K = Kinetic energy
  E_P = Potential energy
  U = Internal energy.

When the value of kinetic and potential energies is quite small, then the internal energy function is U. In that case, internal energy is written as E. This happens in the case of different gases.

2.5 SIGN CONVENTION FOR HEAT AND WORK INTERACTIONS

The most widely used sign conventions for work and heat interaction are shown in Fig. 2.3 and this convention is used in this book too for better understanding of the problems it is associated with. The work done by the system on the surroundings is treated as a +ve quantity and work done on the system is a −ve quantity. In case of heat transfer, it is just the reverse. The heat transfer from the system to the surrounding is treated as a −ve quantity and the same from the surroundings to the system is treated as a +ve quantity. Generally, Q and W cannot be of the same sign.

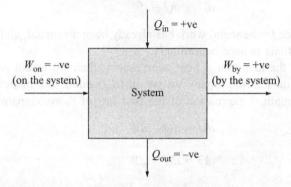

Fig. 2.3 Sign convention for heat–work interaction.

It can be summarized as

Heat: $\qquad\qquad\qquad\qquad\qquad Q_{in} = +ve \quad Q_{out} = -ve$

Work: $\qquad\qquad\qquad\qquad\qquad W_{on} = -ve \quad W_{by} = +ve$

2.6 FIRST LAW OF THERMODYNAMICS FOR CLOSED SYSTEM

The closed system is basically of fixed mass. There is no mass transfer across the system boundary. Only energy can be exchanged between the system and the surroundings. It does appear as heat and work. For a closed system, the total energy of the surroundings is expressed in terms of heat and work interaction.

The first law states that total energy is conserved. The change in internal energy is equal to the amount added by heating minus the amount lost by doing work on the surroundings. The first law can be stated mathematically as

$$\Delta E = Q - W \qquad\qquad\qquad (2.4)$$

where

ΔE = Change in internal energy

Q = Amount of heat added to the system

W = Work done on the system.

In differential form, Eq. (2.4) is given by

$$dE = dQ - dW \tag{2.5}$$

From the definition of total energy in differential form

$$dE = d(KE) + d(PE) + dU \tag{2.6}$$

Comparing Eqs. (2.5) and (2.6), we have

$$dE = d(KE) + d(PE) + dU = dQ - dW \tag{2.7}$$

In most of the cases, kinetic and potential energies are very small. Hence, the equation reduces to

$$dU = dQ - dW \tag{2.8}$$

The sign convention for heat and work has already been discussed, and the sign of the above quantities in the equations is used accordingly.

It is analogous to the idea that if heat were money, then we would say that any change in our savings (dU) is equal to the money we put in (dQ) minus the money we spend.

Hence, the mathematical expression of the first law of thermodynamics is given by

$$dU = dQ - dW$$

or

$$dQ = dU + dW \tag{2.9}$$

EXAMPLE 2.1   The total energy of a typical closed system is given by $E = 50 + 25T + 0.05T^2$ in Joules. The amount of heat absorbed by the system can be expressed as $Q = 4000 + 10T$ in Joules. Estimate the work done during the processes in which temperature rises from 400 Kelvin to 800 Kelvin.

Solution: The amount of work done can easily be calculated by using the first law of thermodynamics. From this, we have

$$Q = \Delta E + W$$

or

$$W = Q - \Delta E$$

$$= \int_{400}^{800} [(4000 + 10T) - (50 + 25T + 0.05T^2)]\ dT$$

$$= \int_{400}^{800} (3950 - 15T - 0.05T^2)\ dT$$

$$= -9486.66 \text{ kJ}$$

EXAMPLE 2.2 A paddle-wheel is employed in a rigid container for stirring a hot fluid to be cooled. The internal energy of the hot fluid is 1000 kJ. During the cooling process, the fluid losses 600 kJ of heat. For this process, the work done by the paddle-wheel on the fluid is 100 kJ. Calculate the final internal energy of the fluid.

Solution: The system is considered to be a closed system. No mass transfer takes place across the system. The tank is rigid. So, the kinetic and potential energies are zero.
Therefore

$$\Delta E = \Delta U + \Delta PE + \Delta KE = \Delta U$$

From the first law of thermodynamics, we have

$$Q = \Delta U + W$$

or

$$\Delta U = Q - W$$

or

$$U_2 - U_1 = Q - W$$

or

$$U_2 - 1000 = -600 \text{ kJ} - (-100 \text{ kJ})$$

or

$$U_2 = 500 \text{ kJ}$$

Hence, the final internal energy is 500 kJ.

EXAMPLE 2.3 A stirrer–container assembly contains a certain amount of fluid. The stirrer performs 3 hp work on the system. The heat developed by stirring is 4000 kJ/h and is transferred to the surroundings. Determine the change in internal energy of the system.

Solution: The work done by the stirrer on the system is given by

$$W = 3 \text{ hp} = 3 \times 745.7 \text{ W} = 2237.1 \quad W = 2237.1 \text{ J/s}$$

The amount of heat transferred to the surroundings can be expressed in terms of J/s as

$$Q = 4000 \text{ KJ/h} = \frac{4000 \times 1000}{3600} = 1,111.1 \text{ J/s}$$

Now, from the first law of thermodynamics

$$Q = \Delta U + W$$

or

$$\Delta U = Q - W$$
$$= [(1,111.11) - (-2237.1)] \text{ J/s}$$
$$= 3348.21 \text{ J/s}$$

Therefore, the change in internal energy of the system would be 1125.99 J/s.

EXAMPLE 2.4 A system consisting of a gas confined in a cylinder undergoes a series of processes shown in Fig. 2.4. During the process A–1–B, 60 kJ of heat is added while it does 35 kJ of work. Then the system follows the process A–2–B, during which 50 kJ of work is performed on the system. Then the system returns to the initial state along the path B–3–A and 70 kJ of work is done on the system. Calculate

(a) The amount of heat flowing into the system during the process A–2–B
(b) the amount of heat transferred between the system and the surroundings during the process B–3–A.

Solution: The internal energy of the process A–1–B can be estimated as (see Fig. 2.4)

$$Q = \Delta U + W$$

or

$$\Delta U = Q - W$$
$$= 60 - 35 = 25 \text{ kJ}$$

Process: A–2–B

$$Q = \Delta U + W$$
$$= 25 - 50 = -25 \text{ kJ}$$

Process: B–3–A

$$Q = \Delta U + W$$
$$= 25 - 70 = -45 \text{ kJ}$$

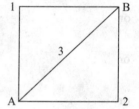

Fig. 2.4 Example 2.4.

Therefore, the amount of heat transferred between the system and the surroundings during process B–3–A is 45 kJ.

EXAMPLE 2.5 A system undergoes a constant pressure process 1–2, during which 100 kJ of work done on the system and 50 kJ of heat as energy is released to the surroundings. Then the system follows a constant volume process 2–3 during which 80 kJ of heat is added to the system. Then the system returns to its initial state along the path 3–1 by an adiabatic process. Calculate the change in internal energy during each process and the work done during the adiabatic process.

Solution: The internal energy of process 1–2 can be calculated as (see Fig. 2.5)

$$\Delta U_{12} = Q_{12} - W_{12}$$
$$= -50 - (-100) = 50 \text{ kJ}$$

For the process 2–3,

$$W = PdV = 0$$

as the process is a constant-volume process, i.e.

$$V = \text{constant} \quad \text{or} \quad dV = 0$$

Hence

$$\Delta U_{23} = Q_{23} - W_{23}$$
$$= 80 - 0 = 80 \text{ kJ}$$

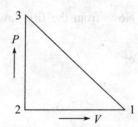

Fig. 2.5 Example 2.5.

For the process 3–1,

Since the process is adiabatic, there is no heat transfer between the system and the surroundings, i.e., $Q = 0$.

Hence

$$\Delta U_{31} = Q_{31} - W_{31}$$
$$= -W_{31}$$

We know, for a cyclic process, the internal energy change is zero.

or

$$\Delta U_{12} + \Delta U_{23} + \Delta U_{31} = 0$$

or

$$\Delta U_{31} = -\Delta U_{12} - \Delta U_{23}$$
$$= -50 - 80 = -130 \text{ kJ}$$

Putting the value of ΔU_{31} in the following equation, we get

$$\Delta U_{31} = -W_{31} \qquad \text{or} \qquad -130 = -W_{31} \qquad \text{or} \qquad W_{31} = 130 \text{ kJ}$$

Hence, the work done during the adiabatic process, $W_{31} = 130$ kJ.

2.7 INTERNAL ENERGY OF AN IDEAL GAS

The internal energy of an ideal gas is a function only of temperature. It was demonstrated experimentally by Joule in 1843. Joule carried out the experiment to find out the change in internal energy of a gas. In his experimental arrangement, shown in Fig. 2.6, two pressure vessels A and B are connected by a pipe and a valve. These vessels are kept immersed in a water bath (or insulator). Initially, vessel A contains air at a high pressure while the vessel B is practically evacuated. A thermometer is used to measure the temperature of the water bath. Now, when the thermal equilibrium is attained, the temperature is recorded. The value is then opened. As a consequence, the gas from vessel A is readily passed into vessel B until the pressures are equalized. The temperature of the water bath is recorded again. It is observed that there is no change in temperature.

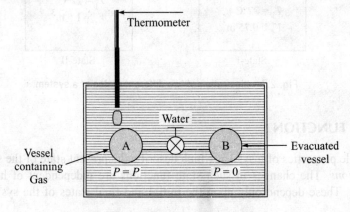

Fig. 2.6 Demonstrating Joule's experiment.

Since the gas is expanded freely against zero pressure (free expansion), no work is done by the gas, i.e., $dW = 0$, and it is assumed that no heat transfer takes place between the water bath and vessels, i.e., $dQ = 0$.

From the first law of thermodynamics, we have

$$dU = dQ - dW = 0$$

or $$U_2 - U_1 = 0$$

or $$U_2 = U_1 \qquad (2.10)$$

Hence, it can be concluded that there is no change in internal energy of an ideal gas undergoing an adiabatic change.

2.8   THERMODYNAMIC STATE

The thermodynamic state of a system is better characterized and described by its properties or variables. In other words, a system is said to be in proper state when the state variables such as temperature, pressure, and volume have definite values.

At a given state of the system, all the variables have a fixed value. Any change in value of even one variable will change the state of the system.

Thus the thermodynamic state refers to the condition of a system in which the system exists. For instance, in a homogeneous system the composition is fixed, and temperature, pressure, and volume are interrelated with each other. Now if the system is considered to obey the ideal gas law $PV = RT$ and two thermodynamic state variables are known, the third variable can be evaluated using R = gas constant.

These two variables among P, V, and T are called *independent variables* and the third one is called the *dependent variable*. The dependent variable depends upon the other two variables. The different states of a system are shown in Fig. 2.7.

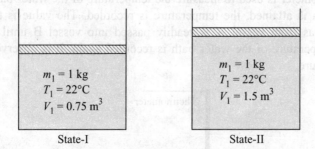

$$m_1 = 1 \text{ kg}$$
$$T_1 = 22°C$$
$$V_1 = 0.75 \text{ m}^3$$

State-I

$$m_1 = 1 \text{ kg}$$
$$T_1 = 22°C$$
$$V_1 = 1.5 \text{ m}^3$$

State-II

Fig. 2.7   Illustration of two different states of a system.

2.9   STATE FUNCTION

The measurable properties of a system which describe the present state of the system are known as *state functions*. The changes in such state functions are independent of how the change is brought about. These depend only upon the initial and final states of the system.

For example, suppose a traveller wants to climb a mountain peak 2 km above his guesthouse. In this case the guesthouse position is the initial state and the mountain peak is the final state. He can go up to the mountain peak by different paths, but the distance remains 2 km. The height cannot be changed by choosing another path. The distance referred to in this case is analogous to functions such as E, H, S, G, T, P, and V. These are state functions.

2.10 ENTHALPY

Enthalpy is a state function and is generally denoted by H. It has a definite value depending on the state of the system. It is an extensive property of the system, and is often convenient in dealing with the system at constant pressure. It is also known as the *heat content function*.

Definition

It is defined as the amount of energy within the system or the substance that is available for conversion into heat. It can be mathematically expressed as

$$H = U + PV \tag{2.11}$$

where

U = Internal energy of the system
P = Absolute pressure of the system
V = Volume of the system.

In other words, it is nothing but the sum of the internal energy and the product of pressure and volume (i.e., work done).

Equation (2.11) can be expressed in differential form as

$$\Delta H = \Delta U + \Delta(PV)$$
$$dH = dU + d(PV) \tag{2.12}$$

On integration, we have

$$\Delta H = \Delta U + P\Delta V \quad (\text{since } P = \text{Constant})$$
$$= \Delta U + P(V_2 - V_1)$$
$$= \Delta U + \Delta nRT \tag{2.13}$$

EXAMPLE 2.6 1 kg of water is vaporized in a container at the constant temperature of 373 K and the constant pressure of 1,01,325.0 N/m². The specific volume of liquid and vapour at these conditions are 1.04×10^{-3} and 1.673 m³/kg respectively. The amount of heat added to the water is 2257 kJ. Calculate the change in internal energy (ΔU) and enthalpy (ΔH).

Solution: From the definition of enthalpy, we have

$$\Delta H = \Delta U + P\Delta V$$

$P\Delta V$, the work done due to expansion, can be expressed as

$$P\Delta V = 1,01,325.0 \text{ N/m}^2 \times (1.673 - 0.00104) \text{ m}^3$$
$$= 169,411.34 \text{ N-m}$$

$$= 169,411.34 \text{ J}$$
$$= 169.411 \text{ kJ}$$

Now, the change in internal energy can be estimated as

$$\Delta U = Q - W = (2257 - 169.411) \text{ kJ}$$
$$= 2087.58 \text{ kJ}$$
$$\Delta H = \Delta U + P\Delta V$$
$$= 2087.58 \text{ kJ} + 169.411 \text{ kJ}$$
$$= 2257 \text{ kJ}$$

EXAMPLE 2.7   Calculate $\Delta U, \Delta H, Q$ and W if 1 mol of an organic liquid is converted reversibly into vapour at 353 K by supplying heat from external source. The expansion of vapour takes place at the pressure of 1 atm. The heat of vaporization and the molecular weight of the liquid are 380 J/g and 78 g/mol respectively.

Solution: The amount of heat supplied for conversion of 1 mol of liquid into vapour

$$Q = \text{Heat of vaporization} \times \text{molecular weight}$$
$$= 380 \text{ J/g} \times 78 \text{ g/mol}$$
$$= 29640 \text{ J/mol}$$

The work done due to expansion during vaporization is given by

$$W = P\Delta V = P(V_v - V_1) \quad \text{(since } V_v \gg V_1)$$
$$= P \ V_v = nRT$$
$$= 1 \times 8.314 \times 353 = 2934.84 \text{ J/mol}$$

The change in internal energy of the system can be calculated by using the first law of thermodynamics as

$$\Delta U = Q - W = (29640 - 2934.84) \text{ J/mol} = 26705.16 \text{ J/mol}$$

The enthalpy change for this process

$$\Delta H = \Delta U + P\Delta V$$
$$= 26705.16 + 2934.84 = 29640 \text{ J/mol}$$

EXAMPLE 2.8   The pressure of a gas is given by $P = \dfrac{10}{V}$, where P is in atmospheres and V is in litres. If the gas expands from 10 to 50 L and undergoes an increase in internal energy of 200 cal. How much heat will be absorbed during the process?

Solution: From the definition of enthalpy, we have

$$dQ = dU + PdV$$

Putting the value of P, we get

$$dQ = dU + 10\left(\frac{dv}{V}\right)$$

On integration, it yields

$$Q_2 - Q_1 = U_2 - U_1 + 10 \ln \frac{V_2}{V_1}$$

$$\Delta Q = \Delta U + 10 \ln \frac{50}{10}$$

$$= \Delta U + 10 \ln 5$$

$$= 200 \text{ cal} + 6.99 \text{ L-atm}$$

$$= 200 \text{ cal} + 6.99 \times 24.2 \text{ cal} \qquad \text{(since 1 L-atm} = 24.2 \text{ cal)}$$

$$= 200 \text{ cal} + 169.16 \text{ cal}$$

$$= 369.16 \text{ cal}$$

2.11 HEAT CAPACITY

Before we define the term *heat capacity*, it is necessary to know what specific heat is. The specific heat of a substance is the amount of heat required to raise the temperature of 1 g of a substance by 1 K.

Now, the amount of heat input required to raise the temperature of 1 mole of the substance by 1 K is known as the *molar heat capacity* or simply the *heat capacity* of the substance. It can be mathematically expressed as

$$dQ = CdT$$

where

dQ = Heat input required
C = Heat capacity
dT = Unit rise in temperature.

When the heat is delivered to a system at constant volume, the system does not perform any work and the required heat input is given by

$$dQ = C_V dT$$

where C_V is the heat capacity at constant volume. It can be written as

$$C_V = \left(\frac{\partial Q}{\partial T} \right)_V \tag{2.14}$$

But for the constant-volume process, V = constant or $dV = 0$.
Hence, from the first law of thermodynamics, we have

$$dQ = dU + dW$$
$$dQ = dU + PdV$$

Putting $dV = 0$ in the preceding equation, we have

$$dQ = dU$$

Hence

$$C_V = \left(\frac{\partial Q}{\partial T} \right)_V = \left(\frac{\partial U}{\partial T} \right)_V \tag{2.15}$$

On the other hand, if the heat is delivered to a system at constant pressure, the system is able to perform work, and the required heat input is given by

$$dQ = C_P dT$$

or

$$C_P = \left(\frac{\partial Q}{\partial T}\right)_P$$

For a constant-pressure process, the first law of thermodynamics becomes

$$dQ = dU + PdV$$
$$dQ = dH \quad (\because H = U + PV)$$

At constant pressure

$$dH = dU + PdV$$

Therefore

$$C_P = \left(\frac{\partial Q}{\partial T}\right)_P = \left(\frac{\partial H}{\partial T}\right)_P \tag{2.16}$$

2.11.1   Relation between C_P and C_V

It is an easy task to establish the relationship between C_P and C_V, if the internal energy is considered to be the function of the volume and the temperature.
Hence

$$U = U(V, T)$$

or

$$dU = \left(\frac{\partial U}{\partial T}\right)_V dT + \left(\frac{\partial U}{\partial V}\right)_T dV \tag{2.17}$$

Differentiating with respect to T at constant pressure

$$\left(\frac{\partial U}{\partial T}\right)_P = \left(\frac{\partial U}{\partial T}\right)_V + \left(\frac{\partial U}{\partial V}\right)_T \left(\frac{\partial U}{\partial T}\right)_P \tag{2.18}$$

Now

$$C_P - C_V = \left(\frac{\partial H}{\partial T}\right)_P - \left(\frac{\partial U}{\partial T}\right)_V$$

Putting $H = U + PV$ into this equation, we have

$$C_P - C_V = \left(\frac{\partial U}{\partial T}\right)_P + P\left(\frac{\partial V}{\partial T}\right)_P - \left(\frac{\partial U}{\partial T}\right)_V \tag{2.19}$$

Putting the value of $\left(\frac{\partial U}{\partial T}\right)_P$, we get

$$C_P - C_V = \left(\frac{\partial U}{\partial T}\right)_V + \left(\frac{\partial U}{\partial V}\right)_T \left(\frac{\partial V}{\partial T}\right)_P + P\left(\frac{\partial V}{\partial T}\right)_P - \left(\frac{\partial U}{\partial T}\right)_V$$

$$= \left[\left(\frac{\partial U}{\partial V} \right)_T + P \right] \left(\frac{\partial V}{\partial T} \right)_P \qquad (2.20)$$

For 1 g-mol of an ideal gas

$$\left(\frac{\partial U}{\partial T} \right)_V = 0$$

Therefore

$$C_P - C_V = P \left(\frac{\partial V}{\partial T} \right)_P$$

$$= P \times \frac{R}{P} \quad \text{(since } PV = RT \quad \text{or} \quad \frac{V}{T} = \frac{R}{P} \text{)}$$

$$= R$$

Hence

$$C_P - C_V = R \qquad (2.21)$$

For ideal gases, molar heat capacities are functions only of temperature and are expressed in polynomial form as

$$C_P = a + bT + cT^2 + \cdots$$

where a, b, and c are constants for a given gas and vary for different substances.

There is another important relation between C_P and C_V. It is given by

$$\frac{C_P}{C_V} = \gamma = \text{Heat capacity ratio}$$

γ can be determined at any temperature if the values of C_P and C_V are evaluated at that temperature. Some values of C_P, C_V and γ for different types of ideal gases are given in Table 2.1.

Table 2.1 C_P, C_V and γ for different types of ideal gases

Gas	Type	$\gamma = \dfrac{C_P}{C_V}$	$C_P = \dfrac{\gamma R}{\gamma - 1}$ (J/mol-K)	$C_V = \dfrac{R}{\gamma - 1}$ (J/mol-K)
He	Monatomic	1.67	20.723	12.409
N_2	Diatomic	1.4	29.099	20.785
CO_2	Polyatomic	1.3	36.027	27.713

EXAMPLE 2.9   In an insulated vessel 1 kg of water ($C_V = 4.78$ kJ/kg-K) is stirred by a mass of 40 kg falling through 25 m. Calculate the temperature rise of water. Assume $g = 9.81$ m/s^2.

Solution: Since the system is thermally insulated, $Q = 0$.

From the first law of thermodynamics, we have
$$dQ = dE + dW$$

or

$$= dU + dW \quad (\text{as } E = U + E_K + E_P \text{ and } E_K = E_P = 0 \text{ for a pure substance})$$

or

$$0 = mC_V dT + (-mgh) \quad [\text{work-done on the system}]$$

$$\Delta T = \frac{mgh}{mC_V} = \frac{9.81 \times 25}{4.78} = 51.30 \text{ K}$$

Therefore, the rise in temperature of water is 51.30 K.

2.12 ADIABATIC PROCESS OR ISOCALORIC PROCESS

An adiabatic process is defined as one in which there is no heat transfer or thermal communication between the system and the surroundings. A good example is the sudden bursting of a cycle tube.

The term *adiabatic* has been derived from the Greek word *adiabators*; it means 'not to be passed'. The adiabatic process does not only mean that the process takes place in an insulated container and there is no scope of transferring heat from the system to the surroundings or vice versa, but it does also mean that the system and the surroundings are of the same temperature. Therefore, no temperature gradient exists between them for heat transfer.

Sometimes, a little confusion is created about the difference between adiabatic and isothermal processes. In case of an adiabatic process, the heat content and temperature of the system can be changed to a small extent due to improper insulation, but for isothermal process the temperature can never be changed by any means.

An adiabatic change process, represented in Fig. 2.8, can be explained by the first law of thermodynamics.

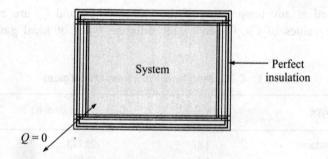

Fig. 2.8 Illustration of an adiabatic system.

Since the system neither receives nor gives out heat to the surroundings, i.e., $dQ = 0$, then from the first law of thermodynamics we have

$$dQ = dU + dW$$

or

$$0 = dU + dW$$

or

$$dU = -dW \tag{2.22}$$

The preceding equation implies that the work done on the system will cause an increase in internal energy and the work done by the system will be at the expense of the internal energy. Since the internal energy is a function of temperature, i.e., $U = U(T)$, the decrease in internal energy results in the lowering of temperature. During the adiabatic expansion of a gas, the system performs work and the final temperature is less than the initial temperature. On the other hand, during compression, work is done on the system and the final temperature increases.

2.13 REVERSIBLE ADIABATIC EXPANSION OF IDEAL GAS

Suppose that 1 mol of an ideal gas enclosed in a vessel is allowed to expand reversibly and adiabatically. Let the initial state of the gas be P_1, V_1, T_1 and the final state be P_2, V_2, T_2. Now, we can calculate the thermodynamic quantities such as work done and internal energy, and can deduce the different relationships between pressure (P), volume (V) and temperature (T).

2.13.1 Adiabatic Work Done (W_{ad})

From the first law of thermodynamics, we have

$$dQ = dU + dW$$

Since the expansion is adiabatic, $dQ = 0$.

On putting this into the preceding equation, it yields

$$dW = -dU$$
$$W_{ad} = dW = -C_V dT = -C_V(T_2 - T_1)$$
$$= C_V(T_1 - T_2) \tag{2.23}$$

For 1 mol of an ideal gas

$$T = \frac{PV}{R}$$

Hence

$$W_{ad} = C_V\left(\frac{P_1 V_1}{R} - \frac{P_2 V_2}{R}\right) = \frac{C_V}{R}(P_1 V_1 - P_2 V_2) \tag{2.24}$$

$$= \frac{P_1 V_1 - P_2 V_2}{\gamma - 1} = \frac{P_2 V_2 - P_1 V_1}{1 - \gamma} \tag{2.25}$$

$$= \frac{R(T_2 - T_1)}{1 - \gamma} \tag{2.26}$$

2.13.2 Internal Energy (U)

The internal energy of the system can be calculated very easily if it is considered to be a function of temperature and volume. Hence

$$U = U(T, V)$$

On partial differentiation

$$\partial U = \left(\frac{\partial U}{\partial T}\right)_V dT + \left(\frac{\partial U}{\partial V}\right)_T dV$$

We know that for an ideal gas

$$\left(\frac{\partial U}{\partial V}\right) = 0$$

Therefore, putting this condition into the preceding equation, we get

$$\partial U = C_V dT \qquad \text{(since by definition } C_V = \left(\frac{\partial U}{\partial T}\right)_V \text{)}$$

Now

$$\Delta U = \int_1^2 \partial U = U_2 - U_1 = \int_{T_1}^{T_2} C_V \, dT$$

or

$$\Delta U = C_V(T_2 - T_1) \tag{2.27}$$

2.13.3 Enthalpy (*H*)

From the definition of enthalpy (*H*)

$$H = U + PV$$

Considering the initial and final states, it can be written as

$$
\begin{aligned}
\Delta H = H_2 - H_1 &= (U_2 + P_2 V_2) - (U_1 + P_1 V_1) \\
&= (U_2 - U_1) + (P_2 V_2 - P_1 V_1) \\
&= \Delta U + R(T_2 - T_1) \\
&= C_V(T_2 - T_1) + R(T_2 - T_1) \\
&= (C_V + R)(T_2 - T_1) \qquad \text{(as } C_P - C_V = R \text{ or } C_P = C_V + R) \tag{2.28}
\end{aligned}
$$

2.13.4 Relation between Temperature and Volume

Under adiabatic condition, the first law of thermodynamics yields

$$dU = -dW = -PdV \tag{i}$$

Again, for ideal gas

$$dU = C_V dT \tag{ii}$$

Comparing Eqs. (i) and (ii), we have

$$C_V dT = -PdV \tag{iii}$$

For 1 mol of an ideal gas

$$PV = RT \qquad \text{or} \qquad P = \frac{RT}{V}$$

Putting the value of *P* in Eq. (iii)

$$C_V dT = -\frac{RT}{V} dV \tag{iv}$$

or

or

$$C_V \int_{T_1}^{T_2} \frac{dT}{T} = -R \int_{V_1}^{V_2} \frac{dV}{V} \qquad \text{(v)}$$

On integration, we get

$$C_V \ln \frac{T_2}{T_1} = -R \ln \frac{V_2}{V_1} = R \ln \frac{V_1}{V_2} \qquad \text{(vi)}$$

or

$$\ln \frac{T_2}{T_1} = \frac{R}{C_V} \ln \frac{V_1}{V_2}$$

$$= \frac{C_P - C_V}{C_V} \ln \frac{V_1}{V_2} \qquad \left(\text{as } C_P - C_V = R \text{ and } \frac{C_P}{C_V} = \gamma = \text{Heat capacity ratio}\right)$$

or

$$\frac{T_2}{T_1} = \left(\frac{V_1}{V_2}\right)^{\gamma-1} \qquad (2.29)$$

or

$$T_2 V_2^{\gamma-1} = T_1 V_1^{\gamma-1} = \text{Constant} \qquad (2.30)$$

It can be generalized as

$$TV^{\gamma-1} = \text{Constant}$$

2.13.5 Relation between Pressure and Volume

Applying ideal gas law, T (temperature) of the above equation can be replaced by PV/R by putting this value in the preceding equation as

$$\frac{PV}{R} V^{\gamma-1} = \text{Constant}$$

or

$$PV^{\gamma} = R \times \text{Constant} = \text{Constant}$$

or

$$PV^{\gamma} = \text{Constant} \qquad (2.31)$$

2.13.6 Relation between Temperature and Pressure

Again using ideal gas law, V of the preceding equation can be replaced by $\dfrac{RT}{P}$, i.e.

$$V = \frac{RT}{P}$$

Putting the value in the previous equation, we get

$$TV^{\gamma-1} = \text{Constant}$$

or

$$T\left(\frac{RT}{P}\right)^{\gamma-1} = \text{Constant}$$

or

$$T^{\gamma}\, P^{1-\gamma} = \frac{\text{Constant}}{R^{\gamma-1}} = \text{Constant}$$

or

$$T^{\gamma}\, P^{1-\gamma} = \text{Constant} \tag{2.32}$$

EXAMPLE 2.10 An ideal gas is compressed adiabatically and reversibly in a piston–cylinder assembly from 30 L to 3 L at 300 K. Calculate the final temperature, given that heat capacity at constant volume, $C_V = 5$ cal/mol.

Solution: The relation between temperature and volume of an ideal gas undergoing adiabatic change is given by

$$T_1 V_1^{\gamma-1} = T_2 V_2^{\gamma-1} = \text{Constant}$$

$$\frac{T_2}{T_1} = \left(\frac{V_1}{V_2}\right)^{\gamma-1}$$

Here

$$T_2 = ?$$
$$T_1 = 300 \text{ K}$$
$$V_1 = 30 \text{ L}$$
$$V_2 = 3 \text{ L}$$

and

$$\gamma = \frac{C_P}{C_V} = \frac{C_V + R}{C_V} = \frac{5+2}{5} = 1.4$$

(since $R = 1.987$ cal/°K $\simeq 2$ cal/°K)

Substituting these values, we get

$$\frac{T_2}{300} = \left(\frac{30}{3}\right)^{\gamma-1} = (10)^{1.4-1}$$

$$T_2 = 753.6 \text{ K}$$

EXAMPLE 2.11 2 mol of an ideal gas was initially at 293 K and 15 atm. The expansion of gas takes place adiabatically when the external pressure is reduced to 5 atm. What will be the final temperature and volume? Also calculate the work done during the process, given that $C_P = 8.58$ cal/mol/degree.

Solution: We are given that $C_P = 8.58$ cal/mol/degree.
Since the gas is behaving ideally, then $R = 1.987$ cal/degree-mol ≈ 2 cal/degree-mol

Putting the values of C_P and R in the following relation, we get

$$C_P - C_V = R$$

or
$$C_V = C_P - R = 8.58 - 2 = 6.58 \text{ cal/degree-mol}$$

Therefore

$$\gamma = \text{heat capacity ratio} = \frac{C_P}{C_V} = \frac{8.58}{6.58} = 1.3$$

Since the gas is ideal, it obeys the ideal gas law, i.e., $PV = nRT$.
Substituting the value

$$P_1 = 15 \text{ atms}$$
$$n = 2$$
$$R = 0.082 \text{ L-atm/degree-K}$$
$$T = 293 \text{ K}$$

in the equation

$$PV = nRT$$

we get

$$V_1 = \frac{nRT}{P_1} = \frac{2 \times 0.082 \times 293}{15} = 3.2 \text{ L}$$

Hence, under adiabatic conditions

$$PV^\gamma = \text{Constant}$$

or

$$P_1 V_1^\gamma = P_2 V_2^\gamma$$

or

$$\left(\frac{V_2}{V_1}\right) = \left(\frac{P_1}{P_2}\right)^{\frac{1}{\gamma}}$$

or

$$\left(\frac{V_2}{3.2}\right) = \left(\frac{15}{5}\right)^{\frac{1}{1.3}}$$

or

$$V_2 = 7.45 \text{ L}$$

Hence, the final volume of the system is 7.45 L.

Now, to determine the final temperature, we have

$$\frac{T_2}{T_1} = \left(\frac{V_1}{V_2}\right)^{\gamma - 1}$$

or

$$\left(\frac{T_2}{293}\right) = \left(\frac{3.2}{7.45}\right)^{1.3 - 1}$$

or

$$T_2 = 227.3 \text{ K} = -46°C$$

Adiabatic work done can be calculated as

$$W = \frac{P_1V_1 - P_2V_2}{\gamma - 1} = \frac{15 \times 3.2 - 5 \times 7.45}{1.3 - 1}$$
$$= 35.83 \text{ L-atm}$$

2.14   CONSTANT-VOLUME (ISOCHORIC) PROCESS

This process is also termed *isovolumetric process* as the volume of the system remains constant during this process. The term *isochoric* has been derived from a Greek word, in which *iso* means equal and *chor* means place. If there is no change in the volume of the system, then the change in other thermodynamic parameters can be calculated with the help of the first law of thermodynamics.

For a common example, a gas is heated in a rigid vessel. Then the pressure and temperature of the gas will increase, but there will be no change in volume.

From the first law of thermodynamics, we have

$$dQ = dU + dW$$
$$dQ = dU + PdV \qquad (2.33)$$

At constant volume, i.e., where V = constant

$$dV = 0$$

Putting the condition in Eq. (2.33), we have

$$dQ = dU$$

For n moles of gas

$$dQ = ndU \qquad (2.34)$$

Hence, the heat absorbed by the system at constant volume goes completely into increasing the internal energy of the system.

2.15   CONSTANT-PRESSURE (ISOBARIC) PROCESS

This is a thermodynamic process in which the pressure of the gas remains constant, i.e., P = constant or $dP = 0$. Suppose that in a piston–cylinder assembly a certain amount of fluid is taken in the cylinder. If an arrangement is made for heating the gas from an external source, then the volume of gas will go on increasing under constant pressure.

Now, on the basis of the above process, the change in enthalpy can be calculated in the following way:

Considering the system under two different states

1. At the initial state

$$H_1 = U_1 + P_1V_1$$

2. At the final state

$$H_2 = U_2 + P_2V_2$$

Hence, change in enthalpy

$$
\begin{aligned}
\Delta H &= H_2 - H_1 \\
&= (U_2 + P_2V_2) - (U_1 + P_1V) \\
&= (U_2 - U_1) + (P_2V_2 - P_1V_1) \\
&= \Delta U + \Delta PV
\end{aligned}
\tag{2.35}
$$

Since the term *enthalpy* is conveniently used to express the process under constant pressure, the above equation can be written as

$$
\begin{aligned}
\Delta H &= \Delta U + P\Delta V \\
&= \Delta U + \Delta W \\
&= \Delta Q
\end{aligned}
$$

Therefore

$$\Delta H = \Delta Q_P \tag{2.36}$$

Hence, the enthalpy change of a system is equal to the change in heat absorbed at constant pressure.

2.16 CONSTANT-TEMPERATURE (ISOTHERMAL) PROCESS

The constant temperature or isothermal process is defined as the process in which the temperature of the system remains constant, i.e., a temperature constancy is maintained throughout the operation. If the gas is allowed to expand at constant temperature, the pressure will decrease during expansion. Hence, if an ideal gas undergoes an expansion at constant temperature, the expansion process is termed *isothermal expansion*, and it will be accompanied by no change in internal energy.

Since the internal energy of an ideal gas is a function only of temperature, i.e., $U = U(T)$, then the imposition of constant temperature implies the constancy of internal energy, i.e. $\Delta U = 0$. Under such conditions, the change in other thermodynamic parameters such as work done, internal energy, and heat interaction can be calculated from the knowledge of the first law of thermodynamics.

From first law of thermodynamics, we have

$$
\begin{aligned}
dQ &= dU + dW \\
dQ &= C_V dT + PdV
\end{aligned}
\tag{2.37}
$$

At constant temperature, i.e., when temperature (T) = constant

$$dT = 0$$

Putting the condition in Eq. (2.37), we have

$$dQ = dW = PdV$$

$$Q = W = \int_1^2 PdV = RT \ln \frac{V_2}{V_1} \qquad \left(\text{as } P = \frac{RT}{V}\right)$$

For n moles of an ideal gas

$$Q = W = nRT \ln \frac{V_2}{V_1} = nRT \ln \frac{P_1}{P_2} \qquad (2.38)$$

Hence, the characteristics of an isothermal process for an ideal gas can be summarized as follows:

- Change in internal energy, $\Delta U = 0$
- Change in enthalpy, $\Delta H = 0$
- $Q = W = nRT \ln \dfrac{V_2}{V_1} = nRT \ln \dfrac{P_1}{P_2}$.

EXAMPLE 2.12 1 kg of air at 50°C expands reversibly and adiabatically to 5 times its original volume. The initial pressure of the air mass was 8 atm. Determine the final pressure, temperature, and work done when the expansion is (i) adiabatic and (ii) isothermal, given that the heat capacity ratio = $\gamma = 1.4$.

Solution: (i) Adiabatic Process

Since the expansion takes place adiabatically, $dQ = 0$.

Final Pressure

Let the original volume be V; then the final volume would be 5 V.

Now, the relation between volume and pressure of a gas undergoing adiabatic process

$$P_1 V_1^{\gamma} = P_2 V_2^{\gamma}$$

$$\frac{P_2}{P_1} = \left(\frac{V_1}{V_2}\right)^{\gamma}$$

$$P_2 = P_1 \left(\frac{V_1}{V_2}\right)^{\gamma} = 8\left(\frac{1}{5}\right)^{1.4} = 0.84 \text{ atm}$$

Hence the final pressure is 0.84 atm.

Final Temperature

From the relation between the temperature and the volume of a gas undergoing an adiabatic process, we have

$$\frac{T_2}{T_1} = \left(\frac{V_1}{V_2}\right)^{\gamma-1}$$

or

$$\frac{T_2}{323} = \left(\frac{1}{5}\right)^{1.4-1}$$

$$T_2 = 170 \text{ K}$$

Hence, the final temperature is 170 K.

Work done

This is given by

$$W_{ad} = \frac{R(T_2 - T_1)}{1 - \gamma} = \frac{0.082(170 - 323)}{1 - 1.4} \quad \text{(since } R = 0.082 \text{ L-atm)}$$

$$= 31.365 \text{ L-atm}$$

(ii) **Isothermal Process**

In an isothermal process, the temperature remains constant, or

$$T_1 = T_2 = T = 323 \text{ K}$$

Final Pressure

From the ideal gas law, we have

$$\frac{P_2 V_2}{T_2} = \frac{P_1 V_1}{T_1}$$

or

$$P_2 V_2 = P_1 V_1 \text{ (as } T_1 = T_2 = T)$$

or

$$P_2 = \frac{P_1 V_1}{V_2} = \frac{8 \times 1}{5} = 1.6 \text{ atm}$$

Work done

In an isothermal process, the work done by the system is given by

$$W = RT_1 \ln \frac{V_2}{V_1} = 0.082 \times 323 \ln 5 = 42.62 \text{ L-atm}$$

EXAMPLE 2.13   5 kg of air is heated from an initial state of 37°C and 101.33 kPa until its temperature reaches 237°C. Calculate ΔU, Q, W, and ΔH for the following processes:

(i)   isochoric process
(ii)   isobaric process.

Air is assumed to be an ideal gas. We are given that

$$C_P = 29.1 \text{ J/mol-K}$$
$$C_V = 20.78 \text{ J/mol-K}$$

Molecular weight of air = 29.

Solution: Since air is assumed to behave ideally, the ideal gas law $PV = nRT$ is applicable to it, and the number of moles of air can be calculated by the relation

$$PV = nRT = \frac{m}{M} RT$$

where

m = Mass of the gas
M = molecular weight.

$$\text{Number of moles} = n = \frac{m}{M} = \frac{5}{29} = 0.1724 \text{ kmol}$$

Volume of air at initial state

$$V_1 = \frac{nR_1T_1}{P_1} = \frac{0.1724 \times 8.314 \times 310}{101.33} = 4.385 \text{ m}^3$$

(a) Isochoric Process

The process is an isochoric one, i.e., volume V = constant. Therefore

$$V_1 = V_2 = 4.385 \text{ m}^3 \quad \text{and} \quad T_2 = 237°C = 510 \text{ K}$$

(i) *Change in internal energy* (ΔU)

$$\begin{aligned}
\Delta U &= nC_V(T_2 - T_1) \\
&= 0.1724 \times 20.785 \times (510 - 310) \\
&= 716.66 \text{ kJ}
\end{aligned}$$

(ii) *Heat Supplied* (Q)

$$\begin{aligned}
Q &= \Delta U + W \\
&= \Delta U + PdV \\
&= \Delta U \quad \text{(as } V = \text{Constant)} \\
&= 716.66 \text{ kJ}
\end{aligned}$$

(iii) *Work done* (W)

$$W = Q - \Delta U = 0$$

(iv) *Change in enthalpy* (ΔH)

$$\begin{aligned}
\Delta H &= \Delta U + P\Delta V \\
&= \Delta U + nR\Delta T \\
&= \Delta U + nR(T_2 - T_1) \\
&= 716.66 + 0.1724 \times 8.314 (510 - 310] \\
&= 1003.32 \text{ kJ}
\end{aligned}$$

(b) Isobaric Process

The process is a constant pressure process, so P is constant.
So

$$P_1 = P_2 = 101.33 \text{ kPa} \qquad T_2 = 510 \text{ K}$$

(i) *Change in enthalpy* (ΔH)

$$\begin{aligned}
\Delta H &= nC_P(T_2 - T_1) \\
&= 0.1724 \times 29.1 \times (510 - 310) \\
&= 1003.36 \text{ kJ}
\end{aligned}$$

(ii) *Heat supplied at constant P* (Q)

$$Q_P = \Delta H_P = 1003.36 \text{ kJ}$$

(iii) *Change in internal energy* (ΔU)

$$\Delta U = \Delta H - P\Delta V$$

$$= \Delta H - nR\Delta T$$
$$= 1003.36 - 0.1724 \times 8.314 \, (510 - 310)$$
$$= 716.69 \text{ kJ}$$

(iv)   *Work done* (W)

$$W = Q - \Delta U$$
$$= 1003.36 - 716.69 \text{ kJ} = 286.67 \text{ kJ}$$

EXAMPLE 2.14   One mol of an ideal gas, used as a working substance in a Carnot cycle, operates initially at 610 K and 10^6 N/m^2 in the compression stage. The gas then expands isothermally to a pressure of 10^5 N/m^2 and adiabatically at 310 K. Determine ΔE, Q and W for each step, and the net work done and the efficiency of the cycle, given that the heat capacity of the gas at constant volume, $C_V = 20.78$ J/mol-K.

Solution: The data given are

$$T_1 = 610 \text{ K}$$
$$T_2 = 310 \text{ K}$$
$$P_1 = 10^6 \text{ N/m}^2$$
$$P_2 = 10^5 \text{ N/m}^2$$

Step I: *Isothermal expansion of ideal gas*
Here

$$P_1 = 10^6 \text{ N/m}^2$$
$$P_2 = 10^5 \text{ N/m}^2$$
$$T_1 = 610 \text{ K}$$

Work done

$$W_1 = 2.303 \, nRT \log \frac{P_1}{P_2}$$

$$= 2.303 \times 1 \times 8.314 \text{ J/K-mol} \times 610 \text{ K} \times \log \frac{10^6 \text{ N/m}^2}{10^5 \text{ N/m}^2}$$

$$= 11{,}679.75 \text{ J/mol}$$

We know that the isothermal expansion of an ideal gas is $\Delta E_1 = 0$.
Therefore, from the mathematical expression of the first law of thermodynamics, we get

$$dQ = dU + dW$$
$$\Delta Q = \Delta E + W$$

The work of expansion is equal to the amount of heat absorbed, i.e.,

$$W_1 = Q_2 = 11{,}679.75 \text{ J/mol}$$

Step II: *Adiabatic expansion of ideal gas*
Since the expansion of gas takes place adiabatically, therefore $Q = 0$.
   Then, from the first law of thermodynamics, we have

$$\Delta E_2 = C_V \, (T_2 - T_1) = 20.78 \, (310 - 610) = -6234 \text{ J/mol} = W_2$$

Step III: *Isothermal compression of ideal gas*
Here

$$T_2 = 310 \text{ K}$$
$$P_1 = 10^5 \text{ N/m}^2$$
$$P_2 = 10^6 \text{ N/m}^2$$

Since the ideal gas undergoes an isothermal process, therefore,

$$\Delta E_3 = 0$$

Work done

$$= W_3 = 2.303 \, nRT_2 \log \frac{P_1}{P_2}$$

$$= 2.303 \times 1 \times 8.314 \text{ J/K-mol} \times 310 \times \log \frac{10^5 \, \text{N/m}^2}{10^6 \, \text{N/m}^2}$$

$$= -5935.6 \text{ J/mol}$$

$$Q = \text{Amount of heat rejected by the system to the surroundings}$$
$$= W = 5935.6 \text{ J/mol}$$

Step IV: Adiabatic compression of gas
Here

$$T_1 = 310 \text{ K}$$
$$T_2 = 610 \text{ K}$$
$$Q = 0$$
$$\Delta E_4 = C_V (T_2 - T_1) = 20.78 \, (610 - 310) \text{ K} = 6234 \text{ J/mol} = W_4$$

Net work done for a complete cycle

$$W = W_1 + W_2 + W_3 + W_4$$
$$= 11679.75 - 6234 - 5935.6 + 6234$$
$$= 5744.15 \text{ J/mol}$$

The efficiency of the cycle is given by

$$\eta = 1 - \frac{T_1}{T_2} = 1 - \frac{310}{610} = 0.492 \approx 0.5$$

EXAMPLE 2.15 Show that for an ideal gas the amount of work done by reversible isothermal expansion is always greater than that by irreversible isothermal expansion.

Solution: We know that for an ideal gas the work done by reversible isothermal expansion

$$W_{\text{rev}} = nRT \ln \frac{v_2}{v_1} = nRT \ln \frac{P_1}{P_2}$$

The relation can be represented as

$$W_{\text{rev}} = nRT \ln \frac{P_1}{P_2} \qquad W_{\text{rev}} > W_{\text{irrev}}$$

For a negligible change, we can write here

$$W_{rev} = -nRT \ln\left(1 - \frac{P_1}{P_2}\right) \qquad \text{(since } \ln(1 - x) \approx -x\text{)}$$

or

$$W_{rev} = nRT\left(\frac{P_1}{P_2} - 1\right)$$

Now, the work is done by irreversible isothermal expansion of an ideal gas. We have

$$W_{irrev} = nRT\left(1 - \frac{P_2}{P_1}\right)$$

Combining the two equations obtained for work done by reversible and irreversible expansion, we get

$$W_{rev} - W_{irrev} = nRT\left(\frac{P_1}{P_2} - 1\right) - nRT\left(1 - \frac{P_2}{P_1}\right)$$

or

$$W_{rev} - W_{irrev} = nRT\left(\frac{P_1}{P_2} + \frac{P_2}{P_1} - 2\right)$$

$$= \frac{nRT}{P_1 P_2}(P_1 - P_2)^2$$

The L.H.S. and R.H.S. of the preceding equation are of positive quantity. Therefore $W_{rev} > W_{irrev}$.

EXAMPLE 2.16 In a frictionless piston–cylinder arrangement, an ideal gas undergoes a compression process from an initial state of 1 bar at 300 K to 10 bars at 300 K. The entire process comprises the following two mechanically reversible processes:

(a) Cooling at constant pressure followed by heating at constant volume
(b) Heating at constant volume followed by cooling at constant pressure.

Calculate Q, ΔU and ΔH for the two processes, given that $C_P = 29.10$ kJ/kmol-K, $C_V = 20.78$ kJ/Kmol-K and volume of the gas at initial state equal to 24.942 m³/kmol.

Solution: Basis: 1 kmol of ideal gas
Initial state: $P_1 = 1$ bar, $T_1 = 300$ K, $V_1 = 24.942$ m³/kmol
Final state: $P_2 = 10$ bar, $T_2 = 300$ K, $V_2 = ?$
The final volume of the gas can be calculated as

$$V_2 = \frac{P_1 V_1}{P_2} = \frac{1 \times 24.942}{10} = 2.494 \text{ m}^3$$

(a) In the first step of the first process, the cooling of gas takes place at constant pressure. Here the volume is reduced appreciably and consequently the temperature decreases as, say T', which is

$$T' = \frac{T_1 V_2}{V_1} = \frac{300 \times 2.494}{24.942} = 30 \text{ K}$$

(i) Heat requirement (Q_P):
Since the process is a constant-pressure process,

$$Q_P = \Delta H_P = C_P \Delta T = C_P(T_2 - T_1) = 29.10 (30 - 300) = -7857 \text{ kJ}$$

(ii) Change in internal energy (ΔU_P):

$$\Delta U = \Delta H - \Delta(PV) = \Delta H - P\Delta V$$
$$= \Delta H - P(V_2 - V_1)$$
$$= -7857 - [(1 \times 100) \text{ kN/m}^2] [2.494 - 24.942) \text{ m}^3]$$
$$= -5612 \text{ kN-m} = -5612 \text{ kJ}$$

[Since, 1 bar = 10^5 Pa = 100 kPa = 100 kN/m^2, 1 kN-m = 1 kJ]

In the second step, the gas is heated at constant volume, i.e.

$$V = \text{constant}$$
$$dV = 0$$

Then from the first law of thermodynamics, we have

$$dQ = dU$$

(i) Heat requirement (Q_V):

$$Q_V = \Delta U_V = C_V \Delta T = C_V(T_2 - T_1) = 20.78 (300\text{-}30) = 5612 \text{ kJ}$$

(ii) Change in internal energy (ΔU_V):

$$\Delta U_V = 5612 \text{ kJ}$$

Since the process comprises two steps, therefore the actual Q and ΔH can be obtained by summing up the respective values for two different steps. Therefore

$$Q = Q_P + Q_V = -7857 + 5612 = -2245 \text{ kJ}$$
$$\Delta U = -5612 \text{ kJ} + 5612 \text{ kJ} = 0$$

(b) In the first step of the second process, the gas is heated at constant volume, i.e., $V =$ constant or $dV = 0$.
The temperature of the air, say T', at the end of the first step can be calculated as

$$T' = \frac{T_1 P_2}{P_1} = \frac{300 \times 10}{1} = 3000 \text{ K}$$

(i) Heat requirement (Q_V):

$$Q_V = \Delta U_V = C_V \Delta T = C_V(T_2 - T_1) = 20.78 (3000 - 300) = 56108 \text{ kJ}$$

(ii) Change in internal energy (ΔU_V):

$$\Delta U_V = 56108 \text{ kJ}$$

In the second step, the gas is cooled at constant pressure of 10 bar.

(i) Heat requirement (Q_P):

$$Q_P = \Delta H_P = C_P \Delta T = C_P(T_2 - T_1) = 29.10 (300 - 3000) = -78570 \text{ kJ}$$

(ii) Change in internal energy (ΔU_P):

$$\Delta U = \Delta H - \Delta(PV) = \Delta H - P\Delta V$$
$$= \Delta H - P(V_2 - V_1)$$
$$= -78570 - [(10 \times 100) \text{ kN/m}^2] [(2.494 - 2.494 - 24.942) \text{ m}^3] = -56122 \text{ kJ}$$

For the two steps, the above quantities can be summed up as

$$Q = Q_P + Q_V = 56108 - 78570 = -22462 \text{ kJ}$$
$$\Delta U = 56108 \text{ kJ} - 56122 \text{ kJ} = -14 \text{ kJ}$$

EXAMPLE 2.17 A spherical balloon of 1 m diameter contains a gas at 120 kPa. The gas inside the balloon is heated until the pressure reaches 360 kPa. During heating the pressure of the gas inside the balloon is proportional to the cube of the diameter of the balloon. Determine the work done by the gas inside the balloon.

Solution: The pressure P of the gas inside the balloon is proportional to the cube of the diameter of the balloon, i.e.

$$P = kD^3$$

where k = proportionality constant.

Now, work done by the gas inside the balloon can be estimated as

$$W = \int P dV = \int kD^3 \cdot d\left(\frac{4}{3}\pi r^3\right)$$

$$= \int kD^3 \cdot d\left(\frac{4}{3}\pi \frac{D^3}{8}\right)$$

$$= \frac{\pi k}{6} \int_{D_1}^{D_2} D^3 d\,(D^3)$$

$$= \frac{\pi k}{12}(D_2^6 - D_1^6)$$

We are given that $P_1 = kD_1^3$ or $k = \dfrac{P_1}{D_1^3} = \dfrac{120 \times 10^3}{1} = 120 \times 10^3$.

Now, we can calculate the final pressure. It is given by

$$P_2 = kD_2^3$$

or

$$360 \times 10^3 = 120 \times 10^3 \times D_2^3$$

or

$$D_2 = 1.442 \text{ m}$$
$$= \text{Diameter of the balloon at the final pressure}$$

Hence, work done by the gas would be

$$W = \frac{120 \times 10^3}{1} \times \frac{\pi}{2} [(1.442^6 - 1)] = 251013.25 \text{ J} = 251.01 \text{ kJ}$$

EXAMPLE 2.18 1 kmol of argon gas confined in a cylinder undergoes a change from an initial condition of 10 bar and 250 K to a final condition of 1 bar and 300 K. The gas follows the equation $PV = RT$. Given that $C_P = 29.10$ kJ/kmol-K and $C_V = 20.78$ kJ/kmol-K, determine the changes in internal energy, enthalpy, heat and work requirement.

Solution: It is well-known that the changes in state variables do not depend on the process or path by which the change is brought about. Therefore, with obvious reason, we can consider the change to be taken place in the following two mechanically reversible processes:

  (i) a constant-volume process in which gas is cooled at final pressure and low temperature; and
  (ii) a constant-pressure process in which gas is heated at final temperature and volume.

Before going to the estimation of change in internal energy and enthalpy, it is necessary to calculate the volume of argon gas at two different states.

Initial condition: $P_1 = 10$ bar and $T_1 = 250$ K
Final condition: $P_2 = 1$ bar and $T_2 = 300$ K

Therefore, the initial volume

$$V_1 = \frac{RT_1}{P_1} = \frac{8.314 \times 250}{1000} = 2.078 \text{ m}^3$$

and the final volume

$$V_2 = \frac{RT_2}{P_2} = \frac{8.314 \times 300}{100} = 24.942 \text{ m}^3$$

Calculation based on first-step process:

In this constant-volume process, the initial pressure of 10 bar is reduced to a final pressure of 1 bar, and consequently the temperature decreases. This temperature is supposed to be T' and it can be determined as

$$T' = \frac{P_2 V_1}{R} = \frac{100 \times 2.078}{8.314} = 25 \text{ K}$$

  (i) Change in internal energy:

$$\Delta U_V = C_V \Delta T = C_V(T - T_1) = 20.78\,(25 - 250) = -4675.5 \text{ kJ}$$

  (ii) Change in enthalpy:

$$\Delta H = \Delta U + \Delta(PV) = \Delta U + V\Delta P$$
$$= -4675.5 + (100 - 1000) = -5575.5 \text{ kJ}$$

  (iii) Work required:
    Since the process is a constant-volume process, V = constant or $dV = 0$. Therefore

$$W = P\int dV = 0$$

  (iv) Heat requirement:
    Putting $W = 0$ into the first law of thermodynamics, we have

$$Q = \Delta U = -4675.5 \text{ kJ}$$

Calculation based on second-step process:

In this process, the gas is heated at constant pressure to the final temperature of $T_2 = 300$ K

(i) Change in enthalpy:

$$\Delta H_P = C_P \Delta T = C_P (T_2 - T_1) = 29.1(300 - 25) = 8002.5 \text{ kJ}$$

(ii) Change in internal energy:

$$\Delta U = \Delta H - \Delta(PV) = \Delta H - P\Delta V = 8002.5(24.942 - 2.078) = -5716.1 \text{ kJ}$$

(iii) Heat requirement:

$$Q = \Delta H = 8002.5 \text{ kJ}$$

(iv) Work required:

From the first law, we have

$$W = Q - \Delta U = 8002.5 - 5716.1 = 2286.4 \text{ kJ}$$

For the above two-step processes, ΔU and ΔH can be obtained as

$$\Delta U = -4675.5 + 5716.1 = 1040.6 \text{ kJ}$$
$$\Delta H = -5575.5 + 8002.5 = 2427 \text{ kJ}$$

2.17 MASS BALANCE FOR OPEN SYSTEM

In order to prepare a mass balance for an open system, the region in Fig. 2.9 is identified only for the analysis of mass interaction. The first law of thermodynamics can be applied to establish a mathematical expression on mass balance.

It is very much clear from Fig. 2.9 that the two fluid input streams m_1 and m_2 enter the control volume and one fluid output stream m_3 leaves the control volume.

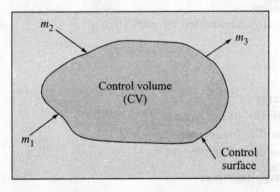

Fig. 2.9 Control volume for the analysis of mass transfer.

On the basis of the sign convention used in this book, the mass balance equation can be represented as

Rate of input − Rate of output = Rate of accumulation

It can be mathematically expressed as

$$(\dot{m}_1 + \dot{m}_2) - \dot{m}_3 = \frac{dm_{CV}}{dt}$$

or

$$\frac{dm_{CV}}{dt} + (\dot{m}_3 - \dot{m}_1 - \dot{m}_2) = 0$$

or

$$\frac{dm_{CV}}{dt} + [\dot{m}_3 - (\dot{m}_1 + \dot{m}_2)] = 0$$

or

$$\frac{dm_{CV}}{dt} + (\dot{m}_{out} - \dot{m}_{in}) = 0 \tag{2.39}$$

In case of consideration of the number of inlets and outlets, summation signs are used. Then Eq. (2.39) can be expressed as

$$\frac{dm_{CV}}{dt} + \sum \dot{m}_{out} - \sum \dot{m}_{in} = 0 \tag{2.40}$$

2.18 ENERGY BALANCE FOR OPEN SYSTEM (OR FIRST LAW OF THERMODYNAMICS TO FLOW PROCESSES)

Like mass balance of an open system based on the law of conservation of mass, the energy balance can also be established for an open system on the basis of the law of conservation of energy.

Consider the fluid stream is flowing into and out of the control volume shown in Fig. 2.10. The total energy associated with transferring of fluid mass comprises internal energy (U), kinetic energy and potential energy.

Hence, the total energy transported by each stream

$$= U + \frac{1}{2}u^2 + Zg$$

where

U = Internal energy

u = Average velocity of the stream

Z = Elevation of the flowing masses

g = Acceleration due to gravity.

So, the rate of energy transport

$$= \dot{m}\left(U + \frac{1}{2}u^2 + Zg\right)$$

It is necessary to consider the flow energy (PV) of the fluid entering or leaving a control volume. This additional form of energy is required to push the fluid mass into or out of the control volume.

This flow energy comes from the work done to induce flow of the fluid. This work is known as *Flow work*.

It is clear from Fig. 2.10 that the amount of heat $\dot{Q}$ is transferred to the control volume and the work done by the fluid is $\dot{W}$.

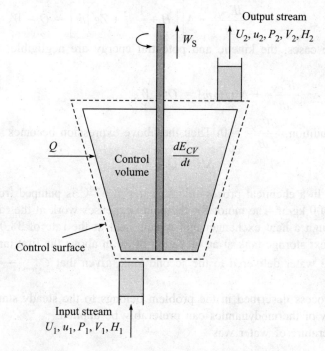

Fig. 2.10   Illustration on energy balance of open system.

So the rate of energy storage within the control volume is given by

$$\frac{dE_{CV}}{dt} = \Delta\left[\left(U + \frac{1}{2}u^2 + Zg\right)\dot{m}\right] + \dot{Q} - \dot{W} \qquad (2.41)$$

The flow work done by the control volume

$$= (PV)_{out} \cdot (\dot{m}_{out})$$

The flow work done on the control volume

$$= (PV)_{in} \cdot (\dot{m}_{in})$$

Net flow work done by the control volume

$$= [(PV)_{out} \cdot (\dot{m}_{out}) - (PV)_{in} \cdot (\dot{m}_{in})]$$
$$= \Delta[(PV)\dot{m}]$$

The net work done by the control volume

$$\dot{W} = \text{Shaft work} + \text{Flow work}$$
$$= W_S + \Delta[(PV)\dot{m}]$$

Applying the first law of thermodynamics, we have

$$\frac{dE_{CV}}{dt} + \Delta\left[\left(U + \frac{1}{2}u^2 + Zg\right)\dot{m}\right] + \Delta[(PV)\dot{m}] = \dot{Q} - \dot{W}_S$$

or

$$\frac{dE_{CV}}{dt} + \Delta\left[\left(H + \frac{1}{2}u^2 + Zg\right)\dot{m}\right] = \dot{Q} - \dot{W}_S \tag{2.42}$$

In most of the cases, the kinetic and potential energy are negligible; then the equation becomes

$$\frac{dE_{CV}}{dt} + \Delta[(H)\dot{m}] = \dot{Q} - \dot{W}_S \tag{2.43}$$

At steady state condition, $\dfrac{dE_{CV}}{dt} = 0$. Then the above expression becomes

$$\Delta H = \dot{Q} - \dot{W}_S \tag{2.44}$$

EXAMPLE 2.19 In a chemical process plant, water at 67°C is pumped from a storage tank at the rate of 20,000 kg/hr. The motor for the pump expenses work at the rate of 1.5 hp. The water passes through a heat exchanger and rejects heat at the rate of 38,000 kJ/min and is delivered to the next storage tank at an elevation of 20 m above the first tank. Determine the temperature of the water delivered to the second tank, given that $C_{P_{H_2O}}$ = 4.2 kJ/kg-K.

Solution: The process described in the problem belongs to the steady state flow process in which the first law of thermodynamics can preferably be applied.
Initially, the temperature of water was

$$T = 67°C = 340 \text{ K}$$

The mass flow rate of water

$$\dot{m} = 20,000 \text{ kg/h}$$

Heat rejected per kg of water

$$Q = \frac{38,000 \times 60}{20,000} = 114 \text{ kJ/kg}$$

As per the sign convention adopted in this book, Q will be of negative sign, i.e.,

$$Q = -114 \text{ kJ/kg}$$

Shaft work expended per kg of water pumped

$$W_S = 1.5 \text{ hp}$$
$$= 0.161 \text{ kJ/kg}$$

Here, W_S will be treated as negative quantity. So

$$W_S = -0.161 \text{ kJ/kg}$$

Potential energy

$$mg\Delta Z = 20 \times 9.81 \times 10^{-3} = 0.1962 \text{ kJ/kg}$$

Kinetic energy of water is assumed to be negligible and $\dfrac{dE_{CV}}{dt} = 0$ for steady flow process. Putting these values in Eq. (2.47), we get

$$\Delta H = Q - W_S - g\Delta Z$$
$$= -114 + 0.161 - 0.1962 = -114.03 \text{ kJ/kg}$$

or

$$H_2 = H_1 - 114.03$$

or

$$H_2 = C_P \Delta T - 114.03$$
$$= C_P(T_2 - T_1) - 114.03$$
$$= 4.2 \ (340 - 273) - 114.03$$
$$= 167.37 \text{ kJ/kg}$$

Now, the temperature of water in the second tank, say T, can be determined as

$$4.2 \ (T - 273) = 167.37$$

Solving for T, we have

$$T = 312.85 \text{ K}$$

SUMMARY

The quantitative relationship between work and heat is established with the help of demonstration of Joule's classical paddle-wheel experiment. The mechanical equivalency of heat leads to the formulation of the first law of thermodynamics. The law of conservation of energy represents the essence of first law. This law has been substantiated with the help of different statements and mathematically expressed for a system undergoing a cyclic process. Internal energy is one of the important parts of first law and a function of temperature for an ideal gas and its change is observed in Joule's experiment. Internal energy is a thermodynamic state function. The sign convention adopted for work and heat mandates that the work done by the system and the heat transferred to the system from the surroundings are considered positive quantities, and the work done on the system and the heat transferred to the surroundings from the system are taken as negative quantities. For a closed system, the first law is analyzed and the mathematical formulation is developed in different simple forms. The state of a system is better described by the state function of enthalpy, which is defined as the summation of the internal energy and the product of P and V. The heat capacities at constant pressure and volume are C_P and C_V respectively, and these play an important role in deducing several important thermodynamic relations. A process is said to be *adiabatic* if no heat transfer takes place between the system and the surroundings. Different relations are established on the basis of adiabatic change of a process. The three important processes—constant-temperature, constant-volume and constant-pressure—are defined and analyzed for knowing their scope for employment in several engineering devices. In the last section, the mass and energy balance are prepared for an open system, based on the law of conservation of mass and energy respectively, and the application of the first law has been emphasized for flow processes.

KEY TERMS

Adiabatic Process   The process in which there is no heat transfer between the system and the surroundings.

Enthalpy   The sum of the internal energy and the work done, or the amount of energy within the system (or the substance) that is available for conversion into heat.

First Law of Thermodynamics   Whenever a system undergoes a cyclic change, the cyclic integral of the heat produced is proportional to the cyclic integral of the work done.

Heat Capacity   The amount of heat input required to raise the temperature of 1 mol of the substance by 1 K.

Internal Energy   A certain amount of energy that every system has within itself, or the energy possessed by a body by virtue of the motion of the molecules of the body and the internal attractive and repulsive forces between the molecules.

Kinetic Energy   The energy possessed by a body due to its motion.

Law of Conservation of Energy   Energy can neither be created nor be destroyed. It can only be transformed from one form to another.

Mechanical Equivalence of Heat   1 BTU = 778 ft-lb.

Non-flow System   A system in which there is no mass transfer across the boundary of the system.

Specific Heat   The specific heat of a substance is the amount of heat required to raise the temperature of 1 g of a substance by 1 K.

State Function   The measurable properties of a system which describe the present state of the system.

Thermodynamic State   A system is said to be in *proper state* when the state variables such as temperature, pressure, and volume have definite values.

Work   A form of energy in transit.

IMPORTANT EQUATIONS

1. The total sum of the energy in the system and the surrounding is a constant value, i.e.

$$E_{\text{System}} + E_{\text{Surroundings}} = \text{Constant}$$

2. The first law of thermodynamics is given by

$$J\oint \partial Q = \oint \partial W \qquad (2.1)$$

3. The total energy E of a thermodynamics system is given by

$$E = E_K + E_P + U$$

where

E_K = Kinetic energy
E_P = Potential energy
U = Internal energy.

4. The mathematical expression of the first law of thermodynamics is given by

$$dQ = dU + dW \qquad (2.9)$$

5. The internal energy of an ideal gas is given by

$$dU = dQ - dW = 0$$

6. The enthalpy of a system is defined by

$$H = U + PV \qquad (2.11)$$

where

U = Internal energy of the system
P = Absolute pressure of the system
V = Volume of the system.

7. The condition of an adiabatic process is given by

$$dQ = 0$$

8. The adiabatic work done is given by

$$W_{ad} = dW = -C_V dT = -C_V(T_2 - T_1)$$
$$= C_V(T_1 - T_2) \qquad (2.23)$$

9. The work done by the adiabatic process for 1 mol of an ideal gas is given by

$$W_{ad} = C_V\left(\frac{P_1V_1}{R} - \frac{P_2V_2}{R}\right) = \frac{C_V}{R}(P_1V_1 - P_2V_2) = \frac{P_1V_1 - P_2V_2}{\gamma - 1} \qquad (2.25)$$

$$= \frac{P_2V_2 - P_1V_1}{1 - \gamma}$$

$$= \frac{R(T_2 - T_1)}{1 - \gamma} \qquad (2.26)$$

10. Internal energy of the system is given by

$$\Delta U = C_V(T_2 - T_1) \qquad (2.27)$$

11. The enthalpy of a thermodynamic system considering the initial and final states is given by

$$\Delta H = H_2 - H_1 = (U_2 + P_2V_2) - (U_1 + P_1V_1) = (C_V + R)(T_2 - T_1) \qquad (2.28)$$

12. The relation between the temperature and the volume under an adiabatic change in a process is given by

$$T_2 V_2^{\gamma-1} = T_1 V_1^{\gamma-1} = \text{Constant} \qquad (2.30)$$

13. The relation between the pressure and the volume under an adiabatic change in a process is given by

$$PV^\gamma = \text{Constant} \qquad (2.31)$$

14. The relation between the temperature and the pressure under an adiabatic change in a process is given by

$$T^\gamma P^{1-\gamma} = \text{Constant} \qquad (2.32)$$

15. The heat capacity of a system at constant volume is given by

$$C_V = \left(\frac{\partial Q}{\partial T}\right)_V = \left(\frac{\partial H}{\partial T}\right)_V$$

16. The heat capacity of a system at constant pressure is given by

$$C_P = \left(\frac{\partial Q}{\partial T}\right)_P = \left(\frac{\partial H}{\partial T}\right)_P$$

17. The relation and difference between C_P and C_V is given by

$$C_P - C_V = R$$

$$C_P - C_V = \left[\left(\frac{\partial U}{\partial V}\right)_T + P\right]\left(\frac{\partial V}{\partial T}\right)_P$$

18. The polynomial form of the molar heat capacities for an ideal gas is given by

$$C_P = a + bT + cT^2 + \cdots$$

where a, b, and c are constants for a given gas and which vary for different substances.

19. The heat capacity ratio is given by

$$\frac{C_P}{C_V} = \gamma$$

20. In a constant-volume process for n moles of gas

$$dQ = ndU \qquad (2.34)$$

21. In a constant-volume process for n moles of gas

$$\Delta H = H_2 - H_1 = \Delta U + \Delta PV \qquad (2.35)$$

EXERCISES

A. Review Questions

1. Explain the importance of the demonstration of Joule's experiment in the formulation of the first law of thermodynamics.

2. Give a proper concept of energy, internal energy, kinetic energy and potential energy.
3. Justify the following statement: "*The first law of thermodynamics is nothing but the law of conservation of energy*".
4. Derive the mathematical expression of the first law of thermodynamics.
5. Explain the application of the first law of thermodynamics to the flow process.
6. What is internal energy? Prove that internal energy is a state function.
7. What is the significance of Joule's experiment in finding out the change in internal energy of an ideal gas?
8. Prove the following statement: '*For an ideal gas, the internal energy is a function of temperature only*'.
9. What is thermodynamic state and what are state functions?
10. Define the term *enthalpy*. How does it relate to the internal energy?
11. State the first law of thermodynamics and mention its importance for a cyclic process.
12. Derive an expression of the work done in a constant-temperature process.
13. Prepare an energy balance for an open system.
14. Give a brief account of constant-volume and constant-pressure processes.
15. What is isothermal expansion? With the help of a neat sketch, substantiate the significance of a porous plug experiment.
16. What do you mean by heat capacity and specific heat?
17. Prepare an energy balance over an isothermal compression system.
18. If $C_P = a + bT + CT^2$, derive a relation to the isobaric mean heat capacity $\dot{Q}$.
19. Prove that $C_P > C_V$, where the notations have their usual meanings.
20. Derive an expression for the work done if an ideal gas undergoes an adiabatic change.

B. Problems

1. The total energy of a typical closed system is given by $E = 25 + 155\,T + 0.07\,T^2$ in Joules. The amount of heat absorbed by the system can be expressed as $Q = 3500 + 9T$ in Joules. Estimate the work done during the processes in which the temperature rises from 350 K to 700 K.
2. The latent heat of vaporization of Freon-11 at 23.6°C and 1 atm is 5960 g-cal/g-mol. Calculate ΔU and ΔH of this process.
3. In a stirrer–container assembly, the stirrer performs 3 hp work on the system containing a certain amount of fluid. The heat developed by stirring is 5000 kJ/hr and is transferred to the surroundings. Determine the change in internal energy of the system.
4. A system consisting of a gas confined in a cylinder undergoes a series of processes shown in Fig. 2.11. During the process A-1-B, 70 kJ of heat is added while it does 45 kJ of work. Then the system follows the process A-2-B, during which 55 kJ of work is performed on the system. How much heat flows into the system during the process A-2-B? Then the system returns to the initial state along the path B-3-A, and 80 kJ of work is done on the system. Calculate the amount of heat transferred between the system and the surroundings during the process B-3-A.

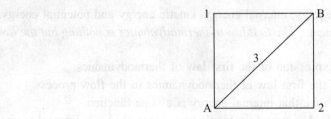

Fig. 2.11   Problem 4.

5. In a constant-volume calorimeter, methane undergoes a compression process. The released heat at 25°C is 3720 cal. What will be the enthalpy change of the process at 25°C?

6. A piston–cylinder assembly containing a gas undergoes a process in which the temperature of the system rises from 100°C to 150°C. The heat transmission per degree rise in temperature is governed by the equation

$$\frac{dQ}{dT} = 1.25 \; \frac{kJ}{K}$$

The work done on the system per degree rise in temperature is given by

$$\frac{dW}{dT} = (5 - 0.25T) \; \frac{kJ}{K}$$

Calculate the change in internal energy of the system during the process.

7. An insulated frictionless piston–cylinder assembly containing 500 g of H_2 gas. The gas is compressed at 27°C adiabatically from 10 L to 5 L. Calculate the final temperature and also ΔU, ΔH, Q, and W, given that $C_V = (3/2)R$.

8. 10 moles of an ideal gas at 37°C are allowed to expand isothermally from an initial pressure of 15 atm to a final pressure of 5 atm against a constant external pressure of 1 atm. Calculate ΔU, ΔH, Q, and W for the process.

9. 0.52 kg air is heated reversibly at constant pressure from an initial state of 37°C and 1 kPa until its volume is doubled. Calculate ΔU, ΔH, Q, and W for the process.

10. A bullet of 3 g flying horizontally at 2 km/s strikes a fixed wooden block ($m = 6$ kg, $C_V = 0.14$ kJ/kg-K) and is embedded in it. Assume that there is no heat loss from the block and that the block does not change its volume. What will be the change in temperature of the block?

11. The pressure of a gas is given by $P = \dfrac{15}{V}$, where P is in atmosphere and V is in litres.

    If the gas expands from 20 to 60 L and undergoes an increase in internal energy of 225 cal. How much heat will be absorbed during the process?

12. 5 moles of an ideal gas was initially at 315 K and 20 atm. The expansion of gas takes place adiabatically when the external pressure is reduced to 7 atm. What will be the final temperature and volume? Also calculate the work done during the process, given that $C_P = 8.58$ cal/mol/°C.

13. 3 kg of air at 45°C expands reversibly and adiabatically to 4 times its original volume. The initial pressure of the air mass was 9 atm. Determine the final pressure, temperature,

and work done when the expansion is (i) adiabatic and (ii) isothermal, given that the heat capacity ratio, $\gamma = 1.4$.

14. In a constant-volume calorimeter, 1 mol of trinitrotoluene (TNT) on explosion produces 3 mol of CO and 2 mol of N_2. When 0.1572 g TNT are exploded at 37°C, the heat evolved is 450 cal. Calculate ΔU and ΔH if 1 mol of TNT explodes at 27°C.

15. In an adiabatic change for an ideal gas, show that the work done in an adiabatic expansion

$$W = \frac{P_1 V_1}{\gamma - 1}\left[1 - \left(\frac{P_2}{P_1}\right)^{\frac{\gamma - 1}{\gamma}}\right]$$

16. Show that for an ideal gas, when volume and enthalpy are separate functions of temperature and pressure

$$C_P - C_V = \left[V + \left(\frac{\partial H}{\partial T}\right)_P \left(\frac{\partial T}{\partial P}\right)_H\right]\left(\frac{\partial P}{\partial T}\right)_V$$

17. A certain quantity of an ideal gas is contained in a cylinder and occupies a volume of 1.0 dm^3 at 3 atm pressure. The gas is transferred by different paths to a final state where it occupies a volume of 3.0 dm^3. The paths are mechanically reversible and the whole process takes place by the following two processes:

    (i) an adiabatic expansion
    (ii) an isothermal expansion to 3.0 dm^3 followed by a change of pressure.

    Calculate the work done of the gas, given that $\gamma = 1.4$.

18. Hydrogen gas is expanded reversibly and adiabatically from a volume of 2.12 dm^3 at a pressure of 4 atm and 32°C until the volume is doubled. Determine the following:

    (i) final temperature and pressure of the gas
    (ii) Q, W, ΔU, and ΔH for the gas.

    Assume that the gas is an ideal one. $C_P = 29.01$ J/kmol and $\gamma = 1.4$.

19. One mol of nitrogen at 25°C and 1 atm is allowed to expand reversibly to a volume of 50 dm^3. If the gas is assumed to be ideal, calculate the final pressure for the following cases:

    (i) isothermal expansion
    (ii) adiabatic expansion

    Take $\gamma = 1.67$.

20. 5 kg of N_2 is heated from an initial state of 37°C and 101.33 kPa until its temperature reaches 237°C. Calculate ΔU, Q, W, and ΔH for the following processes:

    (i) isochoric process
    (ii) isobaric process.

    N_2 is assumed to be an ideal gas. $C_P = 29.10$ J/mol-K, $C_V = 20.78$ J/mol-K, and molecular weight of nitrogen = 28.

21. A horizontal piston–cylinder assembly is placed in a constant temperature bath. The piston slides in the cylinder with negligible friction, and the external force holds it in place

against an initial gas pressure of 12 bar. The initial gas volume is 0.04 m³. The external force is reduced gradually, allowing the gas to expand until its volume doubles. If the gas follows the relation PV^T = Constant, determine

  (i) the work done by the gas
  (ii) the work done if the external force were suddenly reduced to half its initial value instead of being gradually reduced.

22. N_2 is contained in a cylinder of 30 L capacity at 75 atm and 30°C. Suddenly a valve is opened to release N_2 into the atmosphere. As a result, the pressure of the cylinder drops and finally reaches 5 atm. What is the temperature of the gas in the cylinder? Assume that nitrogen behaves ideally and take $\gamma = 1.4$.

23. A rigid vessel containing 5 mol of He gas at 25°C is heated to 225°C. Calculate the heat requirements for the process, given that C_V = 20.78 kJ/ kmol-K. Neglect the heat capacity of the vessel.

24. A spherical balloon of 1 m diameter contains a gas at 150 kPa. The gas inside the balloon is heated until the pressure reaches 600 kPa. During heating the pressure of the gas inside the balloon is proportional to the cube of the diameter of the balloon. Determine the work done by the gas inside the balloon.

25. 5 kmol of Ar gas confined in a cylinder undergoes a change from an initial condition of 20 bar and 350 K to a final condition of 2 bar and 350 K. The gas follows the equation $PV = RT$.
Given that C_P = 29.10 kJ/kmol-K and C_V = 20.78 kJ/kmol-K, determine the change in internal energy, change in enthalpy, heat and work requirement.

26. A thermally insulated cylinder having a frictionless piston contains N_2 gas. The piston is held in place by latches in such a way that it divides the cylinder into two equal halves. Each half of the cylinder contains 30 L of N_2 gas at 2 atm and 25°C. The restraining latches are removed and heat is supplied to the gas to the lower half of the cylinder until the piston compresses the gas on the upper half to 4.5 atm. Take C_P = 29.10 kJ/kmol-K and C_V = 20.78 kJ/mol-K.

  (i)   Calculate the final temperature of the gas on the upper half.
  (ii) Determine the work done on the gas on the upper half.
  (iii) What is the final temperature of the gas on the lower half?
  (iv) How much heat is supplied to the lower half of the cylinder?

Properties of Pure Substances

3

LEARNING OBJECTIVES

After reading this chapter, you will be able to:

- Know the definition of phase and learn about phases of a pure substance
- Discuss the different phase change processes of a pure substance
- Understand the meaning of compressed and saturated liquid, saturated and superheated vapour, saturation temperature and pressure
- Highlight the $P-V-T$ behaviour of pure substances
- Represent the property diagrams such as $P-T$, $T-V$, and $P-V$ for a phase change process
- Follow the mathematical representation of $P-V-T$ behaviour
- Understand the ideal gas equation of state
- Discuss different generalized equations of state for real gases, such as the van der Waals, Redlich–Kwong, Peng–Robinson, Redlich–Kwong–Soave equations
- Understand the concept of compressibility factor and the law of corresponding state
- Use the generalized virial coefficient correlations

This chapter deals with the study of a system comprising a single component, i.e., a system in which only one pure substance is present. The states as specified its properties under which a pure substance may exist in a particular phase and various phase-change processes are discussed. The $P-V-T$ behaviour of a pure substance is illustrated with the help of property diagrams. The equation of state play a dominant role in representing the $P-V-T$ relation of the fluids. For ideal and real gas, the equations of state are elaborately discussed along with the determination of values of various constants. The compressibility factor, which measures the deviation of real gases from ideal gas behaviour is introduced. The law of corresponding states and the concept of acentric factor are nicely presented here to highlight the essence of critical constants and reduced parameter. Some familiar equations of state, such as van der Waals, Kammerlingh–Onnes, Peng–Robinson, Benedict–Webb–Rubin, Redlich–Kwong–Soave have been discussed in this chapter.

3.1 PURE SUBSTANCE

A pure substance is defined as a substance which has fixed chemical composition throughout. Alternatively, it is a substance of constant chemical composition throughout its mass. It is one component system. It may exist in one or more phases; for example, water, helium, nitrogen, and oxygen.

Again, it is well known to us that air has a uniform chemical composition and it is a mixture of 79% N_2 and 21% O_2 by volume. With obvious reason, it should be a pure substance and air is often considered to be a pure substance. But it has been observed that its composition may vary.

Suppose, a certain quantity of air is confined in a cylinder–piston assembly. Now, if the gas is allowed to undergo an expansion or a compression process, its composition does not change. In such a case, the air can be considered as a pure substance. Again, if the air is cooled to sufficiently low temperature where it may be available in both liquid and vapour phases. In this case, the air cannot be treated as pure substance. Because, the composition of the liquid will differ from that of vapour.

A mixture of two or more phases of a pure substance is regarded as pure substance subject to the similarity in chemical composition of all the phases. For example, a mixture of ice and liquid water is a pure substance, because both the phases have the same chemical composition. Again, a mixture of oil and water, or a mixture of liquid and gaseous air, cannot be pure substances.

3.2 PHASES OF A PURE SUBSTANCE

A phase is defined as a physically distinct but homogeneous part of a system separated from other parts by easily identifiable boundary surfaces.

It can be defined in other way that a system which is uniform throughout both in chemical composition and physical state, is called phase. The pure substance can exist in three different principal phases, such as solid, liquid and gas. But a substance may have different phases within the principal phase.

For instance, water can exist in three phases: solid–ice, liquid–water and gas–vapour or steam. Sulphur can exist in two phases: rhombic form (lemon yellow in colour) and monoclinic form (orange in colour); carbon may exist as graphite or diamond in the solid phase; helium has two liquid phases; iron has three solid phases; ice may exist at seven different phases at high pressure, etc.

3.3 PHASE CHANGE PROCESSES OF PURE SUBSTANCES

For a pure substance, the coexistence of two phases in equilibrium is a familiar phenomenon. In this regard, we can cite a few simple and common examples, such as the existence of water as a mixture of liquid and vapour in the boiler of a steam power plant, or the phase change of a refrigerant from liquid to vapour in the evaporator of a refrigerator. Here, we shall discuss the different phase-change processes, considering a common instance.

State I: Compressed or Sub-cooled Liquid

Consider a cylinder–piston assembly containing a liquid water at 25°C and 1 atm pressure. Under these conditions, water exists in the liquid phase and it is known as a *Compressed liquid or Sub-cooled liquid*. The changes take place in a stepwise manner and will be demonstrated with the help of Fig. 3.1 and the relevant graphical representation of temperature versus volume in Fig. 3.2.

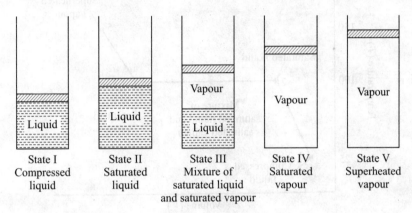

Fig. 3.1 Demonstration of phase-change processes.

A liquid at a temperature below its saturation temperature, at a specific pressure, is called sub-cooled liquid or compressed liquid. The term *compressed liquid* implies that the liquid exists at a pressure which is greater than the saturation pressure at the specified temperature, and is not about to vaporize.

State II: Saturated Liquid

Water is then heated until its temperature reaches to, say, 50°C. Due to increase in temperature, the liquid will expand slightly, which in turn will increase the specific volume. Since the process takes place in the cylinder–piston device, then on expansion the piston will move up slightly. During this process, the pressure inside the cylinder will remain constant at 1 atm. At this condition too, water is compressed liquid as its vaporization is yet to take place.

On further heating, the temperature of water will go on increasing and it reaches to 100°C. At this state, the water is still in liquid phase and further addition of heat will cause some of the liquid to vaporize. It necessarily means that a phase-change will take place from liquid to vapour. At this state, the liquid is known as *saturated liquid.*

Saturated liquid is defined as the liquid which is in equilibrium with its own vapour at a specified temperature or pressure. In other words, a liquid that is about to vaporize is called saturated liquid. At boiling point, a liquid is always saturated. For example, water boils at 100°C and standard atmospheric pressure of 101.32 kPa or 1 atm and produces vapour. Under this condition, water is a saturated liquid.

State III: Saturated Liquid and Vapour

This is the intermittent state at which saturated liquid is in the phase before reaching the saturated vapour state on further application of heat. Actually, this state is midway between liquid and vapour states, when the cylinder contains equal amounts of liquid and vapour. The state is represented by point 3 in the *T–V* diagram in Fig. 3.2. A substance at states between 2 and 4 is referred to as a *saturated liquid–vapour mixture*, as liquid and vapour coexist in equilibrium at this state.

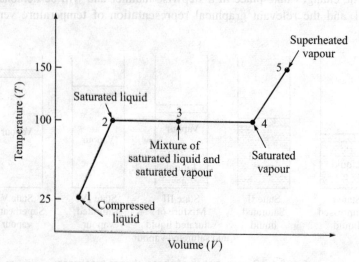

Fig. 3.2 *T–V* diagram for heating process of water at constant pressure.

State IV: Saturated Vapour

Once boiling starts, the temperature will stop rising. During the boiling of the liquid, it is observed that the volume of the liquid increases and a steady decline in the liquid level occurs. This is because of the liquid turning into vapour. The process continues until the last drop of the liquid vaporizes. At this point, the entire cylinder is filled with vapour, i.e., it is vapour-bound. The vapour will have adequate tendency to be condensed under heat loss. In this situation, the vapour is considered to be saturated vapour. Point 4 in the *T–V* diagram (see Fig. 3.2) represents the saturated vapour state.

Saturated vapour is defined as the vapour which is in equilibrium with its own liquid at a specified temperature or pressure. In other words, vapour that is about to condense is called saturated vapour. At the condensing point, the vapour is always saturated.

State V: Superheated Vapour

On further transfer of heat, saturated vapour absorbs heat and it results in an increase of both the temperature and the specific volume. At point 5, the temperature of the vapour reaches, say, 150°C. Now, if some amount of heat is transferred from the vapour, there may be a drop in temperature, but no condensation will take place as the temperature of the vapour remains above 100°C.

Superheated vapour is defined as the vapour existing at a temperature above the saturation temperature at a specified pressure. For instance, steam at 1 atm and 150°C is superheated vapour. Alternatively, the vapour that is not about to condense is termed superheated vapour.

If the entire process is reversed by cooling the water vapour while maintaining the pressure at the same level, the system will return to its original state.

In this chapter, we should know some important definitions. These can be represented as follows:

Saturation Temperature

It is defined as the temperature at which a liquid is in equilibrium with its own vapour at a specified pressure. Alternatively, it can be said that the temperature at which a pure substance changes its phase is known as the saturation temperature. At this temperature, the liquid starts boiling. For example, water boils at 100°C. Hence, 100°C is the saturation temperature of water.

Saturation Pressure

It is defined as the pressure at which a liquid is in equilibrium with its own vapour at a specified temperature. Alternatively, it can be said that the pressure at which a pure substance changes its phase is known as the saturation pressure. At this pressure, liquid starts boiling. For example, water boils at the standard atmospheric pressure of 101.32 kPa or 1 bar or 1 atm. Therefore, this pressure is the saturation pressure of water.

Latent Heat of Fusion

It is defined as the amount of heat absorbed during the melting process. It is equivalent to the amount of energy released during freezing. For example, the latent heat of fusion of water is 333.7 kJ/kg.

Latent Heat of Vaporization

It is defined as the amount of heat absorbed during vaporization process. It is equivalent to the amount of energy released during condensation. For example, the latent heat of vaporization of water is 2257.1 kJ/kg.

Triple Point

The state at which all the three phases—solid, liquid, and vapour—coexist in equilibrium is called the triple point. It can be better explained by the Gibbs phase rule

$$F = C - P + 2$$

where
   F = Degrees of freedom
   C = Number of components
   P = Number of phases.

At the triple point, $C = 1$, $P = 3$. Hence, from the preceding equation, we get

$$F = 1 - 3 + 2 = 0$$

Therefore, the degree of freedom is zero at the triple point.
For example, the triple point of water is at $P = 0.611$ kPa and $T = 0.01$°C.

Critical Point

It is defined as the highest temperature and pressure above which a liquid cannot exist. At the critical point, the liquid and vapour are indistinguishable. The properties of saturated liquid and vapour are identical at critical point.

At the critical point, the temperature and pressure are known as critical temperature and critical pressure respectively.

3.4   *P–V–T* BEHAVIOUR OF PURE SUBSTANCE

The *P–V–T* behaviour of a pure substance can be better described by the property diagrams for phase-change processes. The property diagrams help to study the variations of properties of pure fluids when the change in phase occurs.

The best known property diagrams are:

- *T–V* diagram
- *P–V* diagram
- *P–T* diagram.

3.4.1   *T–V* Diagram

Consider a certain amount of solid ice taken in a piston–cylinder device. If the heat is transferred to the system at constant pressure, say P_1, till the solid converts into vapour, then the nature of the temperature versus volume diagram can be represented by Fig. 3.3.

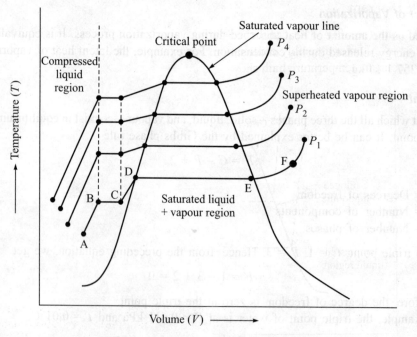

Fig. 3.3   *T–V* diagram of a pure substance.

On application of heat, the temperature and the molar volume of the solid ice increase at constant pressure P_1. This process is represented by the line AB in the T–V diagram.

At the point B, the solid ice starts melting while the temperature remains constant till it reaches point C, where it completely exists as a liquid. The segment BC represents the melting process.

As the heating process continues, the molar volume and temperature of the liquid increase and the system follows the path CD. At point D the liquid starts vaporizing.

Due to the continuous heat transmission to the system, vaporization of liquid continues while its temperature remains constant, and the system follows the path DE. Here, the vaporization process is indicated by the segment DE. At state D, the liquid exists as a saturated liquid. At state E, all the liquid is completely vaporized. The vapour at state E is referred to as saturated vapour.

On further addition of heat, the vapour follows the path EF. If similar experiments are carried out at higher pressures P_2 to P_4 and the results are represented on the same diagram, they will appear as shown in Fig. 3.3.

It is important to note that all the saturated liquid states are connected by a line. The line is called the *saturated liquid line*. Similarly, all the saturated vapour states are connected by a line. The line is called the *saturated vapour line*. These two lines meet at the critical point and form a dome. All the compressed liquid states are situated in the region to the left of the saturated liquid line. The region is known as the *compressed liquid region*. All the superheated vapour states are situated in the region to the right of the saturated vapour line. The region is known as the *superheated vapour region*. All the states that involve both phases in equilibrium are situated under the dome, which is called the *saturated liquid–vapour mixture region*.

3.4.2 P–V Diagram

The P–V diagram (see Fig. 3.4) can be suitably studied by the example given here. Let us consider a piston–cylinder assembly containing liquid water at 10 bar and 150°C. The assembly is placed in a constant temperature bath shown in Fig. 3.5 to maintain the temperature constancy

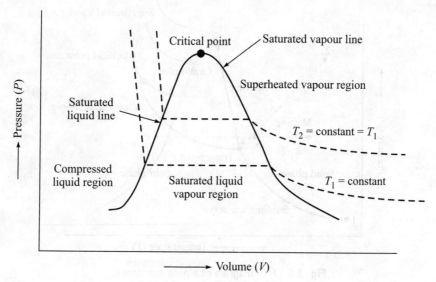

Fig. 3.4 *P–V* diagram of a pure substance.

of the system. Under this condition, water exists as a compressed liquid. Now, if the weights placed on the top of the piston are removed one by one, then the pressure inside the cylinder decreases gradually. To maintain the temperature constancy throughout the entire system, water is allowed to exchange heat with the surroundings. As the pressure decreases, the expansion takes place and the volume of water will increase slightly. At this temperature, when pressure reaches the saturation pressure (for water, P_{Sat} = 475.8 kPa at 150°C), the water will start boiling. The vaporization will take place at constant temperature and pressure due to heat absorption from the surroundings. Thus the total volume increases. Once the last drop of liquid is vaporized, further reduction in pressure results in a further increase in volume. It is necessary to remember that during the phase-change process, no weights are removed, because

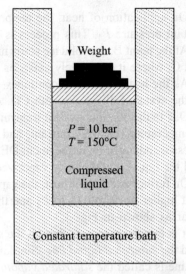

Fig. 3.5 Illustration on piston–cylinder assembly at constant temperature bath.

it may cause the temperature and pressure of the system to drop; then the process would no longer be isothermal.

3.4.3 *P–T* Diagram

The *P–T* diagram of a pure substance is often called the *phase diagram*, because all the three phases—solid, liquid and vapour—are separated from each other by three lines. From the phase diagram, shown in Fig. 3.6, it is very much clear that the sublimation curve 1 separates the solid

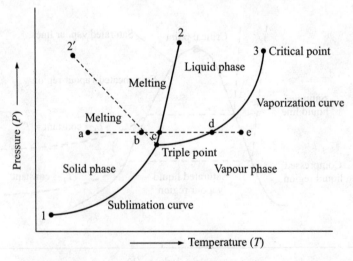

Fig. 3.6 *P–T* diagram of a pure substance.

and vapour phases. Along the sublimation curve, the solid and vapour phases coexist in equilibrium, i.e., solid is in equilibrium with its vapour. Sublimation is the process by which solid vaporizes without melting.

The melting or fusion lines 2 and 2′ separate the solid and liquid phases. Along the melting curve, both solid and liquid phases coexist in equilibrium. Similarly, the vaporization curve 3 separates the vapour and liquid phases. The liquid and vapour phases coexist in equilibrium along the vaporization curve. These three lines meet at the triple point. At this point, all the three phases can exist together in equilibrium. The vaporization line starts at the triple point and terminates at the critical point because there is no physical distinction between liquid and vapour phases.

The T–P diagram can be better described by a common and simple instance. Suppose that an ice block exists at a certain pressure and temperature at point a in Fig. 3.6. On application of heat at constant pressure, the ice block starts melting at point b and changes to liquid. The points b and c conjoin on the fusion line. It is heated further at point c at constant pressure. At constant temperature, the process results in vapour at point d. That is, the vaporization of the liquid starts at point d. After all the liquid gets vaporized, the temperature increases along the line de. Then, on further heating, the substance converts into gas beyond the critical temperature.

3.5 EQUATION OF STATE—MATHEMATICAL REPRESENTATION OF P–V–T BEHAVIOUR

Any equation that relates the pressure, temperature and volume of a substance is known as *equation of state*. Property relations that involve other property of a substance at equilibrium states are also referred to as equations of state. There are several equations of state. The simplest and best known equations of state for substances in the gas phase is the ideal gas equation. In other words, a mathematical expression of the type $f(P,V,T) = 0$ which describes the P–V–T behaviour of a substance is called an *equation of state*.

An equation of state may be solved if any one of three quantities P, V, and T is considered a function of the other two. For example, V, the specific volume of a system, is a function of temperature T and pressure P. Then it can be mathematically expressed as

$$V = f(T, P)$$

or

$$dV = \left(\frac{\partial V}{\partial T}\right)_P dT + \left(\frac{\partial V}{\partial P}\right)_T dP \tag{3.1}$$

This equation is related to the following important properties of a liquid:

Isothermal Compressibility

It is defined as the rate of decrease of volume with respect to pressure at constant temperature per unit volume.

Mathematically

$$\alpha = -\frac{1}{V}\left(\frac{\partial V}{\partial P}\right)_T \tag{3.2}$$

In this regard, we can define another important term *isentropic compressibility*. It is denoted by α. It is defined as the rate of decrease of volume with respect to pressure at constant entropy per unit volume.

It can be mathematically expressed as

$$\alpha = -\frac{1}{V}\left(\frac{\partial V}{\partial P}\right)_S \tag{3.3}$$

Volume Expansivity or Coefficient of Volume Expansion (β)

It is defined as the rate of change of volume with respect to temperature at constant pressure per unit volume.

$$\beta = \frac{1}{V}\left(\frac{\partial V}{\partial T}\right)_P \tag{3.4}$$

From Eq. (3.1), we have

$$dV = \left(\frac{\partial V}{\partial T}\right)_P dT + \left(\frac{\partial V}{\partial P}\right)_T dP$$

Now, Eq. (3.2) can be expressed as

$$\left(\frac{\partial V}{\partial T}\right)_P = -\alpha V \tag{3.5}$$

Equation (3.4) can be expressed as

$$\left(\frac{\partial V}{\partial T}\right)_P = \beta V \tag{3.6}$$

On substitution of Eqs. (3.5) and (3.6) into Eq. (3.1), we get

$$dV = \beta V dT - \alpha V dP$$

or

$$\frac{dV}{V} = \beta dT - \alpha dP \tag{3.7}$$

as liquid β is always +ve (except for water between 0°C and 4°C) and α is +ve. On integration of Eq. (3.7), we get

$$\ln\frac{V_2}{V_1} = \beta(T_2 - T_1) - \alpha(P_2 - P_1) \tag{3.8}$$

3.6 IDEAL GAS EQUATION OF STATE

Ideal gas equation of state is established on the basis of three familiar laws.

These are:

- **Boyle's Law:** It states that the pressure of a gas varies inversely or is inversely proportional to the volume when temperature is constant. It can be mathematically expressed as

$$P \propto \frac{1}{V} \qquad \text{when } T = \text{constant}$$

or
$$PV = K = \text{constant} \tag{3.9}$$

- **Charles' Law:** It states that the molar volume of a gas is directly proportional to the temperature when the pressure is constant. It can be mathematically expressed as

$$V \propto T \quad \text{when } P = \text{constant}$$

or
$$V = K'T \tag{3.10}$$

- **Avogadro's Law:** It states that at normal temperature and pressure, the number of molecules is directly proportional to volume. It can be mathematically expressed as

$$V \propto n \tag{3.11}$$

Combining Eqs. (3.9), (3.10), and (3.11), the ideal gas equation of state can be established well in the following way:

$$V \propto \frac{nT}{P}$$

or
$$PV = nRT \tag{3.12}$$

where

$$R = \text{Gas constant}$$
$$= \text{Constant of proportionality.}$$

For 1 g-mol of an ideal gas, i.e., $n = 1$, Eq. (3.12) becomes

$$PV = RT \tag{3.13}$$

The above equations (3.12) and (3.13) are known as *ideal gas equation of state*. The gases obeying this equation of state are known as ideal gases.

Based on the following two assumptions, the ideal gas equation state is established:

1. The volume occupied by the gas molecules is negligible compared to the volume available for free motion.
2. The molecules are independent of each other and do not exert any attractive forces.

Equation (3.12) can also be written as

$$PV = \frac{m}{M} RT$$

where
m = Mass of the pure substance
M = Molecular weight of the pure substance.

EXAMPLE 3.1 Determine the molar volume of a gas at 37°C and a pressure of 2 bar. Assume that the gas behaves ideally.

Solution: Since the gas behaves as an ideal gas, then the ideal gas equation can be applied to determine the volume of the gas. It is given by

$$PV = RT$$

or

$$V = \frac{RT}{P} \tag{i}$$

Here, $T = (273 + 37)$ K $= 310$ K, $P = 2$ bar. Putting these values in Eq. (i), we get

$$V = \frac{8.314 \times 310}{2 \times 10^5} = 1.28 \times 10^{-2} \text{ m}^3$$

Hence, the molar volume of the gas is 1.28×10^{-2} m^3.

EXAMPLE 3.2   8.0 m^3 of air is enclosed by a frictionless piston at in a cylinder at 300 kPa. The gas undergoes a compression process with no change in temperature and its volume becomes 2.0 m^3. What is the pressure of air after compression? Air is assumed to be an ideal gas.

Solution: We are given that

$$V_1 = 8.0 \text{ m}^3$$
$$P_1 = 300 \text{ kPa}$$
$$V_2 = 2.0 \text{ m}^3$$

Applying the ideal gas equation, we get

$$\frac{P_1 V_1}{T_1} = \frac{P_2 V_2}{T_2}$$

Since the process takes place with no change in temperature, then $T_1 = T_2$. Hence the above equation becomes

$$P_1 V_1 = P_2 V_2$$

or

$$P_2 = \frac{P_1 V_1}{V_2}$$

Substituting the values, we get

$$P_2 = \frac{300 \times 8}{2} = 1200 \text{ kPa}$$

Therefore, the pressure of air after compression is 1200 kPa.

EXAMPLE 3.3   A vessel contains 6 m^3 of air at a pressure of 500 kPa. If one-fifth of the air be removed by an air pump, what will be the pressure of the remaining air, the temperature being constant? We are given that characteristic gas constant, $R = 0.287$ kJ/kg-K.

Solution: Applying the characteristic gas equation to the gas initially, and to the gas which is left in the vessel after one-fifth of the gas has been removed, we get

$$P_1 V_1 = m_1 R T_1$$
$$500 \times 6 = m_1 \times 0.287 \times T_1 \tag{i}$$

Similarly

$$P_2 V_2 = m_2 R T_2$$

$$P_2 \times 6 = m_2 \times 0.287 \times T_2 \tag{ii}$$

But $T_2 = T_1$ and $m_2 = \dfrac{4}{5} m_1$. Therefore, Eq. (ii) becomes

$$P_2 \times 6 = \frac{4}{5} m_1 \times 0.287 \times T_1 \tag{iii}$$

Dividing Eq. (i) by Eq. (iii), we get

$$\frac{500}{P_2} = \frac{5}{4}$$

or

$$P_2 = 400 \text{ kPa}$$

EXAMPLE 3.4   A steel cylinder containing air has a closely fitted piston and a set of stops as shown in Fig. 3.7. The piston is loaded with certain weights. The air inside the cylinder is initially at 450°C and 3 bar. The air is then cooled by transferring heat to the surroundings.

(a)   What is the temperature of air when the piston reaches the stop?
(b)   If the cooling is continued until the temperature reaches 30°C, what is the pressure of air inside the cylinder at this state?

Solution: Refer to Fig. 3.7.

(a)   Since the weight remains the same, therefore, the final pressure is equal to the initial pressure. This is a constant pressure process. Therefore

$$P_1 = P_2 = 3 \text{ bar}$$
$$T_1 = 450°C = (450 + 273) \text{ K} = 723 \text{ K}$$

Volumetric ratio

$$\frac{V_2}{V_1} = \frac{2.5}{2.5 + 2.5} = 0.5$$

Applying

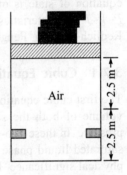

Fig. 3.7   Example 3.4.

$$\frac{P_1 V_1}{T_1} = \frac{P_2 V_2}{T_2}$$

or

$$T_2 = \frac{T_1 V_2}{V_1} = 723 \text{ K} \times 0.5 = 361.5 \text{ K} = 88.5°C$$

Hence the temperature of air is 88.5°C.

(b)   when the piston rests on the stops, the pressure exerted by the weight, air and atmosphere will be different. But there will be no further decrease in volume. This is a constant-volume process. Therefore

$$V_3 = V_2$$
$$T_3 = (273 + 30) \text{ K} = 303 \text{ K}$$

Similarly

$$\frac{P_3 V_3}{T_3} = \frac{P_2 V_2}{T_2}$$

$$P_3 = \frac{T_3 P_2}{T_2} \quad \text{(since } V_2 = V_3\text{)}$$

$$= \frac{3 \times 303}{361.5} = 2.51 \text{ bar}$$

3.7 EQUATION OF STATE FOR REAL GASES

The ideal gas equation is very simple in form and to use, but the limitation of this equation is its incapability of being applied to an appreciable range of temperature and pressure. In order to represent the P–V–T behaviour of both liquid and vapour, it is essential to obtain an equation of state which must encompass a wide range of temperatures and pressures with no limitation. This sort of equation is complicated in form, but is very much capable of providing accurate results. Therefore, these equations are successfully and most widely used in different areas. In this regard, numerous such equations have been proposed to serve the purpose. Van der Waal's equation of state is one such simple equation, which is very much useful to substantiate the P–V–T characteristics of both liquid and vapour behaviours. Other equations of state are Redlich–Kwong, Peng–Robinson, Benedict–Webb–Rubin, and Kammerlingh–Onnes equations.

3.7.1 Cubic Equation of State—van der Waals Equation for Real Gases

The first cubic equation of state was proposed by van der Waals in 1873 to determine the molar volume of both the saturated vapour and saturated liquid phases at given temperature and pressure. In these two-phase regions, three real roots are obtained. The smallest is applicable for saturated liquid phase and the largest is for saturated vapour phase. The intermediate is of no physical significance. For all the cubic equations, P can be explicitly expressed as a function of V and T, i.e., given by $P = f(T, V)$.

Cubic equations have three roots. This implies the following possibilities:

- All three roots are real. While two roots having physical significance are real, one is hypothetical.
- One root is real and the other two are complex. It necessarily means only one real root and so only one phase exists.

The van der Waals equation of state was introduced to explain the P–V–T behaviour of a real gas, as the ideal gas equation of state fails to do the same task. It is given by

$$\left(P + \frac{a}{V^2}\right)(V - b) = RT \tag{3.14}$$

where a and b are the constants.

To rectify the error for neglecting two factors, van der Waals introduced two correction terms in the ideal gas equation, one for pressure and other for volume.

These two factors are

1. Intermolecular force of attraction
2. Volume occupied by the gas molecules

The intermolecular forces cannot be ignored when the two molecules are near each other. Again, the volume of the molecules is not negligible as compared to the volume of the gas. In the van der Waals equation, $\dfrac{a}{V^2}$ accounts for the intermolecular force of attraction and b accounts for the excluded volume or the volume occupied by the gas molecules per unit mass. The pressure correction term, P', was introduced to replace P in the ideal gas equation by $\left(P + \dfrac{a}{V^2}\right)$, and the volume correction term, V', to replace V in the ideal gas equation by $V - b$.

Equation (3.14) can be rearranged as

$$P = \frac{RT}{V - b} - \frac{a}{V^2} \tag{3.15}$$

Here, a and b are the positive constants. When $a = 0$ and $b = 0$, Eq. (3.15) becomes $PV = RT$, i.e., the ideal gas equation is recovered.

Now, for a particular substance, the constants a and b can be determined from the idea of critical constants, such as critical pressure, P_C and critical temperature, T_C. It has been observed from the P–V diagram that the critical isotherm exhibits a horizontal inflexion point at the critical point as shown in Fig. 3.8. Then the first and second derivatives of P with respect to V at the critical point must be zero.

The following conditions must be satisfied:

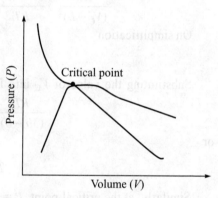

Fig. 3.8 Inflection point of a pure substance at critical state.

$$\left(\frac{\partial P}{\partial V}\right)_{T_C} = 0$$

and

$$\left(\frac{\partial^2 P}{\partial V^2}\right)_{T_C} = 0$$

Differentiating Eq. (3.15) with respect to V at constant temperature, we have

$$\left(\frac{\partial P}{\partial V}\right)_{T} = -\frac{RT}{(V - b)^2} + \frac{2a}{V^3} \tag{3.16}$$

$$\left(\frac{\partial^2 P}{\partial V^2}\right)_{T} = \frac{2RT}{(V - b)^3} - \frac{6a}{V^4} \tag{3.17}$$

At the critical point, $T = T_C$, $\left(\dfrac{\partial P}{\partial V}\right)_{T_C} = 0$ and $\left(\dfrac{\partial^2 P}{\partial V^2}\right)_{T_C} = 0$. Hence, Eq. (3.16) becomes

$$0 = -\frac{RT_C}{(V_C - b)^2} + \frac{2a}{V_C^3}$$

or

$$\frac{RT_C}{(V_C - b)^2} = \frac{2a}{V_C^3} \tag{3.18}$$

and Eq. (3.17) becomes

$$0 = \frac{2RT_C}{(V_C - b)^3} - \frac{6a}{V_C^4}$$

or

$$\frac{2RT_C}{(V_C - b)^3} = \frac{6a}{V_C^4} \tag{3.19}$$

Dividing Eq. (3.18) by Eq. (3.19), we get

$$\frac{RT_C}{(V_C - b)^2} \cdot \frac{(V_C - b)^3}{RT_C} = \frac{2a}{V_C^3} \cdot \frac{V_C^4}{6a}$$

On simplification

$$V_C = 3b$$

Substituting the value of V_C into Eq. (3.18), we get

$$\frac{RT_C}{(3b - b)^2} = \frac{2a}{(3b)^3}$$

or

$$T_C = \frac{8a}{27Rb}$$

Similarly, at the critical point, $P = P_C$, $T = T_C$, and $V = V_C$; then the van der Waals equation can be expressed as

$$P_C = \frac{RT_C}{V_C - b} - \frac{a}{V_C^2} \tag{3.20}$$

Putting the values of T_C and V_C in Eq. (3.20), we get

$$P_C = \frac{R \cdot \dfrac{8a}{27Rb}}{3b - b} - \frac{a}{(3b)^2} = \frac{a}{27b^2}$$

Therefore, at the critical point, the values of three critical constants are

$$V_C = 3b \qquad T_C = \frac{8a}{27Rb} \qquad \text{and} \qquad P_C = \frac{a}{27b^2}$$

From the value of the critical volume, temperature and pressure, the constants a and b of the van der Waals equation can be evaluated.

$$P_C = \frac{a}{27b^2} \qquad \text{or} \qquad a = 27P_C b^2 \tag{3.21}$$

Again

$$T_C = \frac{8a}{27Rb} \qquad \text{or} \qquad a = \frac{27RbT_C}{8} \tag{3.22}$$

Equating Eqs. (3.21) and (3.22), we have

$$27P_Cb^2 = \frac{27RbT_C}{8}$$

or

$$b = \frac{RT_C}{8P_C} \qquad (3.23)$$

Substituting the value of b in Eq. (3.22), we get

$$a = \frac{27RT_C}{8} \cdot \frac{RT_C}{8P_C} = \frac{27R^2T_C^2}{64P_C} \qquad (3.24)$$

Again, the second set of values of a and b can be determined by putting $V_C = 3b$, or

$$b = \frac{V_C}{3}$$

Putting this value in the following equation, we get

$$P_C = \frac{a}{27b^2} = \frac{a}{27 \cdot \left(\dfrac{V_C}{3}\right)^2} = \frac{a}{3V_C^2}$$

or

$$a = 3P_C V_C^2$$

Putting the value of b in the following equation, we get

$$T_C = \frac{8a}{27Rb} = \frac{8a}{27R \cdot \dfrac{V_C}{3}} = \frac{8a}{9RV_C}$$

Hence, the two sets of values of a and b are

1. $a = \dfrac{27R^2T_C^2}{64P_C}$, $b = \dfrac{RT_C}{8P_C}$

2. $a = 3P_C V_C^2 = \dfrac{9RT_C V_C}{8}$, $b = \dfrac{V_C}{3}$

The approximate value of van der Waals constants a and b for some common gases are given in Table 3.1. Sometimes, these values are found to be quite useful to solve the problems within a short span of time.

Table 3.1 Values of van der Waals constants for some common gases

Gas	$a\left(\dfrac{Nm^4}{(kmol)^2}\right)$	$b\left(\dfrac{m^3}{kmol}\right)$
Ammonia	426,295	0.0373
Carbon dioxide	368,127	0.0428
Nitrogen	137,450	0.0378
Oxygen	139,044	0.0317

(Contd.)

Table 3.1   Values of van der Waals constants for some common gases   (*Contd.*)

Gas	$a \left(\dfrac{\text{Nm}^4}{(\text{kmol})^2} \right)$	$b \left(\dfrac{\text{m}^3}{\text{kmol}} \right)$
Hydrogen	24,800	0.0266
Helium	3440	0.0232
Air	137,052	0.0366
Freon-12	1,082,470	0.0998

EXAMPLE 3.5   If a cylinder of volume 0.1 m³ is filled with 1.373 kg of ammonia at 1.95 MPa, determine the temperature at which ammonia exists in the cylinder. Assume that ammonia obeys the van der Waals equation of state. The van der Waals constants a and b for ammonia are 422.546×10^{-3} m³/mol² and 37×10^{-6} m³/mol respectively.

Solution: Here molecular weight of ammonia = 17

Molar mass of ammonia (NH_3) = 17×10^{-3} kg/mol

$$\text{Quantity of ammonia} = 1.373 \text{ kg} = \frac{1.373}{17 \times 10^{-3}} = 80.765 \text{ mol}$$

Molar volume of ammonia

$$V = \frac{0.1}{80.765} = 1.238 \times 10^{-3} \text{ m}^3/\text{mol}$$

Applying the van der Waals equation of state, we get

$$P = \frac{RT}{V - b} - \frac{a}{V^2}$$

or

$$1.95 \times 10^6 = \frac{8.314T}{1.238 \times 10^{-3} - 37 \times 10^{-6}} - \frac{422.546 \times 10^{-3}}{(1.238 \times 10^{-3})^2}$$

or

$$T = 321.5 \text{ K}$$

Hence, the temperature at which ammonia exists in the cylinder is 321.5 K.

3.7.2   Redlich–Kwong Equation of State

These two parameters of equation of state was proposed by Redlich and Kwong in 1949. It is given by

$$P = \frac{RT}{V - b} - \frac{a}{T^{0.5}V(V + b)} \tag{3.25}$$

where a and b are empirical constants.

This equation provides an impressive result at moderate density and the temperature greater than critical temperature.

At the critical point

$$\left(\frac{\partial P}{\partial V} \right)_T = 0 \quad \text{and} \quad \left(\frac{\partial^2 P}{\partial V^2} \right)_T = 0 \tag{3.26}$$

From the knowledge of critical constants of the gas, the values of a and b can be determined in terms of critical temperature and pressure as

$$a = \frac{0.4278 \, R^2 T_C^{2.5}}{P_C} \quad \text{and} \quad b = \frac{0.0867 RT_C}{P_C} \tag{3.27}$$

This equation is used in many engineering calculations.

EXAMPLE 3.6 Calculate the molar volume of methane at 773 K and 15 bar using the following methods:

(a) Ideal gas equation of state

(b) van der Waals equation of state, with $a = 0.2303$ Nm^4/mol^2 and $b = 4.3073.12 \times 10^{-5}$ m^3/mol

(c) Virial equation of state, $Z = 1 + \dfrac{B}{V}$

(d) Redlich–Kwong equation of state, when $T_C = 190.6$ K and $P_C = 45.99$ bar.

For methane, $T_C = 190.6$ K, $P_C = 45.99$ bar, $V_C = 98.6$ cm^3/mol, and $Z_C = 0.286$.

Solution: We are given that $P = 15$ bar $= 15 \times 10^5$ N/m^2 and $T = 773$ K. Therefore

(a) Applying ideal gas equation of state, we have

$$V = \frac{RT}{P} = \frac{8.314 \times 773}{15 \times 10^5} = 4.28 \times 10^{-3} \text{ m}^3/\text{mol}$$

(b) From the van der Waals equation of state, we have

$$P = \frac{RT}{V - b} - \frac{a}{V^2}$$

or

$$\left(P + \frac{a}{V^2}\right)(V - b) - RT = \Delta$$

By trial and error method, the gas volume ▒▒▒▒rmined. First, a value for V is to be assumed. Then it is substituted in the equati▒▒▒▒evaluate the left hand side of the equation. If it is zero, then the value assumed is ▒▒▒rect one. If it is not so, then assume another value and repeat the calculation until the L.H.S as well as Δ becomes zero.

We can safely proceed to use the trial and error method by considering the volume obtained from using the ideal gas equation.

The results are as follows:

Iteration number	Volume of gas (in m^3/mol)	Δ
1.	4.28×10^{-3}	-71.222
2.	4.23×10^{-3}	-146.22
3.	4.30×10^{-3}	-40.77
4.	4.32×10^{-3}	-10.722
5.	4.34×10^{-3}	19.228
6.	4.33×10^{-3}	4.228

It is found from the table that the volume of gas, V, equal to 4.33×10^{-3} is close to zero compared to the other values.

(c) From the virial equation of state, we have

$$Z = 1 + \frac{B}{V}$$

or

$$\frac{PV}{RT} = 1 + \frac{B}{V}$$

or

$$V^2 - \frac{VRT}{P} - \frac{BRT}{P} = 0$$

or

$$V^2 - \frac{8.314 \times 773 V}{15 \times 101325} - \frac{1.3697 \times 10^{-5} \times 8.314 \times 773}{15 \times 101325} = 0$$

or

$$V^2 - 4.229 \times 10^{-3} V - 5.79 \times 10^{-8} = 0$$

Solving the quadratic equation and taking the positive root, we get

$$V = 4.243 \times 10^{-3} \ \text{m}^3/\text{mol}$$

(d) From the Redlich–Kwong equation of state, we have

$$P = \frac{RT}{V - b} - \frac{a}{T^{0.5} V(V + b)}$$

or

$$V^3 - \frac{RT}{P} V^2 - \left(b^2 + \frac{bRT}{P} - \frac{a}{PT^{0.5}} \right) V - \frac{ab}{PT^{0.5}} = 0$$

Now, let us calculate the ░░░░░ constants a and b by the following relation:

$$\frac{0.4278 R^2 T_C^{2.5}}{P_C} = 3.2224 \ \text{Nm}^4/\text{mol}^2$$

$$\frac{0.0867 RT_C}{P_C} = 2.9853 \times 10^{-5} \ \text{m}^3/\text{mol}$$

Substitute the values of a and b in the cubic equation of Redlich–Kwong, we get

$$V^3 - 4.229 \times 10^{-3} V^2 - 5.0898 \times 10^{-8} V - 2.276 \times 10^{-12} = 0$$

or

$$V = 4.241 \times 10^{-3} \ \text{m}^3/\text{mol}$$

The other two roots are imaginary.

3.7.3 Redlich–Kwong–Soave Equation of State

The Redlich–Kwong–Soave equation of state is a two-parameter equation and is the modified form of Redlich–Kwong equation. This modification was proposed by Soave in 1972. The modified form is given by

$$P = \frac{RT}{V - b} - \frac{a\alpha}{V(V + b)} \tag{3.28}$$

where a and b are constants.

Pitzer's acentric factor, ω, is incorporated to this equation to provide the more accurate results, where $\omega = -1.00 - \log\left(\dfrac{P^S}{P_C}\right)_{T_R = 0.7}$

where

P^S = Vapour pressure

P_C = Critical pressure

T_R = Reduced temperature.

This equation is widely used to determine the properties of hydrocarbons.

3.7.4 Peng–Robinson Equation of State

The Peng–Robinson equation of state was proposed in 1976. It is given by

$$P = \frac{RT}{V - b} - \frac{a\alpha}{V(V + b) + b(V - b)} \tag{3.29}$$

where a and b are constants and α is a function of the reduced temperature and the acentric factor.

This equation is widely used to determine the thermodynamic properties of hydrocarbons and gases such as nitrogen and oxygen.

3.7.5 Benedict–Webb–Rubin Equation of State

The Benedict–Webb–Rubin equation of state was introduced in 1940. This equation of state is a complex one as it contains eight parameters. But it has greater accuracy in results than the other equations of state. This model is known as the *multi-parameter model*. This equation is observed to be suitable for finding out the thermodynamic properties of hydrocarbons. Therefore, it is extensively used in petroleum industries.

$$P = \frac{RT}{V} + \frac{B_0 RT - A_0 - \dfrac{C_0}{T^2}}{V^2} + \frac{bRT - a}{V^3} + \frac{a\alpha}{V^6} + \frac{c}{V^3 T^2}\left(1 + \frac{\gamma}{V^2}\right)e^{-\frac{\gamma}{V^2}} \tag{3.30}$$

where A_0, B_0, C_0, a, b, c, α and γ are constants.

3.7.6 Beattie–Bridgeman Equation of State

This equation of state was proposed by Beattie and Bridgeman in 1928. This is the most widely used equation of state. This is a five-constant model. The equation is given by

$$P = \frac{RT}{V^2}\left[V + B_0\left(1 - \frac{b}{V}\right)\right]\left(1 - \frac{c}{VT^3}\right) - A_0\left(1 - \frac{a}{V}\right) \tag{3.31}$$

where

a, b, A_0, B_0 and c are constants
V = Molal volume in L/g-mol
P = Pressure in atm
T = Temperature in K
R = Universal gas constant = 0.082 L-atm/g-mol K.

This equation provides accurate results for substances of low density.

3.7.7 Virial Equation of State

The term *virial* has been derived from the Latin word *vis*, which means *force*. Therefore, the interaction forces between the molecules are taken into consideration in deriving the virial equation.

In 1901, Kammerlingh–Onnes suggested that PV can be expressed in a power series of pressure, as

$$PV = RT(1 + B'P + C'P^2 + D'P^3 + \cdots) \tag{3.32}$$

where B', C' and D' are constants.

It can be more conveniently expressed in terms of volume as

$$Z = \frac{PV}{RT} = 1 + \frac{B}{V} + \frac{C}{V^2} + \frac{D}{V^3} + \cdots \tag{3.33}$$

where

B = Second virial coefficient
C = Third virial coefficient
D = Fourth virial coefficient
Z = Compressibility factor.

This equation is known as the *Virial equation of state*.

These coefficients are functions of the temperature and nature of the gas and play an important role in expressing the deviation from ideal gas behaviour. Of these coefficients, B and C, the second and third virial coefficients, are most important to be considered for substantiating the two-body and three-body interactions respectively.

Equation (3.29) can be represented as limiting to the second term of the power series. We have

$$PV = RT\left(1 + \frac{B}{V}\right)$$

or

$$PV = RT + \frac{BRT}{V} \tag{3.34}$$

From Eq. (3.32), we get

$$PV = RT + B'RTP \tag{3.35}$$

By comparing the Eqs. (3.34) and (3.35), we have

$$\frac{B}{V} = B'P$$

or

$$B = B'VP = B'RT$$

again

$$B' = \frac{B}{RT}$$

Similarly, limiting to the third term of the power series, C' can be determined as

$$\frac{PV}{RT} = 1 + \frac{B}{V} + \frac{C}{V^2} \qquad (3.36)$$

$$P = RT\left(\frac{1}{V} + \frac{B}{V^2} + \frac{C}{V^3}\right)$$

From Eq. (3.32), we get

$$PV = RT(1 + B'P + C'P^2)$$

or

$$Z = 1 + B'P + C'P^2 \qquad (3.37)$$

Putting the value of P in Eq. (3.37), we have

$$Z = 1 + B'RT\left(\frac{1}{V} + \frac{B}{V^2} + \frac{C}{V^3}\right) + C'(RT)^2\left(\frac{1}{V} + \frac{B}{V^2} + \frac{C}{V^3}\right)^2 \qquad (3.38)$$

or, $$Z = 1 + \frac{B'RT}{V} + \frac{BB'RT + C'(RT)^2}{V^2} + \frac{CB'RT + 2C'(RT)^2 B + D'(RT)^3}{V^2} + \cdots \quad (3.39)$$

By comparing the $\frac{1}{V^2}$ term between Eqs. (3.39) and (3.33), we have

$$C = BB'RT + C'(RT)^2$$

or

$$C' = \frac{C - B^2}{(RT)^2} \quad \text{(since } B = B'RT, \; BB'RT = B^2\text{)}$$

Similarly, by comparing the $\frac{1}{V^3}$ term between Eqs. (3.39) and (3.33), we have

$$D = CB'RT + 2C'(RT)^2 B + D'(RT)^3$$

or

$$D' = \frac{D - 3BC + 2B^3}{(RT)^3}$$

EXAMPLE 3.7 Find the second, third and fourth virial coefficients of the van der Waals equation of state. (WBUT, 2008)

Solution: From the van der Waals equation, we get

$$\left(P + \frac{a}{V^2}\right)(V - b) = RT$$

or

$$PV = RT - \frac{a}{V} + Pb + \frac{ab}{V^2} \qquad \text{(i)}$$

The equation can be expressed in terms of P as

$$P = \frac{RT}{V - b} - \frac{a}{V^2}$$

We substitute the value of P in the R.H.S. of equation (i) and put $a = RT = $ constant, because it was observed that PV is constant for a vapour or a gas along an isotherm at the critical point, and this equals RT. Hence, we have

$$Z = 1 + \frac{b}{V - b} - \frac{1}{V} \qquad \text{(ii)}$$

It can be rearranged as

$$Z = 1 + \frac{b}{V}\left(1 - \frac{b}{V}\right)^{-1} - \frac{1}{V} \qquad \text{(iii)}$$

We know that

$$\left(1 - \frac{b}{V}\right)^{-1} = 1 + \frac{b}{V} + \frac{b^2}{V^2} + \cdots \qquad \text{(iv)}$$

Therefore, Eq. (iii) can be simplified as

$$Z = 1 + (b - 1)\frac{1}{V} + \frac{b^2}{V^2} + \frac{b^3}{V^3} + \cdots \qquad \text{(v)}$$

Now, from the virial equation of state, we get

$$Z = \frac{PV}{RT} = 1 + \frac{B}{V} + \frac{C}{V^2} + \frac{D}{V^3} + \cdots \qquad \text{(vi)}$$

Comparing the coefficients of Eqs. (v) and (vi), we have

$$B = b - 1$$
$$C = b^2$$
$$D = b^3$$

EXAMPLE 3.8 Determine the molar volume of *n*-butane at 500 K and 8 MPa by making use of (a) the ideal gas law and (b) the virial equation of state. The virial coefficients of *n*-butane are given by $B = -0.265 \times 10^{-3}$ m³/mol and $C = 0.3025 \times 10^{-7}$ m⁶/mol².

Solution: (a) The ideal gas equation of state gives

$$V = \frac{RT}{P} = \frac{8.314 \times 500}{8 \times 10^6} = 0.5196 \times 10^{-3} \text{ m}^3/\text{mol}$$

(b) The virial equation of state is given by

$$\frac{PV}{RT} = 1 + \frac{B}{V} + \frac{C}{V^2} \qquad \text{(i)}$$

The virial equation of state is pressure-explicit, and hence it is easier to calculate P if V is given. On the other hand, to calculate V from P, it is necessary to solve the equation by trial and error.

Let us assume that $V = 0.219 \times 10^{-3}$ m^3/mol and check whether equation (i) is satisfied or not.

L.H.S. of equation (i)

$$= \frac{PV}{RT} = \frac{8 \times 10^6 \times 0.219 \times 10^{-3}}{8.314 \times 500} = 0.4215$$

R.H.S. of equation (i)

$$= 1 + \frac{B}{V} + \frac{C}{V^2} = 1 + \frac{(-0.265 \times 10^{-3})}{0.219 \times 10^{-3}} + \frac{0.3025 \times 10^{-7}}{(0.219 \times 10^{-3})} = 0.4206$$

Hence, the L.H.S. and R.H.S of Eq. (i) differs only in the 3rd decimal place. So, Eq. (i) is satisfied.

Therefore, $V = 0.219 \times 10^{-3}$ m^3/mol.

EXAMPLE 3.9 Prove that

$$C_P - C_V = \frac{TV\beta^2}{\alpha}$$

where

α = Isothermal compressibility
β = Volume expansivity.

Solution: From the expression of isothermal compressibility, we have

$$\alpha = -\frac{1}{V}\left(\frac{\partial V}{\partial P}\right)_T$$

or

$$\left(\frac{\partial V}{\partial P}\right)_T = -\alpha V \qquad (3.40)$$

Again, volume expansivity is given by

$$\beta = \frac{1}{V}\left(\frac{\partial V}{\partial T}\right)_P$$

or

$$\left(\frac{\partial V}{\partial T}\right)_P = \beta V \qquad (3.41)$$

Putting the values in the following equation, we get

$$C_P - C_V = -T\left(\frac{\partial V}{\partial T}\right)_P^2 \left(\frac{\partial P}{\partial V}\right)_T$$

$$= -T(\beta V)^2 \left(-\frac{1}{\alpha V}\right) \left(\text{since } \left(\frac{\partial P}{\partial V}\right)_T = \frac{1}{\left(\frac{\partial V}{\partial P}\right)_T}\right)$$

$$= \frac{T\beta^2 V^2}{\alpha V}$$

$$= \frac{TV\beta^2}{\alpha}$$

Hence proved.

EXAMPLE 3.10 For liquid acetone at 20°C and 1 bar,

$$\beta = 1.487 \times 10^{-3}/°C \qquad \alpha = 62 \times 10^{-6}/\text{bar} \qquad V = 1.287 \text{ cm}^3/\text{g}$$

Find out

(a) the value of $\left(\dfrac{\partial P}{\partial T}\right)_V$ at 20°C and 1 bar

(b) the pressure generated by heating at constant volume V from 20°C and 1 bar to 30°C.

(c) the change in volume for a change from 20°C and 1 bar to 0°C and 10 bar.

Solution: (a) the value of the derivative $\left(\dfrac{\partial P}{\partial T}\right)_V$ can be determined by the following way, in which volume V = constant or $dV = 0$. We have

$$\frac{dV}{V} = \beta dT - \alpha dP$$

or

$$0 = \beta dT - \alpha dP$$

or

$$\alpha dP = \beta dT$$

or

$$\left(\frac{\partial P}{\partial T}\right)_V = \frac{\beta}{\alpha} = \frac{1.487 \times 10^{-3}}{62 \times 10^{-6}} = 24 \text{ bar/°C}$$

(b) Applying the same equation, we get

$$\left(\frac{\partial P}{\partial T}\right)_V = \frac{\beta}{\alpha}$$

or

$$\frac{\partial P}{\partial T} = \frac{\beta}{\alpha}$$

or

$$\frac{\Delta P}{\Delta T} = \frac{P_2 - P_1}{T_2 - T_1} = \frac{\beta}{\alpha}$$

or

$$P_2 - P_1 = \frac{\beta}{\alpha}(T_2 - T_1) = 24 \times (30 - 20) = 240 \text{ bar}$$

or

$$P_2 = P_1 + 240 \text{ bar} = (1 + 240) = 241 \text{ bar}$$

(c) The change in volume ΔV can be obtained from the following equation:

$$\ln\frac{V_2}{V_1} = \beta(T_2 - T_1) - \alpha(P_2 - P_1)$$

$$= 1.487 \times 10^{-3} (0 - 20) - 62 \times 10^{-6} (10 - 1) = -0.0303$$

or

$$\frac{V_2}{V_1} = 1.0307$$

$$V_2 = 1.0307 \times 1.287 = 1.326 \text{ cm}^3/\text{g}$$

$$\Delta V = V_2 - V_1 = (1.326 - 1.287) \text{ cm}^3/\text{g} = 0.039 \text{ cm}^3/\text{g}$$

3.8 COMPRESSIBILITY FACTOR

The deviation from ideal gas behaviour at a given temperature and pressure can be measured by the introduction of a correction factor called the *compressibility factor (Z)*. It is also the ratio of the volume of the real gas to the volume occupied by the ideal gas at the same temperature and pressure.

The non-ideal behaviour of real gases is more conveniently expressed by the following equation:

$$PV = ZRT$$

or

$$Z = \frac{PV}{RT}$$

or

$$Z = \frac{V}{\dfrac{RT}{P}} = \frac{V_{real}}{V_{ideal}} \tag{3.42}$$

where

Z = Compressibility factor

V_{real} = Volume of real gas

V_{ideal} = Volume occupied by the ideal gas.

The compressibility factor Z depends on the pressure at a given temperature. It has been shown in Fig. 3.9. The value of Z with respect to pressure can be judged with the help of plot of Z vs P.

For an ideal gas, $Z = 1$ at all temperatures and pressures

For a real gas, Z can be different from unity.

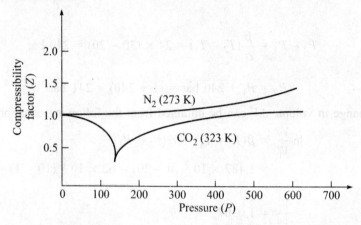

Fig. 3.9 Graphical representation of compressibility factor versus pressure.

The value of Z signifies the departure (extent of deviation) from the ideal gas behaviour. Alternatively, it can also be said that Z is a measure of the deviation for real gases.

A parameter introduced to express the non-ideality of the gases is known as *residual volume*. It is denoted by V^R. This is nothing but the difference between the volume of real gas and the volume predicted by ideal gas. It can be mathematically expressed as

$$V^R = \frac{RT}{P} - \frac{ZRT}{P} = (1 - Z)\frac{RT}{P} \qquad (3.43)$$

3.9 LAW OF CORRESPONDING STATE

It has been experimentally observed that the behaviour of the different gases are not same at same temperature and pressure. But they exhibit the same behaviour at reduced temperature and pressure. *The compressibility factor, Z, for all gases is approximately the same at the same reduced temperature and reduced pressure.* This is known as the *law of corresponding state*. This law was first proposed by van der Waals in 1873.

Reduced temperature (T_r) is defined as the ratio of temperature of the system to the critical temperature. Mathematically

$$T_r = \frac{T}{T_C}$$

where

T = Temperature of the system

T_C = Critical temperature.

Similarly, *reduced pressure* (P_r) is defined as the ratio of pressure of the system to the critical pressure. Mathematically,

$$P_r = \frac{P}{P_C}$$

where

> P = Pressure of the system
>
> P_C = Critical pressure.

The compressibility factors for different gases at reduced temperature and pressure are shown in the plot of Z versus P_r. It is found from Fig. 3.10 that the gases have the same Z value at the same reduced properties. That is, the gases are obeying the law of corresponding state.

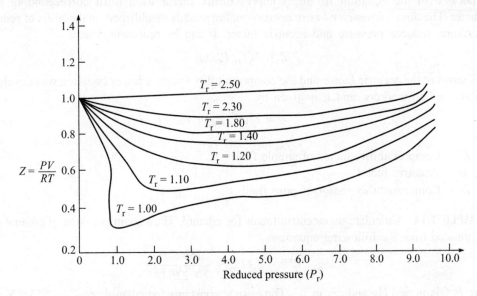

Fig. 3.10   Comparison of Z factors for various gases.

This law can also be stated as: *If two or more substances have the same reduced pressure and temperature, then they will have the same reduced volume.* It can be mathematically represented as

$$Z = f(T_r, P_r)$$

This law is known as the *two-parameter law of corresponding state.*

3.10   ACENTRIC FACTOR

The acentric factor, ω, was introduced by Pitzer and co-workers to measure the deviation of the intermolecular potential of a molecule (pure fluid) from that of a simple spherical molecule (simple fluid). The factor accounts for characteristic of molecular structure. For a pure fluid, it is defined in terms of reduced vapour pressure.

It can be mathematically expressed as

$$\omega = -1 - \log_{10} \left. P_r^{Sat} \right|_{T_r = 0.7} \tag{3.44}$$

Molecules which are spherical with respect to geometry and force field are taken as reference molecules. These are inert fluids such as argon, krypton, xenon, which are also known as *simple fluids*. It was noted from the plot of $\log_{10} P_r^{Sat}$ versus $\dfrac{1}{T_r}$ that $\log_{10} P_r^{Sat} = -1$ at $T_r = 0.7$ for simple fluids for which the acentric factor is defined to be zero (i.e., $\omega = 0$).

It has been experimentally observed that the simple fluids have similar intermolecular potentials and obey the law of corresponding state. Therefore, the value of ω can be determined for any fluid from the knowledge of T_r and P_r. In addition to T_r and P_r, the acentric factor was incorporated in the equation for the compressibility factor as a third corresponding state parameter. The *three-parameter law of corresponding state* is established on the basis of reduced temperature, reduced pressure and acentric factor. It can be represented as

$$Z = Z(T_r, P_r, \omega)$$

In terms of the acentric factor and the compressibility factor, a linear equation was developed by Pitzer and co-workers, and it is given by

$$Z = Z^0(T_r, P_r) + \omega Z^1(T_r, P_r) \qquad (3.45)$$

where

Z^0 = Compressibility factor of simple fluid
ω = Acentric factor
Z^1 = Compressibility factor of pure fluid.

EXAMPLE 3.11 Calculate the acentric factor for ethanol. The vapour pressure of ethanol can be estimated from the following equation:

$$\log_{10} P^{Sat} = 8.1122 - \frac{1592.864}{t + 226.184}$$

where P^{Sat} is in mm Hg and t is in °C. The critical constants for ethanol are $T_C = 513.9$ K and $P_C = 61.48$ bar.

Solution: At $T_r = 0.7$, $T = 0.7 \times 513.9 = 359.73$ K

At this temperature, the value of P^{Sat} is to be determined. The equation is given by

$$\log_{10} P^{Sat} = 8.1122 - \frac{1592.864}{t + 226.184}$$

where,

$$t = 359.73 - 273.15 = 86.58 \text{ °C}$$

or

$$\log_{10} P^{Sat} = 3.01934$$

or

$$P^{Sat} = \frac{1045.53}{760} \times 101325 = 1.394 \times 10^5 \text{ N/m}^2$$

or

$$P_r^{Sat} = \frac{1.394 \times 10^5}{61.48 \times 10^5} = 0.022674$$

or

$$\log P_r^{Sat} = -1.64447$$

Now, putting the value of P_r^{Sat} in the expression for acentric factor, we have

$$\omega = -1 - \log (P_r^{Sat})_{T_r = 0.7}$$
$$= -1 - (-1.64447) = 0.64447$$

SUMMARY

In this chapter, the concept of pure substance and its existence in a particular phase have been discussed. The most important terminology such as saturation temperature, saturation pressure, critical point, sub-cooled liquid, superheated vapour, triple point, latent heat of fusion, latent heat of vaporization, and reduced parameter have clearly been discussed, as the pure substance is better characterized by these parameters. The $P-V-T$ behaviour of a pure substance have been prominently illustrated with the help of property diagrams such as $T-V$, $P-V$ and $P-T$ diagrams. The equation of state helps to make one understand the $P-V-T$ behaviour of the pure substance by establishing the functional relationship between temperature, pressure and volume. The simplest and familiar equation of state is the ideal gas equation of state, which is mathematically expressed as $PV = RT$. Real gases exhibit ideal gas behaviour at relatively low pressure and high temperature. The compressibility factor, Z, which accounts for the deviation from ideal gas behaviour is defined by $Z = \dfrac{PV}{RT}$, where $Z = 1$ for ideal gas and more than unity for real gases.

The law of corresponding states tells us that the compressibility factor is the same for all gases at the same reduced temperature and pressure. The $P-V-T$ relation of fluids for real gases have been elaborately discussed with some best-known equations of state, such as van der Waals, Kammerlingh–Onnes, Peng–Robinson. Benedict–Webb–Rubin, and Redlich–Kwong–Soave. The importance of acentric factor has been presented in detail.

KEY TERMS

Acentric Factor   The third corresponding state parameter introduced in the equation for the compressibility factor.

Avogadro's Law   At normal temperature and pressure, the number of molecules is directly proportional to the volume.

Boyle's Law   The molar volume of a gas varies inversely or is inversely proportional to the pressure when temperature is constant.

Charles' Law   The molar volume of a gas is directly proportional to the temperature when the pressure is constant.

Compressed Liquid   A liquid at a temperature below its saturation temperature, at a specific pressure.

Compressibility Factor (Z)   The correction factor by which the deviation from ideal gas behaviour at a given temperature and pressure can be measured.

Critical Point The highest temperature and pressure above which a liquid cannot exist.

Cubic Equation of State The equation of state which can encompass a wide range of temperature and pressure.

Equation of State Any equation that relates the pressure, temperature and volume of a substance.

Ideal Gas Equation of State The equation obeyed by an ideal gas.

Isothermal Compressibility The rate of decrease of volume with pressure at constant temperature per unit volume.

Latent Heat of Fusion The amount of heat absorbed during the melting process.

Latent Heat of Vaporization The amount of heat absorbed during the vaporization process.

Law of Corresponding State The compressibility factor, Z, for all gases is approximately the same at the same reduced temperature and reduced pressure.

Phase A physically distinct but homogeneous part of a system separated from other parts by boundary surfaces.

Pure Substance A substance which has fixed chemical composition throughout.

Residual Volume A parameter introduced to express the non-ideality of the gases; it is denoted by V^R and is nothing but the difference between the volume of the real gas and the volume predicted by the ideal gas.

Saturated Liquid The liquid which is in equilibrium with its own vapour at a specified temperature or pressure.

Saturated Vapour The vapour which is in equilibrium with its own liquid at a specified temperature or pressure.

Saturation Pressure The pressure at which a liquid is in equilibrium with its own vapour at a specified temperature.

Saturation Temperature The temperature at which a liquid is in equilibrium with its own vapour at a specified pressure.

Superheated Vapour The vapour existing at a temperature above the saturation temperature at a specified pressure.

Triple Point The state at which all the three phases—solid, liquid and vapour—coexist in equilibrium.

Virial equation The equation by which PV of a pure substance can be represented in a power series.

Volume Expansivity The rate of change of volume with respect to temperature at constant pressure per unit volume.

IMPORTANT EQUATIONS

1. Boyle's Law is given by

$$P \propto \frac{1}{V} \quad \text{when } T = \text{Constant}$$

or

$$PV = K = \text{Constant} \tag{3.9}$$

2. Charles' Law is given by

$$V \propto T \quad \text{when } P = \text{Constant}$$

or

$$V = K'T \tag{3.10}$$

3. Avogadro's Law is given by

$$V \propto n \tag{3.11}$$

4. Ideal gas equation of state is given by

$$V \propto \frac{nT}{P}$$

or

$$PV = nRT \tag{3.12}$$

5. Van der Waals equation of state for real gases

$$\left(P + \frac{a}{V^2}\right)(V - b) = RT \tag{3.14}$$

6. Two sets of values of van der Waals constants a and b are given by

1. $a = \dfrac{27R^2 T_C^2}{64 P_C} \qquad b = \dfrac{RT_C}{8 P_C}$

2. $a = 3P_C V_C^2 = \dfrac{9RT_C V_C}{8} \qquad b = \dfrac{V_C}{3}$

7. Redlich–Kwong equation of state is given by

$$P = \frac{RT}{V - b} - \frac{a}{T^{0.5} V(V + b)} \tag{3.25}$$

8. Redlich–Kwong–Soave equation of state is given by

$$P = \frac{RT}{V - b} - \frac{a\alpha}{V(V + b)} \tag{3.28}$$

9. Peng–Robinson equation of state is given by

$$P = \frac{RT}{V - b} - \frac{a\alpha}{V(V + b) + b(V - b)} \tag{3.29}$$

10. Benedict–Webb–Rubin equation of state is given by

$$P = \frac{RT}{V} + \frac{B_0 RT - A_0 - \dfrac{C_0}{T^2}}{V^2} + \frac{bRT - a}{V^3} + \frac{a\alpha}{V^6} + \frac{c}{V^3 T^2}\left(1 + \frac{\gamma}{V^2}\right)e^{-\frac{\gamma}{V^2}} \tag{3.30}$$

where A_0, B_0, C_0, a, b, c, α and γ are constants.

11. Beattie–Bridgeman equation of state is given by

$$P = \frac{RT}{V^2}\left[V + B_0\left(1 - \frac{b}{V}\right)\right]\left(1 - \frac{c}{VT^3}\right) - A_0\left(1 - \frac{a}{V}\right) \tag{3.31}$$

where

a, b, A_0, B_0, c = Constants
V = Molal volume in L/g mol
P = Pressure in atm
T = Temperature in K
R = Universal gas constant = 0.082 L-atm/g-mol-K.

12. Virial equation is given by

$$PV = RT(1 + B'P + C'P^2 + D'P^3 + \cdots) \tag{3.32}$$

where

B', C' and D' are constants.

It can be more conveniently expressed in terms of the volume as

$$Z = \frac{PV}{RT} = 1 + \frac{B}{V} + \frac{C}{V^2} + \frac{D}{V^3} + \cdots \tag{3.33}$$

where

B = Second virial coefficient
C = Third virial coefficient
D = Fourth virial coefficient
Z = Compressibility factor.

13. Value of second virial coefficient is given by

$$B' = \frac{B}{RT}$$

14. Value of third virial coefficient is given by

$$C' = \frac{C - B^2}{(RT)^2}$$

15. Value of fourth virial coefficient is given by

$$D' = \frac{D - 3BC + 2B^3}{(RT)^3}$$

16. Isothermal compressibility is given by

$$\alpha = -\frac{1}{V}\left(\frac{\partial V}{\partial P}\right)_T$$

17. Isentropic compressibility is given by

$$\alpha = -\frac{1}{V}\left(\frac{\partial V}{\partial P}\right)_S$$

18. Volume expansivity or coefficient of volume expansion (β)

$$\beta = \frac{1}{V}\left(\frac{\partial V}{\partial T}\right)_P$$

19. Compressibility factor is given by

$$Z = \frac{PV}{RT}$$

or

$$Z = \frac{V}{\frac{RT}{P}} = \frac{V_{real}}{V_{ideal}} \qquad (3.42)$$

where

Z = Compressibility factor
V_{real} = Volume of real gas
V_{ideal} = Volume occupied by ideal gas.

20. Residual volume is given by

$$V^R = \frac{RT}{P} - \frac{ZRT}{P} = (1 - Z)\frac{RT}{P}$$

$$= \frac{RT}{P} - \frac{ZRT}{P} = (1 - Z)\frac{RT}{P} \qquad (3.43)$$

21. Law of corresponding state is given by

$$Z = f(T_r, P_r)$$

22. Acentric factor is given by

$$\omega = -1 - \log_{10} P_r^{sat}\Big|_{T_r = 0.7} \qquad (3.44)$$

EXERCISES

A. Review Questions

1. What is pure substance? Give some examples to justify the concept.
2. What do you mean by the different phases of a pure substance?
3. What is the difference between compressed liquid and saturated liquid?
4. Give an informatory note on saturation temperature and saturation pressure.
5. Explain the significance of critical point and triple point.
6. Draw a T–V diagram and indicate the saturated liquid and saturated vapour lines.
7. Draw a P–T diagram and show the sublimation and vaporization curves.
8. What is the importance of latent heat of fusion and vaporization?
9. *Any equation that relates to the pressure, temperature and volume is called an equation of state.* Justify the statement.
10. Establish the ideal gas equation of state from Charles' law, Boyle's law and Avogadro's law. Enumerate the limitation of the ideal gas law.
11. What is the necessity of cubic equation of state?
12. How does the van der Waals equation of state play an important role in improving the ideal gas law?

13. Explain the importance and application of Peng–Robinson equation of state.
14. Name some proposed model equations of state for real gases.
15. What is compressibility factor? Explain its necessity for showing the deviation from ideal gas behaviour for real gases.
16. Discuss the importance of generalized compressibility chart.
17. State the law of corresponding state and show that the gases are obeying the law by plotting a graph of Z versus P_R.
18. Write a short note on superheated vapour.
19. What is virial equation of state?
20. Write informatory notes on
    Redlich–Kwong–Soave equation of state
    Benedict–Webb–Rubin equation of state.
21. How do you explain the significance of virial coefficients?
22. What is acentric factor?

B. Problems

1. Determine the molar volume of a gas at 137°C and a pressure of 12 bar. Assume the gas behaves ideally.
2. A spherical balloon with a diameter of 6 m is filled with helium at 20°C and 200 kPa. Determine the mole number and the mass of the helium in the balloon.
3. 12.0 m^3 of air is enclosed by a frictionless piston at in a cylinder at 325 kPa. The gas undergoes a compression process with no change in temperature and its volume becomes 36.0 m^3. What is the pressure of air after compression? Air is assumed to be an ideal gas.
4. Carbon dioxide occupies a tank at 100°C. If the volume of the tank is 0.5 m^3 and the pressure is 500 kPa, determine the mass of the gas in the tank.
5. Nitrogen at 15 bar is used to fill a container of 0.25 m^3. The filling process is very slow and the contents of the tank attain the room temperature of 295 K. How much gas is there in the container?
6. For a gas, $T_C = 304.2$ K and $P_C = 72.8$ atm. Calculate the van der Waals constant for the gas.
7. Prove that

$$\left(\frac{\partial \beta}{\partial P}\right)_T = -\left(\frac{\partial \alpha}{\partial T}\right)_P$$

where
α = Isothermal compressibility
β = Volume expansivity.

8. Calculate the molar volume of ammonia at 350 K and 100 bar using van der Waals equation of state, given that $a = 0.4233$ Nm^4/mol^2 and $b = 3.73 \times 10^{-5}$ m^3/mol.
9. Determine the specific volume of N_2 gas at 10 MPa and 150 K based on (a) ideal gas equation and (b) generalized compressibility factor.

10. Determine the molar volume of ammonia vapour and ammonia liquid at 321.55 K and 1.95 MPa. Ammonia is assumed to follow van der Waals equation of state.

11. A tank of 1 m^3 volume is filled with 5 kg ammonia at 300 K. Determine the pressure exerted by ammonia using the Redlich–Kwong equation.

12. Calculate the mass of ethane contained in a 0.3 m^3 cylinder at 60°C and 130 bar using the virial equation of state $Z = 1 + \dfrac{BP}{RT}$ for gases, and compare it with the ideal gas law.

    For ethane, $T_C = 305.3$ K, $P_C = 48.72$ bar, and $\omega = 0.1$.

13. Calculate the pressure of CO_2 occupying a volume of 0.425 m^3 at 327 K by using the following equation of state:

    (a) Ideal gas equation of state
    (b) Van der Waals equation of state, when $a = 0.364$ Nm^4/mol^2 and $b = 4.267 \times 10^{-5}$ m^3/mol.

14. Calculate the pressure of 1.0 kmol of methane occupying a volume of 0.9 m^3 in a vessel at a constant temperature of 533 K by using (a) ideal gas equation, (b) van der Waals equation of state when $a = 0.4233$ Nm^4/mol^2 and $b = 3.73 \times 10^{-5}$ m^3/mol, and (c) Redlich–Kwong equation of state when $P_C = 123.2$ bar and $T_C = 398$ K.

15. Calculate the molar volume of methane at 672 K and 12 bar using the following methods:

    (a) Ideal gas equation of state
    (b) Van der Waals equation of state, where $a = 0.2303$ Nm^4/mol^2 and $b = 4.3073.12 \times 10^{-5}$
    (c) Virial equation of state

    $$Z = 1 + \frac{B}{V}$$

    (d) Virial equation of state

    $$Z = 1 + \frac{BP}{RT}$$

    (e) Redlich–Kwong equation of state. For methane, $T_C = 190.6$ K, $P_C = 45.99$ bar, $V_C = 98.6$ cm^3/mol, and $Z_C = 0.286$.

16. The virial equation for ethane is given by $PV = RT + BP$. At 0°C, $B = -0.1814$ L/mol. Calculate the volume of 1 mol of ethane at 10 atm, given that van der Waals constant $a = 5.489$ atm-L/mol. Estimate the critical volume of ethane.

17. Determine the compressibility factor of steam at 627 K and 200 kPa using (a) van der Waals equation and (b) Redlich–Kwong equation of state when $P_C = 123.2$ bar and $T_C = 398$ K.

18. A steel cylinder of 6 L capacity contains 500 g of nitrogen. Calculate the temperature to which the cylinder may be heated without the pressure exceeding 50 atm, given that the compressibility factor, $Z = 0.945$.

19. Using the Peng–Robinson equation of state, calculate the volume of a gas contained in a vessel of capacity 0.8 m^3 at 574 K and 18 bar. $T_C = 190.6$ K, $P_C = 45.99$ bar, $V_C = 98.6$ cm^3/mol, $Z_C = 0.286$ and $\omega = 0.0112$.

20. Calculate the acentric factor for ethanol. The vapour pressure of methanol can be estimated from the following equation:

$$\log_{10} P^{\text{Sat}} = 8.1122 - \frac{1592.864}{t + 226.184}$$

where P^{Sat} is in mm Hg and t is in °C. The critical constants for ethanol are $T_C = 508.9$ K and $P_C = 59.48$ bar.

21. Calculate the second virial coefficients for benzene at 350 K and 8 atm, given that $T_C = 560.3$ K, $P_C = 47.89$ bar, and $\omega = 0.2112$.

22. The Berthelot equation of state is given by

$$\left(P + \frac{a}{TV^2} \right)(V - b) = RT$$

where a and b are constants. Show that $a = \dfrac{27R^2T_C^3}{64P_C}$, $b = \dfrac{RT_C}{8P_C}$, $Z_C = \dfrac{3}{8}$ from the knowledge of point of inflexion at critical isotherms.

23. The Dieterici equation of state is given by

$$P(V - b)\exp\left(\frac{a}{RTV} \right) = RT$$

where a and b are constants. Develop the relations to determine the constants a and b in terms of T_C and P_C.

24. Calculate the volume occupied by isopropanol vapour at 200°C and 10 bar by using

   (a) Ideal gas equation of state
   (b) Virial equation of state

$$Z = 1 + \frac{B}{V}$$

   (c) Virial equation of state

$$Z = 1 + \frac{BP}{RT}$$

We are given that $B = -388$ cm^3/mol and $C = -26{,}000$ cm^6/mol^2.

Heat Effects

4

LEARNING OBJECTIVES

After reading this chapter, you will be able to:

- Know about exothermic and endothermic reactions
- Define the meaning of standard heat of reaction, standard heat of formation, and standard heat of combustion
- State and explain Hess's law of constant heat summation with example
- Discuss Hess's law from the thermodynamic point of view
- Understand the concept of adiabatic and theoretical flame temperatures
- Explain the effect of temperature on the enthalpy change of a chemical reaction

Every chemical reaction involves heat transfer for the breaking of the bonds in the reactants and formation of new bonds to give the products. Heat as energy is required to break the bonds and is released during the formation of the bonds. The calculation of heat transfer rate accompanying a chemical reaction is of great concern to all chemical engineers, because the success of a chemical process very much depends upon the proper design of a chemical reactor, the heart of the chemical process plant. This is the foremost duty of a chemical engineer to produce a significant extent of chemical using modern science and technology in a safe and economic way to fulfil the demands of society by proper design of chemical reactors.

The application of principle of thermodynamics makes the procedure of evaluation of heat changes in a chemical reaction simpler without performing tedious experiment. The chapter deals with the analysis of a chemical process from the thermodynamic point of view. The heat effects associated with combustion, formation, and reaction processes are discussed. The essence of Hess's law of constant heat summation in estimating the heat of formation of compound has been substantiated with sufficient examples. The effect of temperature has a great influence on the rate of reaction and the enthalpy change associated with it. It has precisely been explained along with suitable examples.

4.1 EXOTHERMIC AND ENDOTHERMIC REACTIONS

It is our common experience that all the chemical reactions are accompanied by either evolution or absorption of heat. On the basis of this, chemical reactions can be divided into two categories. These are:

(i) Exothermic reaction
(ii) Endothermic reaction.

Exothermic reaction is defined as the reaction in which heat escapes from the system to the surroundings. For example, carbon is oxidized to form CO_2. The reaction is:

$$C + O_2 = CO_2 \qquad \Delta H = -393.5 \text{ kJ/mol}$$

On the other hand, endothermic reaction is defined as the reaction in which heat is absorbed by the system from the surroundings. For example, hydrogen is iodized to form hydrogen iodide. The reaction is:

$$H_2 + I_2 = 2HI \qquad \Delta H = +49.4 \text{ kJ/mol}$$

It is important to remember that if the chemical reaction takes place at constant pressure, then the heat change is denoted by Q_P. Similarly, when the chemical reaction occurs at constant volume, the necessary heat change is denoted by Q_V. From the thermodynamic point of view, the most useful sign convention adopted to indicate that the heat change for a chemical reaction is +ve for heat absorption and −ve for heat evolution.

Now, the magnitude of Q_P and Q_V may be different for the same chemical reaction. For example,

$$C_7H_8 + 9O_2 = 7CO_2 + 4H_2O \qquad \begin{aligned} Q_P &= -3893.9 \text{ kJ/mol} \\ Q_V &= -3888.69 \text{ kJ/mol} \end{aligned}$$

4.2 RELATION BETWEEN Q_P AND Q_V

We know that for a constant pressure process which is mechanically reversible

$$Q_P = \Delta H_P = \Delta U_P + P\Delta V \qquad (4.1)$$

where ΔV is the change in volume of the system.

Again, for a mechanically reversible constant-volume process

$$Q_V = \Delta U_V \qquad (4.2)$$

Subtracting Eq. (4.2) from (4.1), we get

$$Q_P - Q_V = \Delta U_P + P\Delta V - \Delta U_V \qquad (4.3)$$

In case of ideal gas, the change in internal energy, ΔU, is a function of temperature only and is independent of pressure and volume at constant temperature. On the other hand, the difference between ΔU_P and ΔU_V is considered to be negligible compared to other heat effects.

The substitution of this condition, Eq. (4.3) gives

$$Q_P - Q_V = P\Delta V \qquad (4.4)$$

For a gaseous reaction, if n_1 and n_2 are the numbers of reactants and resultant molecules respectively, then

$$P\Delta V = P(n_2 - n_1)\frac{RT}{P} = \Delta nRT \tag{4.5}$$

Hence, Eq. (4.4) becomes

$$Q_P - Q_V = \Delta nRT$$

or

$$Q_P = Q_V + \Delta nRT \tag{4.6}$$

where Δn is the increase in numbers of moles in the reaction.

EXAMPLE 4.1 The heat of combustion of liquid ethanol into CO_2 and liquid water is 327 kcal at constant pressure. The temperature is maintained at 327°C. Calculate Q_V.

Solution: The reaction involved is

$$C_2H_5OH + 3O_2 = 2CO_2 + 3H_2O$$
$$\text{liquid} \quad\quad \text{gas} \quad\quad \text{gas} \quad\quad \text{liquid}$$

Let Δn be the increase in numbers of moles in the reaction = $2 - 3 = -1$.
Then

$$\begin{aligned}
Q_V = Q_P - \Delta nRT &= -327 - ((-1) \times 2 \times 10^{-3} \times 300) \\
&= -327 + (2 \times 10^{-3} \times 300) \\
&= -326.4 \text{ kcal}
\end{aligned}$$

4.3 HESS'S LAW OF CONSTANT HEAT SUMMATION

This law states: *For a given chemical process, the net heat change will be the same, whether the process occurs in one or in several stages.*

For example, carbon and oxygen can directly give carbon dioxide, or these can first form carbon monoxide which may then be oxidized to form carbon dioxide. The net heat changes in the case two will be the same.

(a) $C + O_2 = CO_2$ $\quad\quad\quad Q_P = -393.5$ kJ/mol

(b) $C + \dfrac{1}{2}O_2 = CO$ $\quad\quad Q_P' = -110.61$ kJ/mol

$CO + \dfrac{1}{2}O_2 = CO_2$ $\quad\quad Q_P'' = -282.89$ kJ/mol

$$\overline{C + O_2 = CO_2} \quad\quad\quad Q_P = Q_P' + Q_P''$$

It is observed from the preceding reaction that the heat changes for reaction (a) is exactly equal to the sum of those of the split-up reactions in (b). Hence, Hess's law of constant heat summation is supported well by the preceding reactions.

4.3.1 Thermodynamic Explanation of Hess's Law

Hess's Law of constant heat summation can be better studied and explained by the first law of thermodynamics. If the final and initial states are fixed, then the change in heat-content or energy-content of a system i.e., ΔH_P and ΔU_V would be independent of the path. Hence Q_P and Q_V will be independent of the path, i.e., heat change of a reaction will be the same, whether the reaction takes place in a single step or in several steps.

A unique advantage of this principle is that heat changes in the process can be added or subtracted, and thermo-chemical equations may be treated algebraically. Some examples are cited later to substantiate the usefulness of the preceding principle.

4.4 STANDARD HEAT OF REACTION

The enthalpy or heat change accompanying the chemical reaction is known as the *heat of reaction*. It is denoted by ΔH. It is the difference between the enthalpy of the products and that of the reactants.

$$\Delta H_{reaction} = \Sigma \Delta H_{products} - \Sigma \Delta H^0_{reactants} \tag{4.7}$$

The value $\Delta H_{reaction}$ may either be positive or negative. If $\Delta H_{reaction} = +ve$, the heat is absorbed by the system from the surroundings, i.e., it is an endothermic reaction; and if $\Delta H_{reaction} = -ve$, the heat is liberated from the system to the surroundings, i.e., it is an endothermic reaction.

However, it not possible to determine the absolute value of enthalpy change for a chemical reaction at different conditions of temperature and pressure at which the reaction takes place. In order to calculate the enthalpy change for a chemical reaction, a standard reference state is chosen. In chemical thermodynamics, it is convenient to choose a common standard reference state of 25°C (298 K) and 1 atm, and it is necessary to assume every substance to have zero enthalpy at this standard state.

Now, the standard heat or enthalpy of reaction can be defined as the enthalpy change associated with the reaction in which the reactants as well as the products are at standard states. It is denoted by ΔH^0_T, the superscript 0 denotes the standard state and the subscript T denotes the temperature at which the reaction takes place.

For example

$$COCl_2 + H_2S \rightarrow 2HCl + COS \qquad \Delta H^0_{298} = -78.705 \text{ kJ}$$
$$C_6H_{12}O_6 \text{ (s)} + 6O_2 \text{ (g)} \rightarrow 6CO_2 \text{ (g)} + 6H_2O \text{ (l)} \qquad \Delta H^0_{298} = -28.170 \text{ kJ}$$

EXAMPLE 4.2 Consider the following heat changes in the oxidation of magnesium and iron. Reaction involved:

(i) $Mg + \dfrac{1}{2}O_2 = MgO \qquad \Delta H = -602.06 \text{ kJ/mol}$

(ii) $2Fe + \dfrac{3}{2}O_2 = Fe_2O_3 \qquad \Delta H = -810.14 \text{ kJ/mol}$

Calculate the heat produced in the reaction.

$$3Mg + Fe_2O_3 = 3MgO + 2Fe$$

Solution: Multiplying reaction (i) by 3, we get

(iii) $3Mg + \dfrac{3}{2}O_2 = 3MgO$ $\Delta H = -3 \times 610.01$ kJ/mol

Adding (iii) and (ii), we get

$3Mg + \dfrac{3}{2}O_2 = 3MgO$ $\Delta H = -3 \times 610.01$ kJ/mol

$2Fe + \dfrac{3}{2}O_2 = Fe_2O_3$ $\Delta H = -810.14$ kJ/mol

$3Mg + Fe_2O_3 = 3MgO + 2Fe$ $\Delta H = -1019.9$ kJ/mol

Hence, by applying standard heat of reaction, we can compute the heat of a reaction in the formation of a compound, though the elements would not directly combine to produce the compound.

The heat of reaction at constant pressure, Q_P, is actually the enthalpy change of the process, and it is the difference between the enthalpies of products and reactants.

$$Q_P = \Delta H = H_{products} - H_{reactants}$$

4.5   STANDARD HEAT OF FORMATION

The change in enthalpy associated with the formation of a gram-molecule of a substance from its constituent elements is called its *heat of formation*. When 1 g-mol of HCl is produced from hydrogen and chlorine, 92.36 kJ/mol of heat are evolved. Hence, the heat of formation of HCl is -92.36 kJ/mol.

The heat of formation of a compound is denoted by ΔH_f, where the subscript f denotes the formation reaction. The heat of formation of some common reactions can be tabulated as follows:

Compound	ΔH_f (kJ/mol)
$\dfrac{1}{2}H_2 + \dfrac{1}{2}Cl_2 = HCl$	-92.36
$\dfrac{1}{2}N_2 + \dfrac{3}{2}H_2 = NH_3$	-46.22
$\dfrac{1}{2}N_2 + \dfrac{1}{2}O_2 = NO$	$+90.43$
$2C + H_2 = C_2H_2$	$+226.88$

In order to calculate the change in enthalpy in formation reactions, it is necessary to assume a common reference state. It has been assigned the zero value for the enthalpy of every element at the standard reference state of 1 atm and 25°C (298 K). The enthalpy or heat of formation of a substance at this standard reference state is known as the *standard heat of formation*.

For example

$$C \text{ (s)} + 2H_2 \text{ (g)} + \dfrac{1}{2}O_2 \text{ (g)} \rightarrow CH_3OH \text{ (g)}$$

Here, the standard heat of formation of methanol is -238.64 kJ/mol, i.e.,

$$H^0_{f298} = -238.64 \text{ kJ/mol}.$$

Suppose elements A and B are converted to compound C.

$$A + B \rightarrow C$$

The heat of formation of C,

$$\Delta H_{f(C)} = H_C - H_A - H_B \qquad (4.8)$$

The standard enthalpy is H^0; the superscript 0 indicating standard state. We can then write

$$\Delta H^0_{f(C)} = H^0_C - H^0_A - H^0_B \qquad (4.9)$$

The enthalpies of the pure elements A and B in the standard state are assigned zero value, i.e., $H^0_A = H^0_B = 0$. Then we have

$$\Delta H^0_{f(C)} = H^0_C - H^0_A - H^0_B = H^0_C$$

or

$$H^0_C = \Delta H^0_f \qquad (4.10)$$

This equation implies that the standard enthalpy of a compound is equal to its standard enthalpy of formation.

Consider the reaction

$$P + Q \rightarrow R + S$$

Now the enthalpy of the reaction can be expressed as

$$\begin{aligned}
\Delta H^0_{\text{reaction}} &= H^0_R + H^0_S - H^0_P - H^0_Q \\
&= \Delta H^0_{f(R)} + \Delta H^0_{f(S)} - H^0_{f(P)} - H^0_{f(Q)} \\
&= \Sigma \Delta H^0_{f(\text{products})} - \Sigma \Delta H^0_{f(\text{reactants})}
\end{aligned} \qquad (4.11)$$

For a chemical reaction, the standard heat of formation is nothing but the heat change involved in the transformation of reactants into products. It is independent of the path by which a chemical reaction goes to completion. It depends on the initial states of the reactant molecules and the final states of the product. Hence, for the calculation of H°_{f298} for a reaction, the main chemical reaction can for better results be replaced by a set of reactions in such a way that the sum of the relevant chosen reactions results in the desired reaction.

EXAMPLE 4.3 Calculate the standard heat of formation of CH_4, given the following experimental results at 25°C and 1 atm:

(a) $2H_2 \text{ (g)} + O_2 \text{ (g)} \rightarrow 2H_2O \text{ (l)}$ $\Delta H_1 = -241.8 \times 2 = -483.6$ kJ/gmol H_2

(b) $C \text{ (graphite)} + O_2 \text{ (g)} \rightarrow CO_2 \text{ (g)}$ $\Delta H_2 = -393.51$ kJ/gmol C

(c) $CH_4 \text{ (g)} + 2O_2 \text{ (g)} \rightarrow CO_2 \text{ (g)} + 2H_2O \text{ (l)}$ $\Delta H_3 = -802.36$ kJ/gmol CH_4

Solution: Since the standard heat of formation of methane is to be determined, the possibility of methane formation by the above reaction is to be considered.

Hence, (a) + (b) − (c) gives

$$2H_2\ (g) + O_2\ (g) \rightarrow 2H_2O\ (l)$$
$$C\ (graphite) + O_2\ (g) \rightarrow CO_2\ (g)$$
$$CH_4\ (g) + 2O_2\ (g) \rightarrow CO_2\ (g) + 2H_2O\ (l)$$
$$(-)\qquad\ (-)\qquad\quad (-)\qquad\quad (-)$$

$$C + 2H_2 = CH_4$$
$$\Delta H_{f(CH_4)} = [(-241.8 \times 2) + (-393.51) - (-802.36)]\ \text{kJ/gmol}$$
$$= -74.75\ \text{kJ/gmol}$$

Therefore, the standard heat of formation of CH_4 is 74.75 kJ/gmol.

EXAMPLE 4.4   Calculate the values of ΔH^0_{298} for the following reactions in the transformation of glucose in an organism:

$$C_6H_{12}O_6\ (s) = 2\ C_2H_5OH\ (l) + 2CO_2\ (g) \qquad\qquad (A)$$
$$C_6H_{12}O_6\ (s) + 6O_2\ (g) = 6CO_2\ (g) + 6H_2O\ (l) \qquad\qquad (B)$$

Which of these reactions supplies more energy to the organism?

We are given that the heat of formation of $C_6H_{12}O_6 = -1273.0$ kJ/mol
the heat of formation of $C_2H_5OH\ (l) = -277.6$ kJ/mol
the heat of formation of $CO_2\ (g) = -393.5$ kJ/mol
the heat of formation of $H_2O\ (l) = -285.8$ kJ/mol

Solution: For reaction (A), the heat of reaction can be estimated as

$$\Delta H_{\text{reaction}} = 2 \times (-277.6) + 2 \times (-393.5) - (-1273)$$
$$= 669.2\ \text{kJ}$$

For reaction (B), the heat of reaction can be calculated as

$$\Delta H_{\text{reaction}} = 6 \times (-393.5) + 6 \times (-285.8) - (-1273)$$
$$= -2802.8\ \text{kJ}$$

Hence, reaction (B) supplies more energy to the organism.

EXAMPLE 4.5   On the basis of the data and the chemical reactions given in the following lines, find the heat of formation of $ZnSO_4$ from its constituent elements.

$$Zn + S = ZnS \qquad\qquad \Delta H = -44.0\ \text{kcal/kmol} \qquad (A)$$
$$2ZnS + 3O_2 = 2ZnO + 2SO_2 \qquad\qquad \Delta H = -221.88\ \text{kcal/kmol} \qquad (B)$$
$$2SO_2 + O_2 = 2SO_3 \qquad\qquad \Delta H = -46.88\ \text{kcal/kmol} \qquad (C)$$
$$ZnO + SO_3 = ZnSO_4 \qquad\qquad \Delta H = -55.10\ \text{kcal/kmol} \qquad (D)$$

Solution: On multiplying (A) by 2 and adding Eq. (B), The rearrangement gives

$$2Zn + 2S = 2ZnS \qquad\qquad \Delta H_a = 2 \times (-44.0)\ \text{kcal/kmol}$$
$$2ZnS + 3O_2 = 2ZnO + 2SO_2 \qquad\qquad \Delta H_b = -221.88\ \text{kcal/kmol}$$

$$2Zn + 2S + 3O_2 = 2ZnO + 2SO_2 \quad \Delta H = -309.88 \text{ kcal/kmol} \tag{E}$$

Similarly, on multiplying (C) by 2 and adding Eq. (D), the rearrangement gives

$$2SO_2 + O_2 = 2SO_3 \qquad \Delta H_c = -46.88 \text{ kcal/kmol}$$
$$2ZnO + 2SO_3 = 2ZnSO_4 \qquad \Delta H_d = 2 \times (-55.10) \text{ kcal/kmol}$$

$$2ZnO + 2SO_3 + O_2 = 2ZnSO_4 \qquad \Delta H = -157.08 \text{ kcal/kmol} \tag{F}$$

Now on addition of Eqs. (E) and (F), we get

$$2Zn + 2S + 3O_2 = 2ZnO + 2SO_2 \quad \Delta H = -309.88 \text{ kcal/kmol}$$
$$2ZnO + 2SO_3 + O_2 = 2ZnSO_4 \quad \Delta H = -157.08 \text{ kcal/kmol}$$

$$2Zn + 2S + 4O_2 = 2ZnSO_4 \tag{G}$$

or

$$Zn + S + O_2 = ZnSO_4 \tag{H}$$

Then

$$\begin{aligned}
\Delta H_{f(H_2SO_4)} &= 2\Delta H_a + \Delta H_b + \Delta H_c + 2\Delta H_d \\
&= 2(-44) + (-221.88) + (-46.88) + 2(-55.10) \\
&= (-309.88) + (-157.08) \\
&= -466.96 \text{ kcal/kmol for 2 kmols of ZnSO}_4
\end{aligned}$$

Therefore, for 1 kmol of $ZnSO_4$, it is -233.48 kcal/kmol.

4.6 STANDARD HEAT OF COMBUSTION

The combustion process is considered to be an exothermic process in which heat is liberated by the reaction between the substance (compound fuel) and molecular oxygen. The combustible elements in the substance such as carbon and hydrogen are oxidized to form the product of combustion.

The heat change accompanying the complete combustion of a gram-mole of the substance at a given temperature under normal pressure is called its *heat of combustion*.

The standard heat of combustion is defined as the enthalpy change of a substance undergoing the combustion process at the standard reference state of 25°C and 1 atm. This is always expressed per mole of the substance burnt or oxidized. For example,

$$H_2 \text{ (g)} + \frac{1}{2}O_2 \text{ (g)} \rightarrow H_2O \text{ (l)} \qquad H^0_{H_2,298} = -285.83 \text{ J/mol}$$

The preceding reaction implies that the amount of heat liberated with the combustion of 1 kmol of H_2 to form liquid water is +285.83 kJ. So, the standard enthalpy or heat of combustion of H_2 is -285.83 kJ.

The heat of combustion is also referred to as heating value or calorific value. It is a positive quantity, while the heat of combustion is a negative quantity. The heating value of solids and liquids is usually determined by Boys or Junkers calorimeters.

The heating value of a fuel may be of two kinds:

(1) Higher Heating Value (HHV) or Gross Heating Value (GHV)
(2) Lower Heating Value (LHV) or Net Heating Value (NHV).

After the completion of combustion operation of a fuel, the product is observed to contain some water vapour, which is condensed to liquid water on cooling at the temperature of 25°C. The liquid water then releases the latent heat of condensation. This heat is added to the heat of combustion of a fuel.

The heating value measured with water in the products in the liquid state is known as *higher* or *gross heating value*. The heating value measured when the combustion products contain water in the vapour state is known as *lower* or *net heating value*.

The relation between the two kinds of heating values is

$$HHV \text{ (or GHV)} = LHV \text{ (or NHV)} + (h_{fg})_{25°C} \qquad (4.12)$$

The heating value of solid fuel is expressed in kJ/kg, liquid fuel in kJ/L and gaseous fuel in kJ/m^3.

The heat of combustion of gas is generally determined with the help of a bomb-calorimeter. The values of heat of combustion at 25°C for various compounds are shown in Table 4.1.

Table 4.1 Heat of combustion at 25°C in kcal

Number	Compounds	Heat of combustion in kcal at 25°C
1	Methane	−212.8
2	Ethane	−372.8
3	Ethylene	−337.2
4	Acetylene	−310.6
5	Benzene (l)	−780.9
6	Naphthalene (s)	−1228.2
7	Methanol (l)	−173.6
8	Ethanol (l)	−326.7
9	Cane sugar (s)	−1349.0
10	Phenol (s)	−732.0
11	Aniline (l)	−812.0
12	Benzoic acid (s)	−771.4

EXAMPLE 4.6 Calculate the heat of formation of ammonia. The heats of combustion of ammonia and hydrogen are −90.6 and −68.3 kcal respectively.

Solution: The reactions are

$$2\,NH_3 + 3O = N_2 + 3H_2O \qquad \Delta H = -2 \times 90.6 = -181.2 \qquad (A)$$
$$H_2 + O = H_2O \qquad \Delta H = -68.3 \qquad (B)$$
$$3H_2 + 3O = 3H_2O \qquad \Delta H = -3 \times 68.3 = -204.9 \qquad (C)$$

Subtracting (C) from (A) and re-arranging, we get

$$2\,NH_3 = N_2 + 3H_2 \qquad \Delta H = -181.2 + 204.9 = 23.7 \qquad (D)$$

that is

$$N_2 + 3H_2 = 2\ NH_3 \quad \Delta H = 23.7$$

Hence, the heat of formation of ammonia = $[-204.9 - (-181.2)]/2 = -11.8$ kcal.

4.7 ADIABATIC FLAME TEMPERATURE

If a reaction proceeds without loss or gain of heat and if all the products remain together in a single mass or stream of materials, these products will assume a definite temperature known as the *adiabatic flame temperature* or *adiabatic reaction temperature*.

In an adiabatic combustion process, as shown in Fig. 4.1, if there is no work transfer, no change in kinetic energy, and no change in potential energy, then all energy released during the combustion process increases the temperature of the products. This temperature is known also as the adiabatic combustion temperature or adiabatic flame temperature.

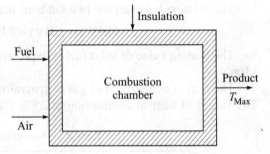

Fig. 4.1 Adiabatic combustion chamber.

Actually, the sudden occurrence of a spontaneous process produces a large amount of heat during the chemical change, and if there is no chance of dissipation of heat to the environment, then it will cause an increase in temperature of the product. This process taking place and going on completion within a short span of time can be treated as an adiabatic process.

It is necessary to define the another important term *theoretical flame temperature*. It is defied as the temperature attained when a fuel is burned in air or oxygen without gain or loss of heat. The maximum adiabatic flame temperature is attained when the fuel is burned with the theoretically required amount of pure oxygen. Sometimes the excess amount of air is supplied to ensure the complete combustion. It does not affect the adiabatic flame temperature to that extent, because a certain amount of heat is lost as it radiates from the surface area of reactor even in case of proper insulation arrangement.

EXAMPLE 4.7 The heat of combustion of acetylene is −310600 cal. If the gas is heated from room temperature (25°C) to a higher temperature. Determine the maximum attainable temperature.

Solution: We are given that the heat of combustion of acetylene is −310600 cal.
   Reaction involved:

$$C_2H_2 + \frac{5}{2}O_2 = 2CO_2 + H_2O \quad \Delta H = 310600 \text{ cal}$$

The gases present in the flame zone after combustion are carbon dioxide, water vapour, and the unreacted nitrogen of the air. Since 5/2 mol of oxygen were required for combustion, nitrogen present would be approximately 10 mol. Hence the composition of the resultant gases would be 2 mol CO_2, 1 mol H_2O vapour and 10 mol N_2.

The heat capacities of these gases approximately are

$$2 \times C_{P_{CO_2}} = 2(6.4 + 10.1 \times 10^{-3}T) = 12.8 + 20.2 \times 10^{-3}T$$
$$C_{P_{H_2O}} = 7.22 + 2.4 \times 10^{-3}T$$
$$10 \times C_{P_{N_2}} = 10(6.45 + 1.4 \times 10^{-3}T) = 64.5 + 14 \times 10^{-3}T$$

$$\sum C_P = 84.52 + 36.6 \times 10^{-3}T$$

The gases have been heated from room temperature 25°C = 273 + 25 = 298 K to a higher temperature T by the heat of combustion.

$$Q = \int_{298}^{T} \sum C_P dT = \int_{298}^{T} (84.52 + 36.6 \times 10^{-3}T) dT$$

or

$$Q = 84.52(T - 298) + 18.3 \times 10^{-3}(T^2 - 298^2) = 310600$$

or

$$84.52T + 18.3 \times 10^{-3}T^2 = 337415$$
$$T = 2566.5 \text{ K}$$

Hence, the maximum attainable temperature is 2566.5 K.

EXAMPLE 4.8 Calculate the theoretical temperature of combustion of ethane with 25% excess air. The average specific heats in kJ/kg-K may be taken as follows:

$$CO_2 = 54.56 \text{ kJ/kmol-K}$$
$$O_2 = 35.20 \text{ kJ/kmol-K}$$
$$\text{Steam} = 43.38 \text{ kJ/ kmol-K}$$
$$N_2 = 33.32 \text{ kJ/kmol-K.}$$

The combustion reaction for ethane is

$$2C_2H_6 \text{ (g)} + 7O_2 \text{ (g)} = 4CO_2 \text{ (g)} + 6H_2O \text{ (g)} \quad \Delta H_{273} = -1560000 \text{ kJ/kmol}$$

Solution: Since the air is 25% in excess of the amount required, the combustion may be written as

$$C_2H_6 \text{ (g)} + \frac{7}{2}O_2 \text{ (g)} = 2CO_2 \text{ (g)} + 3H_2O \text{ (g)}$$

25% excess air is supplied. Hence, the equation would be represented as

$$C_2H_6 \text{ (g)} + 3.5O_2 \text{ (g)} + 0.25 \times 3.5O_2 \text{ (g)} + \left(4.375 \times \frac{79}{21}\right)N_2$$
$$= 2CO_2 \text{ (g)} + 3H_2O \text{ (g)} + 0.875O_2 + 16.46N_2 \quad\quad\quad\quad \text{(A)}$$

since the air contains N_2 = 79% and O_2 = 21%.

Considering reaction (A), the amount of O_2

$$= 3.5 + (3.5 \times 0.25) \text{ mol} = 3.5 + 0.875 = 4.375 \text{ mol}$$

Amount of N_2 present for 4.375 mol of O_2 = 4.375 × 79/21 = 16.46 mol

Let the initial temperature of ethane and air be $0°C$, and the temperature of the products of combustion be $T°C$.

Since, heat liberated by combustion = heat accumulated by combustion products

$$1560000 = \left[2 \times 54.5(C_{B_{CO_2}}) + 3 \times 43.38(C_{P_{H_2O}}) + 0.875 \times 35.2(C_{P_{O_2}}) + 16.46 \times 33.32(C_{P_{N_2}})\right] \times T$$

$$= T \times (109 + 130.14 + 30.8 + 548.44)$$

$$= 818.38T$$

$$T = 1906°C$$

Therefore, the theoretical temperature of combustion is $1906°C$.

4.8   EFFECT OF TEMPERATURE ON HEAT OF REACTION: KIRCHHOFF'S EQUATION

The rate of a chemical reaction is greatly influenced by temperature. Generally, it increases with increase in temperature and results in the formation of product to a significant extent. The change in enthalpy associated with the reaction also depends on temperature. Therefore, in addition to the rapidity of the reaction, chemical engineers and scientists are interested in calculating the enthalpy change of a reaction at higher temperatures, because chemical reactions are very rarely carried out at the standard reference temperature of 298 K (25°C).

The influence of temperature on the enthalpy change of a reaction can be better studied and explained in the following way. Let us consider the reaction

$$aA + bB \rightarrow cC + dD$$

The enthalpy change for the reaction is given by

$$\Delta H = \sum H_{\text{products}} - \sum H_{\text{reactants}}$$

$$= (cH_C + dH_D) - (aH_A + bH_B) \tag{4.13}$$

Differentiating this with respect to temperature at constant pressure, we have

$$\left[\frac{\partial(\Delta H)}{\partial T}\right] = c\left(\frac{\partial H_C}{\partial T}\right)_P + d\left(\frac{\partial H_D}{\partial T}\right)_P - a\left(\frac{\partial H_A}{\partial T}\right)_P - b\left(\frac{\partial H_B}{\partial T}\right)_P$$

$$= cC_{P_C} + dC_{P_D} - aC_{P_A} - bC_{P_B}$$

$$= \sum C_{P_{\text{products}}} - \sum C_{P_{\text{reactants}}} \tag{4.14}$$

or

$$\left[\frac{\partial(\Delta H)}{\partial T}\right]_P = \Delta C_P \tag{4.15}$$

or

$$\int d(\Delta H) = \int \Delta C_P dT \tag{4.16}$$

Integrating between the limits T K and 0 K, we have

$$\Delta H_T = \Delta H_0 + \int_0^T \Delta C_P dT \tag{4.17}$$

where ΔH_0 is the integration constant as well as the enthalpy change at 0 K. Equation (4.17) is called Kirchhoff's equation.

Case I: This is applicable when C_P values of reactants and products are independent of temperature and temperature range is small. For the process occurring at T_1 K, we have

$$\Delta H_{T_1} = \Delta H_0 + \int_0^{T_1} \Delta C_P dT = \Delta H_0 + \Delta C_P T_1$$

Similarly, for the process occurring at T_2 K, we have

$$\Delta H_{T_2} = \Delta H_0 + \int_0^{T_2} \Delta C_P dT = \Delta H_0 + \Delta C_P T_2$$

Hence

$$\Delta H_{T_1} - \Delta H_{T_2} = \Delta C_P (T_1 - T_2) \tag{4.18}$$

Case II: When C_P values change with temperature and temperature range is not small, i.e., if it is necessary to estimate the change of enthalpy of a reaction at higher temperature, then C_P must be expressed as a function of temperature.

It is to be noted that C_P slowly increases with temperature and the influence of temperature is empirically given by a power series in T.

$$C_P = a + bT + cT^2 + \cdots$$

where a, b, and c are constants.

For a process that takes place at T_1 K, Eq. (4.17) can be expressed as

$$\Delta H_{T_1} = \Delta H_0 + \int_0^{T_1} \Delta C_P dT$$

$$= \Delta H_0 + \int_0^{T_1} (C_{P_{\text{products}}} - C_{P_{\text{reactants}}}) dT$$

$$= \Delta H_0 + \int_0^{T_1} [(a_P + b_P T + c_P T^2 + \cdots) - (a_R + b_R T + c_R T^2 + \cdots)] dT$$

$$= \Delta H_0 + \int_0^{T_1} (\alpha + \beta T + \gamma T^2 + \cdots) dT$$

where

$$\alpha = a_P - a_R$$
$$\beta = b_P - b_R$$
$$\gamma = c_P - c_R$$

$$= \Delta H_0 + \left(\alpha T_1 + \frac{\beta}{2} T_1^2 + \frac{\gamma}{3} T_1^3 + \cdots \right) \tag{4.19}$$

Similarly for the process occurring at T_2 K, we have

$$\Delta H_{T_2} = \Delta H_0 + \left(\alpha T_2 + \frac{\beta}{2} T_2^2 + \frac{\gamma}{3} T_2^3 + \cdots\right) \qquad (4.20)$$

Hence

$$\Delta H_{T_1} - \Delta H_{T_2} = \alpha(T_1 - T_2) + \frac{\beta}{2}(T_1^2 - T_2^2) + \frac{\gamma}{3}(T_1^3 - T_2^3) + \cdots$$

EXAMPLE 4.9 The latent heat of fusion of ice at 0°C is 1440 cal/mol and the heat capacity of ice/mole is 6.7 cal. Calculate the latent heat of ice at –20°C.

Solution: We are given that $T_1 = 0°C = 273$ K and $T_2 = -20°C = 253$ K. From Eq. (4.17), we get

$$\Delta H_{T_1} = \Delta H_0 + \int_0^{T_1} \Delta C_P dT$$

$$\Delta H_{273} = \Delta H_0 + \int_0^{273} \Delta C_P dT \qquad (4.21)$$

For the process occurring at 253 K

$$\Delta H_{253} = \Delta H_0 + \int_0^{253} \Delta C_P dT \qquad (4.22)$$

Subtracting Eq. (4.21) from Eq. (4.22), we get

$$\Delta H_{T_2} - \Delta H_{T_1} = C_P(T_2 - T_1)$$

or

$$\Delta H_{253} - \Delta H_{273} = 8.7 \times (253 - 273) = -8.7 \times 20 = -174$$

or

$$\Delta H_{253} = \Delta H_{273} - 174 = 1440 - 174 = 1266 \text{ cal/mol}$$

Therefore, the latent heat of ice at –20°C is 1266 cal/mol.

EXAMPLE 4.10 The heat of formation of ammonia from its constituent elements is 11030 cal/mol at 300 K. What will be its heat of formation at 1273 K?

We are given that $C_{P_{H_2}} = 6.94 - 0.2 \times 10^{-3}T$, $C_{P_{N_2}} = 6.45 + 1.4 \times 10^{-3}T$ and $C_{P_{NH_3}} = 6.2 + 7.8 \times 10^{-3}T - 7.2 \times 10^{-6}T^2$.

Solution: The chemical reaction involved is

$$N_2 + 3H_2 = 2NH_3$$

$$\frac{1}{2}N_2 + \frac{3}{2}H_2 = NH_3$$

The equation can be represented as

$$\Delta H_T - \Delta H_0 = \int_0^T \Delta C_P dT$$

For temperature T_1,

$$\Delta H_{T_1} - \Delta H_0 = \int_0^{T_1} (C_{P_{\text{products}}} - C_{P_{\text{reactants}}})\, dT$$

$$= \int_0^{T_1} \left(C_{P_{\text{NH}_3}} - \frac{1}{2} C_{P_{\text{H}_2}} - \frac{1}{2} C_{P_{\text{N}_2}} \right) dT$$

$$= \int_0^{T_1} \left[(6.2 + 7.3 \times 10^{-3} T - 7.2 \times 10^{-6} T^2) - \frac{1}{2}(6.94 - 0.2 \times 10^{-3} T) - \frac{1}{2}(6.45 + 1.4 \times 10^{-3} T) \right] dT$$

$$= \int_0^{T_1} (-7.435 + 7.4 \times 10^{-3} T - 7.2 \times 10^{-6} T^2)\, dT$$

or

$$\Delta H_{T_1} = \Delta H_0 - 7.435 T_1 + 3.7 \times 10^{-3} T_1^2 - 2.4 \times 10^{-6} T_1^3 \qquad (4.23)$$

Similarly, for temperature T_2,

$$\Delta H_{T_2} = \Delta H_0 - 7.435 T_2 + 3.7 \times 10^{-3} T^2 - 2.4 \times 10^{-6} T_2^3 \qquad (4.24)$$

Subtracting Eq. (4.23) from (4.24), we get

$$\Delta H_{T_2} - \Delta H_{T_1} = -7.435(T_2 - T_1) + 3.7 \times 10^{-3}(T_2^2 - T_1^2) - 2.4 \times 10^{-6}(T_2^3 - T_1^3)$$

$$\Delta H_{1273} - \Delta H_{300} = -7.435(1273 - 300) + 3.7 \times 10^{-3}(1273^2 - 300^2)$$
$$- 2.4 \times 10^{-6}(1273^3 - 300^3)$$
$$= -1091 \text{ cal}$$

$$\Delta H_{1273} = \Delta H_{300} - 1091 = (-11030 - 1091) \text{ cal} = -12121 \text{ cal}$$

EXAMPLE 4.11 A sample of dry flue gas has the following composition by volume: CO_2—13.4%, N_2—80.5% and O_2—6.1%. Calculate the excess air supplied. Assume that the fuel contains no nitrogen. The oxygen and nitrogen must have come from air.

Solution: Basis: 100 m³ of flue gas

$$N_2 \text{ content} = 80.5 \text{ m}^3$$

$$\text{Volume of air supplied} = \frac{80.5}{0.79} = 101.9 \text{ m}^3$$

$$\text{Volume of } O_2 \text{ in air supply} = 101.9 \times 0.21 = 21.4 \text{ m}^3$$

$$\text{Volume of } O_2 \text{ present in the flue gas} = 6.1 \text{ m}^3$$

$$\text{Volume } O_2 \text{ used up in combustion of the fuel} = 21.4 - 6.1 = 15.3 \text{ m}^3$$

$$\text{Percentage of excess air supplied} = \left(\frac{6.1}{15.3} \right) \times 100 = 39.9$$

SUMMARY

The evolution or absorption of heat associated with chemical reaction is an important factor to be taken into consideration for the proper design of a chemical reactor. One of the major responsibilities of a process engineer is to estimate the heat transfer rate accompanying a chemical reaction that takes place in the heart of the chemical process industries, i.e., the chemical reactor. These have been discussed in detail. The change in enthalpy of a chemical reaction can be estimated from various useful thermodynamic relations. Examples have been cited to calculate the standard enthalpy of combustion and formation for different types of chemical reactions. The rate and enthalpy of the chemical reaction are greatly affected by temperature. In addition to this, the definition and importance of adiabatic flame temperature for combustion of a fuel have been discussed.

KEY TERMS

Adiabatic Flame Temperature   A definite temperature attained by the products of a reaction if it proceeds without loss or gain of heat and if all the products of the reaction remain together in a single mass or stream of materials.

Endothermic Reaction   The reaction in which heat is absorbed by the system from the surroundings.

Exothermic Reaction   The reaction in which heat escapes from the system to the surroundings.

Heat of Combustion   The heat change accompanying the complete combustion of a gram-mole of the substance at a given temperature under normal pressure.

Heat of Formation   The enthalpy change occurring when a gram-molecule of a substance is formed from its constituent elements.

Heat of Reaction   The enthalpy or heat change accompanying the chemical reaction.

Hess's Law   For a given chemical process, the net heat change will be the same, whether the process occurs in one or in several stages.

Standard Heat of Combustion   The standard heat of combustion is defined as the enthalpy change of a substance undergoing combustion process at the standard reference state of 25°C and 1 atm.

Standard Heat of Formation   The heat of formation of a substance at standard reference state of 25°C and 1 atm.

Standard Heat of Reaction   The heat change in a reaction at standard reference state of 25°C and 1 atm.

Theoretical Flame Temperature   The temperature attained when a fuel is burned in air or oxygen without gain or loss of heat.

IMPORTANT EQUATIONS

1. For a constant-pressure process

$$Q_P = \Delta H_P = \Delta U_P + P\Delta V \tag{4.1}$$

2. For a mechanically reversible constant-volume process

$$Q_P - Q_V = \Delta U_P + P\Delta V - \Delta UV \tag{4.3}$$

3. For a gaseous reaction, if n_1 and n_2 are the numbers of reactant and product molecules respectively, then

$$P\Delta V = P(n_2 - n_1)\frac{RT}{P} = \Delta nRT \tag{4.5}$$

4. The relation between Q_P and Q_V is given by

$$Q_P - Q_V = \Delta nRT \tag{4.6}$$

5. The enthalpy change of the chemical reaction is the difference between the enthalpy of the products and that of the reactants, or

$$\Delta H_{\text{reaction}} = \sum \Delta H_{\text{products}} - \sum \Delta H^0_{\text{reactants}} \tag{4.7}$$

6. The relation between the two kinds of heating values is given by

$$\text{HHV (or GHV)} = \text{LHV (or NHV)} + (h_{fg})_{25°C} \tag{4.12}$$

7. For the following reaction

$$aA + bB \rightarrow cC + dD$$

The enthalpy change for the reaction is given by

$$\Delta H = \sum H_{\text{products}} - \sum H_{\text{reactants}} \tag{4.13}$$
$$= (cH_C + dH_D) - (aH_A + bH_B)$$

8. Integrating between the limits T K and 0 K

$$\Delta H_T = \Delta H_0 + \int_0^T \Delta C_P dT \tag{4.17}$$

where ΔH_0 is the integration constant as well as the enthalpy change at 0 K.

9. When the C_P values of reactants and products are independent of temperature and the temperature range is small, then for the process occurring at T_1 K

$$\Delta H_{T_1} = \Delta H_0 + \int_0^{T_1} \Delta C_P dT = \Delta H_0 + \Delta C_P T_1$$

Similarly, for the process occurring at T_2 K, we have

$$\Delta H_{T_2} = \Delta H_0 + \int_0^{T_2} \Delta C_P dT = \Delta H_0 + \Delta C_P T_2$$

Hence

$$\Delta H_{T_1} - \Delta H_{T_2} = \Delta C_P (T_1 - T_2) \tag{4.18}$$

10. When C_P values change with temperature, and the temperature range is not small

$$C_P = a + bT + cT^2 + \cdots$$

where a, b, and c are constants.

11. For a process taking place between T_1 K and T_2 K

$$\Delta H_{T_1} - \Delta H_{T_2} = \alpha(T_1 - T_2) + \frac{\beta}{2}(T_1^2 - T_2^2) + \frac{\gamma}{3}(T_1^3 - T_2^3) + \cdots$$

EXERCISES

A. Review Questions

1. Explain the role of thermo-chemistry in the design and analysis of chemical processes and their relevant equipments.
2. State Hess's law of constant heat summation with a suitable example. How can it be explained from the thermodynamic point of view?
3. What is the usefulness of Hess's law of constant heat summation?
4. Write informatory notes on the following:
   (a) Heat of formation   (b) Heat of reaction
   (c) Heat of combustion   (d) Standard heat of reaction
   (e) Standard heat of formation
5. How do you estimate the heat of reaction at unknown temperature with respect to the heat of reaction of the substance at known temperature?
6. Define the term *adiabatic flame temperature*. How can it be determined?
7. What is the importance of theoretical flame temperature?
8. What is excess air? What role does it play for the complete combustion of a fuel?
9. What is the difference between excess air and theoretical air?
10. Name the factors responsible for incomplete combustion even if excess air is supplied to the fuel.
11. Which instrument is used to determine the heat of combustion of a fuel?
12. What is the difference between complete combustion and theoretical combustion?

B. Problems

1. At 17°C, at constant pressure, the heat of combustion of amorphous carbon is 96960 cal and that of CO to CO_2 is 67960 cal. Determine the heat of formation of CO at constant volume.
2. The heat of formation of ammonia and hydrogen fluoride gas are −46.1 kJ and −271.2 kJ respectively. Estimate the heat of the reaction

$$2NH_3 \text{ (g)} + 3F_2 \text{ (g)} = N_2 \text{ (g)} + 6HF \text{ (g)}$$

3. The latent heat of formation of H_2O (g) $= -5800$ cal at 500 K. What will be the heat of formation at 1000 K? We are given that

$$C_{P_{H_2}} = 6.94 - 0.2 \times 10^{-3}T$$
$$C_{P_{O_2}} = 6.15 + 3.2 \times 10^{-3}T$$
$$CP_{H_2O} = 7.25 + 2.28 \times 10^{-3}T$$

4. For the reaction

$$CO + \frac{1}{2}O_2 = O_2 \qquad \Delta H = -67650 \text{ cal at } 25°C$$

find out ΔH of the process at 100°C, given that molal heat capacities $C_{P_{CO}} = 6.97$, $C_{P_{CO_2}} = 8.97$ and $C_{P_{O_2}} = 7.0$.

5. The heat of combustion of n-heptane (l) at constant volume and at 25°C is -1150 kcal/mol. Calculate the ΔH_f^0 for n-heptane, given that $\Delta H_{f_{CO_2(g)}} = -94$ kcal/mol, $\Delta H_{f_{H_2O(l)}} = -68$ kcal/mol.

6. Calculate the theoretical flame temperature for CO when burned with 100% excess air while both the reactants are at 373 K. The heat capacities of CO, O_2, N_2 and CO_2 are 29.33 J/mol-K, 34.83 J/mol-K, 33.03 J/mol-K and 53.59 J/mol-K, respectively. The standard heat of combustion at 298 K is -283178 kJ/mol CO_2.

7. Calculate the standard heat of the following gaseous reaction at 773 K:

$$CO_2 + 4H_2 \rightarrow 2H_2O + CH_4 \qquad \Delta H_{298} = -164.987 \text{ kJ}$$

The specific heat of the components are represented by $C_P = a + bT + CT^2$, where C_P is in J/mol-K and a, b, and c are constants.

	a	$b \times 10^3$	$c \times 10^6$
CO_2	26.75	42.26	-14.25
H_2	26.88	4.35	-0.33
H_2O	29.16	14.49	-2.02
CH_4	13.41	77.03	-18.74

8. Determine the standard heat of the following reaction at 298 K:
   (a) CaC_2 (s) $+ H_2O$ (l) $\rightarrow C_2H_2$ (g) $+ CaO$ (s)
   (b) $4NH_3$ (g) $+ 5O_2$ (g) $\rightarrow 4NO$ (g) $+ 6H_2O$ (g)
   (c) C_2H_4 (g) $+ \frac{1}{2}O_2$ (g) $\rightarrow (CH_2)_2O$ (g)
   (d) H_2S (g) $+ 2H_2O$ (g) $\rightarrow 3H_2$ (g) $+ SO_2$ (g).

9. Calculate the standard enthalpy change of combustion at 298.15 K for C_2H_5OH (l) if H_2O is in the gaseous state.

10. The standard enthalpy change of combustion of acetylene is -1300.48 kJ at 298 K with H_2O in the liquid state. Calculate the standard enthalpy of formation of acetylene.

11. A student pours chloroform into the palm of his hand. As the liquid evaporates, his hands feels cold. Is the process of evaporation of chloroform exothermic or endothermic?

12. The heat of combustion of cyclopropane $(CH_2)_3$ is -2090 kJ/mol. The heat of formation of CO_2 and H_2O are -393.5 and -285.8 kJ/mol, respectively. Calculate (a) the heat of

formation of cyclopropane and (b) the heat of polymerization of cyclopropane to propylene, given that $\Delta H_f \ (CH_2)_3 = 20.5$ kJ/mol.

13. The chemical equation for the water gas reaction between CO and steam is

$$CO \ (g) + H_2O \ (g) \rightarrow CO_2 \ (g) + H_2 \ (g)$$

Determine the enthalpy of reaction at 25°C and 1 atm.

14. Sucrose ($C_{12}H_{22}O_{11}$) burns at 25°C to form CO_2 gas and liquid H_2O, releasing 5,640,000 kJ/kmol of heat according to the equation

$$C_{12}H_{22}O_{11} \ (s) + 12O_2 \ (g) \rightarrow 12CO_2 \ (g) + 11H_2O \ (l) + 5,640,000 \text{ kJ/kmol}$$

Determine the enthalpy of formation of sucrose.

15. Propane gas at room temperature is burned with enough air so that combustion is complete and gases leave the burner at 1400 K. The combustion gas is then mixed with sufficient air so that the resulting gas mixture for drying is at 400 K. How many moles of gas are available for drying per mole of propane burned?

16. H_2 gas is produced by reforming methane via the following reaction:

$$CH_4 + H_2O = 3H_2 + CO$$

The reaction is endothermic and carried out at 1200 K. Methane and steam, each at 800 K and 1 atm, are fed to the reactor in a 1:1 ratio. The reaction essentially goes to completion, and the hydrogen and CO leave at 1200 K. How much heat must be supplied per mole of methane reacted?

17. Calculate the theoretical temperature of combustion of ethane with 25% excess air. The average specific heats in kJ/kg-K may be taken as follows: $CO_2 = 1.24$, $O_2 = 1.10$, Steam $= 2.41$, $N_2 = 1.19$. The combustion reaction for ethane is

$$2C_2H_6 \ (g) + 7O_2 \ (g) = 4CO_2 \ (g) + 6H_2O \ (g) \qquad \Delta H_{273} = -1560 \text{ MJ/kmol}$$

18. A sample of coal has 80% C, 0.5% H_2, 0.5% S, and 14.5% ash. Calculate the theoretical quantity of air necessary for the combustion of 1 kg of coal. Find the composition of the flue gas by weight and by volume if 25% excess air is supplied.

19. Calculate the theoretical flame temperature when hydrogen burns with 400% excess air at 1 atm. The reactant enters at 100°C. The reaction involved is

$$H_2 + \tfrac{1}{2} O_2 = H_2O$$

Second Law of Thermodynamics

5

LEARNING OBJECTIVES

After reading this chapter, you will be able to:

- Discuss the significance and limitations of the first law of thermodynamics
- Represent the basic concepts of heat engine, heat pump and refrigerator, along with the terminology involved in explaining them
- Define the thermal efficiency of heat engine and the COP of heat pump and refrigerator
- Understand the second law of thermodynamics
- State the Kelvin–Planck and Clausius statements of the second law of thermodynamics
- Explain the equivalence of Kelvin–Planck and Clausius statements
- Describe the operational steps of Carnot cycle and state the Carnot theorem
- Discuss the application of the second law of thermodynamics to develop the thermodynamic temperature scale
- Understand the concept of entropy and its changes for different processes
- Deduce the mathematical expression of the second law of thermodynamics
- Understand entropy from the microscopic point of view and its balance for an open system
- State the third law of thermodynamics
- Know about the criterion of irreversibility
- Understand the Clausius inequality

In this chapter, first of all, we have recapitulated the importance of the first law of thermodynamics. Although no process can violate this law but it has certain limitations. These have been substantiated by citing different logical instances. The inadequacy of first law necessitates the introduction of second law of thermodynamics in which impossibility of transferring heat from lower temperature region to higher one without help of any external agent and no possibility for any device to convert the total heat input into net amount of work have been emphasized in Clausius and Kelvin–Planck statements respectively. The equivalence of these two statements has been clearly discussed.

Carnot cycle is an ideal and hypothetical cycle consisting of two isothermal and two adiabatic processes. The working principle is described with the help of illustration. Carnot theorem is discussed. It is well known to us work can be converted so easily into heat but the reverse process cannot occur by itself. It requires a device, which is called the *heat engine*. But this is necessary to remember that all the input heat can be converted into net amount of work. It necessitates to define the term *thermal efficiency* of a heat engine like the coefficient of performance for the refrigerator and heat pump. The second law of thermodynamics plays an important role in developing the thermodynamic temperature scale, which is of paramount importance in understanding the condition of the processes. Like internal energy, entropy is a thermodynamic state function. The change in entropy for several processes is required to be calculated. The term *entropy* has been introduced from the microscopic point of view.

5.1 LIMITATIONS OF FIRST LAW OF THERMODYNAMICS

The first law of thermodynamics basically represents the principle of conservation of energy, which states that energy can neither be created nor be destroyed; only one form of energy can be transformed into another form of energy. It is very much concerned with the change in energy of a process and its quantitative estimation. From the knowledge of first law of thermodynamics, it is very much clear that

(i) Work and heat are mutually interconvertible form of energy;

(ii) Appearance of one form of energy indicates the disappearance of an equivalent amount of another form of energy.

But it is necessary to remember that there are some limitations in the first law of thermodynamics. The first law of thermodynamics does not give any information as to whether the transformation of energy would actually take place or not. If it does, then (i) in which direction, i.e., what would be the direction of energy transformation? and (ii) to what extent of conversion, i.e., what extent of transformation would take place? This inadequacy of the first law of thermodynamics can be proved by the following examples:

1. Consider two bodies brought into thermal contact with each other. If the temperature difference exists between two bodies, then heat energy will flow from one body to another spontaneously. Then from the point of view of the first law, if the certain amount of heat is lost by the first body, exactly the equivalent amount of heat would be gained by the second body obeying law of conservation of energy. The process may take place in reverse order also depending upon the body of higher temperature. But the law does not indicate, firstly, whether the first or second body will give out the heat as energy, i.e., the extent of heat of the two bodies, and secondly, in which direction, i.e., the direction of transferring heat from one body to another. These are the limitations of the first law of thermodynamics. Hence, we must be aware of the temperatures of two bodies to know the direction of flow of heat.

2. Suppose that in an electrolysis process, two electrodes are partially immersed in an electrolytic solution. They are externally well-connected. According to the principle of electrochemistry, there will be a transformation of chemical energy into electrical energy.

But whether the current will pass from the first electrode to the second or vice versa, can not be determined unless we know the potentials of the two electrodes.

3. Consider the combustion of methane in the presence of oxygen. The change is

$$CH_4 + 2O_2 \rightarrow CO_2 + 2H_2O$$

This is a spontaneous reaction. But this is not possible to form CH_4 and O_2 by the recombination of CO_2 and H_2O, because the reaction will never proceed in the reverse direction on its own. But this information could not be obtained from the first law of thermodynamics. To know the direction of changes, one has to know the kinetics of the reaction.

These valid examples support the insufficiency of the first law of thermodynamics to ensure the occurrence of the process and to ascertain the direction and extent of changes of a chemical or physical process. These insufficiencies are remedied by introducing the second law of thermodynamics.

5.2 BASIC CONCEPTS OF HEAT ENGINES, HEAT PUMPS AND REFRIGERATORS

Before going to the detailed discussion of the second law of thermodynamics, it is very convenient to introduce the basic concepts of heat engines, heat pumps and refrigerators, and it is necessary to have an idea about the terminology which is used in explaining the operational methodology of heat engines, heat pumps, and refrigerators in a proper way. The most important terms are:

(i) **Thermal Reservoir:** It is defined as the hypothetical body with large capacity which can supply or absorb a significant amount of thermal energy without undergoing any change in temperature. For instance, atmosphere, oceans, lakes, and rivers are such bodies because they have large capability of storing thermal energies.

(ii) **Working Fluid:** It is defined as the substance which in operation absorbs the heat from a hot reservoir, release the heat to a cold reservoir, and finally returns to its original state completing the cycle.

(iii) **Source:** A thermal reservoir or a body at high temperature from which a definite amount of heat is transferred by the working fluid is called a *source*. Simply put, a reservoir that supplies heat as energy is known as a *heat source*.

(iv) **Sink:** A thermal reservoir or a body at low temperature to which a definite amount of heat is transferred by the working fluid is called a *sink*. In other words, a reservoir that absorbs heat as energy is known a as *heat sink*.

Now it will be advantageous to introduce the basic concepts of heat pump, heat engine and refrigerator with the help of the preceding terms.

5.2.1 Heat Engines

It is our common experience that work can easily be converted into heat. But the reverse process cannot occur by itself. For the reverse process, i.e., transformation of heat into work requires a special device known as a *heat engine*. Alternatively, a heat engine is defined as the device operating in a cycle for the continuous transformation of thermal energy (heat) into work. In a heat engine the heat always flows from a higher temperature region to a lower temperature one and produces a network shown in Fig. 5.2.

For example, consider the mechanical work done by the stirrer or agitator in a chemical reactor. If the agitator is allowed to rotate inside the cylindrical vessel, then first the work is converted into internal energy of the water as demonstrated in Fig. 5.1(a) and (b). This energy may then leave the water as heat. So, work is finally converted into heat. But the reverse is not true, i.e., transferring heat to the water does not cause the agitator to rotate. Hence, from this observation, we can conclude that work can be converted into heat easily but not the other way round. This is possible only with the assistance of a heat engine. So, heat engine is an energy conversion device which continuously converts the energy absorbed as heat into work.

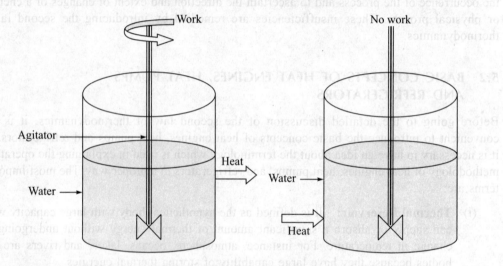

Fig. 5.1   (a) and (b) Demonstration of conversion of work into heat and heat into no work.

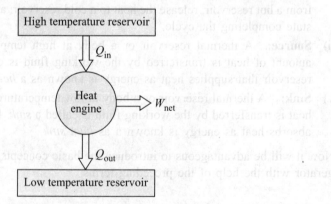

Fig. 5.2   Basic operation in a heat engine.

The common characteristics of a heat engine can be summarized as follows:

1. It absorbs heat from a high temperature source
2. It releases heat into a low temperature sink
3. Heat engine operates on a cycle
4. It delivers the net work.

Thermal Efficiency of Heat Engine

A heat engine is an energy (heat) transformation device. Its performance capability is judged by introducing a factor called the *energy transformation efficiency* or *thermal efficiency*. The thermal efficiency of a heat engine is defined as the ratio of net work output to the total heat input, or

$$\eta = \frac{\text{Net work output}}{\text{Heat input}}$$

$$= \frac{W_{\text{net}}}{Q_{\text{in}}}$$

$$= \frac{Q_{\text{in}} - Q_{\text{out}}}{Q_{\text{in}}} \tag{5.1}$$

This is to be noted that the thermal efficiency of a heat engine is always less than unity and can never be greater than 60 to 70% in actual practice. But as per the first law of thermodynamics, it should be 100%. So, an obvious question may arise: Why can't 100% thermal efficiency be possible. This can be answered with proper justification by using the second law of thermodynamics.

5.2.2   Heat Pumps

The heat pump is defined as the device by which heat flows from a low temperature region to a high temperature one. The objective of a heat pump is to maintain the heated space (a region to be heated) at high temperature by delivering heat. This is accomplished by absorption of heat from a low temperature source, such as a river, the surrounding land, or the cold outside air in winter, and supplying heat to a warmer medium such as a house. It has been depicted in Fig. 5.3.

The heat pump is a reversed heat engine—a device that can be used for heating houses and commercial buildings during winter and cooling them during summer.

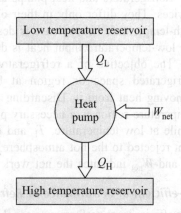

Fig. 5.3   Basic operation of a heat pump.

Co-efficient of Performance of Heat Pump

A heat pump or refrigerator is not an energy transformation device, i.e., heat is not converted into work. To say very clearly, a significant low or high temperature is generated by the expenditure of work. Therefore, the performance of heat pump or refrigerator is expressed in terms of the co-efficient of performance. The performance capability of a heat pump is expressed in terms of the co-efficient of performance, denoted by COP_{HP}. It is given by

$$COP_{HP} = \frac{\text{Heat absorbed}}{\text{Work required}} = \frac{\text{Heating effect}}{\text{Work input}}$$

$$= \frac{Q_H}{Q_H - Q_L} = \frac{Q_H}{W_{net}} \qquad (5.2)$$

5.2.3 Refrigerators

As far as the nature of the heat flow is concerned, whenever a temperature difference exists between two regions, heat flows in the direction of the decreasing temperature, i.e., from higher temperature region to a lower temperature one and the heat transfer process occurs in nature without requiring any devices. But the reverse process, cannot occur by itself. The transfer of heat from a low-temperature medium (source) to a high temperature one (sink) requires some special devices called *refrigerators*.

This mechanism of a refrigerator employs virtually a substance used as working fluid for carrying heat from the source to sink. The fluid is known as *refrigerant*. It is needless to mention that the efficiency of a refrigerator as well as the economy of cold production in the small scale or large scale application area, dominantly depends on the thermo-physical properties and role of a working fluid. The different properties of refrigerants will be discussed later in this chapter.

A refrigerator is schematically represented in Fig. 5.4 to give a transparent idea about its basic working principle as well as direction of heat transfer. Here, Q_L is the amount of heat removed from the refrigerated space.

Refrigerators and heat pumps are basically identical devices. They differ only in their objectives, in that the high-temperature heat-release is desired for heating and the low-temperature input heat is desired for cooling.

The objective of a refrigerator is to maintain the refrigerated space or region at low temperature by removing heat from it. Discarding this heat to a higher temperature region is a necessary part of this operation. While at low temperature, T_L and Q_H is the amount of heat rejected to the hot atmosphere, at high temperature T_H and W_{net} indicates the net work input to the refrigerator.

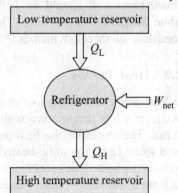

Fig. 5.4 Basic operation in a refrigerator.

Co-efficient of Performance of Refrigerator

The efficiency or performance capability of a refrigerator is expressed in terms of a suitable index known as the *co-efficient of performance*. It is denoted by COP_R.

A refrigerator is said to be most efficient if it produces a maximum amount of desired effect with minimum expenditure of work. The desired effect produced by a refrigerating machine is known as the *refrigerating effect*.

The performance capability of a refrigeration system is judged by the ratio of the refrigerating effect (or the amount of heat removed) to the work required. Actually, this ratio is called the *coefficient of Performance* (COP_R).

Referring to Fig. 5.4, if Q_L is the amount of heat removed at low temperature T_L, and Q_H is the amount of heat rejected at high temperature T_H, then by the first law of thermodynamics the external work required for transferring the heat is

$$W_{net} = Q_H - Q_L$$

The COP of a refrigerator can be expressed as

$$COP_R = \frac{\text{Heat removed}}{\text{Work required}} \quad \text{or} \quad \frac{\text{Refrigerating effect}}{\text{Work input}}$$

$$= \frac{Q_L}{Q_H - Q_L} = \frac{Q_L}{W_{net}} \tag{5.3}$$

It is necessary to remember that in most of the cases, COP of a refrigerator can be greater than unity, i.e., the amount of heat removed from the refrigerated space can be greater than the amount of work input and normally varies between 2 and 7.

5.3 STATEMENTS OF SECOND LAW OF THERMODYNAMICS

The first law is concerned with the quantity of energy and the conversion of energy from one form to another form of energy. The direction and extent of energy transformed cannot be ascertained with the help of the first law of thermodynamics. This insufficiency of the first law is compensated by introducing the second law of thermodynamics, as it guides a process in a definite direction. There are several logically acceptable statements on the second law of thermodynamics for spontaneous processes. These are:

- Heat cannot spontaneously pass from a colder body to a warmer body.
- All spontaneous processes are irreversible and tend to be in equilibrium.
- Every system which is left to itself will, on average, change towards a condition of maximum probability.
- There is a general tendency in nature for energy to pass from a higher-temperature region to a lower-temperature region.
- When an actual process occurs, it is impossible to invent a means of restoring every system concerned to its original condition.
- It is impossible by an engine to derive mechanical work from any portion of matter by cooling it below the temperature of the coldest part of the surroundings.
- It is impossible to construct an engine, working in a cycle, which would produce no effect except the raising of a weight and the cooling of a heat reservoir.

Of the above statements, the Kelvin–Planck and Clausius statements are two classical statements discussed in the following sections precisely with their illustrations.

5.3.1 Kelvin–Planck Statement

The Kelvin–Planck statement is related to the working principle of a heat engine. Heat engine plays a dominant role to substantiate the Kelvin–Planck statement. The basic operations in a heat engine have already been demonstrated in the preceding section. On the basis of working principle and the thermal efficiency of a heat engine, Kelvin–Planck statement of second law of thermodynamics may be expressed as follows:

It is impossible for a heat engine that operates on a cycle to convert the heat absorbed completely into a net amount of work.

It can be stated in another way as:

It is impossible for any device that operates on a cycle to receive heat from a single reservoir and produce a net amount of work.

It is very much clear from the preceding statements that a heat engine essentially requires a minimum of two thermal reservoirs to accomplish work in a cycle.

In a heat engine, a definite amount of heat is transferred from a high-temperature reservoir to a low-temperature reservoir. A part of the heat is converted into useful work and the remaining heat is rejected to the low-temperature reservoir to complete the cycle. It is depicted in Fig. 5.5. Henceforth, it is needless to mention that total heat cannot be converted into work. Therefore, the thermal efficiency of a heat engine can never be 100%. This is a limitation applied to both actual and ideal heat engines.

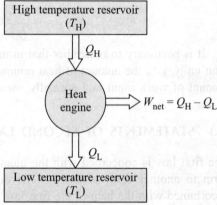

Fig. 5.5 A heat engine obeying Kelvin–Planck statement.

5.3.2 Clausius Statement

The Clausius statement is related to the working principle of a refrigerator or heat pump. We have demonstrated the basic operation in a refrigerator or heat pump. A certain amount of heat is abstracted from the lower-temperature region and released into a higher-temperature region. This process cannot take place spontaneously by itself.

It necessitates the work input. The Clausius statement is led by this observation and may be expressed as:

It is not possible to construct a device that operates on a cycle and produces no effect other than the transfer of heat from a low-temperature body to a high-temperature body.

In other words:

Heat cannot, of itself, pass from a cold body to a hot body.

The Clausius statement is well-supported by Fig. 5.6. The above statement implies that heat cannot be transferred from the body of lower temperature to that of higher

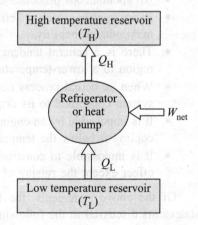

Fig. 5.6 A refrigerator or heat pump obeying Clausius statement.

temperature without work supplied from an external source. A refrigerator cannot be operated if its compressor is not run by the help of an external power source such as an electric motor.

Hence, a spontaneous process of transferring heat from a high-temperature body to a low-temperature body cannot proceed in the reverse direction unless it is externally aided.

5.3.3 Equivalence of Kelvin–Planck and Clausius Statements

The Kelvin–Planck and Clausius statements are equivalent in their importance as well as results. These two classical statements can be significantly treated as the second law of thermodynamics. The equivalence of Kelvin–Planck and Clausius statements can be better judged by showing the violation of one statement leads to violation of other. Any device that violates the Kelvin–Planck statement also violates the Clausius statement. The reverse is also true. This can be demonstrated by Fig. 5.7 and the following discussion.

Consider a combined system, represented by dotted rectangle, that consists of a heat engine and a refrigerator, shown in Fig. 5.7. The system operates between a high-temperature reservoir and a low-temperature reservoir. The heat engine violates the Kelvin–Planck statement by receiving all the heat Q_H from a single thermal reservoir and converting it into useful work W. That is, the thermal efficiency of the heat engine is 100%. But this can never be true. Now, the produced work is supplied to drive a refrigerator which transfers heat Q_L from the low-temperature reservoir and an amount of heat $(Q_L + Q_H)$ to a high-temperature reservoir. The combined system in Fig. 5.7 can be treated as a refrigerator which transfers an amount of heat Q_L [the difference between $(Q_L + Q_H)$ and Q_H] from a lower-temperature reservoir to a higher-temperature reservoir without the help of any external work input. This is a violation of the Clausius statement. Hence, the violation of Kelvin–Planck statement is nothing but the violation of the Clausius statement.

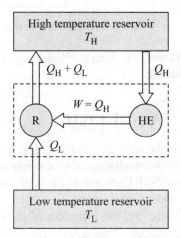

Fig. 5.7 Violation of Kelvin–Planck statement leads to violation of Clausius statement.

Similarly, the violation of the Clausius statement can be shown. Consider again the heat engine–refrigerator combination of Fig. 5.8. The refrigerator is assumed to violate the Clausius statement by absorbing heat Q_L from a low-temperature reservoir and discarding it to a high temperature reservoir without requiring any work input from outside. The heat engine operates between two thermal reservoirs. Let the heat engine absorb an amount of heat Q_H from the high-temperature reservoir at T_H and produce useful work W and release heat Q_L to the low-temperature reservoir. Then the heat received by the heat engine from the high-temperature reservoir will be $Q_H = Q_L + W$.

So the heat engine absorbs the heat $Q_H - Q_L = W$. The engine has not released any amount of heat to the low-temperature reservoir. This does necessarily mean that the heat engine operates with a single reservoir and converts the total heat into work with 100% efficiency. This is a violation of the Kelvin–Planck statement.

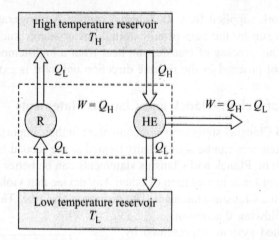

Fig. 5.8 Violation of Clausius statement leads to violation of Kelvin–Planck statement.

Hence, the Kelvin–Planck and Clausius statements of second law of thermodynamics are equivalent.

5.4 CARNOT CYCLE

This is the most efficient and ideal reversible cycle, which can explain the principle of the second law of thermodynamics in a better way. This cycle was first proposed by the French engineer Nicolas Sadi Carnot in 1824. With the help of this cycle, he explained how work is obtained from heat and to what extent.

Therefore, the main aim of the invention was:

(i) To estimate the work obtained from heat during its passage from a higher to a lower temperature

(ii) To obtain the maximum work in a cyclic operation where every step is to be carried out in a reversible fashion.

Carnot cycle is a hypothetical cycle. The heat engine operating on this cycle is known as *Carnot engine*. It is not possible to construct a Carnot engine, as it consists solely of reversible processes in all stages. This is because the reversible cycle cannot be achieved in practice due to the unavoidable irreversibilities associated with it. Though it is not a practical cycle, it serves as a standard of perfection against which the performance of any practical engine operating between high and low temperature can be compared.

Carnot cycle consists of four successive operational steps. These are:

(i) Reversible isothermal expansion
(ii) Reversible adiabatic expansion
(iii) Reversible isothermal compression
(iv) Reversible adiabatic compression.

Consider an adiabatic cylinder–piston assembly as a Carnot heat engine in which 1 mole of an ideal gas has been taken as a working substance. The cylinder head is insulated in such a way, that it may be removed so easily as to bring the cylinder into contact with the reservoir, to communicate thermally between the system and the surroundings. It has been depicted in Fig. 5.9.

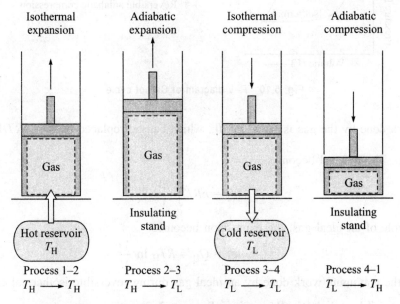

Fig. 5.9 Operation in Carnot cycle.

Step I: Reversible Isothermal Expansion (Process 1–2)

The heat source at temperature T_H is put in contact with the cylinder containing ideal gas. Now the gas is allowed to expand isothermally (as T_H = constant) from the state point 1 to 2 (volume change V_1 to V_2). As a result, the temperature of the gas decreases. Internal work is done by the system on the surroundings. Due to the drop in temperature, a certain amount of heat is transferred from the hot reservoir into the cylinder gas, to maintain the initial temperature T_H. The temperature constancy is maintained at T_H. Hence the gas expands isothermally and reversibly, and finally reaches position 2 along with the piston. This operation is represented diagrammatically by line 1–2 of the P–V diagram in Fig. 5.10.

Since the process is an isothermal expansion of an ideal gas, so the change in internal energy is zero, i.e., $\Delta U = 0$ or $dU = 0$.

The first law of thermodynamics then becomes

$$dQ = dU + dW$$
$$= 0 + dW$$
$$dQ = dW$$

It implies that all the heat added to the system is used to do work. The amount of heat absorbed by the gas is Q_H (corresponding to temperature T_H).

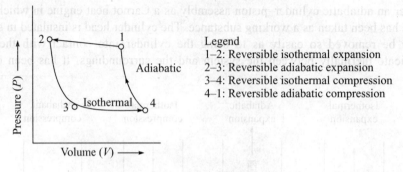

Fig. 5.10 *P–V* diagram of Carnot cycle.

The work done by the gas is $W_1 = \int\limits_{1}^{2} PdV$, which can be replaced by $P = nRT/V$ using the ideal gas law. The integral becomes

$$W_1 = nRT \int\limits_{V_1}^{V_2} \frac{dV}{V}$$

For 1 mole of an ideal gas, the expression becomes

$$W_1 = Q_H = RT_H \ln \frac{V_2}{V_1} \tag{5.4}$$

This is the maximum work done by an ideal gas during reversible isothermal expansion.

Step II: Reversible Adiabatic Expansion (Process 2–3)

At step II the cylinder head, which was a perfect heat conductor and in contact with the heat source, is now insulated properly by replacing the heat source with the insulating stand. So there is no scope of heat transfer between the system and the surroundings. The expansion of gas goes on continuously until its temperature drops from T_H to T_L. So, the gas is allowed to expand adiabatically and reversibly from state point 2 to 3 (volume change V_2 to V_3). The pressure remains uniform throughout the system. This operation is represented diagrammatically by line 2–3 of the *P–V* diagram.

As the expansion is adiabatic, heat absorbed by the gas = 0, i.e., $Q = 0$.
Putting this in the first law of thermodynamics, we have

$$dQ = dU + dW$$
$$0 = dU + dW$$
$$dW = -dU = -C_V (T_L - T_H) = C_V (T_H - T_L)$$

So, the work done by the system for Step II

$$W_2 = C_V (T_H - T_L) \tag{5.5}$$

Step III: Reversible Isothermal Compression (Process 3–4)

The insulation is removed from the cylinder head, and it is brought into contact with the sink at T_L to make it perfect conductor. Now the piston is pushed inward by an external force, doing work on the gas. The gas is now allowed to compress isothermally from state point 3 to 4, which in turn causes the temperature to rise. Due to the rise in temperature, heat is transferred from the cylinder to the sink (cold reservoir) to maintain at temperature T_L by dropping itself. Thus the gas temperature is kept constant at T_L. During this process, work is done on the system. The piston finally reaches state point 4. Pressure and volume change from P_3 and V_3 to P_4 and V_4.

The amount of heat released from the gas during this process is Q_L (corresponding to T_L). Therefore, the reversible work of compression (work done on the system) is given by

$$W_3 = Q_L = -\int_{V_3}^{V_4} P dV = -RT_L \ln \frac{V_4}{V_3} = RT_L \ln \frac{V_3}{V_4} \qquad (5.6)$$

Step IV: Reversible Adiabatic Compression (Process 4–1)

The cylinder head which was in contact with cold reservoir is now insulated properly by replacing the cold reservoir with the insulating stand. The system becomes adiabatic. The gas is compressed reversibly and adiabatically from the state point 4 to 1. The gas returns to its initial state. The temperature rises from T_L to T_H during this reversible adiabatic compression process, which completes the cycle. In other words, the gas is allowed to compress adiabatically and reversibly along 4–1 to bring it back to its original pressure, volume and temperature.

Since the pressure is adiabatic, the amount of heat absorbed by the gas = 0, i.e., $Q = 0$. Putting this in the first law of thermodynamics, we have

$$dQ = dU + dW$$
$$0 = dU + dW$$
$$dW = -dU = -C_V (T_H - T_L)$$

The work done on the system or work of compression is given by

$$W_4 = -[-C_V (T_H - T_L)]$$
$$= C_V (T_H - T_L) \qquad (5.7)$$

The system returns to its original state and the Carnot cycle is completed. The net work done by the gas in the complete cycle is given by

$$W_{cycle} = W_1 + W_2 - W_3 - W_4$$

$$= RT_H \ln \frac{V_2}{V_1} + C_V(T_H - T_L) - RT_L \ln \frac{V_3}{V_4} - C_V(T_H - T_L)$$

$$= RT_H \ln \frac{V_2}{V_1} - RT_L \ln \frac{V_3}{V_4}$$

But considering the adiabatic changes in operations 2 and 4, we have

(i) $T_L V_2^{\gamma-1} = T_H V_3^{\gamma-1}$

(ii) $T_H V_1^{\gamma-1} = T_L V_4^{\gamma-1}$

Dividing the above equations, we get

$$\frac{V_2}{V_1} = \frac{V_3}{V_4}$$

Hence

$$W = Q_H - Q_L = RT_H \ln \frac{V_2}{V_1} - RT_L \ln \frac{V_3}{V_4}$$

$$= RT_H \ln \frac{V_2}{V_1} - RT_L \ln \frac{V_2}{V_1}$$

$$= R \ln \frac{V_2}{V_1}(T_H - T_L) \tag{5.8}$$

5.4.1 Efficiency of Carnot Cycle

Carnot efficiency is the maximum possible efficiency for an engine cycle operating between high and low temperature reservoirs at T_H and T_L respectively. The temperatures in Carnot efficiency expression must be expressed in terms of degrees Kelvin. Temperature in °C in the expression of efficiency causes erroneous results.

The efficiency of the Carnot cycle can be defined as

$$\eta = \frac{\text{Heat converted into work}}{\text{Heat absorbed from the source } (T_H)}$$

$$= \frac{\text{Net work output}}{\text{Total heat input}}$$

$$= \frac{W}{Q_H} = \frac{R \ln \dfrac{V_2}{V_1}(T_H - T_L)}{RT_H \ln \dfrac{V_2}{V_1}}$$

$$= \frac{(T_H - T_L)}{T_H} = 1 - \frac{T_L}{T_H} < 1 \tag{5.9}$$

This equation can be rearranged as

$$\frac{W}{Q_H} = \frac{(T_H - T_L)}{T_H}$$

$$W = Q_H \frac{(T_H - T_L)}{T_H} \tag{5.10}$$

This equation can be generalized as

$$W = Q \frac{\Delta T}{T} \tag{5.11}$$

This relation shows that maximum amount of work can be obtained from the engine operating between the temperatures T_H and T_L. This is the mathematical expression of the second law of thermodynamics.

Hence the efficiency of a Carnot cycle is given by

$$\frac{W}{Q_H} = \frac{Q_H - Q_L}{Q_H} = \frac{T_H - T_L}{T_H} \tag{5.12}$$

It follows that $W < Q_H$, i.e., only a part of the heat absorbed by the system at T_H is transformed into work.

From the expression of efficiency, the following can be concluded:

1. The efficiency of any reversible engine
    (a) depends only on the temperature of the source and the sink
    (b) is independent of the nature of the working substance.

2. Total heat input can be transformed into work only when $T_L = 0$; then the expression becomes $W = Q$. These cannot be achieved in practice. Hence, the heat can not be transformed completely into work.

3. When $T_H = \infty$ or $T_L = 0$, $\eta = 1$; but these temperatures cannot be realized in actual practice. Thus the efficiency of an engine can never be unity.

4. When $T_H = T_L$, $W = 0$, i.e., there will be no work output by the engine operating under isothermal condition.

5. The efficiency increases with increase in upper temperature and constancy in lower temperature.

6. The efficiency increases with decrease in lower temperature and constancy in upper temperature.

EXAMPLE 5.1 A heat engine is working between a source at 550°C and a sink at 27°C. What is the efficiency of the heat engine?

Solution: The theoretical efficiency of a heat engine is given by

$$\eta = \frac{\text{Net work output}}{\text{Total heat input}}$$

$$= \frac{W_{net}}{Q_{in}}$$

$$= \frac{Q_{in} - Q_{out}}{Q_{in}} = \frac{T_H - T_L}{T_H}$$

$$= \frac{(550 + 273) - (27 + 273)}{(550 + 273)}$$

$$= 0.635$$

Hence the theoretical efficiency of the heat engine is 63.5%.

EXAMPLE 5.2 A reversible heat engine receives the heat from a high-temperature source at 810 K and release the same to a low-temperature sink at 300 K. Determine

(a) the thermal efficiency of the heat engine;
(b) the efficiency of the engine if the higher temperature is increased to 1366 K;
(c) the maximum efficiency of the engine, if the lower temperature is increased to 344 K while the higher temperature is kept at 810 K.

Solution: From the expression of thermal efficiency of heat engine, we have

$$\eta = \frac{T_H - T_L}{T_H}$$

(a) Given that $T_H = 810$ K and $T_L = 300$ K, we have

$$\eta = \frac{T_H - T_L}{T_H} = \frac{810 - 300}{810} = 0.629$$

Thus the efficiency of the heat engine is 62.9%.

(b) Given that $T_H = 1366$ K and $T_L = 300$ K, we have

$$\eta = \frac{T_H - T_L}{T_H} = \frac{1366 - 300}{1366} = 0.78$$

Thus the efficiency of the heat engine is 78.0%.

(c) In the same way, we can find out the efficiency of the heat engine as $T_H = 810$ K and $T_L = 344$ K. Therefore,

$$\eta = \frac{T_H - T_L}{T_H} = \frac{810 - 344}{810} = 0.575$$

The efficiency of the heat engine is 57.5%.

EXAMPLE 5.3   A Carnot engine absorbs heat to the tune of 585 kJ/cycle from a hot reservoir at 650°C and discards heat to a cold reservoir at 30°C. Then

(a) What is the theoretical efficiency of the Carnot engine?

(b) What amount would be released to the cold reservoir?

Solution:

(a) We are given that $T_H = 650 + 273 = 923$ K and $T_L = 30 + 273 = 303$ K. By applying Eq. (5.9), we have

$$\eta = \frac{T_H - T_L}{T_H} = \frac{923 - 303}{923} = 0.671$$

(b) From the knowledge of the ratio of heat and temperature between the two regions, we have

$$\frac{Q_H}{T_H} = \frac{Q_L}{T_L}$$

or

$$Q_L = \frac{T_L \, Q_H}{T_H} = \frac{303 \text{ K} \times 585 \text{ kJ}}{923 \text{ K}}$$

$$= 192 \text{ kJ}$$

EXAMPLE 5.4   It is desired to produce a 1 kg ice block from water in a freezer box of a refrigerator at 273 K while the temperature of the environment is 295 K. Given that the latent heat of fusion of ice at 273 K is 335 kJ/kg, determine

(a) the minimum work requirement

(b) the amount of heat released to the surroundings.

Solution:
(a) We know that the coefficient of performance of a refrigerating machine is

$$\text{COP}_\text{R} = \frac{Q_\text{L}}{W_\text{net}} = \frac{T_\text{L}}{T_\text{H} - T_\text{L}}$$

The preceding equation can be rearranged as

$$W_\text{net} = \frac{Q_\text{L}(T_\text{H} - T_\text{L})}{T_\text{L}} = \frac{335 \text{ kJ/kg}(295 - 273)\text{ K}}{273\text{ K}} = 26.99\text{ kJ} = 27\text{ kJ}$$

(b) The amount of heat released can be estimated as

$$W_\text{net} = Q_\text{H} - Q_\text{L}$$

or

$$Q_\text{H} = W_\text{net} + Q_\text{L}$$
$$= (27 + 335)\text{ kJ}$$
$$= 362\text{ kJ}$$

EXAMPLE 5.5 A Carnot engine working between a high-temperature source at 373 K and a low-temperature sink at 275 K receives 50 kJ of heat from a high-temperature region. Determine

(i) the minimum work required
(ii) the efficiency
(iii) the amount of heat released.

Solution:
(i) We know the thermal efficiency of a heat engine is given by

$$\eta_\text{HE} = \frac{W_\text{net}}{Q_\text{H}}$$

$$W_\text{net} = Q_\text{H} \cdot \text{COP}_\text{HE} = Q_\text{H} \cdot \frac{T_\text{H} - T_\text{L}}{T_\text{H}}$$

The minimum work done can be computed by using the following relation:

$$W_\text{net} = \frac{Q_\text{H}(T_\text{H} - T_\text{L})}{T_\text{H}} = \frac{50\text{ kJ}(373 - 275)\text{ K}}{373\text{ K}} = 13.1\text{ kJ}$$

(ii) The efficiency of a heat engine is

$$\eta = \frac{T_\text{H} - T_\text{L}}{T_\text{H}} = \frac{373 - 275}{373} = 0.262$$

(iii) The amount of heat released can be calculated as

$$\eta = \frac{\text{Net work output}}{\text{Total heat input}}$$

$$0.262 = \frac{W_\text{net}}{Q_\text{in}}$$

$$0.262 = \frac{Q_{in} - Q_{out}}{Q_{in}}$$

$$= 1 - \frac{Q_{out}}{Q_{in}}$$

$$Q_{out} = 50 \, (1 - 0.262) = 36.9 \text{ kJ}$$

EXAMPLE 5.6   An inventor claims to have developed a heat engine that receives 7000 J/s of heat from a source at 400°C. The power output of the engine is 5 hp. The temperature of the surroundings is 24°C. It can be treated as a sink for the engine. Can this claim be valid? Justify your answer.

Solution: The net work output can be calculated as

$$W = 5 \times 745.7 = 3728.5 \text{ W}$$

The heat input is, $Q = 7000 \text{ J/s} = 7000 \text{ W}$

The thermal efficiency of the claimed engine is, therefore,

$$= \frac{3728.5 \text{ W}}{7000 \text{ W}} = 0.5326$$

The theoretical efficiency of a Carnot heat engine is

$$\eta = \frac{T_H - T_L}{T_H} = \frac{673 - 297}{673} = 0.558$$

This claim is not valid, because the theoretical efficiency of the Carnot engine is greater than that of the engine claimed by the inventor. Carnot engine is considered the standard for comparison.

5.5   CARNOT THEOREM

It is very much clear from the Kelvin–Planck and Clausius statements of the second law of thermodynamics that the Kelvin–Planck statement refers to a heat engine and the Clausius statement refers to a refrigerator or heat pump. As per the former statement, the efficiency of a heat engine can never be 100% as it cannot convert the heat received completely to useful work, because a heat engine cannot operate with a single reservoir. The latter statement implies that heat cannot flow from a low-temperature reservoir to a high-temperature one by itself, i.e., a refrigerator or a heat pump cannot operate without requiring any work input from an external source. Hence, it can be concluded that the above two statements are related to the performance capability or efficiency of reversible and irreversible heat engines. These are known as the *Carnot Theorems*.

Theorem I: *The efficiency of an irreversible heat engine is always less than the efficiency of a reversible one operating between the same two reservoirs.*

Theorem II: *The efficiencies of all reversible heat engines operating between the same two reservoirs are the same. The efficiency of a heat engine does not depend on the nature of the working medium but only on the temperatures of the reservoirs between which it operates.*

We can prove the above two theorems by demonstrating that the violation of one theorem leads to the violation of the second law of thermodynamics.

Let us prove the first theorem. Suppose two heat engines are operating between the same thermal reservoirs. Among them, one is reversible and other is irreversible. Now, the amount of heat Q_H supplied to the engines is the same. Both of them will produce a certain amount of work, say, W_{rev} and W_{irrev} by reversible and irreversible heat engines respectively. In order to violate the first theorem, the irreversible heat engine is assumed to be more efficient that the reversible one, i.e., $\eta_{irrev} > \eta_{rev}$. So, this relation implies that the irreversible engine will produce more useful work than the reversible heat engine.

Now, consider that the reversible engine operates in the opposite direction with respect to the initial direction. Then it will act as a refrigerator. In this case, the amount of work received by the refrigerator is obviously W_{rev}, which will induce the refrigerator to discard an amount of heat Q_H to the high-temperature reservoir. The irreversible heat engine will receive heat Q_H from the same reservoir. Therefore, no temperature gradient exists in the high-temperature reservoir, and consequently the net exchange of heat for this reservoir is zero. Since this reservoir is of no meaningful use, it could be eliminated from the circuit by delivering the discharged heat Q_H into the irreversible heat engine directly. This operation is shown in Fig. 5.11.

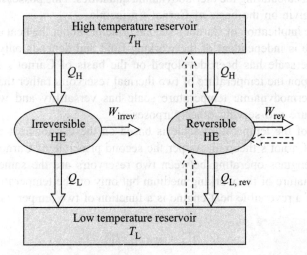

Fig. 5.11 Violation of first Carnot theorem leads to violation of Kelvin–Planck statement of the second law of thermodynamics.

Now, the combined system consisting of a irreversible heat engine and refrigerator, as shown in Fig. 5.12, delivers a net amount of work $W_{irrev} - W_{rev}$ by heat with a single reservoir. This can never be possible. This is a violation of the Kelvin–Planck statement. So, the initial assumption, $\eta_{irrev} > \eta_{rev}$, is absolutely wrong.

Hence, we can conclude that the reversible heat engine is always more efficient than the any heat engine. In the same fashion, we can prove the second Carnot theorem.

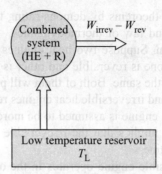

Fig. 5.12 Combined system (heat engine + refrigerator).

5.6 THERMODYNAMIC TEMPERATURE SCALE

The second law of thermodynamics plays a dominant role in the development of the thermodynamic temperature scale. It is widely used to measure the temperature. It is an useful tool that also helps in calculating the thermodynamic quantities. This possesses some universality. It was coined by Kelvin on the basis of Carnot's principle.

According to the implication of Carnot's second principle about the heat engine, the efficiency of a reversible cycle is independent of the working fluid and depends only on the temperature of the reservoir. The scale has been developed on the basis of Carnot's theorem on the heat engine, depending upon the temperature of two thermal reservoirs rather than the working fluid. That is why the thermodynamic temperature scale has versatility and wide acceptability in measuring temperature and serving other purposes.

The derivation of the temperature scale is based on the Carnot heat engine following the second principle of Carnot's theorem. As per the second principle of Carnot, the efficiencies of all reversible heat engines operating between two reservoirs are the same, and the efficiency depends not on the nature of the working medium but only on the temperature of the reservoirs. So, the efficiency of a reversible heat engine is a function of two temperatures T_H and T_L of the reservoirs, i.e.,

$$\eta_{rev} = g\ (T_H,\ T_L) \tag{5.13}$$

From the expression of thermal efficiency of Carnot cycle, we have

$$\eta_{rev} = 1 - \frac{Q_L}{Q_H} \tag{5.14}$$

Combining Eqs. (5.13) and (5.14), we get

$$\eta_{rev} = g\ (T_H,\ T_L) = 1 - \frac{Q_L}{Q_H}$$

$$\frac{Q_H}{Q_L} = f(T_H,\ T_L) \tag{5.15}$$

To develop the temperature scale based on the functional form of $f(T_H,\ T_L)$, consider three reversible heat engines A, B, and C of Fig. 5.13. The first one (engine A) operating between a

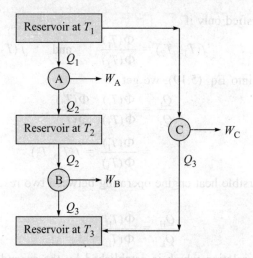

Fig. 5.13 Three reversible heat engines operating between thermal reservoirs.

high-temperature reservoir at T_1 and a low-temperature reservoir at T_2 absorbs the heat Q_1 and release the heat Q_2 by doing work W_A. The second one (engine B) absorbs the heat Q_2 at T_2 and release the heat Q_3 at T_3 operating between the two reservoirs by doing work W_B. The third one (engine C) operating between the two temperatures T_3 and T_1 absorbs the heat Q_1 from the point T_1 and releases the heat Q_3 at T_3 by doing work W_C.

It is to be noted that the amount of heat discarded by the heat engines B and C must be the same, as A and B can be combined into one reversible engine.

By the application of Eq. (5.15) to engine A, we have

$$\frac{Q_1}{Q_2} = f(T_1, T_2) \tag{5.16}$$

For engine B, we have

$$\frac{Q_2}{Q_3} = f(T_2, T_3) \tag{5.17}$$

For engine C, we have

$$\frac{Q_1}{Q_3} = f(T_1, T_3) \tag{5.18}$$

Considering Eqs. (5.16), (5.17) and (5.18), we have

$$\frac{Q_1}{Q_3} = \frac{Q_1}{Q_2} \cdot \frac{Q_2}{Q_3} \tag{5.19}$$

or

$$f(T_1, T_3) = f(T_1, T_2) \cdot f(T_2, T_3) \tag{5.20}$$

From the expression (5.20), it is clear that the left hand side is independent of T_2. So the right hand side must also be independent of T_2.

This condition is satisfied only if

$$f(T_1, T_2) = \frac{\Phi(T_1)}{\Phi(T_2)} \quad \text{and} \quad f(T_2, T_3) = \frac{\Phi(T_2)}{\Phi(T_3)}$$

Putting these values into Eq. (5.19), we get

$$\frac{Q_1}{Q_3} = \frac{\Phi(T_1)}{\Phi(T_2)} \cdot \frac{\Phi(T_2)}{\Phi(T_3)}$$

$$= \frac{\Phi(T_1)}{\Phi(T_3)} = f(T_1, T_3) \tag{5.21}$$

Therefore, for a reversible heat engine operating between two reservoirs at temperature T_H and T_L

$$\frac{Q_H}{Q_L} = \frac{\Phi(T_H)}{\Phi(T_L)} \tag{5.22}$$

This is an important relation which is established by the second law of thermodynamics based on the Carnot cycle on which a reversible heat engine operates. Lord Kelvin took an initiative to define the thermodynamic temperature scale by considering

$$\Phi(T) = T$$

So, Eq. (5.22) becomes

$$\frac{Q_H}{Q_L} = \frac{T_H}{T_L} \tag{5.23}$$

This scale is known as the Thermodynamic Temperature Scale or Kelvin Scale in which the temperature ranges from zero to infinity. This scale is a ratio of two absolute temperatures. But it does not provide any information about the magnitude of Kelvin.

Therefore, to measure the temperature of an object, we use water at its triple point (273.16 K) in the low-temperature reservoir of a reversible heat engine depicted in Fig. 5.14. Suppose the temperature of an object is T. Then to get the temperature of reservoir T, we measure Q_H and Q_L.

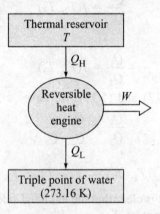

Fig. 5.14 Determination of thermodynamic temperature on Kelvin scale.

Then, T can be determined by

$$T = T_L \frac{Q_H}{Q_L} = 273.16 \text{ K} \frac{Q_H}{Q_L} \tag{5.24}$$

5.7 CONCEPT OF ENTROPY

In 1865, Clausius had invented a new thermodynamic property called *entropy*. The term is derived from a Greek word. As per the Greek word's meaning, the suffix *en* means energy and *trope* means change. It thus summarily means *energy change*. It is a state variable that plays a dominant role in understanding the condition of a thermodynamic system. The application of the second law of thermodynamics to a process in a closed system led to the establishment of entropy as the property of internal energy by the first law of thermodynamics.

Entropy is defined as the measure of disorderliness or randomness and multiplicity of a system. Substances which are highly disordered have high entropies. It denotes the extent to which atoms, molecules or ions are distributed in a disorderly manner in a given region. That is, higher the disorderness, higher is the entropy. It is also a measure of the amount of energy which is unavailable to do work.

It is independent of the path of transformation of the system. This function is called *entropy*, denoted by S. It is the ratio of the reversible heat change to the temperature, i.e.

$$dS = \frac{dQ_{rev}}{T}$$

where dQ_{rev} = reversible heat change and rev implies the reversibility of the system.

Example of Physical Concept of Entropy

1. Suppose a certain quantity of yellowish chlorine gas is unconfined or released at the corner of a room. Then it is observed that the gas spreads in all directions of the room until it is most chaotically distributed over the entire room and the equilibrium is reached. This is a natural process. Never we find opposite to happen in practice. It is needless to mention that this process is an irreversible process and has always a tendency to attain a state of equilibrium. Since it is an irreversible process, it will, thus, lead to an increase in entropy. Here, higher the tendency of the system to be in equilibrium, greater is the disorderness or randomness of the molecules, higher will be the entropy.

2. If a coloured material like anhydrated copper sulphate is placed at the bottom of a bottle which is full of water, the colour will slowly spread from the lower part of the bottle to the upper layer of clear water and after a sufficiently long time, the water will be uniformly coloured. By means of diffusion, the coloured material moves from a region of higher concentration to that of lower concentration. Diffusion occurs as a result of the random thermal movement of molecules. During its movement, a molecule collides with other molecules, changes its speed and direction. The process of diffusion continues until the uniformity of concentration is reached, or in other words, a state of equilibrium has been achieved. This is an irreversible process in which molecules are colliding with each other in a disorderly manner. In order to be in equilibrium, the molecules of the system will be more random or disordered. The entropy also goes on

increasing and reaches its maximum. Hence, it can be concluded here that greater the tendency of the system to be in equilibrium, greater is the disorderness or randomness of the molecules, the more intense will be the entropy.

5.7.1 Entropy—A Thermodynamic State Function

If a reversible heat engine operates between T_H and T_L, then from the mathematical expression of efficiency of a Carnot engine, we have

$$\eta = \frac{W}{Q_H} = \frac{Q_H - Q_L}{Q_H} = \frac{T_H - T_L}{T_H} = 1 - \frac{T_L}{T_H} \qquad (5.25)$$

Suppose the reversible engine absorbs the heat Q_H from the high-temperature reservoir at temperature T_H and releases the heat Q_L to the low-temperature reservoir at temperature T_L; so the net work output

$$W_{net} = Q_H - Q_L$$

Substitution of this into Eq. (5.25) yields

$$\frac{Q_H - Q_L}{Q_H} = 1 - \frac{T_L}{T_H} \qquad (5.26)$$

or

$$1 - \frac{Q_L}{Q_H} = 1 - \frac{T_L}{T_H} \qquad (5.27)$$

or

$$\frac{Q_L}{Q_H} = \frac{T_L}{T_H} \qquad (5.28)$$

or

$$\frac{Q_H}{T_H} = \frac{Q_L}{T_L} \qquad (5.29)$$

As per the sign convention, the heat absorbed by the system should be treated as a positive quantity, i.e., $Q_H = +ve$, and the heat released to the system should be treated as a negative quantity, i.e., $Q_L = -ve$.

Applying this sign convention to Eq. (5.29), we have

$$\frac{Q_H}{T_H} = \frac{-Q_L}{T_L} \qquad (5.30)$$

or

$$\frac{Q_H}{T_H} + \frac{Q_L}{T_L} = 0 \qquad (5.31)$$

Equation (5.31) does necessarily imply that for a reversible heat engine or Carnot engine undergoing a cyclic process, the summation of these 2 quantities—Q_H/T_H and Q_L/T_L—is zero. Hence, for an imaginary Carnot cycle comprising a number of cycles, as shown in the $P–V$ diagram in Fig. 5.15, Eq. (5.31) suggests certain existence of a new thermodynamic property whose change is measured by the quantity Q/T.

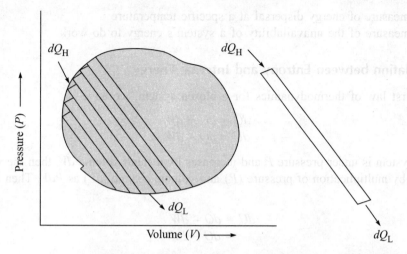

Fig. 5.15   Imaginary Carnot cycle consisting of a number of reversible cycles.

For an infinitesimal reversible change, Eq. (5.31) can be written as

$$\frac{dQ_{H}}{T_{H}} + \frac{dQ_{L}}{T_{L}} = 0 \qquad (5.32)$$

In a Carnot cycle, it is the ratio of the reversible heat change to the temperature, i.e.,

$$\oint \frac{dQ_{rev}}{T} = 0 \qquad (5.33)$$

where dQ_{rev} is the reversible heat change.

For an arbitrary cycle, the above ratio shows the characteristic of a new thermodynamic property. The differential change of this property can be written as

$$dS = \oint \frac{dQ_{rev}}{T} = 0 \qquad (5.34)$$

On substitution of this value of dS into Eq. (5.20), we have

$$\oint dS = 0$$

5.7.2   Entropy—At a Glance

The following expressions can each be considered definitions of entropy:

- A direct measure of the randomness of a system
- A measure of disorder in the universe or of the availability of energy in a system to do work
- A parameter representing the state of disorder of a system at the atomic, ionic, or molecular level
- An index of the tendency of a system towards spontaneous change
- A measure of the partial loss of ability of a system to perform work due to the effects of irreversibility

- A measure of energy dispersal at a specific temperature
- A measure of the unavailability of a system's energy to do work.

5.7.3   Relation between Entropy and Internal Energy

From the first law of thermodynamics for a closed system, we have

$$dQ = dU + dW$$
$$dU = dQ - dW \tag{5.35}$$

If the system is under pressure P and increases by a small volume dV, then the work done is obtained by multiplication of pressure (P) and volume change (dV) as PdV. Then Eq. (5.35) becomes

$$dU = dQ - dW$$
$$= dQ - PdV \tag{5.36}$$

From the definition of entropy established by the second law of thermodynamics, we get

$$dS = \frac{dQ}{T}$$

or

$$dQ = TdS \tag{5.37}$$

Substituting the value of dQ in Eq. (5.35), it yields,

$$dU = TdS - PdV \tag{5.38}$$

This is the *combined form of the first law and the second law of thermodynamics*. This is an useful relation in the deduction of various other thermodynamic relations of different functions.

5.7.4   Relation between Entropy and Enthalpy

From the definition of enthalpy, we know that it is nothing but the summation of the internal energy and the product of pressure and volume (that is, work done).

Hence, it can be expressed as

$$H = U + PV$$
$$dH = dU + PdV + VdP$$
$$= TdS - PdV + PdV + VdP$$
$$= TdS + VdP$$

This is an important relation which is applied to calculate the heat interactions in different processes.

5.8   CALCULATION OF ENTROPY CHANGES

Since entropy is a state function, any change or difference in its values always depends on the initial and final states of the system and is independent of the path of transformation. But entropy change is regardless of whether a system undergoes a reversible or an irreversible process.

Reversibility of a process is essential for the calculation of change in entropy, because we know that the change in entropy experienced by a system undergoing a reversible process (following the initial and final states) is given by

$$\Delta S = S_2 - S_1 = \int_2^1 \frac{dQ_{\text{rev}}}{T} \qquad (5.39)$$

i.e., this equation is applicable only for a reversible process, although no process is associated with the reversibility in practice.

Now, if the system undergoes an irreversible process, then how can the entropy change be calculated? The answer is quite simple. In order to calculate the entropy change in an irreversible process, the process must be replaced by an equivalent reversible process. In other words, an imaginary reversible process must be devised for achieving the same change. Then the summation of entropy change by the imaginary reversible process will give us the value for the actual irreversible process.

Hence, it can be concluded that change in entropy is easily computed by assuming that the actual processes are reversible. This technique will be demonstrated for the following processes.

5.8.1 Entropy Change in Reversible Process

Carnot cycle is a reversible cycle and it comprises four successive operational steps. Every step is carried out in a reversible fashion. Hence, the Carnot cycle can be considered here for the estimation of entropy changes in a reversible process.

In a Carnot cycle, the change in entropy for each step can be computed as follows:

Process I (*Reversible isothermal expansion*):   The change in entropy for the process is given by

$$S_{\text{I}} = \frac{dQ_H}{T_H} \qquad (5.40)$$

where Q_H is the amount of heat received from the source and T_H is the corresponding temperature.

Process II (*Reversible adiabatic expansion*):   Since the process is adiabatic, therefore, $dQ = 0$. Hence, the change in entropy for the process is given by

$$S_{\text{II}} = 0 \qquad (5.41)$$

Process III (*Reversible isothermal compression*):   The change in entropy for the above process is given by

$$S_{\text{III}} = \frac{dQ_L}{T_L} \qquad (5.42)$$

where Q_L is the amount of heat given up to the sink and T_L is the corresponding temperature.

Process IV (*Reversible adiabatic compression*):   Since the process is adiabatic, therefore, $dQ = 0$. Hence, the change in entropy for the process is given by

$$S_{\text{IV}} = 0 \qquad (5.43)$$

Therefore, the net change in entropy for the entire cycle can be represented as

$$\Delta S = S_{\text{I}} + S_{\text{II}} + S_{\text{III}} + S_{\text{IV}} = \frac{dQ_H}{T_H} + 0 + \frac{dQ_L}{T_L} + 0 = \frac{dQ_H}{T_H} + \frac{dQ_L}{T_L} \qquad (5.44)$$

In a Carnot cycle, we know that

$$\frac{dQ_H}{T_H} = \frac{dQ_L}{T_L}$$

or,

$$\frac{dQ_H}{T_H} + \frac{-dQ_L}{T_L} = 0$$

where dQ_H is the heat absorbed by the gas and dQ_L is the heat released from the gas. Hence, $-dQ_L$ means heat absorbed.

So, we may write

$$\frac{dQ_H}{T_H} + \frac{dQ_L}{T_L} = 0 \qquad (5.45)$$

where dQ stands for heat changes, contains own signs.

Substituting Eq. (5.45) into Eq. (5.44), we get

$$\Delta S = \frac{dQ_H}{T_H} + \frac{dQ_L}{T_L} = 0$$

So, for the entire Carnot cycle, heat interactions with respect to temperature can be represented as

$$\oint dS = \oint \frac{dQ}{T} = \frac{dQ_H}{T_H} + \frac{dQ_L}{T_L} = 0 \qquad (5.46)$$

Hence, in any reversible cyclic process, the net entropy-change is zero.

5.8.2 Entropy Change in Irreversible Process

Since the entropy is a state function, the entropy change, ΔS of a system from a given state 1 to a state 2 is always same and does not depend on the path of the change chosen. We know to calculate the entropy for a reversible path and it is given by

$$\Delta S = S_2 - S_1 = \int_1^2 \frac{dQ_r}{T}$$

But in order to calculate the change in entropy of a system undergoing irreversible process, we must imagine the system from state 1 to a state 2 along an 'imagined' or 'hypothetical' path that is reversible and then consider the ratio of the heat increment over the temperature at each point and sum up the quotients.

Let us consider a heat reservoir 'A' which is maintained at temperature T_1. Now if it is allowed to be brought in thermal contact with the second reservoir 'B', being held at a lower temperature T_L, then a quantity of heat 'Q' will flow irreversibly from high temperature to the low-temperature reservoir.

Then, the decrease in entropy of reservoir $A = \dfrac{Q}{T_1}$ the increase in entropy of reservoir

$B = \dfrac{Q}{T_2}$

Therefore, the net entropy-change for the irreversible flow of heat from reservoir A to B is given by

$$\Delta S = \frac{Q}{T_2} - \frac{Q}{T_1} = Q\left(\frac{T_1 - T_2}{T_1 T_2}\right) = + \text{ve quantity}$$

Hence, the irreversible flow of heat leads to an increase in entropy.

5.8.3 Entropy at Phase Change

When the two phases are in equilibrium condition, the reversible phase transition or change can occur. A pure substance under constant pressure and temperature undergoes phase transitions such as vaporization, fusion, allotropic transition, etc. The estimation of the entropy change at phase transition is quite simple and this is done on the basis of the latent heat of phase change and the temperature at which phase change occurs.

Consider a glass of ice water, in which external pressure is constant. Now, the phase transition temperature of ice water, $T_{tr} = 273$ K when ice is in equilibrium with liquid water at 1 atm; and for boiling water, $T_{tr} = 373$ K when liquid water in equilibrium with vapour at 1 atm. In this case, any transfer of heat between the system and the surroundings is reversible, since the two phases in the system are in equilibrium. So a *phase transition is reversible*. It is regardless of how the ice melts (what path it follows); since entropy is a state function, it is independent of the path.

In the preceding case, the change in entropy for a unit mole of pure substance at fixed temperature and pressure under a reversible process is given by the following equations.

For fusion (melting)

$$\Delta S_{\text{fus}} = \frac{\Delta h_{\text{fus}}}{T_{tr}} = \frac{\Delta h_{\text{sf}}}{T_{tr}} = \frac{h_{\text{f}} - h_{\text{s}}}{T_{tr}} = \frac{Q}{T_{tr}} \qquad (5.47)$$

where

ΔS_{fus} = Entropy of fusion (melting)

Δh_{sf} = Enthalpy (latent heat) of fusion.

For vaporization

$$\Delta S_{\text{vap}} = \frac{\Delta h_{\text{vap}}}{T_{tr}} = \frac{\Delta h_{\text{fg}}}{T_{tr}} = \frac{h_{\text{g}} - h_{\text{f}}}{T_{tr}} = \frac{Q}{T_{tr}} \qquad (5.48)$$

where

ΔS_{vap} = Entropy of vaporization

Δh_{fg} = Enthalpy (latent heat) of vaporization.

For sublimation

$$\Delta S_{\text{sub}} = \frac{\Delta h_{\text{sub}}}{T_{tr}} = \frac{\Delta h_{\text{sg}}}{T_{tr}} = \frac{h_{\text{g}} - h_{\text{s}}}{T_{tr}} = \frac{Q}{T_{tr}} \qquad (5.49)$$

where

ΔS_{sub} = Entropy of sublimation

Δh_{sg} = Enthalpy (latent heat) of sublimation.

In the above three cases, T_{tr} is the phase transition temperature and the suffixes s, f, and g denote solid, liquid and gas respectively.

EXAMPLE 5.7 Calculate the change in entropy for the conversion of 1 mol of ice to liquid at 273 K and 1 atm. The latent heat of fusion is 6500 J/mol.

Solution: By using Eq. (5.47), the entropy change for fusion is given by

$$\Delta S_{fus} = \frac{\Delta h_{fus}}{T_{tr}} = \frac{Q}{T_{tr}} = \frac{6500 \text{ J/mol}}{273 \text{ K}} = 23.80 \text{ J/mol-K}$$

5.8.4 Entropy Changes of Ideal Gas

Consider 1 mol of an ideal gas enclosed in a cylinder fitted with a frictionless and movable piston at constant pressure. Suppose the gas is reversibly expanding from V_1 to V_2. As the process is reversible, the pressure of the gas is approximately equal to the external pressure against which the gas is expanding at all stages. Hence, under these conditions, the maximum work done by the gas is equal to PdV. From the first law of thermodynamics for a closed system, we have

$$dQ = dU + dW$$

$$= dU + PdV$$

$$= C_V dT + \frac{RT}{V} dV \quad \text{(for an ideal gas, } dU = C_V dT \text{ and } PV = RT\text{)}$$

Dividing T by both sides, we have

$$\frac{dQ}{T} = C_V \frac{dT}{T} + \frac{RT}{TV} dV$$

$$dS = C_V \frac{dT}{T} + \frac{R}{V} dV \tag{5.50}$$

On integration with integral value, Eq. (5.24) yields

$$S_2 - S_1 = C_V \ln \frac{T_2}{T_1} + R \ln \frac{V_2}{V_1}$$

$$\Delta S = C_V \ln \frac{T_2}{T_1} + R \ln \frac{V_2}{V_1} \tag{5.51}$$

$$= C_V \ln \frac{T_2}{T_1} + R \ln \frac{P_1}{P_2} \tag{5.52}$$

For an ideal gas, $C_P - C_V = R$. On substitution of this value of C_V into Eq. (5.52), we have

$$\Delta S = (C_P - R) \ln \frac{T_2}{T_1} + R \ln \frac{V_2}{V_1} \tag{5.53}$$

Again, for an ideal gas, we know

$$\frac{P_1 V_1}{T_1} = \frac{P_2 V_2}{T_2}$$

or

$$\frac{V_2}{V_1} = \frac{P_1 T_2}{P_2 T_1} \tag{5.54}$$

Putting Eq. (5.54) into (5.55), we obtain

$$\Delta S = C_P \ln \frac{T_2}{T_1} + R \ln \frac{P_1}{P_2} \tag{5.55}$$

Now, we consider the following three cases:

Case 1 When $T_1 = T_2$ i.e., for isothermal change of a process, Eq. (5.55) reduces to

$$(\Delta S)_T = R \ln \frac{P_1}{P_2} = R \ln \frac{V_2}{V_1} \tag{5.56}$$

Case 2 When $P_1 = P_2$ i.e., for isobaric change of a process, Eq. (5.55) reduces to

$$(\Delta S)_P = C_P \ln \frac{T_2}{T_1} \tag{5.57}$$

Case 3 When $V_1 = V_2$ i.e., for isochoric change of a process, Eq. (5.51) reduces to

$$(\Delta S)_V = C_V \ln \frac{T_2}{T_1} \tag{5.58}$$

EXAMPLE 5.8 Calculate the change in entropy when 5 moles of an ideal gas expands from a volume of 5 L to 50 L at 27°C.

Solution: We know that the change in entropy for an isothermal change of an ideal gas is given by

$$(\Delta S)_T = nR \ln \frac{P_1}{P_2} = nR \ln \frac{V_2}{V_1}$$

Here

$$V_1 = 5 \text{ L}$$
$$V_2 = 50 \text{ L}$$
$$n = 5 \text{ mol}$$

Putting the values into the preceding equation, we get

$$(\Delta S)_T = 5 \times 1.987 \times \ln \frac{50}{5} = 22.876 \text{ cal/deg} = 22.876 \text{ eu}$$

EXAMPLE 5.9 Calculate ΔS when 8 mol of an ideal gas are heated from a temperature of 350 K to a temperature of 700 K at constant pressure. Assume that $C_P = \frac{5}{2} R$.

Solution: Gas constant, $R = 8.314$ J/mol-K. The entropy change is

$$(\Delta S)_P = nC_P \ln \frac{T_2}{T_1}$$

$$= 8 \times \frac{5}{2} \times 8.314 \times \ln \frac{700}{350} = 115.25 \text{ J/K}$$

EXAMPLE 5.10 5 mol of an ideal gas is expanded from a temperature of 300 K at 3 bars to 400 K at 12 bars. Assume that $C_P = 26.73$ kJ/mole-K.

Solution: The entropy the process can be calculated by the following expression:

$$\Delta S = C_P \ln \frac{T_2}{T_1} + R \ln \frac{P_1}{P_2}$$

$$= 26.73 \ln \frac{400}{300} + 8.314 \ln \frac{3}{12}$$

$$= 3.837 \text{ kJ/mol-K}$$

5.8.5 Entropy Changes in Mixture of Non-identical Ideal Gases

When two non-identical ideal gases, each at pressure P and temperature T, are brought into contact with each other, they would diffuse and mix up spontaneously. This is an irreversible process and will yield a considerable entropy change. This process can be replaced by an equivalent reversible process, and the change in entropy associated with such mixing of two ideal gases can easily be estimated through the *two-step reversible operations* that take place before and after mixing.

Suppose N_A and N_B are gram-moles of two ideal gases A and B, mixed at constant temperature T and constant pressure P. Then the total mole number of the mixture

$$N = N_A + N_B$$

The mole fractions of gas A and gas B respectively are

$$x_A = \frac{N_A}{N}$$

and

$$x_B = \frac{N_B}{N}$$

Step I: Before mixing

Gases A and B are allowed to undergo an isothermal expansion from an initial pressure P to their partial pressures p_A and p_B in the final mixture. In this step, the expression on their entropy changes can be derived from the first law of thermodynamics.

For gas A

$$dQ = dU + dW$$
$$= C_V dT + P dV$$

Dividing both sides by T, we get

$$\frac{dQ}{T} = \frac{C_V dT}{T} + \frac{P dV}{T}$$

$$dS = C_V \frac{dT}{T} + R \frac{dV}{V} \qquad \text{(since } PV = RT \text{ for 1 mol of an ideal gas)}$$

or

$$dS = C_V \, d \, (\ln T) + R \, d \, (\ln V)$$

or

$$S = C_V \ln T + R \ln V + S^0 \qquad \text{where } S^0 = \text{constant}$$

$$= (C_P - R) \ln T + R \ln \frac{RT}{P} + S^0 \qquad (\text{since } C_P - C_V = R \text{ for 1 mol of ideal gas})$$

$$= C_P \ln T - R \ln T + R \ln R + R \ln T - R \ln P + S^0$$
$$= C_P \ln T - R \ln P + S^0$$

For N_A moles of gas A

$$S_A' = N_A (C_P \ln T - R \ln P + S_A^0) \qquad (5.59)$$

Similarly, for N_B moles of gas B

$$S_B' = N_B (C_P \ln T - R \ln P + S_B^0) \qquad (5.60)$$

Step II: After mixing

The ideal gases are allowed to pass through a semi-permeable membrane which will permit the flow of any one of the two gases A and B.

The total pressure P remains constant, and it comprises partial pressures p_A and p_B, i.e.

$$P = p_A + p_B$$

Hence, the partial pressure of gas A in mixture $= p_A/P = x_A$ and of gas B $= p_B/P = x_B$.

The entropy changes associated with this step are:

For gas A

$$S_A'' = N_A (C_P \ln T - R \ln p_A + S_A^0)$$

For gas B

$$S_B'' = N_B (C_P \ln T - R \ln p_B + S_B^0)$$

Now, the entropy change due to mixing is

$$\Delta S_M = S_A'' + S_B'' - S_A' - S_B'$$

$$= -N_A R \ln \frac{p_A}{P} - N_B R \ln \frac{p_B}{P}$$

$$= -N_A R \ln x_A - N_B R \ln x_B$$

$$= -NR \left(\frac{N_A}{N} \ln x_A + \frac{N_B}{N} \ln x_B \right)$$

$$= -NR (x_A \ln x_A + x_B \ln x_B) \qquad (5.61)$$

$$= -NR \sum x_i \ln x_i$$

For 1 mol of the mixture

$$\Delta S_M = -R \sum x_i \ln x_i \qquad (5.62)$$

where

x_i = Mole fraction of species i
R = Universal gas constant.

EXAMPLE 5.11 Calculate the entropy of 1 kmol of air containing 21% oxygen and 79% nitrogen by volume. These are at the same temperature and pressure.

Solution: The change in entropy is given by

$$\Delta S_M = -NR (x_A \ln x_A - x_B \ln x_B)$$

where

$x_A = 0.21$ for oxygen
$x_B = 0.79$ for nitrogen
$N = 1$ kmol.

Hence, for 1 kmol of air, the entropy change is

$$= -8314 (0.21 \ln 0.21 + 0.79 \ln 0.79)$$
$$= 4.27 \text{ kJ/kmol-K}$$

EXAMPLE 5.12 An ideal gas mixture contains 5.6 L of oxygen and 16.8 L of hydrogen at N.T.P. Calculate the increase in entropy.

Solution:

$$5.6 \text{ L of oxygen} = \frac{5.6}{22.4} = \frac{1}{4} \text{ g-mol} = 0.25 \text{ g-mol}$$

$$16.8 \text{ L of hydrogen} = \frac{16.8}{22.4} = \frac{3}{4} \text{ g-mol} = 0.75 \text{ g-mol}$$

Hence, mole fraction of oxygen $= \frac{1}{4}$ and mole fraction of hydrogen $= \frac{3}{4}$

Total number of moles

$$N = \frac{1}{4} + \frac{3}{4} = 1$$

According to Eq. (5.21), the change in entropy is given by

$$\Delta S_M = -NR \left(x_A \ln x_A - \frac{x_B}{N} \ln x_B \right)$$

$$= -1.987 (0.25 \ln 0.25 + 0.75 \ln 0.75)$$
$$= -1.987 \times 0.56 = 1.112 \text{ cal/K}$$

5.8.6 Entropy Changes with Temperature

Suppose two fluids in a heat exchanger are at different temperatures. They are brought in contact with each other. They would mix up and exchange the heat with each other. If this spontaneous process is allowed to take place for a certain period of time, they will attain a final temperature (which may be termed *intermediate temperature* with respect to the temperatures of hot and cold fluid). This process would of course be accompanied by a change in entropy. The entropy change of each fluid can be evaluated by the following equation:

$$\Delta S = \int_{T_1}^{T_2} \frac{dQ}{T}$$

where

dQ = Heat change
T_1 = Initial temperature
T_2 = Final temperature.

Now, the heat change can be expressed in terms of heat capacity.
For a constant-pressure process, the entropy change will be

$$\Delta S_P = \int_{T_1}^{T_2} \frac{dQ_P}{T} = \int_{T_1}^{T_2} C_P \frac{dT}{T} = C_P \ln \frac{T_2}{T_1} \tag{5.63}$$

For a constant-volume process, the change in entropy is given by

$$\Delta S_V = \int_{T_1}^{T_2} \frac{dQ_V}{T} = \int_{T_1}^{T_2} C_V \frac{dT}{T} = C_V \ln \frac{T_2}{T_1} \tag{5.64}$$

For a given mass of fluid m, the preceding expressions will be

$$\Delta S_P = \int_{T_1}^{T_2} mC_P \frac{dT}{T} = mC_P \ln \frac{T_2}{T_1} \tag{5.65}$$

$$\Delta S_V = m \int_{T_1}^{T_2} C_V \frac{dT}{T} = mC_V \ln \frac{T_2}{T_1} \tag{5.66}$$

Hence the total entropy change of the process can be estimated by summing up the individual changes. The two preceding equations are quite useful in calculating the entropy changes in various processes such as mixing of two fluids and quenching of metal bodies in liquids.

EXAMPLE 5.13 What would be the change in entropy when 80 g of argon is heated from 300 K to 500 K at constant volume. Assume specific heat $C_V = 0.3122$ kJ/kg-K.

Solution: 1 g-mol of argon = 40 g
So

80 g of argon = 2 g-mol
Initial temperature, $T_1 = 300$ K
Final temperature, $T_2 = 500$ K

The entropy change is given by

$$\Delta S_V = nC_V \ln \frac{T_2}{T_1}$$

$$= 2 \times 0.3122 \ln \frac{500}{300} = 0.318 \text{ kJ/K}$$

EXAMPLE 5.14 A process is carried out at constant volume. It is found that the entropy change is 1.0 kJ/kg-K. If $C_V = 0.918$ kJ/kg-K and the lower temperature of the process is 18°C, then determine the upper temperature of the process.

Solution: Initial temperature, $T_1 = 18°C = 291$ K

Let the upper temperature of the process be T.

The entropy change of the process can be estimated by the equation

$$\Delta S_V = \int_{T_1}^{T_2} C_V \frac{dT}{T} = \int_{291}^{T} C_V \frac{dT}{T}$$

or

$$1.0 = 0.918 \ln \frac{T_2}{291}$$

or

$$\frac{T_2}{291} = e^{1.089} = 2.971$$

or

$$T_2 = 864.6 \text{ K} = 591.64°C$$

Therefore, the upper temperature of the process is 591.64°C.

EXAMPLE 5.15   Estimate the change in entropy if 5 kg of water at 350 K is mixed adiabatically with 20 kg of water at 250 K. Assume the specific heat of water is 4.2 kJ/kg-K.

Solution: Suppose the final temperature of the process is T. From the energy balance of the process, we have

$$(5)\ (4.2)\ (350 - T) = (20)\ (4.2)\ (T - 250)$$

On solving the equation, the final temperature (T) obtained is 275 K.

(a) Change in entropy of the hot water

$$\Delta S_1 = mC \int_{350}^{275} \frac{dT}{T} = 5 \times 4.2 \times \ln \frac{275}{350} = -5.083 \text{ kJ/K}$$

(b) Change in entropy of the cold water

$$\Delta S_2 = mC \int_{250}^{275} \frac{dT}{T} = 20 \times 4.2 \times \ln \frac{275}{250} = 8.006 \text{ kJ/K}$$

(c) Total entropy change

$$\Delta S_{total} = \Delta S_1 + \Delta S_2 = -5.083 \text{ kJ/K} + 8.006 \text{ kJ/K} = 2.923 \text{ kJ/K}$$

EXAMPLE 5.16   12 g of a metal is heated from 294 K to 574 K in a constant-pressure process. The melting point of metal is 505 K. We are given that

Latent heat of fusion of the metal = 14.5 cal/g
Specific heat of solid metal = 0.052
Specific heat of liquid metal = 0.062.

Calculate the change in entropy of the process.

Solution: To estimate the change in entropy in the given process, the three processes, i.e., heating of solid metal, fusion of metal, and heating of liquid metal, are to be taken into consideration.

Assume that

Entropy change in heating solid metal $= \Delta S_1$
Entropy change in fusion of metal $= \Delta S_2$
Entropy change in heating liquid metal $= \Delta S_3$

(a) Entropy change in heating 12 g of metal from 294 K to 574 K

$$\Delta S_1 = mC_P \int_{294}^{505} \frac{dT}{T} = 12 \times 0.052 \times \ln \frac{505}{294} = 0.337 \text{ kJ/K}$$

(b) Entropy change in fusion of metal

$$\Delta S_2 = \frac{dQ}{T} = \frac{m\lambda}{T} = \frac{12 \times 4.5}{505} = 0.106 \text{ kJ/K}$$

(c) Entropy change in heating liquid metal from 505 K to 574 K

$$\Delta S_3 = mC_P \int_{505}^{574} \frac{dT}{T} = 12 \times 0.062 \times \ln \frac{574}{505} = 0.094 \text{ kJ/K}$$

(d) Total entropy change

$$\Delta S_{\text{total}} = \Delta S_1 + \Delta S_2 + \Delta S_3$$
$$= 0.337 \text{ kJ/K} + 0.106 \text{ kJ/K} + 0.094 \text{ kJ/K} = 0.537 \text{ kJ/K}$$

EXAMPLE 5.17   Calculate the change in molar entropy of water when it is heated from 137°C to 877°C. The molar specific heat of water, $C_P = 7.25 + 2.28 \times 10^{-3} T$.

Solution: Entropy change in heating water is given by

$$\Delta S = mC_P \int_{410}^{1150} \frac{dT}{T} = \int_{410}^{1150} \frac{7.25 + 0.00228 \, T}{T} dT$$

$$= 7.25 \ln \frac{1150}{410} + 0.0022 \,(1150 - 410)$$

$$= 9.092 \text{ cal/K}$$

EXAMPLE 5.18   A 40 kg block of iron casting at 625 K is dropped into a well-insulated vessel containing 160 kg of water at 276 K. Calculate the entropy change for (a) the iron block, (b) the water and (c) the entire process. Assume that the specific heat of iron is 0.45 kJ/kg-K and that of water is 4.185 kJ/kg-K.

Solution: Let the final temperature of the system be T.
Then, from the energy balance over the process, we have

$$40 \,(0.45) \,(625 - T) = 160 \,(4.185) \,(T - 276)$$

Solving the equation yields $T = 285.13$ K.
Suppose that

Change in entropy of iron casting $= \Delta S_1$
Change in entropy of water $= \Delta S_2$

(a) The change in entropy of the iron casting can be estimated as

$$\Delta S_1 = mC_P \int_{525}^{285.13} \frac{dT}{T} = 40 \times 0.45 \times \ln \frac{285.13}{625}$$

$$= -14.13 \text{ kJ/K}$$

(b) The change in entropy of water is given by

$$\Delta S_2 = mC_P \int_{276}^{285.13} \frac{dT}{T} = 160 \times 4.185 \times \ln \frac{285.13}{276}$$

$$= 21.74 \text{ kJ/K}$$

(c) The total entropy change of the process is

$$\Delta S_{\text{total}} = \Delta S_1 + \Delta S_2$$

$$= -14.13 \text{ kJ/K} + 21.74 \text{ kJ/K}$$

$$= 7.61 \text{ kJ/K}$$

5.9 MATHEMATICAL STATEMENT OF SECOND LAW OF THERMODYNAMICS

We have already discussed the second law of thermodynamics and its importance with different statements and examples. In light of the earlier discussion, the mathematical statement of second law of thermodynamics may be established.

Suppose a heat reservoir at a temperature T_H is brought in contact with a second reservoir at a lower temperature T_L. Let a quantity of heat Q be transferred from the high-temperature to the low-temperature reservoir.

Then

Decrease in entropy of high temperature reservoir $= \dfrac{Q}{T_H}$

Increase in entropy of low temperature reservoir $= \dfrac{Q}{T_L}$

Net change in entropy

$$\Delta S_{\text{total}} = \frac{Q}{T_L} - \frac{Q}{T_H} = Q\left(\frac{T_H - T_L}{T_H T_L}\right) \tag{5.67}$$

In the preceding irreversible heat transfer, the total entropy change ΔS_{total} would be positive if $q = +\text{ve}$ quantity and $T_H > T_L$, i.e., there is existence of finite difference in temperature between two heat reservoirs.

If the process is made reversible, T_H will be infinitesimally higher than T_L, i.e., the magnitude of T_H is just greater than T_L. Then ΔS_{total} approaches zero.

Now, consider a closed adiabatic system. It consists of an irreversible forward process and a reversible return process. The irreversible forward process is represented by the path AB in the P–V diagram in Fig. 5.16. No heat transfer takes place in this process. In this process the system is taken from the initial state A to the final state B. Now, suppose the system is restored to its initial state along the reversible path BA.

Hence, the entropy change for the cycle is

$$\Delta S = S_A - S_B = \int_A^B dS = \int_A^B \frac{dQ_{rev}}{T} \qquad (5.68)$$

Since, the change in internal energy is zero i.e., $\Delta U = 0$.
On substitution of this value into the first law of thermodynamics, we have

$$W = Q_{rev}$$

The preceding equation implies the complete conversion of heat into work by the system. It violates the second law of thermodynamics (Kelvin–Planck statement). Therefore, the second law of thermodynamics requires that Q_{rev} be either negative or zero. For an irreversible cycle, Q_{rev} cannot be zero; it would then be negative and the entropy change from A to B must be positive, or

$$\Delta S_A - \Delta S_B > 0$$

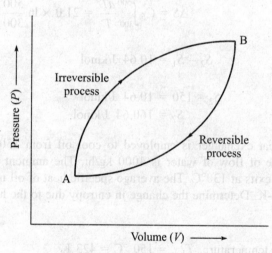

Fig. 5.16   Cycle consisting of irreversible and reversible processes.

So the irreversible process in a closed adiabatic system leads to an increase in entropy. Entropy change must be zero for a reversible process taking place in a closed adiabatic system.

An isolated system is a closed adiabatic system. As no thermal communication exists between the system and the surroundings, the entropy change must be equal to or greater than zero, or

$$\Delta S_{isolated} \geq 0 \qquad (5.69)$$

The isolated system consists of a system and the surroundings. Therefore, the entropy change of an isolated system is the sum of the entropy changes of the system and the surroundings, or

$$\Delta S_{system} + \Delta S_{surroundings} \geq 0 \tag{5.70}$$

$$\Delta S_{total} \geq 0 \tag{5.71}$$

This is the mathematical statement of the second law of thermodynamics, which indicates that the total entropy change associated with the process is positive. In the limiting case of a reversible process, the entropy change is zero.

Equation (5.71) is also known as the *principle of increase in entropy* and can be summarized as follows:

$$\Delta S \begin{cases} > 0 & \text{Irreversible process} \\ = 0 & \text{Reversible process} \end{cases}$$

EXAMPLE 5.19 An ideal gas ($C_P = 21.0$ J/kmol) undergoes a constant-pressure change from 300 K to 500 K. The molar entropy of the gas at 300 K is 150 J/kmol. Find the entropy at 500 K.

Solution: This is a constant-pressure process. Hence, the entropy change can be computed by the following relation:

$$\Delta S = C_P \int_{300}^{500} \frac{dT}{T} = 21.0 \times \ln \frac{500}{300}$$

or

$$S_2 - S_1 = 10.64 \text{ J/kmol}$$

or

$$S_2 - 150 = 10.64 \text{ J/kmol}$$
$$S_2 = 160.64 \text{ J/kmol}$$

EXAMPLE 5.20 A heat exchanger is employed to cool oil from 150°C to 50°C by using ambient water. The rate of flow of water is 4000 kg/hr. The ambient water enters the heat exchanger at 20°C and exits at 130°C. The average specific heat of oil is 2.5 kJ/kg-K and that of water is 4.185 kJ/kg-K. Determine the change in entropy due to the heat exchange process.

Solution: For oil

$$\text{Inlet temperature, } T_{1_{oil}} = 150 \text{ °C} = 423 \text{ K}$$
$$\text{Outlet temperature, } T_{2_{oil}} = 50 \text{ °C} = 323 \text{ K}$$
$$\Delta T_{oil} = 423 \text{ K} - 323 \text{ K} = 100 \text{ K}$$

For water

$$\text{Inlet temperature, } T_{1_{water}} = 20°C = 293 \text{ K}$$
$$\text{Outlet temperature, } T_{2_{water}} = 130°C = 403 \text{ K}$$
$$\Delta T_{water} = 403 \text{ K} - 293 \text{ K} = 110 \text{ K}$$

The mass flow rate of oil, m_{oil}, can be calculated by the enthalpy balance over the process. That is

$$\text{Heat lost by oil} = \text{Heat gained by water}$$
$$m_{oil} C_{P_{oil}} \Delta t_{oil} = m_{water} C_{P_{water}} \Delta t_{water}$$

or

$$m_{oil} \times 2.5 \times 100 = 4000 \times 4.185 \times 110$$

or

$$m_{oil} = 7365.6 \text{ kg/h}$$

(a) Change in entropy of the oil

$$\Delta S_{oil} = m_{oil} C_{P_{oil}} \int_{423}^{323} \frac{dT}{T} = 7365.6 \times 2.5 \times \ln \frac{323}{423} = -4980.9 \text{ kJ/K}$$

(b) Change in entropy of the water

$$\Delta S_{water} = m_{water} C_{P_{water}} \int_{423}^{323} \frac{dT}{T} = 4000 \times 4.185 \times \ln \frac{403}{293} = 5330.91 \text{ kJ/K}$$

(c) Total entropy change

$$\Delta S_{total} = \Delta S_{oil} + \Delta S_{water} = -4980.9 \text{ kJ/K} + 5330.91 \text{ kJ/K} = 350.01 \text{ kJ/K}$$

EXAMPLE 5.21 In the presence of catalyst, vegetable oil is hydrogenated at 250°C by continuously stirring with the help of an agitator for 20 minutes using a 650 W motor. Calculate the change in entropy of the oil.

Solution: Amount of energy supplied as work

$$Q = 650 \times 20 \times 60 = 7,80,000 \text{ J} = 780 \text{ KJ}$$
$$T = 250 + 273 = 523 \text{ K}$$

The change in entropy is

$$\Delta S = \left(\frac{Q}{T} \right)_R = \frac{780}{523} = 1.49 \text{ kJ/K}$$

EXAMPLE 5.22 An insulated tank of volume 2 m³ is divided into two equal compartments by a thin and rigid partition. One compartment contains an ideal gas at 400 K and 300 kPa, while the other is completely evacuated. Now, the partition is suddenly removed and the gases are allowed to mix. The equilibrium is reestablished by equalizing the pressure and the temperature. Estimate the change in entropy of the gas.

Solution: Since the system is well-insulated, there is no scope of transferring heat and work between the system and the surroundings, i.e, $\Delta Q = 0$ and $\Delta W = 0$. Therefore, by the first law of thermodynamics, we have that the change in internal energy, $\Delta U = 0$, as the internal energy of an ideal gas depends only on the temperature. The process is isothermal, so $\Delta T = 0$.

Now, we consider the temperature, pressure and volume of the two compartments before and after the mixing of the ideal gas. We have

Before mixing: $T_1 = 400$ K, $V_1 = 1$ m³, $P_1 = 300$ kPa
After mixing: $T_2 = 400$ K, $V_2 = 2$ m³, $P_2 = 150$ kPa.

Hence, the molar entropy change is given by

$$(\Delta S)_T = nR \ln \frac{P_1}{P_2}$$

where

$$n = \frac{PV}{RT} = \frac{300 \times 1}{8.314 \times 400} = 0.089 \text{ kmol}$$

Hence, on substitution of the value of n into the expression of entropy change, we get

$$(\Delta S)_T = nR \ln \frac{P_1}{P_2} = 0.089 \times 8.314 \times \ln \frac{300}{150} = 0.5128 \text{ kJ/K}$$

Since the process is adiabatic, there is no change in the surroundings due to exchange of heat. It results in almost no change in entropy. Hence, we conclude that the process is irreversible due to some obvious reasons.

EXAMPLE 5.23   Two iron blocks of same size are at distinct temperatures T_1 and T_2, brought in thermal contact with each other. The transfer process is allowed to take place until thermal equilibrium is attained. Suppose, after the attainment of equilibrium, that the blocks are at the final temperature T. Show that the change in entropy of the process is given by

$$\Delta S = C_P \ln \left[\frac{(T_2 - T_1)^2}{4 T_1 T_2} + 1 \right] \qquad \text{(WBUT, 2008)}$$

Solution: Let the two blocks A and B be at temperatures T_1 and T_2 respectively. Suppose $T_2 > T_1$. Then from the heat balance over the process, we have

$$\text{Heat lost by block B} = \text{Heat gained by block A}$$

or

$$C_{P_B}(T_2 - T) = C_{P_A}(T - T_1)$$

or

$$T = \frac{T_1 + T_2}{2} \qquad (\text{since } C_{P_B} = C_{P_A} = C_P)$$

Block B is at higher temperature; so a definite amount of heat will flow from block B to block A. Therefore, the entropy will go on decreasing in block B and increasing in block A.

Decrease in entropy of block B

$$\Delta S_B = C_P \ln \frac{T}{T_2}$$

Increase in entropy of block A

$$\Delta S_A = C_P \ln \frac{T}{T_1}$$

Net entropy change $= \Delta S = \Delta S_A + S_B = C_P \ln \dfrac{T}{T_1} + C_P \ln \dfrac{T}{T_2}$

$$= C_P \left(\ln \frac{T}{T_1} + \ln \frac{T}{T_2} \right) = C_P \left[\ln \frac{T^2}{T_1 T_2} \right] = C_P \left[\ln \frac{(T_1 + T_2)^2}{4 T_1 T_2} \right]$$

$$= C_P \left(\ln \frac{T_1^2 + T_2^2 + 2 T_1 T_2}{4 T_1 T_2} \right) = C_P \left(\ln \frac{T_1^2 + T_2^2 - 2 T_1 T_2 + 4 T_1 T_2}{4 T_1 T_2} \right)$$

$$\Delta S = C_P \ln\left[\frac{(T_2 - T_1)^2}{4T_1T_2} + 1\right]$$

Hence proved.

5.10 ENTROPY BALANCES FOR OPEN SYSTEMS

The entropy balance relation for an open system is analogous to the energy balance relation, and it differs from the energy balance only in that the entropy is not conserved. As per the definition of open system (or control volume), it is a system in which mass and energy can both cross the boundary. Hence, to derive a generalized entropy balance equation for an open system, depicted in Fig. 5.17, in addition to the heat as energy, the mass is also to be taken into consideration. This is because the mass contains entropy as well as energy and the amount of mass is proportional to the entropy and the heat. It is necessary to remember that entropy can be transferred from one system to another by two mechanisms. These are:

1. Heat transmission
2. Mass flow.

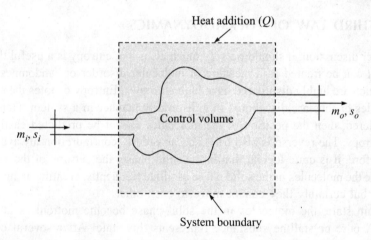

Fig. 5.17 Entropy balance for open system.

Again, it is well-known from the second law of thermodynamics and the principle of increase in entropy that the total entropy change for any process must be positive, and it is zero for any reversible process. Hence the entropy change of a system is equal to the entropy transfer. This is to be considered while making an entropy balance for both the system and the surroundings.

Now, the irreversibilities of a system always lead to an increase in entropy and therefore the term *entropy generation*, denoted by S_G, plays an important role to substantiate the irreversibility of a process.

The basic entropy balance relation can be stated as

Rate of inflow of entropy – Rate of outflow of entropy + Rate of entropy generation = Rate of accumulation entropy

Hence, for Fig. 5.17, considering the reversibility and irreversibility associated with the operation, the entropy balance equation can be expressed as

$$\left(\sum \frac{Q_K}{T_K} + \sum m_i s_i \right) - \sum m_o s_o + S_G = \frac{dS_{CV}}{dT} \tag{5.72}$$

$$\underset{\text{Inflow}}{\qquad\qquad} \underset{\text{Outflow}}{\qquad} \underset{\text{Generation}}{\qquad} \underset{\text{Accumulation}}{\qquad}$$

Now, for a steady-flow process, $\dfrac{dS_{CV}}{dT} = 0$. On substitution of the preceding condition into Eq. (5.72), it reduces to

$$S_G = \sum m_o s_o - \sum m_i s_i - \sum \frac{Q_K}{T_K} \tag{5.73}$$

If there is one entrance and one exit, then the entropy balance equation is given by

$$S_G = m(s_o - s_i) - \sum \frac{Q_K}{T_K} \tag{5.74}$$

For an adiabatic system, the entropy balance relation further reduces to

$$S_G = m(s_o - s_i) \tag{5.75}$$

5.11 THE THIRD LAW OF THERMODYNAMICS

From the earlier discussion, it should be very much clear that entropy is a useful thermodynamic property and it can be treated as a measure of molecular disorder or randomness of a system. Substances which are highly disordered have high entropies. Entropy denotes the extent to which atoms, molecules or ions are distributed in a disorderly manner in a system. Hence if a system is more disordered, then the position of the molecules cannot be predicted easily. It results in increase in entropy. The reverse is also true, i.e., an orderly configuration always results in less entropy. Therefore, it is quite logical that in the solid phase, the entropy of the substance is the lowest, because the molecules in the solid phase oscillate frequently, creating an uncertainty about their position; but certainly they cannot move to each other.

At a certain state, the molecules in the solid phase become motionless at absolute zero temperature. A pure crystalline substance represents this state. After several experiments on measuring the heat capacities of pure crystalline substances at absolute zero temperature (0 K), it is observed that they have identical entropy at 0 K. This experiment has led to the establishment of the third law of thermodynamics, which states: *The absolute entropy of a pure crystalline substance is zero at absolute zero temperature.*

Therefore, the absolute value of the entropy of a substance can be calculated with the help of the third law of thermodynamics. The condition to be obeyed is that the entropy of the substance is zero at $T = 0$ K. For this purpose, the measurement of heat capacities at different temperatures and the latent heats of phase transition are necessary. For instance, the entropy of a substance at T may be estimated by the following relation:

$$S = \int_0^{T_f} \frac{C_{Ps} dT}{T} + \frac{\Delta H_f}{T_f} + \int_{T_f}^{T_b} \frac{C_{Pl} dT}{T} + \frac{\Delta H_v}{T_b} + \int_{T_b}^{T} \frac{C_{Pg} dT}{T} \tag{5.76}$$

where

> the melting point of the substance is T_f
> the boiling point of the substance is T_b
> the specific heat of the solid is C_{Ps}
> the specific heat of the liquid is C_{Pl}
> the specific heat of the gas is C_{Pg}
> the latent heat of fusion is ΔH_f
> the latent heat of vapourization is ΔH_v.

It is necessary to remember that the above relation is a useful tool to measure the heat capacity at cryogenic temperature, i.e., very low temperature, and the third law is confined strictly to pure crystalline substances.

5.12 ENTROPY: MICROSCOPIC POINT OF VIEW

The term *entropy* is generally interpreted in three ways. These are:

(a) from the macroscopic point of view (Classical thermodynamics)
(b) from the microscopic point of view (Statistical thermodynamics)
(c) from the information point of view (Information theory).

So far we have discussed the concept of entropy from the macroscopic point of view. Now we will discuss the statistical explanation of entropy, i.e., from the microscopic viewpoint. Before going to the microscopic interpretation of entropy, it is worth mentioning that the mathematical relationship between entropy and probability, known as *statistical thermodynamics*, deals with probability, the partition function, and thermodynamic properties.

As per the microscopic definition of entropy in statistical thermodynamics, it is the number of microscopic configurations (probabilities) that result in the observed macroscopic explanation of the thermodynamic system, and it can be mathematically expressed as

$$S = k_B \ln \Omega \tag{5.77}$$

where

k_B = Boltzmann's constant = 1.380×10^{-23} J/K

$\quad = \dfrac{R}{N_A}$ (where R is universal gas constant and N_A is Avogadro's number)

Ω = Number of microstates corresponding to the observed thermodynamic macrostates calculated with the help of the multiplicity function.

This definition is considered to be the fundamental definition of entropy wherefrom all the other definitions, based on the macroscopic and information points of view, can be mathematically derived, but not vice versa.

In 1896, the famous physicist Ludwig Boltzmann visualized a way to measure the entropy for a system of atoms and molecules in the gas phase with the help of the preceding mathematical expression.

Equation (5.77) can be proved by the following discussion.

Entropy is a measure of the randomness or disorder in the system. All spontaneous changes of physical process are irreversible and in the direction of probability, and occur with an increase in entropy. Entropy is thus an extensive property function that accounts for the effects of irreversibility in thermodynamic systems.

Consider an insulated rigid container of two non-identical ideal gases A and B, separated into two equal volumes by a heat-conducting partition (such as a semi-permeable membrane), so that the temperature of the gas in each half is the same. Suppose the partition is removed, so that the gas molecules are free to diffuse quickly throughout the total volume. The gas would mix up irreversibly. This process is an adiabatic expansion. After the partition is taken away, only one half of the container is occupied by molecules.

In this case, initially the molecules are not distributed in a random manner throughout the volume of the container, but are agitated into one part. Therefore, the process finally leads to an increase in randomness or disorder as well as an increase in entropy.

It may be illustrated in another way. Suppose we were able to see the gas molecules of the preceding mixing process in distinct colours, say, the molecules of A in red and of B in white. When the partition is withdrawn, the red molecules start to move into the white region and the white molecules into the red region. The gases would mix up. As the process goes on, there would be more and more of the different-coloured molecules in the region. If the process is allowed to go on for a long time, probably we could not be able to pick out individual molecules. Finally, the existence of a uniform pink gas throughout the volume will be observed. This final state is more random or disorderly than the initial state. Thus, higher the randomness or disorder, higher is the number of configurations or thermodynamic probabilities, and so higher will be the entropy.

The entropy change of the preceding system is given by

$$S_{\text{system}} = S_A + S_B \tag{5.78}$$

where

S_A = Entropy of the molecules of A
S_B = Entropy of the molecules of B.

Let Ω_A be the number of configurations or thermodynamic probabilities of A, and Ω_B be the number of configurations or thermodynamic probabilities of B. Since higher the randomness or disorder, higher is the number of configurations or thermodynamic probabilities, the entropy can be considered a universal function of thermodynamic probability. So, we can write for A and B

$$S_A = f(\Omega_A) \quad \text{and} \quad S_B = f(\Omega_B)$$

Hence

$$S_{\text{system}} = f(\Omega_{\text{system}}) = f(\Omega_A \ \Omega_B) \tag{5.79}$$

Putting the preceding values into Eq. (5.23), we have

$$f(\Omega_A \ \Omega_B) = f(\Omega_A) + f(\Omega_B) \tag{5.80}$$

This relation can be written as

$$\ln pq = \ln p + \ln q$$

Therefore, the monotonically increasing function having a single value with the previous property is given by

$$f(\Omega) = k \ln \Omega$$

or

$$S = k \ln \Omega \tag{5.81}$$

where k is Boltzmann constant.

Equation (5.81) was established by Boltzmann. The behaviour of entropy, stated in this equation, can be summarized as follows:

(i) S is maximum when Ω is maximum, which means that the larger number of ways of arranging the molecules in various energy levels leads to greater randomness.

(ii) S is minimum when Ω is minimum; for example, say $\Omega = 1$, $S = 0$. There is no randomness in case of a smaller number of ways of arranging the molecules in various energy levels.

Hence, greater the entropy, greater is the number of ways of arranging the molecules in various energy levels, and also, greater is the disorder or randomness in the system. In essence, the most common interpretation of entropy is that it is a measure of the disorder or randomness in the microscopic system and also of the degree of information or ignorance about a system.

5.13 CRITERION OF IRREVERSIBILITY

We have already discussed in the preceding chapter about the reversibility and irreversibility of the thermodynamic system. A process is said to be reversible if both the system and its surroundings can be restored to their respective original state by reversing the direction of the process. All spontaneous processes are irreversible. They do not proceed in the reverse direction on their own. It is clear to say that the first law of thermodynamics supports the irreversibility of a system and the second law of thermodynamics supports the reversibility, as the Carnot cycle operating on a reversible process can be explained better by the second law of thermodynamics. Therefore, the first law does not deny the possibility of a spontaneous process reversing on its own; the second law, however, does deny the same. That also does necessarily mean that the second law does not deny the possibility of a reversible process proceeding in either direction. In this regard, the criterion for irreversibility of a process can be stated as: if the conditions of a given reversible process lead to the violation of the second law of thermodynamics, then the process is said to be irreversible. In other words, if an irreversible process obeys the significance of the first law of thermodynamics, then the process is definitely an irreversible one. To identify the irreversible process, this criterion can be applied.

5.14 CLAUSIUS INEQUALITY

A reversible heat engine operates between a high-temperature reservoir at T_H and a low-temperature reservoir at T_L. Then the thermal efficiency of the engine is given by

$$\eta = \frac{T_H - T_L}{T_H} = 1 - \frac{T_L}{T_H} \tag{5.82}$$

Suppose the engine is an irreversible one and operates between the same temperature reservoirs. If it receives the heat of Q_H from a high-temperature reservoir at T_H and discards the heat of Q_L to a low-temperature reservoir at T_L, then the efficiency of the heat engine is given by

$$\eta = \frac{W_{net}}{Q_H} = \frac{Q_H - Q_L}{Q_H} = 1 - \frac{Q_L}{Q_H} \qquad (5.83)$$

According to the Carnot theorem, the thermal efficiency of an irreversible heat engine is always less than the efficiency of a reversible one operating between the same two temperature reservoirs.

Using Eqs. (5.82) and (5.83) together, we have

$$1 - \frac{Q_L}{Q_H} < 1 - \frac{T_L}{T_H}$$

or

$$-\frac{Q_L}{Q_H} < -\frac{T_L}{T_H}$$

or

$$\frac{Q_L}{Q_H} > \frac{T_L}{T_H}$$

or

$$\frac{Q_H}{T_H} < \frac{Q_L}{T_L}$$

or

$$\frac{Q_H}{T_H} - \frac{Q_L}{T_L} < 0 \qquad (5.84)$$

As per the sign convention, if we take the amount of heat discarded (Q_L) as –ve, Eq. (5.84) becomes

$$\frac{Q_H}{T_H} + \frac{Q_L}{T_L} < 0 \qquad (5.85)$$

Therefore, in differential form, Eq. (5.85) can be generalized as an integral around an irreversible cycle as follows:

$$\oint \frac{dQ}{T} < 0 \qquad (5.86)$$

For the reversible cycle, we already obtained the relation

$$\oint \frac{dQ}{T} = 0 \qquad (5.87)$$

On combination of Eqs. (5.86) and (5.87), the inevitable result is

$$\oint \frac{dQ}{T} \leq 0 \qquad (5.88)$$

which is the *Clausius inequality*. It states that: *Whenever a system undergoes a cyclic change, the algebraic sum of the ratios of the heat interactions to the temperature at which the heat interaction occurs, over the complete cycle, is less than or equal to zero.*

The inequality is valid for all thermodynamic cycles, reversible or irreversible, including the refrigeration cycles. It was first investigated by the German physicist, R.J.E. Clausius, one of the founders of thermodynamics.

5.15 IRREVERSIBILITY AND LOST WORK

It is true that all spontaneous processes are irreversible and accompanied by an increase in entropy. This is the *principle of increase in entropy*. However, increase in entropy is nothing but the measure of lost work due to the irreversibility of the naturally occurring processes. This statement can be proved by the following discussion:

Suppose a heat reservoir at temperature T_H is brought in contact with the second reservoir at a lower temperature T_L. Let a quantity of heat Q be transferred from the high-temperature reservoir to the low-temperature reservoir.

Then

the decrease in entropy of the high-temperature reservoir $= \dfrac{Q}{T_H}$

the increase in entropy of the low-temperature reservoir $= \dfrac{Q}{T_L}$

Net change in entropy in this irreversible process (see Eq. (5.67))

$$\Delta S_{total} = \frac{Q}{T_L} - \frac{Q}{T_H} = Q \cdot \frac{T_H - T_L}{T_H T_L}$$

Now consider the Carnot engine as a reversible heat engine which operates between these high-temperature and low-temperature reservoirs and receives the same amount of heat Q. We know, the efficiency of the Carnot engine is given by (see Eq. (5.68))

$$\eta = \frac{W_{net}}{Q} = \frac{T_H - T_L}{T_H}$$

where

W_{net} = Net work output
Q = Amount of heat received.

Equation (5.68) can be rearranged as Eq. (5.69)

$$W_{net} = Q \frac{T_H - T_L}{T_H}$$

Equation (5.69) implies that the net amount of work results from a reversible process operating between T_H and T_L. This extent of work was lost in the irreversible process, because a spontaneous process is always associated with unavoidable irreversibilities.

Comparing Eqs. (5.67) and (5.69), we can have the following expression of lost work (see Eq. (5.70)):

$$W_{lost} = T_L (\Delta S)_{total}$$

Hence, the amount of work lost is the product of the total entropy changes in an irreversible process and the temperature of the low-temperature reservoir, and could have been saved for a reversible process, but was lost due to the irreversible process.

EXAMPLE 5.24 Referring to Example 5.18, compute the amount of work lost in the process.

Solution: We can incorporate here the change in entropy of the iron block and water once again.

(a) Change in entropy of iron casting

$$\Delta S_1 = mC_P \int_{625}^{285.13} \frac{dT}{T} = 40 \times 0.45 \times \frac{285.13}{625}$$

$$= -14.13 \text{ kJ/K}$$

(b) Change in entropy of water

$$\Delta S_2 = mC_P \int_{276}^{285.13} \frac{dT}{T} = 160 \times 4.185 \times \frac{285.13}{276}$$

$$= 21.74 \text{ kJ/K}$$

(c) Total entropy change of the process

$$\Delta S_{total} = \Delta S_1 + \Delta S_2$$

$$= -14.13 \text{ kJ/K} + 21.74 \text{ kJ/K}$$

$$= 7.61 \text{ kJ/K}$$

Hence, the work lost can be computed using Eq. (5.63) as

$$W_{lost} = T_L (\Delta S)_{total}$$

$$= 276 \times 7.61 \text{ kJ/K} = 2100.36 \text{ kJ}$$

EXAMPLE 5.25 It is desired to cool a variety of aromatic oil in a heat exchanger from 515 K to 315 K at a rate of 4750 kg/hr. The temperature of the cooling water is 290 K and it is supplied at a rate of 9500 kg/hr. We are given that the average specific heat of the aromatic oil is 3.2 kJ/kg-K, and the average specific heat of water is 4.185 kJ/kg-K. Determine the entropy change of the process and check whether the process is reversible or irreversible.

Solution: To compute the change in entropy of the entire process, the exit temperature must be known. So, we can make an enthalpy balance over the process to meet our purpose. The enthalpy balance gives

$$\underbrace{m_1 C_{P_1} \Delta T_1}_{\text{Oil}} = \underbrace{m_2 C_{P_2} \Delta T_2}_{\text{Water}}$$

$$4750 \times 3.2 \times (515 - 315) = 9500 \times 4.185 \ (T - 290)$$

Solving the preceding equation, we get the exit temperature

$$T = 366.46 \text{ K}$$

- Change in entropy of aromatic oil

$$\Delta S_1 = mC_P \int_{315}^{515} \frac{dT}{T} = 4750 \times 3.2 \times \ln \frac{315}{515}$$

$$= -7488.40 \text{ kJ/K}$$

- Change in entropy of water

$$\Delta S_2 = mC_P \int_{290}^{366.46} \frac{dT}{T} = 9500 \times 4.185 \times \ln \frac{366.46}{290}$$

$$= 9188.42 \text{ kJ/K}$$

- Total change in entropy

$$\Delta S_{total} = \Delta S_1 + \Delta S_2$$

$$= -7488.40 \text{ kJ/K} + 9188.42 \text{ kJ/K} = 1700.02 \text{ kJ/K}$$

It is a +ve quantity. So the process is irreversible.

SUMMARY

The first law of thermodynamics basically represents the law of conservation of energy. It is very much concerned with the different forms of energy and their quantitative estimation of naturally occurring processes. Therefore, no process can violate the first law of thermodynamics, but it has certain limitations. Occurrence of a spontaneous process in their own direction is supported well by the first law, but this law does not give any information about the direction and extent of this process. This inadequacy necessitates the establishment of the second law of thermodynamics. To explain the incompleteness of the second law, certain logical examples have been cited. Of different statements of the second law, only classic statements such as Kelvin–Planck and Clausius statements have been discussed, showing that they are equivalent to each other and the violation of one statement leads to the violation of the other.

The Kelvin–Planck statement refers to a heat engine and the Clausius statement refers to a heat pump or refrigerator. The heat engine is an energy conversion device that operates on a cycle and converts the heat as energy into work. Carnot engine is the most efficient reversible engine that operates on a cycle between two thermal reservoirs and can better explain the principle of transformation of heat into work through the absorption of heat from a high-temperature source and release of heat to a low-temperature sink. The Carnot cycle consists of two isothermal and two adiabatic processes. All the processes are carried out in reversible fashion.

Since the Clausius statement is related to the working principle of the refrigerator or the heat pump, it demonstrates the possibility of attainment of cooling or heating effects other than by transferring heat from a low-temperature region to a high-temperature one. Here the refrigerator or the heat pump operates on the reversed Carnot engine.

However, the efficiency of the Carnot engine is maximum among all engines operating between two temperature reservoirs. The efficiency of a heat engine depends only upon the temperature of the two thermal reservoirs and is independent of the working medium.

The thermal efficiency of a heat engine is given by

$$\eta = \frac{\text{Net work output}}{\text{Heat input}} = \frac{W_{net}}{Q_{in}} = \frac{Q_{in} - Q_{out}}{Q_{in}}$$

The COP of a refrigerator can be expressed as

$$COP_R = \frac{Q_L}{Q_H - Q_L} = \frac{Q_L}{W_{net}}$$

The COP of a heat pump can be expressed as

$$COP_R = \frac{Q_H}{Q_H - Q_L} = \frac{Q_H}{W_{net}}$$

For a reversible heat engine operating between two reservoirs at temperatures T_H and T_L, we have as per the thermodynamic temperature scale

$$\frac{Q_H}{Q_L} = \frac{\Phi(T_H)}{\Phi(T_L)}$$

The second law of thermodynamics leads to the definition of a thermodynamic property known as *entropy*. It is a measure of the randomness or disorder of a system and can be mathematically expressed as

$$dS = \frac{dQ_{rev}}{T}$$

where dQ_{rev} is reversible heat change and rev implies the reversibility of the system.

The estimation of entropy change for different processes has been discussed. The combined form of the first and second laws, a useful relation in the deduction of various other thermodynamic relations of different functions, has been shown.

The Clausius inequality states: *Whenever a system undergoes a cyclic change, the algebraic sum of the ratios of heat interactions to the temperature at which the heat interaction occurs, over the complete cycle is less than or equal to zero.* Mathematically,

$$\oint \frac{dQ}{T} \leq 0$$

The third law of thermodynamics states: *The absolute entropy of a pure crystalline substance is zero at absolute zero temperature.*

The microscopic definition of entropy in statistical thermodynamics is: *It is the number of microscopic configurations (probabilities) that result in the observed macroscopic explanation of the thermodynamic system.* It can be mathematically expressed as

$$S = k_B \ln \Omega \qquad (5.77)$$

where

k_B = Boltzmann's constant = 1.38×10^{-23} J/K

$= \dfrac{R}{N_A}$ (where R = universal gas constant and N_A = Avogadro's number)

Ω = Number of microstates corresponding to the observed thermodynamic macrostates calculated with the help of the multiplicity function

This fundamental relationship between probability and entropy yields a constructive idea: *Greater the entropy, greater is the number of ways of arranging the molecules in various energy levels, and greater is the disorder or randomness in the system.*

In addition to this idea, a presentation of the entropy balance has been made over an open system that provides some important information about entropy generation based on different process conditions.

The entropy balance for open systems is given by

$$\sum \frac{Q_K}{T_K} + \sum m_i s_i - \sum m_o s_o + S_G = \frac{dS_{CV}}{dT}$$

For a steady flow process

$$S_G = \sum m_o s_o - \sum m_i s_i - \sum \frac{Q_K}{T_K}$$

If there is one entrance and one exit, then the entropy balance equation is given by

$$S_G = m(s_o - s_i) - \sum \frac{Q_K}{T_K}$$

KEY TERMS

Carnot Cycle   A reversible cycle proposed by Sadi Carnot in 1824, consisting of two isothermal and two adiabatic processes, and which can be treated as the model cycle to explain the conversion of heat into work.

Clausius Statement   It is not possible to construct any device which could convey heat from a low-temperature region to a high-temperature region without the help of any external agency.

Entropy   A thermodynamic property and state function; it is a measure of the randomness of a system or disorder in the universe.

Heat Engine   A device employed for transformation of heat into work.

Heat Pump   The device by which heat flows from a low-temperature region to a high-temperature one, the objective of a heat pump is to maintain the heated space (a region to be heated) at high temperature by delivering heat.

Isentropic Process   A process carried out at constant entropy.

Isothermal Process   A process carried out at constant temperature.

Kelvin–Planck Statement   It is impossible for any device that operates on a cycle to receive heat from a single reservoir and produce net work.

Principle of Increase in Entropy   The entropy of an isolated system must be equal to or greater than zero.

Refrigerator   A device by which low temperature as well as coldness is produced through the transfer of heat from a low-temperature medium (source) to a high-temperature one (sink).

Sink   A body at low temperature to which a definite amount of heat is transferred by working fluid.

Source A body at high temperature from which a definite amount of heat is transferred by working fluid.

Thermal Efficiency Thermal efficiency of a heat engine is defined as the ratio of the net work output to the total heat input.

Thermal Reservoir The hypothetical body with relatively large capacity which can supply or absorb a significant amount of thermal energy without undergoing any change in temperature.

Thermodynamic Temperature Scale A scale widely used to measure temperature; it is a useful tool that also helps in calculating the thermodynamic quantities and possesses some degree of universality. It was coined by Kelvin on the basis of Carnot's principle.

Third Law of Thermodynamics The absolute entropy of a pure crystalline substance is zero at absolute zero temperature.

Working Fluid The substance which in operation absorbs heat from a hot reservoir, release the heat to a cold reservoir, and returns to its original state on completing the cycle.

IMPORTANT EQUATIONS

1. The thermal efficiency of a heat engine is given by

$$\eta = \frac{\text{Net work output}}{\text{Heat input}} = \frac{W_{\text{net}}}{Q_{\text{in}}} = \frac{Q_{\text{in}} - Q_{\text{out}}}{Q_{\text{in}}} \qquad (5.1)$$

2. The COP of a refrigerator can be expressed as

$$\text{COP}_R = \frac{\text{Heat removed}}{\text{Work required}} \quad \text{or} \quad \frac{\text{Refrigerating effect}}{\text{Work input}}$$

$$= \frac{Q_L}{Q_H - Q_L} = \frac{Q_L}{W_{\text{net}}} \qquad (5.3)$$

3. The efficiency of a Carnot cycle is given by

$$\eta = \frac{W}{Q_H} = \frac{Q_H - Q_L}{Q_H} = \frac{T_H - T_L}{T_H} \qquad (5.12)$$

4. For a reversible heat engine operating between two reservoirs at temperature T_H and T_L, as per the thermodynamic temperature scale

$$\frac{Q_H}{Q_L} = \frac{\Phi(T_H)}{\Phi(T_L)} \qquad (5.22)$$

5. The differential form of entropy is given by

$$dS = \frac{dQ_{\text{rev}}}{T}$$

where dQ_{rev} = reversible heat change and rev implies the reversibility of the system.

6. In a Carnot cycle, the differential change of the property can be written as

$$\oint \frac{dQ_{rev}}{T} = \oint dS = 0$$

7. The combined form of the first law and the second law of thermodynamics is given by

$$dU = TdS - PdV \tag{5.38}$$

8. The relation between entropy and enthalpy is given by

$$dH = TdS + VdP$$

9. The change in entropy at phase transition is given by:
   for fusion (melting)

$$\Delta S_{fus} = \frac{\Delta h_{fus}}{T_{tr}} = \frac{\Delta h_{sf}}{T_{tr}} = \frac{h_f - h_s}{T_{tr}} = \frac{Q}{T_{tr}} \tag{5.47}$$

where
   ΔS_{fus} = Entropy of fusion (melting)
   Δh_{sf} = Enthalpy (latent heat) of fusion.

For vaporization

$$\Delta S_{vap} = \frac{\Delta h_{vap}}{T_{tr}} = \frac{\Delta h_{fg}}{T_{tr}} = \frac{h_g - h_f}{T_{tr}} = \frac{Q}{T_{tr}} \tag{5.48}$$

where
   ΔS_{vap} = Entropy of vaporization
   Δh_{fg} = Enthalpy (latent heat) of vaporization.

For sublimation

$$\Delta S_{sub} = \frac{\Delta h_{sub}}{T_{tr}} = \frac{\Delta h_{sg}}{T_{tr}} = \frac{h_g - h_s}{T_{tr}} = \frac{Q}{T_{tr}} \tag{5.49}$$

where
   ΔS_{sub} = Entropy of sublimation
   Δh_{sg} = Enthalpy (latent heat) of sublimation.

10. The entropy change for ideal gas can be estimated for
   (i) When $T_1 = T_2$, i.e., at isothermal condition

$$(\Delta S)_T = R \ln \frac{P_1}{P_2} = R \ln \frac{V_2}{V_1} \tag{5.56}$$

   (ii) When $P_1 = P_2$, i.e., at isobaric condition

$$(\Delta S)_P = C_P \ln \frac{T_2}{T_1} \tag{5.57}$$

   (iii) When $V_1 = V_2$, i.e., at isochoric condition

$$(\Delta S)_V = C_V \ln \frac{T_2}{T_1} \tag{5.58}$$

11. The entropy changes in a mixture of gases for 1 mol of the mixture

$$\Delta S_M = -R \sum x_i \ln x_i \qquad (5.62)$$

where

x_i = Mole fraction of species i

R = Universal gas constant.

12. The mathematical statement of the second law of thermodynamics for isolated system

$$\Delta S_{\text{isolated}} \geq 0 \qquad (5.69)$$

The entropy change of the total system is the sum of the entropy change of the system and the surroundings, and is given by

$$\Delta S_{\text{system}} + \Delta S_{\text{surroundings}} = 0 \qquad (5.70)$$
$$\Delta S_{\text{total}} \geq 0 \qquad (5.71)$$

13. The principle of increase in entropy can be summarized as follows:

$$\Delta S \begin{cases} > 0 & \text{Irreversible process} \\ = 0 & \text{Reversible process} \end{cases}$$

14. The entropy balance equation for an open system can be expressed as

$$\sum \frac{Q_K}{T_K} + \sum m_i s_i - \sum m_o s_o + S_G = \frac{dS_{CV}}{dT} \qquad (5.72)$$

For a steady-flow process

$$S_G = \sum m_o s_o - \sum m_i s_i - \sum \frac{Q_K}{T_K} \qquad (5.73)$$

If there is one entrance and one exit, then the entropy balance equation is given by

$$S_G = m(s_o - s_i) - \sum \frac{Q_K}{T_K} \qquad (5.74)$$

For an adiabatic system, the entropy balance relation further reduces to

$$S_G = m(s_o - s_i) \qquad (5.75)$$

15. The entropy of a pure crystalline substance at low temperature may be estimated by the following relation:

$$S = \int_0^{T_f} \frac{C_{Ps} dT}{T} + \frac{\Delta H_f}{T_f} + \int_{T_f}^{T_b} \frac{C_{Pl} dT}{T} + \frac{\Delta H_v}{T_b} + \int_{T_b}^{T} \frac{C_{Pg} dT}{T} \qquad (5.76)$$

where

Melting point of the substance is T_f

Boiling point of the substance is T_b

Specific heat of the solid is C_{Ps}

Specific heat of the liquid is C_{Pl}

Specific heat of the gas is C_{Pg}

Latent heat of fusion is ΔH_f

Latent heat of vaporization is ΔH.

16. The microscopic definition of entropy in statistical thermodynamics can be mathematically expressed as

$$S = k_B \ln \Omega \qquad (5.77)$$

where

k_B = Boltzmann's constant = 1.38×10^{-23} J/K

$\qquad = \dfrac{R}{N_A}$ (where R = universal gas constant and N_A = Avogadro's number)

Ω = Number of microstates corresponding to the observed thermodynamic macrostates calculated with the help of the multiplicity function.

17. The Clausius inequality can be expressed as

$$\oint \frac{dQ}{T} \le 0 \qquad (5.88)$$

18. The lost work due to the irreversibility of a spontaneous process is

$$W_{lost} = T_L \, (\Delta S)_{total}$$

EXERCISES

A. Review Questions

1. Explain the importance of the first law of thermodynamics. Does the first law give any information about the direction of a process?
2. What are the limitations of the first law of thermodynamics?
3. State and explain the second law of thermodynamics.
4. Explain the Kelvin–Planck and Clausius statements.
5. Prove the equivalence of the Kelvin–Planck and Clausius statements.
6. Justify: "The violation of the Kelvin–Planck statement is nothing but the violation of the Clausius statement".
7. What is a heat engine? Calculate its thermal efficiency.
8. What are the differences between a heat pump and a refrigerator?
9. What are the common characteristics of a heat engine?
10. Comment on the statement: "Heat cannot be easily converted into work".
11. What is the basic difference in working principle between a heat pump and a heat engine?
12. Explain the working principle of the Carnot cycle with the help of a P–V diagram.
13. Why can the thermal efficiency of a heat engine never be 100%?
14. Is it possible to achieve a reversible heat engine in practice?
15. What is entropy? Give an example.
16. Derive an expression for the change in entropy of the mixture of two non-identical ideal gases.
17. What is thermodynamic temperature scale? How is it established? Explain the importance of the thermodynamic temperature scale.

18. Deduce the relationship between entropy and internal energy.
19. Derive an expression for the change in entropy of an ideal gas.
20. Prove the Clausius inequality.
21. What does the principle of increase in entropy state?
22. State the third law of thermodynamics. Is the third law of thermodynamics applied only to pure crystalline substance?
23. Give an informatory note on the microscopic point of view of entropy.
24. Mention the criterion of irreversibility. How does the irreversible process relate to entropy?
25. With the help of a neat sketch, show the entropy balance for an open system.
26. What is the criterion of an irreversible process?
27. What is the relation between entropy and probability? Give an example.

B. Problems

1. A Carnot cycle operates between 800°C and 150°C. Determine the efficiency of the cycle, the heat released, and the work output if 100 kJ heat enters the cycle.
2. An inventor claims to have developed an engine of thermal efficiency 92%. Check the validity of his claim.
3. An inventor claims to have developed a heat pump with a COP of 4.5. It maintains the cold space at 255 K. The temperature of the environment is 305 K. Check the validity of his claim.
4. A heat engine absorbs 250 kJ of heat from a source at 310 K. The heat engine produces a work to the tune of 65 kJ, discarding 110 kJ of heat to a thermal reservoir at 312 K and 80 kJ of heat to another reservoir at 270 K. Does this engine violate the second law of thermodynamics?
5. A piston–cylinder device contains 1.2 kg of saturated water vapour at 180°C. Heat is transferred to steam. As a result, steam expands reversibly to a final pressure of 700 kPa. Determine the heat transferred and the work done during this process.
6. Determine the change in entropy when 200 g of ice at 0°C is converted into water at the same temperature, given that latent heat of ice = 80 cal/g.
7. Calculate the entropy change when 96 g of methane is heated from 35°C to 200°C at constant volume. Assume $C_V = 1.735$ kJ/kg-K.
8. 7 mol of an ideal gas ($C_V = 5$ cal) was initially at 30°C and 2 atm. The gas was transferred to the state when the temperature is 110°C and pressure 12 atm. Determine the change in entropy of the process.
9. What would be the change in entropy when 2 kg of air is heated from 30°C to 200°C at constant pressure? Assume $C_P = 1.005$ kJ/kg-K.
10. Calculate the increase in entropy when 2.8 L of oxygen are mixed with 19.6 L of hydrogen at N.T.P.
11. 5 kilograms of helium undergoes a constant-volume change from 317 K to 572 K. Determine the entropy change. Assume specific heat, $C_V = 3.115$ kJ/kg-K.

12. A process is carried out at constant pressure. It is found that the entropy change is 0.832 kJ/kg-K. If C_P = 0.846 kJ/kg-K and the upper temperature of the process is 315°C, then determine the lower temperature of the process.

13. Find the change in entropy of H_2 when it is heated from 157°C to 923°C. The molar specific heat of H_2 is given by C_P = 6.94 − 0.2 × $10^{-3}T$.

14. A 30 kg block of steel casting at 635 K is quenched in 135 kg oil at 283 K. If there are no heat losses, determine the change in entropy. Assume that the specific heat of steel is 0.72 kJ/kg-K and that of oil is 2.4 kJ/kg-K.

15. An iron block weighing 15 kg at a temperature of 450°C is in a well-insulated container having 120 kg of water at 5°C. Determine the change in entropy of
    (a) the iron block, (b) the water and (c) the total process. Assume that the specific heat of iron is 0.45 kJ/kg-K and that of water is 4.185 kJ/kg-K.

16. One kg of air is heated at constant volume from 100°C to 400°C. If C_V is 0.7186 kJ/kg-K, determine the change in entropy of the air.

17. 5 moles of water are evaporated at 100°C. Calculate ΔS, given that the latent heat of vaporization = 9720 cal.

18. One kg of air is heated at constant temperature of 30°C by the addition of 190 kJ. Determine the change in entropy of the process.

19. A frictionless piston–cylinder device contains 1 kg saturated liquid at 110°C. The water in the cylinder is heated at 420°C by bringing the assembly into contact with a body until the water is completely converted into saturated vapour at constant pressure. Determine the change in entropy, given that the latent heat of vaporization = 2256.94 kJ/kg-K.

20. In the presence of catalyst, hydrogenation of vegetable oil takes place at 300°C by continuously stirring with the help of an agitator for 30 min using a 700 W motor. Calculate the change in entropy of the oil.

21. An insulated tank of volume 2 m^3 is divided into two equal chambers by a partition. One chamber contains an ideal gas of 500 K and 4 MPa, while the other is completely evacuated. Now, the partition is withdrawn and the gases are allowed to mix. Estimate the change in entropy of the gas.

22. Two perfectly insulated tanks having a volume of 1 m^3 are connected by means of a small pipeline fitted with a valve. One tank contains an ideal gas at 3 bar and 292 K, and the other is completely evacuated. The valve is opened, and the pressure and the temperature are equalized. Determine the change in entropy of the gas.

23. An ideal gas at an initial pressure of 25 atm and 315 K occupies a volume of 2 L. Determine the entropy of the system for a reversible isothermal expansion of an ideal gas to a final volume of 10 L.

24. 5 moles of an ideal gas are allowed to expand from an initial state of 110 dm^3 at 320 K to a final state of 150 dm^3 at 430 K. Estimate the change in entropy for the process, given that heat capacity at constant volume, C_V = (7/2)R.

25. A 25 kg copper block at 373 K is dropped into an insulated tank that contains 80 L of water at 293 K. Determine the final equilibrium temperature and the total change in entropy. Assume that the specific heat of steel is 0.72 kJ/kg-K and that of oil is 2.4 kJ/kg-K.

Thermodynamic Property Relations

6

Some properties such as temperature, pressure, mass, and volume can be measured directly. From the knowledge of these properties, the other properties like density and specific volume can be determined. But it is not so easy task to determine the properties such as internal energy, enthalpy, and entropy, because they cannot be measured directly or by using some simple relations based on easily measurable properties. Hence, it is essential to develop some fundamental relations between the commonly encountered thermodynamic properties and express the properties which cannot be measured directly in terms of easily measurable properties. In light of that necessity, Maxwell's relation (forms the basis of some thermodynamic relations), the Clausius–Clapeyron equation (for the study of phase change processes like vaporization, sublimation, etc.), Gibbs-Helmholtz relation etc. have been discussed clearly. The general expressions for the differential changes in enthalpy, entropy, internal energy, and heat capacity at constant pressure and constant volume are also developed. The expressions for isothermal compressibility and volume expansivity have been developed. The significance of the Joule–Thomson coefficient as a measure of temperature change with pressure at constant enthalpy, along with the inversion temperature for ideal and real gases, is discussed. Fugacity is some sort of idealized pressure because it is exactly equal to the pressure for an ideal gas and will be the property similar to the pressure for a real gas. The effect of temperature and pressure on fugacity and the determination of fugacity for a pure substance have been discussed. Then we calculate the departure functions with the help of the equations of state of a pure substance, such as the virial equation of state and the cubic equation of state.

6.1 THERMODYNAMIC PROPERTIES

Before we go on to derive the several thermodynamic relations among the properties of a pure substance, it is necessary to know the different thermodynamic properties and how they can be classified. A thermodynamic property is an observable physical characteristic which is experimentally measurable and is used to define the state of the system. They can be broadly classified into three groups, namely, reference or primitive, energy, and derived properties. One must remember that a thermodynamic property comprises the following salient features:

1. All the thermodynamic properties of a system should have a definite value in a particular state.
2. The change in the property of a system should solely depend upon the state (initial and final), and be independent of the path followed by the system to reach a definite state.

   ❖ **Reference or Primitive Properties**
   The properties which enable us to define the state of the system are known as reference properties. As the name implies, these properties have absolute values which are measured relative to some arbitrary reference state. For example, temperature, pressure, volume, entropy etc. are reference properties.

   ❖ **Energy Properties**
   These are defined as the properties in which the changes in the thermodynamic functions such as internal energy (U), enthalpy (H), etc. indicate some useful work under certain conditions. For example, Gibbs free energy (G), Helmholtz free energy (A), internal energy (U), enthalpy (H), etc.

   ❖ **Derived or Mathematically Derived Properties**
   These properties are mathematically derived on the basis of the aforementioned reference and energy properties. For instance, Joule–Thomson coefficient (μ), specific heat (C), isothermal compressibility (α), volume expansivity (β), etc.

6.2 MATHEMATICAL PRE-REQUISITES

Based on the state postulate, several expressions are developed in this chapter which express the state of a pure substance by two independent variables. All other variables can be expressed with the help of those two variables.

It can be mathematically expressed as

$$z = z(x, y) \tag{6.1}$$

where x and y are the two independent variables and z is the other variable.

To establish the basic important thermodynamic relations, differential changes of the variables, and the method of partial derivatives are necessary to be taken into consideration. Consider Eq. (6.1), in which the variation of $z(x, y)$ with x when y is held constant is known as the partial derivative of z with respect to x.

Theorem 1: If there is a relation between the variables x, y, and z, then z may be expressed as a function of the other two variables, x and y, i.e.,

$$z = f(x, y)$$

On total differentiation

$$dz = \left(\frac{\partial z}{\partial x}\right)_y dx + \left(\frac{\partial z}{\partial y}\right)_x dy \tag{6.2}$$

If $\left(\dfrac{\partial z}{\partial x}\right)_y = M$ and $\left(\dfrac{\partial z}{\partial y}\right)_x = N$, then $dz = Mdx + Ndy$, where z, M and N are functions of

x and y. Now, on partial differentiation of M with respect to y and N with respect to x, we have

$$\left(\frac{\partial M}{\partial y}\right)_x = \left(\frac{\partial^2 z}{\partial y \partial x}\right) \tag{6.3}$$

and

$$\left(\frac{\partial N}{\partial x}\right)_y = \left(\frac{\partial^2 z}{\partial y \partial x}\right) \tag{6.4}$$

Comparing Eqs. (6.3) and (6.4), we get

$$\left(\frac{\partial M}{\partial y}\right)_x = \left(\frac{\partial N}{\partial x}\right)_y \tag{6.5}$$

This is the condition of a perfect differential.

Theorem 2: If the relation $f(x, y, z) = 0$ exists among three variables x, y and z, then we can write

$$x = x(y, z)$$

The total differential of x becomes

$$dx = \left(\frac{\partial x}{\partial y}\right)_z dy + \left(\frac{\partial x}{\partial z}\right)_y dz \tag{6.6}$$

and

$$z = z(x, y)$$

The total differential of z becomes

$$dz = \left(\frac{\partial z}{\partial x}\right)_y dx + \left(\frac{\partial z}{\partial y}\right)_x dy \qquad (6.7)$$

Eliminating dx by combining Eqs. (6.6) and (6.7), we have

$$dz = \left[\left(\frac{\partial z}{\partial x}\right)_y \left(\frac{\partial x}{\partial y}\right)_z + \left(\frac{\partial z}{\partial y}\right)_x\right] dy + \left(\frac{\partial x}{\partial z}\right)_z \left(\frac{\partial z}{\partial x}\right)_y dz \qquad (6.8)$$

Rearranging it, we have

$$\left[\left(\frac{\partial z}{\partial x}\right)_y \left(\frac{\partial x}{\partial y}\right)_z + \left(\frac{\partial z}{\partial y}\right)_x\right] dy = \left[1 - \left(\frac{\partial x}{\partial z}\right)_y \left(\frac{\partial z}{\partial x}\right)_y\right] dz \qquad (6.9)$$

The variables y and z are independent of each other. Setting the terms in each bracket equal to zero gives

$$\left(\frac{\partial x}{\partial z}\right)_y \left(\frac{\partial z}{\partial x}\right)_y = 1 \qquad (6.10)$$

It can be rewritten as

$$\left(\frac{\partial x}{\partial z}\right)_y = \frac{1}{\left(\dfrac{\partial z}{\partial x}\right)_y} \qquad (6.11)$$

This is the reciprocal relation.

Theorem 3: Consider three variables x, y, and z. Among them, any one may be considered the function of the other two.

So

$$x = f(y, z)$$

or

$$dx = \left(\frac{\partial x}{\partial y}\right)_z dy + \left(\frac{\partial x}{\partial z}\right)_y dz \qquad (6.12)$$

Similarly

$$z = f(x, y)$$

or

$$dz = \left(\frac{\partial z}{\partial x}\right)_y dx + \left(\frac{\partial z}{\partial y}\right)_x dy \qquad (6.13)$$

Substituting the values of dz in Eq. (6.12), we get

$$dx = \left(\frac{\partial x}{\partial y}\right)_z dy + \left(\frac{\partial x}{\partial z}\right)_y \left[\left(\frac{\partial z}{\partial x}\right)_y dx + \left(\frac{\partial z}{\partial y}\right)_x dy\right]$$

or

$$dx = \left[\left(\frac{\partial x}{\partial y}\right)_z + \left(\frac{\partial x}{\partial z}\right)_y \left(\frac{\partial z}{\partial y}\right)_x\right] dy + \left(\frac{\partial x}{\partial z}\right)_y \left(\frac{\partial z}{\partial x}\right)_y dx$$

$$dx = \left[\left(\frac{\partial x}{\partial y}\right)_z + \left(\frac{\partial x}{\partial z}\right)_y \left(\frac{\partial z}{\partial y}\right)_x\right] dy + dx$$

$$\left(\frac{\partial x}{\partial y}\right)_z = -\left(\frac{\partial x}{\partial z}\right)_y \left(\frac{\partial z}{\partial y}\right)_x$$

$$\left(\frac{\partial x}{\partial y}\right)_z \left(\frac{\partial y}{\partial z}\right)_x \left(\frac{\partial z}{\partial x}\right)_y = -1 \qquad\qquad (6.14)$$

This is known as *cyclic relation*. Hence, the relationship between P, V, and T for a pure substance is given by

$$\left(\frac{\partial P}{\partial V}\right)_T \left(\frac{\partial V}{\partial T}\right)_P \left(\frac{\partial T}{\partial P}\right)_V = -1$$

EXAMPLE 6.1 Show that for 1 g-mol of an ideal gas, the equation of state $PV = RT$ is well supported by the cyclic relation $\left(\frac{\partial P}{\partial V}\right)_T \left(\frac{\partial V}{\partial T}\right)_P \left(\frac{\partial T}{\partial P}\right)_V = -1$, assuming that one variable is dependent on the other two.

Solution: Here, pressure (P), molar volume (V) and temperature (T) of the gas are the three variables.

From the ideal gas equation of state, $P = \dfrac{RT}{V}$, we have

$$\left(\frac{\partial P}{\partial V}\right)_T = -\frac{RT}{V^2} \qquad\qquad \text{(i)}$$

$$\left(\frac{\partial V}{\partial T}\right)_P = \frac{R}{P} \qquad\qquad \text{(ii)}$$

$$\left(\frac{\partial T}{\partial P}\right)_V = \frac{V}{R} \qquad\qquad \text{(iii)}$$

Substitute Eqs. (i), (ii) and (iii) into the ideal gas equation of state,

$$\left(\frac{\partial P}{\partial V}\right)_T \left(\frac{\partial V}{\partial T}\right)_P \left(\frac{\partial T}{\partial P}\right)_V = -\frac{RT}{V^2} \cdot \frac{R}{P} \cdot \frac{V}{R} = -\frac{RT}{PV} = -1$$

Hence, the ideal gas equation of state is supported well by the cyclic relation.

6.3 FREE ENERGY FUNCTIONS

The free energy functions are extremely useful functions in studying several physicochemical processes. These are:

(a) Helmholtz free energy, A (work function).
(b) Gibbs' free energy, G (Gibbs' function or Gibbs' energy)

6.3.1   Helmholtz Free Energy (Work Function), A

The Helmholtz free energy of a system is a state function and an extensive property of the system, denoted by A. It is defined by the following equation:

$$A = U - TS \tag{6.15}$$

where

U = Internal energy
T = Absolute temperature
S = Entropy.

Equation (6.15) can be represented as

$$\Delta A = \Delta U - T\Delta S - S\Delta T \tag{6.16}$$

For an isothermal reversible change of the system, Eq. (6.16) is given by

$$\Delta A_T = \Delta U - T\Delta S$$
$$= \Delta U - Q$$
$$= -W_{max}$$
$$-\Delta A_T = W_{max} \tag{6.17}$$

Here, W_{max} is the maximum amount of work involved in the reversible process. Equation (6.17) implies that a decrease in work function, A, at constant temperature gives the maximum amount of reversible work done by the system. This work may be either purely mechanical or partly mechanical and partly external.

Now, if the process is an irreversible one, then the decrease of the work function should not be equivalent to the maximum work done by the system but rather would exceed the output of work.

Mathematically

$$-\Delta A_T > W \tag{6.18}$$

If this process is not isothermal, then the change in work function will not be equivalent to the maximum work.

In that case, Eq. (6.15) can be expressed as

$$dA = dU - TdS - SdT \tag{6.19}$$

Since the process is a reversible one, then $dQ = TdS$ and Eq. (6.19) becomes

$$dA = dU - dQ - SdT$$

or

$$dA = dW - SdT = -PdV - SdT \tag{6.20}$$

For an isochoric process

$$\left(\frac{\partial A}{\partial T} \right)_V = -S \tag{6.21}$$

For an isothermal process

$$\left(\frac{\partial A}{\partial V}\right)_T = -P \tag{6.22}$$

6.3.2 Gibbs' Free Energy (Gibbs Function), G

Gibbs' free energy is a single-valued state function of the thermodynamic system and is an extensive property. Gibbs' free energy of a system is defined as

$$G = H - TS \tag{6.23}$$

where

H = Enthalpy

T = Absolute temperature

S = Entropy.

Equation (6.23) can be represented as

$$G = U + PV - TS \qquad \text{(since enthalpy } H = U + PV\text{)}$$

If a system undergoes an isothermal reversible change at constant pressure, then the equation is given by

$$\Delta G = \Delta U + P\Delta V - V\Delta P - T\Delta S - S\Delta T \tag{6.24}$$

or

$$\Delta G_{T,P} = \Delta U + P\Delta V - T\Delta S$$
$$= \Delta U + P\Delta V - Q$$
$$= P\Delta V - W$$
$$-\Delta G_{T,P} = W - P\Delta V \tag{6.25}$$

where

$P\Delta V$ = Mechanical work involved in the system during transformation

W = Maximum total work output.

$W - P\Delta V$ is the amount of work obtainable from the system for external use other than the mechanical work.

It is very much clear from Eq. (6.25) that the decrease in Gibbs' free energy is equal to the amount of maximum work which a system can do at constant temperature and pressure over and above the mechanical work.

Hence, Gibbs' free energy is such a thermodynamic property of the system whose decrease is the measure of the useful external work available during the transformation of the system. It is to be kept in mind that Eq. (6.25) holds good only for an isothermal change of the process.

Now, Eq. (6.24) can be re-written as

$$dG = dU + PdV + VdP - TdS - SdT \tag{6.26}$$

For a reversible process, we know

$$dU + PdV = dQ = TdS \tag{6.27}$$

Therefore, Eq. (6.26) becomes

$$dG = VdP - SdT \tag{6.28}$$

For an isothermal process

$$\left(\frac{\partial G}{\partial P}\right)_T = V \tag{6.29}$$

For an isobaric process

$$\left(\frac{\partial G}{\partial T}\right)_P = -S \tag{6.30}$$

6.3.3 Variation of Free Energy with Pressure at Constant Temperature

From Eq. (6.28), we have

$$dG = VdP - SdT$$

At constant temperature, $dT = 0$. So the equation reduces to

$$(dG)_T = VdP \tag{6.31}$$

Now, if the system is changed from state 1 to state 2, then

$$\int_1^2 dG = \int_1^2 VdP$$

or

$$G_2 - G_1 = \Delta G = \int_1^2 VdP \tag{6.32}$$

If the system contains n moles of an ideal gas, then V can be represented as $V = \dfrac{nRT}{P}$ and Eq. (6.32) can be represented as

$$G_2 - G_1 = \Delta G = nRT \int_1^2 \frac{dP}{P} = nRT \ln \frac{P_2}{P_1} \tag{6.33}$$

This equation can be applied to estimate Gibbs' free energy changes for a process at known initial pressure (P_1) and final pressure (P_2) at two different states and also to calculate the free energy at one pressure level if the free energy at the other pressure level is known.

6.4 THERMODYNAMIC PROPERTY RELATIONS

For the determination of some thermodynamic properties such as internal energy, enthalpy and entropy which cannot be measured directly, it is necessary to develop some thermodynamic property relations. Examples of such property relations are Maxwell's relation established on the basis of some thermodynamic relations, Clausius–Clapeyron equation for the study of phase change processes like vaporization and sublimation, and Gibbs–Helmholtz relation. These have been discussed in the following section.

6.4.1 Fundamental Property Relations

From the mathematical expression of the first law of thermodynamics, we have

$$dQ = dU + PdV \tag{6.34}$$

Again, we know that the definition of entropy is given by

$$dS = \frac{dQ}{T}$$

or

$$dQ = TdS \tag{6.35}$$

Putting Eq. (6.35) into Eq. (6.34), we get

$$TdS = dU + PdV$$

or

$$dU = TdS - PdV \tag{6.36}$$

This is the *first thermodynamic relation* necessary for establishing Maxwell's relation. For the second, the definition of enthalpy is given by

$$H = U + PV \tag{6.37}$$

Differentiating it, we get

$$dH = dU + PdV + VdP \tag{6.38}$$

Substituting Eq. (6.36) into Eq. (6.38), we get

$$dH = TdS - PdV + PdV + VdP$$
$$dH = TdS + VdP \tag{6.39}$$

This is the *second thermodynamic relation.*
Again, Helmholtz free energy is expressed as

$$A = U - TS$$

Differentiating it, we get

$$dA = dU - TdS - SdT \tag{6.40}$$

Substituting Eq. (6.36) into Eq. (6.40), we get

$$dA = TdS - PdV - TdS - SdT$$
$$dA = -PdV - SdT \tag{6.41}$$

This is the *third thermodynamic relation.*
Gibbs free energy is mathematically given by

$$G = H - TS$$

Differentiating it, we get

$$dG = dH - TdS - SdT \tag{6.42}$$

Substituting Eq. (6.38) into Eq. (6.42), we get

$$dG = TdS + VdP + TdS - SdT$$
$$dG = VdP - SdT \qquad (6.43)$$

This is the *fourth thermodynamic relation.*

6.4.2 Maxwell's Relations

The equations related to the partial derivatives of properties P, V, T, and S of a simple compressible system with respect to each other are known as *Maxwell's relations*. These relations are derived from the definitions of the thermodynamic potentials. The two methods are discussed in the following parts:

Method 1: We have obtained the four thermodynamic relations. These are exact differentials applicable to a pure substance undergoing an infinitesimal reversible process.

1. $dU = TdS - PdV$
2. $dH = TdS + VdP$
3. $dA = -PdV - SdT$
4. $dG = VdP - SdT$.

The relations resemble the following form of the equations. If Z is a function of x and y, i.e., $Z = Z(x, y)$, then the total differential of Z is expressed as

$$dZ = \left(\frac{\partial Z}{\partial x}\right)_y dx + \left(\frac{\partial Z}{\partial y}\right)_x dy \qquad (6.44)$$
$$dZ = Mdx + Ndy$$

where

$$M = \left(\frac{\partial Z}{\partial x}\right)_y \qquad \text{and} \qquad N = \left(\frac{\partial Z}{\partial y}\right)_x$$

Then

$$\left(\frac{\partial M}{\partial y}\right)_x = \left(\frac{\partial^2 Z}{\partial y \partial x}\right) \qquad \text{and} \qquad \left(\frac{\partial N}{\partial x}\right)_y = \left(\frac{\partial^2 Z}{\partial y \partial x}\right)$$

Equating, we get

$$\left(\frac{\partial M}{\partial y}\right)_x = \left(\frac{\partial N}{\partial x}\right)_y \qquad (6.45)$$

By applying the cross-derivative rule, we have

A. $\left(\dfrac{\partial T}{\partial V}\right)_S = -\left(\dfrac{\partial P}{\partial S}\right)_V$ from the relation $dU = TdS - PdV$

B. $\left(\dfrac{\partial T}{\partial P}\right)_S = \left(\dfrac{\partial V}{\partial S}\right)_P$ from the relation $dH = TdS + VdP$

C. $\left(\dfrac{\partial P}{\partial T}\right)_V = \left(\dfrac{\partial S}{\partial V}\right)_T$ from the relation $dA = -PdV - SdT$

D. $\left(\dfrac{\partial V}{\partial T}\right)_P = -\left(\dfrac{\partial S}{\partial P}\right)_T$ from the relation $dG = VdP - SdT$

These four equations A, B, C, and D are the Maxwell's relations referred to previously. They are extensively applied for the determination of change in entropy, which cannot be measured directly by simply measuring the changes in pressure, volume, and temperature. Application of these relations is confined to the simple compressible systems. Maxwell's relations are frequently used to derive useful thermodynamic relations.

Method 2:   Considering the first thermodynamic relation, we get

$$dU = TdS - PdV \tag{6.46}$$

When V = constant, $dV = 0$; then Eq. (6.46) becomes

$$dU = TdS - PdV = TdS \tag{6.47}$$

Differentiating with respect to S at constant V, we get

$$\left(\frac{\partial U}{\partial S}\right)_V = T \tag{6.48}$$

Again, when S = constant, $dS = 0$; then Eq. (6.46) becomes

$$dU = TdS - PdV = -PdV \tag{6.49}$$

Differentiating with respect to V at constant S, we get

$$\left(\frac{\partial U}{\partial V}\right)_S = -P \tag{6.50}$$

Now, differentiating Eq. (6.48) with respect to V at constant S, we get

$$\left(\frac{\partial^2 U}{\partial V \partial S}\right)_V = \left(\frac{\partial T}{\partial V}\right)_S \tag{6.51}$$

Differentiating Eq. (6.50) with respect to S at constant V, we get

$$\left(\frac{\partial^2 U}{\partial V \partial S}\right)_V = -\left(\frac{\partial P}{\partial S}\right)_V \tag{6.52}$$

Comparing Eqs. (6.51) with (6.52), we get

$$\left(\frac{\partial T}{\partial V}\right)_S = -\left(\frac{\partial P}{\partial S}\right)_V \tag{6.53}$$

Equation (6.53) is the first Maxwell's equation. Similarly, we can derive the other three equations.

6.4.3   Clapeyron Equation

The enthalpy change associated with the phase change processes such as vaporization and freezing is determined by the Clapeyron equation with the help of the thermodynamic correlation between pressure, volume, and temperature. These enthalpy changes are known as enthalpy of vaporization, enthalpy of fusion, etc. In the study of a phase change process, the saturation pressure is considered to be the relevant pressure, and it depends only on the temperature and is independent of volume. During a phase change, there is always a possibility of transition of the substance from one phase to another phase by altering any of the variables of the system.

Let us consider two phases in equilibrium in a system at constant temperature and pressure. ΔG_1 and ΔG_2 are the free energies of liquid and vapour respectively. Now, at equilibrium, the change in free energy is zero, i.e.

$$\Delta G = 0$$

or

$$\Delta G_2 - \Delta G_1 = 0$$

or

$$\Delta G_1 = \Delta G_2 \tag{6.54}$$

The change of free energy with change of temperature and pressure is given by

$$dG = VdP - SdT$$

Therefore, the preceding equation becomes

$$V_1 \Delta P - S_1 \Delta T = V_2 \Delta P - S_2 \Delta T$$

$$\Delta P(V_2 - V_1) = \Delta T(S_2 - S_1)$$

$$\frac{\Delta P}{\Delta T} = \frac{S_2 - S_1}{V_2 - V_1} = \frac{\Delta S}{V_2 - V_1}$$

$$\frac{dP}{dT} = \frac{\Delta S}{V_2 - V_1} \tag{6.55}$$

Since the phases are in equilibrium, then for the reversible transition process we know that the change in entropy

$$\Delta S = \frac{Q}{T} = \frac{\Delta H}{T}$$

where ΔH is the change in enthalpy for the reversible transition.

Equation (6.55) becomes

$$\frac{dP}{dT} = \frac{\Delta S}{V_2 - V_1} = \frac{\Delta H}{T(V_2 - V_1)} \tag{6.56}$$

This is known as the *Clapeyron equation.*

The Clapeyron equation may be simplified for vaporization (solid–vapour and liquid–vapour phase change) of a substance by utilizing some approximations. At low pressures, $V_g \gg V_1$ and therefore $V_g = V_1$.

In that case, Eq. (6.56) becomes

$$\frac{dP}{dT} = \frac{\Delta H}{T(V_g - V_1)} = \frac{\Delta H}{TV_g} \tag{6.57}$$

where
V_g = Volume of gas
V_l = Volume of liquid.

Assuming that the vapour behaves like an ideal gas and therefore substituting $V = \dfrac{RT}{P}$ into Eq. (6.57), we get

$$\frac{dP}{dT} = \frac{\Delta H}{TV_g} = \frac{\Delta H}{T\left(\dfrac{RT}{P}\right)} = \frac{P\Delta H}{RT^2} \tag{6.58}$$

or

$$\frac{1}{P}\frac{dP}{dT} = \frac{\Delta H}{RT^2} \tag{6.59}$$

or

$$\frac{d\ln P}{dT} = \frac{\Delta H}{RT^2} \tag{6.60}$$

For small temperature range, the enthalpy change of the process can be treated further as a constant. Then on integration, Eq. (6.60) yields

$$\ln\frac{P_2}{P_1} = \frac{\Delta H}{R}\left(\frac{1}{T_1} - \frac{1}{T_2}\right) \tag{6.61}$$

This equation is called the *Clausius–Clapeyron equation* and is used to determine the variation of pressure with temperature. It may be applied to the sublimation process (solid–vapour phase change process).

EXAMPLE 6.2 Calculate the change in freezing point of water at 0°C per atm change of pressure, given that the heat of fusion of ice is 335 J/g, the density of water is 0.9998 g/cm³, and the density of ice is 0.9168 g/cm³.

Solution: We are given that the density of water, ρ_{water} = 0.9998 g/cm³. So, the volume of water, V_{water} = 1.0002 cm³/g. Similarly, the density of ice, ρ_{ice} = 0.9168 g/cm³ and the volume of ice, V_{ice} = 1.0908 cm³/g.

From Eq. (6.56), we get

$$\frac{dP}{dT} = \frac{\Delta S}{V_2 - V_1} = \frac{\Delta H}{T(V_2 - V_1)}$$

Substituting the values into Eq. (6.58), we have

$$\frac{\Delta P}{\Delta T} = \frac{335617.8 - 161.3}{273 \times (1.0002 - 1.0908)} = -13.543 \text{ J/cm}^3\text{K} = -133.7 \text{ atm/K}$$

$$\frac{\Delta T}{\Delta P} = -0.0075 \text{ K/atm}$$

Thus an increase in pressure of 1 atm lowers the freezing point by 0.0075 K. The negative sign indicates that an increase in pressure causes a decrease in temperature.

EXAMPLE 6.3 Determine the enthalpy of vaporization of water at 150°C, given that the saturation pressure is 361.3 kPa at 140°C and 617.8 kPa at 160°C, and the specific volume at 150°C is 0.3917 m^3/kg.

Solution: From Eq. (6.57), we have

$$\frac{dP}{dT} = \frac{\Delta H}{T(V_g - V_l)} = \frac{\Delta H}{TV_g}$$

It can be rearranged as

$$\frac{\Delta P}{\Delta T} = \frac{\Delta H}{TV_g}$$

Substituting the values, we get

$$\frac{\Delta P}{\Delta T} = \frac{617.8 - 361.3}{(160 - 140)} = \frac{\Delta H}{TV_g}$$

or

$$\Delta H = 12.825 \times TV_g = 12.285 \times (273 + 150) \times 0.3917$$
$$= 2125 \text{ kJ/kg}$$

EXAMPLE 6.4 Determine the saturation pressure of the refrigerant R–134a at –45°C. At 40°C, the latent heat of vaporization is 225.86 kJ/kg and the saturation pressure is 51.25 kPa.

Solution: We are given that $T_1 = 40°C$ and $T_2 = -45°C$. From Eq. (6.61), we get

$$\ln\frac{P_2}{P_1} = \frac{\Delta H}{R}\left(\frac{1}{T_1} - \frac{1}{T_2}\right)$$

or

$$\ln\frac{P_2}{51.25} = \frac{225.86}{0.0815}\left(\frac{1}{223} - \frac{1}{228}\right)$$

or

$$P_2 = 39.48 \text{ kPa}$$

Therefore, the saturation pressure of the refrigerant R–134a is 39.48 kPa at –45°C.

But the actual value obtained from the other source is 39.15 kPa. Thus the percentage of error is 1. So the calculated value is quite acceptable.

EXAMPLE 6.5 The temperature dependence of vapour pressure of an organic compound is given by

$$\log P = -\frac{834.13}{T} + 1.75 \log T - 8.375 \times 10^{-3}T + 5.324$$

Estimate the enthalpy of vaporization at its boiling point, –103.9°C.

Solution: From Eq. (6.60), we get

$$\frac{d \ln P}{dT} = \frac{\Delta H}{RT^2}$$

The vapour pressure of the compound is given by

$$\log P = -\frac{834.13}{T} + 1.75 \log T - 8.375 \times 10^{-3} T + 5.324$$

Dividing both sides of the equation by 2.303, we have

$$\frac{1}{2.303} \log P = -\frac{834.13}{T} + \frac{1.75 \log T - 8.375 \times 10^{-3} T + 5.324}{2.303}$$

Differentiating it with respect to T, we get

$$\frac{d \ln P}{dT} = \frac{2.303 \times 834.13}{T^2} + \frac{1.75}{T} - 2.303 \times 8.375 \times 10^{-3} T + 5.324$$

or

$$\frac{d \ln P}{dT} = \frac{\Delta H}{RT^2} = \frac{2.303 \times 834.13}{T^2} + \frac{1.75}{T} - 2.303 \times 8.375 \times 10^{-3}$$

or

$$\Delta H = R \times 2.303 \times 834.13 + 1.75RT + 2.303 \times 8.375 \times 10^{-3} \times RT^2$$

where $R = 8.314$ J/mol-K and $T = 169.25$ K.

Substituting the values of R and T, the enthalpy of vaporization is found to be

$$\Delta H = 13.84 \text{ kJ/mol}$$

6.4.4 Gibbs–Helmholtz Equation

The Gibbs–Helmholtz equation indicates the temperature dependence of free energy.
From Eq. (6.20), we get

$$dA = -PdV - SdT$$

Then for an isochoric process

$$\left(\frac{\partial A}{\partial T} \right)_V = -S$$

Equation (6.28) gives

$$dG = VdP - SdT$$

Then for an isobaric process

$$\left(\frac{\partial G}{\partial T} \right)_P = -S$$

From the definition of work function, we know that

$$A = U - TS$$

Substitution of $\left(\dfrac{\partial A}{\partial T}\right)_V = -S$ into the equation $A = U - TS$ gives

$$A = U + T\left(\frac{\partial A}{\partial T}\right)_V \tag{6.62}$$

Similarly, the definition of work function provides

$$G = H - TS$$

Substitution of $\left(\dfrac{\partial G}{\partial T}\right)_P = -S$ into the equation $G = H - TS$ gives

$$G = H + T\left(\frac{\partial G}{\partial T}\right)_P \tag{6.63}$$

Equations (6.62) and (6.63) are known as *Gibbs–Helmholtz equations*. These may be rearranged as

$$\left(\frac{\partial A}{\partial T}\right)_V = \frac{A - U}{T} \tag{6.64}$$

$$\left(\frac{\partial G}{\partial T}\right)_P = \frac{G - H}{T} \tag{6.65}$$

respectively.

To find out the differential of G/T with respect to temperature, we know that

$$\left[\frac{\partial}{\partial T}\left(\frac{G}{T}\right)\right]_P = \frac{1}{T}\left(\frac{\partial G}{\partial T}\right)_P - G \cdot \frac{1}{T^2}$$

$$= -\frac{S}{T} - \frac{G}{T^2} = -\frac{G + TS}{T^2}$$

or

$$\left[\frac{\partial}{\partial T}\left(\frac{G}{T}\right)\right]_P = -\frac{H}{T^2} \qquad \text{(since } G = H - TS\text{)} \tag{6.66}$$

Similarly

$$\left[\frac{\partial}{\partial T}\left(\frac{A}{T}\right)\right]_P = -\frac{U}{T^2} \tag{6.67}$$

Again, as the process takes place at constant temperature, we have:

(i) At initial state

$$G_1 = H_1 - TS_1 \tag{6.68}$$

(ii) At final state

$$G_2 = H_2 - TS_2 \tag{6.69}$$

Subtracting Eq. (6.69) from Eq. (6.68), we get

$$\Delta G = \Delta H - T\Delta S \tag{6.70}$$

Now

$$\Delta S = S_2 - S_1 = -\left[\left(\frac{\partial G_2}{\partial T}\right)_P - \left(\frac{\partial G_1}{\partial T}\right)_P\right] = -\left(\frac{\partial(\Delta G)}{\partial T}\right)_P \qquad (6.71)$$

Substituting this into Eq. (6.70), we get

$$\Delta G = \Delta H + T\left(\frac{\partial(\Delta G)}{\partial T}\right)_P \qquad (6.72)$$

It can be rearranged as

$$\left(\frac{\partial(\Delta G)}{\partial T}\right)_P = \frac{\Delta G - \Delta H}{T} \qquad (6.73)$$

Similarly

$$\left(\frac{\partial(\Delta A)}{\partial T}\right)_P = \frac{\Delta A - \Delta U}{T} \qquad (6.74)$$

Equations (6.73) and (6.74) are other useful forms applied to different physico-chemical processes.

6.4.5 General Equations for Differential Changes in Internal Energy

If internal energy U is considered to be the function of T and V, i.e., $U = f(T, V)$, then the total differential of U is given by

$$dU = \left(\frac{\partial U}{\partial T}\right)_V dT + \left(\frac{\partial U}{\partial V}\right)_T dV \qquad (6.75)$$

From the definition of C_V, we get

$$C_V = \left(\frac{\partial U}{\partial T}\right)_V \qquad (6.76)$$

Substituting Eq. (6.76) into (6.75), we get

$$dU = C_V dT + \left(\frac{\partial U}{\partial V}\right)_T dV \qquad (6.77)$$

Now, if entropy S is considered to be the function of temperature and volume, i.e., $S = f(T, V)$, then its total differential is given by

$$dS = \left(\frac{\partial S}{\partial T}\right)_V dT + \left(\frac{\partial S}{\partial V}\right)_T dV \qquad (6.78)$$

According to the first law of thermodynamics, the differential form of conservation of energy for a closed system containing a compressible substance for an internally reversible process can be expressed as

$$dU = dQ - dW \qquad (6.79)$$

But $dQ = TdS$ and $dW = PdV$; thus Eq. (6.79) becomes

$$dU = TdS - PdV \qquad (6.80)$$

Substituting Eq. (6.78) into (6.80), it yields

$$dU = T\left(\frac{\partial S}{\partial T}\right)_V dT + \left[T\left(\frac{\partial S}{\partial V}\right)_T - P\right]dV \qquad (6.81)$$

Equating the coefficients of dT and dV in Eqs. (6.77) and (6.81) gives

$$\left(\frac{\partial S}{\partial T}\right)_V = \frac{C_V}{T} \qquad (6.82)$$

$$\left(\frac{\partial U}{\partial V}\right)_T = \left[T\left(\frac{\partial S}{\partial V}\right)_T - P\right] \qquad (6.83)$$

From the third Maxwell's relation, we have $\left(\frac{\partial S}{\partial V}\right)_T = \left(\frac{\partial P}{\partial T}\right)_V$. So Eq. (6.83) gives

$$\left(\frac{\partial U}{\partial V}\right)_T = \left[T\left(\frac{\partial P}{\partial T}\right)_V - P\right] \qquad (6.84)$$

Putting Eq. (6.84) into (6.77) to yield

$$dU = C_V dT + \left[T\left(\frac{\partial P}{\partial T}\right)_V - P\right]dV \qquad (6.85)$$

This is a useful expression on change in internal energy for a simple compressible substance.

6.4.6 General Equations for Differential Changes in Enthalpy

From the second thermodynamic relation (see Eq. (6.39)), we have

$$dH = TdS + VdP$$

We choose enthalpy to be a function of T and P, i.e., $H = f(T, P)$, then the total differential of H is given by

$$dH = \left(\frac{\partial H}{\partial T}\right)_P dT + \left(\frac{\partial H}{\partial P}\right)_T dP \qquad (6.86)$$

Using the definition of $C_P \left[= \left(\frac{\partial H}{\partial T}\right)_P\right]$, we get

$$dH = C_P dT + \left(\frac{\partial H}{\partial P}\right)_T dP \qquad (6.87)$$

Now, if entropy S is considered to be the function of T and P, i.e., $S = f(T, P)$, then its total differential is given by

$$dS = \left(\frac{\partial S}{\partial T}\right)_P dT + \left(\frac{\partial S}{\partial P}\right)_T dP \qquad (6.88)$$

Substituting Eq. (6.88) into (6.39) gives

$$dH = T\left(\frac{\partial S}{\partial T}\right)_P dT + \left[V + T\left(\frac{\partial S}{\partial P}\right)_T\right] dP \qquad (6.89)$$

Comparing Eqs. (6.87) and (6.89), we get

$$\left(\frac{\partial S}{\partial T}\right)_P = \frac{C_P}{T}$$

and

$$\left(\frac{\partial H}{\partial P}\right)_T = \left[V + T\left(\frac{\partial S}{\partial P}\right)_T\right] \qquad (6.90)$$

Using the fourth Maxwell's relation $\left(\frac{\partial V}{\partial T}\right)_P = -\left(\frac{\partial S}{\partial P}\right)_T$, we get

$$\left(\frac{\partial H}{\partial P}\right)_T = \left[V - T\left(\frac{\partial V}{\partial T}\right)_P\right] \qquad (6.91)$$

Putting Eq. (6.91) into (6.87) yields

$$dH = C_P dT + \left[V - T\left(\frac{\partial V}{\partial T}\right)_P\right] dP \qquad (6.92)$$

This is a useful expression on change in enthalpy generally applied to a simple compressible substance.

6.4.7 General Equations for Differential Changes in Entropy

Consider entropy S as a function of temperature and volume, i.e., $S = f(T, V)$, then its total differential is given by (see Eq. (6.78))

$$dS = \left(\frac{\partial S}{\partial T}\right)_V dT + \left(\frac{\partial S}{\partial V}\right)_T dV$$

Using the relation $\left(\frac{\partial S}{\partial T}\right)_V = \frac{C_V}{T}$ and the third Maxwell's relation $\left(\frac{\partial S}{\partial V}\right)_T = \left(\frac{\partial P}{\partial T}\right)_V$, we get

$$dS = \frac{C_V}{T} dT + \left(\frac{\partial P}{\partial T}\right)_V dV \qquad (6.93)$$

Now, if entropy S is considered to be the function of T and P, i.e., $S = f(T, P)$, then its total differential is given by (see Eq. (6.88))

$$dS = \left(\frac{\partial S}{\partial T}\right)_P dT + \left(\frac{\partial S}{\partial P}\right)_T dP$$

Using the relation $\left(\dfrac{\partial S}{\partial T}\right)_P = \dfrac{C_P}{T}$ and the fourth Maxwell's relation $\left(\dfrac{\partial V}{\partial T}\right)_P = -\left(\dfrac{\partial S}{\partial P}\right)_T$, we have

$$dS = \frac{C_P}{T}dT - \left(\frac{\partial V}{\partial T}\right)_P dP \qquad (6.94)$$

6.4.8 *TdS* Equations

Let S be imagined as a function of T and P, then Eq. (6.88) can be written as

$$dS = \left(\frac{\partial S}{\partial T}\right)_P dT + \left(\frac{\partial S}{\partial P}\right)_T dP$$

Then by making use of $\left(\dfrac{\partial S}{\partial T}\right)_P = \dfrac{C_P}{T}$ and the fourth Maxwell's relation $\left(\dfrac{\partial V}{\partial T}\right)_P = -\left(\dfrac{\partial S}{\partial P}\right)_T$, and multiplying both sides of the equation by T, we get

$$TdS = C_P dT - T\left(\frac{\partial V}{\partial T}\right)_P dP \qquad (6.95)$$

Equation (6.95) is known as the *first TdS equation*.
Again, if S is a function of T and V, then from Eq. (6.78), we get

$$dS = \left(\frac{\partial S}{\partial T}\right)_V dT + \left(\frac{\partial S}{\partial V}\right)_T dV$$

Then by making use of $\left(\dfrac{\partial S}{\partial T}\right)_V = \dfrac{C_V}{T}$, the third Maxwell's relation $\left(\dfrac{\partial S}{\partial V}\right)_T = \left(\dfrac{\partial P}{\partial T}\right)_V$ and multiplying both sides of the equation by T, we get

$$TdS = C_V dT + T\left(\frac{\partial P}{\partial T}\right)_V dV \qquad (6.96)$$

Equation (6.96) is known as the *second TdS equation*.

6.4.9 Heat Capacity Relations

The heat capacity at constant pressure can mathematically be expressed as

$$C_P = \left(\frac{\partial H}{\partial T}\right)_P$$

or

$$dH = C_P dT$$

or

$$C_P = \frac{dH}{dT} = \frac{d(U + PV)}{dT} \qquad \text{(since enthalpy } H = U + PV)$$

$$= \frac{dU}{dT} + \frac{d(PV)}{dT} = \frac{dU}{dT} + \frac{d(RT)}{dT}$$

$$= \frac{dU}{dT} + R = C_V + R \qquad (dU = C_V dT)$$

or

$$C_P - C_V = R$$

From Eq. (6.94), we have

$$dS = \frac{C_P}{T} dT - \left(\frac{\partial V}{\partial T}\right)_P dP$$

Dividing both sides by dT and keeping V constant, we have

$$\left(\frac{\partial S}{\partial T}\right)_V = \frac{C_P}{T} - \left(\frac{\partial V}{\partial T}\right)_P \left(\frac{\partial P}{\partial T}\right)_V \qquad (6.97)$$

We know that $\left(\frac{\partial S}{\partial T}\right)_V = \frac{C_V}{T}$. Putting it into Eq. (6.97) yields

$$\frac{C_V}{T} = \frac{C_P}{T} - \left(\frac{\partial V}{\partial T}\right)_P \left(\frac{\partial P}{\partial T}\right)_V \qquad (6.98)$$

Rearranging it, we get

$$C_P - C_V = T\left(\frac{\partial V}{\partial T}\right)_P \left(\frac{\partial P}{\partial T}\right)_V \qquad (6.99)$$

From the cyclic relation between P, V, and T, we know that

$$\left(\frac{\partial P}{\partial V}\right)_T \left(\frac{\partial V}{\partial T}\right)_P \left(\frac{\partial T}{\partial P}\right)_V = -1$$

wherefrom we can get

$$\left(\frac{\partial P}{\partial T}\right)_V = -\left(\frac{\partial V}{\partial T}\right)_P \left(\frac{\partial P}{\partial V}\right)_T$$

On substitution of this condition, Eq. (6.99) yields

$$C_P - C_V = -T\left(\frac{\partial V}{\partial T}\right)_P^2 \left(\frac{\partial P}{\partial V}\right)_T \qquad (6.100)$$

Sometimes it is more useful to find out the relation based on the influence of pressure and volume on C_P and C_V. In order to obtain those relations, we consider the relation obtained in Eq. (6.94).

$$dS = \frac{C_P}{T} dT - \left(\frac{\partial V}{\partial T}\right)_P dP$$

For an isothermal change of the process, when $T =$ constant and $dT = 0$, Eq. (6.94) becomes on differentiation with respect to P

$$\left(\frac{\partial S}{\partial P}\right)_T = -\left(\frac{\partial V}{\partial T}\right)_P \qquad (6.101)$$

For the isobaric process, when P = constant and dP = 0, Eq. (6.94) becomes on differentiation with respect to T

$$\left(\frac{\partial S}{\partial T}\right)_P = \frac{C_P}{T} \qquad (6.102)$$

Differentiating Eq. (6.101) with respect to T at constant P and Eq. (6.102) with respect to P at constant T, we get

$$\frac{\partial^2 S}{\partial T \partial P} = -\left(\frac{\partial^2 V}{\partial T^2}\right)_P \qquad (6.103)$$

$$\frac{\partial^2 S}{\partial P \partial T} = \frac{1}{T}\left(\frac{\partial C_P}{\partial P}\right)_T \qquad (6.104)$$

Comparing Eqs. (6.103) and (6.104), we have

$$\left(\frac{\partial C_P}{\partial P}\right)_T = -T\left(\frac{\partial^2 V}{\partial T^2}\right)_P \qquad (6.105)$$

or

$$\left(\frac{\partial C_P}{\partial V}\right)_T \left(\frac{\partial V}{\partial P}\right)_T = -T\left(\frac{\partial^2 V}{\partial T^2}\right)_P$$

or

$$\left(\frac{\partial C_P}{\partial V}\right)_T = -T\left(\frac{\partial^2 V}{\partial T^2}\right)_P \left(\frac{\partial P}{\partial V}\right)_T \qquad (6.106)$$

Similarly, to find out the expression for the effect of pressure and volume on C_V, we consider the relation $dS = \dfrac{C_V}{T}\, dT + \left(\dfrac{\partial P}{\partial T}\right)_V dV$ and in the same fashion, we get

$$\left(\frac{\partial C_V}{\partial V}\right)_T = T\left(\frac{\partial^2 P}{\partial T^2}\right)_V \qquad (6.107)$$

and

$$\left(\frac{\partial C_V}{\partial P}\right)_T = T\left(\frac{\partial^2 P}{\partial T^2}\right)_V \left(\frac{\partial V}{\partial P}\right)_T \qquad (6.108)$$

EXAMPLE 6.6   A gas obeys the equation of state $V = \dfrac{RT}{P} - \dfrac{C}{T^3}$. Find out the variation of C_P with pressure at constant temperature.

Solution: From Eq. (6.104), we get

$$\left(\frac{\partial C_P}{\partial P}\right)_T = -T\left(\frac{\partial^2 V}{\partial T^2}\right)_P$$

Now, differentiating the equation $V = \dfrac{RT}{P} - \dfrac{C}{T^3}$ with respect to temperature at constant pressure, we get

$$\left(\frac{\partial V}{\partial T}\right)_P = \frac{R}{P} + \frac{3C}{T^4}$$

On further differentiation, we get

$$\left(\frac{\partial^2 V}{\partial T^2}\right)_P = -\frac{12C}{T^5}$$

Substituting the values into Eq. (6.104), we have

$$\left(\frac{\partial C_P}{\partial P}\right)_T = -T\left(-\frac{12C}{T^5}\right) = \frac{12C}{T^4}$$

EXAMPLE 6.7 Derive an expression to calculate the change in enthalpy and entropy of a real gas undergoing an isothermal compression and obeying the following equation of state:

$$V = \frac{RT}{P} + b - \frac{a}{RT}$$

Solution: From Eq. (6.92), we get

$$dH = C_P dT + \left[V - T\left(\frac{\partial V}{\partial T}\right)_P\right]dP$$

Since the gas undergoes an isothermal process, then T = constant and $dT = 0$. Substituting this condition into Eq. (6.92), we have

$$dH = \left[V - T\left(\frac{\partial V}{\partial T}\right)_P\right]dP$$

On integration, we get

$$\Delta H = H_2 - H_1 = \int\left[V - T\left(\frac{\partial V}{\partial T}\right)_P\right]dP$$

After differentiation, the equation of state becomes

$$\left(\frac{\partial V}{\partial T}\right)_P = \frac{R}{P} + \frac{a}{RT^2}$$

Thus

$$V - \left(\frac{\partial V}{\partial T}\right)_P = \frac{RT}{P} + b - \frac{a}{RT} - T\left(\frac{R}{P} + \frac{a}{RT^2}\right) = b - \frac{2a}{RT}$$

Hence

$$\Delta H = H_2 - H_1 = \int_{P_1}^{P_2}\left[V - T\left(\frac{\partial V}{\partial T}\right)_P\right]dP = \left(b - \frac{2a}{RT}\right)(P_2 - P_1)$$

Again, we know from Eq. (6.94)

$$dS = \frac{C_P}{T} dT - \left(\frac{\partial V}{\partial T}\right)_P dP$$

For isothermal compression, the preceding equation reduces to

$$dS = -\left(\frac{\partial V}{\partial T}\right)_P dP$$

The change in entropy would be

$$\Delta S = S_2 - S_1 = -\int \left(\frac{\partial V}{\partial T}\right)_P dP = -\int \left(\frac{R}{P} + \frac{a}{RT^2}\right) dP = -R \ln \frac{P_2}{P_1} - \frac{a(P_2 - P_1)}{RT^2}$$

6.4.10 Isothermal Compressibility and Volume Expansivity

These two important thermodynamic properties applied to a pure substance are already defined with mathematical expression in Chapter 3. *Isothermal compressibility* is defined as the rate of decrease of volume with respect to pressure at constant temperature per unit volume.

Mathematically (see Eq. (3.2))

$$\alpha = -\frac{1}{V} \left(\frac{\partial V}{\partial P}\right)_T$$

and *volume expansivity* or *coefficient of volume expansion* (β) is defined as the rate of change of volume with respect to temperature at constant pressure per unit volume (see Eq. (3.4)).

$$\beta = \frac{1}{V} \left(\frac{\partial V}{\partial T}\right)_P$$

From Eq. (3.1), we have

$$dV = \left(\frac{\partial V}{\partial T}\right)_P dT + \left(\frac{\partial V}{\partial P}\right)_T dP$$

Now, Eq. (3.2) can be expressed as

$$\left(\frac{\partial V}{\partial P}\right)_T = -\alpha V$$

Equation (3.4) can be expressed as (see Eq. (3.6))

$$\left(\frac{\partial V}{\partial T}\right)_P = \beta V$$

On substitution of Eqs. (3.5) and (3.6) into Eq. (3.1), we get

$$dV = \beta V dT - \alpha V dP$$

or

$$\frac{dV}{V} = \beta dT - \alpha dP \quad \text{(see Eq. (3.7))}$$

From the fourth Maxwell's relation, we get

$$\left(\frac{\partial S}{\partial P}\right)_T = -\left(\frac{\partial V}{\partial T}\right)_P = -\beta V \tag{6.109}$$

Again, from (6.91), we get

$$\left(\frac{\partial H}{\partial P}\right)_T = \left[V - T\left(\frac{\partial V}{\partial T}\right)_P\right]$$

$$= V - T\beta V = V(1 - \beta T) \tag{6.110}$$

Substituting Eq. (6.110) into Eq. (6.87), we have

$$dH = C_P dT + \left(\frac{\partial H}{\partial P}\right)_T dP$$

$$= C_P dT + V(1 - \beta T)dP$$

EXAMPLE 6.8 Prove that $dS = \dfrac{C_P}{T}dT - \beta V dP$.

Solution: If S is considered to be a function of T and P, then from the first TdS equation, we get (see Eq. (6.95))

$$TdS = C_P dT - T\left(\frac{\partial V}{\partial T}\right)_P dP$$

or

$$dS = \frac{C_P}{T}dT - \left(\frac{\partial V}{\partial T}\right)_P dP \quad \text{(see Eq. (6.94))}$$

We know that the volume expansivity β is defined as (see Eq. (3.4))

$$\beta = \frac{1}{V}\left(\frac{\partial V}{\partial T}\right)_P$$

or

$$\left(\frac{\partial V}{\partial T}\right)_P = \beta V \quad \text{(see Eq. (3.6))}$$

Substituting Eq. (3.6) into Eq. (6.94), we have

$$dS = \frac{C_P}{T}dT - \left(\frac{\partial V}{\partial T}\right)_P dP = \frac{C_P}{T}dT - \beta V dP$$

EXAMPLE 6.9 Prove that $TdS = C_V dT + \dfrac{T\beta}{\alpha}dV$, following the method of partial differentials, where the symbols have their usual meanings.

Solution: If S is a function of T and V, then from the second TdS equation (see Eq. (6.96)), we get

$$TdS = C_V dT + T\left(\frac{\partial P}{\partial T}\right)_V dV$$

We know that the isothermal compressibility α and volume expansivity β are defined as

$$\alpha = -\frac{1}{V}\left(\frac{\partial V}{\partial P}\right)_T \qquad \text{and} \qquad \beta = \frac{1}{V}\left(\frac{\partial V}{\partial T}\right)_P$$

Now

$$\frac{\beta}{\alpha} = \frac{\dfrac{1}{V}\left(\dfrac{\partial V}{\partial T}\right)_P}{-\dfrac{1}{V}\left(\dfrac{\partial V}{\partial P}\right)_T} = -\frac{\left(\dfrac{\partial V}{\partial T}\right)_P}{\left(\dfrac{\partial V}{\partial P}\right)_T} = -\left(\frac{\partial V}{\partial T}\right)_P\left(\frac{\partial P}{\partial V}\right)_T$$

From the cyclic relation among the three variables P, V, and T, we have

$$\left(\frac{\partial P}{\partial V}\right)_T\left(\frac{\partial V}{\partial T}\right)_P\left(\frac{\partial T}{\partial P}\right)_V = -1$$

Using this relation, we get

$$\frac{\beta}{\alpha} = -\left(\frac{\partial V}{\partial T}\right)_P\left(\frac{\partial P}{\partial V}\right)_T = \frac{1}{\left(\dfrac{\partial T}{\partial P}\right)_V} = \left(\frac{\partial P}{\partial T}\right)_V$$

Substituting it into Eq. (6.96), we get

$$TdS = C_V dT + \frac{T\beta}{\alpha}dV$$

EXAMPLE 6.10 Prove that $C_P - C_V = \dfrac{TV\beta^2}{\alpha}$

where
   α = Isothermal compressibility
   β = Volume expansivity. $\hfill$ (WBUT, 2007)

Solution: Isothermal compressibility

$$\alpha = -\frac{1}{V}\left(\frac{\partial V}{\partial P}\right)_T$$

Volume expansivity

$$\beta = \frac{1}{V}\left(\frac{\partial V}{\partial T}\right)_P$$

We know from Eq. (6.99) that

$$C_P - C_V = -T\left(\frac{\partial V}{\partial T}\right)_P^2\left(\frac{\partial P}{\partial V}\right)_T$$

$$C_P - C_V = -T(\beta V)^2 \cdot \left(-\frac{1}{\alpha V}\right)$$

$$= \frac{T\beta^2 V^2}{\alpha V} = \frac{TV\beta^2}{\alpha}$$

EXAMPLE 6.11 Show that $\left(\dfrac{\partial U}{\partial P}\right)_T = (\alpha P - \beta T)V$.

Solution: From the definition of enthalpy, we know that $H = U + PV$.
On differentiation, we get

$$dU = dH - PdV - VdP$$

Differentiating again with respect to P at constant T, we get

$$\left(\frac{\partial U}{\partial P}\right)_T = \left(\frac{\partial H}{\partial P}\right)_T - P\left(\frac{\partial V}{\partial P}\right)_T - V$$

Substituting Eq. (6.109) into the preceding equation, we have

$$\left(\frac{\partial U}{\partial P}\right)_T = V(1 - \beta T) - P(-\alpha V) - V = (\alpha P - \beta T)V$$

EXAMPLE 6.12 For a pure substance, if internal energy is considered to be the function of any two of the independent variables P, V, and T, then derive the following relations:

(a) $\left(\dfrac{\partial U}{\partial V}\right)_T = \dfrac{T\beta}{\alpha} - P$

(b) $\left(\dfrac{\partial U}{\partial P}\right)_V = C_V\dfrac{\alpha}{\beta}$ (WBUT, 2008)

Solution:

(a) The first thermodynamic relation is given by $dU = TdS - PdV$.
Differentiating it with respect to V at constant T, we get

$$\left(\frac{\partial U}{\partial V}\right)_T = T\left(\frac{\partial S}{\partial V}\right)_T - P = T\left(\frac{\partial P}{\partial T}\right)_V - P \quad \text{(using Maxwell's relation)}$$

From the cyclic relation among the three variables P, V, and T, we have

$$\left(\frac{\partial P}{\partial V}\right)_T\left(\frac{\partial V}{\partial T}\right)_P\left(\frac{\partial T}{\partial P}\right)_V = -1$$

which can be rearranged as

$$\left(\frac{\partial P}{\partial T}\right)_V = -\left(\frac{\partial V}{\partial T}\right)_P\left(\frac{\partial P}{\partial V}\right)_T$$

Substituting this relation, we have

$$\left(\frac{\partial U}{\partial V}\right)_T = \left(\frac{\partial P}{\partial T}\right)_V - P = -T\left(\frac{\partial V}{\partial T}\right)_P\left(\frac{\partial P}{\partial V}\right)_T - P = \frac{T\beta}{\alpha} - P$$

$$\left(\text{since } \alpha = -\frac{1}{V}\left(\frac{\partial V}{\partial P}\right)_T \text{ and } \beta = \frac{1}{V}\left(\frac{\partial V}{\partial T}\right)_P\right)$$

(b) Differentiating $dU = TdS - PdV$ with respect to P at constant V, we get

$$\left(\frac{\partial U}{\partial P}\right)_V = T\left(\frac{\partial S}{\partial P}\right)_V$$

Using Maxwell's relation, we obtain

$$\left(\frac{\partial U}{\partial P}\right)_V = T\left(\frac{\partial S}{\partial P}\right)_V = -T\left(\frac{\partial V}{\partial T}\right)_S$$

From the cyclic relation among the three variables S, T, and V, we have

$$\left(\frac{\partial S}{\partial T}\right)_V \left(\frac{\partial T}{\partial V}\right)_S \left(\frac{\partial V}{\partial S}\right)_T = -1$$

or

$$\left(\frac{\partial S}{\partial T}\right)_V \left(\frac{\partial V}{\partial S}\right)_T = -\left(\frac{\partial V}{\partial T}\right)_S$$

Substituting this, we get

$$\left(\frac{\partial U}{\partial P}\right)_V = -T\left(\frac{\partial V}{\partial T}\right)_S = T\left(\frac{\partial S}{\partial T}\right)_V \left(\frac{\partial V}{\partial S}\right)_T \qquad \text{(i)}$$

Now, by dividing $dU = TdS - PdV$ by dT, keeping V constant, we can have

$$\left(\frac{\partial U}{\partial T}\right)_V = T\left(\frac{\partial S}{\partial T}\right)_V$$

wherefrom

$$C_V = T\left(\frac{\partial S}{\partial T}\right)_V$$

Maxwell's relation gives $\left(\frac{\partial S}{\partial V}\right)_T = \left(\frac{\partial P}{\partial T}\right)_V$ and we have got $\left(\frac{\partial P}{\partial T}\right)_V = \dfrac{\beta}{\alpha}$ earlier.
Therefore Eq. (i) yields

$$\left(\frac{\partial U}{\partial P}\right)_V = T\left(\frac{\partial S}{\partial T}\right)_V \left(\frac{\partial V}{\partial S}\right)_T = T \cdot \frac{C_V}{T}\left(\frac{\partial T}{\partial P}\right)_V = C_V \frac{\alpha}{\beta}$$

EXAMPLE 6.13 For a pure substance, derive the relation $\left(\dfrac{\partial H}{\partial V}\right)_T = \dfrac{(T\beta - 1)}{\alpha}$.

Solution: We know that

$$dH = TdS + VdP$$

Differentiating it with respect to V at constant T, we get

$$\left(\frac{\partial H}{\partial V}\right)_T = T\left(\frac{\partial S}{\partial V}\right)_T + V\left(\frac{\partial P}{\partial V}\right)_T$$

Using Maxwell's relation $\left(\dfrac{\partial S}{\partial V}\right)_T = \left(\dfrac{\partial P}{\partial T}\right)_V$, we obtain

$$\left(\frac{\partial H}{\partial V}\right)_T = T\left(\frac{\partial P}{\partial T}\right)_V + V\left(\frac{\partial P}{\partial V}\right)_T \tag{i}$$

We have obtained the relation $\left(\dfrac{\partial P}{\partial T}\right)_V = \dfrac{\beta}{\alpha}$. Substituting it into Eq. (i), we get

$$\left(\frac{\partial H}{\partial V}\right)_T = T\left(\frac{\partial P}{\partial T}\right)_V + V\left(\frac{\partial P}{\partial V}\right)_T = \frac{T\beta}{\alpha} - V\cdot\frac{1}{\alpha V} = \frac{(T\beta-1)}{\alpha} \quad \left(\text{as } \alpha = \frac{1}{V}\left(\frac{\partial V}{\partial P}\right)_T\right)$$

EXAMPLE 6.14 Calculate the difference between C_P and C_V for a copper block having isothermal compressibility $\alpha = 0.837 \times 10^{-11}$ m²/N and volume expansivity $\beta = 54.2 \times 10^{-6}$ K⁻¹ at 227°C, given that the specific volume of copper is 7.115×10^{-3} m³/kmol. If C_P for copper is 26.15 J/mol-K at this temperature, then what percentage of error would be made in C_V if C_P is assumed to be equal to C_V?

Solution: We know that

$$C_P - C_V = \frac{TV\beta^2}{\alpha}$$

Putting the values, we get

$$C_P - C_V = \frac{TV\beta^2}{\alpha}$$

$$= \frac{500 \times (7.115 \times 10^{-3}) \times (54.2 \times 10^{-6})^2}{0.837 \times 10^{-11}}$$

$$= 1248.5 \text{ J/kmol-K} = 1.2485 \text{ J/mol-K}$$

We are given that $C_P = 26.15$ J/mol-K. Then $C_V = 26.15$ J/mol-K $- 1.25$ J/mol-K $= 24.90$ J/mol-K

Hence, percentage of error $= \dfrac{1.249}{24.9} \times 100 = 0.05 = 5\%$.

EXAMPLE 6.15 Prove that $C_P - C_V = R$ for an ideal gas.

Solution: For an ideal gas, we know that

$$C_P - C_V = -T\left(\frac{\partial V}{\partial T}\right)_P^2 \left(\frac{\partial P}{\partial V}\right)_T$$

The ideal gas obeys the equation $PV = RT$. Therefore, on differentiation with respect to T at constant P, we get

$$\left(\frac{\partial V}{\partial T}\right)_P = \frac{R}{P}$$

Again, we obtain $\left(\dfrac{\partial P}{\partial V}\right)_T = -\dfrac{RT}{V^2}$ after differentiating it with respect to volume at constant T. Thus

$$C_P - C_V = T\left(\frac{\partial V}{\partial T}\right)_P^2 \left(\frac{\partial P}{\partial V}\right)_T = -T\left(\frac{R}{P}\right)^2 \left(-\frac{RT}{V^2}\right) = R \qquad (\text{as } PV = RT)$$

EXAMPLE 6.16 Show that for a van der Waals gas, C_V is a function only of temperature.

Solution: A van der Waals gas follows the equation of state that is given by

$$P = \frac{RT}{V - b} - \frac{a}{V^2}$$

which gives on differentiation

$$\left(\frac{\partial P}{\partial T}\right)_V = \frac{R}{V - b} \qquad \text{and} \qquad \left(\frac{\partial^2 P}{\partial T^2}\right)_V = 0$$

Now, we know that

$$\left(\frac{\partial C_V}{\partial V}\right)_T = T\left(\frac{\partial^2 P}{\partial T^2}\right)_V$$

After putting the values, we get

$$\left(\frac{\partial C_V}{\partial V}\right)_T = T\left(\frac{\partial^2 P}{\partial T^2}\right)_V = 0$$

It can be represented as

$$\left(\frac{\partial C_V}{\partial P}\right)_T = \left(\frac{\partial C_V}{\partial V}\right)_T \left(\frac{\partial V}{\partial P}\right)_T = 0$$

Hence, C_V is independent of pressure and volume and is a function only of temperature.

EXAMPLE 6.17 Show that $\left(\dfrac{\partial \beta}{\partial P}\right)_T = -\left(\dfrac{\partial \alpha}{\partial T}\right)_P$.

Solution: If volume is considered to be a function of temperature and pressure, i.e., $V = f(T, P)$, then on total differentiation we get

$$dV = \left(\frac{\partial V}{\partial T}\right)_P dT + \left(\frac{\partial V}{\partial P}\right)_T dP$$

Dividing both sides, by V we obtain

$$\frac{dV}{V} = \frac{1}{V}\left(\frac{\partial V}{\partial T}\right)_P dT + \frac{1}{V}\left(\frac{\partial V}{\partial P}\right)_T dP$$

$$= \beta dT - \alpha dP$$

or

$$dV = \beta V dT - \alpha V dP$$

Since the property volume is an exact differential, then it must satisfy the following condition:

$$\left[\frac{\partial}{\partial P}(\beta V)\right]_T = -\left[\frac{\partial}{\partial T}(-\alpha V)\right]_P$$

or

$$\left(\frac{\partial \beta}{\partial P}\right)_T = -\left(\frac{\partial \alpha}{\partial T}\right)_P$$

EXAMPLE 6.18 Show that the C_V of an ideal gas does not depend upon a specific volume.

Solution: For an ideal gas, we know that

$$\left(\frac{\partial C_V}{\partial V}\right)_T = T\left(\frac{\partial^2 P}{\partial T^2}\right)_V$$

Again, for 1 mol of an ideal gas, $P = \dfrac{RT}{V}$, wherefrom we get on differentiation with respect to T at constant V

$$\left(\frac{\partial P}{\partial T}\right)_V = \frac{R}{V}$$

or

$$\left(\frac{\partial^2 P}{\partial T^2}\right)_V = 0$$

Using Eq. (6.106), we obtain

$$\left(\frac{\partial C_V}{\partial V}\right)_T = T\left(\frac{\partial^2 P}{\partial T^2}\right)_V = 0$$

Therefore, the specific heat at constant volume of an ideal gas is independent of the specific volume.

6.4.11 Joule–Thomson Coefficient

In Chapter 7, we have discussed the throttling process along with the significance of the Joule–Thomson coefficient, μ. If a compressed gas is allowed to pass slowly through a flow-restricting device like a porous barrier plug or throttling valve, it undergoes a significant pressure drop. As a result, it experiences a change in temperature which varies in magnitude and sign. This involves either heating up or cooling down depending upon the nature of the gas. The process accompanied by a change in temperature resulting from a reduction in pressure is known as the *throttling process*. But it has been observed severally that there is a decrease in temperature of the gas. This phenomenon is known as *Joule–Thomson effect*.

Joule–Thomson coefficient can be defined as the rate of change in temperature with respect to pressure at constant enthalpy. It can mathematically be expressed as

$$\mu_{JT} = \left(\frac{\partial T}{\partial P}\right)_H \tag{6.111}$$

In Eq. (6.111), when $\mu = +ve$ there will be a cooling effect, and when $\mu = -ve$ there will be a heating effect. Hence if the gas is passed through the porous plug at ordinary (lower) temperature and lower pressure, gases will be cooled down. In most of the cases, this is what happens. But in case of helium and hydrogen at room temperature, there will be a heating effect instead of cooling effect at ambient temperature. This can be explained in the following way.

Let H be imagined as a function of temperature and pressure, i.e.

$$H = f(T, P)$$

$$dH = \left(\frac{\partial H}{\partial T}\right)_P dT + \left(\frac{\partial H}{\partial P}\right)_T dP \tag{6.112}$$

In isenthalpic change, $dH = 0$. Hence Eq. (6.112) becomes

$$\left(\frac{\partial H}{\partial T}\right)_P dT = -\left(\frac{\partial H}{\partial P}\right)_T dP$$

or

$$\left(\frac{\partial T}{\partial P}\right)_H = -\frac{\left(\dfrac{\partial H}{\partial P}\right)_T}{\left(\dfrac{\partial H}{\partial T}\right)_P}$$

or

$$\mu_{JT} = -\frac{1}{C_P}\left(\frac{\partial H}{\partial P}\right)_T \tag{6.113}$$

Again, we know

$$\left(\frac{\partial H}{\partial P}\right)_T = V - T\left(\frac{\partial V}{\partial T}\right)_P \tag{6.114}$$

From compressibility factor

$$Z = \frac{PV}{RT}$$

or

$$V = \frac{ZRT}{P}$$

This factor is used to measure the deviation from ideal gas behaviour, as H_2 and He are real gases whose behaviour deviates from that of ideal gases.

Putting the value of V in Eq. (6.114), we get

$$\left(\frac{\partial H}{\partial P}\right)_T = \frac{ZRT}{P} - T\left(\frac{\partial V}{\partial T}\right)_P$$

$$= \frac{ZRT}{P} - T\left[\frac{\partial(ZRT/P)}{\partial T}\right]_P$$

$$= \frac{ZRT}{P} - T\left[\frac{\partial}{\partial T}\left(\frac{ZRT}{P}\right)\right]_P$$

$$= \frac{ZRT}{P} - T\left[\frac{R}{P}\left\{T\left(\frac{\partial Z}{\partial T}\right)_P + Z\right\}\right]$$

$$= \frac{ZRT}{P} - T\left[\frac{RT}{P}\left(\frac{\partial Z}{\partial T}\right)_P + \frac{RZ}{P}\right]$$

or

$$\left(\frac{\partial H}{\partial P}\right)_T = \frac{ZRT}{P} - T\left[\frac{RT}{P}\left(\frac{\partial Z}{\partial T}\right)_P + \frac{RZ}{P}\right]$$

or

$$\left(\frac{\partial H}{\partial P}\right)_T = -\frac{RT^2}{P}\left(\frac{\partial Z}{\partial T}\right)_P \tag{6.115}$$

Substituting Eq. (6.115) into Eq. (6.113), we have

$$\mu_{JT} = -\frac{1}{C_P} \cdot -\frac{RT^2}{P}\left(\frac{\partial Z}{\partial T}\right)_P = \frac{RT^2}{C_P P}\left(\frac{\partial Z}{\partial T}\right)_P$$

Equation (6.113) can be rearranged as

$$\mu_{JT} = -\frac{1}{C_P}\left(\frac{\partial H}{\partial P}\right)_T = -\frac{1}{C_P}\left(\frac{\partial U}{\partial P}\right)_T - \frac{1}{C_P}\left(\frac{\partial (PV)}{\partial P}\right)_T \quad \text{(since } H = U + PV) \tag{6.116}$$

In case of an ideal gas, both the terms of the equation would be zero separately, as $\left(\frac{\partial U}{\partial P}\right)_T = 0$ and $\left(\frac{\partial (PV)}{\partial P}\right)_T = 0$. But for a real gas, it will be different and Eq. (6.116) can be represented as

$$\mu_{JT} = -\frac{1}{C_P}\left(\frac{\partial U}{\partial V}\right)_T\left(\frac{\partial V}{\partial P}\right)_T - \frac{1}{C_P}\left(\frac{\partial (PV)}{\partial P}\right)_T \tag{6.117}$$

$\left(\frac{\partial U}{\partial P}\right)_T = -$ve quantity for a real gas. Then the first term of Eq. (6.117) is +ve. The sign of the

term $\left(\frac{\partial (PV)}{\partial P}\right)_T$ may be +ve or −ve. But it is sure that at low temperature and pressure,

$\left(\frac{\partial (PV)}{\partial P}\right)_T$ is −ve. Then the second term is +ve. Thus, on substitution of these conditions into

Eq. (6.117), μ_{JT} wil have a +ve value. Hence, there will be a cooling effect. But it is needless

to mention that at high temperature and pressure, $\left(\frac{\partial (PV)}{\partial P}\right)_T = +$ve. As a result, $\mu_{JT} = -$ve, which

means a heating effect. For helium and hydrogen, $\mu_{JT} = -$ve only because of higher temperatures and pressures, i.e., heating effect.

For every gas there is a temperature where $\mu_{JT} = 0$, i.e., neither heating nor cooling of the gas passing through the porous plug occurs. This temperature is known as the *inversion temperature*.

The Joule–Thomson coefficient is the slope of the isenthalpic lines, and these are shown on the P–T diagram in Fig. 6.1. The line that passes through the points of zero slope or zero Joule-Thomson coefficient is called the *inversion line*, and the temperature at a point where a constant enthalpy line intersects the inversion line is called the *inversion temperature*. The point at which $\mu_{JT} = 0$ is called the *inversion point*.

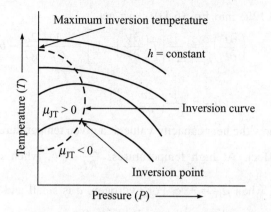

Fig. 6.1 Illustration of Joule–Thomson inversion curve.

The inversion temperature of a gas can be discussed suitably if the entropy is considered to be the function of temperature and pressure. So

$$S = f(T, P)$$

or

$$dS = C_P \frac{dT}{T} - \left(\frac{\partial V}{\partial T}\right)_P dP$$

$$\left(\text{since } \left(\frac{\partial S}{\partial P}\right)_T = -\left(\frac{\partial V}{\partial T}\right)_P \text{ from Maxwell's relation}\right)$$

Again, we know that $dH = TdS + VdP$. Putting the value of dS in this equation, we get

$$dH = T\left[\frac{C_P}{T}dT - \left(\frac{\partial V}{\partial T}\right)_P dP\right] + VdP \qquad (6.118)$$

Since it is an isenthalpic change, $dH = 0$. Hence Eq. (6.118) becomes

$$C_P dT = \left[T\left(\frac{\partial V}{\partial T}\right)_P - V\right]dP$$

$$\left(\frac{\partial T}{\partial P}\right)_H = \frac{1}{C_P}\left[T\left(\frac{\partial V}{\partial T}\right)_P - V\right] \qquad (6.119)$$

For a van der Waals gas, we know

$$\left(P + \frac{a}{V^2}\right)(V - b) = RT$$

$$\left(\frac{\partial V}{\partial T}\right)_P = \frac{R}{\left(P - \frac{a}{V^2}\right)} = \frac{R}{\left(P + \frac{a}{V^2}\right) - \frac{2a}{V^2}} = \frac{R}{\left(\frac{RT}{V - b}\right) - \frac{2a}{V^2}}$$

or

$$T\left(\frac{\partial V}{\partial T}\right)_P - V = \frac{RT}{\left(\frac{RT}{V - b}\right) - \frac{2a}{V^2}} - V \approx \frac{2a}{RT} - b \qquad (6.120)$$

Substituting Eq. (6.120) into Eq. (6.119), we get

$$\left(\frac{\partial T}{\partial P}\right)_H = \frac{1}{C_P}\left[T\left(\frac{\partial V}{\partial T}\right) - V\right] = \frac{1}{C_P}\left(\frac{2a}{RT} - b\right)$$

or

$$\mu_{JT} = \frac{1}{C_P}\left(\frac{2a}{RT} - b\right) \tag{6.121}$$

Therefore, if we know the heat capacity value at a given temperature, then we can determine the Joule–Thomson effect. At high temperatures, $\frac{2a}{RT} < b$, when $\mu_{JT} = -$ve, and at low temperatures, $\frac{2a}{RT} > b$, when $\mu_{JT} = +$ve. For hydrogen, a is small and $\frac{2a}{RT}$ is less than b even for ordinary temperatures; so there will be a heating effect.

Now, if the inversion temperature is denoted by T_i, then on substituting $\mu_{JT} = 0$ into Eq. (6.121) we have

$$\frac{2a}{RT_i} - b = 0$$

or

$$T_i = \frac{2a}{Rb} \tag{6.122}$$

The inversion of some common gases are shown in Table 6.1.

Table 6.1 Inversion temperature of some common gases

Gas	Oxygen	Helium	Hydrogen	Argon	Nitrogen	Carbon dioxide	Air	Neon
Inversion Temperature (K)	893	40	202	723	621	−1500	603	270

EXAMPLE 6.19 The van der Waals constants for carbon dioxide are $a = 3.59$ L^2-atm/mol^2, $b = 0.043$ L/mol. What is the inversion temperature of the gas?

Solution: From Eq. (6.122), we obtain the expression for inversion temperature as

$$T_i = \frac{2a}{Rb}$$

On substitution of the values, we get

$$T_i = \frac{2a}{Rb} = \frac{2 \times 3.59 \text{ L}^2 - \text{atm/mol}^2}{0.082 \text{ L-atm/mol} - \text{K} \times 0.043 \text{ L/mol}}$$

$$= 2036.3 \text{ K}$$

EXAMPLE 6.20 Show that the Joule–Thomson coefficient (μ_{JT}) of an ideal gas is zero.
(WBUT, 2007)

Solution: Consider that the entropy S is a function of temperature and pressure, i.e.

$$S = f(T, P)$$

On total differentiation, we get

$$dS = \left(\frac{\partial S}{\partial T}\right)_P dT + \left(\frac{\partial S}{\partial P}\right)_T dP \qquad \text{(i)}$$

$$= \frac{C_P}{T} dT - \left(\frac{\partial V}{\partial T}\right)_P dP \qquad \left(\text{since } \left(\frac{\partial S}{\partial P}\right)_T = -\left(\frac{\partial V}{\partial T}\right)_P\right)$$

Substituting Eq. (i) into the following thermodynamic relation, we get

$$dH = TdS + VdP$$

$$= T\left[\frac{C_P}{T} dT - \left(\frac{\partial V}{\partial T}\right)_P dp\right] + VdP \qquad \text{(ii)}$$

Since, the Joule–Thomson process is an isenthalpic process, then $dH = 0$. Thus, Eq. (ii) becomes

$$C_P dT = T\left(\frac{\partial V}{\partial T}\right)_P dP - VdP$$

$$= \left[T\left(\frac{\partial V}{\partial T}\right)_P - V\right] dP$$

$$\left(\frac{\partial T}{\partial P}\right)_H = \frac{1}{C_P}\left[T\left(\frac{\partial V}{\partial T}\right)_P - V\right]$$

$$\mu_{JT} = \frac{1}{C_P}\left(T \times \frac{R}{P} - V\right) \qquad \left(\text{for ideal gas, } PV = RT \text{ or } \left(\frac{\partial V}{\partial T}\right)_P = \frac{R}{P}\right)$$

$$= \frac{1}{C_P}\left(\frac{RT}{P} - V\right) = \frac{1}{C_P}\left(\frac{PV}{P} - V\right) \qquad (\because \text{ since } PV = RT)$$

$$= \frac{1}{C_P}(V - V) = 0$$

EXAMPLE 6.21 If the three variables H, T, and P are related as $f(H, T, P) = 0$, then show that $\mu_{JT} = \frac{v}{C_P}(T\beta - 1)$.

Solution: If $f(H, T, P) = 0$, then from the cyclic relation we get

$$\left(\frac{\partial H}{\partial T}\right)_P \left(\frac{\partial T}{\partial P}\right)_H \left(\frac{\partial P}{\partial H}\right)_T = -1$$

It can be rearranged as

$$\left(\frac{\partial T}{\partial P}\right)_H = -\frac{\left(\frac{\partial H}{\partial P}\right)_T}{\left(\frac{\partial H}{\partial T}\right)_P} \qquad \text{(i)}$$

We know that $dH = TdS + VdP$; differentiating it with respect to P at constant T, we have

$$\left(\frac{\partial H}{\partial P}\right)_T = T\left(\frac{\partial S}{\partial P}\right)_T + V$$

Using the fourth Maxwell's relation $\left(\dfrac{\partial V}{\partial T}\right)_P = -\left(\dfrac{\partial S}{\partial P}\right)_T$, we get

$$\left(\frac{\partial H}{\partial P}\right)_T = T\left(\frac{\partial S}{\partial P}\right)_T + V = -T\left(\frac{\partial V}{\partial T}\right)_P + V = -\beta V T + V = -V(\beta T - 1)$$

Substitution of this into Eq. (i) yields

$$\mu_{JT} = \left(\frac{\partial T}{\partial P}\right)_H = -\frac{\left(\dfrac{\partial H}{\partial P}\right)_T}{\left(\dfrac{\partial H}{\partial T}\right)_P} = \frac{V(T\beta - 1)}{C_P}$$

6.5   RESIDUAL PROPERTY

Residual properties or departure functions are useful to calculate the thermodynamic property of real fluids when the P–V–T data of the substance is unavailable or insufficient. Residual property is defined as the difference between the thermodynamic property of a substance (real fluid) and that of an ideal gas at the specified temperature and pressure. Common residual properties include those of enthalpy, entropy, and internal energy. It can schematically be represented as

Thermodynamic property of a substance at given temperature and pressure (M)	−	Thermodynamic property of an ideal gas at same temperature and pressure (M^{ig})	=	Residual property of a substance at given temperature and pressure (M^R)

where M is the molar value of extensive thermodynamic properties like V, U, H, S, and G, of a real substance at T and P, M^{ig} is that of an ideal gas at the same condition, and M^R is the residual property.

Hence, a residual property gives the difference between the real state, at a finite volume or non-zero pressure and temperature, and the ideal state, usually at zero pressure or infinite volume and temperature.

For example, the *residual enthalpy* or *enthalpy departure* of a substance can be expressed accordingly as $H^R = H - H^{ig}$, the residual internal energy as $U^R = U - U^{ig}$, and the Gibbs free energy as $G^R = G - G^{ig}$.

Residual properties are computed by integrating a function which depends on an equation of state and its derivative.

6.6   EVALUATION OF RESIDUAL PROPERTIES FROM EQUATION OF STATE

This involves determination of the residual internal energy, enthalpy, entropy, and Gibbs function from equation of state.

6.6.1   Residual Internal Energy Relation from Equation of State

Let us calculate the change in internal energy of a real gas from an initial state of T_1, V_1 to the final state T_2, V_2. Since the internal energy is a state function, therefore change in internal energy is independent of the path. In order to estimate ΔU, we can consider a pathway comprising the following three steps:

1. Calculate ΔU at constant temperature T_1, describing how the gas behaves ideally while the volume changes from V_1 to ∞, i.e., initial state $T_1, V_1 \rightarrow$ final state T_1, ∞.
2. Calculate ΔU for ideal gas while temperature changes from T_1 to T_2 at constant volume, i.e., initial state $T_1, \infty \rightarrow$ final state T_2, ∞.
3. Temperature remains unchanged at T_2, volume changed to V_2, i.e., initial state $T_1, \infty \rightarrow$ final state T_2, ∞.

Now, for the evaluation of change in internal energy, we can consider $U = U(T, V)$. Then the total differential of U is given by

$$\Delta U = \left(\frac{\partial U}{\partial T}\right)_V dT + \left(\frac{\partial U}{\partial V}\right)_T dV \tag{6.123}$$

Integrating Eq. (6.123) with limits V_1, T_1 to V_2, T_2, we get

$$U_2 - U_1 = \int_{T_1}^{T_2} \left(\frac{\partial U}{\partial T}\right)_V dT + \int_{V_1}^{V_2} \left(\frac{\partial U}{\partial V}\right)_T dV \tag{6.124}$$

We know

$$dU = TdS - PdV$$

or

$$\left(\frac{\partial U}{\partial V}\right)_T = T\left(\frac{\partial S}{\partial V}\right)_T - P = T\left(\frac{\partial P}{\partial T}\right)_V - P \quad \text{(using Maxwell's relation)}$$

Then Eq. (6.123) becomes

$$\Delta U = \int_{T_1}^{T_2} C_V dT + \int_{V_1}^{V_2} \left[T\left(\frac{\partial P}{\partial T}\right)_V - P \right] dV \tag{6.125}$$

The total internal energy change from T_1, V_1 to T_2, V_2 is given by

$$\Delta U = \Delta U_1 + \Delta U_2 + \Delta U_3$$

which is the summation of internal energies in steps 1, 2 and 3.

Now

$$\Delta U_1 = \int_{T_1, V_1}^{T_1, \infty} C_V dT + \int_{T_1, V_1}^{\infty} \left[T\left(\frac{\partial P}{\partial T}\right)_V - P \right] dV$$

$$\Delta U_2 = \int_{T_1, \infty}^{T_2, \infty} C_V dT + \int_{T_1, \infty}^{T_2, \infty} \left[T\left(\frac{\partial P}{\partial T}\right)_V - P \right] dV = \int_{T_1}^{T_2} C_V^0 dT$$

$$\Delta U_3 = \int_{T_2, \infty}^{T_2, V_2} C_V dT + \int_{T_2, \infty}^{T_2, V_2} \left[T\left(\frac{\partial P}{\partial T}\right)_V - P \right] dV$$

On substitution, we get

$$\Delta U = \int_{T_1, V_1}^{T_1, \infty} C_V dT + \int_{T_1, V_1}^{\infty} \left[T \left(\frac{\partial P}{\partial T} \right)_V - P \right] dV + \int_{T_1}^{T_2} C_V dT + \int_{T_2, \infty}^{T_2, V_2} C_V dT + \int_{T_2, \infty}^{T_2, V_2} \left[T \left(\frac{\partial P}{\partial T} \right)_V - P \right] dV \quad (6.126)$$

At constant temperature

$$\int_{T_1, V_1}^{T_1, \infty} C_V dT = \int_{T_2, \infty}^{T_2, V_2} C_V dT = 0$$

Hence, Eq. (6.126) becomes

$$\Delta U = \int_{T_1}^{T_2} C_V^0 dT + \int_{T_2, \infty}^{T_2, V_2} \left[T \left(\frac{\partial P}{\partial T} \right)_V - P \right] dV - \int_{T_1, \infty}^{T_1, V_1} \left[T \left(\frac{\partial P}{\partial T} \right)_V - P \right] dV \quad (6.127)$$

For ideal gas

$$\int_{T_1}^{T_2} C_V^0 dT = 0$$

Then the residual internal energy is given by

$$U^R = \int_{\infty}^{V} \left[T \left(\frac{\partial P}{\partial T} \right)_V - P \right] dV \quad (6.128)$$

6.6.2 Residual Enthalpy Relation from Equation of State

Like internal energy, change in enthalpy of a real gas can also be estimated when the gas changes in enthalpy takes place from an initial state T_1, P_1 to the final state T_2, P_2. Since enthalpy is a state function, therefore change in enthalpy does not depend on the path. In order to calculate ΔH, we can take a convenient path by which the real gas reaches a final state.

Now for the determination of enthalpy change, we can consider $H = H(T, P)$. Then the total differential of H is given by

$$\Delta H = \left(\frac{\partial H}{\partial T} \right)_P dT + \left(\frac{\partial H}{\partial P} \right)_T dP \quad (6.129)$$

Using this equation, we get the generalized enthalpy change equation

$$\Delta H = \int_{T_1}^{T_2} C_P dT + \int_{P_1}^{P_2} \left[V - T \left(\frac{\partial V}{\partial T} \right)_P \right] dP \quad (6.130)$$

Integrating Eq. (6.129) within the limits P_1, T_1 and P_2, T_2 by two different paths, we get

$$H_2 - H_1 = \int_{T_1}^{T_2} \left(\frac{\partial H}{\partial T} \right)_{P_1} dT + \int_{P_1}^{P_2} \left(\frac{\partial H}{\partial P} \right)_{T_2} dP \quad (6.131)$$

$$H_2 - H_1 = \int_{P_1}^{P_2} \left(\frac{\partial H}{\partial P} \right)_{T_1} dP + \int_{T_1}^{T_2} \left(\frac{\partial H}{\partial T} \right)_{P_2} dT \quad (6.132)$$

For the consideration of two paths, one can face a genuine problem for the term $\left(\dfrac{\partial H}{\partial T}\right)_P$, i.e., heat capacity at constant pressure (C_P). To avoid such a problem of C_P calculation, we devise an equivalent pathway of three steps to estimate the change in enthalpy from T_1, P_1 to T_2, P_2. These are:

Step 1: Initial state $T_1, P_1 \rightarrow$ Final state $T_1, 0$;
Step 2: Initial state $T_1, 0 \rightarrow$ Final state $T_2, 0$;
Step 3: Initial state $T_2, 0 \rightarrow$ Final state T_2, P_2.

Considering the enthalpy change equation, we get

$$\Delta H = \int\limits_{T_1,P_1}^{T_1,0} C_P dT + \int\limits_{T_1,P_1}^{T_1,0}\left[V - T\left(\frac{\partial V}{\partial T}\right)_P\right]dP + \int\limits_{T_1}^{T_2} C_P^0 dT + \int\limits_{T_2,0}^{T_2,P_2} C_P dT + \int\limits_{T_2,0}^{T_2,P_2}\left[V - T\left(\frac{\partial V}{\partial T}\right)_P\right]dP$$

At constant temperature

$$\int\limits_{T_1,P_1}^{T_1,0} C_P dT = \int\limits_{T_2,0}^{T_2,P_2} C_P dT = 0 \tag{6.133}$$

Therefore, Eq. (6.133) reduces to

$$\Delta H = \int\limits_{T_1}^{T_2} C_P^0 dT + \int\limits_{T_2,0}^{T_2,P_2}\left[V - T\left(\frac{\partial V}{\partial T}\right)_P\right] - \int\limits_{T_1,0}^{T_1,P_1}\left[V - T\left(\frac{\partial V}{\partial T}\right)_P\right]dP \tag{6.134}$$

$$\Delta H = \int\limits_{T_1}^{T_2} C_P^0 dT + H_2^R - H_1^R$$

For ideal gas

$$\int\limits_{T_1}^{T_2} C_P^0 dT = 0$$

Then the residual enthalpy is given by

$$H^R = \int\limits_0^P \left[V - T\left(\frac{\partial V}{\partial T}\right)_P\right]dP \tag{6.135}$$

Again the residual enthalpy can be calculated when enthalpy is considered to be the function of T and V. From the definition of enthalpy, we have

$$H = U + PV$$

Accordingly

$$H^R = U^R + (PV)^R = U^R + [PV - (PV)^{ig}] = U^R + ZRT - RT = U^R + RT(Z - 1) \tag{6.136}$$

We obtained from Eq. (6.128)

$$U^R = \int\limits_\infty^V \left[T\left(\frac{\partial P}{\partial T}\right)_V - P\right]dV$$

Substituting the value of U^R into Eq. (6.136), we get

$$H^R = \int_0^P \left[V - T\left(\frac{\partial V}{\partial T}\right)_P \right] dP + RT(Z - 1) \tag{6.137}$$

6.6.3 Residual Entropy Relation from Equation of State

We can estimate the change in entropy in the same fashion as we did in case of internal energy and enthalpy. Let us consider the real gas change from initial state T_1, P_1 to the final state T_2, P_2.

Now for the determination of entropy change, we can consider

$$\Delta S = C_P \frac{dT}{T} - \left(\frac{\partial V}{\partial T}\right)_P dP = C_P \frac{dT}{T} - R\frac{dP}{P}$$

For an ideal gas, it can be expressed as

$$S_2^0 - S_1^0 = C_P^0 \int_{T_1}^{T_2} \frac{dT}{T} - R \ln \frac{P_2}{P_1} \tag{6.138}$$

Here, we will evaluate the change in entropy in a little different way as

$$S_2 - S_1 = - \int_{P_1}^{P=0} \left(\frac{\partial V}{\partial T}\right)_{P,T_1} dP + \int_{T_1}^{T_2} \left(\frac{C_P}{T}\right)_{P=0} dT - \int_{P=0}^{P_2} \left(\frac{\partial V}{\partial T}\right)_{P,T_2} dP \tag{6.139}$$

For an ideal gas, substituting the relation (6.138), we get

$$(S_2 - S_2^0) - (S_1 - S_1^0) = - \int_{P=0}^{P_2} \left[\left(\frac{\partial V}{\partial T}\right)_{P_1} - \frac{R}{P}\right]_{T_2} dP + \int_{P=0}^{P_2} \left[\left(\frac{\partial V}{\partial T}\right)_P - \frac{R}{P}\right]_{T_1} dP \tag{6.140}$$

Equation (6.140) can also be written as

$$(S - S^0)_{T,P} = \int_{P=0}^{P} \left[\left(\frac{\partial S}{\partial P}\right)_T + \frac{R}{P}\right] dP = \int_{P=0}^{P} \left[\frac{R}{P} - \left(\frac{\partial V}{\partial T}\right)_P\right] dP \tag{6.141}$$

where

   S = Entropy of a real gas at given temperature and pressure
   S^0 = Entropy of a gas behaving ideally at the same temperature and pressure.

Then the residual entropy is given by

$$S^R = \int_{P=0}^{P} \left[\frac{R}{P} - \left(\frac{\partial V}{\partial T}\right)_P\right] dP \tag{6.142}$$

Now, suppose it is required to estimate the change in entropy when it changes from an initial state T_1, V_1, to the final state T_2, V_2. In that case, Eq. (M) can be expressed as

$$(S - S^0)_{T,P} = - \int_{P=0}^{P} \left[\left(\frac{\partial V}{\partial T}\right) - \frac{R}{P}\right]_T dP = \int_{V=\infty}^{V} \left(\frac{\partial P}{\partial T}\right)_V dV + \int_{P=0}^{P} R\frac{dP}{P} \tag{6.143}$$

If the pressure is expressed in terms of molar volume, then

$$(S - S_0)_{T,P} = \int\limits_{V=\infty}^{V} \left(\frac{\partial P}{\partial T}\right)_V dV + \int \left[R\frac{d(PV)}{PV} - R\frac{dV}{V} \right]$$

$$= \int\limits_{V=\infty}^{V} \left(\frac{\partial P}{\partial T}\right)_V dV + R\int\limits_{PV=RT}^{PV} d\ln(PV) - \int\limits_{V=\infty}^{V} R\frac{dV}{V}$$

$$= R\ln\left(\frac{PV}{RT}\right) + \int\limits_{V=\infty}^{V} \left[\left(\frac{\partial P}{\partial T}\right)_V - \frac{R}{V}\right] dV = R\ln Z + \int\limits_{V=\infty}^{V} \left[\left(\frac{\partial P}{\partial T}\right)_V - \frac{R}{V}\right] dV \quad (6.144)$$

6.6.4 Residual Gibbs Function Relation from Equation of State

The fundamental property relation for a gas or vapour of constant composition may be written as (see Eq. (6.28))

$$dG = VdP - SdT \qquad \text{when } G = G\,(T,\,P)$$

At constant temperature, the preceding equation reduces to

$$dG = VdP \tag{6.145}$$

If Eq. (6.145) is applied to an ideal gas at isothermal condition, then

$$dG^{ig} = V^{ig}dP \tag{6.146}$$

Subtracting Eq. (6.145) from Eq. (6.146), we get

$$dG - dG^{ig} = (V - V^{ig})\,dP$$

or

$$dG^{R} = V^{R}dP \tag{6.147}$$

Dividing both sides of Eq. (6.147) by RT and integrating from zero pressure to arbitrary pressure P gives

$$\int\limits_{0}^{P} \frac{dG^{R}}{RT} = \int \frac{V^{R}}{RT}\,dP \tag{6.148}$$

or

$$\left(\frac{G^{R}}{RT}\right)_{P=P} - \left(\frac{G^{R}}{RT}\right)_{P=0} = \int\limits_{0}^{P} \frac{Z-1}{P}\,dP \tag{6.149}$$

where residual volume $V^{R} = V - V^{ig} = \dfrac{ZRT}{P} - \dfrac{RT}{P} = (Z-1)\dfrac{RT}{P}$.

At zero pressure, the homogeneous fluid behaves like an ideal gas and thus the residual property will be zero, i.e., in Eq. (6.149), $\left(\dfrac{G^{R}}{RT}\right) = 0$. Hence, Eq. (6.149) becomes

$$\frac{G^R}{RT} = \int_0^P (Z - 1)\frac{dP}{P} \tag{6.150}$$

Now, consider Gibbs free energy is a function of T and V, i.e., $G = G(T, V)$, then the residual Gibbs free energy can be estimated in the following way:

We know that for a real gas

$$PV = ZRT \tag{6.151}$$

On differentiation, the Eq. (6.151) yields at constant temperature

$$PdV + VdP = RTdZ$$

Dividing both sides by RTZ, we get

$$\frac{dP}{P} = \frac{dZ}{Z} - \frac{dV}{V} \tag{6.152}$$

Substituting Eq. (6.152) into (6.150), we get

$$\frac{G^R}{RT} = \int (Z - 1)\left(\frac{dZ}{Z} - \frac{dV}{V}\right) \quad \text{at constant temperature}$$

$$\frac{G^R}{RT} = \int_1^Z \frac{Z-1}{Z}dZ - \int_\infty^V \frac{(Z-1)}{V}dV \tag{6.153}$$

Here the pressure limit $P = 0$ to $P = P$ corresponds to the compressibility factor as $Z = 1$ to $Z = Z$ and consequently the volume limit $V = \infty$ to $V = 0$. Then Eq. (6.153) becomes

$$\frac{G^R}{RT} = Z - 1 - \ln Z - \int_\infty^V \frac{Z-1}{V}dV \tag{6.154}$$

For the second term of Eq. (6.153), put $Z = \frac{PV}{RT}$. So

$$\frac{G^R}{RT} = Z - 1 - \ln Z - \int_\infty^V \frac{PV - RT}{RTV}dV$$

$$\frac{G^R}{RT} = Z - 1 - \ln Z + \frac{1}{RT}\int_\infty^V \frac{RT - PV}{V}dV$$

$$\frac{G^R}{RT} = Z - 1 - \ln Z + \frac{1}{RT}\int_\infty^V \left(\frac{RT}{V} - P\right)dV \tag{6.155}$$

6.7 RESIDUAL PROPERTIES FROM VIRIAL EQUATION OF STATE

The residual properties of pure substances can be computed with the help of the virial equation of state. From the knowledge of the compressibility factor and the virial equation of state, we obtain

$$Z = \frac{PV}{RT} = 1 + \frac{BP}{RT}$$

or

$$Z - 1 = \frac{BP}{RT} \qquad (6.156)$$

We know that for a real gas at low to moderate pressure

$$V = \frac{ZRT}{P} = \frac{RT}{P}\left(1 + \frac{BP}{RT}\right) = B + \frac{RT}{P} \qquad (6.157)$$

Differentiating it with respect to T at constant P, we get

$$\left(\frac{\partial V}{\partial T}\right)_P = \frac{\partial B}{\partial T} + \frac{R}{P} \qquad (6.158)$$

Now, we know from the virial equation of state that the second virial coefficient can be represented as

$$\left(\frac{BP_C}{RT_C}\right) = B^0 + \omega B^1$$

or

$$B = \frac{RT_C}{P_C}(B^0 + \omega B^1) \qquad (6.159)$$

Differentiating Eq. (6.159) with respect to T, we obtain

$$\frac{dB}{dT} = \frac{RT_C}{P_C}\left(\frac{dB^0}{dT} + \omega\frac{dB^1}{dT}\right) \qquad (6.160)$$

where

$$B^0 = 0.083 - \frac{0.422}{T_r^{1.6}} \Rightarrow \frac{dB^0}{dT_r} = \frac{0.675}{T_r^{2.6}}$$

and

$$B^1 = 0.139 - \frac{0.172}{T^{4.2}} \Rightarrow \frac{dB^1}{dT_r} = \frac{0.722}{T_r^{5.2}}$$

From Eqs. (6.135) and (6.142), we have

$$H^R = \int_0^P \left[V - T\left(\frac{\partial V}{\partial T}\right)_P\right]dP$$

$$S^R = \int_{P=0}^P \left[\frac{R}{P} - \left(\frac{\partial V}{\partial T}\right)_P\right]dP$$

Substituting Eqs. (6.157) and (6.158) into Eqs. (6.135) and (6.142), we get

$$H^R = \left(BP - PT\frac{dB}{dT}\right) \qquad (6.161)$$

$$S^R = -P\frac{dB}{dT} \qquad (6.162)$$

Substituting all the possible values, we get

$$\frac{H^R}{RT_C} = P_r\left[\left(0.083 - \frac{1.097}{T_r^{1.6}}\right) + \omega\left(0.139 - \frac{0.894}{T_r^{4.2}}\right)\right] \tag{6.163}$$

$$\frac{S^R}{R} = -P_r\left[\frac{0.675}{T_r^{2.6}} + \omega\left(\frac{0.722}{T_r^{5.2}}\right)\right] \tag{6.164}$$

It is necessary to remember that residual properties estimated on the basis of the virial equation of state are valid only in the low pressure region.

EXAMPLE 6.22 Using the virial equation of state, estimate the residual enthalpy and entropy for propane at 60°C and 2.5 bar, given that $T_C = 370$ K, $P_C = 42.57$ bar, and $\omega = 0.153$.

Solution: From the given data, we can obtain
Reduced temperature

$$T_r = \frac{333.15}{370} = 0.9$$

Reduced pressure

$$P_r = \frac{2.5}{42.57} = 5.873 \times 10^{-2}$$

From Eq. (6.163), we get

$$\frac{H^R}{RT_C} = P_r\left[\left(0.083 - \frac{1.097}{T_r^{1.6}}\right) + \omega\left(0.139 - \frac{0.894}{T_r^{4.2}}\right)\right]$$

or

$$\frac{H^R}{RT_C} = 5.873 \times 10^{-2}\left[\left(0.083 - \frac{1.097}{(0.9)^{1.6}}\right) + \omega\left(0.139 - \frac{0.894}{(0.9)^{4.2}}\right)\right]$$

or

$$H^R = RT_C \times (-8.263 \times 10^{-2}) = 8.314 \times 370 = -0.254 \text{ kJ/mol}$$

Hence, the residual enthalpy for propane is -0.254 J/mol.

Now, to calculate the residual entropy, we apply Eq. (6.164)

$$\frac{S^R}{R} = -P_r\left[\frac{0.675}{T_r^{2.6}} + \omega\left(\frac{0.722}{T_r^{5.2}}\right)\right]$$

or

$$\frac{S^R}{R} = -5.873 \times 10^{-2}\left(\frac{0.675}{(0.9)^{2.6}} + \frac{0.153 \times 0.722}{(0.9)^{5.2}}\right)$$

or

$$S^R = -6.335 \times 10^{-2} \times 8.314 = -0.527 \text{ J/mol-K}$$

Therefore, the residual entropy for propane is estimated as -0.527 J/mol-K.

6.8 RESIDUAL PROPERTIES FROM CUBIC EQUATION OF STATE

6.8.1 Residual Internal Energy Relation from Cubic Equation of State

We consider the van der Waals equation as a cubic equation of state in order to estimate the residual internal energy of a real gas. It is given by

$$P = \frac{RT}{V - b} - \frac{a}{V^2} \tag{6.165}$$

Differentiating it with respect to T at constant V, we get

$$\left(\frac{\partial P}{\partial T} \right)_V = \frac{R}{V - b} \tag{6.166}$$

Thus

$$T \left(\frac{\partial P}{\partial T} \right)_V - P = \frac{RT}{V - b} - P = \frac{RT}{V - b} - \left(\frac{RT}{V - b} - \frac{a}{V^2} \right) = \frac{a}{V^2} \tag{6.167}$$

We obtained earlier that the residual internal energy relation is

$$U^R = \int\limits_{\infty}^{V} \left[T \left(\frac{\partial P}{\partial T} \right)_V - P \right] dV \tag{6.168}$$

Substituting Eq. (6.167) into (6.168), we have

$$U^R = \int\limits_{\infty}^{V} \frac{a}{V^2} \, dV = -\frac{a}{V} \tag{6.169}$$

6.8.2 Residual Entropy Relation from Cubic Equation of State

The van der Waals equation of state is given by

$$P = \frac{RT}{V - b} - \frac{a}{V^2}$$

The residual entropy is obtained as (see Eq. (6.144))

$$S^R = R \ln Z + \int\limits_{V=\infty}^{V} \left[\left(\frac{\partial P}{\partial T} \right)_V - \frac{R}{V} \right] dV$$

Accordingly, we get from the van der Waals equation of state that

$$\left(\frac{\partial P}{\partial T} \right)_V = \frac{R}{V - b}$$

and

$$\left(\frac{\partial P}{\partial T} \right)_V - \frac{R}{V} = \frac{R}{V - b} - \frac{R}{V} \tag{6.170}$$

Substitution of Eq. (6.170) into (6.144) yields

$$S^R = R \ln Z + \int\limits_{V=\infty}^{V} \left[\left(\frac{\partial P}{\partial T} \right)_V - \frac{R}{V} \right] dV = R \ln Z + \int\limits_{V=\infty}^{V} \left(\frac{R}{V - b} - \frac{R}{V} \right) dV \tag{6.171}$$

On integration with integral limits $V = \infty$ and $V = V$, we have

$$S^R = R \ln Z + R \ln \frac{V - b}{V} \qquad (6.172)$$

Putting $Z = \dfrac{PV}{RT}$ into Eq. (6.172), we get

$$S^R = R \ln \frac{P(V - b)}{RT} \qquad (6.173)$$

6.8.3 Residual Enthalpy Relation from Cubic Equation of State

The residual enthalpy relation can also be established from the cubic equation of state. We got the relation for residual enthalpy (see Eq. (6.136)) that can further be represented as

$$H^R = U^R + [RT(Z - 1)]$$

We obtained the residual internal energy relation from the van der Waals equation (see Eq. (6.169)) that

$$U^R = -\frac{a}{V}$$

Therefore, substitution of Eq. (6.169) into (6.136) yields

$$H^R = [RT(Z - 1)] - \frac{a}{V} = \left[RT\left(\frac{PV}{RT} - 1 \right) \right] - \frac{a}{V} = PV - RT - \frac{a}{V} \qquad (6.174)$$

We know that the van der Waals equation of state is given by

$$P = \frac{RT}{V - b} - \frac{a}{V^2}$$

Multiplying both sides of the equation by $\dfrac{V}{RT}$, we get

$$\frac{PV}{RT} = \frac{V}{V - b} - \frac{a}{RTV}$$

$$\Rightarrow \qquad Z = \frac{V}{V - b} - \frac{a}{RTV}$$

The expression can easily be developed for $Z - 1$ as

$$Z - 1 = \frac{b}{V - b} - \frac{a}{RTV} \qquad (6.175)$$

Substitution of $Z - 1$ into Eq. (6.174) yields

$$H^R = \frac{bRT}{V - b} - \frac{2a}{V} \qquad (6.176)$$

This is the expression of residual enthalpy from the cubic equation of state.

EXAMPLE 6.23 Estimate the residual entropy, enthalpy and internal energy at 298 K and 10 bar for nitrogen obeying the van der Waals equation of state, given that $T_C = 126.2$ K and $P_C = 34.0$ bar.

Solution: From the given data, we can determine the van der Waals constants a and b by using the following relation:

$$a = \frac{27R^2 T_C^2}{64 P_C} = 1.39 \text{ Pa-m}^6/\text{mol}^2$$

$$b = \frac{RT_C}{8 P_C} = 0.039 \times 10^{-3} \text{ m}^3/\text{mol}$$

The residual entropy can be estimated with the help of Eq. (6.173) as

$$S^R = R \ln \frac{P(V-b)}{RT}$$

$$S^R = 8.314 \ln \left[\frac{10 \times 10^5 (2.425 \times 10^{-3} - 0.039 \times 10^{-3})}{8.314 \times 298} \right] = -0.3134 \text{ J/mol–K}$$

Similarly, the residual enthalpy can be estimated by Eq. (6.174) as

$$H^R = PV - RT - \frac{a}{V}$$

$$H^R = \left[\{(10 \times 10^5)(2.425 \times 10^{-3})\} - 8.314 \times 298 - \frac{1.39}{2.425 \times 10^{-3}} \right]$$

$$= -625.765 \text{ J/mol}$$

The residual internal energy would be

$$U^R = -\frac{a}{V} = -\frac{1.39}{2.425 \times 10^{-3}} = -573.195 \text{ J/mol}$$

6.9   FUGACITY AND FUGACITY COEFFICIENT

As the compressibility factor measures the deviation from ideal gas behaviour in terms of P, V and T, fugacity measures the deviation from ideal gas behaviour in terms of chemical potential. The concept of fugacity, a new device to express the chemical potential (μ) or the molar Gibbs free energy (G) of a real gas, was introduced by the American chemist Gilbert N. Lewis in 1901. The term *fugacity* is literally defined as the *e* tendency to escape or flee' from one state to another. Lower the fugacity of a phase, lower is the tendency to the state. Fugacity is another way of representing the behaviour of real gases.

Fugacity is an intensive property. It is a sort of idealized pressure. It has the same dimension as pressure, usually atm or bar. It is preferably used in the study of phase. It is also useful for multi-component equilibrium involving any combination of solid, liquid and gas equilibria. It is a helpful engineering tool to predict the final phase and reaction state of multi-component mixtures at various temperatures and pressures without performing an experiment of a single substance. When a system approaches ideal gas state at very low pressure, the chemical potential approaches negative infinity, which is very much undesirable for the purpose of mathematical modelling. Under these conditions, fugacity approaches zero.

For a pure fluid, the change in the Gibbs free energy is given by

$$dG = VdP - SdT$$

For an isothermal change, it becomes

$$dG = VdP$$

For one mole of an ideal gas undergoing the same change in temperature and pressure

$$dG^{ig} = V^{ig}dP = \frac{RT}{P}dP = RTd \ln P \qquad (6.177)$$

For a real gas at constant temperature, the change in the Gibbs free energy is given by

$$dG = RTd \ln f \qquad (6.178)$$

where f stands for fugacity.

Integrating Eq. (6.178), we get

$$G_2 - G_1 = RT \ln \frac{f_2}{f_1} \qquad (6.179)$$

Subtracting Eq. (6.178) from (6.177), we have

$$d(G - G^{ig}) = RTd \ln f - RTd \ln P = RTd \ln \frac{f}{P} \qquad (6.180)$$

By definition, $G - G^{ig} = G^R$ is the residual function.

For an ideal gas, fugacity is exactly equal to pressure, i.e., $\frac{f}{P} = 1$. But for a non-ideal gas,
fugacity will not be equal to pressure. It will be a similar property. Again, at low pressure, the
real gases behave ideally. Then fugacity may be replaced by pressure. Hence, fugacity can be
mathematically expressed as

$$\lim_{P \to 0} \frac{f}{P} = 1 \quad \text{or} \quad \frac{f}{P} = 1 \quad \text{as } P \to 0$$

The ratio of fugacity to pressure is known as the *fugacity coefficient*. It is denoted by ϕ.
It is dimensionless and is independent of the nature of the gas. It can be expressed as $\frac{f}{P} = \phi$.

Therefore, Eq. (6.180) becomes

$$dG^R = RTd \ln \phi \qquad (6.181)$$

Integration of Eq. (6.181) gives

$$G - G^{ig} = G^R = RT \ln \phi + C \qquad (6.182)$$

where C is the integration constant, which is a function of temperature.

For an ideal gas, since $G^R = 0$ and $\phi = 1$, we get $C = 0$. Thus Eq. (6.182) would be

$$G^R = RT \ln \phi = RT \ln \frac{f}{P} \qquad (6.183a)$$

or

$$\frac{G^R}{RT} = \ln \phi = \sum x_i \ln \bar{\phi} \qquad (6.183b)$$

6.9.1 Effect of Temperature and Pressure on Fugacity

From Eq. (6.183a and b), we get

$$G^R = RT \ln \phi = RT \ln \frac{f}{P}$$

This equation can also be expressed as

$$G - G^0 = RT \ln \phi = RT \ln \frac{f}{f^0} \qquad (6.184)$$

where G^0 and f^0 are respectively the molar free energy and the fugacity of a pure substance. At very low pressure, when the gas is assumed to behave as an ideal one, then

$$\frac{G}{T} - \frac{G^0}{T} = RT \ln \frac{f}{f^0}$$

Differentiating it with respect to temperature at constant pressure, we get

$$\left[\frac{\partial}{\partial T} \left(\frac{G}{T} \right) \right]_P - \left[\frac{\partial}{\partial T} \left(\frac{G^0}{T} \right) \right]_P = R \left[\left(\frac{\partial \ln f}{\partial T} \right)_P - \left(\frac{\partial \ln f^0}{\partial T} \right)_P \right] \qquad (6.185)$$

Since $f^0 = P$, i.e., fugacity is equal to pressure for an ideal gas, then $\left(\dfrac{\partial \ln f^0}{\partial T} \right)_P = 0$. Then Eq. (6.185) reduces to

$$\left[\frac{\partial}{\partial T} \left(\frac{G}{T} \right) \right]_P - \left[\frac{\partial}{\partial T} \left(\frac{G^0}{T} \right) \right]_P = R \left(\frac{\partial \ln f}{\partial T} \right)_P \qquad (6.186)$$

We know that the Gibbs function $G = H - TS$. From Eq. (6.66), we obtain

$$\left[\frac{\partial}{\partial T} \left(\frac{G}{T} \right) \right]_P = -\frac{H}{T^2}$$

Therefore, its substitution into Eq. (6.186) yields

$$\left(\frac{\partial \ln f}{\partial T} \right)_P = \frac{H^0 - H}{RT^2} \qquad (6.187)$$

This equation implies the influence of temperature on fugacity at constant pressure. Now, we can derive another expression which indicates the effect of pressure on fugacity at constant temperature.

For a pure substance, the change in the Gibbs free energy is given by

$$dG = VdP - SdT$$

For an isothermal change of the process, it becomes

$$dG = VdP \qquad \text{(at constant temperature)}$$

Again, we know that

$$dG = VdP = RTd \ln f$$

or

$$\left(\frac{\partial \ln f}{\partial P} \right)_T = \frac{V}{RT}$$

6.9.2 Fugacity Coefficient from Compressibility Factor

The compressibility factor can be used to predict the fugacity coefficient at low pressure. For a real gas, it is defined as

$$Z = \frac{PV}{RT} = 1 + \frac{BP}{RT}$$

or

$$Z - 1 = \frac{BP}{RT} \tag{6.188}$$

We know that for a real gas at low to moderate pressure

$$V = \frac{ZRT}{P} = \frac{RT}{P}\left(1 + \frac{BP}{RT}\right) = B + \frac{RT}{P} \tag{6.189}$$

We know

$$dG = VdP = RTd \ln f \tag{6.190}$$

For an ideal gas, it becomes

$$dG^0 = VdP = \frac{RT}{P}dP = RTd \ln P \tag{6.191}$$

Subtracting Eq. (6.191) from Eq. (6.190), we get

$$dG - dG^0 = RTd \ln \frac{f}{P} = V - \frac{RT}{P} \tag{6.192a}$$

Since for an ideal gas, fugacity is exactly equal to pressure, i.e., $\frac{f}{P} = 1$, as $P \rightarrow 0$, then

$$RT \ln \frac{f}{P} = \int_0^P \left(V - \frac{RT}{P}\right) dP$$

or

$$\ln \frac{f}{P} = \int_0^P \frac{Z-1}{P} dP \quad \left(\text{as } Z = \frac{PV}{RT}\right) \tag{6.192b}$$

The value of Z with respect to pressure can be obtained from the compressibility charts. Therefore, the fugacity coefficient f can easily be computed.

6.9.3 Fugacity Coefficient from Virial Equation of State

The fugacity coefficient can be estimated from the virial coefficient of correlation. It is well known to us that at low pressure the compressibility factor is given by

$$Z = \frac{PV}{RT} = 1 + \frac{BP}{RT}$$

It can also be expressed as

$$Z = 1 + \frac{BP}{RT} = 1 + \left(\frac{BP_C}{RT_C}\right)\frac{P_r}{T_r}$$

or

$$Z - 1 = \left(\frac{BP_C}{RT_C}\right)\frac{P_r}{T_r} \tag{6.193}$$

Now, we know from the virial equation of state that the second virial coefficient B can be represented as

$$\frac{BP_C}{RT_C} = B^0 + \omega B^1$$

Hence, Eq. (6.193) can be expressed as

$$Z - 1 = \left(\frac{BP_C}{RT_C}\right)\frac{P_r}{T_r} = (B^0 + \omega B^1)\left(\frac{P_r}{T_r}\right) \tag{6.194}$$

Again we know that

$$\ln \phi = \ln\frac{f}{P} = \int_0^P (Z - 1)\frac{dP}{P} \tag{6.195}$$

wherefrom we can write

$$\ln \phi = \ln\frac{f}{P} = \int_0^P (Z - 1)\frac{dP}{P} = \int_0^P \left(\frac{BP}{RT}\right)\frac{dP}{P} = \frac{BP}{RT} \tag{6.196}$$

or

$$f = Pe^{\frac{BP}{RT}} \tag{6.197}$$

On integrating between the limits $P = P_r$ and $P = 0$, Eq. (6.195) becomes

$$\ln \phi = \ln\frac{f}{P} = \int_0^{P_r} (Z - 1)\frac{dP_r}{P_r} = \int_0^{P_r} (B^0 + \omega B^1)\frac{dP_r}{T_r} = (B^0 + \omega B^1)\frac{P_r}{T_r} \tag{6.198}$$

where $B^0 = 0.083 - \dfrac{0.422}{T_r^{1.6}}$ and $B^1 = 0.139 - \dfrac{0.172}{T_r^{4.2}}$.

EXAMPLE 6.24 Estimate the fugacity of iso-butane at 15 atm and 87°C using the compressibility factor correlation $Z = 1 + \dfrac{BP}{RT}$, given that the second virial coefficient, $B = -4.28 \times 10^{-4}\,\text{m}^3/\text{mol}$.

Solution: The given compressibility factor correlation $Z = 1 + \dfrac{BP}{RT}$ can be rearranged as

$$Z - 1 = \frac{BP}{RT}$$

or

$$\frac{Z-1}{P} = \frac{B}{RT}$$

Now from Eq. (6.192b), we obtain

$$\ln \frac{f}{P} = \int_0^P \frac{Z-1}{P} \, dP$$

On substitution of the value $Z - 1$, the preceding equation yields

$$\ln \frac{f}{P} = \int_0^P \frac{Z-1}{P} \, dP = \int_0^P \frac{B}{RT} \, dP = \frac{B}{RT} (P - 0) = \frac{BP}{RT}$$

Substituting the values, we get

$$\ln \frac{f}{P} = \frac{BP}{RT} = \frac{-4.28 \times 10^{-4} \times 15 \times 10^5}{8.314 \times 360} = -0.214$$

or

$$\frac{f}{P} = 0.807$$

$$f = 15 \times 0.807 = 12.10 \text{ atm}$$

Therefore, the fugacity of iso-butane is estimated as 12.10 atm.

6.9.4 Fugacity Coefficient by Equation of State

At constant temperature, the fundamental property relation is given by

$$dG = VdP = RTd \ln f$$

Integrating within the integral limits $P = P$, $P = P^0$ and $f = f, f = f^0$, we have

$$\ln \frac{f}{f^0} = \frac{1}{RT} \int_{P^0}^P VdP \tag{6.199}$$

The equation of state is available and of greater importance in the form of $V = V(T, P)$, also called the *pressure-explicit form*, in which the integral on the right hand side can be evaluated. But it is necessary to introduce the *volume-explicit form*, i.e., $P = P(T, V)$ to estimate the fugacity coefficient, because the volume-explicit form can be measured directly. Some equations like the van der Waals equation can be easily solved by this technique. The only requirement is that the integral has to be determined with the help of integration by parts.

$$\int VdP = PV - \int PdV \tag{6.200}$$

$$\int_{P^0}^P VdP = PV - P^0V^0 - \int_{V^0}^V PdV \tag{6.201}$$

Substituting Eq. (6.201) into Eq. (6.199), we get

$$\ln \frac{f}{f^0} = \frac{1}{RT} (PV - P^0V^0) - \frac{1}{RT} \int_{V^0}^V PdV \tag{6.202}$$

$$\ln \frac{f}{f^0} = (Z - 1) - \frac{1}{RT} \int_{V^0}^{V} P dV \quad \left(\text{as } Z = \frac{PV}{RT} \text{ and } P^0 V^0 = RT \text{ for ideal gas}\right) \quad (6.203)$$

This equation can be applied to calculate the fugacity coefficient of a real gas from the equation of state.

SUMMARY

Before understanding the essence of thermodynamic property relations, it is necessary to define and classify the thermodynamic properties. The mathematical preliminaries have been discussed to enable us to establish the thermodynamic relation and prove the desired expressions. The determination of immeasurable thermodynamic properties by measurable ones such as temperature, pressure, volume, etc. has been discussed. The fundamental property relations have been derived on the basis of the first law and the second law of thermodynamics, and the Gibbs and Helmholtz free energy functions. With their help, some important relations such as Maxwell's relation and Gibbs–Helmholtz relations have been deduced. For the study of phase change in a system such as freezing and vaporization, Clausius-Clapeyron equation—a correlation between temperature and pressure—is quite useful. The Joule–Thomson coefficient plays an important role in explaining the possibility of attainment of cooling or heating effect by a compressed gas passing through the porous plug. The significance of inversion temperature has been substantiated with the help of the inversion curve for different gases. The concept of fugacity has been introduced to discuss the deviation of a pure substance from ideal behaviour. The influence of pressure and temperature on fugacity has been discussed with some interesting examples. Another factor, i.e., the residual properties such as residual enthalpy and residual entropy are useful to calculate the thermodynamic property of real fluids when the *PVT* data of the substance is unavailable or insufficient. They are derived from the equation of state, compressibility factor, virial coefficient correlation, cubic equation of state, etc.

KEY TERMS

Clapeyron Equation The equation that determines the enthalpy change associated with phase change processes such as vaporization and freezing with the help of the thermodynamic correlation between pressure, volume, and temperature.

Derived or Mathematically Derived Properties These properties are mathematically derived on the basis of the above reference and energy properties, for example, Joule–Thomson coefficient.

Energy Properties The properties in which the changes in the thermodynamic functions such as internal energy (U) and enthalpy (H) indicate some useful work under certain conditions; examples are Gibbs free energy (G), Helmholtz free energy (H), internal energy (U), and enthalpy (H).

Fugacity The tendency to escape or flee from one state to another; lower the fugacity of a phase, lower is the tendency towards the state.

Gibbs Free Energy (Gibbs Function) A single-valued state function of the thermodynamic system and an extensive property. Gibbs free energy of a system is defined as

$$G = H - TS$$

Helmholtz Free Energy (Work Function) A state function and an extensive property of the system, denoted by A. It is defined by the following equation:

$$A = U - TS$$

Inversion Temperature For every gas, there is a temperature where $\mu = 0$, when neither heating nor cooling of the gas passing through the porous plug occurs. This temperature is known as the *inversion temperature.*

Isothermal Compressibility The rate of decrease of volume with respect to pressure at constant temperature per unit volume.

Joule–Thomson Coefficient The rate of change in temperature with respect to pressure at constant enthalpy.

Reference or Primitive Properties The properties which enable us to define the state of the system; they have absolute values which are measured relative to some arbitrary reference state, for example, temperature, pressure, volume, and entropy.

Residual Property The difference between the thermodynamic property of a substance (real fluid) and that of an ideal gas at the specified temperature and pressure.

Thermodynamic Diagram A graph which shows a set of thermodynamic properties for a particular substance in terms of independent variables; it is widely used in the analysis of thermodynamic processes.

Throttling Process The process accompanied by a change in temperature resulting from a reduction in pressure.

Volume Expansivity The rate of change of volume with respect to temperature at constant pressure per unit volume.

IMPORTANT EQUATIONS

1. The cyclic relation between P, V and T for a pure substance is given by

$$\left(\frac{\partial P}{\partial V}\right)_T \left(\frac{\partial V}{\partial T}\right)_P \left(\frac{\partial T}{\partial P}\right)_V = -1$$

2. The Helmholtz free energy (work function) is defined by the following equation:

$$A = U - TS \tag{6.15}$$

3. Gibbs free energy is defined as

$$G = H - TS \tag{6.23}$$

4. The first thermodynamic (fundamental property) relation can be expressed as

$$dU = TdS - PdV \qquad (6.36)$$

5. The second thermodynamic relation is given by

$$dH = TdS + VdP \qquad (6.39)$$

6. The third thermodynamic relation is represented by

$$dA = -PdV - SdT \qquad (6.41)$$

7. The fourth thermodynamic relation is given by

$$dG = VdP - SdT \qquad (6.43)$$

8. Maxwell's relations are given by

(i) $\left(\dfrac{\partial T}{\partial V}\right)_S = -\left(\dfrac{\partial P}{\partial S}\right)_V$
(ii) $\left(\dfrac{\partial T}{\partial P}\right)_S = \left(\dfrac{\partial V}{\partial S}\right)_P$

(iii) $\left(\dfrac{\partial P}{\partial T}\right)_V = \left(\dfrac{\partial S}{\partial V}\right)_T$
(iv) $\left(\dfrac{\partial V}{\partial T}\right)_P = -\left(\dfrac{\partial S}{\partial P}\right)_T$

9. The Clapeyron equation for the study of phase changes is given by

$$\frac{dP}{dT} = \frac{\Delta S}{V_2 - V_1} = \frac{\Delta H}{T(V_2 - V_1)} \qquad (6.56)$$

10. The Clausius–Clapeyron equation used to determine the variation of pressure with temperature in a phase change process is given by

$$\ln\frac{P_2}{P_1} = \frac{\Delta H}{R}\left(\frac{1}{T_1} - \frac{1}{T_2}\right) \qquad (6.61)$$

11. The Gibbs–Helmholtz equations are given by

(i) $A = U + T\left(\dfrac{\partial A}{\partial T}\right)_V$ \qquad (6.62)

(ii) $G = H + T\left(\dfrac{\partial G}{\partial T}\right)_P$ \qquad (6.63)

12. The TdS equations are given by

(i) $TdS = C_P dT - T\left(\dfrac{\partial V}{\partial T}\right)_P dP$ \qquad (6.95)

(ii) $TdS = C_V dT + T\left(\dfrac{\partial P}{\partial T}\right)_V dV$ \qquad (6.96)

13. The Joule–Thomson coefficient is given by

$$\mu_{JT} = \left(\frac{\partial T}{\partial P}\right)_H \qquad (6.111)$$

14. Fugacity can be mathematically expressed as

$$\lim_{P \to 0} \frac{f}{P} = 1 \quad \text{or} \quad \frac{f}{P} = 1 \quad \text{as } P \to 0$$

15. The fugacity coefficient is denoted by ϕ and can be expressed as

$$\frac{f}{P} = \phi$$

EXERCISES

A. Review Questions

1. What is thermodynamic property? How can it be classified? Give an example for each kind of thermodynamic property.
2. Establish the cyclic relation between P, V, and T.
3. What are the fundamental property relations? Derive the four thermodynamic relations.
4. Derive Maxwell's relation among thermodynamic properties.
5. What are free energy functions? Classify them. Mention the importance of free energy functions in the analysis of thermodynamic processes.
6. For an isothermal reversible change of the system

$$-\Delta A_T = W_{\text{max}}$$

   explain the significance of the preceding equation in the light of Helmholtz free energy.
7. In the study of the phase change process of a pure substance, derive the Clapeyron equation.
8. What assumptions are involved in the Clausius–Clapeyron equation?
9. Derive the following relation:

$$\ln \frac{P_2}{P_1} = \frac{\Delta H}{R} \left(\frac{1}{T_1} - \frac{1}{T_2} \right)$$

10. Establish the Gibbs–Helmholtz equations.
11. Prove that

$$dH = C_P dT + \left[V - T \left(\frac{\partial V}{\partial T} \right)_P \right] dP$$

12. Deduce two TdS equations and mention why TdS equations are so useful.
13. Define the terms *isothermal compressibility* and *volume expansivity*.
14. What is Joule–Thomson coefficient? How does it relate to the heating or cooling effect of a gas passing through a porous plug?
15. With the help of a neat sketch of the porous plug experimental assembly, prove that Joule–Thomson expansion is an isenthalpic process.
16. Justify the statement with mathematical expression:
    "For all the gases, the positive and negative values of the Joule–Thomson coefficient do not indicate the attainment of cooling effect and heating effect respectively."

17. What is inversion temperature? Mention the importance of this temperature in explaining the heating or cooling effect of a gas with the help of the Joule–Thomson inversion curve.

18. Prove that C_P and C_V of an ideal gas depend only on temperature.

19. What is a residual property? Define residual enthalpy, entropy, and internal energy.

20. Derive the relations for the estimation of residual enthalpy, entropy, and Gibbs free energy from the following relations:

   (a) Ideal gas equation of state
   (b) Compressibility factor
   (c) Virial coefficient
   (d) Cubic equation of state.

21. What is fugacity and fugacity coefficient? How are they related? How does it play an important role in explaining the deviation from ideal gas behaviour?

22. Explain some important methods for the estimation of the fugacity coefficient of a pure substance.

23. What is thermodynamic diagram? How can it be categorized? What is its importance? How is the thermodynamic diagram constructed?

B. Problems

1. Using the ideal gas equation of state, verify (a) the cyclic relation and (b) the reciprocity relation at constant P.

2. If $f(S, T, P) = 0$, then by following the cyclic relation, prove that

$$\left(\frac{\partial S}{\partial T}\right)_P \left(\frac{\partial T}{\partial P}\right)_S \left(\frac{\partial P}{\partial S}\right)_T = -1$$

   One variable is assumed to be dependent on the other two.

3. Applying the Clapeyron equation, estimate the enthalpy of vaporization of the refrigerant R–134a at 40°C, and compare it with the tabulated value.

4. Using the Clapeyron equation, estimate the enthalpy of vaporization of refrigerant steam at 300 kPa and compare it with the tabulated value.

5. If the internal energy of a substance is considered to be a function of temperature and volume, then show that

$$dU = C_V dT + \left(\frac{T\beta}{\alpha} - P\right) dV$$

6. Derive the following relations for a pure substance:

   (a) $\left(\dfrac{\partial H}{\partial T}\right)_V = C_V + \dfrac{\beta V}{\alpha}$

   (b) $\left(\dfrac{\partial H}{\partial P}\right)_T = V(1 - T\beta)$

   (c) $\left(\dfrac{\partial H}{\partial P}\right)_V = \dfrac{C_V \alpha}{\beta} + V$

   (d) $\left(\dfrac{\partial S}{\partial P}\right)_T = T - \dfrac{1}{\beta}$

   (e) $\left(\dfrac{\partial H}{\partial S}\right)_V = T\left(1 + \dfrac{V\beta}{C_V d}\right)$

   (f) $\left(\dfrac{\partial T}{\partial S}\right)_U = \dfrac{T}{C_V}\left(1 - \dfrac{T\beta}{P\alpha}\right)$

7. Show that for a van der Waals gas

$$\alpha = \frac{V-b}{PV - \dfrac{a}{V} + \dfrac{2ab}{V^2}} \quad \text{and} \quad \beta = \frac{R}{PV - \dfrac{a}{V} + \dfrac{2ab}{V^2}}$$

8. For the equation of state $Z = 1 + \dfrac{B}{V}$, show that

$$\alpha = \frac{1}{P + \dfrac{BRT}{V^2}} \quad \text{and} \quad \beta = \frac{1 + \dfrac{B}{V} + \dfrac{T}{V}\dfrac{dB}{dT}}{T\left(1 + \dfrac{2B}{V}\right)}$$

9. For a pure substance, if the internal energy is considered to be the function of any two of the independent variables P, V, and T, then show that

(a) $\left(\dfrac{\partial U}{\partial V}\right)_P = \dfrac{C_P}{V\beta} - P$

(b) $\left(\dfrac{\partial U}{\partial T}\right)_P = C_P - PV\beta$

10. Prove that

(a) $\left(\dfrac{\partial C_P}{\partial P}\right)_T = -T\left(\dfrac{\partial^2 V}{\partial T^2}\right)_P$

(b) $\left(\dfrac{\partial C_V}{\partial V}\right)_T = T\left(\dfrac{\partial^2 P}{\partial T^2}\right)_V$

11. For a van der Waals gas, show that

(a) $dU = C_V dT + \dfrac{a}{V^2} dV$

(b) $dS = C_V \dfrac{dT}{T} + \dfrac{R dV}{V-b}$

12. A gas obeys the equation of state $V = \dfrac{RT}{P} - \dfrac{C}{T^2} + \dfrac{D}{T^3}$. Find out the variation C_P at constant temperature.

13. For a van der Waals gas, show that

$$C_P - C_V = \frac{TR^2 V^3}{RTV^3 - 2a(V-b)^2}$$

14. Show that

(a) $\left(\dfrac{\partial P}{\partial T}\right)_V = \dfrac{\beta}{\alpha}$

(b) $\left(\dfrac{\partial H}{\partial S}\right)_T = T - \dfrac{1}{\beta}$

15. Prove that for a van der Waals gas

$$\left(\frac{\partial C_V}{\partial V}\right)_T = 0$$

16. Show that the C_V of an ideal gas is independent of the specific volume.

17. For a gas, if enthalpy is considered to be the function of temperature and pressure, then show that the Joule–Thomson coefficient is

$$\mu_{JT} = -\frac{1}{C_P}\left(\frac{\partial H}{\partial P}\right)_T$$

18. Show that

$$\mu_{JT} = \frac{RT^2}{C_P P}\left(\frac{\partial Z}{\partial T}\right)_P$$

where Z is the compressibility factor.

(WBUT, 2006)

19. For a gas obeying the van der Waals equation of state, prove that the inversion temperature is $T = \dfrac{2a}{Rb}$, where a and b are van der Waals constants.

20. Derive an expression for the fugacity coefficient of a gas obeying the van der Waals equation of state.

21. Using Maxwell's relation, prove that

$$\left(\frac{\partial S}{\partial V}\right)_T = \left(\frac{\partial P}{\partial T}\right)_V = \frac{\beta}{\alpha}$$

22. Calculate ΔA and ΔG when 1 mol of an ideal gas is allowed to expand isothermally at 300 K from a pressure of 100 atm to 1 atm.

23. Show that for an ideal gas, $\left(\dfrac{\partial E}{\partial V}\right)_T = 0$, and for a van der Waals gas, $\left(\dfrac{\partial E}{\partial V}\right)_T = \dfrac{an^2}{V^2}$.

24. Using the fundamental property relation $G = H - TS$, show that

$$\left(\frac{\partial H}{\partial P}\right)_T = V - T\left(\frac{\partial V}{\partial T}\right)_P = V(1 - \alpha T)$$

25. Show that

(a) $\left(\dfrac{\partial H}{\partial P}\right)_T = \left[\dfrac{\partial(V/T)}{\partial(1/T)}\right]_P$ (b) $\dfrac{\partial}{\partial T}\left(\dfrac{\Delta G}{T}\right) = -\dfrac{\Delta H}{T^2}$

26. The heat of vaporization of ether is 25.98 kJ/mol at its boiling point, 34.5°C.

   (a) Calculate the rate of change of vapour pressure with temperature $\dfrac{dP}{dT}$ at the boiling point.

   (b) What is the boiling point at 750 mm?

   (c) Estimate the vapour pressure at 36°C.

27. If the temperature dependence of vapour pressure of an organic compound is given by

$$\log P = -\frac{1246.038}{T + 221.354} + 6.95$$

then calculate the enthalpy of vaporization at 298 K.

28. The van der Waals constants for nitrogen, $a = 1.39$ L^2-atm/mol^2, $b = 0.039$ L/mol. What is the inversion temperature of the gas?

29. Show that the internal energy of an ideal gas is a function only of temperature.

30. For a gas obeying the equation of state $V = B + \dfrac{RT}{P}$, the Joule–Thomson coefficient is given by

$$\mu_{JT} = -\frac{1}{C_P}\left(T\frac{dB}{dT} - B\right)$$

31. Calculate the residual enthalpy and entropy for carbon dioxide at 393 K and 12 MPa using any equation of state.

32. Estimate the residual enthalpy and entropy for n-octane at 60°C and 5 bar using the virial equation of state, given that $T_C = 569.4$ K, $P_C = 24.97$ bar, and $\omega = 0.398$.

33. Using the generalized virial coefficient of correlation, estimate the residual enthalpy and entropy for ethylene at 339.7 K and 1 bar, given that $T_C = 283$ K, $P_C = 51.17$ bar, and $\omega = 0.089$.

34. Calculate the residual enthalpy and entropy for propane at 312 K and 2 MPa using the van der Waals equation of state, given that $a = 0.877$ Pa(m^3/mol)^2, $b = 0.84 \times 10^{-4}$ m^3/mol.

35. Estimate the fugacity of a gas obeying the virial equation of state at 100°C and 50 atm, given that the virial coefficient, $B = -73$ cm^3/mol.

36. Estimate the fugacity of carbon monoxide at 50 bar and 200 bar, if the following data are applicable at 273 K:

P (in bar)	25	50	100	200	400
Z	0.9890	0.9792	0.9741	1.0196	1.2482

37. Derive an expression to calculate the change in enthalpy and entropy of a real gas obeying the following equation of state along an isothermal path between the initial and final pressures P_1 and P_2 respectively:

$$V = \frac{RT}{P} + b - \frac{a}{RT}$$

38. Show that $\left(\frac{\partial C_P}{\partial P}\right)_T = \frac{2a}{T^2}$ for a gas under isenthalpic condition obeying the equation of state

$$V = \frac{RT}{P} + b - \frac{a}{RT}$$

39. For a gas which obeys the equation of state $\left(P + \frac{a}{V^2}\right)V = RT$, prove that the Joule–Thomson coefficient is

$$\mu_{JT} = \frac{2aRT}{C_P V^2\left(P^2 - \frac{a^2}{v^4}\right)}$$

40. Show that $\left(\frac{\partial C_P}{\partial P}\right)_T = \frac{6B}{T^3}$ for a gas obeying the equation of state $V = \frac{RT}{P} + A - \frac{B}{T^2}$.

    [**Hint:** We know that $\left(\frac{\partial C_P}{\partial P}\right)_T = -T\left(\frac{\partial^2 V}{\partial T^2}\right)_P$. Find out $\left(\frac{\partial V}{\partial T}\right)_P$ and $\left(\frac{\partial^2 V}{\partial T^2}\right)_P$, and substitute.]

Thermodynamics to Flow Processes   7

The chapter deals with the application of thermodynamics to different steady-flow processes such as throttling, compression, and exchanging of heat, as well as unsteady-flow processes such as charging of rigid vessels from pipelines and discharging of fluid from tanks. The devices pertaining to the steady-flow processes such as nozzles, diffusers, compressors, heat exchangers, and throttling valves have been analyzed with the mass and energy balance for the control volume, depending upon the law of conservation of energy and mass. Hence, the chapter starts with the discussion on the continuity equation and Bernoulli's equation as the basis of mass and energy analysis of the control volume of the process. We extend our task towards the analysis and solution of some typical problems related to the uniform and transient flow processes. We also apply the conservation of mass and energy principles to analyze the control volume of unsteady-flow processes.

7.1 STEADY-FLOW PROCESSES AND DEVICES

Before we start our discussion, it is essential to have a clear understanding of the meaning of the term *steady* or *uniform*. *Steady* means no change with time and *uniform* means no change with location over a particular region. Here, the steady-flow processes imply the processes in which a fluid flows through an open system under steady-state condition. The flow rate of mass does not vary with time. Most of the continuous thermodynamic processes with control volume, such as throttling, compression, refrigeration, pumping, and mixing belong to steady-flow processes. The devices relevant to the process are known as *steady-flow devices*. They operate for a long span of time steadily. The common examples of such devices are compressor, heat exchanger, nozzle, diffuser, pump, and turbine. In this section, we discuss the basis of control volume analysis of some steady-flow processes based on the law of conservation of mass and energy, along with the illustration of some common steady-flow engineering devices for their mass and energy balance.

7.1.1 Continuity Equation and Mass Analysis of Control Volume

The continuity equation basically represents the law of conservation of mass. For a control volume, shown in Fig. 7.1, the macroscopic balance can be represented as

Rate of accumulation of mass within the control volume
= Rate of inflow of mass − Rate of outflow of mass

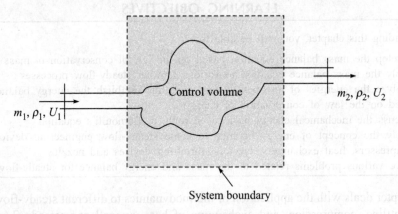

Fig. 7.1 Mass balance for a control volume.

It can be mathematically expressed as

$$\Delta \dot{m}_{CV} = \dot{m}_{in} - \dot{m}_{out} \qquad (7.1)$$

In the rate form

$$\frac{dm_{CV}}{dt} = \sum \dot{m}_{in} - \sum \dot{m}_{out} \qquad (7.2)$$

Here, the summation sign implies that all the inlets and outlets are to be taken into consideration.

Equation (7.2) can be rearranged as

$$\frac{dm_{CV}}{dt} + \left[\sum \dot{m}_{out} - \sum \dot{m}_{in} \right] = 0 \qquad (7.3)$$

Referring to Fig. 7.1, consider the mass of a substance m flowing into or out of the control volume through a cross-sectional area A during a time interval dt. If the average velocity of fluid is U, then the preceding expression becomes

$$\frac{dm_{CV}}{dt} + \Delta(\rho UA) = 0 \tag{7.4}$$

Now, in case of mass balance for a steady-flow process, the total amount of mass in the control volume remains unchanged with time. It necessarily means no accumulation of mass within the control volume, i.e., $\dfrac{dm_{CV}}{dt} = 0$.

On substitution of this condition into the preceding equation, we have

$$\Delta(\rho UA) = 0 \tag{7.5}$$

Considering a single stream consisting of only one inlet and one outlet in a device such as pump, compressor or nozzle, the equation can be written as

$$\rho_1 U_1 A_1 = \rho_2 U_2 A_2 \tag{7.6}$$

Here, subscripts 1 and 2 denote the inlet and outlet condition of the stream respectively. In case of incompressible fluid, $\rho_1 = \rho_2$. Then Eq. (7.6) becomes

$$U_1 A_1 = U_2 A_2$$

EXAMPLE 7.1   The inlet and outlet diameters of a pipe are 15 cm and 20 cm respectively. If the velocity of water flowing through the pipe at the inlet is 7 m/s, find the rate of flow through the pipe. Determine also the velocity of water at the outlet.

Solution: Refer to Fig. 7.2.
At the inlet

> Diameter, $d_1 = 15$ cm $= 0.15$ m
>
> Area, $A_1 = \dfrac{\pi}{4}d^2 = \dfrac{\pi}{4}(0.15)^2 = 0.01767$ m^2.

At the outlet

> Diameter, $d_2 = 20$ cm $= 0.2$ m
>
> Area, $A_2 = \dfrac{\pi}{4}d^2 = \dfrac{\pi}{4}(0.2)^2 = 0.0314$ m^2.

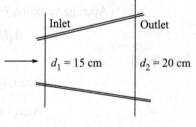

Fig. 7.2   Example 7.1.

The rate of flow through the pipe can be estimated as

$$Q = A_1 U_1 = 0.01767 \text{ m}^2 \times 7 \text{ m/s} = 0.1236 \text{ m}^3/\text{s}$$

Using continuity equation, we have

$$\rho_1 U_1 A_1 = \rho_2 U_2 A_2$$

For water

$$\rho_1 = \rho_2 = \rho$$

Hence the preceding equation reduces to

$$A_1 U_1 = A_2 U_2$$

Therefore

$$U_2 = \frac{A_1 U_1}{A_2} = \frac{0.0176 \times 7}{0.0314} = 3.936 \text{ m/s}$$

EXAMPLE 7.2 Water flowing through a pipe of 20 cm diameter. The flow splits into two parts and passes through pipes of diameters 15 cm and 10 cm respectively. Find the discharge of the 10 cm diameter pipe, if the average velocity of water flowing through this pipe is 3 m/s. Determine also the velocity in the 10 cm diameter pipe if the average velocity in the 15 cm diameter pipe is 2.5 m/s.

Solution: Refer to Fig. 7.3.

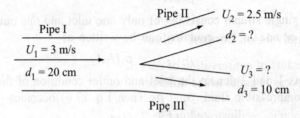

Fig. 7.3   Example 7.2.

For pipe I

Diameter, $d_1 = 20$ cm $= 0.2$ m

Area, $A_1 = \frac{\pi}{4}d^2 = \frac{\pi}{4}(0.2)^2 = 0.0314 \text{ m}^2$

Average velocity, $U_1 = 3$ m/s

$A_1 U_1 = 0.0942 \text{ m}^3/\text{s}$

For pipe II

Diameter, $d_2 = 15$ cm $= 0.15$ m

Area, $A_2 = \frac{\pi}{4}d^2 = \frac{\pi}{4}(0.15)^2 = 0.0176 \text{ m}^2$

Average velocity, $U_2 = 2.5$ m/s

$A_2 U_2 = 0.044 \text{ m}^3/\text{s}$

For pipe III

Diameter, $d_3 = 10$ cm $= 0.10$ m

Area, $A_3 = \frac{\pi}{4}d^2 = \frac{\pi}{4}(0.10)^2 = 0.0078 \text{ m}^2$

(i)   Let the discharge through pipe 1 be Q_1. It is given by

$$Q_1 = A_1 \times V_1 = 0.0314 \times 3 = 0.0942 \text{ m}^3/\text{s}$$

(ii)   To determine the velocity in the 10 cm diameter pipe, the continuity equation can be used as

$$Q_1 = Q_2 + Q_1$$

or

$$A_1U_1 = A_2U_2 + A_3U_3$$

or

$$0.0942 = 0.044 + 0.0078 \ U_3$$

or

$$U_3 = 6.43 \ \text{m/s}$$

7.1.2   Energy Analysis of Control Volume and Bernoulli's Equation

This is based on the law of conservation of energy, i.e., the net change in the total energy of the system during a process is equal to the difference between the total energy entering and leaving the system.

For a control volume, as illustrated in Fig. 7.4, the basis of energy balance can be expressed as

$$\text{Rate of accumulation of energy within the control volume}$$
$$= \text{Rate of inflow of energy} - \text{Rate of outflow of energy} \qquad (7.7)$$

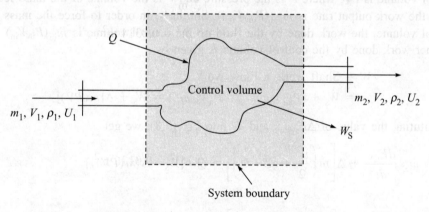

Fig. 7.4   Energy balance for a control volume.

Now, if the rate of heat transferred to the system is $\dot{Q}$ and the work produced by the system at a rate is $\dot{W}$, then these heat inputs and work outputs are to be taken into consideration for energy balance. Hence Eq. (7.7) can be mathematically represented as

$$\frac{dE_{\text{CV}}}{dt} = (\Delta\dot{E}_{\text{in}} + \dot{Q}) - (\dot{E}_{\text{out}} + \dot{W})$$

or

$$\frac{dE_{\text{CV}}}{dt} = (\dot{E}_{\text{in}} - \dot{E}_{\text{out}}) + \dot{Q} - \dot{W}$$

or

$$\frac{dE_{CV}}{dt} + (\dot{E}_{out} - \dot{E}_{in}) = \dot{Q} - \dot{W}$$

or

$$\frac{dE_{CV}}{dt} + \Delta \dot{E}_{system} = \dot{Q} - \dot{W} \tag{7.8}$$

where

$\Delta \dot{E}_{system}$ = Change in total energy of the system during a process

= Change in internal energy + change in kinetic energy + change in potential energy

= $\Delta U + \Delta K.E. + \Delta P.E.$

$$= \Delta \left[\dot{m} \left(\frac{u^2}{2} + gZ + U \right) \right] \tag{7.9}$$

Here

m = Mass flow rate
Z = Height above a datum level
u = Average velocity.

The work output rate $\dot{W}$ essentially consists of two parts. These are shaft work and flow work. The shaft work done by the fluid in the control volume is W_S. The flow work done by the control volume is PV, where P is the pressure and V is the volume of the mass leaving. In this case, the work output rate is $\dot{m}_{out}(P_{out}V_{out})$. Similarly, in order to force the mass $\dot{m}_{in}$ into the control volume, the work done by the fluid on the control volume is $\dot{m}_{in}(P_{in}V_{in})$.

The net work done by the control volume is given by

$$\dot{W} = \text{Shaft work} + \text{Flow work}$$

$$= \dot{W}_S + (P_{out}V_{out}\dot{m}_{out} - P_{in}V_{in}\dot{m}_{in}) = \dot{W}_S + \Delta[\dot{m}(PV)] \tag{7.10}$$

Substituting the value of ΔE_{system} and $\dot{W}$ into Eq. (7.8), we get

$$\frac{dE_{CV}}{dt} + \Delta \left[\dot{m} \left(\frac{u^2}{2} + gZ + U \right) \right] = \dot{Q} - [\dot{W}_S + \Delta\{\dot{m}(PV)\}]$$

or

$$\frac{dE_{CV}}{dt} + \Delta \left[\dot{m} \left(\frac{u^2}{2} + gZ + H \right) \right] = \dot{Q} - \dot{W}_S \quad (\text{since enthalpy, } H = U + PV)$$

For a steady flow process, the total energy content of the control volume remains constant. Thus, the rate of change of the total energy with respect to time is zero, i.e., $\dfrac{dE_{CV}}{dt} = 0$.

Therefore, the equation of energy balance for a general steady flow system reduces to

$$\Delta \left[\dot{m} \left(\frac{u^2}{2} + gZ + H \right) \right] = \dot{Q} - \dot{W}_S \tag{7.11}$$

This is known as the *steady-flow energy balance equation.*

In a single-stream system, where, only one stream enters and leaves the control volume, and the condition is $\dot{m}_{in} = \dot{m}_{out} = \dot{m}$. In this case the steady flow energy balance equation on unit mass basis becomes

$$\Delta\left[\left(\frac{u^2}{2} + gZ + H\right)\right] = \dot{Q} - \dot{W}_S \qquad (7.12)$$

where

$$Q = \frac{\dot{Q}}{\dot{m}} = \text{Heat transfer per unit mass of the fluid}$$

$$W_S = \frac{\dot{W}_S}{\dot{m}} = \text{Work done per unit mass of the fluid.}$$

For the flowing stream experiencing a negligible change in kinetic and potential energy, Eq. (7.12) reduces to

$$\Delta \dot{m} H = \dot{Q} - \dot{W}_S$$

or

$$\Delta H = \frac{\dot{Q}}{\dot{m}} - \frac{\dot{W}_S}{\dot{m}}$$

or

$$\Delta H = Q - W_S \qquad (7.13)$$

If the friction factor of a fluid flow is taken into consideration, the mechanical energy balance equation can easily be derived with the help of Eq. (7.12), as the friction concerned with the flow of a fluid degrades the mechanical energy through its transformation into heat.

It can be expressed from the definition of enthalpy as

$$H = U + PV$$

On total differentiation, we get

$$dH = dU + PdV + VdP \qquad (7.14)$$

Since the work done by the fluid, $dW = PdV$ for a reversible process, Eq. (7.14) becomes

$$dH = dU + dW + VdP \qquad (7.15)$$

From the first law of thermodynamics, we have

$$dQ = dU + dW$$

Putting the value of dQ into Eq. (7.15), we get

$$dH = dQ + VdP$$

Integrating, we find

$$\Delta H = Q + \int_{P_1}^{P_2} VdP \qquad (7.16)$$

Substituting Eq. (7.16) into Eq. (7.12), it reduces to

$$-W_S = \frac{\Delta U^2}{2} + g\Delta Z + \int_{P_1}^{P_2} VdP$$

When the changes in kinetic and potential energies of the fluid are negligible, it reduces further to

$$W_S = -\int_{P_1}^{P_2} VdP \tag{7.17}$$

Assuming the reversibility of the processes, the preceding equations are derived. But in practice, we must keep track of the irreversibilities associated with the work lost by the process. Incorporating the irreversibility of the process, we get the form as

$$\frac{\Delta U^2}{2} + g\Delta Z + \int_{P_1}^{P_2} VdP + W_S + W_F = 0 \tag{7.18}$$

where W_F is the lost work due to friction.

Equation (7.18) is known as the *mechanical energy balance equation.*

On the basis of this balance equation, Bernoulli's equation can be derived. Consider that the above fluid is an incompressible one, whose density ρ is independent of the change in temperature or pressure, and which is flowing through a conduit. It involves no work interaction and irreversibility such as friction. Hence, the shaft work term is zero, i.e., $W_S = 0$ and the friction factor related to the lost work, $W_F = 0$.

Since the fluid is incompressible, its volume V remains constant during the process and is a function of ρ. Therefore

$$\int_{P_1}^{P_2} VdP = V\Delta P = \frac{\Delta P}{\rho}$$

Hence, the preceding mechanical energy balance equation can be reduced to

$$\frac{\Delta u^2}{2} + gZ + \frac{\Delta P}{\rho} = 0 \tag{7.19}$$

or

$$\frac{u^2}{2} + gZ + \frac{P}{\rho} = \text{Constant} \tag{7.20}$$

or

$$\text{Kinetic energy + Potential energy + Pressure energy = Constant}$$

This is known as *Bernoulli's equation* and is applicable to the fluid which is

(a) Incompressible
(b) Non-viscous and
(c) Not exchanging work with the surroundings.

EXAMPLE 7.3 A pipe, through which water is flowing, has diameters 30 cm and 15 cm at cross-sections 1 and 2 respectively. The discharge of the pipe is 40 L/s. The cross-section 1 is 8 m and cross-section 2 is 6 m above the reference level. If the pressure at section 1 is 5 bars, determine the pressure at section 2.

Solution: Referring to Fig. 7.5, we have:

Section 1:

Diameter, $d_1 = 30$ cm $= 0.3$ m

$$\text{Area, } A_1 = \frac{\pi}{4}d_1^2 = \frac{\pi}{4}(0.3)^2 = 0.0706 \text{ m}^2$$

$P_1 = 5$ bar $= 5 \times 10^5$ N/m^2

$Z_1 = 8$ m

Section 2:

Diameter, $d_2 = 15$ cm $= 0.15$ m

$$\text{Area } A_2 = \frac{\pi}{4}d_2^2 = \frac{\pi}{4}(0.15)^2 = 0.0176 \text{ m}^2$$

$P_2 = ?$

$Z_2 = 6$ m

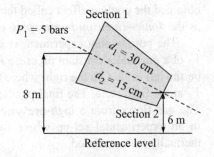

Fig. 7.5 Example 7.3.

We are given that the discharge, $Q = 40$ L/s $= 0.04$ m^3/s.

Here, to find the average velocity at sections 1 and 2, the continuity equation can be used as

$$A_1 U_1 = A_2 U_2 = Q$$

Hence

$$U_1 = \frac{Q}{A_1} = \frac{0.04}{0.07} = 0.571 \text{ m/s}$$

$$U_2 = \frac{Q}{A_2} = \frac{0.04}{0.017} \text{ } 2.352 \text{ m/s}$$

Applying Bernoulli's equation at sections 1 and 2, we get

$$\frac{U_1^2}{2g} + Z_1 \frac{P_1}{\rho g} = \frac{U_2^2}{2g} + Z_2 + \frac{P_2}{\rho g}$$

or

$$\frac{(0.571)^2}{2 \times 9.81} + 8 + \frac{5 \times 10^5}{1000 \times 9.81} = \frac{(2.352)^2}{2 \times 9.81} + 6 + \frac{P_2}{1000 \times 9.81}$$

or

$$P_2 = 513092.43 \text{ N/m}^2 = 5.13 \times 10^5 \text{ N/m}^2 = 5.13 \text{ bar}$$

7.1.3 Throttling Device

When a compressed gas expands adiabatically and slowly through a flow-restricting device like a porous plug or throttling valve, it undergoes a significant pressure drop without doing any work. As a result, it experiences a change in temperature which varies in magnitude and sign. This involves either heating up or cooling down, depending upon the nature of the gas.

The process accompanied by a change in temperature resulting from a reduction in pressure is known as the *throttling process*. The validity of this process principle was tested first by Joule and Thomson in 1862 by conducting an experiment which involved forcing a gas through a porous plug. They observed the change in temperature produced by free expansion of gas and obtained the cooling effect called the Joule–Thomson effect. The experiment is popularly known as the *Joule–Thomson porous plug experiment*.

The porous plug experiment is schematically represented by Fig. 7.6. Suppose the volume V_1 of a gas under constant pressure P_1 is allowed to pass through a porous plug from the region on the left to that on the right where the constant pressure is P_2. The volume of the gas becomes V_2 at the final state. The final position of the two pistons is shown by dotted lines in the figure. The gas flows from a high-pressure region to a low-pressure one. The process is carried out in an experimental set-up where two frictionless pistons move slowly. The entire system is thermally well-insulated.

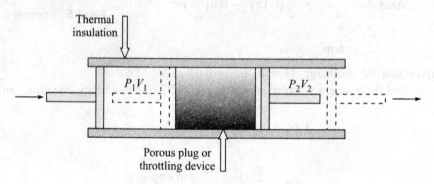

Fig. 7.6 Demonstration of operation in a throttling device.

So

The work done by the piston in the left chamber $= W_1 = P_1V_1$
The work done by the piston in the right chamber $= W_2 = P_2V_2$

Since the Joule–Thomson expansion is carried out adiabatically, i.e., $dQ = 0$, we have from the first law of thermodynamics,

$$dQ = dU + dW$$

or

$$0 = (U_2 - U_1) + (W_2 - W_1)$$

or

$$U_2 - U_1 = -(P_2V_2 - P_1V_1) = P_1V_1 - P_2V_2$$

or

$$U_2 + P_2V_2 = U_1 + P_1V_1 \text{ (by definition, enthalpy } H = U + PV)$$

or

$$H_2 - H_1 = 0$$

or

$$\Delta H = 0 \qquad (7.21)$$

i.e., there is no change in heat content or enthalpy in the adiabatic expansion of a gas through a porous plug. Thus, Joule–Thomson expansion is an isenthalpic expansion. Whether a gas will produce a heating effect or a cooling effect during throttling expansion can be better explained by using the *Joule–Thomson coefficient*. It is symbolized by μ_{JT} and is defined as the rate of change in temperature with respect to pressure at constant enthalpy. Mathematically, it can be expressed as

$$\mu_{JT} = \left(\frac{\partial T}{\partial P}\right)_H$$

μ_{JT} = +ve implies a cooling effect, i.e., drop in temperature, as ΔP = –ve; ΔT = –ve and
μ_{JT} = –ve, implies a heating effect, i.e., rise in temperature, as ΔP = +ve; and ΔT = +ve and
μ_{JT} = 0, implies no temperature change, i.e., neither heating nor cooling effect. The temperature at which no temperature change occurs is known as the *inversion temperature*.

Joule–Thomson expansion is employed for the liquefaction of gases as well as for the production of low temperature.

Thus, the characteristics of the Joule–Thomson coefficient can be summarized as follows:

$$\mu_{JT} \begin{cases} = +\text{ve} = \text{cooling effect, i.e., temperature decreases} \\ = -\text{ve} = \text{heating effect, i.e., temperature increases} \\ = 0 = \text{no change in temperature.} \end{cases}$$

7.1.4 Compressor

A compressor is a mechanical device that increases the pressure of a gas by reducing its volume and finally increases the temperature of gas. Alternatively, a compressor is capable of compressing the gas to very high pressure. Compressors can be classified depending upon the way they work. They are:

1. Reciprocating compressor
2. Dynamic or turbo or centrifugal compressor.

Reciprocating Compressors

They employ pistons to push gas to high pressure and are suitable for use in the following cases:

❖ Low flow rates
❖ High compression ratios, i.e., to achieve large pressure difference.

Dynamic Compressor

They use a set of rotating blades to add velocity and pressure to fluid. They operate at high speed and are driven by steam or gas turbines or electric motors. They are preferable in the following areas:

❖ High flow rate
❖ Low compression ratio.

Now, we derive the equations by which we can estimate how much energy is required for different steady-state compression processes.

Adiabatic Compression

When a gas is allowed to undergo an adiabatic compression, as represented by Fig. 7.7, its temperature increases. Since the temperature change is accompanied by a change in the specific volume, the work required to compress the gas also changes. The changes in kinetic energy and potential energy are small. Hence, the energy balance for steady-state compression gives

$$\Delta\left(\frac{U^2}{2} + gZ + H\right) = Q - W_S$$

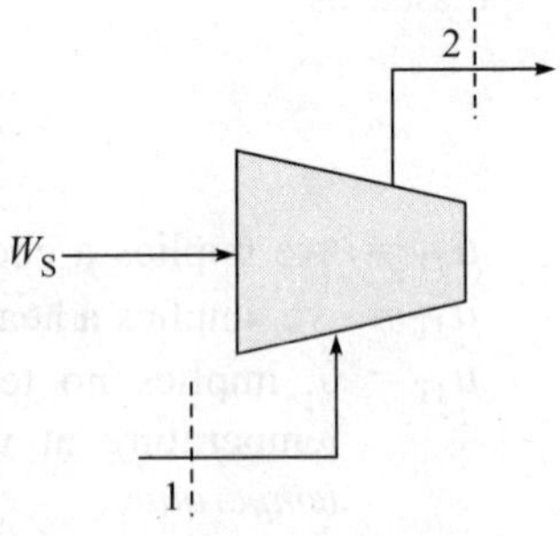

For adiabatic compression, $Q = 0$, and since the changes in kinetic and potential energy are small, the velocity and static heads may be neglected.

Fig. 7.7 Steady-state adiabatic compression process.

Substituting the above conditions, we get

$$W_S = -\Delta H$$

If there is no heat transfer to or from the gas being compressed, the minimum requirement of shaft work for compression of gas is known as *isentropic work*, and it is given by

$$W_{iso} = -\Delta H_S$$

where ΔH_S is change in enthalpy for isentropic compression.

Deviation from ideal behaviour of a gas must be accounted for by introducing an isentropic compressor efficiency as

$$\eta_{compressor} = \frac{W_{iso}}{W_{ac}} = \frac{\text{Isentropic work}}{\text{Actual work}} \qquad (7.22)$$

A compressor without internal cooling can be assumed to be adiabatic and reversible. In this case, the shaft work requirement can be written as

$$W_{iso} = -\int_{P_1}^{P_2} V dP \qquad (7.23)$$

For isentropic compression of an ideal gas, we know

$$PV^\gamma = \text{Constant}$$

$$PV^\gamma = P_1 V_1^\gamma = \text{Constant}$$

or

$$\frac{V^\gamma}{V_1^\gamma} = \left(\frac{P_1}{P}\right)^{1/\gamma}$$

or

$$V = V_1 \left(\frac{P_1}{P}\right)^{1/\gamma} \tag{7.24}$$

Putting the value of V into Eq. (7.5), we get

$$W_{\text{iso}} = -V_1 \int_{P_1}^{P_2} \left(\frac{P_1}{P}\right)^{1/\gamma} \tag{7.25}$$

$$= \frac{\gamma}{\gamma - 1} P_1 V_1 \left[1 - \left(\frac{P_2}{P_1}\right)^{\frac{\gamma-1}{\gamma}}\right] \tag{7.26}$$

$$= \frac{\gamma}{\gamma - 1} RT_1 \left[1 - \left(\frac{P_2}{P_1}\right)^{\frac{\gamma-1}{\gamma}}\right] \tag{7.27}$$

For ideal gas, the work requirement for the compression can be estimated.

Suppose T_1 and T_2 are the initial and final temperatures of ideal gas respectively. Then the work required is given by

$$W_S = -\Delta H = -C_P(T_2 - T_1) = C_P(T_1 - T_2) \tag{7.28}$$

Isothermal Compression

In an isothermal compression, the temperature of gas is maintained as constant during the entire operation. All heat of compression is removed at the instant it is created, i.e., sufficient cooling is provided to make the process isothermal. Since it is an isothermal compression, it will be accompanied by no change in internal energy or enthalpy for an ideal gas, i.e., $\Delta U = 0$ and $\Delta H = 0$.

Putting into the first law of thermodynamics, we have

$$dQ = dW \tag{7.29}$$

Considering the work of a mechanically reversible process, $dW = PdV$, and substituting, we get

$$dQ = dW = PdV \tag{7.30}$$

On replacing P by RT/V, Eq. (7.30) can be written as

$$dW = RT \frac{dV}{V} \tag{7.31}$$

On integration, Eq. (7.31) yields

$$W = RT \ln \frac{V_2}{V_1} = RT \ln \frac{P_1}{P_2} \tag{7.32}$$

This is the minimum work requirement for compressing an ideal gas from a given state to another state.

For a given compression ratio and suction condition, the work requirement in isothermal compression is less than that for adiabatic compression. The only reason is that cooling is useful in compressors.

Isothermal efficiency is defined as the ratio of isothermal work to actual work, i.e.,

$$\eta_{\text{iso}} = \frac{W_{\text{iso}}}{W_{\text{actual}}} \tag{7.33}$$

Multistage Compression

When it is not possible to achieve a higher compression ratio, multistage compression with inter-stage cooling technique is employed. The huge amount of heat of compression is removed by intercooling in such a way that the compressed gas is cooled between each stage by passing it through an intercooler. The chief advantage is that temperature constancy (or isothermal condition) can be maintained and thereby the requirement of work for compression can be reduced.

This can be demonstrated in a proper way with the help of Fig. 7.8 that if we compare the work input requirement for isothermal, adiabatic and multistage compressions. Suppose that an ideal gas is allowed to undergo a compression in a two-stage compressor where the series of compression processes are executed between the limiting pressure levels P_1 and P_2. All the processes are carried out in an internally reversible way. Now, in the first stage, the gas is compressed from an initial pressure P_1 to an intermediate pressure P_2 and cooled at constant pressure to the initial temperature T_1. In the second stage, the compression of gas takes place from the intermediate pressure P_2 to final pressure P_3.

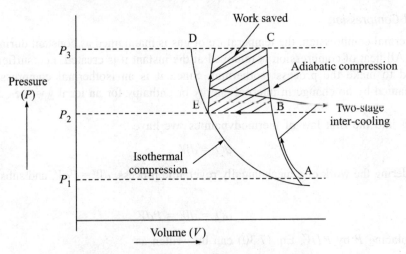

Fig. 7.8 Multistage compression with inter-cooling.

The area ABCDEA in the $P-V$ diagram represents the amount of work required and the shaded area indicates the amount of work saved using two-stage compression with inter-stage cooling.

For a multistage compression, the total work input is the sum of inputs for individual stages of compression.

Accordingly, the total work requirement for a two-stage compressor operating between P_1 and P_3 with an intermediate pressure P_2 is given by

$$W_{Comp,Total} = W_{Comp,Stage\,I} + W_{Comp,Stage\,II}$$

$$= \frac{\gamma R T_1}{\gamma - 1}\left[\left(\frac{P_2}{P_1}\right)^{\frac{\gamma-1}{\gamma}} - 1\right] + \frac{\gamma R T_1}{\gamma - 1}\left[\left(\frac{P_3}{P_2}\right)^{\frac{\gamma-1}{\gamma}} - 1\right] \quad (7.34)$$

Now, it is necessary to find out the value of P_2 because it minimizes the work of compression during two-stage compression. P_2 can be determined by differentiating Eq. (7.15) with respect to P_2 and equating the derivative to zero.

After simplification, Eq. (7.15) yields

$$P_2 = (P_1 P_3)^{1/2}$$

or

$$\frac{P_2}{P_1} = \frac{P_3}{P_2} \quad (7.35)$$

The above relation implies that the pressure ratio across individual stages of the compressor must be the same for the minimization of compression for a two-stage compressor.

The compression ratio in Eq. (7.35) that minimizes the total work is such that each stage has an identical ratio.

So, Eq. (7.35) can be generalized for n stages as

$$\frac{P_2}{P_1} = \frac{P_3}{P_2} = \cdots = \frac{P_{n+1}}{P_n} = \left(\frac{P_{n+1}}{P_n}\right)^{1/n} \quad (7.36)$$

EXAMPLE 7.4 Air at 100 kPa and 320 K is to be compressed steadily to 600 kPa and 430 K in a reversible compressor. The mass flow rate of the air is 0.03 kg/s and the heat losses during the process are estimated to be 15 kJ/kg. Neglecting the changes in kinetic and potential energy and assuming air to be an ideal gas, estimate the power requirement of the compressor.

Solution: The energy balance around the compressor can be expressed as

$$\frac{dE_{system}}{dt} = \dot{E}_{in} - \dot{E}_{out}$$

Since it is a steady-flow process and no change occurs in kinetic and potential energy, $\dfrac{dE_{system}}{dt} = 0$ and the equation reduces to

$$\dot{E}_{in} = \dot{E}_{out}$$

or

$$W_{in} + \dot{m} H_1 = \dot{Q}_{out} + \dot{m} H_2$$

or

$$W_{in} = \dot{m} Q_{out} + \dot{m}(H_2 - H_1) \quad \left(\text{since } Q_{out} = \frac{\dot{Q}_{out}}{\dot{m}}\right)$$

or

$$W_{in} = \dot{m}[Q_{out} + (H_2 - H_1)]$$

Here

$\dot{m}$ = Mass flow rate of air = 0.03 kg/s

$\dot{W}_{in}$ = ?

Q_{out} = 15 kJ/kg.

From enthalpy chart for air

H_2 = Enthalpy of air at 430 K = 431.43 kJ/kg

H_1 = Enthalpy of air at 320 K = 320.20 kJ/kg.

Substituting the values, we get the power requirement of the compressor as

$$\dot{W}_{in} = 0.03 \text{ kg/s} \times [15 \text{ kJ/kg} + (431.43 - 320.29) \text{ kJ/kg}$$
$$= 3.78 \text{ kJ/s} = 3.78 \text{ kW}$$

EXAMPLE 7.5 CO_2 enters an adiabatic compressor at 100 kPa and 250 K at a rate of 0.1 m³/s and leaves at 500 kPa. CO_2 is assumed to behave as an ideal gas. Determine the work of compression per unit mass for

(a) Reversible adiabatic compression when $\gamma = 1.4$
(b) Isothermal compression
(c) Single-stage compression when $\gamma = 1.3$.

Solution:

(a) The work for compressing an ideal gas is given by

$$W_{ad} = C_P(T_1 - T_2) = \frac{\gamma R}{\gamma - 1}(T_1 - T_2)$$

Here

$\gamma = 1.4$

$R = 8.314$

$T_1 = 250$ K

$T_2 = ?$

The final temperature T_2 of an ideal gas undergoing adiabatic compression can be determined by the following relation:

$$\frac{T_2}{T_1} = \left(\frac{P_2}{P_1}\right)^{\frac{\gamma}{\gamma - 1}}$$

where

$P_1 = 100$ kPa

$P_2 = 500$ kPa

or

$$T_2 = 250\left(\frac{500}{100}\right)^{\frac{1.4}{1.4 - 1}} = 395.25 \text{ K}$$

Substituting these values, the work of compression in an adiabatic process can be estimated as

$$W_{ad} = \frac{\gamma R}{\gamma - 1}(T_1 - T_2) = \frac{1.4 \times 8.314}{1.4 - 1}(250 - 395.25) = -4226.62 \text{ J/mol}$$
$$= -4226.62/\text{molecular weight of } CO_2 (=44)$$
$$= -96.05 \text{ J/g}$$

(b) Work done by isothermal compression of an ideal gas is given by

$$W_{iso} = -RT \ln \frac{P_2}{P_1} = -8.314 \times 250 \times \ln \frac{500}{100} = -3345.21 \text{ J/mol} = -76.02 \text{ J/g}$$

(c) The work done in single-stage compression is

$$W_{single\text{-}stage} = \frac{\gamma RT_1}{\gamma - 1}\left[1 - \left(\frac{P_2}{P_1}\right)^{\frac{\gamma-1}{\gamma}}\right] = \frac{\gamma P_1 V_1}{\gamma - 1}\left[1 - \left(\frac{P_2}{P_1}\right)^{\frac{\gamma-1}{\gamma}}\right]$$
$$= \frac{1.3 \times 100 \times 0.1}{1.3 - 1}\left[1 - \left(\frac{500}{100}\right)^{\frac{1.3-1}{1.3}}\right]$$
$$= -19.41 \text{ kW}$$

7.1.5 Ejectors or Jet Pumps

It is a pump-like device without any moving part or piston that uses high-velocity steam as a motive fluid to entrain and compress vapours or gases or a second fluid stream. This creates an appreciable vacuum in any vessel or chamber to the suction line of the device.

An ejector, as illustrated in Fig. 7.9, consists of the following main components:

1. Motive fluid inlet nozzle—through which a motive fluid (generally high-pressure steam) is accelerated;

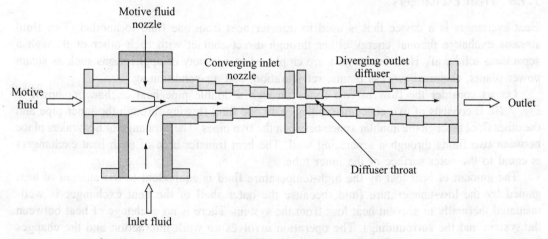

Fig. 7.9 Illustration of a modern ejector.

2. Second fluid inlet—through which a second fluid (gases or vapours) enters
3. Converging inlet nozzle—where the intermixing of the second and motive fluids occurs with further acceleration of the second fluid and deceleration of the motive fluid;
4. Diverging outlet diffuser—for deceleration of the mixed fluid;
5. Compressed fluid outlet—through which compressed fluid is discharged.

The steam jet ejector operates on Bernoulli's principle. The high-pressure motive fluid is passed through an inlet nozzle; after expansion, it is converted into a high-velocity jet at the converging nozzle, where it creates a low pressure. Consequently, the low pressure draws the second fluid (gas or vapour) into the convergent–divergent nozzle, where the second fluid and the motive fluid mix up. Now, the pressure energy of the inlet fluid is converted to kinetic energy in the form of velocity head at the diffuser throat.

In the diverging section of the diffuser the mixed fluid decelerates, and as a result the kinetic energy of the mixed stream is converted back to pressure energy. The mixture is discharged through the outlet of the ejector to the surroundings.

In this regard, it is important to consider two key parameters which are extremely useful for the design of a steam jet ejector. These are compression ratio and entrainment ratio of ejectors.

Compression ratio is defined as the ratio of the discharge pressure of the ejector to the inlet pressure of the suction vapour or gas.

Entrainment ratio is defined as the ratio of the amount of motive fluid required to entrain and compress a given amount of the second fluid.

The materials for construction of an ejector are generally carbon steel, stainless steel, titanium, PTFE, and carbon. Ejectors require easy maintenance and they are lower in cost. Steam ejectors are widely used in distillation, evaporation, and refrigeration to maintain sub-atmospheric pressure. It is also in greater use in aircraft. Special care must be taken to handle corrosive gases because they may lead to mechanical damage of the ejector. Ejectors also find application in railway locomotives for creating vacuum to operate the brakes.

7.1.6 Heat Exchangers

Heat exchanger is a device that is used to transfer heat from one fluid to another. Two fluid streams exchange thermal energy either through direct contact with each other or through a separating solid wall. Heat exchangers are employed in a variety of applications such as steam power plants, chemical process plants, refrigeration, and air-conditioning.

Let us consider the flow of two fluids through a double pipe heat exchanger shown in Fig. 7.10. It consists of two concentric pipes, with one fluid flowing through the inner pipe and the other fluid through the annular space between the two pipes. The exchange of heat takes place between two fluids through a separating wall. The heat transfer area of such heat exchangers is equal to the outer surface of the inner tube.

The amount of heat lost by the high-temperature fluid is equivalent to the amount of heat gained by the low-temperature fluid, because the outer shell of the heat exchanger is well-insulated thermally to prevent heat loss from the system. There is no exchange of heat between the system and the surroundings. The operation involves no work interaction and the changes in kinetic energy and potential energy are negligible.

We can analyze the flow process in two possible ways:

1. By assuming the whole system as control volume, as shown in Fig. 7.10
2. By assuming one fluid as control volume, as shown in Fig. 7.11.

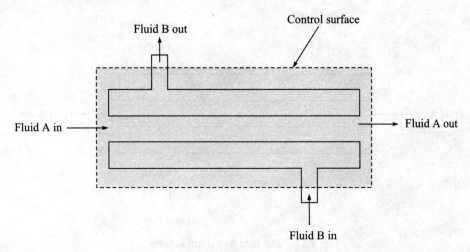

Fig. 7.10 Entire heat exchanger as control volume.

For the first case, there is no heat across the control surface because the outer wall of the system is well-insulated and the control surface is heat-preventive within the layer. There is no transfer of work across the control surface, but the mass crosses. So, neglecting the heat losses, the energy balance equation can be written as

$$\dot{E}_{in} - \dot{E}_{out} = \frac{dE_{CV}}{dt} \tag{7.37}$$

For a steady-state flow process

$$\frac{dE_{CV}}{dt} = 0$$

Hence, Eq. (7.37) becomes

$$\dot{E}_{in} = \dot{E}_{out} \tag{7.38}$$

Inlet stream:

$$\dot{E}_{in} = \dot{m}_{A_{in}} h_{A_{in}} + \dot{m}_{B_{in}} h_{B_{in}} \tag{7.39}$$

Outlet stream:

$$\dot{E}_{out} = \dot{m}_{A_{out}} h_{A_{out}} + \dot{m}_{B_{out}} h_{B_{out}} \tag{7.40}$$

Substituting Eqs. (7.39) and (7.40) into Eq. (7.38), we get

$$\dot{m}_{A_{in}} h_{A_{in}} + \dot{m}_{B_{in}} h_{B_{in}} = \dot{m}_{A_{out}} h_{A_{out}} + \dot{m}_{B_{out}} h_{B_{out}}$$

or

$$\dot{m}_A (h_{A_{in}} - h_{A_{out}}) = \dot{m}_B (h_{B_{out}} - h_{B_{in}}) \tag{7.41}$$

or

$$-\dot{Q}_{A_{in}} = \dot{Q}_{B_{out}}$$

For the second case, all the considerations mentioned in the first case will remain the same for the second case as shown in Fig. 7.11, except for the consideration of heat transfer, because the amount of heat transferred from one fluid to another is not zero.

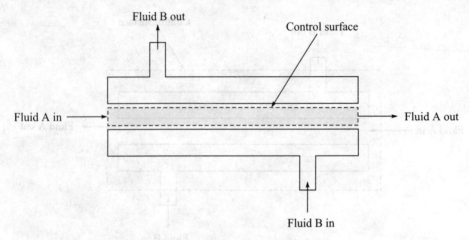

Fig. 7.11 One fluid as control volume.

So, the rate of heat transfer Q_A around fluid A is given by

$$-\dot{Q}_{A_{in}} = \dot{m}_A(h_{A_{in}} - h_{A_{out}}) = \dot{Q}_{B_{out}}$$

and that around fluid B is given by

$$\dot{Q}_{B_{out}} = \dot{m}_B(h_{B_{out}} - h_{B_{in}})$$

EXAMPLE 7.6 A shell and tube heat exchanger is used to cool lubricating oil by water at the rate of 180 kg/min. The oil enters the heat exchanger at 353 K and leaves at 308 K. The specific heat of oil is 3553 kJ/kmol-K. The cooling water enters at 288 K and leaves at 305 K. The specific heat of water is 4.18 kJ/kmol-K. Assuming the heat losses are negligible, estimate the flow rate of water required.

Solution: From the heat balance around the heat exchanger for oil, we have

$$\dot{Q}_{oil} = \dot{m}_{oil} h_{oil} = \dot{m}_{oil} C_{P_{oil}} \Delta t_{oil}$$
$$= 180 \text{ kg/min} \times 3.553 \text{ kJ/kmol-K} \times (308 - 353) \text{ K}$$
$$= -28779.3 \text{ kJ/min}$$

This is the amount of heat transferred from oil to water. Hence, in the same way, the heat balance gives for water

$$\dot{Q}_{water} = \dot{m}_{water} h_{water} = \dot{m}_{water} C_{P_{water}} \Delta t_{water}$$
$$-28779.3 = \dot{m}_{water} \times 4.18 \text{ kJ/kmol-K} \times (305 - 288) \text{ K}$$
$$\dot{m}_{water} = 405 \text{ kg/min}$$

EXAMPLE 7.7   Steam is desired to cool by water in a condenser. Steam enters the condenser at 50 kPa and 50°C with a flow rate of 10 kg/min and leaves at 30°C. The cooling water flows inside the tubes at 15 kPa and 15°C and leaves at 25°C. Determine (a) the mass flow rate of cooling water required and (b) the heat transfer rate in the condenser.

Solution: (a) The mass and energy balance of the above flow system gives

$$\dot{m}_{\text{steam}}(H_1 - H_2) = \dot{m}_{\text{water}}(H_4 - H_3)$$

It is necessary to know the values of enthalpy corresponding to the temperature and that can be obtained from the steam table as

$$H_1 \text{ at } 50°C = 2645.9 \text{ kJ/kg}$$
$$H_2 \text{ at } 30°C = 768.2 \text{ kJ/kg}$$
$$H_3 \text{ at } 15°C = 62.982 \text{ kJ/kg}$$
$$H_4 \text{ at } 25°C = 104.83 \text{ kJ/kg}$$

Substituting, we find

$$10 \text{ kg/min} \times (2645.9 - 768.2) = \dot{m}_{\text{water}}(104.83 - 62.982)$$

or

$$\dot{m}_{\text{water}} = 448.69 \text{ kg/min}$$

(b) The rate of heat transfer can be estimated as

$$\dot{Q} = \dot{m}_{\text{water}}(H_4 - H_3)$$
$$= 448.69 \times (104.83 - 62.982)$$
$$= 18776.77 \text{ kJ/min}$$

7.1.7 Nozzles and Diffusers

A nozzle is a device or contrivance that increases the velocity of a flowing fluid at the expense of pressure drop. It converts the mechanical energy to kinetic energy. The nozzle was developed by the Swedish inventor Gustaf de Laval in the 19th century. Its cross-sectional area changes in the direction of flow and decreases at the throat of the nozzle for subsonic flow and increases for supersonic flow. The operation in a nozzle is accompanied by an increase in kinetic energy.

A diffuser is a device that increases the pressure of a fluid by decreasing its velocity. It converts the kinetic energy to pressure energy. In a diffuser, the variation of cross-sectional area with subsonic and supersonic flow is just the opposite of that of a nozzle. So, the duties of a nozzle and a diffuser are opposite to each other.

Nozzles and diffusers are widely used in rockets, turbines, ejectors, space vehicles, etc.

For the analysis of fluid flow through a nozzle and diffuser, the following assumptions are taken into account:

❖ The flow of fluid is isentropic. As a result, the flow is reversible and adiabatic.
❖ The flow is frictionless ($F = 0$) and has no dissipative losses.
❖ The change in potential energy is zero, i.e., ΔP.E. = 0.
❖ There is no work interaction, i.e., $W = 0$.

❖ There is negligible heat transfer, i.e., $Q = 0$.
❖ The fluid flow behaviour is compressible.

Suppose that a compressible fluid is flowing through a variable cross-sectional area of the conduit. In this case, the continuity equation becomes

$$d(\rho u A) = 0$$

or

$$(\rho A)du + ud(\rho A) = 0$$

or

$$du = -\frac{u}{\rho A} d(\rho A) \tag{7.42}$$

For the isentropic and reversible flows, the energy balance equation without friction, work interaction, and potential change can be reduced to

$$udu + vdP = 0$$

or

$$udu + \frac{dP}{\rho} = 0$$

or

$$dP = -\rho u du \tag{7.43}$$

Substituting Eq. (7.42) into (7.43), we get

$$dP = \frac{u^2}{A} d(\rho A) = \frac{u^2}{A}(\rho dA + A d\rho)$$

or

$$\frac{dP}{d\rho} = \frac{\rho u^2}{A}\left(\frac{dA}{d\rho}\right) + u^2 \tag{7.44}$$

or

$$\frac{dA}{A} = \frac{\delta \rho}{\rho u^2} - \frac{dP}{\rho}$$

or

$$\frac{dA}{A} = \frac{\delta \rho}{\rho u^2}\left(1 - u^2 \frac{dP}{d\rho}\right)$$

For the isentropic flow, $\dfrac{dP}{d\rho} = c^2$, where c is the speed of sound in the fluid.

Putting $\dfrac{dP}{d\rho} = c^2$ into Eq. (7.44), we get

$$\frac{dA}{A} = \frac{dP}{\rho u^2}\left(1 - \frac{u^2}{c^2}\right)$$

or

$$\frac{dA}{A} = \frac{dP}{\rho u^2} (1 - M^2) \qquad (7.45)$$

where $M = \dfrac{u}{c}$ = Mach number.

From Eq. (7.45), it can be concluded that for the nozzle, $dP < 0$, and

(a) $dA < 0$, when $M < 1$, the flow is subsonic ($v < c$) and the convergent nozzle is required;

(b) $dA > 0$, when $M > 1$, the flow is supersonic ($v > c$) and the divergent nozzle is required; and

(c) $dA = 0$, when $M = 1$, and at the throat of the nozzle the subsonic flow of fluid changes to supersonic flow as convergence changes into divergence. Here, the flow velocity of fluid is equal to the local velocity of sound, i.e., $v = c$.

These three cases are corroborated in Fig. 7.12

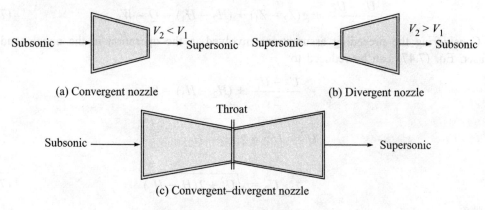

(a) Convergent nozzle (b) Divergent nozzle

(c) Convergent–divergent nozzle

Fig. 7.12   Various kinds of nozzles.

For diffusers, $dP > 0$, and

(a) $dA > 0$, when $M < 1$, the flow is subsonic ($v < c$), and the divergent diffuser is required;

(b) $dA < 0$, when $M > 1$, the flow is supersonic ($v > c$), and the convergent diffuser is required; and

(c) $dA = 0$, when $M = 1$, the supersonic flow of fluid changes to subsonic flow as divergence changes into convergence.

These cases are corroborated in Fig. 7.13

Inlet and Exit Velocity of Stream

In order to calculate the velocity of the fluid stream at entry and exit for nozzle and diffuser, the energy balance equation is given by

$$\frac{\Delta U^2}{2} + g\Delta Z + \Delta H = Q - W_S \qquad (7.46)$$

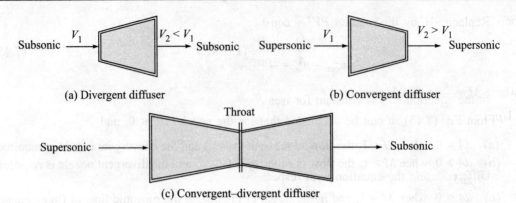

(a) Divergent diffuser   (b) Convergent diffuser

(c) Convergent–divergent diffuser

Fig. 7.13 Various kinds of diffusers.

Equation (7.46) can be written as

$$\frac{U_2^2 - U_1^2}{2} + g(Z_2 - Z_1) + (H_2 - H_1) = Q - W_S \tag{7.47}$$

Considering the preceding assumptions involved in the operation in the nozzle and the diffuser, Eq. (7.47) can be reduced to

$$\frac{U_2^2 - U_1^2}{2} + (H_2 - H_1) = 0$$

or

$$U_2^2 - U_1^2 = 2(H_1 - H_2)$$

or

$$U_2 = \sqrt{U_1^2 + 2(H_1 - H_2)} \tag{7.48}$$

Critical Pressure Ratio

As an isentropic and reversible flow of a fluid behaves ideally with constant heat capacities, the relationship between velocity and pressure in a nozzle can be expressed by the following relations already established:

$$u\,du + V\,dP = 0$$

or

$$u\,du = -V\,dP \tag{7.49}$$

Now, we integrate Eq. (7.49) with the inlet and outlet conditions of the nozzle. The inlet condition is $u = u_{in}$ when $P = P_{in}$ and the outlet condition is $u = u_{out}$ when $P = P_{out}$. Since the process is a reversible isentropic flow of an ideal gas, the pressure–volume relation is

$$PV^\gamma = \text{Constant}$$

Integrating Eq. (7.49), we get

$$u_{out}^2 - u_{in}^2 = -2 \int_{P_{in}}^{P_{out}} V\,dP \tag{7.50}$$

Replacing V by the relation $PV^\gamma = $ constant and integrating further, we get

$$u_{out}^2 - u_{in}^2 = -\frac{2\gamma P_{in} V_{in}}{\gamma - 1}\left[1 - \left(\frac{P_{out}}{P_{in}}\right)^{\frac{\gamma-1}{\gamma}}\right] \qquad (7.51)$$

In the equation PV^γ is constant for isentropic flow of an ideal fluid, V can be replaced by $1/\rho$, i.e.

$$PV^\gamma = \frac{P}{\rho^\gamma} = \text{Constant}$$

Differentiating the equation with respect to ρ at constant entropy S, we get

$$\left(\frac{\partial P}{\partial \rho}\right)_S = \gamma \cdot \frac{P}{\rho} = \gamma PV \qquad (7.52)$$

Equation (7.52) can be solved for the pressure ratio of $\frac{P_{out}}{P_{in}}$. The condition is that the magnitude of U_{out} is maximum and equivalent to the speed of sound.

It can be mathematically expressed as

$$u_{out}^2 = c^2 = \left(\frac{\partial P}{\partial \rho}\right)_S \qquad (7.53)$$

Comparing Eqs. (7.52) and (7.53), we get

$$u_{out}^2 = \gamma PV \qquad (7.54)$$

This is the maximum velocity of fluid attained at the throat. It may increase when the pressure at the outlet to a nozzle decreases. The velocity of the fluid becomes sonic at the throat. The pressure at which this condition is achieved is known as *critical pressure*. It is denoted by P_{cr}. At this critical pressure ratio, $u_{out} = u_{throat}$.

Substituting the above condition in Eq. (7.51), we have

$$\frac{P_{out}}{P_{in}} = \left(\frac{2}{\gamma+1}\right)^{\frac{\gamma}{\gamma-1}} \qquad (7.55)$$

where

$\frac{P_{out}}{P_{in}}$ = Critical pressure ratio

γ = Heat capacity ratio = $\frac{C_P}{C_V}$.

Equation (7.55) can be expressed in terms of Mach number, M, as

$$\frac{P_{out}}{P_{in}} = \frac{1}{\left(1 + \frac{\gamma-1}{2}M^2\right)^{\frac{\gamma}{\gamma-1}}} \qquad (7.56)$$

EXAMPLE 7.8 A nozzle, to which steam at 500 kPa and 623 K is entering at the rate of 12 kg/s and leaving at 500 kPa and 523 K, is fitted to a long pipe. The amount of heat loss to the environment is calculated to be 120 kW. If the velocity of steam at the entrance of the nozzle is assumed to be negligible, then find the velocity at the outlet. We are given that the enthalpy of steam at 500 kPa and 623K is 3168 kJ/kg and the enthalpy of steam at 50 kPa and 523 K is 2976 kJ/kg.

Solution: For the determination of the velocity of steam at the outlet, we consider the energy balance equation. It is given by

$$\frac{\Delta u^2}{2} + g\Delta Z + \Delta H = Q - W_S \tag{7.57}$$

Considering the negligible changes in potential energy and no work done, i.e., $\Delta Z = 0$ and $W_S = 0$, the above equation becomes

$$\frac{\Delta u^2}{2} + \Delta H = Q \tag{7.58}$$

Here

Change in enthalpy $(\Delta H) = (2976 - 3168)$ kJ/kg $= -192$ kJ/kg

Amount of heat loss, $Q = -120$ kW $= -120$ kJ/s

Since

$$Q = \frac{\dot{Q}}{\dot{m}} = \frac{-120 \, \text{kJ/s}}{12 \, \text{kg/s}} = -10 \, \text{kJ/kg}$$

therefore on substituting the values of Q and ΔH into Eq. (7.23), we get

$$\frac{\Delta u^2}{2} + (-192) = -10$$

or

$$\frac{\Delta u^2}{2} = 182 \, \text{kJ/kg}$$

or

$$\frac{u_2^2 - u_1^2}{2} = 182 \, \text{kJ/kg} = 182000 \, \text{J/kg}$$

Neglecting the inlet velocity of steam, we have

$$u_2^2 = 2 \times 182000 \, \text{J/kg}$$
$$u_2 = 603.3 \, \text{m/s}$$

EXAMPLE 7.9 An ideal gas enters a high-velocity nozzle which operates at 1000 kPa and 600 K. The inlet velocity of the gas at the nozzle is 50 m/s. Assume that the heat capacity ratio, $\gamma = 1.4$ and the molecular weight of the gas $= 17$. Determine (i) the critical pressure ratio at the throat and (ii) the velocity at the nozzle outlet for a Mach number of 2 and the discharge pressure.

Solution:

(i) Using Eq. (7.36), we get the critical pressure ratio as

$$\frac{P_{cri}}{P_{in}} = \left(\frac{2}{\gamma+1}\right)^{\frac{\gamma}{\gamma-1}}$$

or

$$\frac{P_{cri}}{P_{in}} = \left(\frac{2}{1.4+1}\right)^{\frac{1.4}{1.4-1}} = 0.52$$

(ii) The velocity at the nozzle outlet can be estimated by the equation as

$$u_{out}^2 - u_{in}^2 = -\frac{2\gamma P_{in} V_{in}}{\gamma-1}\left[1-\left(\frac{P_{out}}{P_{in}}\right)^{\frac{\gamma-1}{\gamma}}\right]$$

Since the gas is an ideal one, so $P_{in}V_{in}$ can be calculated as

$$P_{in}V_{in} = \frac{RT_{in}}{M} = \frac{8314 \times 600}{17} = 293435.29 \text{ m}^2$$

Substituting this value into the preceding equation, we get

$$u_{out}^2 = u_{in}^2 + \frac{2\gamma P_{in} V_{in}}{\gamma-1}\left[1-\left(\frac{P_{out}}{P_{in}}\right)^{\frac{\gamma-1}{\gamma}}\right]$$

or

$$u_{throat}^2 = 50^2 + \frac{2 \times 1.4 \times 293435.29}{1.4-1}\left[1-(0.52)^{\frac{1.4-1}{1.4}}\right]$$

or

$$u_{throat} = 594.76 \text{ m/s}$$

When the Mach number is 2, the discharge velocity at the throat is given by

$$u_{out} = 2u_{throat} = 2 \times 594.76 \text{ m/s} = 1189.52 \text{ m/s}$$

EXAMPLE 7.10 Steam at 800 kPa and 773 K enters a nozzle with an enthalpy of 3480 kJ/kg and leaves at 100 kPa and 573 K with an enthalpy of 3074 kJ/kg.

(a) If the initial enthalpy of the steam is negligible, then determine the final velocity.
(b) If the initial velocity is 40 m/s then what is the final velocity of steam?

Solution:

(a) We know that the velocity of fluid steam at exit of the nozzle is given by

$$U_2 = \sqrt{U_1^2 + 2(H_1 - H_2)}$$

We are given that

$H_1 = 3480 \text{ kJ/kg}$
$H_2 = 3074 \text{ kJ/kg}.$

On neglecting the initial velocity, the preceding equation reduces to

$$U_2 = \sqrt{2(H_1 - H_2)}$$

Putting the values of H_1 and H_2, we get

$$U_2 = \sqrt{2(H_1 - H_2)} = \sqrt{2(3480 - 3074)} = 901.11 \text{ m/s}$$

(b) We are given that $U_1 = 40$ m/s. Hence the final velocity can be estimated as

$$U_2 = \sqrt{U_1^2 + 2(H_1 - H_2)}$$

$$= \sqrt{40^2 + 2(3480 - 3074)}$$

$$= 1553.06 \text{ m/s}$$

EXAMPLE 7.11 Exhaust steam at 100 kPa and 200°C enters the subsonic diffuser of a jet engine steadily with a velocity of 190 m/s. The inlet area of the diffuser is 2000 cm². The steam leaves the diffuser with a velocity of 70 m/s. The pressure increases to 200 kPa. The heat losses from the diffuser to the surroundings is estimated to be 100 kW. Determine

(a) The mass flow rate of the steam
(b) The temperature of the steam leaving the diffuser
(c) The area of the diffuser outlet

given that $V_1 = 2.172$ m³/kg and $H_1 = 2875.3$ kJ/kg.

Solution:

(a) The mass flow rate can be calculated by using the following relation:

$$\dot{m} = \rho_1 U_1 A_1 = \frac{U_1 A_1}{V_1}$$

We are given that $U_1 = 190$ m/s

$$U_2 = 70 \text{ m/s}$$
$$A_1 = 2000 \text{ cm}^2 = 0.2 \text{ m}^2$$
$$V_1 = 2.172 \text{ m}^3/\text{kg}$$
$$H_1 = 2875.3 \text{ kJ/kg.}$$

Therefore, the mass flow rate of steam is

$$\dot{m} = \rho_1 U_1 A_1 = \frac{U_1 A_1}{V_1} = \frac{190 \times 0.2}{2.172} = 17.49 \text{ kg/s}$$

(b) The temperature of the steam leaving the diffuser can also be obtained from the steam table, if we know the amount of heat transferred per unit of steam to the surroundings. The amount of heat transferred per unit of steam is given by

$$Q = \frac{\dot{Q}}{\dot{m}} = \frac{100 \text{ kJ/s}}{17.49 \text{ kg/s}} = 5.717 \text{ kJ/kg}$$

The enthalpy of steam at the diffuser outlet can be estimated as

$$H_2 = Q + H_1 + \frac{U_1^2 - U_2^2}{2}$$

$$= 5.717 + 2875.3 + \frac{190^2 - 70^2}{2} = 18481.01 \text{ kJ/kg}$$

The temperature of steam corresponding to this enthalpy and pressure of 200 kPa obtained from the steam table is $T_2 = 393.38$ K and $V_2 = 1.1123$ m^3/kg.

(c) The outlet area of the steam diffuser is calculated as

$$A_2 = \frac{V_2 \, \dot{m}}{U_2} = \frac{1.123 \times 17.49}{70} = 0.280 \text{ m}^2$$

7.2 UNSTEADY-FLOW PROCESSES AND DEVICES

We have discussed the definition and importance of the steady-flow process. The condition of the unsteady-flow process is just opposite to that of the steady-flow process. This implies that during the process the fluid flows through a control volume under unsteady state condition. The term *unsteady* means that change occurs within a control volume with time. The unsteady process is also known as a *transient process*.

In the following section, the mass and energy balance equation for a control volume under unsteady-state condition has been presented with the help of the first law of thermodynamics and the continuity equation. Some common unsteady-flow processes are: charging or filling of a tank from supply lines, discharging or emptying of a tank, and inflating balloons.

7.2.1 Control Volume Analysis of Unsteady Flow Processes

During an unsteady-flow process, the flow rate of the fluid varies with time. It has been shown in control volume analysis for charging and discharging of a rigid tank illustrated in Fig. 7.14(a) and (b).

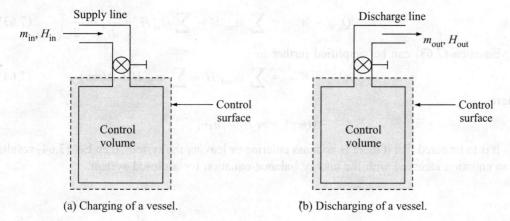

(a) Charging of a vessel.   (b) Discharging of a vessel.

Fig. 7.14

Let us consider the entire tank as a control volume. Since the analysis of the common unsteady-flow process is not an easy task because of the possibility of change in the properties of mass of the inlet and outlet streams during the process, therefore to analyze the filling and emptying process, some important factors are to be taken into account. These are:

(a) The unsteady flow processes are to be treated as uniform flow processes.
(b) The flow of fluid at inlet and outlet is uniform.
(c) The properties of mass of streams entering and leaving the control volume is constant with respect to time.

Hence, from the continuity equation, the mass balance for the uniform flow system can be expressed as

$$\frac{dm_{CV}}{dt} = \sum \dot{m}_{in} - \sum \dot{m}_{out} \tag{7.59}$$

On integration between the initial and final states in the control volume respectively, Eq. (7.59) yields

$$(m_2 - m_1)_{CV} = \sum \dot{m}_{in} - \sum \dot{m}_{out} \tag{7.60}$$

From the energy balance equation obtained from the first law of thermodynamics, we get

$$\frac{dE_{CV}}{dt} + \Delta\left[\dot{m}\left(\frac{u^2}{2} + gZ + H\right)\right] = \dot{Q} - \dot{W}_s \tag{7.61}$$

For the control volume of a uniform flow process, Eq. (7.61) can be written as

$$\dot{Q}_{CV} + \sum \dot{m}_{in}\left(H + \frac{u^2}{2} + gZ\right)_{in}$$

$$= \frac{d}{dt}\left[m\left(H + \frac{u^2}{2} + gZ\right)\right] + \sum \dot{m}_{out}\left(H + \frac{u^2}{2} + gZ\right)_{out} + \dot{W}_{CV} \tag{7.62}$$

For negligible changes in kinetic and potential energy associated with the control volume, Eq. (7.62) reduces to

$$\dot{Q}_{CV} - \dot{W}_{CV} = \sum \dot{m}_{out} H - \sum \dot{m}_{in} H + \frac{d}{dt}\left(\dot{m}\frac{u^2}{2}\right)_{CV} \tag{7.63}$$

Equation (7.63) can be simplified further to

$$Q_{CV} - W_{CV} = \sum m_{out} H - \sum m_{in} H + \Delta(mu)_{CV} \tag{7.64}$$

where

$$\Delta(mu) = m_2 u_2 - m_1 u_1$$

It is to be noted that if there is no mass entering or leaving the system, then Eq. (7.64) results in an equation identical with the energy balance equation for a closed system.

SUMMARY

From the thermodynamic point of view, different steady- and unsteady-flow processes have been analyzed along with the mass and energy balance on the basis of the law of conservation of mass and energy. The basis of the mass and energy balance is expressed as

Rate of input – Rate of output = Rate of accumulation

The mass balance of a flow process in rate form is represented as

$$\dot{m}_{in} - \dot{m}_{out} = \frac{dm_{system}}{dt}$$

It enables the establishment of the continuity equation for the amount of mass flowing through a cross-section per unit time, and is given by

$$\dot{m} = \rho_1 U_1 A_1 = \rho_1 U_1 A_1$$

Similarly, the energy balance equation has been derived from the first law of thermodynamics as well as the law of conservation of energy, and it is in rate form given by

$$\dot{E}_{in} - \dot{E}_{out} = \frac{dE_{system}}{dt}$$

Steady flow process is defined as the process in which the fluid flowing through the control volume is independent of time. Bernoulli's equation and continuity equation have been applied to steady-flow processes such as throttling, compression, and heat exchange operation. The mass and energy balance for a steady flow process are represented as

$$\Delta\left[\dot{m}\left(\frac{u^2}{2} + gZ + H\right)\right] = \dot{Q} - \dot{W}_S$$

The application of thermodynamics is extended to the unsteady-flow processes that are dependent on time, such as charging of a tank, emptying of a tank, and inflating balloons. The mass and energy balance for such a process are expressed as

$$Q_{CV} - W_{CV} = \Sigma m_{out} H - \Sigma m_{in} H + \Delta(mu)_{CV}$$

Hence, the concept of application of thermodynamics to flow processes would thus definitely help engineers to develop and simplify the processes as well as to improve the efficiency of relevant devices.

KEY TERMS

Bernoulli's Equation   A special and restrictive form of the energy equation widely used in fluid mechanics.

Compressor   A device that is capable of compressing the gas to very high pressure.

Continuity Equation   Equation representing the law of conservation of mass in a steady flow system.

Control Volume   A properly selected region in space. It is often called *open system*.

Diffuser A device that increases the pressure of a fluid by decreasing its velocity and so converts kinetic energy to pressure energy.

Ejector A pump-like device, without any moving part or piston, that uses high-velocity steam as a motive fluid to entrain and compress vapours, gases, or a second fluid stream.

Heat Exchanger A device that is used to transfer heat from one fluid to another, in which two fluid streams exchange the thermal energy either through direct contact with each other or through a separating solid wall.

Law of Conservation of Energy Energy can neither be created nor be destroyed. It can only be transformed from one form to another.

Law of Conservation of Mass The net mass transfer during a time interval is equal to the change in the total mass within the control volume.

Nozzle A static device that increases the velocity of a flowing fluid at the expense of pressure drop and so converts mechanical energy to kinetic energy.

Steady Process The process which does not change with time.

Steady-flow Devices The engineering devices which are relevant to the steady-flow processes.

Steady-flow Process The process in which the flow rate of fluid mass through an open system does not vary with time.

Throttling Process A process accompanied by a change in temperature resulting from a reduction in pressure is known as *throttling process*.

Unsteady Process The process which is dependent on time.

IMPORTANT EQUATIONS

1. Continuity equation is given by
$$\rho_1 U_1 A_1 = \rho_2 U_2 A_2 \tag{7.6}$$

2. Energy balance equation for a general steady flow system is given by
$$\Delta \left[\dot{m} \left(\frac{u^2}{2} + gZ + H \right) \right] = \dot{Q} - \dot{W}_S \tag{7.11}$$

3. Steady-flow energy balance equation for a single stream system is given by
$$\Delta \left[\left(\frac{u^2}{2} + gZ + H \right) \right] = \dot{Q} - \dot{W}_S \tag{7.12}$$

4. Steady-flow energy balance equation for a flowing stream experiencing a negligible change in kinetic and potential energy is given by
$$\Delta H = Q - W_S \tag{7.13}$$

5. Mechanical energy balance equation is given by
$$\frac{\Delta U^2}{2} + g\Delta Z + \int_{P_1}^{P_2} V dP + W_S + W_F = 0 \tag{7.18}$$

6. Bernoulli's equation is given by

$$\frac{u^2}{2} + gZ + \frac{P}{\rho} = \text{Constant} \qquad (7.20)$$

7. Joule–Thomson coefficient is given by

$$\mu_{JT} = \left(\frac{\partial T}{\partial P}\right)_H$$

8. Minimum work requirement for compressing an ideal gas from a given state to another state is given by

$$W = RT \ln \frac{V_2}{V_1} = RT \ln \frac{P_1}{P_2} \qquad (7.32)$$

9. Total work requirement for a two-stage compressor operating between P_1 and P_3 with an intermediate pressure P_2 is given by

$$W_{\text{Comp, Total}} = W_{\text{Comp, Stage I}} + W_{\text{Comp, Stage II}}$$

$$= \frac{\gamma RT_1}{\gamma - 1}\left[\left(\frac{P_2}{P_1}\right)^{\frac{\gamma-1}{\gamma}} - 1\right] + \frac{\gamma RT_1}{\gamma - 1}\left[\left(\frac{P_3}{P_2}\right)^{\frac{\gamma-1}{\gamma}} - 1\right] \qquad (7.34)$$

10. Energy balance equation for heat exchanger is given by

$$\dot{m}_A (h_{A_{in}} - h_{A_{out}}) = \dot{m}_B (h_{B_{out}} - h_{B_{in}}) \qquad (7.41)$$

11. For isentropic flow of fluid through nozzle, the relationship between cross-sectional area and velocity of the fluid is given by

$$\frac{dA}{A} = \frac{dP}{\rho u^2}\left(1 - \frac{u^2}{c^2}\right)$$

12. Velocity of fluid stream at entry and exit for nozzle and diffuser is given by

$$U_2 = \sqrt{U_1^2 + 2(H_1 - H_2)} \qquad (7.48)$$

13. The energy balance equation for unsteady-state flow process is given by

$$\frac{dE_{CV}}{dt} = \Delta\left[\dot{m}\left(\frac{u^2}{2} + gZ + H\right)\right] = \dot{Q} - \dot{W}_S \qquad (7.61)$$

14. For control volume of uniform flow process

$$\dot{Q}_{CV} + \sum \dot{m}_{in}\left(H + \frac{u^2}{2} + gZ\right)_{in} = \frac{d}{dt}\left[m\left(H + \frac{u^2}{2} + gZ\right)\right]_{CV}$$

$$+ \sum \dot{m}_{out}\left(H + \frac{u^2}{2} + gZ\right)_{out} + \dot{W}_{CV} \qquad (7.62)$$

15. For negligible change in kinetic and potential energy

$$\dot{Q}_{CV} - \dot{W}_{CV} = \sum \dot{m}_{out} H - \sum \dot{m}_{in} H + \frac{d}{dt}\left(\dot{m}\frac{u^2}{2}\right)_{CV} \qquad (7.63)$$

EXERCISES

A. Review Questions

1. How can you establish the continuity equation?
2. What are the limitations of the continuity equation?
3. Prove: 'Bernoulli's equation is a restrictive form of energy equation'.
4. What are the assumption made for the establishment of Bernoulli's equation?
5. Establish the mechanical energy balance equation starting from the law of conservation of energy for a control volume.
6. Explain the important role of a throttling device for cold production.
7. With the help of a neat sketch, substantiate the significance of a porous plug experiment.
8. When can heating effect instead of cooling effect be produced by a throttling device?
9. Prepare an energy balance over an isothermal compression system.
10. Deduce the expression on the work of compression in a two stage compression system.
11. Derive an expression on the work done in an adiabatic reversible compression process.
12. What is the significance of inter-stage cooling in multistage compression?
13. Develop the energy balance equation for (i) the whole heat exchanger as a control volume and (ii) only one fluid as a control volume.
14. Explain the operation methodology of a nozzle. Mention the application areas of a nozzle and a diffuser.
15. What is the working principle of an ejector? Mention the uses of ejectors.
16. Prepare a mass and energy balance for the unsteady flow process.
17. What is Mach number? How does it play an important role in establishing the relation between the velocity and the cross-sectional area for a nozzle?

B. Problems

1. The inlet and outlet diameters of a conduit are 10 cm and 15 cm respectively. If the velocity of water flowing through the conduit at the inlet is 5 m/s, find the rate of flow through the conduit. Also calculate the velocity of water at the outlet.
2. Water flows through a pipe of 30 cm diameter. The flow splits into two parts and passes through pipes of diameters 25 cm and 15 cm respectively. Find the discharge of the pipe of 15 cm diameter, if the average velocity of water flowing through this pipe is 5 m/s. Determine also the velocity in the pipe of 15 cm diameter if the average velocity in the pipe of 25 cm diameter is 3.5 m/s.
3. A pipe, through which water is flowing, has diameters 30 cm and 15 cm at cross-sections 1 and 2 respectively. The discharge velocity of the pipe is 40 L/s. Cross-section 1 is 8 m above and cross-section 2 is 6 m above the reference level. If the pressure at cross-section 1 is 5 bar, determine the pressure at cross-section 2.
4. A shell and tube heat exchanger is used to cool lubricating oil by water at the rate of 120 kg/min. The oil enters the heat exchanger at 343 K and leaves at 298 K. The specific heat of oil is 3202 kJ/kmol-K. The cooling water enters at 288 K and leaves at 300 K.

The specific heat of water is 4.18 kJ/kmol-K. Assuming the heat losses are negligible, estimate the flow rate of water required.

5. A nozzle, through which steam at 500 kPa and 623 K is entering at the rate of 12 kg/s and leaving at 500 kPa and 523K, is fitted to a long pipe. The amount of heat loss to the environment is calculated to be 120 kW. If the velocity of steam at the entrance of the nozzle is assumed to be negligible, then find the velocity at the outlet, given that the enthalpy of steam at 500 kPa and 623 K is 3168 kJ/kg and the enthalpy of steam at 50 kPa and 523 K is 2976 kJ/kg.

6. Steam at 800 kPa and 773 K enters a nozzle with an enthalpy of 3480 kJ/kg and leaves at 100 kPa and 573 K with an enthalpy of 3074 kJ/kg.

   (a) If the initial enthalpy of the steam is negligible, determine the final velocity.
   (b) If the initial velocity is 40 m/s, what is the final velocity of the steam?

7. Exhaust steam at 100 kPa and 200°C enters the subsonic diffuser of a jet engine steadily with a velocity of 190 m/s. The inlet area of the diffuser is 200 cm^2. The steam leaves the diffuser a the velocity of 70 m/s. The pressure increases to 200 kPa. The heat losses from the diffuser to the surroundings is estimated to be 100 kW. Determine

   (a) The mass flow rate of steam
   (b) The temperature of the steam leaving the diffuser
   (c) The area of the diffuser outlet.

8. A single-stage compressor is used to compress 1000 m^3/hr of air at 100 kPa and 300 K to 600 kPa. The isentropic efficiency of the compressor is 80%. Estimate the power input to the compressor. Take $\gamma = 1.3$.

9. Air enters an adiabatic compressor at 100 kPa and 300 K at a rate of 0.5 m^3/s and leaves at 600 kPa. Air is assumed to behave as an ideal gas. Determine the work of compression per unit mass for

   (a) Reversible adiabatic compression when $\gamma = 1.4$
   (b) Isothermal compression
   (c) Single-stage compression. Take $\gamma = 1.3$.

10. Air enters an adiabatic compressor at 150 kPa and 350 K at a rate of 0.8 m^3/s and leaves at 700 kPa and 500 K. Neglecting the changes in kinetic and potential energy, estimate the power input to the compressor.

11. CO_2 at 100 kPa and 330 K is to be compressed steadily in an adiabatic compressor to 500 kPa and 450 K. The mass flow rate of air is 0.05 kg/s. Assuming that the changes in the kinetic and potential energies are negligible and CO_2 is assumed to behave as an ideal gas, compute the power requirement of the compressor.

12. A two-stage compressor is used to compress CO_2 at a rate of 500 m^3/hr from an initial state of 100 kPa and 300 K to an exit pressure of 700 kPa. Determine the power required to run the compressor. Take $\gamma = 1.3$.

Refrigeration and Liquefaction Processes

8

LEARNING OBJECTIVES

After reading this chapter, you will be able to:

- Understand basic ideas about the refrigerating machine and heat pump, and derive their performance indexes
- Demonstrate the operation of a refrigerator based on the reversed Carnot cycle
- Analyze the vapour–compression, vapour–absorption and air–refrigeration cycles and evaluate their performance capability
- Review the factors associated with the selection of a refrigerant
- Perform a comparative study between refrigeration systems
- Enumerate the applications of refrigeration
- Understand the different liquefaction processes
- Evaluate various ideas about the application of liquefaction

---

In the earlier days, when people had widespread belief that nothing could be colder than ice, natural ice at 0°C was used for refrigeration purposes. This was quite inadequate to meet larger requirements. The field of refrigeration on a large scale was first developed in the nineteenth century. By the end of the nineteenth century, mechanical refrigeration had become a practical reality for the production of appreciable low temperature as we know it today.

Refrigeration is a significant area of application of thermodynamics, as the process works on the principle of the second law of thermodynamics. It is the process in which transfer of heat takes place from a low-temperature region to a relatively high-temperature one. It requires some external work input. The device that substantiates it and produces the cooling effect is known as *refrigerator* or *refrigerating machine*. The most conventional refrigeration system includes the vapour–compression refrigeration cycle.

In the vapour–compression refrigeration cycle, the refrigerant—the working media—carries heat from source (low-temperature region) to sink (high-temperature region), and undergoes a series of operations such as vaporization, condensation, expansion, and compression. Another well-known refrigeration system is vapour–absorption refrigeration, where the refrigerant is dissolved in a liquid before it is compressed. The other refrigeration system, discussed in this chapter, is air or gas refrigeration system, in which the refrigerants remain in the gaseous phase throughout.

Following several international regulatory activities on the protection of the environment, such as Montreal Protocol, Vienna Conference, and Kyoto Protocol, it is necessary to review critically the desirable properties of refrigerants, as these factors play an important role in selecting a promising refrigerant. In addition to the refrigeration systems, the concept of liquefaction process by vaporization, free expansion, and isentropic expansion of liquid have been highlighted.

8.1 REFRIGERATION

Refrigeration may be defined as the process by which heat energy is abstracted from a region of lower temperature and discarded (delivered) to a region of relatively higher temperature by expending mechanical work. Refrigeration is an indispensable and versatile technique in various emerging fields and multi-dimensional areas of application such as food, beverages, medicine, blood preservation, ice manufacturing, medical treatment, and transportation. This process is well-known for its utilization in chemical process industries in manufacturing synthetic rubber, textiles, plastics, etc. It also finds significant industrial applications in the petroleum industry for the purification of lubricating oil, separation of volatile hydrocarbons, separation of gasoline from natural gas, etc.

Apart from its extensive use in the air-conditioning of buildings and in domestic refrigerators to provide a comfortable and healthy living environment, refrigeration is also employed in the maintenance of low temperature in the manufacture of rayon, photographic films, and in cooling car cabins.

The large-scale commercial application of refrigeration principle includes the liquefaction process for the production of pure gases such as O_2 and N_2 from air. The term *refrigeration* implies the maintenance and production of temperature below that of the surroundings in a given space or substance. This is made possible by the absorption of heat at a low temperature level and release of the same at a high temperature level. This process is usually accomplished by evaporation of liquid having low saturation temperature at the pressure of evaporation and returning of the vapour to its original liquid state for re-evaporation to continue the process. The complete cycle consists of four operational steps—compression of the refrigerant, condensation of the vapour into liquid, expansion of the liquid, and finally the evaporation of the liquid refrigerant.

8.1.1 Refrigerators and Heat Pumps

It is well known to us from our common experience and as per the nature of the heat flow that whenever temperature difference exists between two regions, heat flows in the direction of decreasing temperature, i.e., from a higher-temperature region to a lower-temperature one, and the heat transfer process occurs in nature without requiring any devices. But the reverse process cannot occur by itself. The transfer of heat from a low-temperature medium (source) to a high-temperature one (sink) requires some special devices called *refrigerators*.

This mechanism of a refrigerator employs virtually a substance used as working fluid for carrying heat from the source to the sink. The fluid is known as *refrigerant*. It is needless to mention that the efficiency of a refrigerator, as well as the economy of cold production in the small-scale or large-scale application area, dominantly depends on the thermo-physical properties and role of a working fluid. The different properties of refrigerants will be discussed later in this chapter.

A refrigerator is schematically represented in Fig. 8.1(a) to give a transparent idea about its basic working principle as well as the direction of heat transfer. Here, Q_L is the amount of heat removed from the refrigerated space at low temperature T_L, Q_H is the amount of heat released to the hot atmosphere at high temperature T_H, and W_{net} indicates the net work input to the refrigerator.

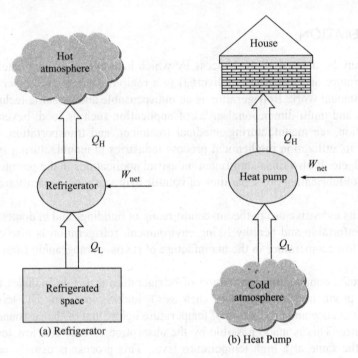

Fig. 8.1 Refrigerator and heat pump.

Now, the heat pump is another device by which heat flows from a low-temperature region to a high-temperature one. Refrigerators and heat pumps are basically the same device, they only differ in their objectives, in that the high-temperature output—heat is desired for heating and the low-temperature input heat is desired for cooling.

The objective of a refrigerator is to maintain the refrigerated space or region at low temperature by removing heat from it. Discarding this heat to a high-temperature region is a necessary part of this operation. On the other hand, the objective of a heat pump is to maintain the heated space (a region to be heated) at high temperature by delivering heat. This is accomplished by absorption of heat from a low-temperature source such as a river, or the surrounding land, or the cold outside air in winter, and supplying heat to a warmer medium such as a house. It has been depicted in Fig. 8.1(b).

The heat pump is a reversed heat engine and a device which can be used for heating houses and commercial buildings during the winter and cooling them during summer.

During winter, the liquid refrigerant is allowed to evaporate in coils kept in the outside air. Heat is absorbed from the low-temperature source and the liquid gets vaporized during this stage. The vapour is then compressed to such a pressure level that it can be condensed at a high temperature. In the condenser, the heat is transferred to cooling water or air, which is used for heating houses. During summer, the heat pump can also be used as an air-conditioning unit by reversing only the flow of the refrigerant. It is necessary to remember that the heat pumps and air-conditioners have the same mechanical components. Therefore, one system can be used as a heat pump in winter and an air-conditioner in summer. This is done by incorporating a master valve in the cycle, which can reverse the flow of the refrigerant. This would ensure the utilization of one system instead of two separate systems, which would be economical for heating and cooling of a house. Here the heat is absorbed from the buildings and released to the outside air. The twofold use of a heat pump is represented in Fig. 8.2(a) and (b).

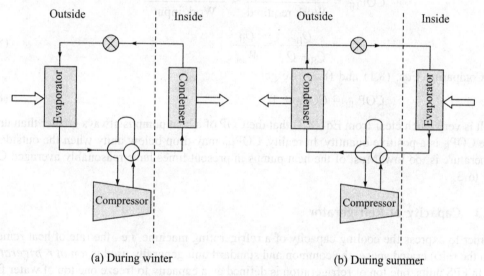

Fig. 8.2 Dual operation of heat pump.

8.1.2 Co-efficient of Performance

A refrigerator is said to be most efficient if it produces a maximum amount of desired effect with a minimum expenditure of work. The desired effect produced by a refrigerating machine is known as *refrigerating effect*.

The efficiency or performance capability of a refrigeration system is judged by the ratio between the refrigerating effects, or the amount of heat removed and the work required.

This is expressed in terms of a suitable index known as the *co-efficient of performance*. It is denoted by COP_R.

Referring to Fig. 8.1(a), if Q_L is the amount of heat removed at the low temperature T_L and Q_H is the amount of heat released at high temperature T_H, then by the first law of thermodynamics the external work required for transferring the heat is

$$W_{net} = Q_H - Q_L$$

and the COP_R of a refrigerator can be expressed as

$$COP_R = \frac{\text{Heat removed}}{\text{Work required}} \quad \text{or} \quad \frac{\text{Refrigerating effect}}{\text{Work input}}$$

$$= \frac{Q_L}{Q_H - Q_L} = \frac{Q_L}{W_{net}} \tag{8.1}$$

It is necessary to remember that in most of the cases, the COP of a refrigerator can be greater than unity, i.e., the amount of heat removed from the refrigerated space can be greater than the amount of work input and normally varies between 2 and 7.

The performance capability of a heat pump is also expressed in terms of the co-efficient of performance, denoted by COP_{HP}. It is given by

$$COP_{HP} = \frac{\text{Heat absorbed}}{\text{Work required}} = \frac{\text{Heating effect}}{\text{Work input}}$$

$$= \frac{Q_H}{Q_H - Q_L} = \frac{Q_H}{W_{net}} \tag{8.2}$$

Comparing Eqs. (8.1) and (8.2), we get

$$COP_{HP} = COP_R + 1 \tag{8.3}$$

It is very much clear from Eq. (8.3) that the COP of a heat pump is always greater than unity, since COP_R is a positive quantity. In reality, COP_{HP} may drop below unity when the outside air temperature is too low. Most of the heat pumps in present times have reasonably averaged COP of 2 to 3.

8.1.3 Capacity of Refrigerator

In order to express the cooling capacity of a refrigerating machine, i.e., the rate of heat removal from the refrigerated space, the common and standard unit generally used is *ton of refrigeration*.

In FPS units, one ton of refrigeration is defined as a capacity to freeze one ton of water from 0°C (32 F) in 24 hours. It is the withdrawal of heat at a rate of 200 BTU/min. The latent heat of ice is 144 BTU per lb.

$$\text{Ton of refrigeration} = \frac{2000 \text{ lb} \times 144 \text{ BTU/lb}}{24 \times 60} = 200 \text{ BTU/min} \tag{8.4}$$

In MKS units, it is defined to be equal to cooling at the rate of 72,000 kcal per 24 hours, or 300 kcal per hour, or 50 kcal per minute, i.e.

$$1 \text{ ton of refrigeration} = \frac{72,000 \text{ kcal}}{24 \times 60} = 50 \text{ kcal/min} \tag{8.5}$$

$$= 200 \text{ BTU/min}$$

$$= 200 \times 1.055 \text{ kJ/min (since 1 BTU = 1.055 kJ)}$$

$$= \frac{200 \times 1.055}{60} \text{ kJ/s}$$

$$= 3.517 \text{ kW} \tag{8.6}$$

It is to be noted that a one-ton refrigerating machine will not produce one ton of ice. Ice-making capacity is defined as the ability of a refrigerating system to make ice, beginning with water at room temperature, which is usually higher than 0°C (273 K). The ice-making capacity for a given machine is always less than its refrigerating capacity, because of the necessity of pre-cooling the water and sub-cooling the ice below the freezing point, in addition to actual freezing. So, the ice-making capacity of a machine should not be confused with its capacity.

8.1.4 Carnot Refrigeration Cycle

Carnot cycle is a reversible cycle. Since it is a reversible cycle, all processes that constitute the Carnot cycle as well as the direction of all heat and work interactions are reversible. Hence, a cycle that operates in the counterclockwise or reverse order on a T–S diagram is known as a *reversed Carnot cycle*.

A refrigerator that operates on the reversed Carnot cycle is known as a Carnot refrigerator. Its scheme is shown in Fig. 8.3. A reversed Carnot engine or Carnot refrigerator is a thermodynamically ideal isothermal-source refrigerator. The performance of all refrigeration systems are judged with respect to the Carnot refrigeration cycle working between the same limits of temperature. The ideal refrigerator consists of two reversible isothermal (constant-temperature) processes, in which Q_L, the amount of heat, is absorbed at the lower temperature T_L and heat Q_H is released at the higher temperature T_H, and two reversible adiabatic/isentropic (constant-entropy) processes.

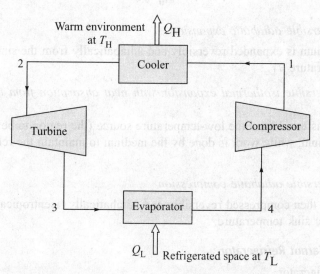

Fig. 8.3 Schematic diagram of Carnot refrigeration cycle.

The steps in the operation of a Carnot refrigerator can be discussed with the help of T–S and P–V diagrams, shown in Fig. 8.4(a) and (b), respectively.

The processes involved in the Carnot refrigerator are as follows:

Process 1–2 *Reversible isothermal compression*
The working medium is compressed, while energy is released to the sink (high-temperature region at T_H) to maintain the refrigerant temperature at a constant level.

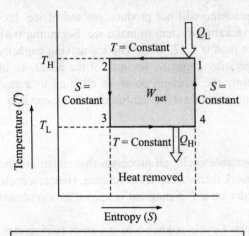

Legend
1–2: Isothermal compression with heat rejection
2–3: Adiabatic expansion
3–4: Isothermal expansion with heat absorption
4–1: Adiabatic compression

(a) T–S diagram of Carnot refrigeration cycle.

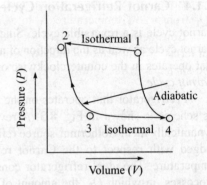

(b) P–V diagram of Carnot refrigeration cycle.

Fig. 8.4

Process 2–3 *Reversible adiabatic expansion*

The working medium is expanded reversibly and adiabatically from the sink temperature T_H to the source temperature T_L.

Process 3–4 *Reversible isothermal expansion with heat absorption from the low-temperature source*

Energy is transferred from the low-temperature source (the region to be cooled) at T_L to the refrigeration medium, while work is done by the medium to maintain the refrigerant at constant temperature.

Process 4–1 *Reversible adiabatic compression*

The refrigerant is then compressed reversibly and adiabatically (isentropically) from the source temperature to the sink temperature.

Performance of Carnot Refrigerator

For a Carnot refrigerator

$$\text{Heat absorbed} = Q_H = T_H (S_2 - S_1)$$
$$\text{Heat removed} = Q_L = T_L (S_2 - S_1)$$
$$\text{Work done} = W_{net} = Q_H - Q_L = T_H (S_2 - S_1) - T_L (S_2 - S_1)$$
$$\text{COP}_{\text{Carnot refrigerator}} = \frac{Q_L}{W_{net}} = \frac{Q_L}{Q_H - Q_L} = \frac{T_L (S_2 - S_1)}{T_H (S_2 - S_1) - T_L (S_2 - S_1)} \tag{8.7}$$

or

$$\frac{Q_L}{W_{net}} = \frac{T_L}{T_H - T_L} \tag{8.8}$$

or

$$W_{net} = Q_L \frac{T_L}{T_H - T_L} \tag{8.9}$$

Equation (8.9) gives a clear idea about the minimum work required for conveying the heat Q_L from a low temperature T_L to a high temperature T_H.

Considering Eqs. (8.7), (8.8), and (8.9) involved in understanding and expressing efficiency in terms of heat removal and work done, the characteristics of a Carnot refrigeration cycle can be summarized as follows:

- No refrigeration cycle can have a higher COP than a reversible cycle operating between the same temperatures.
- The COP of a Carnot refrigeration cycle is a function only of the upper and lower temperatures of the cycle, and it is true that the reversed Carnot cycle is the most efficient refrigeration cycle operating between these two specified temperature levels.
- Equation (8.5) indicates that the coefficient of performance of a reversed Carnot cycle increases as the difference between the upper and lower temperatures decreases.
- The value of T_L is more effective than that of T_H on the COP of a cycle.
- The efficiency of a Carnot refrigeration cycle does not depend on the working fluid.
- It is not possible to construct a refrigerating machine which could convey heat from a low-temperature to a high-temperature region with a lower expenditure of work than that given by Eq. (8.9).
- In practice, the COP of an actual refrigeration cycle is always less than that of an ideal one.

Hence it can be concluded that the Carnot cycle cannot be executed in actual devices. This is not a realistic model for the refrigeration cycle. This can be treated as a standard against which the actual refrigeration cycles are compared.

Limitations of Reversed Carnot Cycle

Carnot cycle is a completely reversible cycle. It consists of two reversible-isothermal and two reversible-isentropic processes. It does necessarily mean that all the processes are carried out in reversible fashion. But, in practice, the reversible process cannot be executed since it requires infinite time. Thus, in all real processes it is difficult to avoid the irreversibilities, and one has to reckon with the finite time available for the execution of these processes.

Among the four reversible processes in a reversible Carnot cycle, the two isothermal heat transfer processes are easily achievable in practice. The processes, i.e., isothermal heat absorption and isothermal heat rejection take place in the evaporator and the condenser respectively. These processes require infinite time, i.e., they have to be executed very slowly, for which significant heat transfer area is required. In spite of this, two isothermal processes are not difficult to achieve, but the processes of adiabatic compression and adiabatic expansion cannot be executed in a practical refrigeration cycle. These have some limitations as discussed in the following lines:

Problems on Compression

Firstly, the compression process in the reversible Carnot cycle involves the wet compression of the liquid–vapour mixture. In practice, when a reciprocating compressor is used, wet compression is not found suitable. After evaporation of refrigerant, it enters the compressor. Due to the appreciable speed of the compressor, the piston reciprocates so shortly that the vapour-bound liquid may remain in the cylinder, which may lead to mechanical damage to the compressor valves and even to the cylinder.

Secondly, the vapour-bound liquid droplets of refrigerant may wash away the lubricating oil in the compressor, which contributes to wearing of the cylinder.

Problems on Expansion

Process 2–3 represents the adiabatic expansion of high-moisture content or liquid refrigerant in a turbine. This process would be executed with infinite speed. Such speed fluctuations in actual devices are not acceptable. This is an impracticality in expanding a liquid or highly wet vapour against a fast-moving piston in a cylinder.

This problem can be overcome if the turbine in a Carnot cycle is replaced by a throttle valve or a very long narrow-bore capillary tube, which substantiates the necessary pressure drop from the condenser pressure to the evaporator pressure and maintains the required mass flow rate.

EXAMPLE 8.1 A Carnot refrigerator is used to maintain a food compartment at $-4°C$. Heat is released from the compartment to a room at $30°C$. If 35 kW heat is removed, determine

(a) The coefficient of performance of the cycle;
(b) The power required;
(c) The rate of heat released to the room.

Solution: Here

Temperature at which heat is absorbed, $T_L = -4°C = 269$ K
Temperature at which heat is removed, $T_H = 30°C = 303$ K.

(a) The coefficient of performance of a Carnot refrigerating machine is given by

$$COP_{\text{Carnot refrigerator}} = \frac{T_L}{T_H - T_L} = \frac{269}{303 - 269} = \frac{269}{34} = 7.91 \approx 8$$

(b) The power requirement of the refrigerator is given by

$$W_{\text{net}} = \frac{Q_L}{COP_{\text{Carnot refrigerator}}} = \frac{30 \text{ kW}}{8} = 3.75 \text{ kW}$$

(c) The rate of heat rejection to the room can be estimated by

$$Q_H = W_{\text{net}} + Q_L \quad (\text{since } W_{\text{net}} = Q_H - Q_L)$$
$$= (30 + 3.75) \text{ kW}$$
$$= 33.75 \text{ kW}$$

EXAMPLE 8.2 (a) A one-ton Carnot refrigerating machine is used to keep a refrigerated space at the temperature of $-10°C$, while the environment is at $45°C$. Determine the power consumption of the machine.

(b) If the machine is used to maintain a freezer box at the temperature of $-20°C$ while the temperature of environment is $45°C$, then how much cooling effect could it produce?
Assume that the power consumption in both the cases are same.

Solution:

(a) Here

Temperature at which heat is absorbed, $T_L = -10°C = 263$ K
Temperature at which heat is removed, $T_H = 45°C = 318$ K
The coefficient of performance of a Carnot refrigerating machine is given by

$$\text{COP}_{\text{Carnot refrigerator}} = \frac{T_L}{T_H - T_L} = \frac{263}{318 - 263} = \frac{263}{55} = 4.78$$

The power consumption of the refrigerating machine is given by

$$W_{\text{net}} = \frac{Q_L}{\text{COP}_{\text{Carnot refrigerator}}} = \frac{1 \text{ ton}}{4.78} = 0.735 \text{ kW}$$

(b) Here, $T_L = -20°C = 253$ K and $T_H = 45°C = 318$ K.

The coefficient of performance of a Carnot refrigerating machine is given by

$$\text{COP}_{\text{Carnot refrigerator}} = \frac{T_L}{T_H - T_L} = \frac{253}{318 - 253} = \frac{253}{65} = 3.89$$

Cooling effect produced, $Q_L = W_{\text{net}} \text{ COP}_{\text{Carnot refrigerator}}$

$$= 0.735 \text{ kW} \times 3.89 = 2.859 \text{ kW}$$

EXAMPLE 8.3 Assume that the Carnot refrigerating machine in Example 8.2 is of unknown capacity. Considering both parts (a) and (b), what would be the increase in the percentage of work input required for the freezer box over the refrigerated space for the same extent of refrigerating effect?

Solution: For refrigerated space:
The co-efficient of performance of the Carnot refrigerator obtained for the first case is 4.78. Now the power consumption would be

$$W_{\text{net, in (1)}} = \frac{Q_L}{\text{COP}_{\text{Carnot refrigerator}}} = \frac{Q_L}{4.78} = 0.209 Q_L$$

For freezer box:
The co-efficient of performance of the Carnot refrigerator obtained for the second case is 3.89. Now the power consumption would be

$$W_{\text{net(2)}} = \frac{Q_L}{\text{COP}_{\text{Carnot refrigerator}}} = \frac{Q_L}{3.89} = 0.257 Q_L$$

Hence, the increase in percentage of work input $= \dfrac{0.257\, Q_L - 0.209\, Q_L}{0.209\, Q_L} \times 100$

$$= 22.96\%$$

EXAMPLE 8.4   A Carnot heat pump is used to maintain a temperature of 24°C inside a building when the outside temperature is 0°C. The amount of heat supplied by the heat pump is 25 kW. Estimate the co-efficient of performance and the power input of the heat pump.

Solution: The COP of a heat pump is given by

$$COP_{\text{Carnot heat pump}} = \frac{T_H}{T_H - T_L} = \frac{297}{297 - 273} = \frac{297}{24} = 12.375$$

The power input can be estimated as

$$W_{\text{net}} = \frac{Q_H}{COP_{\text{Carnot heat pump}}} = \frac{25\ \text{kW}}{12.375} = 2.02$$

EXAMPLE 8.5   In peak winter, it is desired to heat a house with the help of a heat pump when the temperature of outside air is −2°C. If the inside of the house is maintained at 20°C and the house loses 80,000 kJ/h, what is the minimum power input required?

Solution: We know from the ratio of heat and temperature of the two regions that

$$\frac{Q_H}{T_H} = \frac{Q_L}{T_L}$$

$$Q_L = \frac{Q_H}{T_H} T_L = \frac{80,000 \times 271}{293} = 73993.17\ \text{kJ/h}$$

The power input required is

$$W_{\text{net}} = Q_H - Q_L = (80,000 - 73993.17)\ \text{kJ/h}$$
$$= 6006.83\ \text{kJ/h} = 1.668\ \text{kW}$$

8.1.5   Vapour–Compression Refrigeration Cycle

The difficulties associated with the Carnot refrigeration cycle can be successfully eliminated by introducing the vapour–compression refrigeration cycle through the incorporation of two modifications—the complete evaporation of the working fluid before entering into the compressor and replacement of the turbine by a throttling device such as expansion valve or capillary tube.

The vapour–compression cycle is used in most of the domestic refrigerators for cooling bodies to a temperature of −20°C (253 K). These widespread refrigeration systems use low-boiling liquid, for instance, ammonia, freon, and sulphurous anhydride, as refrigerant at low pressures (preferably close to the atmospheric pressure). The refrigerant is sealed in an airtight and leak-proof mechanism and circulated through the system.

The *compression cycle* is provided this name because it involves the compression of the refrigerant by the compressor, which permits the transfer of heat energy. The refrigerant absorbs heat from one place and releases it to another place. The cycle is based on the fact that the vapour pressure of a fluid varies with temperature, and therefore a boiling liquid can be made to absorb heat at one temperature and reject heat at a higher temperature by compressing and condensing the vapour thus generated.

Most of the common domestic and industrial refrigerators work on this cycle. The requirements of fundamental operation of a vapour–compression cycle can be divided into four simple parts:

(i)   Compression
(ii) Condensation
(iii) Expansion
(iv) Vaporization.

These can be explained by the arrangement of the apparatus, which appears in Fig. 8.5. The schematic diagram of a simple vapour–compression cycle is shown in Fig. 8.5 and the corresponding $T–S$ diagram in Fig. 8.6.

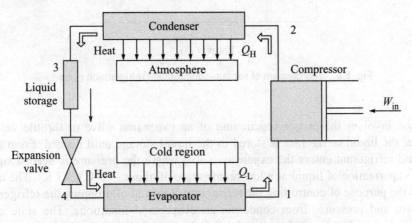

Fig. 8.5   Schematic diagram of vapour–compression system.

Compression

This operational step employs a compressor. The function of the compressor is twofold. One function is that of withdrawing the fluid from the evaporator at a rate that is sufficient to maintain the necessary reduced pressure and temperature in the evaporator. The other is that of compressing and delivering the fluid, at a temperature which is adequately above that of the atmosphere or of the region or substance to which the fluid must next discard its load of energy. Compression of vapour requires the least work when performed isentropically. The state change is represented by line 1–2 in Fig. 8.5 and in the $T–S$ diagram in Fig. 8.6, which show isentropic (or adiabatic) compression from saturated vapour to condenser pressure.

Condensation

The high-pressure refrigerant vapour enters the condenser, and heat is removed from it to change the superheated vapour to saturated or sub-cooled liquid. The state change is represented by line 2–3 in Fig. 8.5 and in the $T–S$ diagram of Fig. 8.6, which show reversible rejection of heat at constant pressure (de-superheating and condensation).

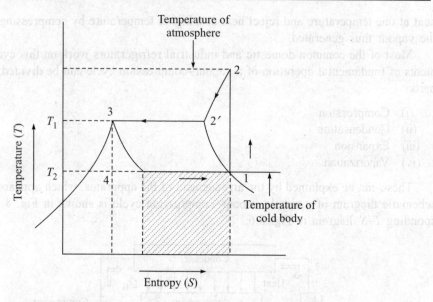

Fig. 8.6   *T–S* diagram of vapour–compression refrigeration cycle.

Expansion

This operation involves the proper functioning of an expansion valve or throttle valve. After condensation, the liquid refrigerant is stored in the liquid storage until needed. From the liquid storage, liquid refrigerant enters the expansion valve where the pressure is needed sufficiently to allow the vaporization of liquid at a low temperature of about $-10°C$ (263 K). The expansion valve serves the purpose of controlling the refrigerant flow and of dropping the refrigerant, both in temperature and pressure, from condenser to evaporator conditions. The state change is represented by line 3–4 in Fig. 8.5 and in the *T–S* diagram in Fig. 8.6, which show irreversible expansion from saturated (or sub-cooled liquid) to the evaporator.

Vaporization

The low-pressure and low-temperature refrigerant, in its liquid state and leaving from the expansion valve changes into vapour as it absorbs a considerable amount of heat from the region or substance which is desired to refrigerate in the evaporator. The vapour leaving the evaporator may be wet, dry-saturated, or superheated; depending on this the process is called *dry* or *wet* *compression*. The state change is represented by line 4–1 in Fig. 8.5 and in the *T–S* diagram in Fig. 8.6, which show reversible addition of heat at constant pressure from evaporating to saturated vapour.

The pressure–enthalpy diagram is very frequently used and is more helpful than the temperature–entropy diagram in the analysis of a vapour–compression refrigeration cycle, because the *P–H* diagram directly shows the enthalpy required for different processes. The *P–H* diagram is shown in Fig. 8.7. In the *P–H* diagram the three processes—evaporation, condensation and expansion—are represented by straight lines. The heat delivered to the evaporator and the heat discarded from the condenser are proportional to the length of the corresponding process curves.

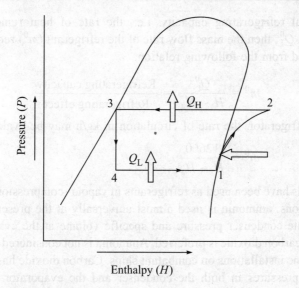

Fig. 8.7 *P–H* diagram of vapour–compression refrigeration cycle.

Performance of Vapour–Compression Cycle

With the help of the *P–H* diagram, the performance of the several phases of the cycle, both for the actual operation and for the ideal cycle, per unit mass of the refrigerant vapour circulated in the system has been calculated in the following way:

At constant pressure

$$dQ = dH$$

For the compressor

$$W = (H_2 - H_1) \text{ kcal/kg} \qquad \text{for isentropic compression}$$

For the condenser

$$Q_H = (H_2 - H_3) \text{ kcal/kg}$$

For the expansion valve

$$H_3 = H_4$$

For the evaporator

$$Q_L = (H_1 - H_4) \text{ kcal/kg}$$

Refrigerating effect $= (H_1 - H_4)$ kcal/kg

$$\text{COP}_{\text{vapour-compression cycle}} = \frac{\text{Refrigerating effect}}{\text{Work input}} = \frac{Q_L}{W_{\text{net}}} \qquad (8.10)$$

$$= \frac{H_1 - H_4}{H_2 - H_1} \qquad (8.11)$$

Now, if the total refrigerating capacity, i.e., the rate of heat removal from the low-temperature source is Q_L^0, then the mass flow rate of the refrigerant (m^0) required for circulation can be easily obtained from the following relation:

$$m^0 = \frac{Q_L^0}{H_1 - H_4} = \frac{\text{Refrigerating capacity}}{\text{Refrigerating effect}} \tag{8.12}$$

For a one-ton refrigerator, the rate of circulation in kg/h may be evaluated by

$$m^0 = \frac{12660}{H_1 - H_4} \tag{8.13}$$

A variety of fluids have been used as refrigerants in vapour–compression systems. For larger commercial installations, ammonia is used almost universally at the present time. It offers the advantage of moderate condenser pressure and specific volume at the evaporator pressure. In marine installations, carbon dioxide is preferred. Ammonia is not considered desirable on account of its toxicity in marine installations on combatant ships. Carbon dioxide has the disadvantage of requiring very high pressures in both the condenser and the evaporator. For the small self-contained household installation, SO_2 vapour is very popular. It offers the advantage of moderate pressure, but requires considerably more compressor displacement on account of fairly high specific volume at the evaporator pressure. Freon (fluoro-and-chlorine derivatives of hydro-carbon), ethyl chloride and propane are also used.

EXAMPLE 8.6 A refrigerant using refrigerant R-12 as the working fluid operates on vapour-compression cycle between 273 K and 313 K. Compute the following:

(i) COP of the vapour–compression cycle
(ii) Refrigerating effect
(iii) COP of an ideal Carnot refrigerator
(iv) Work of compression

Given that

Enthalpy of saturated vapour at 273 K, $H_1 = 187$ kJ/kg

Enthalpy of saturated liquid at 313 K, $H_3 = H_4 = 74$ kJ/kg

Enthalpy of superheated vapour at 273 K, $H_2 = 204$ kJ/kg.

Solution: From the given data, we have

$$H_1 = 187 \text{ kJ/kg}$$
$$H_2 = 204 \text{ kJ/kg}$$
$$H_3 = H_4 = 74 \text{ kJ/kg}$$

(i) The coefficient of performance of the vapour-compression cycle is given by

$$COP_{\text{vapour-compression cycle}} = \frac{\text{Refrigerating effect}}{\text{Work input}} = \frac{Q_L}{W_{\text{net}}}$$

$$= \frac{H_1 - H_4}{H_2 - H_1} = \frac{187 - 74}{204 - 187}$$

$$= \frac{113}{17} = 6.64$$

(ii) Refrigerating effect can be obtained by

$$H_1 - H_4 = 187 - 74 = 113 \text{ kJ/kg}$$

(iii) The COP of an ideal Carnot refrigerator is given by

$$\text{COP}_{\text{Carnot refrigerator}} = \frac{T_L}{T_H - T_L} = \frac{273}{313 - 273} = \frac{273}{40} = 6.82$$

(iv) The work of compression can be estimated as

$$H_2 - H_1 = 204 - 187 = 17 \text{ kJ/kg}$$

EXAMPLE 8.7 A vapour–compression refrigeration cycle in which the refrigerant HFC-134a enters the compressor as superheated vapour at 0.18 MPa and $-10°C$ at a rate of 0.06 kg/s and leaves at 1.0 MPa and 45°C. The refrigerant is cooled in the condenser to 29°C and 0.75 MPa and is throttled to 0.2 MPa.
Determine

(a) The amount of heat removed from the cold space
(b) The power input required
(c) The COP of the refrigeration cycle

Given that
Enthalpy of superheated vapour at $-10°C$ and 0.18 MPa = 245.16 kJ/kg
Enthalpy of superheated vapour at 45°C and 1.0 MPa = 277.22 kJ/kg
Enthalpy of saturated liquid at 29°C and 0.75 MPa = 92.22 kJ/kg.

Solution: With reference to the *T–S* diagram, we have

$$H_1 = 245.16 \text{ kJ/kg}$$
$$H_2 = 277.22 \text{ kJ/kg}$$
$$H_3 = H_4 = 92.22 \text{ kJ/kg}$$

(a) The amount of heat removed from the cold space can be obtained by the equation

$$Q_L = m^0 (H_1 - H_4)$$
$$= 0.06 \text{ kg/s} (245.16 - 92.22) \text{ kJ/kg}$$
$$= 9.17 \text{ kW}$$

(b) The power input required is given by

$$W_{\text{net}} = m^0 (H_2 - H_1) \text{ kcal/kg}$$
$$= 0.06 \text{ kg/s} (277.22 - 245.16) \text{ kJ/kg}$$
$$= 1.92 \text{ kW}$$

(c) COP of the refrigeration cycle is given by

$$\text{COP} = \frac{Q_L}{W_{\text{net}}} = \frac{9.17 \text{ kW}}{1.92 \text{ kW}} = 4.77$$

EXAMPLE 8.8 A vapour–compression refrigeration system using CFC-12 (Freon) rated at 5 tons is employed in a chemical manufacturing plant to maintain the temperature of evaporator and condenser at $-10°C$ and $35°C$ respectively. The isentropic efficiency of compressor is reported to be 85%.

Determine

(a) Mass flow rate of the refrigerant
(b) Power consumption of the compressor
(c) Amount of heat rejected in the condenser
(d) Difference in COP between vapour–compression and Carnot cycle

Given that

Enthalpy of saturated vapour at 263 K, $H_1 = 183.2$ kJ/kg
Enthalpy of saturated liquid at 308 K, $H_3 = H_4 = 69.5$ kJ/kg
Enthalpy of superheated vapour, $H_2 = 208.3$ kJ/kg.

Solution:

(a) The mass flow rate of the refrigerant can be estimated by the relation

$$m^0(H_1 - H_4) = Q_L^0 = \text{Refrigerating capacity}$$

or

$$m^0 = \frac{Q_L^0}{H_1 - H_4} = \frac{5 \text{ ton}}{183.2 - 69.5}$$

$$= \frac{5 \times 3.516 \text{ kW}}{113.7 \text{ kJ/kg}}$$

$$= 0.1546 \text{ kg/s}$$

(b) Power consumption of the compressor is given by

$$W_{net} = \frac{m^0 W}{\eta_{iso-comp}}$$

where

$$m^0 = \text{Mass flow rate of refrigerant} = 0.1546 \text{ kg/s}$$
$$W = H_2 - H_1 = 208.3 \text{ kJ/kg} - 183.2 \text{ kJ/kg} = 25.1 \text{ kJ/kg}$$
$$\eta_{iso-comp} = \text{Isentropic efficiency of compressor} = 0.85.$$

Putting these values in the preceding equation, we get

$$W_{net} = 3.65 \text{ kW}$$

(c) The amount of heat rejected in the condenser is given by

$$Q_H = Q_L^0 + W$$
$$= 5 \text{ ton} + 3.65 \text{ kW}$$
$$= (5 \times 3.516) \text{ kW} + 3.65 \text{ kW}$$
$$= 21.23 \text{ kW}$$

(d) The coefficient of performance of the vapour–compression cycle can be evaluated by

$$\text{COP}_{\text{vapour-compression cycle}} = \frac{Q_L}{W} = \frac{H_1 - H_4}{H_2 - H_1} = \frac{113.7}{25.1} = 4.53$$

The coefficient of performance of the Carnot cycle is given by

$$COP_{Carnot\ cycle} = \frac{T_L}{T_H - T_L} = \frac{263}{308 - 263} = 5.84$$

Hence, the relative coefficient of performance is defined by

$$COP_{relative} = \frac{\text{Actual coefficient of performance}}{\text{Theoretical coefficient of performance}} = \frac{4.53}{5.84} = 0.78$$

8.1.6 Absorption Refrigeration Cycle

Energy is required in the refrigeration system for passing the refrigerant from the lower temperature and pressure of the cycle to the higher temperature and pressure, and this is usually supplied as work at the compressor. Instead of mechanical energy supply through a power-driven compressor in effecting this, arrangement may be made through a direct supply of heat as energy from a high-temperature source. Several typical forms and arrangements of apparatus have been devised for this purpose, all of which may be designated uniquely as absorption–refrigeration systems.

The performance of an absorption system depends upon the use of two substances which have great affinity for each other, and which can be easily separated by the application of heat energy. The principal combinations are sulphuric acid (H_2SO_4) and water (H_2O), ammonia (NH_3) and water (H_2O), lithium bromide ($LiBr_2$) and water (H_2O). In practical use for lower temperatures, NH_3 is employed as the refrigerant and water as the solvent and adsorbent for the system. Various other adaptations of the absorption system employ other adsorbents such as anhydrous ammonium nitrate and silica gel.

The components of a typical absorption-refrigeration system are shown diagrammatically in Fig. 8.8. It is easily noticeable from the figure that the main difference between a vapour–compression system and an absorption system is that the compressor of the vapour–compression system has been replaced by a joint assembly of the distinctive items such as absorber, liquid pump, heat exchanger, and generator. The dashed block of Fig. 8.8 represents the similarity between the two circuits by the usual condenser, expansion valve and evaporator.

The low-pressure refrigerant vapour from the evaporator passes to the absorber, and it is absorbed there by a weak solution of water and ammonia known as *weak aqua*. This absorption process has the characteristics of emitting energy; therefore, it is necessary to circulate the cooling water through coils in the absorber in order to maintain sufficiently low temperature to remove the heat of the solution evolved there and to increase the absorption capacity of water, because water absorbs less ammonia at high temperature. Thus, weak aqua is permitted to build up in order to ensure adequate ammonia concentration and eventually becomes strong aqua. It is then drawn by the liquid pump from the absorber and delivered through the heat exchanger to the generator. In the generator, the strong aqua is heated by supplying heat energy, usually by a low-pressure steam coil, and thereby the temperature is raised sufficiently in order that the ammonia is eliminated subsequently from the strong aqua, and is thus delivered as ammonia vapour to the condenser, where it is condensed.

The weak aqua (containing low concentration of refrigerant) returns to the absorber through the heat exchanger with a restriction which maintains the pressure differential between the high and low sides of the system. Here the function of the heat exchanger is one of warming the strong aqua, which is routed from the absorber to the generator, by cooling of the weak aqua which is returning from the generator to the absorber.

After this, the high-pressure liquid ammonia is passed through a throttle valve to the evaporator, which absorbs the ammonia's latent heat, thus producing low temperature in the range of –60°C (213 K).

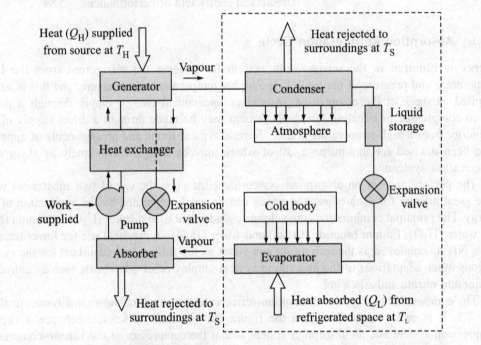

Fig. 8.8 Schematic diagram of a vapour–absorption refrigeration unit.

Performance of Absorption Refrigeration Cycle

The coefficient of performance of an absorption refrigeration cycle is given by:

$$\text{Coefficient of performance} = \frac{\text{Desired output}}{\text{Required Input}}$$

$$\text{COP}_{\text{absorption refrigeration}} = \frac{\text{Refrigeration produced}}{\text{Energy input}} \tag{8.14}$$

$$= \frac{Q_L}{Q_H - W} \cong \frac{Q_L}{Q_H} \tag{8.15}$$

It will be advantageous to determine the significant COP of an absorption refrigeration cycle if it is assumed to be a completely reversible one, which does necessarily mean that there is no irreversibility in the cycle. So, the efficiency of the absorption refrigeration cycle should be judged with respect to the reversed Carnot cycle. In that case, it would be possible for the system to be a reversible one if the heat from the source (Q_H) were transferred to a Carnot heat engine, and the work done by the heat engine were supplied to a Carnot refrigerator to discard the heat from the refrigerated space. Hence the work output of the heat engine is given by

$$W = \eta_{\text{Heat engine}} \, Q_H$$

or

$$W = \frac{T_H - T_S}{T_H} Q_H \tag{8.16}$$

where

$\eta_{\text{Heat engine}}$ = Efficiency of heat engine
Q_H = Amount of heat supplied from the source
T_H = Temperature at which heat is supplied
T_S = Temperature at which heat is rejected to the surroundings.

The amount of heat absorbed, Q_L at T_L, can be obtained from the knowledge of COP of the Carnot refrigerator and the work done by the Carnot heat engine. It is given by

$$Q_L = \text{COP}_{\text{Carnot refrigerator}} \cdot W$$

$$= \frac{T_L}{T_S - T_L} \cdot W \tag{8.17}$$

Equation (8.17) can be re-arranged as

$$W = \frac{T_S - T_L}{T_L} Q_L \tag{8.18}$$

Equating Eqs. (8.16) and (8.18), we have

$$\frac{T_H - T_S}{T_H} Q_H = \frac{T_S - T_L}{T_L} Q_L \tag{8.19}$$

or

$$\frac{Q_L}{Q_H} = \frac{T_L}{T_S - T_L} \cdot \frac{T_H - T_S}{T_H} \tag{8.20}$$

Comparing Eq. (8.20) with Eq. (8.15), we get

$$\text{COP}_{\text{absorption cycle}} = \frac{T_L}{T_S - T_L} \cdot \frac{T_H - T_S}{T_H} \tag{8.21}$$

The coefficient of performance of an absorption refrigeration cycle is normally less than 1, as it is less efficient than the vapour–compression cycle.

EXAMPLE 8.9 An inventor claims to have developed an absorption refrigerating machine that receives heat from a source at 125°C and maintains the refrigerated space at −5°C and have a COP of 2. If the environment temperature is 28°C, can this claim be valid? Justify your answer.

Solution: For the determination of COP of absorption refrigerating machine, we can take help of Eq. (8.21)

$$\text{COP}_{\text{absorption cycle}} = \frac{T_L}{T_S - T_L} \cdot \frac{T_H - T_S}{T_H}$$

where

T_H = Temperature at which heat received = 125°C = 398 K
T_L = Temperature of the refrigerated space = –5°C = 268 K
T_S = Temperature of the environment or surrounding = 28°C = 301 K.

On substitution of these values in the above equation, we get

$$\text{COP}_{\text{absorption cycle}} = \frac{T_L}{T_S - T_L} \cdot \frac{T_H - T_S}{T_H} = \frac{268}{301 - 268} \cdot \frac{398 - 301}{398}$$

$$= 8.121 \times 0.243 = 1.97 \approx 2$$

Hence, the claim of the inventor is valid and reasonable.

8.1.7 Air Refrigeration Cycle

The refrigeration cycle in which air is used as a refrigerant without applying any conventional chlorinated refrigerants like CFC-12 is known as the *air refrigeration cycle*. The major examples are Stirling refrigeration cycle and Bell–Coleman refrigeration system. With these refrigeration systems, very low temperatures can be attained.

Air absorbs the heat from a low-temperature system and discharges the same to a high-temperature system. As air does not change its phase throughout the cycle, the heat carrying capacity per kg of air is very small compared with that of the vapour absorbing machine. The Air Standard Refrigeration system may function depending upon any one of the following two cycles:

(a) *Open Cycle*

In this cycle, the working fluid, i.e., air comes directly to the contact with the products which it is desired to refrigerate in the cold chamber.

(b) *Closed System or Dense Air Cycle*

In this cycle the air passes through the refrigerating coils placed in the chamber (sink), and there extracts the heat energy from the cold region without coming in direct contact with the things kept inside. Its temperature is increased above the atmospheric temperature by compressing it in a compressor, and it is further cooled in the heat exchanger up to atmospheric temperature and is expanded in the expander, lowering its temperature. It again enters the refrigerator (sink) for absorbing the heat, and the cycle is repeated.

The following is a discussion on the Bell–Coleman refrigerator which operates on the working principle of the air standard refrigeration cycle.

Bell–Coleman Refrigeration Cycle

The Bell–Coleman refrigeration cycle is a perfectly reversed closed-loop cycle. It is well known to us that the compression and expansion processes in the Carnot cycle are isothermal, whereas those in the Bell–Colem cycle are isobaric. In fact, in the reversed Bell–Coleman cycle, the energy is absorbed by the refrigerant at variable temperature instead of a constant temperature as in the Carnot cycle.

The arrangement of a reversed Bell–Coleman cycle using air as a working fluid is shown diagrammatically in Fig. 8.9 and the ideal state changes of the cycle are presented in the *T–S* and *P–V* diagrams of Fig. 8.10(a) and (b) respectively.

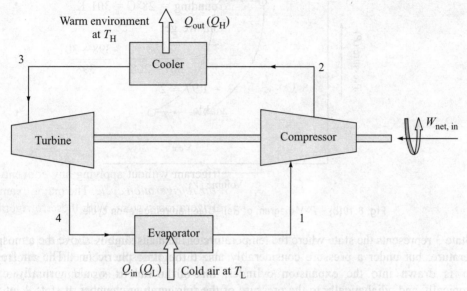

Fig. 8.9 Schematic diagram of Bell–Coleman refrigeration cycle.

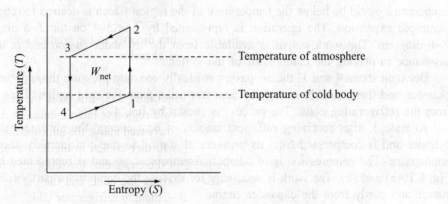

Fig. 8.10(a) *T–S* diagram of Bell–Coleman refrigeration cycle.

This cycle essentially consists of the following four internally reversible processes:

1–2: Isentropic compression
2–3: Isobaric heat rejection
3–4: Isentropic expansion
4–1: Isobaric heat addition

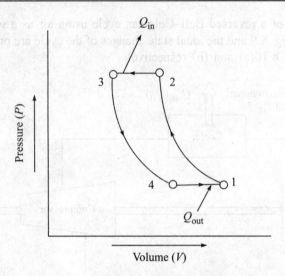

Fig. 8.10(b) *P–V* diagram of Bell–Coleman refrigeration cycle.

State 3 represents the state where the temperature of the air is slightly above the atmospheric temperature, but under a pressure considerably above the atmospheric one. The air from the cooler is drawn into the expansion cylinder, through which it would virtually expand isentropically and adiabatically to the pressure of the refrigerating chamber at state 4, at which state the pressure would commonly exceed considerably the atmospheric pressure, but the temperature would be below the temperature of the region which is desired to refrigerate during isentropic expansion. The operation is represented by line 3–4 on the *T–S* diagram and the *P–V* diagram. The work output is available from the expander engine and is transferred for assistance in driving the compressor of the system.

Between states 4 and 1, the air passes at ideally constant pressure through the refrigerating chamber and there absorbs energy by heat transition from the cold region. Then it is removed from the refrigerating coils. The process is shown by line 4–1.

At state 1, after receiving sufficient amount of heat energy, the air enters the compressor cylinder and is compressed from its pressure at state 3 to one considerably above that of the atmosphere. The compression is of adiabatic/isentropic type and is represented by line 1–2 in Fig. 8.10(a) and (b). The work is necessary for driving the compressor partly from an external source and partly from the expander engine.

At state 2, the hot air is finally exhausted into the cooler, where it emits energy as heat to the circulating water and is re-cooled to the original condition in state 1. The operation is shown by line 2–3. It thus completes one cycle.

Performance of Bell–Coleman Cycle Considering the ideal conditions of constant-pressure cooling and warming, and isentropic expansion and compression, we can calculate the co-efficient of performance of the cycle with the assistance of the *T–S* diagram.

Basis: *m* kg of air

Now

Heat extracted from the cold chamber = $Q_L = Q_{in} = mC_P(T_1 - T_4) = H_1 - H_4$

$$\text{Heat rejected to the cooler} = Q_H = Q_{out} = mC_P(T_2 - T_3) = H_2 - H_3$$

Since

$$\text{Heat rejected} = \text{Work done} + \text{Refrigerating effect (Heat extracted)}$$

So

$$
\begin{aligned}
\text{Work done} &= \text{Heat rejected } (Q_{out}) - \text{Heat extracted } (Q_{in}) \\
&= mC_P(T_2 - T_3) - mC_P(T_1 - T_4) \\
&= mC_P[(T_2 - T_3) - (T_1 - T_4)] \quad (8.22)
\end{aligned}
$$

$$
COP_{\text{Bell–Coleman cycle}} = \frac{\text{Heat extracted}}{\text{Work done}}
$$

$$
= \frac{mC_P(T_1 - T_4)}{mC_P(T_2 - T_3) - mC_P(T_1 - T_4)} \quad (8.23)
$$

For the isentropic process 1–2, we have

$$
\frac{T_2}{T_1} = \left(\frac{P_2}{P_1}\right)^{\frac{(\gamma-1)}{\gamma}} = (R_{ip})^{\frac{(\gamma-1)}{\gamma}} \quad (8.24)
$$

where R_{ip} is the isentropic pressure ratio.

For the isentropic process 3–4, we have

$$
\frac{T_3}{T_4} = \left(\frac{P_3}{P_4}\right)^{\frac{(\gamma-1)}{\gamma}} = (R_{ip})^{\frac{(\gamma-1)}{\gamma}} \quad (8.25)
$$

Comparing Eqs. (8.24) and (8.25), we get

$$
\frac{T_2}{T_1} = \frac{T_3}{T_4} = (R_{ip})^{\frac{(\gamma-1)}{\gamma}} \quad (8.26)
$$

Considering the earlier equation, we can write

$$
\frac{T_2}{T_1} = \frac{T_3}{T_4}
$$

or

$$
\frac{T_1}{T_4} = \frac{T_2}{T_3} \quad (8.27)
$$

By the componendo et dividendo rule, Eq. (8.27) can be written as

$$
\frac{T_1 - T_4}{T_2 - T_3} = (R_{ip})^{\frac{(\gamma-1)}{\gamma}}
$$

or

$$
T_1 - T_4 = (T_2 - T_3)(R_{ip})^{\frac{(\gamma-1)}{\gamma}} \quad (8.28)
$$

Substituting the value of $T_1 - T_4$ in Eq. (8.23), we get

$$
\text{Coefficient of performance} = \frac{T_1 - T_4}{(T_2 - T_3) - (T_1 - T_4)}
$$

$$\text{COP}_{\text{Bell–Coleman cycle}} = \left(R_{ip}^{\frac{(\gamma-1)}{\gamma}} - 1 \right)^{-1} \tag{8.29}$$

$$= \frac{T_1}{T_2 - T_1} \tag{8.30}$$

With the assistance of the *T–S* diagram and the expression of COP, we can see very much clearly that at a given capacity, the performance of the Bell–Coleman refrigeration system may be impressive if the mass flow rate of air is more rapid. Better performance can be achieved in this cycle when the working fluid behaves ideally.

Advantages of Closed System Over Open System

1. In an open system, the working medium absorbs moisture from the freezer box or the space which is to be cooled. During expansion, the moisture may freeze and clog the valves.
2. In an open system, the expansion can be carried only up to the atmospheric pressure, but for a closed system there is no such restriction.
3. In a closed system the suction to the compressor may be at high pressure. The sizes of the expander and the compressor can be kept within reasonable limits by using dense air.

Due to the ready availability of air, the refrigeration system was used widely in various fields for several years, such as in freezing meat and in marine installations. But now it is practically abandoned, partly because of its relatively poor efficiency and partly because of operating difficulties arising from the freezing of any moisture in the air.

Research and development in air refrigeration systems has recently generated a lot of interest due to advancements in aeronautics and the idea of applying air-conditioning systems to aeroplanes. As high-pressure air is already available in an aeroplane, it has been applied recently to aircraft systems with low equipment weight.

EXAMPLE 8.10 An air refrigeration system of 5 tons cooling capacity is employed to maintain the temperature of a refrigerated space at $-20°C$ (253 K) when the temperature of the surroundings is $30°C$ (303 K). The pressure ratio of the compressor is 4. The specific heat of air may be taken as 1.008 kJ/kg and $\gamma = 1.4$.

Estimate

(a) The coefficient of performance of the system
(b) The mass flow rate of air as working medium
(c) The work of compression
(d) The work of expansion
(e) The net work of the system.

Solution: (a) By the definition of the coefficient of performance of the air refrigeration system, we have

$$\text{COP}_{\text{air refrigeration cycle}} = \frac{1}{\left(\dfrac{P_2}{P_1} \right)^{\frac{(\gamma-1)}{\gamma}} - 1} = \frac{T_1}{T_2 - T_1}$$

We are given that

T_1, temperature of the working fluid leaving the evaporator = 253 K

T_2, temperature of the working fluid leaving the cooler = 303 K.

Now, to find out the temperature T_2, using Eq. (8.24), we have

$$T_2 = T_1 \left(\frac{P_2}{P_1} \right)^{\frac{(\gamma-1)}{\gamma}}$$

$$= 253 \, (4)^{\frac{1.4-1}{1.4}} = 375.58 \text{ K}$$

We can also determine T_4 in the same fashion.

$$T_3 = T_4 \left(\frac{P_3}{P_4} \right)^{\frac{(\gamma-1)}{\gamma}}$$

$$303 = T_4 (4)^{\frac{1.4-1}{1.4}}$$

$$T_4 = 204.1 \text{ K}$$

Hence, the COP of the air refrigeration system is

$$\text{COP}_{\text{air refrigeration cycle}} = \frac{T_1}{T_2 - T_1} = \frac{253}{375.58 - 253} = 2.06$$

(b) The mass flow rate of the refrigerant air can be computed by making the energy balance around the refrigerating chamber in the system, i.e.

$$m^0 C_P \, (T_1 - T_4) = 5 \times 12{,}660$$

$$m^0 = \frac{5 \times 12{,}660}{1.008 \times (253 - 204.1)} = \frac{63{,}300}{49.29} = 1284.23 \text{ kg/h}$$

(c) The work of compression can be calculated by the relation

$$W_{\text{compression}} = m^0 C_P \, (T_2 - T_3)$$
$$= 1284.23 \text{ kg/h} \times 1.008 \text{ kJ/kg-K} \times (375.58 - 303)$$
$$= 93{,}955 \text{ kJ/h} = 26.09 \text{ kW}$$

(d) The work of expansion can be estimated as

$$W_{\text{expansion}} = m^0 C_P \, (T_1 - T_4)$$
$$= 1284.23 \text{ kg/h} \times 1.008 \text{ kJ/kg-K} \times (253 - 204.1)$$
$$= 63{,}301.2 \text{ kJ/h} = 17.58 \text{ kW}$$

(e) The net work of the system is given by

$$W_{\text{net}} = W_{\text{compression}} - W_{\text{expansion}}$$
$$= 26.09 \text{ kW} - 17.58 \text{ kW}$$
$$= 8.51 \text{ kW}$$

8.1.8 Comparative Study of Carnot, Vapour–Compression, Absorption and Air Refrigeration Systems

Comparison between Carnot and Vapour–Compression Cycles

1. In case of the Carnot refrigeration cycle, the work of compression is less than that for a vapour–compression cycle. This is due to constant pressure de-superheating in case of the vapour–compression cycle as against constant temperature superheating for the Carnot cycle.

2. The Carnot cycle produces some useful work during isentropic expansion, but a vapour–compression cycle produces no work during the throttling process.

3. In a vapour–compression cycle there is a loss of cooling effect due to the throttling expansion, which results in an increase in entropy; but in the Carnot cycle, it is not so due to isentropic expansion.

Difference between Vapour–Compression and Vapour–Absorption Refrigeration Systems

The main points of difference between a vapour–compression system and vapour–absorption refrigeration system are enumerated in Table 8.1. The various pertinent aspects are capacity, mode of energy supply, space requirements for installation, performance, nature of operation, feeding of refrigerant, etc.

Table 8.1 Comparative study between vapour–compression and vapour–absorption refrigeration systems

		Vapour–compression system	Vapour–absorption system
1.	Place of working	The system is used where mechanical energy is available.	The system is used even at remote places with any available source of thermal energy.
2.	Coefficient of performance	COP of the system is poor at partial loads.	The COP of the system is not affected by load variation.
3.	Extent of input energy	The extent of supplied energy is low. It is about half to one-fourth of the refrigerating effect.	The extent of input energy is high. It is about one and a half times the refrigerating effect.
4.	Ease of charging	The charging of the refrigerant is quite simple.	The charging of the refrigerant is difficult.
5.	Mode of energy input	Mechanical energy is supplied as input energy.	Thermal energy is supplied as input energy.
6.	Installation space	Space requirements for installation of the system are relatively higher.	Space requirements for installation favours the system more than the compression units.
7.	Chances of leakage	More chances of leakage of the refrigerant from the system are there.	No chance of leakage of refrigerant is there.
8.	Creation of noise	Due to the presence of moving parts of the compressor in the circuit, system has more wear and tear as well as noise.	Due to the presence of a minimum number of moving parts, the system is quiet in operation.

Advantages and Disadvantages of Vapour Refrigeration System over Air Refrigeration System

The advantages and disadvantages of the two systems are discussed in the following lines.

Advantages. These are:

1. As the working cycle of the vapour refrigeration system is similar to the Carnot cycle, the COP is quite high. The COP of the vapour cycle lies between 3 and 4, whereas for the air cycle it is always less than 1.
2. The running cost of the vapour refrigeration system is only one-fifth of the air refrigeration system when used at the ground level.
3. As heat is carried away by the latent heat of vapour, the amount of liquid circulated is less per ton of refrigeration; therefore, the size of the evaporator is similar for the same refrigerating effect.
4. Just by adjusting the throttle valve of the same unit, the required temperature of the evaporator can be attained.

Disadvantages. These are:

1. The first investment cost is high.
2. There is a risk of leakage of the refrigerant, leading to ozone layer depletion in the stratosphere.

8.1.9 Selection of Right Refrigerant

The mathematical expression of the COP of an ideal refrigeration cycle implies that the efficiency of a refrigerating machine does not depend on the refrigerant. It is needless to mention that every operational step of an ideal cycle (Carnot refrigeration cycle) is carried out in reversible fashion. But the unavoidable internal irreversibility in a practical vapour-compression refrigeration cycle causes the co-efficient of performance to depend on the working fluid and the nature of the refrigerant. Although no refrigerant has been found to meet all the requirements, the selection of a promising refrigerant is necessary. This is because the COP depends upon the specific application in terms of the refrigerating capacity, cooling effect required, types of compressor used, etc.

In 1850, ethyl-ether was first used as a commercial refrigerant in the vapour–compression refrigeration system. Then ammonia, carbon dioxide, methyl chloride, sulphur dioxide, butane, ethane, propane, and iso-butane had been choices for refrigerants. These substances have been in use literally from cooling technology's infant stage. Unfortunately, accelerated technical developments in this field throughout the world, starting from the last century (1932 onwards), have produced several environmental problems. This is due to the introduction of various man-made synthetic chemicals such as CFCs (Chlorofluorocarbons), HCFCs (Hydro-chlorofluorocarbons), HFCs (Hydrofluorocarbons), etc. These are non-toxic, non-flammable, chemically inert, and extremely stable molecules which are not normally degraded by the usual processes. They do not dissolve in rain. Their atmospheric lifetime is very long, i.e., they can reside in the atmosphere for a long time.

Chlorofluorinated refrigerants were first developed as efficient and safe refrigerant for the household refrigerating machine in 1931. Since then, CFCs have been manufactured for a wide variety of uses as blowing agents in urethane foams, spray propellants in aerospace, cleaning agents for electronic circuit board, etc. From the CFC family of refrigerants, CFC-12 (dichlorodifluoromethane) was selected as most suitable for commercial utilization. CFC-12 is known by the trade name *Freon* and is designated as R-12 according to the nomenclature of the

American Society of Heating, Refrigerating and Air-conditioning Engineers (ASHRAE). The low cost and versatility of CFCs made them refrigerants of choice.

Nowadays, these varieties of halogenated compounds are most widely used as conventional refrigerants due to their excellent thermodynamic properties in producing adequate cold. But they have a strong negative impact on the environment, contributing to ozone layer depletion and global warming. From the point of view of environmental protection, and the global restrictions imposed through various international regulatory activities such as the Montreal Protocol of 1987 on the phasing out of CFCs and the Kyoto Protocol of 1997 on the reduction of emission of greenhouse gases, these refrigerants must be phased out. It was established in 1974 by Mario Molina and F. Sherwood Rowland that chlorine in CFCs is the main culprit atom, strongly responsible for the destruction of the ozone layer as one chlorine atom can destroy 1,00,000 ozone molecules. As a result, penetration of harmful UV rays to the earth's surface increases, which in turn causes skin cancer, eye cataract, disruption of food chain, destruction of marine food web, etc. Due to the presence of hydrogen bonds, HCFCs are not so much responsible for ozone layer degradation. HFCs, which contain no chlorine, do not cause ozone crisis. These species also play a dominant role in global warming by absorbing the infra-red rays. In this alarming situation, the replacement of man-made chemicals by proper alternatives becomes an urgent necessity. The significant drawbacks of halogenated refrigerants could be compensated for by going back to the earlier times of natural refrigerants, as well as by resorting to the various eco-friendly alternative techniques.

Apart from the ozone depletion and global warming potential of anthropogenic synthetic refrigerants, there are many factors that must be taken into consideration while choosing a good refrigerant. So, the desirable properties that a promising refrigerant should possess can be highlighted in the following way:

1. Evaporator and Condenser Pressures

The operating pressure in the condenser should not be high. If it is, the materials used in the construction of evaporators and condensers should be good enough to withstand pressure, which in turn will result in high initial cost of the system. On the other hand, the evaporator pressures should preferably be positive, i.e., above atmospheric pressures, so as to prevent the leakage of air and moisture into the refrigeration system. Table 8.2 shows evaporator pressure at $-15°C$ (258 K) and condenser pressure at 29°C (302 K) for several refrigerants.

Table 8.2 Evaporator and condenser pressures

Refrigerant	Evaporator pressure at $-15°C$ (in kgf/cm^2)	Condenser pressure at 290°C (in kgf/cm^2)
Ammonia	2.34	11.5
Carbon dioxide	23.7	71.2
Refrigerant 11	0.2055	1.2855
Refrigerant 12	1.8	7.32
Refrigerant 22	3.03	12.26
Refrigerant 113	0.0704	0.5527

It is very much clear from Table 8.2 that carbon dioxide operates at extremely high pressure, thus requiring strong metal for compressors and piping vessels, while refrigerant 11 and refrigerant

113 operate below atmospheric pressures where some equipment is required to purge air from the system.

2. Boiling Point

The boiling point of a refrigerant should be appreciably lower than the temperature levels at which the refrigerator works; otherwise there is a chance of moisture and air leakage into the system.

3. Freezing Point

The freezing point of the refrigerant should be appreciably below the operating temperature of system, so as to avoid clogging the pipes. The freezing temperatures of important refrigerants are given in Table 8.3.

Table 8.3 Freezing temperature of refrigerants

Refrigerant	Freezing temperature in °C
Ammonia	−77.8 (195.2 K)
Carbon dioxide	−56.7 (216.3 K)
Refrigerant 11	−111 (162.0 K)
Refrigerant 12	−157.8 (115.2 K)
Refrigerant 22	−160 (113.0 K)
Refrigerant 113	−35 (238.0 K)
Water	0 (273.0 K)

4. Critical Temperature and Pressure

A refrigerant should have the critical temperature and pressure well above the operating temperature and pressure of the system. If the system operates above the critical temperature, then the condensation of the gas becomes impossible after compression at high pressure.

5. Latent Heat

Evaporation of the liquid is an important step in the refrigeration cycle which produces cooling; therefore the latent heat of refrigerant should be as large as possible. Also, the weight of the refrigerant to be circulated in the system will be less if its latent heat is high. This reduces the initial cost of the refrigerant. The size of the system will also be small, which further reduces the initial cost.

6. Specific Volume

The specific volume of the refrigerant vapour (which a compressor is required to pump) roughly determines the size of the compressor. For a reciprocating compressor, a low suction value of the volume pumped is normally desirable, permitting a small displacement. In case of centrifugal compressors, a high suction value of the volume is desired. Then the efficiency of the compressor can be increased.

7. Coefficient of Performance

The coefficient of performance is a factor of paramount importance in selecting a refrigerant. Table 8.4 shows the approximated COP of some refrigerants. This can give us an idea about how to choose a refrigerant.

Table 8.4 Freezing temperature of refrigerants

Refrigerants	Coefficient of performance
Refrigerant 11	5.09
Refrigerant 133	4.92
Ammonia	4.76
Refrigerant 12	4.70
Refrigerant 22	4.66
Carbon dioxide	2.56

8. Flammability and Explosiveness

A good refrigerant must be non-flammable and non-explosive. In the refrigeration system, to avoid the danger of explosion or fire hazard during high compression, the refrigerants should not have flammability and explosiveness even when mixed with air or oil. Hydrocarbons such as propane, ethane and butane are highly flammable and explosive. Ammonia is explosive when mixed with air in concentrations of 16 to 25 per cent (ammonia) by volume. The halogenated hydrocarbons are considered to be non-flammable.

9. Toxicity

Toxicity is one of the major factors in selecting a refrigerant. A toxic refrigerant is always injurious to human beings when mixed in small percentages with air. For instance, ammonia causes toxicity.

10. Corrosiveness

This factor is very important while selecting a refrigerant. Certain metals are attacked by refrigerants. For example, ammonia reacts with copper, brass, or other cuprous alloys in the presence of water. Therefore, in the ammonia system the common materials used are iron and steel. The halogenated hydrocarbons may react with zinc but not with copper, aluminium, iron, and steel. The Freon group does not react with steel, copper, brass, zinc, tin, or aluminium, but is corrosive to magnesium and same other metals.

11. Leakage Tendency and Detectability

The leakage tendency and detectability of a refrigerant is an important factor to be taken into consideration. The leakage tendency should be as minimal as possible, and leakage should be easily detected to avoid the loss of refrigerant from the system. Ammonia leakage is easily detectable by its odour. Halogenated hydrocarbon refrigerants are odourless. So the leakage of such refrigerants cannot be detected so easily. In case of leakage of the refrigerant from the system, the refrigerant will come in contact with the refrigerated products.

12. Viscosity and Thermal Conductivity

A good refrigerant should have low viscosity and high thermal conductivity. These parameters always enhance the better transmission of heat from one temperature zone to another.

13. Reaction with Oil

The refrigerant should not react with the lubricating oil of the compressor. Secondly, the miscibility of the oil (i.e., the tendency of the refrigerant to be dissolved in oil and vice versa) is quite important, as some oil should be carried out of the compressor crank case with the hot refrigerant vapour to lubricate the pistons and discharge valves properly.

14. Cost and Availability

If a refrigerant is relatively cheap, then it would always be a refrigerant of choice. This factor plays an important role in case of bigger refrigeration units. For instance, natural refrigerants (which occur in the biosphere or environment) such as air, CO_2, NH_3, and water are available and cheap. But the halogenated refrigerants such as CFC-12, HCFC-22 and HFC-134a are costly and do not have relatively easy availability.

15. Ozone Depletion Potential (ODP)

Nowadays, it is an environmental issue, and is the index by which the capability of a man-made synthetic chemical, used as refrigerant, to deplete the ozone layer is measured. The halogenated refrigerants have ozone depletion potential while the natural refrigerants (hydrocarbon, air, CO_2, NH_3, and water) do not. Hence, a good refrigerant should have no ozone depletion potential. Table 8.5 shows the ODP of some refrigerants.

Table 8.5 Ozone depletion potential of some common refrigerants

Refrigerants	Ozone depletion potential
CO_2	0
NH_3	0
Propane	0
CFC-12	0.82
HCFC-22	0.055
HFC-134a	0

16. Global Warming Potential (GWP)

It is also an index that indicates the ability of a gas to absorb the infra-red rays. The gases are greenhouse gases such as CO_2, SO_2, CH_4, perfluorocarbons, chlorofluorocarbons, hydro-fluorocabons, and hydrochlorofluorocarbons, which is responsible for global warming. Therefore the refrigerant should have low or negligible GWP. Table 8.6 shows the GWP of some refrigerants.

Table 8.6 Ozone depletion potential of some common refrigerants

Refrigerants	Global warming potential
CO_2	1
NH_3	<1
Prpopane	20
CFC-12	8100
HCFC-22	1500
HFC-134a	1300

8.2 LIQUEFACTION PROCESS

The liquefaction techniques of gases have widespread utility in producing significant cold at cryogenic temperature (appreciably below $-100°C$). The application of liquefaction includes the separation of oxygen and nitrogen from air, preparation of propellants for rockets, and production of liquid propane as domestic fuel. Liquefaction finds its application in liquefying nitrogen, which is used in activities such as studying the materials at low temperature. Liquid hydrogen and helium are also the products of liquefaction process used extensively in space research.

Liquefaction is defined as the process resulting from cooling of a gas to a temperature in the regions of two phases. These can be achieved by the following three methods:

1. Expansion of gas through a work-producing device (isentropic expansion);
2. Joule–Thomson expansion (isenthalpic expansion);
3. By exchanging heat at constant pressure.

It is needless to mention that the critical point is an important factor to be taken into consideration for the liquefaction of a gas. It is the highest temperature and pressure above which a liquid cannot exist. The critical temperatures of the most widely used liquefied gases—helium, hydrogen and nitrogen—are $-268°C$, $-240°C$, and $-147°C$ respectively. These gases cannot exist as liquids at normal atmospheric condition. The necessity of liquefaction involves the production of low temperatures down to these critical temperatures, as it is not possible to attain the low temperature of these magnitudes by simple refrigeration technique.

The following is a discussion on the different liquefaction processes for the generation of cooling effect.

8.2.1 Isentropic Expansion of Gas through Work-Producing Device

The isentropic expansion of the gas through a work-producing device such as an expansion engine is an important method of producing low temperature. This expansion is reversible and adiabatic, known as *isentropic expansion*. We can define the isentropic expansion coefficient μ_S, which expresses the rate of change of temperature with respect to pressure at constant entropy

$$\mu_S = \left(\frac{\partial T}{\partial P} \right)_S \tag{8.31}$$

The isentropic expansion of a gas through an expander always results in temperature decrease. In this method, energy is removed from the gas in the form of external work. For this reason, this method of cold production is sometimes called the *external work method*. At the other end, the expansion through an expansion valve may or may not result in a temperature decrease, and in this method the energy is not removed from the gas, but the molecules are moved further apart under the influence of intermolecular forces; so this method is called the *internal work method*.

This process of isentropic cooling has a major disadvantage due to the fact that if the temperature comes down to a great extent, the lubricating oil used for the moving parts of the expansion engine becomes hardened. Hence, the availability of a suitable lubricating medium, which is not hardened at low temperature, must be ensured for smooth running of such machines. Claude eliminated this difficulty by using hydrocarbons like petroleum ether as lubricants. These lubricants do not get hardened at very low temperature, even at −160°C. Then there is no difficulty at −140°C, which is the critical temperature of air.

In 1902, the French chemist Georges Claude developed the first successful expansion engine illustrated in Fig. 8.11 for the liquefaction of air. Here, cooling was produced in such a fashion that when a highly compressed gas is allowed to expand adiabatically by doing work against external load, it becomes cooled. The cooling of the gas takes place due to the fact that energy necessary for the work done quickly by the gas is extracted from the molecular kinetic energy of the gas itself. The entropy remains constant during such a process.

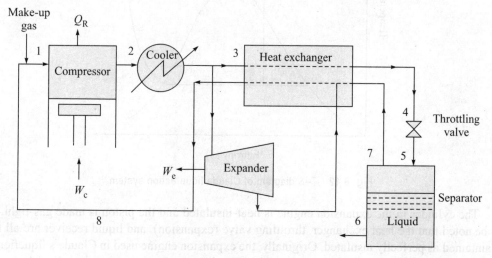

Fig. 8.11 Claude liquefaction system.

In Claude's apparatus, air is compressed to about 40 atm and cooled to room temperature, after which it passes through the first countercurrent heat exchanger. A portion of the air (approximately 80%) is diverted from the mainstream and then passed through the expansion engine (or an expander), which must be suitably braked by coupling it to the compressor or by making it drive an electric generator to ensure that external work is done. The expansion in the engine takes place under adiabatic condition and the work is performed at the expense of the internal energy of the compressed gas. There is consequently a very drastic reduction in temperature. The cooled outgoing gas is reunited with the return stream and passed through the low-pressure side of the second heat exchanger, also called the *liquefier*, and lowers the temperature of the other part of the gas, flowing inward through the second heat exchanger below its critical temperature. Due to this, the expanded stream of gas loses its coldness and returns to the compressor. The remainder of the high-pressure gas (about 20%) is liquefied in the high-pressure tubes of this exchanger under the combined influence of high pressure and low temperature, expanded through an expansion valve (throttling valve), and then accumulated in the liquid receiver. The cold vapour from the liquid receiver is returned through the heat exchangers to cool the incoming gas and recycled back to the compressor as make-up gas. The *T–S* diagram of the Claude liquefaction system is shown in Fig. 8.12.

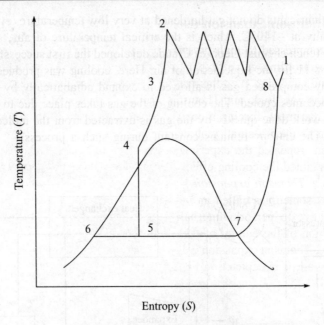

Fig. 8.12 T–S diagram of Claude liquefaction system.

The cylinder in the expansion engine is heat-insulated and the piston is made gas-tight. It is to be noted that the heat exchanger, throttling valve (expansion), and liquid receiver are all to be maintained as perfectly insulated. Originally, the expansion engine used in Claude's liquefier was of the reciprocating type; but in later times, improved types of expansion turbine have been used.

An expansion valve is still required in Claude's system in expanding the two-phase mixture (liquid and vapour), because the expander cannot withstand much liquid in an actual system. Moreover, the liquids have much lower compressibility than the gases; therefore, if liquids were formed in the cylinder of an expansion engine (positive displacement type), high momentary stress would result.

In case of the Claude liquefaction process, under isentropic condition the fraction of gas liquefied in the throttling valve can be computed by the following energy balance equation over the heat exchanger, turbine and the separator along with the fraction (x) of high-pressure gas expanded isentropically:

$$W_e = \psi H_6 + (1 - \psi)H_8 - H_3 \qquad (8.32)$$

where

W_e = Work produced by the turbine under adiabatic condition

$$= x\,(H_7 - H_3) \qquad (8.33)$$

ψ = Fraction of gas liquefied in the throttling valve.

Combining Eqs. (8.32) and (8.33), we get

$$\psi = \frac{H_8 - H_3}{H_8 - H_6} + x\frac{H_3 - H_7}{H_8 - H_6} \qquad (8.34)$$

8.2.2 Joule–Thomson Expansion (Isenthalpic Expansion)

When a gas is compressed to high pressure under adiabatic condition and allowed to expand slowly through a throttling valve (expansion valve), it undergoes a temperature change which varies in magnitude and sign (i.e., this may be either a warming up or cooling down) according to the nature of the gas. Joule and Thomson first tested the validity of this principle in 1862 by conducting an experiment, forcing a compressed gas past a restraining orifice such as porous plug of porcelain or even a small pinhole. Later, Lord Kelvin (whose previous name was Sir William Thomson) repeated the experiment, observed the change in temperature produced by expansion, and attained the cooling effect called *Joule–Thomson effect*. The relevant expansion is known as *Joule–Thomson expansion*.

This effect is sometimes called an *internal work process*, because after expansion the gas molecules will have to perform internal work against their mutual forces of attraction by expending a portion of the energy of the gas itself, due to which the gas will be cooled. The principle of Joule–Thomson expansion was used by Linde in Germany and by Hampson in Great Britain to liquefy air. It is depicted in Fig. 8.13.

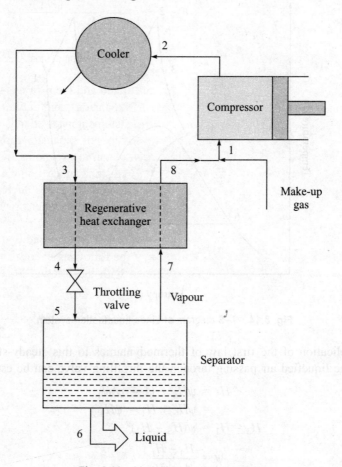

Fig. 8.13 Linde liquefaction system.

The effect of change in temperature due to change in pressure at constant enthalpy is represented by the Joule–Thomson coefficient, defined by

$$\mu_{JT} = \left(\frac{\partial T}{\partial P}\right)_H \tag{8.35}$$

For an increase in temperature during expansion, the Joule–Thomson coefficient (μ_{JT}) is negative; for a decrease in temperature, it is positive.

In the gas liquefaction system shown in Fig. 8.13, the make-up gas and recycle gas are allowed to compress isothermally at state 1. Then the compressed gas is passed on to a cooler. The gas is cooled here by cooling water or ambient air at state 2. Then the high-pressure gas is cooled further in a regenerative counter-current heat exchanger by unliquefied low-pressure gas from the previous cycle to state 3, and it is passed through a throttling valve to state 4. Then the liquid is collected as the final product, which is a mixture of liquid and vapour, at state 5. The liquid is again passed through the heat exchanger at stage 7 and remixed with the make-up gas at state 8. The entire process employed in the system is depicted in Fig. 8.14.

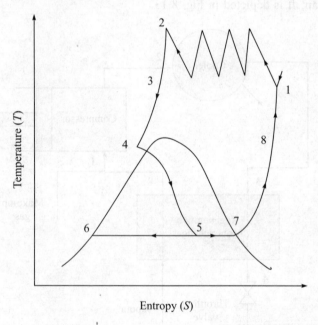

Fig. 8.14 T–S diagram of Linde liquefaction system.

By the application of the first law of thermodynamics to this steady-state operation, the fraction ψ of the liquefied air passing through the throttling valve can be estimated as

$$H_3 = \psi H_6 + (1 - \psi)H_1$$
$$= \psi H_6 + H_1 - \psi H_1$$
$$H_3 - H_1 = \psi(H_6 - H_1)$$

$$\psi = \frac{H_3 - H_1}{H_6 - H_1} \tag{8.36}$$

Comparison between Joule–Thomson and Isentropic Cooling

Now, there are some differences between Joule–Thomson cooling and isentropic cooling. These are discussed in the following lines.

Firstly, in case of Joule–Thomson cooling, the initial temperature of the gas should be below its inversion temperature, whereas for isentropic cooling the effect can be produced for any initial temperature.

Secondly, in Joule–Thomson cooling, enthalpy remains constant. But for isentropic cooling, entropy remains constant.

Thirdly, Joule–Thomson cooling occurs due to the internal work done by the gas against the attractive forces existing among the molecules of a real gas, whereas isentropic cooling is due to the external work done by the gas against the loading agency.

Fourthly, Joule–Thomson cooling effect is smaller than isentropic cooling effect, except when close to critical temperature and/or with very high compression-expansion ratio.

Fifthly, Joule–Thomson cooling takes place only in the case of real gases, while isentropic cooling takes place for all gases.

Sixthly, in isentropic cooling a part of the work of compression is recovered out of the work done by the reciprocating or turbine expander.

8.2.3   Exchanging Heat at Constant Pressure

This method is of greater theoretical and practical importance. Low-temperature refrigeration could be achieved by isothermally compressing and subsequently boiling a liquid under reduced pressure adiabatically. The latent heat required for the evaporation of the liquid (for conversion from the liquid to the gaseous state) comes from itself, thereby producing cold. This results in gradual loss of heat by the liquid whose temperature falls considerably, often as low as 'triple point'. With the help of this method it has been possible to bring the temperature down to −200°C. The working of the domestic and industrial refrigerators and air-conditioners is mainly based on this principle. The cold produced by this technique depends on the nature of the liquid used, its latent heat of vaporization, the rate of vaporization of the liquid, and the amount of heat insulation provided to the system.

SUMMARY

The thermodynamics of various conventional refrigeration systems as well as some special refrigeration systems and liquefaction systems have been discussed. The basic working principle of the refrigerator and the heat pump have substantiated the transmission of heat between high- and low-temperature thermal reservoirs. The circuit components and operational methodology of some refrigerators have been highlighted. Comparative advantages and disadvantages of different refrigeration systems have also been critically presented. While studying the various properties of conventional refrigerants, it is observed that some refrigerants have promising and excellent thermodynamic properties in case of application in industrial refrigeration and air-conditioning systems, but these refrigerants have many significant environmental drawbacks such as ozone depletion potential, global warming potential, inflammability, and toxicity. As there is tremendous international pressure to limit the use of conventional refrigerants, particularly chlorofluorocarbons, in order to avoid further damage to the ozone layer, finding alternative viable and eco-friendly refrigerants is the need of the hour.

The liquefaction process has generated strong interest for the generation of very low temperatures (cryogenic temperatures), which are in great use in space technology, research on materials, household applications, etc. The cooling of a gas by allowing it to expand isentropically by doing external work and cooling by Joule–Thomson expansion through internal work done are well-supported by Claude and Linde liquefaction systems respectively. Some relevant and useful problems have been worked out to corroborate the theoretical background and practicability of the different systems.

KEY TERMS

Chlorofluorocarbons (CFCs) Halogenated compounds having excellent thermodynamic properties and used as efficient refrigerants, but responsible for ozone layer depletion and global warming.

Coefficient of Performance (COP) for Heat Pump The ratio of the energy delivered to the work input.

Coefficient of Performance (COP) for Refrigerator The ratio of the refrigerating effect to the work input.

External Work Method Method in which cold is produced under isentropic expansion of a gas and energy is removed from the gas in the form of external work.

Heat Pump A reversed heat engine whose object is to keep an area warm.

Internal Work Method The method in which energy is not removed from a gas, but the molecules are moved further apart under the influence of intermolecular forces.

Joule–Thomson Coefficient Coefficient representing the effect of change in temperature due to change in pressure at constant enthalpy.

Liquefaction The process by which a gas is liquefied below its critical temperature.

Refrigerant The working fluid in a refrigeration cycle.

Refrigerating Effect The removal of heat from an area by the use of mechanical energy.

Refrigeration A process of producing and maintaining temperatures below that of the surroundings in a given space or substance.

Refrigerator A device used to produce cooling effect by transferring heat from low-temperature to high-temperature regions.

Ton of Refrigeration Standard unit generally used to express the cooling capacity of a refrigerator. It is equivalent to 50 kcal/min or 3.517 kW.

Vapour–Absorption Refrigeration The use of a heat source to replace the compressor in a vapour–compression system.

Vapour–Compression Refrigeration Refrigeration produced by compression of vapour of working fluid.

IMPORTANT EQUATIONS

1. COP of refrigerator is given by

$$\text{COP}_R = \frac{\text{Heat removed}}{\text{Work required}} = \frac{\text{Refrigerating effect}}{\text{Work input}} = \frac{Q_L}{Q_H - Q_L} = \frac{Q_L}{W_{net}} \qquad (8.1)$$

2. COP of heat pump is given by

$$\text{COP}_{HP} = \frac{\text{Heat absorbed}}{\text{Work required}} = \frac{\text{Heating effect}}{\text{Work input}} = \frac{Q_H}{Q_H - Q_L} = \frac{Q_H}{W_{net}} \qquad (8.2)$$

3. $$\text{COP}_{HP} = \text{COP}_R + 1 \qquad (8.3)$$

4. $$1 \text{ ton of refrigeration} = \frac{2000 \text{ lb} \times 144 \text{ BTU/lb}}{24 \times 60} = 200 \text{ BTU/min} \qquad (8.4)$$

5. $$1 \text{ ton of refrigeration} = \frac{72{,}000 \text{ kcal}}{24 \times 60} = 50 \text{ kcal/min} \qquad (8.5)$$

6. 1 ton of refrigeration in kilowatts = 200 BTU/min

$$= 200 \times 1.055 \text{ kJ/min} \quad (\text{since } 1 \text{ BTU} = 1.055 \text{ kJ})$$

$$= \frac{200 \times 1.055}{60} \text{ kJ/s}$$

$$= 3.517 \text{ kW} \qquad (8.6)$$

7. $$\text{COP}_{\text{Carnot refrigerator}} = \frac{Q_L}{W_{net}} = \frac{Q_L}{Q_H - Q_L} = \frac{T_L(S_2 - S_1)}{T_H(S_2 - S_1) - T_L(S_2 - S_1)} \qquad (8.7)$$

or

$$\frac{Q_L}{W_{net}} = \frac{T_L}{T_H - T_L} \qquad (8.8)$$

or

$$W_{net} = Q_L \frac{T_L}{T_H - T_L} \qquad (8.9)$$

8. $$\text{COP}_{\text{vapour–compression cycle}} = \frac{\text{Refrigerating effect}}{\text{Work input}} = \frac{Q_L}{W_{net}} \qquad (8.10)$$

$$= \frac{H_1 - H_4}{H_2 - H_1} \qquad (8.11)$$

9. Mass flow rate of the refrigerant can be expressed as

$$m^0 = \frac{Q^0 L}{H_1 - H_4} = \frac{\text{Refrigerating capacity}}{\text{Refrigerating effect}} \qquad (8.12)$$

10. For a one-ton refrigerator, the rate of circulation in kg/h is given by

$$m^0 = \frac{12660}{H_1 - H_4} \qquad (8.13)$$

11. $COP_{absorption\ refrigeration} = \dfrac{\text{Refrigeration produced}}{\text{Energy input}}$ (8.14)

$$= \dfrac{Q_L}{Q_H + W} \cong \dfrac{Q_L}{Q_H} \qquad (8.15)$$

$$= \dfrac{T_L}{T_S - T_L} \cdot \dfrac{T_H - T_S}{T_H} \qquad (8.21)$$

12. $COP_{Bell-Coleman\ refrigeration} = \dfrac{\text{Heat extracted}}{\text{Work done}}$

$$= \left(R_{ip}^{\frac{\gamma-1}{\gamma}} - 1 \right)^{-1} \qquad (8.29)$$

$$= \dfrac{T_1}{T_2 - T_1} \qquad (8.30)$$

13. The isentropic expansion coefficient is defined by

$$\mu_S = \left(\dfrac{\partial T}{\partial P} \right)_S \qquad (8.31)$$

14. The Joule–Thomson coefficient is defined by

$$\mu_{JT} = \left(\dfrac{\partial T}{\partial P} \right)_H \qquad (835)$$

EXERCISES

A. Review Questions

1. How do you judge the performance capability of a refrigerator?
2. Define the term *ton of refrigeration*. Can it be expressed in kilowatts?
3. What is a heat pump? How can it be used for heating a room during winter?
4. Explain the dual operation of a heat pump during winter and summer.
5. Prove that $COP_{HP} = COP_R + 1$.
6. Can the COP of a refrigerator be greater than unity? Why?
7. How does a refrigerator differ from a heat pump?
8. Justify the statement: *A refrigerating machine works on the principle of the second law of thermodynamics.*
9. Name the different operational steps carried out in a Carnot cycle.
10. With the help of *T–S* diagram, explain the functioning of a Carnot refrigerator.
11. With the help of a neat sketch, explain the working principle of a vapour–compression refrigeration cycle and find its COP.
12. Name some refrigerants used widely in the vapour–compression cycle.
13. What are the important factors to be taken into account for selecting a promising refrigerant in the present scenario of replacement of CFCs?
14. What desirable properties should a refrigerant possess?

15. Explain the operation that takes place in a vapour–absorption cycle.
16. Provide a proper comparative discussion of between vapour–compression and vapour–absorption cycles.
17. Enumerate the various application areas of refrigeration.
18. What are the advantages and disadvantages of the air refrigeration cycle?
19. What are the different processes employed for liquefying a gas?
20. Compare the Linde process with the Claude process for air liquefaction.
21. Mention the usefulness of liquefaction processes.
22. Why are the CFCs still in major use despite having a strong negative impact on the environment?

B. Problems

1. A Carnot refrigerator is used to keep a food compartment at $-2°C$. Heat is rejected from the compartment to a room at $35°C$. If 40 kW heat is removed, determine
   (a) The coefficient of performance of the cycle
   (b) The power required
   (c) The rate of heat rejection to the room.

2. To maintain a freezer box at $-40°C$ on a summer day when the ambient temperature is $27°C$, heat is removed at the rate of 1.25 kW. What is the maximum possible coefficient of performance of the freezer, and what is the minimum power that must be supplied to the freezer?

3. A Carnot refrigerator used for air-conditioning operates between a low temperature of $5°C$ and a high temperature of $45°C$. Now, if it is used for maintaining foodstuffs in a cold storage operating at a low temperature of $-15°C$, and a high temperature of $45°C$ as before, what percentage increase in work input would be required for the food refrigeration unit over the air-conditioning unit for producing the same quantity of cooling?

4. A Carnot refrigerator is used to keep a freezer at $-2°C$. If the room to which the heat is rejected is at $35°C$, calculate for a heat removal rate of 2 kW
   (a) The COP of the cycle
   (b) The power required
   (c) The rate of heat rejected to the room.

5. An inventor claims to have developed a refrigerator, that maintains the refrigerated space at $1°C$ while operating in a room where the temperature is $30°C$, and that has a COP of 11. Is this claim reasonable?

6. (a) A one-ton Carnot refrigerator maintains a cold space at a temperature of $-15°C$, while the surroundings are at $40°C$. Determine the power consumption of the refrigerator.
   (b) If the same refrigerator is used as a freezer maintaining a temperature of $-25°C$, while the surroundings maintain at $40°C$, how much refrigeration will it produce? Assume the same power consumption as calculated in part (a).

7. A large auditorium is desired to be maintained at $20°C$ in both summer and winter seasons. In winter, the temperature of the environment is $4°C$ while the same in summer is $38°C$. The energy loss through the barrier walls of the auditorium exposed to the

atmosphere is at the rate of 22 kJ/s. If the same system is used as a refrigerator in summer and as a heat pump in winter, determine the power requirement to maintain the auditorium at 20°C in summer as well as in winter.

8. In winter, when the outside temperature is 0°C, a room is to be maintained at 25°C. The heat loss from the room is estimated to be 5 kW. Two alternative strategies to keep the room temperature constant are: (a) direct electrical heating and (b) reversing the room air-conditioner. Which is the better proposition?

9. A heat pump is used to maintain a temperature of 20°C inside a building when the outside temperature is −2°C. The amount of heat supplied by the heat pump is 35 kW. Estimate the coefficient of performance and the power input of the heat pump.

10. A Carnot heat pump supplies 25 kW of heat at 25°C when the environment temperature is 0°C. Calculate the COP and the power input.

11. Consider an ideal gas refrigeration cycle using air as the working fluid which is used to maintain a cold space at −15°C while discarding the heat to the environment at 30°C. The pressure ratio of the compressor is 4. Take $C_P = 1.007$ kJ/kg and $\gamma = 1.4$. Determine

    (a) The coefficient of performance
    (b) The cooling rate for a mass flow rate of 0.06 kg/s.

12. A vapour–compression refrigeration unit using steam as the working fluid operates between 306 K and 280 K. The enthalpy of saturated vapour at 280 K is 2514.20 kJ/kg and the enthalpy of saturated liquid at 306 K is 137.82 kJ/kg. Calculate the COP of the cycle.

13. A vapour–compression refrigeration system based on CFC-12 operates between an evaporator temperature of 245 K and a condenser temperature of 310 K. The power input to the compressor is 55 kW. Determine the cooling capacity and the COP of the system, given that

    Enthalpy of saturated vapour at 245 K = 176.7 kJ/kg
    Enthalpy of saturated liquid at 306 K = 71.23 kJ/kg
    Enthalpy of superheated vapour = 204.2 kJ/kg.

14. Determine the COP of a vapour–compression refrigeration system that uses Freon-12 as working fluid and operates between a condensation temperature of 45°C and an evaporator temperature of −20°C. Also determine the theoretical power consumption if the refrigerating capacity of the system is 300 W.

15. An ammonia refrigeration plant is to operate between a saturated liquid at 313 K at the condenser outlet and a saturated vapour at 283 K at the evaporator outlet. If the capacity of the refrigeration plant is desired to the tune of 6 tons, compute

    (a) The coefficient of performance
    (b) The work of compression
    (c) The refrigerating effect
    given that

    Enthalpy of saturated vapour = 1360.7 kJ/kg
    Enthalpy of saturated liquid at 306 K = 172.2 kJ/kg
    Enthalpy of superheated vapour = 1512.2 kJ/kg.

16. A vapour–compression refrigerating machine uses HFC-134a as the working fluid. The fluid enters the compressor as superheated vapour at 0.18 MPa and −10°C and leaves at a rate of 0.9 MPa and 50°C. The mass flow rate of the working fluid is 0.06 kg/s. The working fluid is cooled in the condenser at 29°C and 0.75 MPa and it is throttled to 0.20 MPa. Determine the rate of heat removal from the cold space and the COP, given that

    Enthalpy of saturated vapour = 245.16 kJ/kg
    Enthalpy of saturated liquid = 92.22 kJ/kg
    Enthalpy of superheated vapour = 274.17 kJ/kg.

17. An absorption refrigeration system receives heat from a source at 120°C and maintains the space which is to be cooled at 1°C. The temperature of the surroundings is 26°C. What is the maximum coefficient of performance that could be obtained from the system?

18. In a remote place, the heat energy is supplied to an absorption refrigeration system from a solar pond at 125°C at the rate of 4.5×10^5 kJ/h. The environment is at 28°C and the cold space is maintained at −5°C. Determine the rate at which heat is rejected from the low-temperature space.

19. An ideal air refrigeration cycle is operated at the upper temperature of 35°C and the lower temperature of −20°C. The COP of the cycle is 3. Determine the work of the expander, the work of the compressor, and the net work of the cycle. Also estimate the mass flow rate of air. Take $C_P = 1.007$ kJ/kg and $\gamma = 1.4$.

20. It is planned to maintain a cold storage at −10°C by employing an air refrigeration system which absorbs 900 kJ/min. The temperature of ambient water is 20°C after compression at 492.5 kPa and is later expanded to 101.3 kPa. Air is assumed to behave as an ideal gas. Calculate the COP and the minimum power requirement. Take $C_P = 1.006$ kJ/kg-K and $\gamma = 1.4$.

21. An air refrigeration cycle has a pressure ratio of 5. Air enters the compressor at 12°C and the turbine at 48°C. The mass flow rate of air is 0.08 kg/s. Take $C_P = 1.007$ kJ/kg-K and $\gamma = 1.4$. Determine

    (a) The actual power input coefficient of performance
    (b) The coefficient of performance.

22. A Bell–Coleman refrigeration cycle operates between the pressures of 50 kPa and 250 kPa. Air enters the compressor at 288 K and the turbine at 313 K. The air flow rate is 0.08 kg/s. Take $C_P = 1.008$ kJ/kg-K and $\gamma = 1.4$. Determine

    (a) The power requirement
    (b) The work of expansion
    (c) The coefficient of performance.

23. The Brayton refrigeration cycle works on the principle of air standard refrigeration and uses air as the working fluid. The pressure ratio of the compressor is 5. At the compressor inlet, the pressure is 100 kPa and the temperature is 308 K. Determine the work of compression, the work of expansion, and the coefficient of performance. Take $C_P = 1.008$ kJ/kg-K and $\gamma = 1.4$.

Solution Thermodynamics: Properties

9

The chapter deals with the study of the proper application of thermodynamics to gas mixtures and liquid solutions. In the chemical process industries, most of the operations such as distillation, absorption, adsorption, leaching, extraction, and drying involve transfer of components from one phase to another at different temperatures and pressures. The compositions and properties of the solutions and gas mixtures play an important role in this regard. These are discussed suitably in this chapter. First, we define the term *partial molar property* and its different forms. The chemical potential is a fundamental new property that enables us to treat the phase equilibrium, and we explain the influence of temperature and pressure on it. Activity and activity coefficient, another significant property, measures the deviation from ideal behaviour of chemical substances in a mixture. The temperature and pressure dependence form of this property is shown in simplified form. The change in partial molar properties with composition at constant temperature and pressure can be explained well by the Gibbs–Duhem equation. We derive the most common and useful forms of this equation. Likewise, the importance of fugacity of components in mixtures is discussed. Due to mixing, there must be a change in the measurable properties of various components of the system. It has been explained with the help of the ideal gas mixture model developed by Gibbs. Another useful topic is that of excess properties that facilitate differentiation between the real and ideal behaviours of the mixture. We discuss the potential role of the ideal solution model with Lewis–Randall rule, and also the significance of Henry's law for dilute solutions and Raoult's law for ideal solutions.

9.1 PARTIAL MOLAR PROPERTIES

Partial molar properties play an important role in describing the multicomponent system. It is defined as the property of a component which measures the contribution of the component present in a mixture to the aggregate property of the mixture. Consider any extensive thermodynamic property X (volume, enthalpy, free energy, or heat capacity) of a single-phase, single-component system, which is a function of two independent intensive properties— temperature and pressure— and the size of the system. Then the partial molar property $\overline{X}_i$ of the component i by definition is

$$\overline{X}_i = \left(\frac{\partial X}{\partial n_i} \right)_{T,P,n_j} \tag{9.1}$$

where

n_i = Number of moles of component i in the mixture

n_j = Number of moles of component other than component i in the mixture.

Thus, the partial molar volume, internal energy, enthalpy, and entropy can be expressed as

$$\overline{V}_i = \left(\frac{\partial V}{\partial n_i} \right)_{T,P,n_j} \qquad \overline{U}_i = \left(\frac{\partial U}{\partial n_i} \right)_{T,P,n_j} \qquad \overline{H}_i = \left(\frac{\partial H}{\partial n_i} \right)_{T,P,n_j} \qquad \overline{S}_i = \left(\frac{\partial S}{\partial n_i} \right)_{T,P,n_j}$$

respectively.

Let us consider a multicomponent system in which any extensive property X is a function of the temperature, pressure, and composition of the mixture. Then for a binary system the property X may be expressed as

$$X = f(T, P, n_1, n_2, \ldots)$$

If there is an infinitesimal change in the temperature, pressure, and amount of each component, then the total change in X is given by

$$dX = \left(\frac{\partial X}{\partial P} \right)_{T,n_1,n_2} dP + \left(\frac{\partial X}{\partial T} \right)_{P,n_1,n_2} dT + \left(\frac{\partial X}{\partial n_1} \right)_{P,T,n_2} dn_1 + \left(\frac{\partial G}{\partial n_2} \right)_{P,T,n_1} dn_2 + \cdots \tag{9.2}$$

At constant temperature and pressure, Eq. (9.2) becomes

$$dX_{P,T} = \left(\frac{\partial X}{\partial n_1} \right)_{P,T,n_2} dn_1 + \left(\frac{\partial X}{\partial n_2} \right)_{P,T,n_1} dn_2 + \cdots \tag{9.3}$$

$$dX_{P,T} = \overline{X}_1\, dn_1 + \overline{X}_2\, dn_2 + \cdots \tag{9.4}$$

Here

$$\left(\frac{\partial X}{\partial n_1}\right)_{P,T,n_2} = \overline{X}_1 = \text{Partial molar property of component 1}$$

and

$$\left(\frac{\partial X}{\partial n_2}\right)_{P,T,n_1} = \overline{X}_2 = \text{Partial molar property of component 2}$$

Imagine that at constant temperature and pressure, the system is enlarged in such a way that the mole fraction of each component remains constant, i.e., y_1 and y_2 are changed but the corresponding mole fractions y_1 and y_2 are not changed.

Then, if $n_1 + n_2 = n$

for component 1: mole fraction $y_1 = \dfrac{n_1}{n}$

and

for component 2: mole fraction $y_2 = \dfrac{n_2}{n}$

Therefore

$$dn_1 = y_1 dn \quad \text{and} \quad dn_2 = y_2 dn$$

Substituting it into Eq. (9.4), we get

$$dY_{P,T} = (\overline{Y}_1 x_1 + \overline{Y}_2 x_2 + \cdots)dn \tag{9.5}$$

or

$$Y = \sum n_i \overline{Y}_i \tag{9.6}$$

This equation shows that the extensive property X can be expressed as a weighted sum of the partial molar properties $\overline{X}_i$. Several additional relations involving partial molar properties are developed later in this chapter. For understanding this concept, let us take up the following illustration.

Consider that 1 mol of pure ethanol is added to 1 mol of pure water. This is not an ideal solution. The volume of 1 mol of pure ethanol is 58.0 ml and that of 1 mol of pure water is 18.0 ml. However, mixing of 1 mol of pure ethanol and 1 mol of pure water does not result in (58.0 + 18.0) ml = 76.0 ml of solution; rather the result is 74.3 ml. It is the partial molar volume that is additive or extensive. When the mole fraction is 0.5, the partial molar volume of ethanol is 57.4 ml and that of water is 16.9 ml.

Now, we can calculate the volume of the solution:

$$\begin{aligned} V &= n_1 V_1 + n_2 V_2 \\ &= 1 \text{ mol} \times 57.4 \text{ ml/mol} + 1 \text{ mol} \times 16.9 \text{ ml/mol} \\ &= 74.3 \text{ ml} \end{aligned}$$

Note that the values just cited for the partial molar volume of ethanol and water are only for a particular concentration; in this case, the mole fraction is equal to 0.5 and applies only to the water–ethanol system.

9.1.1 Evaluation of Partial Molar Properties

The partial molar properties of the components in a solution may be evaluated by both analytical and graphical methods.

Analytical Method

In case of analytical methods, when the extensive thermodynamic properties such as volume and internal energy are available, the partial molar properties can be obtained by partial differentiation of extensive properties with respect to the number of moles of the components. For instance, partial molar volume

$$\overline{V}_i = \left(\frac{\partial V}{\partial n_i} \right)_{T,P,n_j} \tag{9.7}$$

Graphical Method

Graphical determination of partial molar properties also has major importance in thermodynamics. When suitable data are available, a simple graphical procedure known as the *method of tangential intercepts* can be used to evaluate partial molar properties (see Fig. 9.1). In principle, the method can be applied for any extensive property. To understand this method, let us consider the volume of a binary system consisting of two components 1 and 2. For this system, the molar volume can be expressed as

$$V = n_1 \overline{V}_1 + n_2 \overline{V}_2 \tag{9.8}$$

where

$\overline{V}_1$ and $\overline{V}_2$ are the partial molar volumes of components 1 and 2
n_1 and n_2 are the gram-moles of components 1 and 2 respectively.
In terms of mole fractions, we get

$$V = y_1 \overline{V}_1 + y_2 \overline{V}_2 \tag{9.9}$$

where

y_1 = Mole fraction of component 1
y_2 = Mole fraction of component 2.

Since for a binary solution, $y_1 + y_2 = 1$, we put $y_1 = 1 - y_2$ into Eq. (9.9). Then it yields

$$V = (1 - y_2)\overline{V}_1 + y_2 \overline{V}_2 \tag{9.10}$$

or

$$V = \overline{V}_1 + y_2(\overline{V}_2 - \overline{V}_1) \tag{9.11}$$

Differentiating it with respect to y_2, we get

$$\frac{dV}{dy_2} = \overline{V}_2 - \overline{V}_1 \tag{9.12}$$

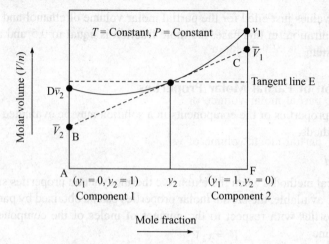

Fig. 9.1 Determination of partial molar volume using method of tangential intercepts.

Substituting Eq. (9.12) into (9.11), we obtain

$$\overline{V}_1 = V - y_2 \frac{dV}{dy_2} \tag{9.13}$$

From Eqs. (9.11) and (9.13), we get

$$\overline{V}_2 = V + (1 - y_2) \frac{dV}{dy_2} \tag{9.14}$$

where $\dfrac{dV}{dy_2}$ is the slope of the curve at any point.

This equation provides the basis for the method of intercepts. In Fig. 9.1, V/n, i.e., the molar volume of the mixture is plotted as a function of y_2, the mole fraction of one of the components at constant T and P.

At $y_1 = 1$, or $y_2 = 0$, the intercept is

$$AB = AD - BD$$

or

$$\overline{V}_1 = V - y_2 \frac{dV}{dy_2}$$

Similarly, at $y_2 = 1$, or $y_1 = 0$, for the intercept

$$CF = CE + EF$$

or

$$\overline{V}_2 = V + (1 - y_2) \frac{dV}{dy_2}$$

At a specified value for y_2, a tangent to the curve is shown in Fig. 9.1. When extrapolated, the tangent line intersects the axis on the left at $\overline{V}_1$ and the axis on the right at $\overline{V}_2$. These values for the partial molar volumes correspond to the particular specifications for T, P, and y_2. At fixed temperature and pressure, $\overline{V}_1$ and $\overline{V}_2$ vary with y_2 and are not equal to the molar specific volumes of pure species 1 and pure species 2, denoted in Fig. 9.1 as $\overline{v}_1$ and $\overline{v}_2$ respectively.

EXAMPLE 9.1   Calculate the partial molar volume of water in a 50 mol per cent ethanol–water solution in which the partial molar volume of ethanol is 52.37×10^{-6} m^3/mol. Given that the density of the mixture is 800.21 kg/m^3.

Solution: Assume that

$$\text{partial molar volume of ethanol (component 1)} = \overline{V}_1$$

and

$$\text{partial molar volume of water (component 2)} = \overline{V}_2$$

Given:

$$\overline{V}_1 = 52.37 \times 10^{-6} \text{ m}^3/\text{mol}$$
$$\overline{V}_2 = ?$$

From Eq. (9.9), we have

$$V = y_1 \overline{V}_1 + y_2 \overline{V}_2$$

Molar mass of ethanol, $M_1 = 46 \times 10^{-3}$ kg/mol
Molar mass of water, $M_2 = 18 \times 10^{-3}$ kg/mol
Mole fraction of ethanol = $y_1 = 0.5$
Mole fraction of water = $y_2 = 0.5$.

Therefore, the molar mass of a 50 mol per cent ethanol–water mixture can be estimated as

$$M = M_1 + M_2 = [(46 \times 10^{-3} \text{ kg/mol}) \times 0.5] + [(18 \times 10^{-3} \text{ kg/mol}) \times 0.5] = 32 \times 10^{-3} \text{ kg/mol}$$

Molar volume of the mixture is given by

$$V = \frac{\text{Mass}}{\text{Density}} = \frac{32 \times 10^{-3}}{800.21} = 39.9 \times 10^{-6} \text{ m}^3/\text{mol}$$

We know that the molar volume of the mixture = $V = y_1 \overline{V}_1 + y_2 \overline{V}_2$

or

$$39.9 \times 10^{-6} = 0.5 \times 52.37 \times 10^{-6} + 0.5 \times \overline{V}_2$$

or

$$\overline{V}_2 = 27.43 \times 10^{-6} \text{ m}^3/\text{mol}$$

Hence, the partial molar volume of water, $\overline{V}_2$, is 27.43×10^{-6} m^3/mol.

EXAMPLE 9.2   In order to prepare 2.0 m^3 of alcohol–water solution, alcohol of mole fraction $X_1 = 0.40$ is required to be mixed with water at 25°C. Determine the volumes of alcohol and water needed to prepare the mixture. Given that
   Partial molar volume of alcohol = 38.3×10^{-6} m^3/mol
   Partial molar volume of water = 17.2×10^{-6} m^3/mol
   Molar volume of alcohol = 39.21×10^{-6} m^3/mol
   Molar volume of water = 18×10^{-6} m^3/mol

Solution: By applying Eq. (9.9), the molar volume of the solution can be estimated as

$$V = y_1 \overline{V}_1 + y_2 \overline{V}_2 = 0.4 \times 38.3 \times 10^{-6} + 0.6 \times 17.2 \times 10^{-6}$$
$$= 25.64 \times 10^{-6} \text{ m}^3/\text{mol}$$

Number of moles of solution required $= 2/(25.64 \times 10^{-6} \text{ m}^3/\text{mol}) = 7.8 \times 10^4$
Number of moles of alcohol $= 0.40 \times 7.8 \times 10^4 = 3.12 \times 10^4$
Number of moles of water $= 0.60 \times 7.8 \times 10^4 = 4.68 \times 10^4$
Volume of alcohol needed $= n_1 V_1 = 3.12 \times 10^4 \times 39.21 \times 10^{-6} = 1.223 \text{ m}^3$
Volume of water needed $= n_2 V_2 = 4.68 \times 10^4 \times 18 \times 10^{-6} = 0.8424 \text{ m}^3$

EXAMPLE 9.3 A student experimenter decides to convert 2000 cm³ of laboratory alcohol containing 96% ethanol and 4% water by mass into vodka having a composition of 56% ethanol and 44% water by mass.
Data given:
At 25°C and 1 atm

In 96% ethanol: $\overline{V}$ (water) $= 0.816 \text{ cm}^3/\text{g}$ $\overline{V}$ (ethanol) $= 1.273 \text{ cm}^3/\text{g}$
In 56% ethanol: $\overline{V}$ (water) $= 0.953 \text{ cm}^3/\text{g}$ $\overline{V}$ (ethanol) $= 1.243 \text{ cm}^3/\text{g}$

The density of water at 25°C and 1 atm is 0.997 cm³/g
  (i) How much water should be added to 2000 cm³ of laboratory alcohol?
  (ii) Calculate the volume of vodka after conversion.

Solution: From the relation

$$V = y_1 \overline{V}_1 + y_2 \overline{V}_2$$

the volume of laboratory alcohol can be calculated as

$$V_a = [(0.96 \times 1.273) + (0.04 \times 0.816)] = 1.255 \text{ cm}^3/\text{g}$$

Mass of 2000 cm³ of laboratory alcohol $= \dfrac{2000}{1.255} = 1594 \text{ g}$

Part I: Let M_w be the mass of water added to the laboratory alcohol. A material balance on ethanol gives

$$1594 \times 0.96 = (M + 1594) \times 0.56$$

Solving for M_w, we get

$$M_w = 1138 \text{ g} = \text{Mass of water}$$
$$\text{Volume of water added} = \frac{\text{Mass}}{\text{Density}} = \frac{1138}{0.997} = 1142 \text{ cm}^3$$

Part II: Mass of vodka obtained $= 1594 \text{ g} + 1138 \text{ g} = 2732 \text{ g}$
Volume of 56% ethanol $= V_v = [(0.56 \times 1.243) + (0.44 \times 0.953)] = 1.115 \text{ cm}^3/\text{g}$
Volume of vodka obtained after conversion $= 1.115 \times 2732 = 3046 \text{ cm}^3$.

EXAMPLE 9.4 In a binary liquid system, the enthalpy of species 1 and 2 at constant temperature and pressure is represented by the following equation:

$$H = 400x_1 + 600x_2 + x_1 x_2 (40x_1 + 20x_2) \quad \text{where } H \text{ is in J/mol}$$

Determine the expressions for $\overline{H}_1$ and $\overline{H}_2$ as functions of x_1, and the numerical values for the pure species enthalpies H_1 and H_2.

Solution: The enthalpy equation is given by

$$H = 400x_1 + 600x_2 + x_1x_2(40x_1 + 20x_2) \tag{i}$$

Since $x_1 + x_2 = 1$, we replace x_2 of Eq. (i) by $1 - x_1$ to get

$$H = 400x_1 + 600 (1 - x_1) + x_1(1 - x_1) [40x_1 + 20(1 - x_1)]$$
$$= 600 - 180x_1 - 20x_1^3$$

Differentiating it with respect to x_1, we get

$$\frac{dH}{dx_1} = -180 - 60x_1^2 \tag{ii}$$

Following Eq. (9.13) after substitution of $dy_2 = -dy_1$, we have in terms of x

$$\overline{H}_1 = H + x_2 \frac{dH}{dx_1} \tag{iii}$$

Substituting Eqs. (i) and (ii) into (iii), we have

$$\overline{H}_1 = 600 - 180x_1 - 20x_1^3 - 180x_2 - 60x_1^2 x_2 \tag{iv}$$

We replace x_2 by $1 - x_1$ and simplify to get

$$\overline{H}_1 = 420 - 60x_1^2 + 40x_1^3 \tag{v}$$

Similarly following Eq. (9.13) after substitution of $x_1 = 1 - x_2$ and $dy_2 = -dy_1$, we have in terms of x

$$\overline{H}_2 = H - x_1 \frac{dH}{dx_1} \tag{vi}$$

On substitution of H and Eq. (ii) into Eq. (vi), we get

$$\overline{H}_2 = H - x_1\frac{dH}{dx_1} = 600 - 180x_1 - 20x_1^3 + 180x_1 + 60x_1^3$$
$$= 600 + 40x_1^3 \tag{vii}$$

Numerical values for the pure-species enthalpy H_1:

We put $H_1 = \overline{H}_1$ when $x_1 = 1$ and $x_2 = 0$ in Eq. (v) to obtain $H_1 = 400$ J/mol. (We can obtain the same value from Eq. (iv).)

Numerical values for the pure-species enthalpy H_2:

We put $H_2 = \overline{H}_2$ when $x_1 = 0$ and $x_2 = 1$ in Eq. (vii) to obtain $H_2 = 600$ J/mol.

EXAMPLE 9.5 The molar volume of a binary mixture consisting of components 1 and 2 at 298 K and 1 atm is represented by the following empirical relation:

$$V = 52.36 - 32.64y_2 - 42.98y_2^2 + 58.77y_2^3 - 23.45y_2^4$$

where V is in cm^3/mol and y_2 is the mole fraction of component 2. Develop the expressions for the partial molar volumes for both the components.

Solution: Differentiating the equation of molar volume, we get

$$V = 52.36 - 32.64y_2 - 42.98y_2^2 + 58.77y_2^3 - 23.45y_2^4$$

or

$$\left(\frac{\partial V}{\partial y_2}\right) = -32.64 - 85.96y_2 + 176.31y_2^2 - 93.80y_2^3 \tag{i}$$

Following Eq. (9.13), we get

$$\overline{V}_1 = V - y_2\frac{dV}{dy_2} \tag{ii}$$

Therefore, Eq. (ii) can be expressed as

$$V - y_2\frac{dV}{dy_2} = (52.36 - 32.64y_2 - 42.98y_2^2 + 58.77y_2^3 - 23.45y_2^4)$$

$$-y_2(-32.64 - 85.96y_2 + 176.31y_2^2 - 93.80y_2^3)$$

or

$$\overline{V}_1 = 52.36 + 42.98y_2^2 - 117.54y_2^3 + 70.35y_2^4 \tag{iii}$$

Similarly, we obtain the expression for $\overline{V}_2$ as

$$\overline{V}_2 = V + y_1\frac{dV}{dy_2} = V + (1 - y_2)\frac{dV}{dy_2}$$

$$= V + \frac{dV}{dy_2} - y_2\frac{dV}{dy_2} = \left(V - y_2\frac{dV}{dy_2}\right) + \frac{dV}{dy_2} = \overline{V}_1 + \frac{dV}{dy_2} \tag{iv}$$

Substituting Eqs. (i) and (iii) into (iv), we get

$$\overline{V}_2 = 12.06 + 7.37y_1^2 - 70.06y_1^3 + 70.35y_1^4$$

EXAMPLE 9.6 The partial molar volume of water (1) in its mixture with methanol (2) at 25°C and 1 atm can be approximated by:

$$\overline{V}_1 = 18.1 + ax_2^2$$

where $a = -3.2$ cm^3/mol.

Develop an expression for the partial molar volume of methanol at the same condition, given that V_2 at 25°C and 1 atm is 40.7 cm^3/mol. Determine also the partial molar volume of methanol at infinite dilution.

Solution: We know that at constant temperature and pressure, the Gibbs–Duhem relation is

$$x_1d\overline{V}_1 + x_2d\overline{V}_2 = 0$$

or

$$d\overline{V}_2 = -\frac{x_1}{x_2}d\overline{V}_1 \tag{i}$$

Differentiating the equation $\overline{V}_1 = 18.1 + ax_2^2$, we get

$$\frac{d\overline{V}_1}{dx_2} = 2ax_2 \tag{ii}$$

Substituting Eq. (ii) into (i), we obtain

$$d\overline{V}_2 = -\frac{x_1}{x_2}d\overline{V}_1 = -\frac{x_1}{x_2} \cdot 2ax_2 \cdot dx_2 \tag{iii}$$

Since $x_1 + x_2 = 1$, we have on putting $x_1 = 1 - x_2$ into Eq. (iii)

$$d\overline{V}_2 = -2a(1 - x_2)dx_2 \qquad \text{(iv)}$$

Now, when $x_2 = 1$, $\overline{V}_2 = V_2$. On integrating Eq. (iv), we get

$$\int_{V_2}^{\overline{V}_2} d\overline{V}_2 = -2a \int_1^{x_2} (1 - x_2)\,dx_2$$

or

$$\overline{V}_2 - V_2 = a(1 - x_2)^2 = ax_1^2 \qquad \text{(v)}$$

Therefore, after substituting the values of V_2 and a, we get

$$\overline{V}_2 = 40.7 - 3.2x_1^2 \qquad \text{(vi)}$$

Now, when $x_2 = 1$, $\overline{V}_2 = V_2$, and when $x_2 = 0$, $\overline{V}_2 = \overline{V}_2^{\infty} = 37.5$ cm^3/mol. Hence the partial molar volume of component 2 at infinite dilution is 37.5 cm^3/mol.

EXAMPLE 9.7   The molar volumes of a binary solution at 25°C are measured as given below:

X_1	0	0.2	0.4	0.6	0.8	1.0
$V \times 10^6$ (m^3/mol)	20.0	21.5	24.0	27.4	32.0	40.0

Using the methods of tangential intercept, calculate the partial molar volumes of components 1 and 2 at

(a) $X_1 = 0.5$
(b) $X_1 = 0.75$.

Solution: Following the data table and the problem, molar volume is plotted against mole fractions as shown in Fig. 9.2. A tangent is drawn to the curve at $X_1 = 0.5$.

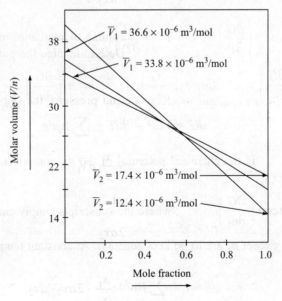

Fig. 9.2 Example 9.7.

When $X_1 = 0.5$, $X_2 = 0.5$. The intercept at $X_2 = 0$ gives $\overline{V}_1 = 33.8 \times 1^{-6}$ m^3/mol and the intercept at $X_2 = 1$ gives $\overline{V}_2 = 17 \times 10^{-6}$ m^3/mol.

Similarly, when $X_1 = 0.75$ or $X_2 = 0.25$, we get

$$\overline{V}_1 = 36.6 \times 10^{-6} \text{ m}^3/\text{mol} \quad \text{and} \quad \overline{V}_2 = 12.4 \times 10^{-6} \text{ m}^3/\text{mol}$$

9.2 CHEMICAL POTENTIAL

The chemical potential of a component is an useful thermodynamic property. It is symbolized by μ. It is defined as the partial derivative of the thermodynamic potentials H, A and G of the ith component of a multicomponent system. Like temperature and pressure, the chemical potential μ_i is an *intensive* property. It is nothing but the rate of increase in Gibbs free energy per mole of the component i added. This quantity plays a central role in the criteria for both chemical and phase equilibria, because of its importance in the study of multicomponent systems. It can be mathematically expressed as

$$\mu_i = \left(\frac{\partial G}{\partial n_i} \right)_{P,T,n_j} \tag{9.15}$$

Consider a thermodynamic system containing n_1, n_2, and n_3 moles of three different components. Its free energy (G) is a function of T, P and the number of moles of the three components. Then the free energy of the system can be expressed as

$$G = f(P, T, n_1, n_2, n_3, \ldots)$$

The total differential will be given by

$$dG = \left(\frac{\partial G}{\partial P} \right)_{T,n_i} dP + \left(\frac{\partial G}{\partial T} \right)_{P,n_i} dT + \left(\frac{\partial G}{\partial n_1} \right)_{P,T,n_2,n_3} dn_1 + \left(\frac{\partial G}{\partial n_2} \right)_{P,T,n_1,n_3} dn_2 + \left(\frac{\partial G}{\partial n_3} \right)_{P,T,n_1,n_2} dn_3 + \cdots$$

We know

$$\left(\frac{\partial G}{\partial P} \right)_{T,n_1,n_2} = V \qquad \left(\frac{\partial G}{\partial T} \right)_{P,n_1,n_2} = -S$$

$$dG = VdP - SdT + \left(\frac{\partial G}{\partial n_1} \right)_{P,T,n_2,n_3} dn_1 + \left(\frac{\partial G}{\partial n_2} \right)_{P,T,n_1,n_3} dn_2 + \left(\frac{\partial G}{\partial n_3} \right)_{P,T,n_1,n_2} dn_3 + \cdots \tag{9.16}$$

$$dG = VdP - SdT + \sum \mu_i dn \tag{9.17}$$

Here, $\mu_i = \left(\frac{\partial G}{\partial n_i} \right)_{P,T,n_j}$ is the chemical potential of the ith component, where $j \neq 1$.

Consider the third term $\left(\frac{\partial G}{\partial n_1} \right)_{P,T,n_2,n_3}$, where the subscripts simply emphasize that P, T and all other mole numbers except n_1 are to be kept constant. At constant temperature and pressure, Eq. (9.17) becomes

$$dG_{P,T} = \sum \mu_i dn_i \tag{9.18}$$

or

$$G_{P,T} = \sum \mu_i n_i \qquad (9.19)$$

For a system consisting of two phases (a, b) in equilibrium, Eq. (9.19) can be written as

$$G_{P,T} = \sum_i \mu_i^a n_i^a + \sum_i \mu_i^b n_i^b = 0 \qquad (9.20)$$

or

$$dG_{P,T} = \sum_i \mu_i^a dn_i^a + \sum_i \mu_i^b dn_i^b = 0 \qquad (9.21)$$

Since $dn_i = dn_i^a + dn_i^b = 0$, then $dn_i^a = -dn_i^b$. Substituting this into Eq. (9.21), we get

$$\sum_i (\mu_i^a - \mu_i^b) dn_i^a = 0 \qquad (9.22)$$

or

$$\mu_i^a = \mu_i^b \qquad (9.23)$$

or

$$\overline{G}_i^a = \overline{G}_i^b \qquad (9.24)$$

Hence one can draw the following conclusion: *At the same temperature and pressure, chemical potential or partial molar free energy of a component in every phase must be the same under equilibrium conditions.*

It is clear from Eq. (9.19) that for a pure substance, $G = \mu n$ or $\mu = \dfrac{G}{n}$, i.e. the chemical potential is the free energy per mole of the substance.

It is necessary to remember that the chemical potential or partial molar free energy is independent of the masses, but depends upon the composition of the system. This implies that the free energy G of the system increases per mole of the added substance, but the increase depends upon the composition. For example, if one mole of H_2O is added to 1 kg of pure water and 1 kg of sugar solution separately, then it is observed that the increase in G will not be exactly the same.

Chemical potential is important when studying the system of reacting substances. It is independent of the microscopic behaviour of the system, i.e., the properties of the constituent species.

To understand the term *chemical potential*, we can cite an example. Consider one mole of methane and 2 moles of oxygen. If a flame is brought near this mixture, then the following reaction will take place:

$$CH_4 + 2O_2 \rightarrow CO_2 + 2H_2O + Q$$

The reaction will be accompanied by the evolution of certain amount of thermal energy. This energy comes from the difference in chemical potential between CH_4 and O_2 on the one hand (higher potential) and CO_2 and H_2O on the other hand (lower potential). The total energy released can be expressed as

$$Q = \mu(CH_4) + 2\mu(O_2) - \mu(CO_2) - 2\mu(H_2O)$$

Similar examples can be found in the case of batteries, where chemical energy is converted into electrical energy.

In terms of Helmholtz free energy, the chemical potential is

$$\mu_i = \left(\frac{\partial A}{\partial n_i}\right)_{P,T,n_j} \quad \text{where } j \neq 1$$

In terms of enthalpy, the chemical potential is

$$\mu_i = \left(\frac{\partial H}{\partial n_i}\right)_{P,T,n_j} \quad \text{where } j \neq 1$$

9.2.1 Influence of Pressure on Chemical Potential

The chemical potential of the ith component of a system is defined as

$$\mu_i = \left(\frac{\partial G}{\partial n_i}\right)_{P,T,n_j}$$

Differentiating it with respect to P at constant T and composition, we get

$$\left(\frac{\partial \mu_i}{\partial P}\right)_{T,n_i} = \frac{\partial}{\partial P}\left[\left(\frac{\partial G}{\partial n_i}\right)_{P,T,n_j}\right]_{T,n_i} = \frac{\partial^2 G}{\partial P \partial n_i} \tag{9.25}$$

Again we know that

$$\left(\frac{\partial G}{\partial P}\right)_{T,n_i} = V$$

On differentiation with respect to n_i, it yields

$$\frac{\partial^2 G}{\partial n_i \partial P} = \left(\frac{\partial V}{\partial n_i}\right)_{T,P,n_j} = \bar{V}_i \tag{9.26}$$

Since dG is an exact differential and

$$\frac{\partial^2 G}{\partial P \partial n_i} = \frac{\partial^2 G}{\partial n_i \partial P}$$

then, comparing Eqs. (9.25) and (9.26), we get

$$\left(\frac{\partial \mu_i}{\partial P}\right)_{T,n_i} = \bar{V}_i \tag{9.27}$$

9.2.2 Influence of Temperature on Chemical Potential

By definition, the chemical potential of the ith component is

$$\mu_i = \left(\frac{\partial G}{\partial n_i}\right)_{P,T,n_j}$$

Differentiating it with respect to T at constant P and composition, we get

$$\left(\frac{\partial \mu_i}{\partial T}\right)_{P,\,n_i} = \frac{\partial^2 G}{\partial T \partial n_i} \tag{9.28}$$

We know that

$$\left(\frac{\partial G}{\partial T}\right)_{P,\,n_i} = -S$$

On differentiation with respect to n_i, it yields

$$\frac{\partial^2 G}{\partial n_i \partial T} = -\left(\frac{\partial S}{\partial n_i}\right)_{T,P,\,n_j} = -\bar{S}_i \tag{9.29}$$

Since G is a state function and dG is an exact differential, and

$$\frac{\partial^2 G}{\partial T \partial n_i} = \frac{\partial^2 G}{\partial n_i \partial T}$$

then, comparing Eqs. (9.28) and (9.29), we get

$$\left(\frac{\partial \mu_i}{\partial T}\right)_{P,\,n_i} = -\bar{S}_i = \text{Partial molar entropy of component } i \tag{9.30}$$

From Eq. (9.30), it is clear that the chemical potential will decrease with increase of temperature, as entropy is always a positive quantity. The rate of change of chemical potential with temperature is different for the gases, liquids, and solids, as

$$S^g > S^l > S^s \tag{9.31}$$

The effect of temperature on chemical potential can also be expressed in the form of the Gibbs–Helmholtz equation. We obtain

$$G = H - TS$$

Differentiating it with respect to n_i, keeping T, P and n_j constant, we get

$$\left(\frac{\partial G}{\partial n_i}\right)_{T,P,\,n_j} = \left(\frac{\partial H}{\partial n_i}\right)_{T,P,\,n_j} - T\left(\frac{\partial S}{\partial n_i}\right)_{T,P,\,n_j} \tag{9.32}$$

or

$$\bar{G}_i = \bar{H}_i - T\bar{S}_i \quad \text{as} \quad \left(\frac{\partial G}{\partial n_i}\right)_{T,P,\,n_j} = \mu_j = \bar{G}_i \tag{9.33}$$

Substituting Eq. (9.30) into Eq. (9.33), we get

$$\bar{G}_i = \bar{H}_i - T\left[-\left(\frac{\partial \mu_i}{\partial T}\right)_{P,\,n_i}\right] = \bar{H}_i + T\left(\frac{\partial \mu_i}{\partial T}\right)_{P,\,n_i}$$

or

$$\left[\frac{\partial\left(\frac{\mu}{T}\right)}{\partial T}\right] = -\frac{\bar{H}_i}{T^2} \tag{9.34}$$

9.3 ACTIVITY AND ACTIVITY COEFFICIENT

Activity

Activity is a measure of *effective concentration* of a component. It results from interaction between the molecules in a non-ideal gas or solution. It is dimensionless. Activity of a component depends on temperature, pressure, and composition. The activity of the component i is defined as the ratio of fugacity of the component i in the solution to the fugacity of component i in the standard state or reference state (existing as a pure substance at the temperature and pressure). It is denoted by a.

Mathematically

$$a_i = \frac{f_i}{f_i^0}$$

Activity for Gaseous Mixture: The activity for the species i in the gaseous mixture can be expressed as

$$a_i = \gamma_i \cdot x_i \cdot \frac{P}{P^0} = \frac{f_i}{P^0} \qquad (9.35)$$

where

γ_i = Activity coefficient of component i

x_i = Mole fraction of component i in the gaseous mixture

P^0 = Standard atmospheric pressure = 1

f_i = Fugacity of the gas = 1 for ideal gas.

Activity for Compound in Liquid: It can be represented as

$$a_i = \gamma_i \frac{C_i}{C^0} \qquad (9.36)$$

where

γ_i = Activity coefficient of component i

C_i = Concentration of component i

C^0 = Concentration of component i in the standard state = 1 mol/L.

Activity Coefficient

Activity coefficient is a measure of the deviation from ideal behaviour of chemical substances in a mixture. In an ideal mixture, the interaction between each pair of chemical components is the same. As a result, the properties of the mixtures can be expressed directly in terms of concentrations or partial pressures. Deviations from ideality are accommodated by modifying the concentration by an activity coefficient. Analogously, expressions involving gases can be adjusted for non-ideality by modifying the partial pressure by a fugacity coefficient.

By definition, activity coefficient is the ratio of activity to the number of moles or mole fractions of component i in a solution. The numerical value of the activity coefficient is dependent on both the standard state and the units of expression for composition. For an ideal solution, activity coefficient is considered to be unity, i.e., $\gamma_i = 1$. It can be expressed as

$$\gamma_i = \frac{a_i}{n_i} = \frac{f_i}{f_i^0 n_i} = \frac{\dfrac{f_i}{n}}{\dfrac{f_i^0 n_i}{n}} = \frac{\bar{f}_i}{f_i^0 x_i} \qquad (9.37)$$

where

a_i = Activity of component i referred to the pure component

f_i = Fugacity of component i

f_i^0 = Concentration of component i in the standard state = 1 mol/L.

For ideal behaviour of pure component at the same pressure, $f_i^0 = f_i$. Hence, the equation yields

$$\gamma_i = \frac{\bar{f}_i}{f_i^0 x_i} = \frac{\bar{f}_i}{f_i x_i} = \frac{a_i}{x_i} \qquad (9.38)$$

9.3.1 Temperature Dependence of Activity Coefficient

For component i in the solution, Eq. (6.186) on influence of temperature on fugacity at constant pressure becomes

$$\left(\frac{\partial \ln f_i}{\partial T} \right)_P = \frac{H_i^0 - H_i}{RT^2}$$

or

$$\left(\frac{\partial \ln \bar{f}_i}{\partial T} \right)_P = \frac{H_i^0 - \bar{H}_i}{RT^2} \qquad (9.39)$$

This is the temperature dependence of the activity coefficient of component i at constant pressure.

Subtracting Eq. (6.187) from (9.39), we have

$$\left[\frac{\partial \ln \left(\dfrac{\bar{f}_i}{f_i} \right)}{\partial T} \right]_P = \frac{H_i - \bar{H}_i}{RT^2} \qquad (9.40)$$

We know from the definition of activity coefficient that $\gamma_i = \dfrac{\bar{f}_i}{f_i^0 x_i}$. Putting this value into Eq. (9.40), we obtain

$$\left[\frac{\partial \ln (\gamma_i x_i)}{\partial T} \right]_P = \frac{H_i - \bar{H}_i}{RT^2} \qquad (9.41)$$

Since the mole fraction of the component i, i.e., x_i does not depend on temperature, it can be neglected. Then the expression becomes

$$\left(\frac{\partial \ln \gamma_i}{\partial T} \right)_P = \frac{H_i - \bar{H}_i}{RT^2} \qquad (9.42)$$

9.3.2 Pressure Dependence of Activity Coefficient

The variation of fugacity of component i in the solution with pressure at constant temperature can be expressed following the equation $\left(\dfrac{\partial \ln f_i}{\partial P} \right)_T = \dfrac{V_i}{RT}$ as

$$\left(\frac{\partial \ln \bar{f}_i}{\partial P}\right)_T = \frac{\bar{V}_i}{RT} \tag{9.43}$$

On subtraction, we get

$$\left(\frac{\partial \ln \bar{f}_i}{\partial P}\right)_T - \left(\frac{\partial \ln f_i}{\partial P}\right)_T = \frac{\bar{V}_i}{RT} - \frac{V_i}{RT}$$

$$\left[\frac{\partial \ln\left(\dfrac{\bar{f}_i}{f_i}\right)}{\partial P}\right]_T = \frac{\bar{V}_i - V_i}{RT} \tag{9.44}$$

Putting $\gamma_i = \dfrac{\bar{f}_i}{f_i^0 x_i}$ into Eq. (9.44), we obtain

$$\left(\frac{\partial \ln(\gamma_i x_i)}{\partial P}\right)_T = \frac{\bar{V}_i - V_i}{RT} \tag{9.45}$$

Since the mole fraction of the component i, i.e., x_i does not depend on pressure, it can be neglected. Then the expression becomes

$$\left(\frac{\partial \ln \gamma_i}{\partial P}\right)_T = \frac{\bar{V}_i - V_i}{RT} \tag{9.46}$$

This equation predicts the pressure dependence of the activity coefficient of the component in the solution.

9.4 GIBBS–DUHEM EQUATION

The change in partial molar properties with composition at constant temperature and pressure can be explained well by the Gibbs–Duhem equation. Of the partial molar properties, the partial molar Gibbs function is particularly useful in describing the behaviour of mixtures and solutions.

Since the free energy (G) of a system is a state function, it depends upon T, P and V. Now, if the system is of fixed composition, then it can be written as $G = f(P, T)$. But in case of an open system where the quantities of the components vary, the free energy will depend upon the quantities of the components.

For a multicomponent system, the free energy (G) is a function of T, P and the quantities of the components.

Consider a binary system which contains n_1 and n_2 moles of components A_1 and A_2 respectively. Then the free energy of the system can be expressed as

$$G = f(P, T, n_1, n_2)$$

If the variables undergo change, then on total differentiation

$$dG = \left(\frac{\partial G}{\partial P}\right)_{T, n_1, n_2} dP + \left(\frac{\partial G}{\partial T}\right)_{P, n_1, n_2} dT + \left(\frac{\partial G}{\partial n_1}\right)_{P, T, n_2} dn_1 + \left(\frac{\partial G}{\partial n_2}\right)_{P, T, n_1} dn_2 + \cdots \tag{9.47}$$

$$dG = V dP - S dT + \left(\frac{\partial G}{\partial n_1}\right)_{P, T, n_2} dn_1 + \left(\frac{\partial G}{\partial n_2}\right)_{P, T, n_1} dn_2 + \cdots \tag{9.48}$$

Here

$$\left(\frac{\partial G}{\partial n_1}\right)_{P,T,n_2} = \text{Partial molar free energy of component } A_1$$

$$\left(\frac{\partial G}{\partial n_2}\right)_{P,T,n_1} = \text{Partial molar free energy of component } A_2.$$

At constant temperature and pressure, Eq. (9.48) becomes

$$dG_{P,T} = \left(\frac{\partial G}{\partial n_1}\right)_{P,T,n_2} dn_1 + \left(\frac{\partial G}{\partial n_2}\right)_{P,T,n_1} dn_2 + \cdots \tag{9.49}$$

$$dG_{P,T} = \mu_1 dn_1 + \mu_2 dn_2 + \cdots \sum \mu_i dn_i \tag{9.50}$$

where $\mu_i = \left(\dfrac{\partial G}{\partial n_i}\right)_{P,T,n_j}$ is the chemical potential of component i.

Equation (9.50) can be written as

$$G_{P,T} = \mu_1 n_1 + \mu_2 n_2 + \cdots + \mu_i n_i + \cdots = \sum \mu_i n_i \tag{9.51}$$

The total derivative of G gives us

$$(dG)_{P,T} = (\mu_1 dn_1 + \mu_2 dn_2 + \cdots) + (n_1 d\mu_1 + n_2 d\mu_2 + \cdots) \tag{9.52}$$

Subtracting Eq. (9.51) from Eq. (9.52) gives

$$n_1 d\mu_1 + n_2 d\mu_2 + \cdots = 0$$

or

$$\sum n_i d\mu_i = 0 \tag{9.53}$$

On dividing by n, Eq. (9.53) yields in terms of mole fraction

$$\sum x_i d\mu_i = 0 \tag{9.54}$$

where $x_i = \dfrac{n_i}{n} = \text{Mole fraction of component } i$.

Equations (9.51), (9.53), and (9.54) are commonly known as *Gibbs–Duhem equations*. For a binary solution containing two components, Eq. (9.54) can be expressed as

$$x_1 d\mu_1 + x_2 d\mu_2 = 0 \tag{9.55}$$

We know that for a solution of component i, the chemical potential is expressed in terms of fugacity as

$$d\mu_i = RTd \ln f_i \tag{9.56}$$

Substituting Eq. (9.56) in (9.55) and dividing by dx_1 gives us

$$\frac{x_1 (RTd \ln f_1)}{dx_1} + \frac{x_2 (RTd \ln f_2)}{dx_1} = 0 \tag{9.57}$$

Since it is a binary solution in a closed system and the number of moles are independent, then

$$dx_1 + dx_2 = 0 \quad \text{or} \quad dx_1 = -dx_2$$

Applying this relation, we get from Eq. (9.57)

$$x_1 \frac{d \ln \bar{f}_1}{dx_1} = x_2 \frac{d \ln \bar{f}_2}{dx_2} \tag{9.58}$$

This relationship can also be expressed in terms of activity and activity coefficient, and would be helpful when fugacities are difficult to measure.

On putting $a_i = \dfrac{\bar{f}_i}{f_i^0} \Rightarrow \bar{f}_i = a_i f_i^0$, Eq. (9.58) yields

$$x_1 \frac{d \ln(a_1 f_1^0)}{dx_1} = x_2 \frac{d \ln(a_2 f_2^0)}{dx_2} \tag{9.59}$$

Since f_i^0 is constant and independent of the composition of the solution in the standard state, then $\dfrac{\partial \ln f_i^0}{\partial x_i} = 0$. Therefore, Eq. (9.59) becomes

$$x_1 \frac{d \ln a_1}{dx_1} = x_2 \frac{d \ln a_2}{dx_2} \tag{9.60}$$

This is another form of *Gibbs–Duhem equation in terms of activity.*
In order to express Eq. (9.60) in terms of activity coefficient, put $a_i = \gamma_i x_i$. Then

$$x_1 \frac{d \ln \gamma_1 x_1}{dx_1} = x_2 \frac{d \ln \gamma_2 x_2}{dx_2} \tag{9.61}$$

$$x_1 \frac{d \ln \gamma_1}{dx_1} + x_1 \frac{d \ln x_1}{dx_1} = x_2 \frac{d \ln \gamma_2}{dx_2} + x_2 \frac{d \ln x_2}{dx_2} \tag{9.62}$$

Here

$$x_1 \left(\frac{\partial \ln x_1}{\partial x_1} \right) = \frac{x_1 \partial x_1}{x_1 \partial x_1} = 1 \quad \text{and similarly} \quad x_2 \left(\frac{\partial \ln x_2}{\partial x_2} \right) = 1$$

Thus Eq. (9.62) reduces to

$$x_1 \frac{d \ln \gamma_1}{dx_1} = x_2 \frac{d \ln \gamma_2}{dx_2} \tag{9.63a}$$

or

$$x_1 d \ln \gamma_1 + x_2 d \ln \gamma_2 = 0 \quad (\text{as } dx_1 = -dx_2) \tag{9.63b}$$

This is a useful form of the Gibbs–Duhem equation in terms of activity coefficient.

The physical significance of the Gibbs–Duhem equation is that the chemical potential of one component in a solution cannot be varied independently from the chemical potentials of the other components of the solution.

EXAMPLE 9.8 The activity coefficient of component 1 in a binary solution is represented by

$$\ln \gamma_1 = ax_2^2 + bx_2^3 + cx_2^4$$

where a, b, and c are constants independent of concentrations. Obtain an expression for γ_2 in terms of x_1.

Solution: We are given that

$$\ln \gamma_1 = ax_2^2 + bx_2^3 + cx_2^4$$

Differentiating both sides, we get

$$d \ln \gamma_1 = 2ax_2 dx_2 + 3bx_2^2 dx_2 + 4cx_2^3 dx_2 \tag{i}$$

Multiplying both sides by x_1, we have

$$x_1 d \ln \gamma_1 = 2ax_1 x_2 dx_2 + 3bx_1 x_2^2 dx_2 + 4cx_1 x_2^3 dx_2 \tag{ii}$$

From the Gibbs–Duhem equation (9.63b), we get

$$x_2 d \ln \gamma_2 = -x_1 d \ln \gamma_1 \tag{iii}$$

Thus, Eq. (iii) can be expressed as

$$x_2 d \ln \gamma_2 = -2ax_1 x_2 dx_2 - 3bx_1 x_2^2 dx_2 - 4cx_1 x_2^3 dx_2$$

or

$$d \ln \gamma_2 = -2ax_1 dx_2 - 3bx_1 x_2 dx_2 - 4cx_1 x_2^2 dx_2 \quad (\text{since } x_1 + x_2 = 1)$$

or

$$d \ln \gamma_2 = 2ax_1 dx_2 + 3bx_1 (1 - x_1) dx_1 + 4cx_1 (1 - x_1)^2 dx_1$$

Integrating it from $x_1 = 0$ (at which $x_2 = 1$, $\gamma_2 = 1$, $\ln \gamma_2 = 0$) to $x_1 = x_1$ (at which $\gamma_2 = \ln \gamma_2$)

$$\int_0^{\ln \gamma_2} d \ln \gamma_2 = (2a + 3b + 4c) \int_0^{x_1} x_1 dx_1 + 3bx_1 - (3b + 8c) \int_0^{x_1} x_1^2 dx_1 + 4c \int_0^{x_1} x_1^3 dx_1$$

or

$$\ln \gamma_2 = (2a + 3b + 4c) \left(\frac{x_1^2}{2} \right) - (3b + 8c) \left(\frac{x_1^3}{3} \right) + 4c \left(\frac{x_1^4}{4} \right)$$

or

$$\ln \gamma_2 = \left(a + \frac{3b}{2} + 2c \right) x_1^2 - \left(b + \frac{8c}{3} \right) x_1^3 + cx_1^4$$

EXAMPLE 9.9 For a binary system, if the activity coefficient of component is $\ln \gamma_1 = ax_2^2$, then derive the expression for component 2.

Solution: From the Gibbs–Duhem relation, we get

$$x_1 d \ln \gamma_1 + x_2 d \ln \gamma_2 = 0$$

or

$$x_2 d \ln \gamma_2 = -x_1 d \ln \gamma_1 \tag{i}$$

Differentiating the equation $\ln \gamma_1 = ax_2^2$, we have

$$d \ln \gamma_1 = 2ax_2 dx_2$$

Substituting it into Eq. (i), we get

$$x_2 d \ln \gamma_2 = -x_1 d \ln \gamma_1 = -2ax_1 x_2 dx_2$$

or

$$d \ln \gamma_2 = -x_1 d \ln \gamma_1 = -2ax_1 dx_2 = 2ax_1 dx_1 \qquad \text{(Since } dx_2 = -dx_1) \quad \text{(ii)}$$

Integrating Eq. (ii), we get

$$\int_0^{\ln \gamma_2} d \ln \gamma_2 = 2a \int_0^{x_1} x_1 dx_1$$

or

$$\ln \gamma_2 = 2a \frac{x_1^2}{2} = ax_1^2$$

9.5 FUGACITY OF COMPONENT IN MIXTURE

In Chapter 6, the term *fugacity* was defined for a pure substance, and it can be represented further as

$$dG = RTd \ln f \quad \text{at constant temperature}$$

Now, the fugacity of a component i in a solution can be expressed as

$$d\overline{G}_i = RTd \ln \bar{f}_i \tag{9.64}$$

At the same condition, i.e., on variation of pressure of component i and at constant temperature, the equation becomes on integration

$$\overline{G}_i - G_i^0 = RT \ln\left(\frac{\bar{f}_i}{f_i^0}\right) \tag{9.65}$$

where f_i^0 is the fugacity of the component i in the standard state in which it behaves ideally.

For an ideal gas, fugacity is exactly equal to pressure, i.e., $\dfrac{f}{P} = 1$. But since a non-ideal gas at low pressure behaves ideally, fugacity may be replaced by pressure. This can be mathematically expressed as

$$\lim_{P \to 0} \frac{f}{P} = 1 \quad \text{or} \quad \frac{f}{P} = 1 \quad \text{as} \quad P \to 0$$

Here, for an ideal gas, $f_i^0 = p_i$, i.e., fugacity is equal to partial pressure. Analogously, for component i in a mixture, it would therefore be

$$\lim_{P \to 0} \frac{f_i^0}{p_i} = 1 \quad \text{or} \quad \frac{f_i^0}{y_i P} = 1 \quad \text{as} \quad P \to 0$$

where

$$p_i = y_i P,$$

in which

p_i = Partial pressure
P = Total pressure
y_i = Mole fraction.

Substituting this condition into Eq. (9.65), we get

$$\overline{G_i} - G_i^0 = RT \ln \left(\frac{\overline{f_i}}{f_i^0} \right) = RT \ln \left(\frac{\overline{f_i}}{y_i P} \right) = RT \ln \overline{\phi_i} \qquad (9.66)$$

where ϕ_i^0 is the fugacity coefficient of component i in the mixture.

EXAMPLE 9.10 Estimate the fugacity of a gaseous mixture consisting of 30% component 1 and 70% component 2 by mole, given that at 100°C and 50 bar, the fugacity coefficients of components 1 and 2 are 0.7 and 0.85.

Solution: We know that for a binary gas mixture obeying the virial equation of state, the fugacity coefficient can be represented by

$$\ln \phi = \sum x_i \ln \overline{\phi} \qquad (i)$$

For components 1 and 2, it can be expressed as

$$\ln \phi = x_1 \ln \overline{\phi_1} + x_2 \ln \overline{\phi_2} \qquad (ii)$$

We are given that

x_1 = Mole fraction of component 1 in the mixture = 0.3
x_2 = Mole fraction of component 2 in the mixture = 0.7
$\overline{\phi_1}$ = Fugacity coefficient of component 1 = 0.7
$\overline{\phi_2}$ = Fugacity coefficient of component 2 = 0.85.

Substituting the values in Eq. (ii), we get

$$\ln \phi = x_1 \ln \overline{\phi_1} + x_2 \ln \overline{\phi_2} = 0.3 \ln 0.7 + 0.7 \ln 0.85$$

or

$$\ln \phi = -0.22$$

or

$$\phi = 0.8025$$

We know that the fugacity coefficient, $\phi = \dfrac{f}{P}$. Therefore, fugacity at 50 bar

$$f = P\phi = 50 \times 0.8025 = 40.125 \text{ bar}$$

EXAMPLE 9.11 Estimate the fugacity of a ternary gas mixture consisting of 30 mol per cent of hydrogen (component 1), 25 mol per cent of nitrogen (component 2), and 45 mol per cent of oxygen (component 3), given that at 150°C and 60 bar, the fugacity coefficients of components 1, 2 and 3 are 0.7, 0.85 and 0.75 respectively.

Solution: We know that a gas mixture obeying the virial equation of state, the fugacity coefficient of the mixture can be represented by

$$\ln \phi = \sum x_i \ln \overline{\phi}$$

For components 1, 2 and 3, it can be expressed as

$$\ln \phi = (x_1 \ln \overline{\phi}_1 + x_2 \ln \overline{\phi}_2 + x_3 \ln \overline{\phi}_3) \tag{i}$$

We are given that $x_1 = 0.3$, $x_2 = 0.25$ and $x_3 = 0.45$

$\overline{\phi}_1$ = Fugacity coefficient of component 1 = 0.7

$\overline{\phi}_2$ = Fugacity coefficient of component 2 = 0.85

$\overline{\phi}_3$ = Fugacity coefficient of component 3 = 0.75.

Substituting the values in Eq. (i), we get

$$\ln \phi = x_1 \ln \overline{\phi}_1 + x_2 \ln \overline{\phi}_2 + x_3 \ln \overline{\phi}_3$$
$$= 0.3 \ln 0.7 + 0.25 \ln 0.85 + 0.45 \ln 0.75$$

or

$$\ln \phi = -0.276$$

or

$$\phi = \frac{f}{P} = 0.7588$$

or

$$f = P \times 0.7588 = 60 \times 0.7588 = 45.53 \text{ bar}$$

9.6 FUGACITY OF LIQUIDS AND SOLIDS

Let us consider a system in which saturated liquid and saturated vapour are in equilibrium with each other at temperature T and saturation pressure P^{Sat}. For such a system, the molar free energy of vapour and liquid in equilibrium are the same.

Therefore we can write

$$G^l = G^v \tag{9.67}$$

Again, from the relationship between molar Gibbs free energy and fugacity, we have

$$dG = RT d \ln f$$

Integrating between the saturated liquid and vapour states, we get

$$\int_{G^l}^{G^v} dG = \int_{f^l}^{f^v} RT d \ln f$$

or

$$G^v - G^l = RT \ln \frac{f^v}{f^l} \tag{9.68}$$

Substituting Eq. (9.67) into (9.68), we get

$$RT \ln \frac{f^v}{f^l} = 0 \quad \text{or} \quad f^v = f^l \tag{9.69}$$

Similarly, as a solid is in equilibrium with its own vapour

$$f^s = f^v \tag{9.70}$$

From Eqs. (9.69) and (9.70), we may write

$$f^v = f^l = f^s \tag{9.71}$$

where the superscripts v, l and s denote the vapour, liquid and solid phases respectively.

When it is necessary to estimate the fugacity of a liquid at T and the pressure varies from the saturation pressure to the higher pressure in which the liquid is supposed to be compressed, then we integrate the pressure-dependence equation of fugacity

$$dG = VdP = RTd(\ln f)$$

At this stage, the liquid is assumed to be incompressible and V^l = constant, or

$$V \int_{P^{Sat}}^{P} dP = RT \int_{f^{Sat}}^{f^l} d\ln f$$

or

$$V^l(P - P^{Sat}) = RT \ln \frac{f^l}{f^{Sat}} \tag{9.72}$$

or

$$\frac{f^l}{f^{Sat}} = \exp \frac{V^l(P - P^{Sat})}{RT} \tag{9.73}$$

or

$$f^l = P^{Sat} \phi^{Sat} \exp \frac{V^l(P - P^{Sat})}{RT} \tag{9.74}$$

where

$$\phi^{Sat} = \text{Fugacity coefficient in saturated condition} = \frac{f^{Sat}}{P^{Sat}}$$

$\exp \dfrac{V^l(P - P^{Sat})}{RT}$ = Poynting pressure correction factor to measure the increase in fugacity of a liquid at a pressure higher than the saturation pressure.

Thus we can estimate the fugacity of solids as well as liquids with the help of the preceding expressions.

EXAMPLE 9.12 The saturation pressure of liquid water at 372.12 K is 100 kPa. Estimate the fugacity of liquid water at 372.12 K and 300 kPa, given that the specific volume of liquid water at 372.12 K is 1.043×10^{-3} m^3/kg.

Solution: Using Eq. (9.73) restated here, the fugacity of liquid can be estimated.

$$\frac{f^l}{f^{Sat}} = \exp \frac{V^l(P - P^{Sat})}{RT}$$

Here

V^l = Molar volume of liquid water

= Specific volume × Molecular weight

= $1.043 \times 10^{-3} \times 18$

= 18.774×10^{-6} m^3/mol

Now, vapour is assumed to behave like an ideal gas. Therefore, in saturated condition, fugacity is equal to pressure, i.e.

$$P^{Sat} = f^{Sat} = 100 \text{ kPa} = 1 \text{ bar}$$

Substituting it into Eq. (9.73), we get

$$\frac{f^l}{f^{Sat}} = \exp \frac{V^l(P - P^{Sat})}{RT} = \exp \frac{18.774 \times 10^{-6}[(3-1) \times 10^5]}{8.314 \times 372.12} = \exp 0.00121$$

$$f^l = 1.0012 \text{ bar (since } f^{Sat} = 1 \text{ bar)}$$

Hence the fugacity of liquid water is 1.0012 bar.

EXAMPLE 9.13 Calculate the fugacity of liquid butadiene at 313 K and 10 bar. The saturation pressure of butadiene at 313 K is 4.2 bar. The molar volume and saturated fugacity of liquid butadiene at 313 K are 90.45×10^{-6} m^3/mol and 4.12 bar respectively.

Solution: We are given that V^l = Molar volume of liquid butadiene = 90.45×10^{-6} m^3/mol

$$f^{Sat} = 4.12 \text{ bar}$$
$$P = 10 \text{ bar}$$
$$P^{Sat} = 4.12 \text{ bar}$$
$$T = 313 \text{ K}$$
$$V^l = 90.45 \times 10^{-6} \text{ m}^3\text{/mol}$$

Applying Eq. (9.73) and putting the preceding values into this equation, we have

$$\frac{f^l}{f^{Sat}} = \exp \frac{V^l(P - P^{Sat})}{RT} = \exp \frac{90.45 \times 10^{-6}[(10 - 4.2) \times 10^5]}{8.314 \times 313} = \exp 0.0201$$

or

$$\frac{f^l}{4.12} = \exp 0.0201 = 1.02$$

or

$$f^l = 1.02 \times 4.12 = 4.202 \text{ bar}$$

Therefore, the fugacity of liquid butadiene at 313 K and 10 bar is 4.202 bar.

EXAMPLE 9.14 A gas obeys the equation of state $P(V - b) = RT$. For this gas, $b = 0.0391$ dm^3/mol. Calculate the fugacity and fugacity coefficient for the gas at 1000°C and 1000 atm.

(WBUT, 2007)

Solution: We know that

$$RT \ln \left(\frac{f}{P}\right) = \int_0^P (V_{real} - V_{id}) \, dP \qquad \text{(i)}$$

Here, for real gas

$$V_{real} = \frac{RT}{P} + b = \frac{RT}{P} + 0.0391$$

for ideal gas

$$V_{id} = \frac{RT}{P}$$

Now, substituting $V_{id} = \frac{RT}{P}$ into Eq. (i), we obtain

$$RT \ln\left(\frac{f}{P}\right) = \int_0^P (V_{real} - V_{id}) dP = \int_0^P \left[\left(\frac{RT}{P} + 0.0391\right) - \left(\frac{RT}{P}\right)\right] dP$$

or

$$RT \ln\left(\frac{f}{P}\right) = 0.0391 \int_0^P dP = 0.0391[P]_0^{1000} = 39.1$$

or

$$\ln\left(\frac{f}{P}\right) = \frac{39.1}{0.08 \times 1273} = 0.374$$

or

$$\phi = \frac{f}{P} = e^{0.74} = 1.453 \text{ atm}$$

or

$$f = P \times 1.453 = 1000 \times 1.453 = 1453 \text{ atm}$$

Therefore, the fugacity and fugacity coefficient of the gas are 1.453 atm and 1453 atm respectively.

Alternative method: The equation of state obeying the gas is given by

$$P(V - b) = RT$$

or

$$\frac{PV}{RT} = 1 + \frac{Pb}{RT}$$

$$\Rightarrow \qquad Z = 1 + \frac{Pb}{RT}$$

Now using the expression for fugacity, we get

$$\ln\frac{f}{P} = \int_0^P \left(\frac{Z-1}{P}\right) dP$$

$$\Rightarrow \qquad \ln\phi = \int_0^P \left(\frac{1 + Pb - 1}{P(RT)}\right) dP$$

On integration, we get

$$\ln \phi = \frac{Pb}{RT} = \frac{1000 \times 0.0391}{0.082 \times 1273} = 0.3743$$

or

$$\phi = e^{0.3743} = 1.454$$

Again

$$\frac{f}{P} = \frac{f}{1000} = 1.454$$

$$\Rightarrow \qquad f = 1454 \text{ atm}$$

Hence, the fugacity and the fugacity coefficient of the gas are 1.453 atm and 1453 atm respectively.

EXAMPLE 9.15 Estimate the fugacity of liquid acetone at 110°C and 275 bar. At 110°C the vapour pressure of acetone is 4.360 bar and the molar volume of liquid acetone is 73 cm³/mol.

(WBUT, 2006)

Solution: We are given that

$$V^l = \text{Molar volume of liquid acetone} = 73 \times 10^{-6} \text{ m}^3/\text{mol}$$
$$P = 275 \text{ bar}$$
$$P^{\text{Sat}} = 4.360 \text{ bar}$$
$$T = 383 \text{ K}.$$

Now, acetone vapour is assumed to behave like an ideal gas. Then, at saturated condition, fugacity is equal to pressure, i.e.

$$P^{\text{Sat}} = f^{\text{Sat}} = 4.360 \text{ bar}$$

Applying Eq. (9.73) and putting the preceding values into this, we have

$$\frac{f^l}{f^{\text{Sat}}} = \exp \frac{V^l(P - P^{\text{Sat}})}{RT}$$

$$= \exp \frac{73 \times 10^{-6}[(275 - 4.360) \times 10^5]}{8.314 \times 383}$$

$$= \exp 0.6204$$

or

$$\frac{f^l}{4.360} = \exp 0.6204 = 1.859$$

or

$$f^l = 1.859 \times 4.360 = 8.105 \text{ bar}$$

Therefore, the fugacity of liquid butadiene at 313 K and 10 bar is 8.105 bar.

9.7 IDEAL GAS MIXTURE MODEL: GIBBS THEOREM

Suppose an ideal gas mixture comprising different gases is formed in a container at temperature T and pressure P. All the ideal gases, whether pure or mixed, have the same molar volume at the same T and P. Even the partial molar volume of an ideal gas is the same as its value in the pure state, and this is equal to $\dfrac{RT}{P}$. According to Gibbs theorem:

> *A partial molar property (other than volume) of a constituent species in an ideal gas*
> *mixture is equal to the corresponding molar property of the species as a pure ideal gas,*
> *at the mixture temperature but at a pressure equal to its partial pressure in the mixture.*

It has been observed that when two or more substances are mixed, then there must be a change in the measurable properties of the various components of the system. The extent of change is not identical for different systems. Considering the following basis, we can estimate the change in thermodynamic properties due to mixing. The basis is:

Change in properties on mixing = Property of the mixture after mixing – Property of the mixture before mixing

If M is any extensive molar property of the solution, M_i is a molar property of component i and x_i is the mole fraction, then we can write

$$M = \sum x_i M_i \tag{9.75}$$

This equation cannot be applied for an ideal or non-ideal solution, if the property change of mixing (ΔM) is not taken into consideration. Hence, due to this consideration Eq. (9.75) yields

$$M = \sum x_i M_i + \Delta M \tag{9.76}$$

where ΔM is the difference between the property of the solution and the sum of the pure components.

At standard state, Eq. (9.76) can be expressed as

$$\Delta M = M - \sum x_i M_i^0 \tag{9.77}$$

From the concept of partial molar property, we get

$$M = \sum x_i \overline{M}_i \tag{9.78}$$

From Eqs. (9.77) and (9.78), we have

$$\Delta M = \sum x_i (\overline{M}_i - M_i^0) \tag{9.79}$$

Similarly, the free energy change of mixing can be written as

$$\Delta G = \sum x_i (\overline{G}_i - G_i^0) \tag{9.80}$$

9.7.1 Entropy Change of Mixing

Let us consider a vessel consisting of two compartments formed by one removable partition. Every compartment contains a known amount of distinct gases as shown in Fig. 9.3. The two gases are at the same temperature and pressure.

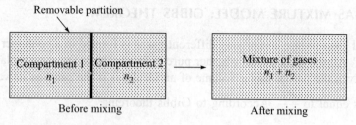

Fig. 9.3 Mixing of gases at constant temperature and pressure.

Now, when the partition is withdrawn and they are brought into contact with each other, they would diffuse and mix up spontaneously. This process will yield a considerable entropy change, and the change in entropy due to mixing can be calculated by two step processes—entropy before mixing ($S_{initial}$) and after mixing (S_{final}).

$$S_{initial} = n_1 S_1^* + n_2 S_2^* = \sum n_i S_i^* \tag{9.81}$$

where S_1^* is the molar entropy of the pure component i.

$$S_{final} = n_1 \bar{S}_1 + n_2 \bar{S}_2 = \sum n_i \bar{S}_i \tag{9.82}$$

where $\bar{S}_1$ is the partial molar entropy of component i.

$$\Delta S_{mixing} = S_{final} - S_{initial} = \sum n_i \bar{S}_i - \sum n_i S_i^* \tag{9.83}$$

The molar free energy of a gas in the mixture is given by

$$\mu_i = \mu_i^* + RT \ln x_i \tag{9.84}$$

Differentiating it with respect to T, keeping other parameters constant, we get

$$\left(\frac{\partial \mu_i}{\partial T}\right)_{P, n_i} = \left(\frac{\partial \mu_i^*}{\partial T}\right)_{P, n_i} + R \ln x_i \tag{9.85}$$

We know that

$$\left(\frac{\partial \mu_i}{\partial T}\right)_{P, n_i} = -\bar{S}_i$$

and similarly

$$\left(\frac{\partial \mu_i^*}{\partial T}\right)_{P, n_i} = -S_i^*$$

Substituting these into Eq. (9.85), we obtain

$$\bar{S}_i = S_i^* - R \ln x_i \tag{9.86}$$

Putting Eq. (9.86) into (9.83), we get

$$\Delta S_{mixing} = S_{final} - S_{initial} = \sum n_i (S_i^* - R \ln x_i) - \sum n_i S_i^* \tag{9.87}$$

or

$$\Delta S_{mixing} = \sum n_i S_i^* - R \sum n_i \ln x_i - \sum n_i S_i^* = -nR \sum \frac{n_i}{n} \ln x_i = -nR \sum x_i \ln x_i \tag{9.88}$$

where x_i is the mole fraction of component i.

EXAMPLE 9.16 Estimate the entropy change of mixing when 2.8 L of oxygen and 19.6 L of hydrogen at 1 atm and 25°C are mixed to prepare a gaseous mixture.

Solution: The change in entropy on mixing of gases is given by Eq. (9.88)

$$\Delta S_{\text{mix}} = - nR \sum x_i \ln x_i$$

For two gases, the preceding equation can be expressed as

$$\Delta S_{\text{mix}} = -nR(x_1 \ln x_1 + x_2 \ln x_2)$$

Here

$$\text{Number of moles of oxygen gas} = n_1 = \frac{2.8}{22.4} = \frac{1}{8} \text{ g-mol}$$

$$\text{Number of moles of hydrogen gas} = n_2 = \frac{19.6}{22.4} = \frac{7}{8} \text{ g-mol}$$

$$\text{Total moles} = n = n_1 + n_2 = \frac{1}{8} + \frac{7}{8} = 1 \text{ g-mol}$$

$$\text{Mole fraction of oxygen} = x_1 = \frac{1/8}{1} = \frac{1}{8}$$

$$\text{Mole fraction of hydrogen} = x_2 = \frac{7/8}{1} = \frac{7}{8}$$

Hence, the entropy change of mixing $= \Delta S_{\text{mix}} = -nR(x_1 \ln x_1 + x_2 \ln x_2)$

$$= -1 \times 1.987 \left(\frac{1}{8} \ln \frac{1}{8} + \frac{7}{8} \ln \frac{7}{8} \right)$$

$$= 0.754 \text{ cal/mol-K}$$

9.7.2 Gibbs Free Energy Change of Mixing

Applying the expression for change in entropy due to mixing, we can estimate the Gibbs free energy change of mixing of ideal gases.

From the definition of Gibbs free energy (G), we have

$$G = H - TS$$

For an isothermal change

$$\Delta G = \Delta H - T\Delta S$$

Considering the mixing of gases, we may write

$$\Delta G_{\text{mix}} = \Delta H_{\text{mix}} - T\Delta S_{\text{mix}} \tag{9.89}$$

Substituting $\Delta S_{\text{mix}} = -nR \sum x_i \ln x_i$ in Eq. (9.89), we get

$$\Delta G_{\text{mix}} = \Delta H_{\text{mix}} + RT \left(n \sum x_i \ln x_i \right) \tag{9.90}$$

For ideal gases, $\Delta H_{mixing} = 0$, as the initial and final temperatures are identical and enthalpy is a function of temperature only. Hence, Eq. (9.90) becomes

$$\Delta G_{mixing} = nRT\left(\sum x_i \ln x_i\right) \tag{9.91}$$

or

$$\frac{\Delta G_{mixing}}{RT} = n\left(\sum x_i \ln x_i\right) \tag{9.92}$$

EXAMPLE 9.17 Derive an expression for the free energy change of mixing when two ideal gases of n_1 and n_2 moles respectively are mixed isothermally but at different pressures.

Solution: Let the initial molar free energies of gas 1 be μ_1^* and of gas 2 be μ_2^*, and the final molar energies (after mixing) be μ_1 and μ_2 respectively.
Before mixing:

$$G_{initial} = n_1\mu_1^* + n_2\mu_2^* \tag{i}$$

After mixing:

$$G_{final} = n_1\mu_1 + n_2\mu_2 \tag{ii}$$

We know that for a pure ideal gas, the chemical potential of the ith component at constant temperature under a pressure P^* is given by

$$\mu_i^* = \mu_i^0 + RT \ln P^* \tag{iii}$$

On substitution of Eq. (iii) into (i), it yields

$$G_{initial} = n_1(\mu_i^0 + RT \ln P_1^*) + n_2(\mu_i^0 + RT \ln P_2^*) \tag{iv}$$

Again, on mixing of the gases under pressure P, the chemical potential at partial pressure p_i will be

$$\mu_i = \mu_i^0 + RT \ln p_i \tag{v}$$

Hence, Eq. (ii) can be expressed as

$$G_{final} = n_1(\mu_1^0 + RT \ln p_1) + n_2(\mu_2^0 + RT \ln p_2) \tag{vi}$$

Therefore, the free energy change of mixing is given by

$$\begin{aligned}
\Delta G_{mixing} &= G_{final} - G_{initial} \\
&= (n_1\mu_1^0 + n_1RT \ln p_1 + n_2\mu_2^0 + n_2RT \ln p_2) - (n_1\mu_i^0 + n_1RT \ln P_1^* + n_2\mu_2^0 + n_2RT \ln P_2^*) \\
&= RT(n_1 \ln p_1 + n_2 \ln p_2) - RT(n_1 \ln P_1^* + n_2 \ln P_2^*) \\
&= RT\, n_1 \ln \frac{p_1}{P_1^*} + n_2 \ln \frac{p_2}{P_2^*} \tag{vii}
\end{aligned}$$

9.7.3 Enthalpy Change of Mixing

Consider the relation for mixing of ideal gases under isothermal and isobaric condition

$$\Delta G_{mixing} = \Delta H_{mixing} - T\Delta S_{mixing}$$

Substituting the values of ΔG_{mixing} and ΔS_{mixing}, we get

$$\Delta H_{mixing} = 0$$

Therefore, there will be no enthalpy change when ideal gases are mixed at constant temperature and pressure.

EXAMPLE 9.18 At 25°C and 1 atm, 0.7 moles of helium were mixed with 0.3 moles of argon. Calculate the free energy and enthalpy change of mixing. Assume that the gases behave ideally.

Solution: From Eq. (9.92), we have

$$\frac{\Delta G_{mixing}}{RT} = n\left(\sum x_i \ln x_i\right) \tag{i}$$

For two gases, Eq. (i) can be expressed as

$$\Delta G_{mixing} = nRT(x_1 \ln x_1 + x_2 \ln x_2)$$

Mole fraction of helium

$$x_1 = \frac{n_1}{n} = \frac{0.7}{1} = 0.7$$

Mole fraction of argon

$$x_2 = \frac{n_2}{n} = \frac{0.3}{1} = 0.3$$

Substituting these into Eq. (i), we get

$$\begin{aligned}\Delta G_{mixing} &= nRT(x_1 \ln x_1 + x_2 \ln x_2)\\ &= 1 \times 8.314 \times 298\,(0.7 \ln 0.7 + 0.3 \ln 0.3) = -1511.79 \text{ J}\end{aligned}$$

Enthalpy change of mixing

$$\Delta H_{mixing} = 0 \quad \text{(since the gases are ideal)}$$

EXAMPLE 9.19 A 20 L vessel is divided into two compartments with the help of a removable partition. The first compartment contains 12 L of hydrogen and the second compartment contains 10 L of nitrogen. Now, the partition is withdrawn and the gases are allowed to mix up isothermally at 1 atm and 298 K. Estimate ΔG_{mixing}, ΔH_{mixing} and ΔS_{mixing}.

Solution: We know that

$$\Delta G_{mixing} = RT\left(n_1 \ln\frac{p_1}{p_1^*} + n_2 \ln\frac{p_2}{p_2^*}\right)$$

Here

$$n_1 = \text{Mole number of hydrogen} = \frac{P_1 V_1}{RT} = \frac{1 \times 12}{0.082 \times 298} = 0.491$$

$$n_2 = \text{Mole number of nitrogen} = \frac{P_2 V_2}{RT} = \frac{1 \times 10}{0.082 \times 298} = 0.409.$$

The final pressure achieved by the gas mixture is

$$P_{final} = \frac{nRT}{V} = \frac{(n_1 + n_2)RT}{V} = \frac{(0.491 + 0.409) \times 0.082 \times 298}{20} = 1.09 \text{ atm}$$

Partial pressure of hydrogen = $p_1 = P_{\text{final}} \times x_1 = 1.09 \times 0.491 = 0.535$ atm

Partial pressure of nitrogen = $p_2 = P_{\text{final}} \times x_2 = 1.09 \times 0.409 = 0.445$ atm

Therefore, the total free energy change of mixing would be

$$\Delta G_{\text{mixing}} = 0.082 \times 298 \left(0.49 \ln \frac{0.535}{1} + 0.409 \ln \frac{0.445}{1} \right)$$

$$= -15.59 \text{ J}$$

$$\Delta S_{\text{mixing}} = -\frac{\Delta G_{\text{mixing}}}{T} = -\frac{-15.59}{298} = 0.023 \text{ J/K}$$

9.7.4 Volume Change of Mixing

By analogy the change in volume due to mixing shown in Fig. 9.4 can be calculated by two step processes—volume before mixing (V_{initial}) and after mixing (V_{final}).

$$V_{\text{initial}} = n_1 V_1^* + n_2 V_2^* = \sum n_i V_i^* \tag{9.93}$$

where V_i^* is the molar volume of the pure component i.

$$V_{\text{final}} = n_1 \overline{V}_1 + n_2 \overline{V}_2 = \sum n_i \overline{V}_i \tag{9.94}$$

where $\overline{V}_i$ is the partial molar volume of component i.

$$\Delta V_{\text{mixing}} = V_{\text{final}} - V_{\text{initial}} = \sum n_i \overline{V}_i - \sum n_i V_i^* \tag{9.95}$$

$T = \text{Constant}, P = \text{Constant}$

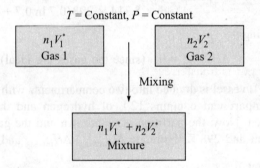

Fig. 9.4 Mixing of gases 1 and 2 at constant temperature and pressure.

The molar volume of each compartment for an ideal gas in the mixture is the same and is given by $\overline{V}_i = \dfrac{RT}{P}$. Again, for a pure gas at same T and P, the molar volume is $V_i^* = \dfrac{RT}{P}$. On substituting this, we get from Eq. (9.95)

$$\Delta V_{\text{mixing}} = V_{\text{final}} - V_{\text{initial}} = \sum n_i \overline{V}_i - \sum n_i V_i^* = \sum n_i \frac{RT}{P} - \sum n_i \frac{RT}{P} = 0 \tag{9.96}$$

Hence, it can be concluded that there is no change in volume when ideal gases are mixed isothermally and isobarically.

9.8 EXCESS PROPERTY OF MIXTURE

Excess property is defined as the difference between the actual property of the solution and the value it would have as an ideal solution at the same temperature, pressure, and composition. It can be schematically represented as

Actual property of the solution at given temperature and pressure (M)	$-$	Property it would have as an ideal gas at the same temperature and pressure (M^{id})	$=$	Excess property of the solution at given temperature and pressure (M^E)

where M is the molar value of an extensive thermodynamic property like V, U, H, S, or G of a real substance at T and P, M^{id} is that of an ideal gas at the same condition, and M^E is the excess property.

For example, the *excess enthalpy* of a solution can be expressed accordingly as

$$H^E = H - H^{id} = \Delta H_{mixing} - \Delta H^{id}_{mixing} = \Delta H_{mixing} \text{ (since for ideal solution, } \Delta H^{id}_{mixing} = 0)$$

the excess internal energy as

$$U^E = U - U^{id} = \Delta U_{mixing} - \Delta U^{id}_{mixing} = \Delta U_{mixing} \quad (\text{as } \Delta U^{id}_{mixing} = 0)$$

the excess Gibbs free energy as

$$G^E = G - G^{id} = \Delta G_{mixing} - \Delta G^{id}_{mixing}$$

and the excess molar volume as

$$V^E = V - V^{id} = \Delta V_{mixing} - \Delta V^{id}_{mixing} = \Delta V_{mixing} \quad (\text{since } \Delta V^{id}_{mixing} = 0]$$

9.8.1 Excess Free Energy of Component in Mixture

When 1 mol of component i is transferred from its standard state of unit fugacity to a solution at the same temperature and pressure, the change in free energy of the solution due to mixing is given by

$$\overline{G}_i - G_i^0 = RT \ln \frac{\overline{f}_i}{f_i^0} \tag{9.97}$$

For an ideal solution, the fugacity of the component i is $\overline{f}_i = x_i f_i$. Substituting this into Eq. (9.97), we get

$$\overline{G}_i^{id} - G_i^0 = RT \ln \frac{\overline{f}_i}{f_i^0} = RT \ln \frac{x_i f_i}{f_i} = RT \ln x_i \tag{9.98}$$

Subtracting Eq. (9.98) from Eq. (9.97), we get

$$\overline{G}_i - \overline{G}_i^{id} = RT \ln \frac{\overline{f}_i}{f_i^0} - RT \ln x_i \tag{9.99}$$

or

$$\overline{G}_i^E = RT \ln \frac{\overline{f}_i}{x_i f_i^0} = RT \ln \gamma_i \tag{9.100}$$

For component i in the solution, the excess free energy can also be expressed as

$$G^E = \sum x_i \overline{G}_i^E \qquad (9.101)$$

where $\overline{G}_i^E$ is the partial molar excess property of component i.

Substituting $\overline{G}_i^E = RT \ln \gamma_i$ into Eq. (9.101), we obtain

$$G^E = RT \sum x_i \ln \gamma_i \qquad \text{for non-ideal solution}$$

or

$$\frac{G^E}{RT} = \sum x_i \ln \gamma_i \qquad (9.102)$$

For binary solution, Eq. (9.102) can be rearranged as

$$\frac{G^E}{RT} = x_1 \ln \gamma_1 + x_2 \ln \gamma_2 \qquad (9.103)$$

In terms of the total excess free energy, the partial molar excess free energy can also be expressed in the following form:

$$\frac{\overline{G}^E}{RT} = \left[\frac{\partial (n_i G^E)}{\partial n_i} \right]_{T, P, n_j} = \ln \gamma_i \qquad (9.104)$$

Considering Eqs. (9.102) and (9.104), we have

$$\frac{G^E}{RT} = \sum x_i \frac{\overline{G}^E}{RT} = \sum x_i \ln \gamma_i \qquad (9.105)$$

EXAMPLE 9.20 The excess Gibbs free energy of a binary liquid at T and P is given by

$$\frac{G^E}{RT} = (-2.6x_1 - 1.8x_2)x_1x_2$$

(a) Find expressions for γ_1 and γ_2.
(b) Show that the expression satisfies the Gibbs–Duhem equation.

Solution: (a) We are given that

$$\frac{G^E}{RT} = (-2.6x_1 - 1.8x_2)x_1x_2$$

$$= 2.6x_1^2x_2 - 1.8x_1x_2^2$$

In terms of number of moles $\left(\text{by } x_1 = \dfrac{n_1}{n} \text{ and } x_2 = \dfrac{n_2}{n} \right)$, we get

$$\frac{G^E}{RT} = -2.6 \frac{n_1^2 n_2}{n^3} - 1.8 \frac{n_1 n_2^2}{n^3}$$

or

$$\frac{nG^E}{RT} = -2.6 \frac{n_1^2 n_2}{n^2} - 1.8 \frac{n_1 n_2^2}{n^2} \qquad (i)$$

Differentiating with respect to n_1 at constant temperature and n_2, we obtain

$$\left[\frac{\partial(nG^E/RT)}{\partial n_1} \right]_{T,\,n_2} = -2.6n_2\left(\frac{2n_1}{n^2} - \frac{2n_1^2}{n^3} \right) - 1.8n_2^2\left(\frac{1}{n^2} - \frac{2n_1}{n^3} \right)$$

or

$$\ln \gamma_1 = -5.2x_1x_2 + 5.2x_1^2x_2 - 1.8x_2^2 + 3.6x_1x_2^2$$

Replacing x_1 by $(1 - x_2)$, we get

$$\ln \gamma_1 = -5.2x_2(1 - x_2) + 5.2x_2(1 - x_2)^2 - 1.8x_2^2 + 3.6x_2^2(1 - x_2)$$

or

$$\ln \gamma_1 = -3.4x_2^2 + 1.6x_2^3$$

Similarly, differentiating with respect to n_2 at constant temperature and n_1, we have

$$\left[\frac{\partial(nG^E/RT)}{\partial n_2} \right]_{T,\,n_1} = -2.6n_1^2\left(\frac{1}{n^2} - \frac{2n_2}{n^3} \right) - 1.8n_1\left(\frac{2n_2}{n^2} - \frac{2n_2^2}{n^3} \right)$$

or

$$\ln \gamma_2 = -2.6x_1^2 + 5.2x_1^2x_2 - 3.6x_1x_2 + 3.6x_1x_2^2$$

Replacing x_2 by $(1 - x_1)$, we get

$$\ln \gamma_2 = -2.6x_1^2 + 5.2x_1^2(1 - x_1) - 3.6x_1(1 - x_1) + 3.6x_1(1 - x_1)^2$$

or

$$\ln \gamma_2 = -x_1^2 - 1.6x_1^3$$

(b) We know that the Gibbs–Duhem equation in terms of activity coefficient is given by

$$x_1 d \ln \gamma_1 + x_2 d \ln \gamma_2 = 0$$

We have

$$\ln \gamma_1 = -3.4x_2^2 + 1.6x_2^3$$

or

$$d \ln \gamma_1 = (-6.8x_2 + 4.8x_2^2)dx_2$$

and

$$\ln \gamma_2 = -x_1^2 - 1.6x_1^3$$

or

$$d \ln \gamma_2 = (-2x_1 - 4.8x_1^2)dx_1$$

Hence

$$\begin{aligned}
x_1 d \ln \gamma_1 + x_2 d \ln \gamma_2 &= x_1 (-6.8x_2 + 4.8x_2^2)dx_2 + x_2(-2x_1 - 4.8x_1^2)dx_1 \\
&= (-6.8x_1x_2 + 4.8x_1x_2^2)dx_2 + (-2x_1x_2 - 4.8x_1^2x_2)dx_1 \quad \text{(as } dx_1 = -dx_2) \\
&= (-6.8x_1x_2 + 4.8x_1x_2^2 + 2x_1x_2 + 4.8x_1^2x_2)dx_2 \\
&= (-4.8x_1x_2 + 4.8x_1x_2(x_1 + x_2))dx_2 \\
&= (-4.8x_1x_2 + 4.8x_1x_2)dx_2 = 0 \quad \text{(since } x_1 + x_2 = 1)
\end{aligned}$$

Hence the expression satisfies the Gibbs–Duhem equation.

It can be verified by the relation of excess Gibbs free energy, which is given by

$$\frac{G^E}{RT} = x_1 \ln \gamma_1 + x_2 \ln \gamma_2 \qquad \text{(ii)}$$

Substituting the values of $\ln \gamma_1$ and $\ln \gamma_2$ in Eq. (ii), we get the actual expression

$$\frac{G^E}{RT} = (-2.6x_1 - 1.8x_2)x_1 x_2$$

9.9 IDEAL SOLUTION MODEL: LEWIS–RANDALL RULE

The evaluation of fugacities of the components in a mixture is not a very easy task. In order to simplify the evaluation methodology, the mixture is modelled as an ideal solution, and the devised model is known as the *ideal solution model*.

An ideal solution is a mixture for which

$$\bar{f}_i = x_i f_i \qquad (9.106)$$

Equation (9.106) is known as the *Lewis–Randall rule*, and it states: *The fugacity of each component in an ideal solution is equal to the product of its mole fraction and the fugacity of the pure component at the same temperature, pressure, and composition of the mixture.* It can also be stated as: *The fugacity of each component in an ideal solution is proportional to the mole fraction of that pure component in the solution.*

Here, $\bar{f}_i$ is the fugacity of the component i in an ideal solution, x_i is the mole fraction, and f_i the fugacity of the pure component at the same temperature and pressure.

The relation shown in Eq. (9.106) can be established in the following way. We know that for an ideal solution of component i

$$\Delta G_{mix}^{id} = RT \sum x_i \ln x_i \qquad (9.107)$$

We obtained the relation earlier from the knowledge of property change of mixing that

$$\Delta M = M - \sum x_i M_i^0 = M - \sum x_i M_i \qquad (9.108)$$

For Gibb's free energy, Eq. (9.107) can be expressed as

$$G - \sum x_i G_i = RT \sum x_i \ln x_i \qquad (9.109)$$

or

$$G = \sum x_i G_i + RT \sum x_i \ln x_i = (x_1 G_1 + x_2 G_2 + \dots) + RT(x_1 \ln x_1 + x_2 \ln x_2 + \dots) \qquad (9.110)$$

In terms of mole

$$G = \left(\frac{n_1}{n} G_1 + \frac{n_2}{n} G_2 + \cdots\right) + RT\left(\frac{n_1}{n} \ln + \frac{n_1}{n} + \frac{n_2}{n} \ln \frac{n_2}{n} + \cdots\right) \qquad (9.111)$$

Multiplying both sides by n, we get

$$nG = (n_1 G_1 + n_2 G_2 + \cdots) + RT\left(n_1 \ln \frac{n_1}{n} + n_2 \ln \frac{n_2}{n} + \cdots\right) \qquad (9.112)$$

Differentiating Eq. (9.112) with respect to n_1 at constant T, P, N_2, we get

$$\left[\frac{\partial(nG)}{\partial n_1}\right]_{T,P,n_2} = G_1 + RT\left[\ln \frac{n_1}{n} + \frac{n_1 n}{n_1}\left(\frac{1}{n} - \frac{n_1}{n^2}\right) + \frac{n_2 n}{n_2} n_2\left(-\frac{1}{n^2}\right) + \cdots\right] \qquad (9.113)$$

$$\overline{G}_1 = G_1 + RT \left(\ln x_1 + 1 - x_1 - x_2 \cdots\right) \tag{9.114}$$

$$\overline{G}_i = G_1 + RT \left(\ln x_1 + 1 - \sum x_i\right) \tag{9.115}$$

Since $\sum x_i = 1$, Eq. (9.115) becomes

$$\overline{G}_1 = G_1 + RT \ln x_1 \tag{9.116}$$

From the expression of fugacity of a component in the mixture, we have

$$\overline{G}_i - G_i = RT \ln \left(\frac{\overline{f}_i}{f_i}\right) \tag{9.117}$$

From Eqs. (9.116) and (9.117), we get

$$\frac{\overline{f}_i}{f_i} = x_i \tag{9.118}$$

$$\overline{f}_i = x_i f_i \tag{9.119}$$

For the ideal gaseous solution, fugacity is equal to partial pressure and therefore Eq. (9.119) becomes

$$\overline{f}_i = p_i = x_i f_i = P x_i \tag{9.120}$$

The Lewis–Randall rule is applicable at low pressures when the gas mixture behaves as an ideal solution. Despite its simplicity in application, the rule can not be applied in all cases, as such a rule provides an erroneous result at moderate to high pressures when the physical properties of the components are not the same. As per the definition of an ideal solution, the following characteristics are exhibited:

❖ The partial molar volume of each component in an ideal solution is equal to the molar volume of the corresponding pure component at the same temperature and pressure. That is

$$\overline{V}_i = V_i \tag{9.121}$$

where

$\overline{V}_i$ = Partial molar volume of pure component i
V_i = Molar volume of pure component i in the ideal solution.

Hence, it can be concluded that there is no volume change on mixing pure components to form an ideal solution.

With the help of Eq. (9.121), the volume of an ideal solution can be expressed as

$$V = \sum_{i=1}^{J} n_i \overline{V}_i = \sum_{i=1}^{J} n_i V_i = \sum_{i=1}^{J} V_i \tag{9.122}$$

where V_i is the volume that the pure component i would occupy when at the temperature and pressure of the mixture.

❖ Similarly, it can be shown that the partial molar internal energy of each component in an ideal solution is equal to the molar internal energy of the corresponding pure component at the same temperature and pressure.

$$\overline{U}_i = U_i \quad \text{taking help of } U = \sum_{i=1}^{J} n_i \overline{U}_i = \sum_{i=1}^{J} n_i U_i = \sum_{i=1}^{J} U_i$$

For enthalpy

$$\overline{H}_i = H_i \quad \text{with } H = \sum_{i=1}^{J} n_i \overline{H}_i = \sum_{i=1}^{J} n_i H_i = \sum_{i=1}^{J} H_i$$

The Lewis–Randall rule requires that the fugacity of mixture component i be evaluated, in terms of the fugacity of the pure component i, at the same temperature and pressure as the mixture and in the same state of aggregation. For example, if the mixture were a gas at T, p, then f_i would be determined for pure i at T, p and as a gas. However, at certain temperatures and pressures of interest, a component of a gaseous mixture may, as a pure substance, be a liquid or solid. We can consider as an example of this an air–water vapour mixture at 20°C (68°F) and 1 atm. At this temperature and pressure, water exists not as vapour but as a liquid.

9.10 RAOULT'S LAW AND IDEAL SOLUTION

We have defined the ideal solution in the preceding section as the solution in which the partial molar volume of each component in an ideal solution is equal to the molar volume of the corresponding pure component at the same temperature and pressure. That is, $\overline{V}_i = V_i$, where $\overline{V}_i$ is the partial molar volume of the pure component i and V_i is the molar volume of the pure component i in the ideal solution. It does necessarily mean that there will be no change in volume on mixing of pure components for the formation of an ideal solution. The ideal solution obeys the expression $\bar{f}_i = x_i f_i$ given by the Lewis–Randall rule. Raoult's law states: *The vapour pressure (P_i) over a solution is the product of the vapour pressure (P_i^{Sat}) of the pure solvent and the mole fraction (y_i) of the solvent in the solution.*

Mathematically

$$\overline{P}_i = y_i P_i^{Sat} \tag{9.123}$$

The expression is found to be quite satisfactory for dilute solutions.

Raoult's law holds good for the volatile components which behave ideally. In that case, $P_i = p_i$, i.e., the vapour pressure of the component will be the partial pressure of the solvent in the vapour phase.

Hence, Raoult's law can be restated as: *The partial pressure of the component in the vapour phase is the product of the vapour pressure (P_i^{Sat}) of the pure solvent and the mole fraction (y_i) of the solvent in the solution.*

Mathematically

$$\overline{p}_i = y_i P_i^{Sat} \tag{9.124}$$

Consider an ideal solution for which the chemical potential of component i is given by

$$\mu_i = \mu_i^0 + RT \ln x_i \tag{9.125}$$

At equilibrium, the chemical potential of the component i in the liquid solution must be the same as that in the vapour phase. Hence

$$\mu_{i,1} = \mu_{i,v} \tag{9.126}$$

Assuming the liquid behaves like an ideal solution, the preceding equation can be represented as

$$\mu_{i,1}^0 + RT \ln x_i = \mu_{i,v}^0 + RT \ln \frac{\bar{f}_i}{f_i^0} \qquad (9.127)$$

For an ideal gas, the fugacity of the vapour component i is

$$f_i^0 = P_i^0$$

Substituting this into Eq. (9.127), we get

$$\mu_{i,1}^0 + RT \ln x_i = \mu_{i,v}^0 + RT \ln \frac{\bar{f}_i}{P_i^0} \qquad (9.128)$$

The corresponding equation for the pure component i in equilibrium with its pure vapour is

$$\mu_{i,1}^0 = \mu_{i,v}^0 + RT \ln \frac{f_i^0}{P_i^0} \qquad (\text{since } x_i = x_1 + x_2 = 1) \quad (9.129)$$

Subtracting Eq. (9.129) from Eq. (9.128), we get

$$RT \ln x_i = RT \ln \frac{\bar{f}_i}{f_i^0} \qquad (9.130)$$

$$\bar{f}_i = x_i f_i^0 \qquad (9.131)$$

Since the vapour behaves ideally, fugacity can be replaced by pressure, as

$$\bar{p}_i = x_i P_i^0 = x_i P_i^{\text{Sat}} \qquad (9.132)$$

Raoult's law is applicable to the system at low to moderate pressure. All ideal solutions obey this law while the vapour phase is in equilibrium with the liquid solution. It applies to solutions having low concentration.

9.11 HENRY'S LAW AND DILUTE SOLUTION

The law formulated by William Henry in 1803 states: *The fugacity or partial pressure of the solute is directly proportional to the concentration of the solute in the solution.* A dilute solution consists of the main constituent, the solvent, and one or more solutes which are the diluted species. The law can be mathematically expressed as

$$\bar{f}_i = y_i H_i \qquad (9.133)$$

where

$\bar{f}_i$ = Fugacity of the component i in a non-ideal solution

y_i = Concentration of the solute in the solution

H_i = Henry's law constant that depends on temperature and pressure. Its unit is atm/mol or Pa/mol. Some values of H_i include that at 23°C for oxygen in water (42900 atm/mol) and for carbon dioxide in water (1560 atm/mol).

Equation (9.133) can also be represented as

$$\bar{p}_i = y_i H_i \qquad (9.134)$$

The fugacity of the component i, $\bar{f}_i$ is replaced by partial pressure $\bar{p}_i$ of the solute over the solution. It is necessary to remember that Henry's law is a limiting law that can be applied only for significantly dilute solutions. The range of concentration in which it applies becomes narrower as the system diverges from non-ideal behaviour. The law is not applicable for solutions where the solvent does react chemically with the gas being dissolved. For example, carbon dioxide reacts and forms hydrated carbon dioxide and then carbonic acid (H_2CO_3) with water.

In terms of comparison, Henry's law is applicable for the solute, while the Lewis–Randall rule is applicable for the solvent over the solution, considering that the mole fraction of the solute tends to zero and for the solvent it tends to 1. Henry's law has been found to be Satisfactorily obeyed at low pressures and where the gas and the solution behave ideally.

Consider a binary solution of acetone and water. A small amount of water is added to a large volume of acetone in a closed vessel. Then the partial pressure of water will be given by Henry's law and that of acetone will be given by Raoult's law. Again, if a few drops of acetone are added to a large volume of water in a closed vessel, then the ratio of the partial pressures of the two components will be the opposite. The relationship between Henry's law and Raoult's law is shown in Fig. 9.5.

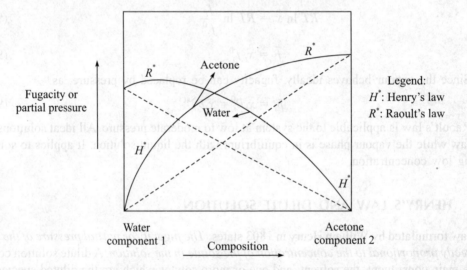

Fig. 9.5 Demonstration of interconnection between Henry's law and Raoult's law on acetone–water system.

Henry's law is considered to be a good approximation when the molecular species of the solute in the gas and in the solution phase are the same. Deviations from the law occur when the gas molecules in the solution undergo dissociation or association. For example, dissociation of ammonia does not obey Henry's law. The law is also not applicable when gases or their solutions behave non-ideally, which is often the case when the solubility is high.

EXAMPLE 9.21 Show that in a binary solution, if the solute obeys Henry's law, the solvent obeys the Lewis–Randall rule.

(WBUT, 2007)

Solution: Consider a binary solution comprising component 1 (solute) and component 2 (solvent). In terms of fugacities, the Gibbs–Duhem equation (9.58) can be expressed as

$$x_1 \frac{d \ln \bar{f_1}}{dx_1} = x_2 \frac{d \ln \bar{f_2}}{dx_2} \tag{i}$$

For the binary solution, since $dx_1 + dx_2 = 0$, then Eq. (i) becomes

$$x_1 d \ln \bar{f_1} + x_2 d \ln \bar{f_2} = 0 \tag{ii}$$

If the solute (component 1) obeys Henry's law, then we can write

$$\bar{f_1} = H_1 x_1$$

(where H_1 = Henry's constant)

or

$$d \ln \bar{f_1} = d \ln(H_1 x_1) = \frac{H_1 dx_1}{H_1 x_1} = \frac{dx_1}{x_1} \tag{iii}$$

Substituting it into Eq. (ii), we obtain

$$x_1 d \ln \bar{f_1} + x_2 d \ln \bar{f_2} = 0$$

or

$$\frac{x_1 dx_1}{x_1} + x_2 d \ln \bar{f_2} = 0$$

or

$$d \ln \bar{f_2} = -\frac{dx_1}{x_2} = \frac{dx_2}{x_2} = d \ln x_2 \quad (\text{as } dx_1 = -dx_2) \tag{iv}$$

Integrating Eq. (iv), we get

$$\int_{f_2}^{\bar{f_2}} d \ln \bar{f_2} = \int_{x_2=1}^{x_2} d \ln x_2$$

or

$$\bar{f_2} = H_2 x_2$$

which is the Lewis–Randall rule for component 2 (solvent).

SUMMARY

The chapter starts with a discussion of the importance of different useful fundamental properties such as partial molar property, chemical potential, fugacity, activity, and excess property of liquid solutions and gas mixtures. Chemical engineers deal with the design and drawing of the process equipment frequently used in chemical industries. The equipment includes distillation column, absorber, adsorber, cooling tower, drier, and extractor, in which mixing and separation take place. The process engineers must be aware of the composition and the properties of the species undergoing the changes brought about, because the transfer of species occurring from one phase to another phase requires complete knowledge of their composition and properties. In this regard, the definition and application of chemical potential, activity coefficient, and fugacity, along with the influence of temperature and pressure on them, have been discussed. The change in property that takes place due to mixing has been substantiated with the help of the ideal gas mixture model developed by Gibbs. We have explained the ideal solution model with the Lewis–Randall rule. The concept of excess property has been introduced to differentiate the ideal and real behaviours of the mixture. Finally, Raoult's and Henry's laws, which are necessary for knowing about the significance of ideal solution and dilute solution respectively, have been discussed.

KEY TERMS

Activity Coefficient   A measure of the deviation from ideal behaviour of chemical substances in a mixture.

Activity   A measure of 'effective concentration' of a component, resulting from interaction between molecules in a non-ideal gas or solution, and depending on temperature, pressure, and composition; the activity of component i is defined as the ratio of its fugacity in the solution to its fugacity in the standard or reference state.

Chemical Potential   The partial derivative of the thermodynamic potentials H, A and G of the ith component of a multicomponent system.

Excess Property   The difference between the actual property of the solution and the value it would have as an ideal solution at the same temperature, pressure, and composition.

Gibbs Theorem   In an ideal gas mixture, each component has its own private properties, and other components present in the mixture have no influence on the properties of any particular component.

Henry's Law   The fugacity or partial pressure of the solute is directly proportional to the concentration of the solute in the solution.

Lewis–Randall Rule   The fugacity of each component in an ideal solution is proportional to the mole fraction of that pure component in the solution.

Partial Molar Property   The property of a component which measures the contribution of the component present in a mixture to the aggregate property of the mixture.

Raoult's Law   The vapour pressure (P_i) over a solution is the product of the vapour pressure (P_i^{Sat}) of the pure solvent and the mole fraction (y_i) of the solvent in the solution.

IMPORTANT EQUATIONS

1. Partial molar property $\overline{X}_i$ of component i by definition is given by

$$\overline{X}_i = \left(\frac{\partial X}{\partial n_i} \right)_{T, P, n_j} \tag{9.1}$$

where

n_i = Number of moles of component i in the mixture

n_j = Number of moles of component other than component i in the mixture.

2. The partial molar property for a multi-component system of constant temperature and pressure in given by

$$dX_{P,T} = \overline{X}_1 \, dn_1 + \overline{X}_2 \, dn_2 + \cdots \tag{9.4}$$

where

$$\overline{X}_1 = \left(\frac{\partial X}{\partial n_1}\right)_{P,T,n_2} = \text{Partial molar property of component 1}$$

$$\overline{X}_2 = \left(\frac{\partial X}{\partial n_2}\right)_{P,T,n_1} = \text{Partial molar property of component 2.}$$

3. Chemical potential of component i can be expressed as

$$\mu_i = \left(\frac{\partial G}{\partial n_i}\right)_{P,T,n_j} \tag{9.15}$$

4. At constant temperature and pressure

$$dG_{P,T} = \sum \mu_i \, dn_i \tag{9.18}$$

5. At equilibrium, the chemical potentials of a component in two phases are given by

$$\mu_i^a = \mu_i^b \tag{9.23}$$

6. For a pure substance

$$G = \mu n$$

7. The effect of temperature on chemical potential is given by

$$\left(\frac{\partial \mu_i}{\partial P}\right)_{T,n_i} = \overline{V}_i \tag{9.27}$$

8. The pressure dependence of chemical potential is given by

$$\left[\frac{\partial(\mu/T)}{\partial T}\right] = -\frac{\overline{H}_i}{T^2} \tag{9.34}$$

9. The activity of component i is represented by

$$a_i = \frac{f_i}{f_i^0}$$

10. The activity coefficient of component i is expressed as

$$\gamma_i = \frac{a_i}{n_i} = \frac{f_i}{f_i^0 n_i} = \frac{f_i/n}{f_i^0 n_i/n} = \frac{\overline{f}_i}{f_i^0 x_i} \tag{9.37}$$

where

a_i = Activity of component i (referred to as the pure component)

f_i = Fugacity of component i

f_i^0 = Concentration of component i in standard state (= 1 mol/L).

11. The temperature-dependent form of activity coefficient is given by

$$\left(\frac{\partial \ln \gamma_i}{\partial T}\right)_P = \frac{H_i - \overline{H}_i}{RT^2} \tag{9.42}$$

12. The pressure-dependent form of activity coefficient is given by

$$\left(\frac{\partial \ln \gamma_i}{\partial P}\right)_T = \frac{\bar{V}_i - V_i}{RT} \tag{9.46}$$

13. The Gibbs–Duhem equations in different forms are given by

(a) $G_{P,T} = \mu_1 n_1 + \mu_2 n_2 + \cdots + \mu_i n_i + \cdots = \sum \mu_i n_i$ (9.51)

(b) $n_1 d\mu_1 + n_2 d\mu_2 + \cdots = 0$ or $\sum n_i d\mu_i = 0$ (9.53)

(c) $\sum x_i d\mu_i = 0$ (9.54)

(d) $x_1 \dfrac{d \ln a_1}{dx_1} = x_2 \dfrac{d \ln a_2}{dx_2}$ (9.60)

(e) $x_1 \dfrac{d \ln \gamma_1}{dx_1} = x_2 \dfrac{d \ln \gamma_2}{dx_2}$ (9.63a)

14. The fugacity of a component in a mixture is given by

$$\bar{G}_i - G_i^0 = RT \ln \frac{\bar{f}_i}{f_i^0} = RT \ln \frac{\bar{f}_i}{y_i P} = RT \ln \bar{\phi}_i \tag{9.66}$$

15. The Gibbs equation for the ideal gas mixture model is given by

$$\Delta M = \sum x_i (\bar{M}_i - M_i^0) \tag{9.79}$$

16. The free energy change of mixing is given by

$$\Delta G = \sum x_i (\bar{G}_i - G_i^0) \tag{9.80}$$

17. The excess free energy change of mixing is given by

$$\frac{G^E}{RT} = \sum x_i \ln \gamma_i \tag{9.102}$$

18. The Lewis–Randall equation for the ideal solution model is given by

$$\bar{f}_i = x_i f_i \tag{9.106}$$

19. Raoult's law is given by

$$P_i = y_i P_i^{Sat} \tag{9.123}$$

20. Henry's law is given by

$$\bar{f}_i = y_i H_i \tag{9.133}$$

EXERCISES

A. Review Questions

1. What is partial molar property? What is its significance in describing a multicomponent system? How do you differentiate between molar volume and partial molar volume?

2. Show that for an ideal gas the partial molar volume in a mixture of two gases is equal to $\dfrac{RT}{P}$ and this is true for all the components of a mixture of ideal gases.

3. Describe the 'method of intercept' to determine the partial molar properties of the components in a binary mixture.

4. What do you mean by chemical potential? Show that the variation of the chemical potential of a component i with pressure is given by $d\mu_i = \overline{V}_i\, dP$.

5. Define chemical potential in terms of internal energy, enthalpy, work function, and Gibbs function.

6. Which of the following derivatives are partial molar properties?

   (a) $\left(\dfrac{\partial V}{\partial n_i}\right)_{T,P,n_j}$ (b) $\left(\dfrac{\partial P}{\partial n_i}\right)_{T,V,n_j}$

   (c) $\left(\dfrac{\partial G}{\partial n_i}\right)_{V,P,n_j}$ (d) $\left(\dfrac{\partial T}{\partial n_i}\right)_{V,P,n_j}$

   (e) $\left(\dfrac{\partial H}{\partial n_i}\right)_{V,P,n_j}$ (f) $\left(\dfrac{\partial H}{\partial n_i}\right)_{S,P,n_j}$

7. Derive an expression for the change in chemical potential with pressure for an ideal gas.

8. Prove the following statement: *At the same temperature and pressure, the chemical potential or partial molar free energy of a component in every phase must be the same under equilibrium conditions.*

9. Explain the influence of temperature and pressure on chemical potential.

10. What are activity and activity coefficient? How does activity coefficient relate to excess property?

11. Deduce the Gibbs–Duhem relation for a multicomponent system.

12. Derive an expression for the fugacity of a component in a mixture.

13. What do you mean by fugacity of solid and liquids?

14. What is the significance of the Poynting pressure correction factor?

15. In the light of Gibbs theorem, discuss the importance of the ideal gas mixture model.

16. What do you mean by property change of mixing? Derive an expression for the free energy change due to mixing in a solution.

17. Define the term 'excess property'. Give an informatory note on the excess free energy of a component in a mixture.

18. Explain the 'ideal solution model' with the help of the Lewis–Randall rule.

19. State Raoult's law. Show that it is a simplified form of the Lewis–Randall rule.

20. Explain Henry's law and show that Raoult's law is a special case of Henry's law.

21. Show that in a binary solution, if the solute obeys Henry's law, the solvent obeys the Lewis–Randall rule.

B. Problems

1. Calculate the partial molar volume of water in a 70 mol per cent ethanol–water solution, in which the partial molar volume of ethanol is $61.23 \times 10^{-6} \, \text{m}^3/\text{mol}$, given that the density of the mixture is $901.57 \, \text{kg/m}^3$.

2. It is desired to prepare $2.5 \, \text{m}^3$ of a 60 mol per cent methanol–water solution. Determine the volumes of methanol and water required to be mixed at ambient temperature, given that the partial molar volumes of methanol and water are $58.3 \times 10^{-6} \, \text{m}^3/\text{mol}$ and $17.2 \times 10^{-6} \, \text{m}^3/\text{mol}$ respectively. The density of the methanol is $782.51 \, \text{kg/m}^3$.

3. In order to prepare $3.0 \, \text{m}^3$ of an alcohol–water solution, alcohol of mole fraction $X_1 = 0.60$ is required to be mixed with water at 25°C. Determine the volumes of alcohol and water needed to prepare the mixture, given that
    Partial molar volume of methanol $= 57.11 \times 10^{-6} \, \text{m}^3/\text{mol}$
    Partial molar volume of water $= 17.5 \times 10^{-6} \, \text{m}^3/\text{mol}$
    Molar volume of alcohol $= 56.21 \times 10^{-6} \, \text{m}^3/\text{mol}$
    Molar volume of water $= 18 \times 10^{-6} \, \text{m}^3/\text{mol}$.

4. Acetone and chloroform are mixed to prepare 1.0 kg of a solution in which the mole fraction of chloroform (x_2) is 0.4690, the partial molar volume of chloroform is $80.235 \, \text{cm}^3/\text{mol}$, and that of acetone is $74.166 \, \text{cm}^3/\text{mol}$. Calculate the volume of the solution.

5. It is desired to prepare $2000 \, \text{cm}^3$ of an antifreeze solution consisting of 30 mol per cent methanol in water. What volumes of pure methanol and pure water at 25°C must be mixed to form the required volume of antifreeze, also at 25°C? The partial molar volumes of methanol and water in a 30 mol per cent methanol solution and their pure species molar volumes, both at 25°C, are:

    Methanol (1): $\overline{V}_1 = 38.632 \, \text{cm}^3/\text{mol}$ $V_1 = 40.727 \, \text{cm}^3/\text{mol}$

    Water (2): $\overline{V}_2 = 17.765 \, \text{cm}^3/\text{mol}$ $V_2 = 18.068 \, \text{cm}^3/\text{mol}$

6. If the molar density of a binary mixture is given by the empirical expression

    $$\rho = a_0 + a_1 x_1 + a_2 x_1^2$$

    Find the corresponding expressions for $\overline{V}_1$ and $\overline{V}_2$.

7. The molar volume of a binary mixture containing benzene and cyclohexane is represented by the empirical relation

    $$V = 109.4 \times 10^{-6} - 16.8 \times 10^{-6}x - 2.64 \times 10^{-6}x^2$$

    where V is in cm^3/mol and x is the mole fraction of component 1. Find the expressions for the partial molar volumes of both the components.

8. The molar volume of a binary mixture is given by

    $$V = y_1 V_1 + y_2 V_2 + y_1 y_2[A + B(y_1 - y_2)]$$

    Determine the partial molar volumes of components 1 and 2.

9. Prove that the chemical potential of the ith component in an ideal gas mixture is given by

    $$\mu_i = \mu_i^0 + RT \ln \frac{p_i}{p^0}$$

10. Estimate the fugacity of liquid n-octane at 427.85 K and 1.0 MPa. The saturation pressure of n-octane at 427.85 K is 0.215 MPa.

11. Calculate the fugacity of a mixture of 60 mol per cent carbon dioxide and 40 mol per cent argon at 600 K and 60 bar, assuming that the mixture behaves like a van der Waals gas.

12. What is the change in entropy when 0.7 m^3 of CO_2 and 0.3 m^3 of N_2, each at 1 bar and 25°C, blend to form a gas mixture at the same conditions? Assume that both are ideal gases.

13. One mole of helium at 100°C is mixed with 0.5 mol of neon at 0°C. Calculate ΔS_{mix} for this mixture if the mixing is isobaric ($P = 1$ atm) and the gases are ideal. We are given that $C_{P,He} = C_{P,Ne} = 2.5R$.

14. Estimate the entropy change of mixing when 0.75 m^3 of hydrogen and 0.25 m^3 of nitrogen at 1 atm and 25°C are mixed to prepare a gaseous mixture.

15. Calculate ΔS_{mixing} for the formation of one mol of a mixture on mixing nitrogen and oxygen in the volume ratio of 4:1.

16. Calculate ΔG_{mixing} and ΔS_{mixing} per litre of mixture containing 15% nitrogen, 55% hydrogen and 30% ammonia at S.T.P.

17. The molar volume of a binary liquid mixture at T and P is given by

$$V = 120x_1 + 70x_2 + (15x_1 + 8x_2)x_1x_2$$

   (a) Find the expressions for the partial molar volume of species 1 and 2;
   (b) Show that the expression satisfies the Gibbs–Duhem equation.

18. If the molar density of a binary mixture is given by the empirical expression

$$\rho = a_0 + a_1x_1 + a_2x_1^2$$

Find the corresponding expressions for $\overline{V}_1$ and $\overline{V}_2$.

19. A vessel, divided into two parts by a partition, contains 4 mol of nitrogen gas at 75°C and 30 bar on one side and 2.5 mol of argon gas at 130°C and 20 bar on the other. If the partition is removed and the gases mix adiabatically and completely, what is the change in entropy? Assume nitrogen to be an ideal gas with $C_V = 5R/2$ and argon to be an ideal gas with $C_V = 3R/2$.

20. The excess enthalpy of a binary solution mixture is given by

$$H^E = x_1x_2(40x_1 + 20x_2) \text{ J/mol}$$

Determine the expressions for $\overline{H}_1^E$ and $\overline{H}_2^E$.

21. The excess free energy of a binary solution mixture is given by

$$\frac{G^E}{RT} = -3x_1x_2(0.4x_1 + 0.5x_2) \text{ J/mol}$$

Find the expressions for γ_1 and γ_2.

22. The excess free energy of a binary solution mixture is given by

$$\frac{G^E}{RT} = x_1x_2[A + B(x_1 - x_2)] \text{ J/mol}$$

Find the expressions for γ_1 and γ_2.

23. The activity coefficients of two components of a solution are given by

$$\ln \gamma_1 = Ax_2^2 + Bx_2^2(3x_1 - x_2) \quad \text{and} \quad \ln \gamma_2 = Ax_1^2 + Bx_1^2(x_1 - 3x_2)$$

Find expressions for $\dfrac{G^E}{RT}$ and show that the equations satisfy the Gibbs–Duhem relation.

24. For a binary system, the influence of composition on property M can be represented by

$$M = x_1M_1 + x_2M_2 + Ax_1x_2$$

where M_1 and M_2 are the values of M for pure chemical species 1 and 2 respectively, and A is a parameter independent of composition. Develop expressions for $\overline{M}_1$ and $\overline{M}_2$.

25. Show that

$$M^E = M^R - \sum_i x_i M_i^R$$

26. For a ternary system, the influence of composition on property M can be represented by

$$M = x_1M_1 + x_2M_2 + x_3M_3 + Cx_1x_2x_3$$

where M_1, M_2 and M_3 are the values of M for pure chemical species 1 and 2 respectively, and A is a parameter independent of composition. Develop expressions for $\overline{M}_1$, $\overline{M}_2$ and $\overline{M}_3$.

27. Show that

$$\gamma_i = \frac{\hat{\phi}_i}{\phi_i}$$

28. For the benzene–chloroform system, H^E at 25°C can be expressed as

$$H^E = x_1x_2[-1690 - 330(x_1 - x_2) + 85(x_1 - x_2)^2]$$

Calculate H^E for $x_1 = 0.1, 0.5$, and 1.0.

29. The volume of a mixture of two organic liquids 1 and 2 is given by

$$V = 110 - 17x_1 - 2.5x_1^2$$

where V is the volume in m^3/mol at 1 bar and 300 K. Find the expressions for $\overline{V}_1$ and $\overline{V}_2$.

30. The fugacity of component 1 in a binary liquid mixture consisting of components 1 and 2 at 298 K and 20 bar is given by $\hat{f}_1 = 50x_1 - 80x_1^2 + 40x_1^3$, where $\hat{f}_1$ is in bar. Determine

(a) Fugacity f_1 of pure component 1;
(b) Fugacity coefficient ϕ_1;
(c) Henry's law constant K_1;
(d) Activity coefficient γ_1.

31. The molar volumes of a binary solution at 25°C are measured as follows:

X_1	0	0.2	0.4	0.6	0.8	1.0
$V \times 10^6$ (m^3/mol)	20.0	21.5	24.0	27.4	32.0	40.0

Using the method of tangential intercept, calculate the partial molar volumes of components 1 and 2 at $X_1 = 0.5$ and $X_1 = 0.75$.

32. Estimate the fugacity of a ternary gas mixture consisting of 25 mol per cent hydrogen (component 1), 35 mol per cent of nitrogen (component 2), and 40 mol per cent of argon (component 3). We are given that at 150°C and 60 bar, the fugacity coefficients of components 1, 2, and 3 are 0.7, 0.85, and 0.90 respectively.

33. The saturation pressure of liquid water at 393.38 K is 200 kPa. Estimate the fugacity of liquid water at 372.12 K and 300 kPa. We are given that the specific volume of liquid water at 372.12 K is $1.061 \times 10^{-3} \, \text{m}^3/\text{kg}$.

34. At 25°C and 1 atm, 0.7 mol of methane were mixed with 0.3 mol of oxygen. Calculate the free energy and enthalpy change of mixing. Assume that the gases are behaving ideally.

35. Ammonia gas obeys the equation of state $A(V - b) = RT$. For this gas, $b = 0.037 \, \text{dm}^3/\text{mol}$. Calculate the fugacity and $\phi = 1.08$ for the gas at 0°C and 50 atm.

(**Ans.** $f = 54 \, \text{atm}, \ \phi = 1.08$)

Vapour–Liquid Equilibrium

10

The operations most commonly performed in the chemical, petroleum, and allied industries and encountered by the process engineers involve distillation, absorption, adsorption, leaching, extraction, drying, etc. These operations involve transfer of components from one phase to another at different temperatures and pressures. More precisely, these are concerned with the coexistence of vapour and liquid phases. The design and analysis of these operations requires the knowledge of phase equilibrium and an availability of vapour–liquid equilibrium (VLE) data. Most of the times, accurate data are not available and so assignments are completed with fragmentary data. Therefore, an engineer often needs to use important correlations to check the correctness of the data. Such correlations are presented and derived in this chapter with the help of the principle of phase equilibrium. For the estimation of activity coefficients at low to moderate pressure,

Wohl's equation, van Laar equation, Margules equation, and UNIFAC method are discussed. Some simple methods are discussed to test the thermodynamic consistency of VLE data. In addition, criteria for phase equilibrium of single and multiple components are discussed. Diagrams are efficient tools to understand the processes clearly. Various diagrams—boiling point diagram, P–x–y, P–T diagram, etc.—have been elaborately discussed. Raoult's law is valid for a system at low to moderate pressure. But the deviation from Raoult's law measures the non-ideality of the solution. Different types of azeotropes are shown in graphical presentations along with the principle of its formation. A stepwise methodology for deducing the dew point, bubble point and flash has been discussed.

10.1 CRITERIA OF EQUILIBRIUM

Thermodynamics is very much concerned with the equilibrium state of the system. The term *equilibrium* is used to imply a state of balance of a system. Equilibrium is a static condition in which no changes occur in the macroscopic properties of a system with time. Consider that a system is undergoing a change in a given state. If there is no change in the state of the system, then the system is considered to be in a state of thermodynamic equilibrium. Hence, a system is said to be in the equilibrium state if the system has no tendency to undergo any further change, i.e., the properties like pressure, temperature, and composition are uniform in magnitude throughout the system.

It is needless to mention that in the equilibrium state there are no unbalanced driving forces within the system, and so the system cannot exchange energy in the form of heat as well as work with the surroundings.

The Clausius inequality plays an important role in establishing the thermodynamic criteria of equilibrium, and it can be substantiated by the following discussion.

Consider that a system is placed in the surroundings and exchanging heat with the reservoir at temperature T and producing work as shown in Fig. 10.1.

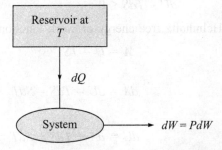

Fig. 10.1 Demonstration on exchanging of heat between system and reservoir.

It can be written for either a finite or an infinitesimal change.
From the first law of thermodynamics

$$dQ = dU + dW = dU + PdV \tag{10.1}$$

From the second law of thermodynamics

$$dS \geq \frac{dQ}{T} \quad \text{or} \quad dQ - TdS \leq 0 \tag{10.2}$$

Combining Eq. (10.1) with (10.2), we have

$$dU + PdV - TdS \leq 0 \tag{10.3}$$

This inequality applies to every incremental change of the system between equilibrium states, and it dictates the direction of change that leads towards equilibrium. This is used for developing several equilibrium criteria under various constraints. Each set of constraints considered corresponds to a physically realistic or commonly encountered situation. The criteria are:

Case I: Isolated System

By definition, an isolated system is incapable of exchanging heat, mass, or work with its surroundings. Therefore, $dQ = 0$ and $dW = 0$. Substituting these, we get, from the first law of thermodynamics (Eq. (10.1)), $dU = 0$. Since it is an isolated system, it would be of constant volume, i.e., $dV = 0$.

Hence, from Eq. (10.3), we have

$$dS_{U,V} \leq 0 \tag{10.4}$$

This equation necessarily implies that entropy can never decrease for any process occurring in a system at constant U and V. This is constant for a reversible process and increases for an irreversible process, which always leads the system toward an equilibrium state. Entropy will have maximum value at equilibrium, when it will not be possible for any further spontaneous process to occur.

Case II: System at Constant T and V

For a closed system undergoing a process at constant temperature and volume, the equation of inequality becomes

$$dU - TdS \leq 0 \tag{10.5}$$

Now, the definition of Helmholtz free energy or work function is given by

$$A = U - TS$$

or

$$dA = dU - TdS - SdT$$

or

$$dU = dA + TdS + SdT \tag{10.6}$$

Comparing it with Eq. (10.3), we have

$$dA \leq -PdV - SdT \tag{10.7}$$

For the process occurring at constant T and P, Eq. (10.7) becomes

$$dA_{T,V} \leq 0 \tag{10.8}$$

The above equation indicates that the spontaneous process taking place at constant T and V is accompanied by decrease in work function or Helmholtz free energy at equilibrium.

Case III: System at Constant T and P

By definition, Gibbs free energy or Gibbs function is

$$G = H - TS$$

or

$$dG = dH - TdS - SdT \tag{10.9}$$

From the definition of enthalpy

$$H = U + PV$$

or

$$dH = dU + PdV + VdP \tag{10.10}$$

Substituting Eq. (10.10) into (10.9), we get

$$dG = dU + PdV + VdP - TdS - SdT$$

or

$$dU = dG - PdV - VdP + TdS + SdT \tag{10.11}$$

Comparing it with Eq. (10.3), we have

$$dG \leq VdP - SdT \tag{10.12}$$

At constant temperature and pressure, Eq. (10.12) becomes

$$dG_{T,P} \leq 0 \tag{10.13}$$

Equation (10.13) implies that for the spontaneous process occurring at constant T and P, Gibbs free energy is minimum at equilibrium.

10.2 CRITERION FOR PHASE EQUILIBRIUM

We have discussed so far the criteria of thermodynamic equilibrium of a system. It enables us to determine the criterion of phase equilibrium. Consider a closed binary system comprising two phases—vapour and liquid. If the system is in thermal and mechanical equilibrium, then another important equilibrium is phase equilibrium, which needs to be taken into consideration because the individual phases are free to exchange mass and energy between them. For instance, a small amount of water left in a glass evaporates, a wet T-shirt hanging in an open area eventually dries, and the aftershave in an open bottle quickly disappears. These examples suggest that there is a driving force between the two phases of a substance that forces the mass to transform from one phase to another phase. The magnitude of this force depends on the relative concentrations of the two phases. At constant temperature and pressure, when two components are mixed well, the state differentials can occur at equilibrium without producing any change in Gibbs free energy. It leads to a more general criterion of equilibrium for a multicomponent system, for which

$$dG = -SdT + VdP + \sum_i G_i dn_i \tag{10.14}$$

At constant temperature, pressure and composition, Eq. (10.14) becomes

$$dG_{T,P} = 0 \tag{10.15}$$

This is the actual criterion of phase equilibrium.

For the case of vapour–liquid equilibrium, Gibbs free energy can be expressed as

$$G = G^v + G^l$$

or

$$dG = dG^v + dG^l = 0 \qquad (10.16)$$

where

G^v = Gibbs free energy of saturated vapour phase

G^l = Gibbs free energy of saturated liquid phase.

10.3 PHASE RULE FOR NON-REACTING SYSTEMS AND DUHEM'S THEOREM

A single-component two-phase system may exist in equilibrium at different temperatures or pressures. However, once the temperature is fixed, the system is locked in equilibrium state and all intensive properties of each phase (except their relative amounts) are fixed. The masses of the phases do not belong to the phase-rule variables, because they have no influence on the intensive state of the system. Temperature, pressure, and composition are considered phase rule variables. Therefore, a single-component two-phase system has one independent property, which may be either temperature or pressure.

The number of independent variables associated with a multi-component, multi-phase non-reacting system is given by Gibbs phase rule developed in 1875, and it can be expressed as

$$F = C - P + 2 \qquad (10.17)$$

where

F = Number of degrees of freedom of the systems or number of independent variables

C = Number of components

P = Number of phases.

2 is included in the equation to account for the fact that the two intensive variables T and P must be specified to describe the state of equilibrium.

For a single-component two-phase system, as shown in Fig. 10.2, $C = 1$ and $P = 2$. Therefore, the number of degrees of freedom of the systems or the number of independent variables, $F = 1 - 2 + 2 = 1$. That is, one independent intensive property is required to be specified. Thus, at the triple point, $F = 1 - 3 + 2 = 0$. It implies that none of the properties of a pure substance at the triple point can be varied.

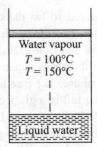

Fig. 10.2 Demonstration of single-component two-phase system having one independent variable.

We can derive Eq. (10.17) from the knowledge of criterion of phase equilibrium. Let us consider a multi-component system comprising C components and P phases. So far as the composition variables are concerned, in every phase the C components would have C different concentrations. But in a closed system, if we arbitrarily choose concentration of $C - 1$ components, the last one is fixed by the remainder. Hence, there will be $C - 1$ compositions in every phase.

For, all the P phases, the number of composition variables = $P(C - 1)$

In addition to this, two variables—temperature and pressure—are there.

Thus, the total number of variables

$$= P(C - 1) + 2$$

Now, we consider the relation between the phase variables. From the phase equilibrium relation, we gather that the chemical potential of component i is the same in all the phases a, b, and c.

$$\mu_i^a = \mu_i^b = \mu_i^c$$

Therefore, for $P - 1$ phases and C components, the total number of phase equilibrium relations would be $(P - 1)C$. So, the degrees of freedom of the system is given by

$$F = P(C - 1) + 2 - C(P - 1) = C - P + 2$$

Duhem's theorem is another kind of rule analogous to the Gibbs phase rule, but it deals also with the extensive property of the system. It applies to a closed system at equilibrium for which the extensive state as well as the intensive state of the system is fixed. The state of such a system is known as *completely determined state* and is characterized by both intensive $[2 + (C - 1)]$ and extensive (C) phase-rule variables. Here, the extensive variables are concerned with the masses or mole numbers. Hence, the total number of variables can be expressed as

$$2 + (C - 1)P + P = 2 + CP \qquad (10.18)$$

where the second term $(C - 1)P$ represents intensive variables and the third term P represents extensive variables (mass).

Since the theorem is applied to a closed system formed from specified amounts of the chemical species, a material balance can be represented for each of the C chemical species. The number of independent phase-equilibrium equations need to be taken into account with C species.

Consequently, the total number of equations

$$= (P - 1)C + C = PC \qquad (10.19)$$

Then, the difference between the number of variables and the number of equations is given by using Eqs. (10.18) and (10.19) as

$$(2 + CP) - PC = 2 \qquad (10.20)$$

Based on Eq. (10.20), Duhem's theorem states: *For any closed system formed initially from a given mass of prescribed chemical species, the equilibrium state is completely determined when any two independent variables are fixed.*

Equation (10.20) implies that only two independent variables (either extensive or intensive) are required to be specified to define the state of the system. Hence, the conclusions from the phase rule can be summarized as follows:

- The number of independent intensive variables is provided by the phase rule.
- When $F = 1$, at least one of the two variables must be extensive.
- When $F = 0$, both must be extensive.

10.4 PHASE EQUILIBRIUM FOR SINGLE-COMPONENT SYSTEM

The equilibrium criterion for a system consisting of two phases of a pure substance such as water is easily developed by considering a mixture of saturated liquid and saturated vapour in equilibrium at a constant temperature and pressure, as shown in Fig. 10.2.

The total Gibbs function of this mixture based on the criterion of equilibrium to the system is given by

$$dG = dG^v + dG^l \tag{10.21}$$

$$= G^v \, dn^v + G^l \, dn^l = 0 \tag{10.22}$$

(as $dG = dG^v + dG^l = \sum_i \overline{G_i} \, dn_i = G^v \, dn^v + G^l \, dn^l$ at constant temperature and pressure)

where

G^v = Gibbs function of saturated vapour phase
G^l = Gibbs function of saturated liquid phase.

Since $dn^v + dn^l = 0$, then $dn^v = -dn^l$. Substituting this into Eq. (10.22), we get

$$(G^l - G^v) \, dn^l = 0$$

or

$$G^v = G^l \tag{10.23}$$

Equation (10.23) does necessarily mean that the two phases of a pure substance are in equilibrium when the Gibbs free energy component in them are the same.

Similar conclusions can be drawn for two other different phases having one component. Consider a closed container consisting of saturated vapour and saturated liquid. If the process is allowed to take place at constant temperature and pressure, then the mass will be transferred from one phase to another phase. Therefore, the change in Gibbs free energy for this reversible process would be

$$G^v = H^v - T\Delta S^v \tag{10.24}$$

and

$$G^l = H^l - T\Delta S^l \tag{10.25}$$

Since the process occurs at constant pressure, we put $\Delta H = Q = T\Delta S$ into Eqs. (10.24) and (10.25) and compare them to obtain

$$G^v = G^l$$

Hence, it is supported well by Gibbs function equation. It can be expressed in terms of fugacity as

$$f^v = f^l \tag{10.26}$$

as Gibbs free energy is related to the fugacity of the pure component by

$$G = G^0 + RT \ln f \tag{10.27}$$

where G^0 is constant.

The above criteria can also be applied to more than two phases.

EXAMPLE 10.1 Applying the criterion for equilibrium, derive the Clausius–Clapeyron equation.

Solution: The Clausius–Clapeyron equation can be derived by applying equilibrium criterion. We know that for a binary mixture comprising two components P and Q at equilibrium, the differential form of Gibbs free energy is given by

$$dG^P = dG^Q \tag{i}$$

From the thermodynamic fundamental relation, we have

$$dG = VdP - SdT \tag{ii}$$

Hence, Eq. (i) can be expressed as

$$V^P\, dP - S^P\, dT = V^Q dP - S^Q dT$$

or

$$(V^P - V^Q)dP = (S^P - S^Q)dT$$

or

$$\frac{dP}{dT} = \frac{S^P - S^Q}{V^P - V^Q} = \frac{\Delta S}{\Delta V} \tag{iii}$$

For the components P and Q, we may also write in the following form of Gibbs energy:

$$G^P = G^Q$$

or

$$G^P = H^P - TS^P$$

and similarly

$$G^Q = H^Q - TS^Q$$

So

$$H^P - TS^P = H^Q - TS^Q$$

$$\Delta H = H^P - H^Q = T(S^P - S^Q) = T\Delta S \tag{iv}$$

Substituting Eq. (iv) into (i), we get

$$\frac{dP}{dT} = \frac{\Delta S}{\Delta V} = \frac{\Delta H}{T\Delta V} = \frac{\Delta H}{T(V_\text{g} - V_1)} \tag{v}$$

Since $V_\text{g} \gg V_1$ and vapour is assumed to behave ideally, Eq. (v) yields

$$\frac{dP}{dT} = \frac{\Delta H}{TV_\text{g}} = \frac{\Delta H}{T\left(\dfrac{RT}{P}\right)} = \frac{P\Delta H}{RT^2} \tag{vi}$$

or

$$\frac{1}{P}\frac{dP}{dT} = \frac{\Delta H}{RT^2} \tag{vii}$$

or

$$\frac{d \ln P}{dT} = \frac{\Delta H}{RT^2} \tag{viii}$$

For small temperature ranges, the enthalpy change of the process can be treated further as a constant. Then on integration, Eq. (viii) yields

$$\ln \frac{P_2}{P_1} = \frac{\Delta H}{R} \left(\frac{1}{T_1} - \frac{1}{T_2} \right) \tag{ix}$$

This equation is called the *Clausius–Clapeyron equation* and is used to determine the variation of pressure with temperature.

10.5 PHASE EQUILIBRIUM FOR MULTI-COMPONENT SYSTEM

In engineering practice, we come across many multi-component systems which have two or more phases. A multi-component multi-phase system at a given temperature and pressure will be in equilibrium when there is no driving force between the different phases of each component. Therefore, for phase equilibrium, the specific Gibbs function or chemical potential of each component must be the same in all phases.

For a closed heterogeneous system consisting of two phases (a, b) in equilibrium, the change in free energy can be written as

$$dG = VdP - SdT + \sum_i \mu_i dn_i \tag{10.28}$$

At constant temperature and pressure, Eq. (10.28) becomes

$$dG = \sum_i \mu_i dn_i \tag{10.29}$$

From the criterion for phase equilibrium, we have

$$dG = 0 \tag{10.30}$$

Comparing Eq. (10.30) with (10.29), we get

$$\sum_i \mu_i dn_i = 0$$

Hence, we may write

$$G_{P,T} = \sum_i \mu_i^a n_i^a + \sum_i \mu_i^b n_i^b = 0 \tag{10.31}$$

or

$$dG_{P,T} = \sum_i \mu_i^a dn_i^a + \sum_i \mu_i^b dn_i^b = 0 \tag{10.32}$$

Since $dn_i = dn_i^a + dn_i^b = 0$, then $dn_i^a = -dn_i^b$. Substituting this into Eq. (10.32), we get

$$\sum_i (\mu_i^a - \mu_i^b) dn_i^a = 0 \tag{10.33}$$

or

$$\mu_i^a = \mu_i^b = \cdots = \mu_i^\pi \qquad (i = 1, 2, 3, \ldots, N) \tag{10.34}$$

Hence, it can be concluded that *for multiple phases at the same temperature and pressure, the chemical potential of a component in every phase must be the same under equilibrium conditions.*

Chemical potential is related to fugacity of a component as

$$\mu_i = G_i + RT \ln \frac{|\hat{f}_i|}{f_i}$$

Therefore, the criterion for equilibrium would be

$$\bar{f}_i^{\,a} = \bar{f}_i^{\,b} = \cdots = \bar{f}_i^{\,\pi} \quad (i = 1, 2, 3, \ldots, N) \tag{10.35}$$

10.6 VAPOUR–LIQUID EQUILIBRIUM DIAGRAM FOR BINARY MIXTURE

Vapour–liquid equilibrium is a condition where either a liquid is in equilibrium with its vapour or a state of coexistence of liquid and vapour phases exists where the rate of evaporation (liquid changing to vapour) is equal to the rate of condensation (vapour changing to liquid).

Let us consider a binary liquid mixture of benzene and toluene existing in two phases shown in Fig. 10.3. In accordance with the phase rule, the degree of freedom or the number of independent variables of the system would be 2 as per the expression $F = C - P + 2 = 2 - 2 + 2 = 2$. Since we know that temperature, pressure, and composition (x, y) of a binary system are considered phase rule

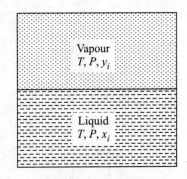

Fig. 10.3 Schematic representation of vapour–liquid equilibrium for benzene–toluene system.

variables, any two of them can be independently varied. For the benzene–toluene system, the volatility of benzene (1) is considered to be more than that of toluene (2). That is, the vapour pressure of benzene is higher than that of toluene. On heating, the liquid (x_1) is converted into vapour (y_1), and the vapour is enriched with greater amounts of volatile component 1. If the process takes place at constant pressure, the pressure is fixed and the liquid phase composition (x_1) can be varied independently as it is given. Therefore, two degrees of freedom are used. The remaining properties are temperature and vapour phase composition (y_1), which can be altered. Hence, at constant pressure, temperature versus composition (vapour) plot is obtained. Similarly, at constant temperature, pressure versus composition (vapour) plot, and at constant temperature or pressure, vapour phase composition versus liquid phase composition can be obtained.

Therefore, the VLE of a binary system can be better represented by the following three kinds of diagrams:

- Temperature–Composition $(T–x–y)$ diagram;
- Pressure–Composition $(P–x–y)$ diagram;
- Pressure–Temperature $(P–T)$ diagram.

10.6.1 Temperature–Composition (*T–x–y*) Diagram

The vapour–liquid equilibrium of a binary mixture at constant pressure can be represented on a *T–x–y* diagram in which temperature (along the ordinate) is plotted against composition of liquid and vapour (along the abscissa). This diagram is known as the *boiling point diagram* and is shown in Fig. 10.4. It is used to understand how the equilibrium values change with temperature. The boiling point diagram for a benzene–toluene system at a total pressure of 101.32 kPa is depicted in the figure. It can be seen from the plot that benzene, which boils at 80.1°C, is more volatile than toluene, which boils at 110.6°C. The upper curve is the saturated vapour curve, which is called the *dew point curve*, and the lower curve is the saturated liquid curve, known as the *bubble point curve*. These two curves meet where the mixture becomes purely one component, where $x_1 = 0$ (and $x_2 = 1$, pure component 2) or $x_1 = 1$ (and $x_2 = 0$, pure component 1). The temperatures at those two points correspond to the boiling points of the two pure components.

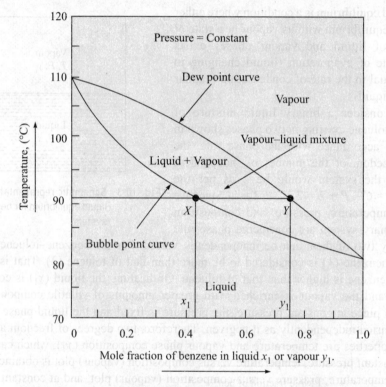

Fig. 10.4 Temperature–composition diagram of binary VLE mixture.

The region bounded by these two curves is called the *two-phase region*, where the mixture is partly liquid and partly vapour. The region below the bubble point curve is known as *sub-cooled region* (sub-cooled liquid) and the region above the dew point curve is *superheated region* (superheated vapour). Any point above the upper curve represents a mixture which consists entirely of vapour and any point below the lower curve represents a completely liquid mixture. Point *x*, on the lower curve, represents a completely liquid mixture of composition $x_1 = 0.42$, which will start to boil at 90°C, and the composition of the vapour in equilibrium is $y_1 = 0.75$, represented by point y, that is, $y_1 > x_1$. So the points x and y, which lie on the same horizontal line, represent equilibrium compositions at a temperature of 95°C. For systems which follow Raoult's law, the boiling point diagram can be drawn from the pure component vapour pressure data. These two curves meet where the mixture becomes purely one component, where $x_1 = 0$ (and $x_2 = 1$, pure component 2) or $x_1 = 1$ (and $x_2 = 0$, pure component 1). The temperatures at those two points correspond to the boiling points of the two pure components.

10.6.2 Pressure–Composition (*P–x–y*) Diagram

At constant temperature the vapour–liquid equilibrium of a binary mixture can be illustrated on a *P–x–y* diagram (see Fig. 10.5) with pressure along the ordinate and the composition of liquid and vapour along the abscissa. Considering again the benzene–toluene system, benzene (species 1) is the lighter or more volatile of the two, so at saturated condition the vapour pressure of benzene is higher than that of toluene (species 2), i.e., $P_1^{Sat} > P_2^{Sat}$.

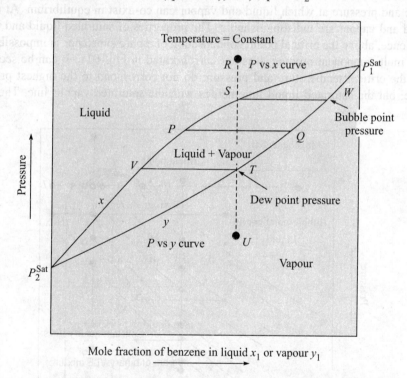

Fig. 10.5 Pressure–composition diagram of binary VLE mixture.

The upper curve is the saturated liquid curve, which is called the *P versus x curve*, and the lower curve is the saturated vapour curve and is known as the *P versus y curve*. The space between the two curves is the region of coexistence of both liquid and vapour phases is called the *two-phase region*. Here the mixture is partly saturated liquid and partly saturated vapour. The vapour is richer in the more volatile component, i.e., benzene. Consequently, at any pressure,

$y_1 > x_1$. In Fig. 10.5, the horizontal lines such as PQ, VT and SW are called *tie lines*. They connect the vapour and liquid phases in equilibrium. Now, on lowering the pressure at constant temperature, the liquid mixture starts forming the vapour bubble and therefore the pressure at which the bubble formation takes place is known as the *bubble point pressure*. The point S in the line $RSTU$ represents the bubble point pressure. The condition of the binary vapour–liquid equilibrium mixture is represented by the line $RSTU$. On further reduction in pressure, the entire liquid is gradually transformed into vapour and a drop of liquid is left. This pressure is called the *dew point pressure*, represented by the point T. The pressure is reduced further and at the point U, the vapour is converted into superheated vapour.

10.6.3 Pressure–Temperature (*P–T*) Diagram

The pressure–temperature diagram for a vapour–liquid equilibrium of a binary mixture of given composition is shown in Fig. 10.6, in which pressure along the ordinate is plotted against temperature along the abscissa. For a pure substance, the critical point is defined as the highest temperature and pressure at which liquid and vapour can co-exist in equilibrium. At the critical point, liquid and vapour are indistinguishable. The properties of saturated liquid and vapour are identical. Hence, above the critical point, condensation of a pure substance is impossible. But for a binary or multicomponent mixture, it is not corroborated in Fig. 10.6. It can be seen from the figure that the critical temperature and pressure do not correspond to the highest pressure and temperature, but the saturated liquid line merges with the saturated vapour line. Therefore, for

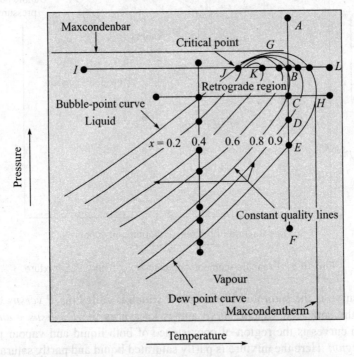

Fig. 10.6 Pressure–temperature diagram of binary VLE mixture.

a binary or multi-component mixture, the critical point can be defined as the point at which liquid and vapour phases cannot be distinguished.

Some important terms used in the graphical representation can be categorically defined as follows:

Maxcondenbar: It is defined as the maximum pressure on the phase boundary envelope of the pressure–temperature diagram of a VLE binary or multi-component mixture. It is represented by the point G in Fig. 10.6.

Maxcondentherm: It is the maximum temperature on the phase boundary envelope of the pressure–temperature diagram of a VLE binary or multi-component mixture. The point H represents the maxcondentherm point.

Retrograde condensation: It is defined as the condensation that takes place in the region bounded by the maxcondenbar, maxcondentherm and critical points.

Dew-point curve: It is the saturated vapour curve in the VLE diagram.

Bubble-point curve: It is the saturated liquid curve in the VLE diagram.

To explain the P–T diagram of a VLE mixture, consider a system in which the pressure is reduced at vapour phase, which is represented by the point A. On reduction of pressure, the condensation starts at point B. As the process proceeds along the line AF in Fig. 10.6, an appreciable amount of liquid is formed till the system reaches the point D. Then it starts getting converted into vapour and the last drop of liquid vaporizes at point E. At constant temperature, this process is known as *retrograde dew formation*. Now, if the temperature is increased gradually at point I, then the condensation starts at point J. The process continues to a certain extent, but then instead of liquid formation, vaporization starts and the process goes on till the system reaches point L, where the last drop of the liquid disappears through vaporization. The system undergoing the processes follows the path IL. This process also belongs to retrograde dew formation.

Now, if the system is considered to be at point L and if the temperature over the system is decreased isobarically, then the system will follow the path LI and there will be formation of bubble initially. On further reduction of temperature, the amount of vapour is increased; but as the process continues the amount goes on decreasing, and when the system reaches finally the point I towards the end, the last bubble disappears. This process is referred to as *retrograde bubble formation*.

10.7 VAPOUR–LIQUID EQUILIBRIUM: RAOULT'S LAW

In the preceding chapter, we have discussed the ideal solution and application of Raoult's law. Here, an attempt is made to substantiate the importance of Raoult's law for the binary vapour–liquid equilibrium solution comprising two components as well as multi-component systems elaborately.

A. For Liquid Phase Composition

Consider a binary solution consisting of two components 1 and 2. Then, according to Raoult's law

$$y_1 P = x_1 P_1^{Sat} \quad \text{for component 1}$$

or

$$y_1 = \frac{x_1 P_1^{Sat}}{P} \tag{10.36}$$

$$y_2 P = x_2 P_2^{Sat} \quad \text{for component 2}$$

or

$$y_2 = \frac{x_2 P_2^{Sat}}{P} \tag{10.37}$$

Since $\sum y_i = y_1 + y_2 = 1$, then after substitution of y_1 and y_2, we get

$$y_1 + y_2 = \frac{x_1 P_1^{Sat}}{P} + \frac{x_2 P_2^{Sat}}{P} = 1$$

or

$$P = x_1 P_1^{Sat} + x_2 P_2^{Sat} \tag{10.38}$$

At the given temperature, the saturated vapour pressures of components 1 and 2, i.e., P_1^{Sat} and P_2^{Sat} are determined by Antoine equations. Apart from binary systems, these equations are also applicable to multi-component systems.

B. For Vapour Phase Composition

In the same fashion, we can determine the composition of the vapour phase for a binary system with the help of Raoult's law.

We know that
for component 1

$$y_1 P = x_1 P_1^{Sat}$$

or

$$x_1 = \frac{y_1 P}{P_1^{Sat}} \tag{10.39}$$

for component 2

$$y_2 P = x_2 P_2^{Sat}$$

or

$$x_2 = \frac{y_2 P}{P_2^{Sat}} \tag{10.40}$$

Since $\sum x_i = x_1 + x_2 = 1$, then after substitution of x_1 and x_2, we get

$$x_1 + x_2 = \frac{y_1 P}{P_1^{Sat}} + \frac{y_2 P}{P_2^{Sat}} = 1$$

or

$$\frac{1}{P} = \frac{y_1}{P_1^{Sat}} + \frac{y_2}{P_2^{Sat}} \tag{10.41}$$

Here also, P_1^{Sat} and P_2^{Sat} are generally calculated with the help of Antoine equations. Apart from binary systems, these equations are also applicable to multi-component systems.

10.8 MODIFIED RAOULT'S LAW FOR VAPOUR–LIQUID EQUILIBRIUM

When the pressure is significantly low, the fugacity coefficient of the components in a liquid mixture is considered to be some sort of idealized pressure. That is, at low pressure, the saturated vapour pressure of the components in the solution is equal to the fugacity of the same. The mathematical expression developed by Raoult's law differs from the incorporation of the activity coefficient to the modified Raoult's law. Hence, the modified form of Raoult's law can be expressed as

$$y_i P = x_i \gamma_i P_i^{Sat} \tag{10.42}$$

A. For Liquid Phase Composition

For a binary solution consisting of two components 1 and 2, the modified Raoult's law is given by

for component 1

$$y_1 P = x_1 \gamma_1 P_1^{Sat}$$

or

$$y_1 = \frac{x_1 \gamma_1 P_1^{Sat}}{P} \tag{10.43}$$

for component 2

$$y_2 P = x_2 \gamma_2 P_2^{Sat}$$

or

$$y_2 = \frac{x_2 \gamma_2 P_2^{Sat}}{P} \tag{10.44}$$

Therefore

$$P = x_1 \gamma_1 P_1^{Sat} + x_2 \gamma_2 P_2^{Sat} \tag{10.45}$$

B. For Vapour Phase Composition

Similarly, we can determine the composition of the vapour phase for a binary system with the help of the modified Raoult's law.

We know that

for component 1

$$y_1 P = x_1 \gamma_1 P_1^{Sat}$$

or

$$x_1 = \frac{y_1 P}{\gamma_1 P_1^{Sat}} \tag{10.46}$$

for component 2

$$y_2 P = x_2 \gamma_2 P_2^{Sat}$$

or

$$x_2 = \frac{y_2 P}{\gamma_2 P_2^{Sat}} \tag{10.47}$$

Hence

$$\frac{1}{P} = \frac{y_1}{\gamma_1 P_1^{Sat}} + \frac{y_2}{\gamma_2 P_2^{Sat}} \tag{10.48}$$

EXAMPLE 10.2   Deduce the Gibbs–Duhem–Margules equation for a binary solution starting from Raoult's law.

Solution: For a binary vapour–liquid equilibrium mixture consisting of components 1 and 2, the Gibbs–Duhem equation can be written as

$$x_1 d\mu_1 + x_2 d\mu_2 = 0 \tag{i}$$

Now, if the chemical potential of any component is a function of temperature, pressure and mole fraction, then

$$\mu_i = \mu(T, P, x_i)$$

or

$$d\mu_i = \left(\frac{\partial \mu_i}{\partial T}\right)_{P, x_i} dT + \left(\frac{\partial \mu_i}{\partial P}\right)_{T, x_i} dP + \left(\frac{\partial \mu_i}{\partial x_i}\right)_{P, T} dx_i \tag{ii}$$

At constant temperature and pressure, Eq. (ii) becomes

$$d\mu_i = \left(\frac{\partial \mu_i}{\partial x_i}\right)_{P, T} dx_i$$

so

$$d\mu_1 = \left(\frac{\partial \mu_1}{\partial x_1}\right)_{P, T} dx_1 \quad \text{and} \quad d\mu_2 = \left(\frac{\partial \mu_2}{\partial x_2}\right)_{P, T} dx_2$$

Substituting the values of $d\mu_1$ and $d\mu_2$ into Eq. (i), we get

$$x_1\left(\frac{\partial \mu_1}{\partial x_1}\right)_{T, P} dx_1 + x_2\left(\frac{\partial \mu_2}{\partial x_2}\right)_{T, P} dx_2 = 0 \tag{iii}$$

The chemical potential of a component i is expressed in terms of fugacity as

$$\mu_i = \mu_i^0 + RT \ln f_i$$

On differentiation with respect to x_i at constant T and P, we get

$$\left(\frac{\partial \mu_i}{\partial x_i}\right)_{T, P} = RT\left(\frac{\partial \ln f_i}{\partial x_i}\right)_{T, P} \tag{iv}$$

Substituting this into Eq. (iii) for components 1 and 2, we get

$$RT\left[x_1\left(\frac{\partial \ln f_1}{\partial x_1}\right)dx_1 + x_2\left(\frac{\partial \ln f_2}{\partial x_2}\right)dx_2\right] = 0 \tag{v}$$

Since for a binary solution, $x_1 + x_2 = 1$ and consequently $dx_1 = -dx_2$, then

$$x_1\left(\frac{\partial \ln f_1}{\partial x_1}\right) - x_2\left(\frac{\partial \ln f_2}{\partial x_2}\right) = 0$$

or

$$\frac{d \ln f_1}{d \ln x_1} = \frac{d \ln f_2}{d \ln x_2}$$

or

$$\frac{x_1}{f_1} \frac{df_1}{dx_1} = \frac{x_2}{f_2} \frac{df_2}{dx_2} \qquad \text{(vi)}$$

This equation is known as the *Gibbs–Duhem–Margules equation*.

EXAMPLE 10.3 A binary liquid mixture consists of 60 mol per cent ethylene and 40 mol per cent propylene. At 423 K, the vapour pressure of ethylene and propylene are 15.2 atm and 9.8 atm respectively. Calculate the total pressure and equilibrium composition of the vapour phase. Assume that the mixture behaves like an ideal solution.

Solution: We know that for ideal solution

$$P = x_1 P_1^{Sat} + x_2 P_2^{Sat} \qquad \text{(i)}$$

Here

$$x_1 = 0.6$$
$$x_2 = 0.4$$
$$P_1^{Sat} = 15.2 \text{ atm}$$
$$P_2^{Sat} = 9.8 \text{ atm}$$

Substituting the values into Eq. (i), we get

$$P = x_1 P_1^{Sat} + x_2 P_2^{Sat} = 0.6 \times 15.2 + 0.4 \times 9.8$$
$$= 13.04 \text{ atm}$$

Hence, the total pressure is 13.04 atm.

Now, the composition of the mixture can be calculated as

$$y_1 = \frac{x_1 P_1^{Sat}}{P} = \frac{0.6 \times 15.2}{13.04} = 0.699 = 0.7$$

and

$$y_2 = \frac{x_2 P_2^{Sat}}{P} = \frac{0.4 \times 9.8}{13.04} = 0.3$$

EXAMPLE 10.4 The pure component vapour pressure of two organic liquids X and Y by Antoine equations are given by

$$\ln P_1^{Sat} = 14.35 - \frac{2942}{T + 220}$$

and

$$P_2^{Sat} = 14.25 - \frac{2960}{T + 210}$$

where P_1^{Sat} and P_2^{Sat} are in kPa and T in °C.

Calculate the composition of liquid and vapour in equilibrium at 77°C and 75 kPa.

Solution: We are given that

$$T = 77°C$$

Hence

$$\ln P_1^{Sat} = 14.35 - \frac{2942}{T + 220} = 14.35 - \frac{2942}{77 + 220} = 4.45$$

or

$$P_1^{Sat} = 85.62 \text{ kPa}$$

and

$$\ln P_2^{Sat} = 14.25 - \frac{2960}{T + 210} = 14.25 - \frac{2960}{77 + 210} = 3.94$$

or

$$P_2^{Sat} = 51.41 \text{ kPa}$$

We know that the total pressure

$$P = x_1 P_1^{Sat} + x_2 P_2^{Sat}$$

or

$$75 = 85.62 x_1 + 51.41 x_2$$

or

$$75 = 85.62 x_1 + 51.41(1 - x_1) = 51.41 + 34.21 x_1$$

or

$$x_1 = 0.69$$

Therefore

$$y_1 = \frac{x_1 P_1^{Sat}}{P} = \frac{0.69 \times 85.62}{75} = 0.787$$

EXAMPLE 10.5 The vapour pressure of acetone, acetonitrile and nitromethane can be represented by Antoine equations as

$$\ln P_1^{Sat} = 14.3916 - \frac{2795.82}{T + 230.0}$$

$$\ln P_2^{Sat} = 14.2724 - \frac{2945.47}{T + 224.0}$$

and

$$\ln P_3^{Sat} = 14.2043 - \frac{2972.64}{T + 209.0}$$

where P_1^{Sat}, P_2^{Sat} and P_3^{Sat} are in kPa and T is in °C.

Assuming that the system follows Raoult's law, calculate:
(a) P and y_1 at $T = 75°C$, $x_1 = 0.30$, $x_2 = 0.40$
(b) P and x_1 at $T = 80°C$, $y_1 = 0.45$, $y_2 = 0.35$.

Solution: (a) We are given that $T = 75°C$.
Hence

$$\ln P_1^{Sat} = 14.3916 - \frac{2795.82}{T + 230.0} = 14.3916 - \frac{2795.82}{75 + 230.0}$$

or

$$P_1^{Sat} = 185.86 \text{ kPa}$$

$$\ln P_2^{Sat} = 14.2724 - \frac{2945.47}{T + 224.0} = 14.2724 - \frac{2945.47}{75 + 224.0} = 4.421$$

or

$$P_2^{Sat} = 83.17 \text{ kPa}$$

and

$$\ln P_3^{Sat} = 14.2043 - \frac{2972.64}{T + 209.0} = 14.2043 - \frac{2972.64}{75 + 209.0} = 3.737$$

or

$$P_3^{Sat} = 41.97 \text{ kPa}$$

We are given that $x_1 = 0.30$ and $x_2 = 0.40$, so that $x_3 = 0.3$.
We know that the total pressure

$$P = x_1 P_1^{Sat} + x_2 P_2^{Sat} + x_3 P_3^{Sat}$$

or

$$P = 185.86 \times 0.3 + 83.17 \times 0.4 + 41.97 \times 0.3 = 101.61 \text{ kPa}$$

Therefore

$$y_1 = \frac{x_1 P_1^{Sat}}{P} = \frac{0.3 \times 185.86}{101.61} = 0.5487$$

$$y_2 = \frac{x_2 P_2^{Sat}}{P} = \frac{0.4 \times 83.17}{101.61} = 0.3274$$

and

$$y_3 = \frac{x_3 P_3^{Sat}}{P} = \frac{0.3 \times 41.97}{101.61} = 0.1239$$

(b) We are given that at $T = 80°C$, $y_1 = 0.45$, $y_2 = 0.35$ and hence $y_3 = 0.20$.

$$\ln P_1^{Sat} = 14.3916 - \frac{2795.82}{80 + 230.0} \qquad \text{or} \qquad P_1^{Sat} = 215.47 \text{ kPa}$$

$$\ln P_2^{Sat} = 14.2724 - \frac{2945.47}{80 + 224.0} \qquad \text{or} \qquad P_2^{Sat} = 97.84 \text{ kPa}$$

$$\ln P_3^{Sat} = 14.2043 - \frac{2972.64}{80 + 209.0} \qquad \text{or} \qquad P_3^{Sat} = 50.32 \text{ kPa}$$

At vapour phase, for component 1

$$y_1 P = x_1 P_1^{Sat}$$

or

$$x_1 = \frac{y_1 P}{P_1^{Sat}}$$

Similarly, for component 2

$$x_2 = \frac{y_2 P}{P_2^{Sat}}$$

and for component 3

$$x_3 = \frac{y_3 P}{P_3^{Sat}}$$

Since $x_1 + x_2 + x_3 = 1$, then

$$\frac{y_1 P}{P_1^{Sat}} + \frac{y_2 P}{P_2^{Sat}} + \frac{y_3 P}{P_3^{Sat}} = 1$$

or

$$P = \frac{1}{\dfrac{y_1}{P_1^{Sat}} + \dfrac{y_2}{P_2^{Sat}} + \dfrac{y_3}{P_3^{Sat}}} = \frac{1}{\dfrac{0.45}{215.47} + \dfrac{0.35}{97.84} + \dfrac{0.20}{50.32}} = 125 \text{ kPa}$$

$$x_1 = \frac{y_1 P}{P_1^{Sat}} = \frac{0.45 \times 125}{215.47} = 0.2610$$

$$x_2 = \frac{y_2 P}{P_2^{Sat}} = \frac{0.35 \times 125}{97.84} = 0.4471$$

and

$$x_3 = \frac{y_3 P}{P_3^{Sat}} = \frac{0.20 \times 125}{50.32} = 0.4968$$

EXAMPLE 10.6 For the system methanol (1)–methyl acetate (2), the activity coefficients for components 1 and 2 are represented by

$$\ln \gamma_1 = Ax_2^2 \quad \text{and} \quad \ln \gamma_2 = Ax_1^2$$

where
$$A = 2.771 - 0.00523T.$$

The vapour pressures of the components are given by the Antoine equation

$$\ln P_1^{Sat} = 16.5915 - \frac{3{,}643.31}{T - 33.424} \quad \text{and} \quad \ln P_2^{Sat} = 14.2532 - \frac{2{,}665.54}{T - 53.424}$$

where P is in kPa and T in K.

Calculate P and y_i at $T = 318.15$ K and $x_1 = 0.25$.

Solution: At $T = 318.15$ K, P_1^{Sat} and P_2^{Sat} are calculated using the Antoine equation as

$$P_1^{Sat} = 44.51 \text{ kPa} \quad \text{and} \quad P_2^{Sat} = 65.64 \text{ kPa}$$
$$A = 2.771 - 0.00523T = [2.771 - (0.00523)(318.15)] = 1.107$$

The activity coefficients are calculated for components 1 and 2 in the following way:

$$\ln \gamma_1 = Ax_2^2 = 1.107 \times (0.75)^2 = 0.6226 \quad \text{or} \quad \gamma_1 = 1.864$$

$$\ln \gamma_2 = Ax_1^2 = 1.107 \times (0.25)^2 = 0.0691 \quad \text{or} \quad \gamma_2 = 1.072$$

The total pressure is given by

$$P = x_1\gamma_1P_1^{Sat} + x_2\gamma_2P_2^{Sat} = (0.25 \times 1.864 \times 44.51) + (0.75 \times 1.072 \times 65.64) = 73.50 \text{ kPa}$$

The vapour phase composition is

$$y_1 = \frac{x_1\gamma_1P_1^{Sat}}{P} = \frac{0.25 \times 1.864 \times 44.51}{73.50} = 0.282$$

$$y_2 = \frac{x_2\gamma_2P_2^{Sat}}{P} = \frac{0.75 \times 1.072 \times 65.64}{73.50} = 0.718$$

10.9 DEVIATIONS FROM RAOULT'S LAW: NON-IDEAL SOLUTIONS

According to Raoult's law, the total vapour pressure of an ideal solution varies linearly with its molar composition. Another important criterion of an ideal liquid mixture is that when a component of the mixture is brought into contact with other components, mixing does not cause any change in the average intermolecular forces. Consequently, there is no change in internal energy or volume due to mixing. Therefore, there would be no absorption or evolution of heat. But in case of a non-ideal solution, there is change in internal energy on mixing of components and there will be either evolution or absorption of heat. These are two non-ideal behaviours of a mixture that can substantiate the deviations from Raoult's law. These deviations can usually be categorized into three types and are discussed in the following paragraphs:

Type I

The system in which the total vapour pressure is continuous and at any phase the system is intermediate between the vapour pressures of the two components. For instance, benzene–toluene, carbon tetrachloride–benzene, and various other systems show this sort of behaviour. For type I, the total pressure and partial vapour pressure are plotted against the composition of the mixture in Fig. 10.7. The dashed lines represent the behaviour of the pure component obeying Raoult's law and the continuous lines show the results of experimental measurements, i.e., the deviation from Raoult's law. In this diagram, the vapour phase composition and liquid phase composition curves are both continuous and the concentrations are different. The liquid and its vapour are in equilibrium with each other. The vapour is richer in more volatile components.

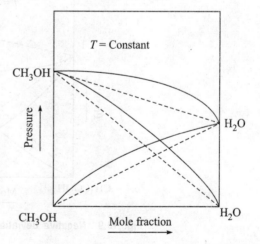

Fig. 10.7 Deviation of CH_3OH–H_2O system.

Type II

This is the system in which the deviations from Raoult's law are positive, but the vapour pressure rises to a maximum and then falls with change in composition. The vapour pressure exerted by the mixture at a given temperature and composition is higher than the estimated value using Raoult's law. The positive deviation from Raoult's law is shown in Fig. 10.8. The mixing of components is accompanied by absorption of heat, i.e., the enthalpy of mixing is positive. The pressure versus composition plot shows that the vapour phase composition curve touches the liquid composition curve at the maximum or minimum points.

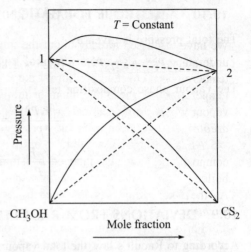

Fig. 10.8 Positive deviation of CH_3OH–CS_2 system.

Type III

This is the system in which the deviations from Raoult's law are negative, when the total vapour pressure exerted by the mixture at a given temperature and composition is lower than the calculated value obtained from using Raoult's law. The systems exhibiting such behaviour are found in toluene–acetic acid, acetone–chloroform, water–nitric acid, pyridine–acetic acid, etc. The negative deviation from Raoult's law is shown in Fig. 10.9. The mixing of components is accompanied by evolution of heat, i.e., the enthalpy of mixing is negative.

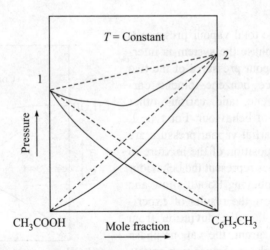

Fig. 10.9 Negative deviation of CH_3COOH–$C_6H_5CH_3$ system.

The diagrams shown in Figures 10.8 and 10.9 for types II and III respectively comprises two parts. One is between component 1 of the ideal solution and a solution of maximum or minimum vapour pressure, and the other is between component 2 of ideal solution and a solution of maximum or minimum vapour pressure.

10.10   AZEOTROPE FORMATION

We have discussed already about the importance of the boiling point diagram. It enables us to understand how the compositions of liquid and vapour change in equilibrium with respect to temperature at constant pressure. Like *P–x* diagram, *T–x–y* also shows a maximum or minimum temperature. At these maximum or minimum boiling points, the composition of the liquid and vapour phases are identical. Such a vapour–liquid mixture is called an azeotrope. The term *azeotrope* stems from a Greek word meaning *boiling without changing*.

An azeotrope is a mixture of two or more pure components in such a ratio that its composition cannot be changed by simple distillation. This is because when an azeotrope is boiled, the resulting vapour has the same ratio of constituents as the original mixture of liquids. As the composition is unchanged by boiling, azeotropes are also known as *constant boiling mixtures*.

For a binary mixture comprising components A and B, azeotropes are formed when there is a large positive deviation from ideality and the vapour pressures of the components A and B are not much different.

Each azeotrope has a characteristic boiling point. The boiling point of an azeotrope is either less than the boiling points of any of its constituents (a positive azeotrope), or greater than the boiling point of any of its constituents (a negative azeotrope).

Azeotropes consisting of two constituents, such as the two preceding examples, are called *binary* azeotropes. Those consisting of three constituents are called *ternary* azeotropes. Azeotropes of more than three constituents are also known.

10.10.1   Types of Azeotropes

Azeotropes can be categorized into two classes. These are:

- **Maximum-boiling Azeotropes**   If the deviation from ideal behaviour is negative and large, the curve in the plot of the total pressure against the liquid and vapour compositions at a constant temperature passes through a minimum at the azeotropic point. An azeotrope that exhibits such behaviour is known as a *maximum boiling azeotrope* (shown in Fig. 10.10) because the boiling point temperature at the azeotropic point is maximum if the total pressure is held constant.

    An example of maximum boiling azeotrope or negative azeotrope is nitric acid at a concentration of 20.2% hydrochloric acid and 79.8% water (by weight). HCl boils at $-84°C$ and water at $100°C$, but the azeotrope boils at $110°C$, which is higher than either of its constituents. Indeed, $110°C$ is the maximum temperature at which any hydrochloric acid solution can boil. It is generally true that a negative azeotrope boils at a higher temperature than for any other ratio of its constituents. Negative azeotropes are also called *maximum boiling mixtures*, i.e. they show negative deviation from Raoult's law ln $\gamma_i < 0$.

- **Minimum-boiling Azeotropes**   If the deviation from ideal behaviour is positive and sufficiently large, and the vapour pressures of the components are not much different, the curve in the plot of the total pressure against the liquid and vapour compositions at

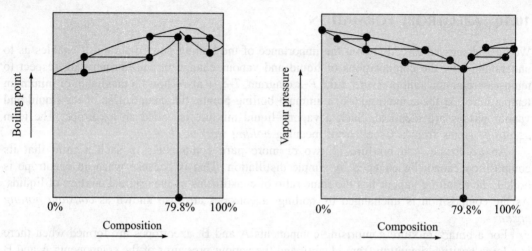

Fig. 10.10   (a) *T–x–y* diagram; (b) *P–x–y* diagram for maximum boiling azeotrope.

a constant temperature passes through a maximum at the azeotropic point. An azeotrope that shows such behaviour is known as *minimum boiling azeotrope*. It is depicted in Fig. 10.11.

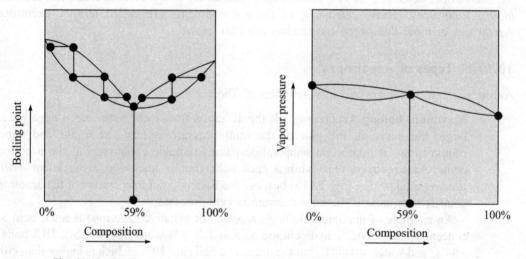

Fig. 10.11   (a) *T–x–y* diagram; (b) *P–x–y* diagram for minimum boiling azeotrope.

A well-known example of positive azeotrope is 59% pyridine and 41% water (by weight). Pyridine boils at 92°C and water boils at 100°C, but the azeotrope boils at 78.1°C, which is lower than either of its constituents. Indeed, 78.1°C is the minimum temperature at which any pyridine–water solution can boil. It is generally true that a positive azeotrope boils at a lower temperature than for any other ratio of its constituents. Positive azeotropes are also called *minimum boiling mixtures*, i.e., positive deviation from Raoult's law ln $\gamma_i > 0$.

EXAMPLE 10.7 For a binary system, the excess Gibbs free energy of components 1 and 2 at 30°C is given by $\dfrac{G^E}{RT} = 0.625x_1x_2$. The vapour pressures of components 1 and 2 are given by

$$\ln P_1^{\text{Sat}} = 13.71 - \frac{3800}{T} \quad \text{and} \quad \ln P_2^{\text{Sat}} = 14.01 - \frac{3800}{T}$$

where P_1^{Sat} and P_2^{Sat} are in bar and T is in K. Calculate the composition of the azeotrope.

Solution: We know that at the azeotropic point

$$P = \gamma_1 P_1^{\text{Sat}} = \gamma_2 P_2^{\text{Sat}}$$

or

$$\frac{\gamma_1}{\gamma_2} = \frac{P_2^{\text{Sat}}}{P_1^{\text{Sat}}}$$

Taking natural logarithm of both sides, we get

$$\ln \gamma_1 - \ln \gamma_2 = \ln P_2^{\text{Sat}} - \ln P_1^{\text{Sat}} = 14.01 - 13.71 = 0.3 \qquad \text{(i)}$$

Now, if $\dfrac{G^E}{RT} = Ax_1x_2$, then $\ln \gamma_1 = Ax_2^2$ and $\ln \gamma_2 = Ax_1^2$.

Comparing it with the given equation and substituting this condition into the equation, we get

$$Ax_2^2 - Ax_1^2 = A(x_2^2 - x_1^2) = 0.625(x_2^2 - x_1^2) = 0.3$$

or

$$0.625\,[(x_2 - x_1)\,(x_2 + x_1)] = 0.3 \qquad \text{(ii)}$$

Since for a binary solution, $x_1 + x_2 = 1$, Eq. (ii) becomes

$$0.625\,(x_2 - x_1) = 0.625(1 - 2x_1) = 0.3 \qquad (\text{as } x_2 = 1 - x_1)$$

or

$$(1 - 2x_1) = \frac{0.3}{0.625} = 0.48$$

or

$$x_1 = 0.26$$

Therefore, the azeotropic composition at 30°C is $x_1 = 0.26$.

10.11 VAPOUR–LIQUID EQUILIBRIUM AT LOW PRESSURE: EXCESS GIBBS FREE ENERGY MODEL

The fugacity coefficient of a component in a liquid mixture can be determined by using the cubic equation of state and applying the concept of residual properties. Similarly, excess properties, another important concept, can be applied to serve the purpose. In this section, various excess Gibbs free energy models for binary solutions have been discussed. The advantage in using the models is that at low to moderate pressure, the excess Gibbs free energy is not a function of pressure, while in general it is a function of temperature, pressure and composition. With the help of these models the activity coefficient can be determined, which in turn helps to find out the fugacity as well as the fugacity coefficient of the component in the solution.

It is necessary to remember that the activity coefficient γ of a pure component is equal to unity.

We know that for a binary solution

$$\frac{G^E}{RT} = g(x_1, x_2)$$

When $x_1 = 1$ and $x_2 = 0$, then the expression becomes

$$\frac{G^E}{RT} = 0 \quad \text{and} \quad G^E = 0$$

Again, when $x_2 = 1$ and $x_1 = 0$, we get $G^E = 0$.

10.11.1   Wohl's Equation

For a binary liquid solution, taking into consideration the two-body and three-body interactions between the molecules, the equation for expressing the excess free energy was developed by Wohl in terms of composition, effective molal volume, and effective volumetric fraction of the separate components. Wohl's three-suffix equation is given by

$$\frac{G^E}{RT(q_1 x_1 + q_2 x_2)} = 2a_{12}z_1z_2 + 3a_{112}z_1^2 z_2 + 3a_{122}z_1 z_2^2 \tag{10.49}$$

where

$\quad\quad x_1, x_2$ = Mole fractions of components 1 and 2;
$\quad\quad q_1, q_2$ = Effective molal volume;
$\quad\quad z_1, z_2$ = Effective volume fraction;
a_{12}, a_{112}, a_{122} = Empirical constants.

The first term on the right-hand side of Eq. (10.49) represents the contribution to the excess free energy due to the two-body interactions of molecules of unlike components, and it is designated as the *two-suffix term*. Similarly, the second and third terms represent the contribution due to the three-body interactions of molecules of unlike components and are designated as *three-suffix terms*.

Since $z_1 + z_2 = 1$, then on substitution into (10.49), it yields

$$\frac{G^E}{RT} = \left(x_1 + \frac{q_2}{q_1} x_2 \right) z_1 z_2 [z_1 q_1 (2a_{12} + 3a_{112}) + z_2 q_1 (2a_{12} + 3a_{112})] \tag{10.50}$$

or

$$\frac{G^E}{RT} = \left(x_1 + \frac{q_2}{q_1} x_2 \right) z_1 z_2 \left(z_1 + \frac{q_1}{q_2} B + z_2 A \right) \tag{10.51}$$

where $\quad\quad A = q_1(2a_{12} + 3a_{122})$ and $B = q_2(2a_{12} + 3a_{112})$.

The activity coefficients can be expressed from Eq. (10.50) as

$$\ln \gamma_1 = z_2^2 \left[A + 2 \left(B \frac{q_1}{q_2} - A \right) z_1 \right] \tag{10.52}$$

$$\ln \gamma_2 = z_1^2 \left[B + 2 \left(A \frac{q_2}{q_1} - B \right) z_2 \right]$$ (10.53)

Equations (10.52) and (10.53) are known as *three-suffix Wohl's equations*. They involve three constants as well as adjustable parameters A, B, and $\frac{q_1}{q_2}$, which are characteristic of each binary system.

Solving for z_1 and z_2, we get

$$z_1 = \frac{x_1}{x_1 + x_2 \left(\dfrac{q_2}{q_1} \right)} \qquad \text{and} \qquad z_2 = \frac{x_2}{x_2 + x_1 \left(\dfrac{q_1}{q_2} \right)}$$

10.11.2 Margules Equation

A. Two Parameters (Three-Suffix)

When $\frac{q_1}{q_2} = 1$, Eqs. (10.52) and (10.53) can be reduced to

$$\ln \gamma_1 = x_2^2[A + 2(B - A)x_1] = (2B - A)x_2^2 + 2(A - B)x_2^3$$ (10.54)
$$\ln \gamma_2 = x_1^2[B + 2(A - B)x_2] = (2A - B)x_1^2 + 2(B - A)x_1^3$$ (10.55)

Equations (10.54) and (10.55) are known as *three-suffix Margules equations*.

The VLE data of binary systems like acetone–chloroform, and chloroform–methanol can be better represented by three-suffix Margules equations.

B. One Parameter (Two-Suffix)

It is observed that in the preceding equations, constant $A = \ln \gamma_1$ at $x_1 = 0$ and $B = \ln \gamma_2$ at $x_2 = 0$.

Now, if $A = B$ in Eqs. (10.54) and (10.55), the simplest form of Margules equations can be obtained as

$$\ln \gamma_1 = Ax_2^2$$ (10.56)
$$\ln \gamma_2 = Ax_1^2$$ (10.57)

Equations (10.56) and (10.57) are known as *two-suffix Margules equations*.

EXAMPLE 10.8 For a liquid mixture following the two-parameter three-suffix Margules equation, the activity coefficients are represented by

$$\ln \gamma_1 = x_2^2[A + 2(B - A)x_1]$$

and

$$\ln \gamma_2 = x_1^2[A + 2(A - B)x_2]$$

Determine the expression for the excess Gibbs free energy of the liquid mixture.

Solution: From the knowledge and expression of excess free energy, we may write

$$\frac{G^E}{RT} = x_1 \ln \gamma_1 + x_2 \ln \gamma_2$$

Now, putting the given expressions into the preceding equation, we get

$$\frac{G^E}{RT} = x_1 x_2^2 [A + 2(B - A)x_1] + x_2 x_1^2 [B + 2(A - B)x_2]$$

or

$$\frac{G^E}{RTx_1x_2} = x_2 [A + 2(B - A)x_1] + x_1 [B + 2(A - B)x_2]$$

or

$$\frac{G^E}{RTx_1x_2} = Ax_2 + 2(B - A)x_1x_2 + Bx_1 + 2(A - B)x_2x_1$$

or

$$\frac{G^E}{RTx_1x_2} = Ax_2 + Bx_1$$

10.11.3   van Laar Equation

When $\dfrac{q_1}{q_2} = \dfrac{A}{B}$, Eqs. (10.52) and (10.53) reduce to the form of the two-suffix equations developed by van Laar and are given by

$$\ln \gamma_1 = Az_2^2 = \frac{Ax_2^2}{\left[\left(\dfrac{A}{B}\right)x_1 + x_2\right]^2} = \frac{A}{\left(1 + \dfrac{Ax_1}{Bx_2}\right)^2} = A\left(1 + \frac{Ax_1}{Bx_2}\right)^{-2} \qquad (10.58)$$

$$\ln \gamma_2 = Bz_1^2 = \frac{Bx_1^2}{\left[x_1 + \left(\dfrac{B}{A}\right)x_2\right]^2} = \frac{B}{\left(1 + \dfrac{Bx_2}{Ax_1}\right)^2} = B\left(1 + \frac{Bx_2}{Ax_1}\right)^{-2} \qquad (10.59)$$

where A and B are the adjustable parameters which can be determined from the experimental values of $\ln \gamma_1$ and $\ln \gamma_2$ as

$$A = \ln \gamma_1 \left(1 + \frac{x_2 \ln \gamma_2}{x_1 \ln \gamma_1}\right)^2 \qquad (10.60)$$

$$B = \ln \gamma_2 \left(1 + \frac{x_1 \ln \gamma_1}{x_2 \ln \gamma_2}\right)^2 \qquad (10.61)$$

It may be noted that for both the Margules and the van Laar equations, A is the terminal value of $\ln \gamma_1$ at $x_1 = 0$ and B is the terminal value of $\ln \gamma_2$ at $x_2 = 0$.

van Laar equation is a simple form containing only two adjustable parameters. Therefore, it is widely used to determine the activity coefficient for the system comprising relatively simple and preferably non-polar liquid systems such as benzene–iso-octane and n-propanol–water systems.

EXAMPLE 10.9 An azeotrope consists of 42.0 mol per cent acetone (1) and 58.0 mol per cent methanol at 760 mm Hg and 313 K. At 313 K, the vapour pressure of acetone and methanol are 786 mm Hg and 551 mm Hg respectively. Calculate the van Laar constants.

Solution: At given mole fractions, the activity coefficients of acetone and methanol can be estimated as
for component 1 (acetone)

$$\gamma_1 = \frac{P}{P_1^{Sat}} = \frac{760}{786} = 0.966$$

and
for component 2 (methanol)

$$\gamma_2 = \frac{P}{P_2^{Sat}} = \frac{760}{551} = 1.379$$

We are given that $x_1 = 0.42$ and $x_2 = 0.58$. Then substituting these values, we get

$$A = \ln \gamma_1 \left(1 + \frac{x_2 \ln \gamma_2}{x_1 \ln \gamma_1} \right)^2 = \ln 0.966 \left(1 + \frac{0.58 \ln 1.379}{0.42 \ln 0.966} \right)^2 = -4.838$$

$$B = \ln \gamma_2 \left(1 + \frac{x_1 \ln \gamma_1}{x_2 \ln \gamma_2} \right)^2 = \ln 1.379 \left(1 + \frac{0.42 \ln 0.966}{0.58 \ln 1.379} \right)^2 = 0.274$$

EXAMPLE 10.10 The azeotrope of the benzene–cyclohexane system has a composition of 53.2 mol per cent benzene with a boiling point of 350.6 K at 101.3 kPa. At this temperature, the vapour pressure of benzene (1) is 100.59 kPa and the vapour pressure of cyclohexane (2) is 99.27 kPa. Determine the activity coefficients for the solution containing 10 mol per cent benzene.

Solution: We are given that $P = 101.3$ KPa. The activity coefficients of acetone and methanol can be estimated as
for component 1 (benzene)

$$\gamma_1 = \frac{P}{P_1^{Sat}} = \frac{101.3}{100.59} = 1.007$$

and
for component 2 (cyclohexane)

$$\gamma_2 = \frac{P}{P_2^{Sat}} = \frac{101.3}{99.27} = 1.020$$

Now, $x_1 = 0.532$ and $x_2 = 0.468$. Substituting these values into the expression of the van Laar equation for adjustable parameters, we get

$$A = \ln \gamma_1 \left(1 + \frac{x_2 \ln \gamma_2}{x_1 \ln \gamma_1} \right)^2 = \ln 1.007 \left(1 + \frac{0.468 \ln 1.02}{0.532 \ln 1.007} \right)^2 = 0.111$$

and

$$B = \ln \gamma_2 \left(1 + \frac{x_1 \ln \gamma_1}{x_2 \ln \gamma_2} \right)^2 = \ln 1.02 \left(1 + \frac{0.532 \ln 1.007}{0.468 \ln 1.02} \right)^2 = 0.0352$$

For the solution containing 10 mol per cent benzene, $x_1 = 0.10$ and $x_2 = 0.90$. The activity coefficients can be computed as

$$\ln \gamma_1 = \frac{A}{\left(1 + \dfrac{Ax_1}{Bx_2}\right)^2} = \frac{0.111}{\left(1 + \dfrac{0.111 \times 0.10}{0.0352 \times 0.90}\right)} = 0.0609$$

or

$$\gamma_1 = 1.062$$

$$\ln \gamma_2 = \frac{B}{\left(1 + \dfrac{Bx_2}{Ax_1}\right)^2} = \frac{0.0352}{\left(1 + \dfrac{0.0352 \times 0.9}{0.111 \times 0.1}\right)^2} = 0.0023$$

or

$$\gamma_2 = 1.002$$

EXAMPLE 10.11 Under atmospheric condition, the acetone–chloroform azeotrope boils at 64.6°C and contains 33.5 mol per cent acetone. The vapour pressures of acetone and chloroform at this temperature are 995 mm Hg and 885 mm Hg respectively. Calculate the composition of the vapour at this temperature in equilibrium with a liquid analyzing 11.1 mol per cent acetone. Apply the van Laar equation of the following form:

$$\ln \gamma_1 = \frac{Ax_2^2}{\left(\dfrac{A}{B}x_1 + x_2\right)^2} \quad \text{and} \quad \ln \gamma_2 = \frac{Bx_1^2}{\left(x_1 + \dfrac{B}{A}x_2\right)^2}$$

(WBUT, 2004)

Solution: Under atmospheric pressure, i.e., $P = 760$ mm Hg, the activity coefficients are for acetone

$$\gamma_1 = \frac{P}{P_1^{\text{Sat}}} = \frac{760}{995} = 0.763$$

and
for chloroform

$$\gamma_2 = \frac{P}{P_2^{\text{Sat}}} = \frac{760}{885} = 0.858$$

We are given that $x_1 = 0.335$ and $x_2 = 0.665$. Then substituting these values into the expression of the van Laar equation for adjustable parameters, we get

$$A = \ln \gamma_1 \left(1 + \frac{x_2 \ln \gamma_2}{x_1 \ln \gamma_1}\right)^2 = \ln 0.763 \left(1 + \frac{0.665 \ln 0.858}{0.335 \ln 0.763}\right)^2 = -1.218$$

and

$$B = \ln \gamma_2 \left(1 + \frac{x_1 \ln \gamma_1}{x_2 \ln \gamma_2}\right)^2 = \ln 0.858 \left(1 + \frac{0.335 \ln 0.763}{0.665 \ln 0.858}\right)^2 = -0.5464$$

For the solution which contains 11.1 mol per cent acetone, $x_1 = 0.111$ and $x_2 = 0.889$. The activity coefficients here can be estimated as

$$\ln \gamma_1 = \frac{Ax_2^2}{\left(\dfrac{A}{B}x_1 + x_2\right)^2} = \frac{(-1.218)(0.889)^2}{\left[\left(\dfrac{-1.218}{-0.5464}\right) \times 0.111 + 0.889\right]^2} = -0.7462$$

or

$$\gamma_1 = 0.4741$$

$$\ln \gamma_2 = \frac{Bx_1^2}{\left(x_1 + \dfrac{B}{A}x_2\right)^2} = \frac{(-0.5464)(0.111)^2}{\left[0.111 + \left(\dfrac{-0.5464}{-1.218}\right) \times 0.889\right]^2} = -0.0257$$

or

$$\gamma_2 = 0.9746$$

The composition of the vapour at 64.6°C can be calculated as

$$y_1 = \frac{1}{1 + \dfrac{\gamma_2 x_2 P_2^{Sat}}{\gamma_1 x_1 P_1^{Sat}}} = \frac{1}{1 + \dfrac{0.9746 \times 0.889 \times 885}{0.4741 \times 0.111 \times 995}} = 0.0639$$

$$y_2 = 1 - 0.0639 = 0.9361$$

At equilibrium, the composition vapour with a liquid analyzing 11.1 mol per cent acetone contains acetone (6.39%) and chloroform (93.61%).

EXAMPLE 10.12 The activity coefficients in a binary system are represented by $\ln \gamma_1 = Ax_2^2$ and $\ln \gamma_2 = Ax_1^2$. Show that if the system forms an azeotrope, the azeotropic composition is given by $x_1 = \dfrac{1}{2}\left(1 + \dfrac{1}{A} \ln \dfrac{P_1^{Sat}}{P_2^{Sat}}\right)$

Solution: We are given that
activity coefficient for component 1

$$\ln \gamma_1 = Ax_2^2$$

and
activity coefficient for component 2

$$\ln \gamma_2 = Ax_1^2$$

We know that

$$y_i P = \gamma_i x_i P_i^{Sat} \tag{i}$$

At azeotropic point, i.e., the point at which the azeotrope forms and the liquid and vapour phase compositions are the same, $y_i = x_i$. Thus, Eq. (i) reduces to

$$P = \gamma_i P_i^{Sat}$$

Now, the activity coefficients for components 1 and 2 at azeotropic point can be expressed as

$$P = \gamma_1 P_1^{Sat} = \gamma_2 P_2^{Sat}$$

or

$$\frac{\gamma_1}{\gamma_2} = \frac{P_2^{Sat}}{P_1^{Sat}}$$

Taking natural logarithm of both sides, we get

$$\ln \gamma_1 - \ln \gamma_2 = \ln P_2^{Sat} - \ln P_1^{Sat} \tag{ii}$$

Substituting the given conditions yields

$$Ax_2^2 - Ax_1^2 = \ln P_2^{Sat} - \ln P_1^{Sat}$$

or

$$A[(x_2 - x_1)(x_2 + x_1)] = \ln P_2^{Sat} - \ln P_1^{Sat} \tag{iii}$$

Since for a binary solution, $x_1 + x_2 = 1$, Eq. (iii) becomes

$$A(x_2 - x_1) = A(1 - 2x_1) = \ln P_2^{Sat} - \ln P_1^{Sat} \quad \text{(as } x_2 = 1 - x_1\text{)}$$

or

$$A(2x_1 - 1) = \ln P_1^{Sat} - \ln P_2^{Sat} = \ln \frac{P_1^{Sat}}{P_2^{Sat}}$$

or

$$2x_1 - 1 = \frac{1}{A} \ln \frac{P_1^{Sat}}{P_2^{Sat}}$$

or

$$x_1 = \frac{1}{2} \left(1 + \frac{1}{A} \ln \frac{P_1^{Sat}}{P_2^{Sat}} \right)$$

Therefore, the azeotropic composition is shown to be

$$x_1 = \frac{1}{2} \left(1 + \frac{1}{A} \ln \frac{P_1^{Sat}}{P_2^{Sat}} \right)$$

10.11.4 Local Composition Model: Wilson and NRTL Equations

The concept of local composition was introduced by G.M. Wilson in 1964, with the development of a model in which it was proposed that within a liquid solution the local composition is different from the overall mixture composition or bulk composition. The local composition is incorporated to the model to account for the nonrandom molecular orientations resulting from differences in molecular size and intermolecular forces.

Consider the liquid structure with molecular orientation for two cases. For case 1, the liquid structure has a central molecule of type 1 and case 2 has a central molecule of type 2, where x_{11} is the number of molecules of type 1 in the vicinity of 1 divided by the total number of surrounding molecules, x_{21} is the number of molecules of type 2 in the vicinity of 2 divided by the total number of surrounding molecules, x_{12} is the number of molecules of type 1 in the vicinity of 2 divided by the total number of surrounding molecules, and x_{22} is the number of molecules of type 2 in the vicinity of 2 divided by the total number of surrounding molecules.

For a binary solution, the ratio of local compositions around molecule 1 is given by

$$\frac{x_{11}}{x_{21}} = \frac{x_1 \exp\left(-\dfrac{G_{11}}{RT}\right)}{x_2 \exp\left(-\dfrac{G_{21}}{RT}\right)} \qquad \text{in which } x_{11} + x_{21} = 1 \qquad (10.62)$$

Around molecule 2, the corresponding ratio is given by

$$\frac{x_{12}}{x_{22}} = \frac{x_1 \exp\left(-\dfrac{G_{12}}{RT}\right)}{x_2 \exp\left(-\dfrac{G_{22}}{RT}\right)} \qquad \text{in which } x_{12} + x_{22} = 1 \qquad (10.63)$$

In Fig. 10.12, the bulk compositions for the two types are the same ($x_1 = x_2 = 0.5$). But $x_{11} = \dfrac{2}{6}$, $x_{21} = \dfrac{4}{6}$, $x_{12} = \dfrac{4}{6}$ and $x_{22} = \dfrac{2}{6}$. It does necessarily mean that molecule 1 is surrounded by more of 2 and molecule 2 is surrounded by more of 1. Hence, the local compositions of 1 and 2 are different, but the bulk composition in this case are the same. For x_{11} and x_{21}, the molecule at the centre is 1, and for x_{12} and x_{22} the molecule at the centre is 2.

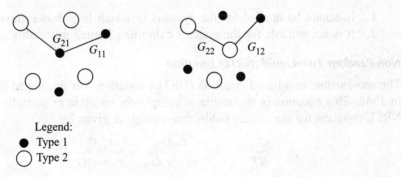

Legend:
● Type 1
○ Type 2

Fig. 10.12 Liquid structure showing interaction energies.

The local composition model necessitates the development of a few equations such as the Wilson equation and the NRTL equation. These are discussed in the following paragraphs.

Wilson Equation

The equation for the excess Gibbs free energy of a binary solution was also developed by Wilson, and that can be expressed as

$$\frac{G^{\mathrm{E}}}{RT} = -x_1 \ln (x_1 + x_2\Lambda_{12}) - x_2 \ln (x_2 + x_1\Lambda_{21}) \qquad (10.64)$$

where Λ_{12} and Λ_{21} are two adjustable parameters, known as the *Wilson parameters*. These are related to pure component molar volumes and characteristic energy differences, and are given by

$$\Lambda_{12} = \frac{V_2}{V_1} \exp\left(-\frac{\lambda_{12} - \lambda_{11}}{RT}\right) = \frac{V_2}{V_1} \exp\left(-\frac{a_{12}}{RT}\right) \tag{10.65}$$

$$\Lambda_{21} = \frac{V_1}{V_2} \exp\left(-\frac{\lambda_{12} - \lambda_{22}}{RT}\right) = \frac{V_1}{V_2} \exp\left(-\frac{a_{21}}{RT}\right) \tag{10.66}$$

where, V_1 and V_2 are molar volumes of components 1 and 2 at temperature T, λ's are the energies of interaction between the molecules of components 1 and 2, and a_{ij} is a constant for a given pair of components independent of temperature and composition.

The activity coefficients derived from this excess Gibbs free energy equation are given by

$$\ln \gamma_1 = -\ln (x_1 + \Lambda_{12} x_2) + x_2 \left(\frac{\Lambda_{12}}{x_1 + \Lambda_{12} x_2} - \frac{\Lambda_{21}}{\Lambda_{21} x_1 + x_2} \right) \tag{10.67}$$

$$\ln \gamma_2 = -\ln (x_2 + \Lambda_{21} x_1) + x_1 \left(\frac{\Lambda_{12}}{x_1 + \Lambda_{12} x_2} - \frac{\Lambda_{21}}{\Lambda_{21} x_1 + x_2} \right) \tag{10.68}$$

The Wilson equation is found to be suitable for application to several completely miscible binary solutions and to provide a better representation of VLE of this sort of solution mixture. It is also preferable for solutions of polar components in non-polar solvents.

Advantage: Wilson equation predicts the influence of temperature on activity coefficient.

Disadvantages: The equation has two main disadvantages. These are:

1. It cannot be applied for the systems in which $\ln \gamma$ shows maxima or minima.
2. It is not suitable for the systems exhibiting limited miscibility.

Non-Random Two-Liquid (NRTL) Equation

The non-random two-liquid equation (NRTL) equation was developed by Renon and Prausnitz in 1968. This equation is applicable to completely miscible or partially miscible systems. The NRTL equation for the excess Gibbs free energy is given by

$$\frac{G^{E}}{RT} = x_1 x_2 \left(\frac{\tau_{21} G_{21}}{x_1 + x_2 G_{21}} + \frac{\tau_{12} G_{12}}{x_2 + x_1 G_{12}} \right) \tag{10.69}$$

where

$$\tau_{12} = \frac{G_{12} - G_{22}}{RT} = \frac{b_{12}}{RT} \quad \text{and} \quad \tau_{21} = \frac{G_{21} - G_{11}}{RT} = \frac{b_{21}}{RT}$$

and

$$G_{12} = \exp(-\alpha_{12} \tau_{12}) \quad \text{and} \quad G_{21} = \exp(-\alpha_{12} \tau_{21})$$

Here α_{12} is incorporated to the equation for considering the randomness of the solution. If $\alpha_{12} = 0$, then the mixture is completely random and it reduces to the Margules equation. The value of α_{12} varies from 0.20 to 0.57.

The activity coefficients derived from this excess Gibbs free energy equation are given by

$$\ln \gamma_1 = x_2^2 \left[\tau_{21} \left(\frac{G_{21}}{x_1 + x_2 G_{21}} \right)^2 + \frac{\tau_{12} G_{12}}{(x_2 + x_1 G_{12})^2} \right] \tag{10.70}$$

$$\ln \gamma_2 = x_1^2 \left[\tau_{12} \left(\frac{G_{12}}{x_2 + x_1 G_{12}} \right)^2 + \frac{\tau_{21} G_{21}}{(x_1 + x_2 G_{21})^2} \right] \qquad (10.71)$$

The NRTL equation is not advantageous to that extent over the van Laar and Margules equations. For strongly non-ideal solutions and partially miscible solutions, the equation provides a good representation.

EXAMPLE 10.13   Using the Wilson and NRTL equations, estimate the activity coefficients of the components of a binary system consisting of iso-butanol (1) and iso-propanol (2) at 50°C and $x_1 = 0.3$. At this temperature, the molar volumes of the components are $V_1 = 65.2$ cm^3/mol and $V_2 = 15.34$ cm^3/mol.
Data given:
for Wilson equation

$$a_{12} = 300.55 \text{ cal/mol}$$
$$a_{21} = 1520.32 \text{ cal/mol}$$

for NRTL equation

$$b_{12} = 685.21 \text{ cal/mol}$$
$$b_{21} = 1210.21 \text{ cal/mol}$$
$$\alpha = 0.552$$

Solution: The required steps are as follows:
A. Estimation of activity coefficients using Wilson equation:
The Wilson parameters are given by

$$\Lambda_{12} = \frac{V_2}{V_1} \exp\left(-\frac{\lambda_{12} - \lambda_{11}}{RT} \right) = \frac{V_2}{V_1} \exp\left(-\frac{a_{12}}{RT} \right) \qquad (i)$$

$$\Lambda_{21} = \frac{V_1}{V_2} \exp\left(-\frac{\lambda_{12} - \lambda_{22}}{RT} \right) = \frac{V_1}{V_2} \exp\left(-\frac{a_{21}}{RT} \right) \qquad (ii)$$

Substituting the values, we get

$$\Lambda_{12} = \frac{V_2}{V_1} \exp\left(-\frac{a_{12}}{RT} \right) = \frac{15.34}{65.2} \exp\left(-\frac{300.55}{2 \times 323} \right) = 0.147$$

$$\Lambda_{21} = \frac{V_1}{V_2} \exp\left(-\frac{a_{21}}{RT} \right) = \frac{65.2}{15.34} \exp\left(-\frac{1520.32}{2 \times 323} \right) = 0.403$$

On substitution of two adjustable parameters, the activity coefficients can be estimated as

$$\ln \gamma_1 = -\ln(x_1 + \Lambda_{12} x_2) + x_2 \left(\frac{\Lambda_{12}}{x_1 + \Lambda_{12} x_2} - \frac{\Lambda_{21}}{\Lambda_{21} x_1 + x_2} \right)$$

$$= -\ln[0.3 + (0.147 \times 0.7)] + 0.7 \left[\frac{0.147}{0.3 + (0.147 \times 0.7)} - \frac{0.403}{(0.403 \times 0.3) + 0.7} \right] = 0.8208$$

$$\gamma_1 = 2.27$$

$$\ln \gamma_2 = -\ln(x_2 + \Lambda_{21}x_1) - x_1\left(\frac{\Lambda_{12}}{x_1 + \Lambda_{12}x_2} - \frac{\Lambda_{21}}{\Lambda_{21}x_1 + x_2}\right)$$

$$= -\ln[0.7 + (0.403 \times 0.3)] - 0.3\left[\frac{0.147}{0.3 + (0.147 \times 0.7)} - \frac{0.403}{0.7 + (0.403 \times 0.3)}\right] = 0.223$$

$$\gamma_2 = 1.25$$

B. Estimation of activity coefficients using NRTL equation:

$$\ln \gamma_1 = x_2^2\left[\tau_{21}\left(\frac{G_{21}}{x_1 + x_2 G_{21}}\right)^2 + \frac{\tau_{12}G_{12}}{(x_2 + x_1 G_{12})^2}\right]$$

$$\ln \gamma_2 = x_1^2\left[\tau_{12}\left(\frac{G_{12}}{x_2 + x_1 G_{12}}\right)^2 + \frac{\tau_{21}G_{21}}{(x_1 + x_2 G_{21})^2}\right]$$

Let us find out the values of the different parameters.

$$\tau_{12} = \frac{G_{12} - G_{22}}{RT} = \frac{b_{12}}{RT} = \frac{685.21}{2 \times 323} = 1.06$$

$$\tau_{21} = \frac{G_{21} - G_{11}}{RT} = \frac{b_{21}}{RT} = \frac{1210.21}{2 \times 323} = 1.87$$

Hence

$$G_{12} = \exp(-\alpha_{12}\tau_{12}) = \exp[-(0.552)(1.06)] = 0.557$$
$$G_{21} = \exp(-\alpha_{12}\tau_{21}) = \exp[-(0.552)(1.87)] = 0.357$$

Substituting the values, we get

$$\ln \gamma_1 = (0.7)^2\left[1.87\left\{\frac{0.357}{0.3 + (0.7 \times 0.357)}\right\}^2 + \frac{1.06 \times 0.557}{\{0.7 + (0.3 \times 0.557)\}^2}\right] = 0.772$$

$$\gamma_1 = 2.164$$

$$\ln \gamma_2 = (0.3)^2\left[1.06\left\{\frac{0.557}{0.7 + (0.37 \times 0.557)}\right\}^2 + \frac{1.87 \times 0.357}{\{0.3 + (0.7 \times 0.357\}^2}\right] = 0.234$$

$$\gamma_2 = 1.263$$

Therefore, the activity coefficients are $\gamma_1 = 2.164$ and $\gamma_2 = 1.263$.

10.11.5 Universal Quasi-Chemical Equation (UNIQUAC)

In order to express the excess Gibbs free energy of a binary solution, the UNIQUAC model was developed by Abrams and Prausnitz in 1975. The equation consists of two parts, namely the combinatorial part and the residual part. The combinatorial part is concerned with the composition, size and shape of the constituent molecules. This part comprises only the pure component properties. The residual part is concerned with the intermolecular forces and it involves two adjustable parameters.

The UNIQUAC equation, substituting $\dfrac{G^E}{RT} = G$, is given by

$$G = G^C \text{ (combinatorial)} + G^R \text{ (residual)}$$

$$G = \underbrace{\left[x_1 \ln \frac{\phi_1}{x_1} + x_2 \ln \frac{\phi_2}{x_2} + \frac{z}{2} \left(q_1 x_1 \ln \frac{\theta_1}{\phi_1} + q_2 x_2 \ln \frac{\theta_2}{\phi_2} \right) \right]}_{\text{Combinatorial part}}$$

$$+ \underbrace{[-q_1 x_1 \ln (\theta_1 + \theta_2 \tau_{21}) - q_2 x_2 \ln (\theta_2 + \theta_1 \tau_{12})]}_{\text{Residual part}} \qquad (10.72)$$

where

θ_i = Area fraction of component $i = \dfrac{x_i q_i}{\displaystyle\sum_j x_j q_j}$

ϕ_i = Volume fraction of component $i = \dfrac{x_i r_i}{\displaystyle\sum_j x_j r_j}$

q_i = Surface area parameter of component i

τ_{ji} = Adjustable parameter $= \exp\left[-\dfrac{(u_{ji} - u_{ii})}{RT} \right] = \exp\left(-\dfrac{a_{ji}}{T} \right)$

u_{ij} = Average interaction energy for the interaction of molecules of component i with the molecules of component j

z = Coordination number (usually taken as 10).

The activity coefficients derived from this excess Gibbs free energy equation are given by

$$\ln \gamma_i = \ln \gamma_i^C \text{ (combinatorial)} + \ln \gamma_i^R \text{ (residual)} \qquad (10.73)$$

$$\ln \gamma_i^C = \ln \frac{\phi_i}{x_i} + \frac{z}{2} q_i \ln \frac{\theta_i}{\phi_i} + l_i - \frac{\phi_i}{x_i} \sum_j x_j l_j \qquad (10.74)$$

$$\ln \gamma_i^R = q_i \left[1 - \ln \left(\sum_j \theta_j \tau_{ji} \right) - \sum_j \frac{\theta_j \psi_{ij}}{\displaystyle\sum_k \theta_k \psi_{kj}} \right] \qquad (10.75)$$

where

$$l_i = \frac{z}{2} (r_i - q_i) - (r_i - 1). \qquad (10.76)$$

In Eq. (10.76), r_i and q_i can be estimated by using the following relations:

$$r_i = \sum_k v_k R_k \qquad (10.77)$$

$$q_i = \sum_k v_k Q_k \qquad (10.78)$$

The UNIQUAC equation is extensively used for estimating the activity coefficient of the components of numerous commonly handled liquid mixtures.

Advantages

The advantages of the UNIQUAC equation are:

1. The equation is suitable for solutions having both small and large molecules.
2. It is simple as it consists of only two adjustable parameters.
3. It is also applicable to a good number of non-polar solutions such as aldehydes, ketones, and alcohols.
4. The constant as in the equation are less dependent on temperature compared to the other equations.

Disadvantage

1. This is mathematically more complicated than the other methods.

10.11.6 Universal Functional Activity Coefficient (UNIFAC) Method

Universal functional activity coefficient method is an efficient method by which the activity coefficients of the components in a solution mixture can be computed with the assistance of the group contribution technique. Like the UNIQUAC method, this method comprises of two parts. These are the combinatorial part, that depends on the volume and the surface area of the molecule, and the residual part, which is dependent on the interaction energies between the molecules. This method is analogous to the UNIQUAC method in estimating the different parameters such as R_k, Q_k, r_i, and q_i. Even in determining the values of γ_i^C and $\ln \gamma_{i,}^R$ the same equation can be applied; but for $\ln \gamma_{i,}^C$ pure component data is required.

The UNIFAC method depends on the concept that a liquid solution may be considered a solution of the structural units from which the molecules are formed, rather than a solution of the molecules. The structural units are called *subgroups*. R_k, the relative volume, and relative surface area, Q_k, are the properties of the subgroups.

The activity coefficients in the UNIFAC method are derived on the basis of the UNIQUAC method, and are given by

$$\ln \gamma_i = \ln \gamma_i^C \text{ (combinatorial)} + \ln \gamma_i^R \text{ (residual)} \tag{10.79}$$

$$\ln \gamma_i^C = \ln \frac{\phi_i}{x_i} + \frac{z}{2} q_i \ln \frac{\theta_i}{\phi_i} + l_i - \frac{\phi_i}{x_i} \sum_j x_j l_j \tag{10.80}$$

$$\ln \gamma_i^R = \sum_k v_k^i (\ln \Gamma_k - \ln \Gamma_k^i) \tag{10.81}$$

where

$$\ln \Gamma_k = Q_k \left[1 - \ln \left(\sum_m \theta_m \psi_{mk} \right) - \sum_m \frac{\theta_m \psi_{mk}}{\sum_n \theta_n \psi_{nm}} \right] \tag{10.82}$$

$$\theta_m = \frac{Q_m X_m}{\sum_n Q_n X_n} = \text{Surface area fraction of group } m$$

$$\psi_{mn} = \exp \left(-\frac{u_{mn} - u_{nn}}{RT} \right) = \exp \left(-\frac{a_{mn}}{T} \right)$$

X_m = Mole fraction of group m in the mixture

$$a_{mn} = \text{Group interaction parameter}$$
$$u_{mn} = \text{Measure of interaction energy between groups } m \text{ and } n$$
$$\ln \Gamma_k = \text{Group residual activity coefficient.}$$

Advantage: The chief advantage of the UNIFAC method is that a relatively small number of subgroups combine to form a very large number of molecules.

10.12 VAPOUR–LIQUID EQUILIBRIUM AT HIGH PRESSURE

In order to estimate the vapour–liquid equilibrium data at high pressure, it is necessary to consider the vaporization or vapour–liquid equilibrium constant K, because the tendency of given components of the solution to the partition between the vapour and liquid phases is accounted for by this K-factor or K-value. It measures the *lightness* of the constituent component, i.e., the tendency of the K-factor is to favour the vapour phase. This can be mathematically expressed as

$$K = \frac{y_i}{x_i} \tag{10.83}$$

where

y_i = Mole fraction of pure component i in vapour phase y

x_i = Mole fraction of pure component i in liquid phase.

We know that at low to moderate pressure, the fugacity of the pure component i in the liquid and vapour phases are equal. This is nothing but the criterion for phase equilibrium, and is given by

$$\bar{f}_i^{\,l} = \bar{f}_i^{\,v} \qquad (i = 1, 2, 3, \ldots, N) \tag{10.84}$$

Again, it is well-known to us that at low pressure the vapour phase can be assumed to behave as an ideal gas. Therefore, $\bar{\phi}_i^{\,v} = \bar{\phi}_i^{\,l} = 1$.

The Poynting pressure correction factor for component i can be written as

$$\bar{f}_i^{\,l} = P_i^{Sat} \phi_i^{Sat} \exp \frac{V_i^l (P - P_i^{Sat})}{RT}$$

$$\gamma_x x_i P_i^{Sat} = P y_i \frac{\bar{\phi}_i^{\,v}}{\phi_i^{Sat}} \exp \frac{V_i^l (P - P_i^{Sat})}{RT} \tag{10.85}$$

At low pressure, $\exp \dfrac{V^l (P - P_i^{Sat})}{RT}$ is also considered to be negligible, and it is equal to unity.

Hence Eq. (10.85) reduces to

$$\gamma_x x_i P_i^{Sat} = P y_i \tag{10.86}$$

Equation (10.84) can be expressed as

$$\bar{\phi}_i^{\,l} x_i P = \bar{\phi}_i^{\,v} y_i P \qquad \left(\text{as } \bar{\phi}_i = \frac{\bar{f}_i}{y_i P} \right) \tag{10.87}$$

$$K_i = \frac{y_i}{x_i} = \frac{\overline{\phi}_i^{\,l}}{\overline{\phi}_i^{\,v}} \tag{10.88}$$

Since at low to moderate pressure, $\overline{\phi}_i^{\,v} = \overline{\phi}_i^{\,l} = 1$, then for pure component i, Eq. (10.88) becomes $K_i = 1$.

Equation (10.88) implies that the K-factor for pure component i can be estimated with help of properties of pure component. It is necessary to remember that the intermolecular forces between the components of light hydrocarbons is significantly weak. Hence, it can also be assumed to behave like an ideal substance.

- When $K_i > 1$, component i exhibits a higher concentration in the vapour phase and can be considered the lighter component.
- When $K_i < 1$, component i shows a higher concentration in the liquid phase and is considered the heavier component.

For vapour and liquid phases in equilibrium, Raoult's law can be quantitatively expressed as

$$y_i P = x_i P_i^{Sat} \tag{10.89}$$

or

$$\frac{y_i}{x_i} = \frac{P_i^{Sat}}{P} \tag{10.90}$$

Hence, the K-factor for Raoult's law is given by

$$K_i = \frac{P_i^{Sat}}{P} \tag{10.91}$$

With respect to Eq. (10.86), the K-factor for modified Raoult's law is given by

$$K_i = \frac{y_i}{x_i} = \gamma_i \frac{P_i^{Sat}}{P} \tag{10.92}$$

For ideal gas, the K-factor is a function of temperature and pressure and independent of the composition of the vapour and liquid phases.

The monographs for the K-values of lighter hydrocarbon systems at low temperature range have been shown in Figures 10.10 and 10.11. These values enable us to calculate the dew point and bubble point of components. A comparative study with equations of K-values for pure component and mixture is shown in Table 10.1.

Table 10.1 K-values for pure component and mixture

Pure component	Mixture
$y_i P = x_i P_i^{Sat}$	$y_i P = x_i \gamma_i P_i^{Sat}$
or $\dfrac{y_i}{x_i} = \dfrac{P_i^{Sat}}{P}$	or $\dfrac{y_i}{x_i} = \gamma_x \dfrac{P_i^{Sat}}{P}$
$K_i = \dfrac{y_i}{x_i} = \dfrac{P_i^{Sat}}{P}$	$K_i = \dfrac{y_i}{x_i} = \gamma_x \dfrac{P_i^{Sat}}{P}$

For binary mixtures, the ratio of the *K*-values for the two components is called the *relative volatility* and is denoted by

$$\alpha = \frac{K_i}{K_j} = \frac{\dfrac{y_i}{x_i}}{\dfrac{y_j}{x_j}} \qquad (10.93)$$

which is a measure of the relative ease or difficulty of separating the two components. Large-scale industrial distillation is rarely undertaken if the relative volatility is less than 1.05, with the volatile component being *i* and the less volatile component being *j*. *K*-values are widely used in the design calculations of continuous distillation columns for distilling multi-component mixtures.

10.12.1 Methodology for Bubble Point Calculations

Before going on to the discussion of the methodology for the estimation of bubble point, it is important to define the term properly. Bubble point is defined as the point at which the first bubble of vapour is formed upon heating a liquid consisting of two or more components at given pressure.

At the bubble point, the composition of the liquid is the same as that of the entire system. We know that

$$K_i = \frac{y_i}{x_i}$$

or

$$y_i = K_i x_i$$

or

$$\sum y_i = \sum K_i x_i \qquad (10.94)$$

where

y_i = Mole fraction of component *i* in the vapour phase and $\sum y_i = 1$

x_i = Mole fraction of component *i* in the liquid phase and $\sum x_i = 1$.

Then, at the given mole fraction of one component in liquid or vapour phase, the mole fraction of another can be obtained by the preceding relation. At constant temperature, assuming the bubble point pressure, we get K_1 and K_2 from the nomographs shown in Figures 10.13 and 10.14 and substitute the values in the equation $\sum y_i = \sum K_i x_i$. We verify whether the summation of the mole fractions of the component in a particular phase is 1. If not so, then we assume another bubble point pressure and repeat the procedure.

Stepwise Methodology at a Glance

- **Step I:** For a given *T* or *P*, with the other not given, assume a value for the missing one.
- **Step II:** At given x_1, find x_2. For example, if $x_1 = 0.45$, then $x_2 = 0.55$.

- **Step III:** Get K_1 and K_2 values from the nomograph with respect to known T or P and assumed T or P.
- **Step IV:** Using the relation $y_i = K_i x_i$, get y_1 and y_2.
- **Step V:** Sum up the values of y_1 and y_2 and verify that it equals 1 or not.
- **Step VI:** If $y_1 + y_2 \neq 1$, then assume another P or T and repeat the procedure until the summation becomes 1. The satisfying T or P is the actual bubble point value.

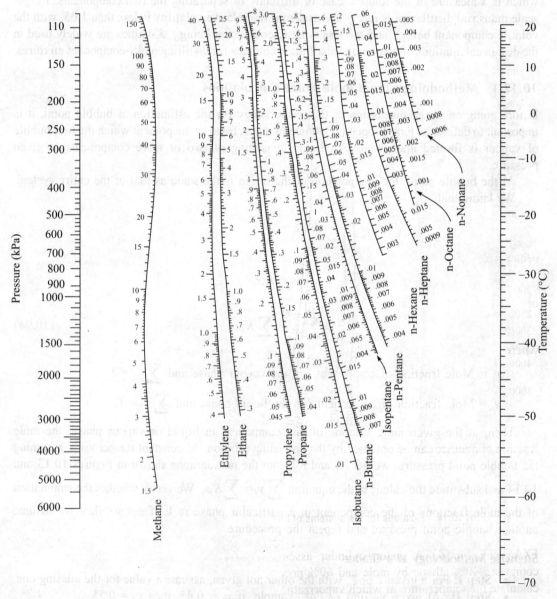

Fig. 10.13 K-values for the systems of light hydrocarbons at low temperature range.

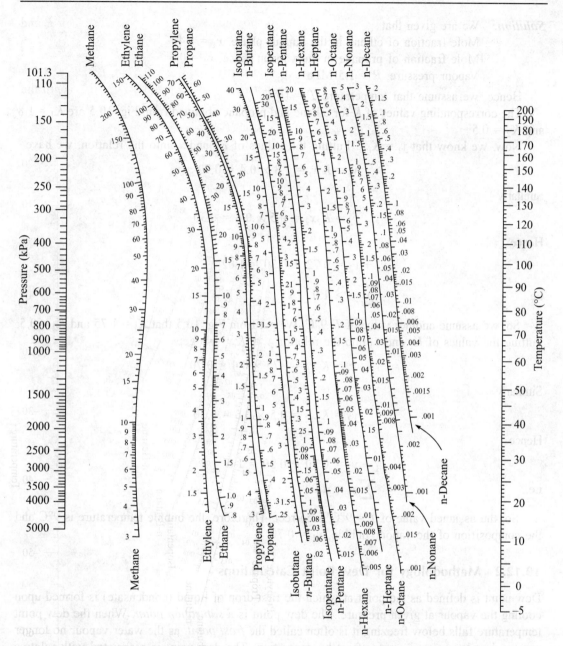

Fig. 10.14   K-values for the systems of light hydrocarbons at high temperature range.

EXAMPLE 10.14   A piston–cylinder assembly contains a binary liquid mixture which comprises 40% ethane by mole and 60% propane by mole at 1.5 MPa. Applying K-method, calculate the temperature at which vaporization begins and the composition of the first vapour bubble formed.

Solution: We are given that

Mole fraction of ethane in the vapour phase, $x_1 = 0.4$

Mole fraction of propane in the vapour phase, $x_2 = 0.6$

Vapour pressure, $P = 1.5$ MPa.

Hence, we assume that temperature $T = 10°C$.

The corresponding values of K for ethane and propane obtained from Fig. 10.5 are $K_1 = 1.8$ and $K_2 = 0.5$.

Now, we know that $y_i = K_i x_i$. Putting the values of K_1 and x_1 into the relation, we have

$$y_i = K_i x_i = 1.8 \times 0.4 = 0.72$$

Similarly

$$y_2 = K_2 x_2 = 0.5 \times 0.6 = 0.30$$

Hence

$$y_1 + y_2 = 0.72 + 0.30 = 1.02$$

i.e.,

$$\sum y_i > 1.0.$$

So, we assume another value of T, 9°C. We get from Fig. 10.5 that $K_1 = 1.75$ and $K_2 = 0.5$. Putting the values of K_1 and x_1 into the relation $y_i = K_i x_i$, we get

$$y_i = K_i x_i = 1.75 \times 0.4 = 0.70$$

Similarly

$$y_2 = K_2 x_2 = 0.5 \times 0.6 = 0.30$$

Hence

$$y_1 + y_2 = 0.70 + 0.30 = 1.0$$

i.e.,

$$\sum y_i = 1.0.$$

So, the assumed value of $T = 9°C$ is correct. Therefore, the bubble temperature is 9°C and the composition of the vapour bubble, $y_1 = 0.70$.

10.12.2   Methodology for Dew Point Calculations

Dew point is defined as the point at which the first drop of liquid (condensate) is formed upon cooling the vapour at given pressure. The dew point is a *saturation point*. When the dew point temperature falls below freezing it is often called the *frost point*, as the water vapour no longer creates dew but instead creates frost by deposition. The dew point is associated with relative humidity. A high relative humidity indicates that the dew point is closer to the current air temperature; if the relative humidity is 100%, the dew point is equal to the current temperature.

At a constant dew point, an increase in temperature will lead to a decrease in relative humidity. At a given pressure, the dew point is independent of temperature and it indicates that the mole fraction of water vapour in the air, and therefore determines the specific humidity of the air. The dew point is an important statistic for general aviation pilots, as it is used to calculate the likelihood of carburettor icing and fog, and to estimate the height of the cloud base.

We know that

$$K_i = \frac{y_i}{x_i}$$

or

$$x_i = \frac{y_i}{K_i}$$

or

$$\sum x_i = \sum \frac{y_i}{K_i} \tag{10.95}$$

The dew point temperature or pressure can be computed using the same method as the one employed for the calculation of bubble point.

Stepwise Procedure

This consists of the following steps:

- **Step I:** For a given T or P, with the other not given, assume the value for the missing one.
- **Step II:** At given y_1, find y_2. For example if $y_1 = 0.4$, then $y_2 = 0.6$.
- **Step III:** Get K_1- and K_2-values from the nomograph with respect to known T or P and assumed T or P.
- **Step IV:** Using the relation $x_1 = \dfrac{y_1}{K_1}$ and $x_2 = \dfrac{y_2}{K_2}$, get x_1 and x_2.
- **Step V:** Sum up the values of x_1 and x_2 and verify whether it equals 1 or not.
- **Step VI:** If $x_1 + x_2 \neq 1$, then assume another P or T and repeat the procedure until the summation becomes 1. The corresponding T or P is the actual bubble point value.

For single-component mixtures the bubble point and the dew point are the same and are referred to as the *boiling point*. Bubble point along with the dew point are useful sources of data in designing different kinds of equipment used in process industries, such as distillation systems.

EXAMPLE 10.15 A vapour mixture of 20 mol per cent methane, 30 mol per cent ethane and 50 mol per cent propane is available at 30°C. Using K-factors, determine the pressure at which condensation begins if the mixture is isothermally compressed and also calculate the composition of the first drop of liquid forms.

Solution: We are given that

$$y_1 = 0.2, \; y_2 = 0.3, \; y_3 = 0.5, \text{ and } T = 30°C.$$

We assume that pressure, $P = 2.0$ MPa. The corresponding values of K for ethane and propane obtained from Fig. 10.5 are $K_1 = 8.5$, $K_2 = 2.0$ and $K_3 = 0.68$.

Now, we know that

$$x_i = \frac{y_i}{K_i}$$

Putting the values of K_1 and x_1 into the relation, we have

$$x_1 = \frac{y_1}{K_1} = \frac{0.2}{8.5} = 0.0235$$

Similarly

$$x_2 = \frac{y_2}{K_2} = \frac{0.3}{2.0} = 0.15$$

$$x_3 = \frac{y_3}{K_3} = \frac{0.5}{0.7} = 0.714$$

Hence

$$x_1 + x_2 + x_3 = 0.0235 + 0.15 + 0.714 = 0.887$$

i.e.,

$$\sum x_i < 1.0.$$

So, we assume another value of $P = 2.15$ MPa at 30°C. We get from Fig. 10.5 that $K_1 = 8.1$, $K_2 = 1.82$ and $K_3 = 0.62$.

Put the values of K_1 and y_1 into the relation $x_i = \dfrac{y_i}{K_i}$, we get

$$x_1 = \frac{y_1}{K_1} = \frac{0.2}{8.1} = 0.0247$$

Similarly

$$x_2 = \frac{y_2}{K_2} = \frac{0.3}{1.82} = 0.1648 \quad \text{and} \quad x_3 = \frac{y_3}{K_3} = \frac{0.5}{0.62} = 0.8065$$

Hence

$$x_1 + x_2 + x_3 = 0.0247 + 0.1648 + 0.8065 = 0.996 = 1$$

i.e.,

$$\sum y_i = 1.0.$$

So, the assumed value of $P = 2.15$ MPa is correct. Therefore, the dew pressure is 2.15 MPa and the composition of the liquid drop, $x_1 = 0.0247$ and $x_2 = 0.1648$.

EXAMPLE 10.16 For a mixture of 10 mol per cent methane, 20 mol per cent ethane and 70 mol per cent propane at 10°C. Determine

(a) The dew point pressure.
(b) The bubble point pressure.

Get K-values from Fig. 10.5.

Solution: (a) Calculation of dew point pressure

We are given that $y_1 = 0.1$, $y_2 = 0.2$ and $y_3 = 0.7$ and $T = 10$°C. We assume that the pressures $P_1 = 690$ kPa, $P_2 = 1035$ kPa, and $P_3 = 870$ kPa. Then the corresponding values of K for ethane and propane, obtained from Fig. 10.5, can be presented as follows:

Species	y_i	$P_1 = 690$ kPa		$P_2 = 1035$ kPa		$P_3 = 870$ kPa	
		K_i	$\dfrac{y_i}{K_i}$	K_i	$\dfrac{y_i}{K_i}$	K_i	$\dfrac{y_i}{K_i}$
Methane	0.10	20.0	0.005	13.2	0.008	16.0	0.006
Ethane	0.20	3.25	0.062	2.25	0.089	2.65	0.075
Propane	0.70	0.92	0.761	0.65	1.077	0.762	0.919
		$\sum \dfrac{y_i}{K_i} = 0.828$		$\sum \dfrac{y_i}{K_i} = 1.174$		$\sum \dfrac{y_i}{K_i} = 1000$	

The assumed value $P = 870$ kPa is correct as it satisfies the dew point equation. Hence, the dew pressure is obtained as 870 kPa. The composition of the dew drop is the estimated value listed in the last column.

(b) Calculation of bubble point pressure

When the system is almost condensed, it is at its bubble point and the given mole fractions of x_i. Again, using trial and error method, we can determine the bubble point pressure by assuming that the pressure for which the corresponding K-values obtained from Fig. 10.5 satisfy the bubble point equation.

Species	y_i	$P_1 = 2622$ kPa		$P_2 = 2760$ kPa		$P_3 = 2656$ kPa	
		K_i	$K_i x_i$	K_i	$K_i x_i$	K_i	$K_i x_i$
Methane	0.10	5.60	0.560	5.25	0.525	5.49	0.549
Ethane	0.20	1.11	0.222	1.07	0.214	1.10	0.220
Propane	0.70	0.335	0.235	0.32	0.224	0.33	0.231
		$\sum K_i x_i = 1.017$		$\sum K_i x_i = 0.963$		$\sum K_i x_i = 1.000$	

Assumed value $P = 2656$ kPa is correct because it satisfies the bubble point equation. Hence, the bubble point pressure is 2656 kPa. The composition of the vapour bubble is the estimated value shown in the last column.

10.12.3 Flash Calculations

Flash calculation is an important application of vapour–liquid equilibrium, in which estimation of mole fractions of components in the liquid and vapour phases at given temperature, pressure and mole fraction of the component in the feed stream is done. Consider that a binary mixture consisting of components A and B flows through a flash vaporizer at the rate of F mol/h as illustrated in Fig. 10.15. After vaporization, the mixture is separated into vapour and liquid streams at equilibrium.

Then the total material balance for the system gives

$$F = V + L$$

where

 F = Total number of moles of feed
 V = Number of moles of vapour stream
 L = Number of moles of liquid stream.

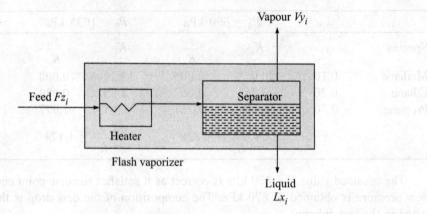

Fig. 10.15 Demonstration of flash distillation.

Material balance for the components gives

$$Fz_i = Lx_i + Vy_i \qquad (10.96)$$

where

z_i = Mole fraction of component i in feed stream

x_i = Mole fraction of component i in liquid phase

y_i = Mole fraction of component i in vapour phase.

We know from the basic concept of the vapour–liquid equilibrium that $y_i = K_i x_i$. Substituting $y_i = K_i x_i$ to replace y_i from Eq. (10.96), we get

$$Fz_i = Lx_i + VK_i x_i = x_i(L + VK_i)$$

$$x_i = \frac{Fz_i}{(L + VK_i)} = \frac{F}{V} \frac{z_i}{\left(\dfrac{L}{V} + K_i \right)} \qquad (10.97)$$

As $\sum x_i = 1$, we have

$$\sum x_i = \frac{F}{V} \frac{z_i}{\left(\dfrac{L}{V} + K_i \right)} = 1 \qquad (10.98)$$

Again, if x_i is replaced by substituting $x_i = \dfrac{y_i}{K_i}$ from Eq. (10.96), we get

$$Fz_i = Lx_i + Vy_i = L\frac{y_i}{K_i} + Vy_i = y_i\left(\frac{L}{K_i} + V \right) \qquad (10.99)$$

$$y_i = \frac{Fz_i}{\left(\dfrac{L}{K_i} + V \right)} = \frac{F}{V} \frac{z_i}{\left(\dfrac{L}{VK_i} + 1 \right)} \qquad (10.100)$$

Since $\sum y_i = 1$, we get

$$\sum y_i = \sum \frac{F}{V} \frac{z_i}{\left(\dfrac{L}{VK_i} + 1\right)} = 1 \qquad (10.101)$$

To determine x_i, y_i, $\dfrac{L}{V}$, etc., the iterative procedure can be employed.

EXAMPLE 10.17 The system acetone (1)–acetonitrile (2)–nitro-methane (3) at 80°C and 100 kPa has the overall composition $z_1 = 0.45$, $z_2 = 0.35$, and $z_3 = 0.20$. Assuming that Raoult's law is appropriate to this system, determine L, V, x_i and y_i. The vapour pressure of pure species are given by

$$P_1^{Sat} = 195.75 \text{ kPa}$$
$$P_2^{Sat} = 97.84 \text{ kPa}$$

and

$$P_3^{Sat} = 50.32 \text{ kPa}.$$

Solution: *Bubble Point Calculation*
We know that the bubble point pressure can calculated as

$$P_{bubble} = x_1 P_1^{Sat} + x_2 P_2^{Sat} + x_3 P_3^{Sat}$$

or

$$P = 0.45 \times 195.75 + 0.35 \times 97.84 + 0.2 \times 50.32 = 132.40 \text{ kPa}$$

Dew Point Calculation

We can assume that $z_i = y_i$, because the given pressure lies between bubble point and dew point pressures. We have

$$P_{dew} = \frac{1}{\dfrac{y_1}{P_1^{Sat}} + \dfrac{y_2}{P_2^{Sat}} + \dfrac{y_3}{P_3^{Sat}}} = \frac{1}{\dfrac{0.45}{195.75} + \dfrac{0.35}{97.84} + \dfrac{0.20}{50.32}} = 101.52 \text{ kPa}$$

It is known that $K_i^{Sat} = \dfrac{P_i^{Sat}}{P}$. So, after substituting the relevant values, we get

$$K_1^{Sat} = \frac{P_1^{Sat}}{P} = \frac{195.75}{101.52} = 1.928$$

$$K_2^{Sat} = \frac{P_2^{Sat}}{P} = \frac{97.84}{101.52} = 0.963$$

and

$$K_3^{Sat} = \frac{P_3^{Sat}}{P} = \frac{50.32}{101.52} = 0.4956$$

Now for the binary system containing components A and B and flowing through a flash vaporizer at the rate of F mol/h, the overall balance for the system is given by

$$L + V = 1 \qquad (i)$$

Material balance for the components gives

$$Fz_i = Lx_i + Vy_i \qquad \text{(ii)}$$

where

z_i = Mole fraction of component i in feed stream
x_i = Mole fraction of component i in liquid phase
y_i = Mole fraction of component i in vapour phase.

For 1 mole of feed, Eq. (ii) becomes

$$z_i = Lx_i + Vy_i$$

or

$$z_i = (1 - V)x_i + Vy_i \qquad \text{(since } L + V = 1) \qquad \text{(iii)}$$

Substituting $x_i = \dfrac{y_i}{K}$ into Eq. (iii), we have

$$\sum y_i = \sum \frac{z_i K_i}{1 + V(K_i - 1)} = 1 \qquad \text{(iv)}$$

Substituting the values of z_i and K_i, we get

$$\frac{z_1 K_1}{1 + V(K_1 - 1)} + \frac{z_2 K_2}{1 + V(K_2 - 1)} + \frac{z_3 K_3}{1 + V(K_3 - 1)} = 1$$

or

$$\frac{0.45 \times 1.928}{1 + V(1.928 - 1)} + \frac{0.35 \times 0.963}{1 + V(0.963 - 1)} + \frac{0.20 \times 0.495}{1 + V(0.495 - 1)} = 1$$

$$V = 0.782 \text{ mol}$$

Therefore

$$L = 1 - V = 1 - 0.782 = 0.218 \text{ mol}$$

$$y_1 = \frac{z_1 K_1}{1 + V(K_1 - 1)} = \frac{0.45 \times 1.928}{1 + 0.782(1.928 - 1)} = 0.502$$

Similarly, $y_2 = 0.331$ and $y_3 = 0.149$. Then, by using the relation $x_i = \dfrac{y_i}{K_i}$, we get

$$x_1 = \frac{y_1}{K_1} = \frac{0.502}{1.928} = 0.2603$$

$$x_2 = \frac{y_2}{K_2} = \frac{0.331}{0.963} = 0.3437$$

and

$$x_3 = \frac{y_3}{K_3} = \frac{0.149}{0.495} = 0.3010$$

EXAMPLE 10.18 For the benzene–toluene system, $z_1 = 0.81$ at 60°C and 70 kPa. The vapour pressure of the components are given by the Antoine equation $\ln P = A - \dfrac{B}{T + C}$, where T is in °C and P is in kPa. Antoine constants for the components are as follows:

Compound	A	B	C
Benzene	14.2321	2773.61	220.13
Toluene	15.0198	3102.64	220.02

Determine the moles and composition of the vapour and liquid streams V, L and x, y respectively.

Solution: At 60°C, the saturated vapour pressures for components 1 and 2 respectively can be calculated by using the Antoine equation as P_1^{Sat} (benzene) = 76.02 kPa and P_2^{Sat} (toluene) = 51.40 kPa. We know that

$$P = x_1 P_1^{Sat} + x_2 P_2^{Sat}$$

Substituting the values, we get

$$70 = x_1 \times 76.02 + x_2 \times 51.40 = x_1 \times 76.02 + (1 - x_1) \times 51.40$$

or

$$x_1 = 0.7554$$

Now

$$y_1 = \frac{x_1 P_1^{Sat}}{P} = \frac{0.7554 \times 76.02}{70} = 0.8203$$

Hence, the liquid and vapour phase compositions are $x_1 = 0.7554$ and $y_1 = 0.8203$.

The basis is 1 mol of feed stream. The moles of liquid and vapour streams can be estimated by making a component balance as

$$Fz_i = Lx_i + Vy_i = Lx_i + (1 - L)y_i$$

or

$$z_i = Lx_i + (1 - L)y_i$$

Substituting the values, we get

$$0.81 = L \times 0.7554 + (1 - L) \times 0.8203$$

or

$$L = 0.312 \text{ mol}$$

Therefore, $V = 1 - L = 1 - 0.312 = 0.688$ mol.

10.13 THERMODYNAMIC CONSISTENCY TEST OF VLE DATA

For the design of equipment related to process industries and laboratories, design engineers necessarily require results of good accuracy. Therefore, it is important to test the thermodynamic consistency of the vapour–liquid equilibrium data. The test of VLE data is usually made on the basis of Gibbs–Duhem equation in terms of activity coefficients. The data can be obtained from pressure–temperature–mole fractions (P–T–x–y) diagram. If the experimental data are inconsistent with Gibbs–Duhem equation, the data are incorrect due to the systematic error.

There are several tests available to check the consistency of VLE data. A few important tests are discussed here.

Before going on to consider the different methods for the test of vapour–liquid equilibrium data, we need to recall the Gibbs–Duhem equation. We know that for a binary system at constant pressure and temperature, the Gibbs–Duhem equation in terms of activity coefficient is given by

$$x_1 \frac{d \ln \gamma_1}{dx_1} = x_2 \frac{d \ln \gamma_2}{dx_2} \qquad (10.102)$$

or

$$x_1 \frac{d \ln \gamma_1}{dx_1} = (1 - x_1) \frac{d \ln \gamma_2}{dx_1} \qquad (10.103)$$

Method I: This is a simple method used to test the consistency of VLE data. In this method, we plot the experimentally determined values of γ against the mole fractions of the components in liquid and vapour phases, i.e., γ_1 versus x_1 and $\ln \gamma_2$ versus x_2. Then we find out the slope for satisfying Eq. (10.102). Although it is not so easy to determine the slope correctly, this method is generally employed for checking how much the available data are correct. We put the data into Eq. (10.102) and check the result for L.H.S. and R.H.S of the equation, whether they come with positive quantity or not.

The nature of the curve for the plot $\ln \gamma_1$ versus x_1 is shown in Fig. 10.16. Figures having this sort of graphical nature with tangent drawn show the thermodynamic consistency of the data.

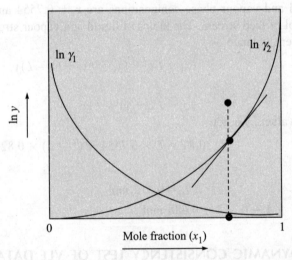

Fig. 10.16 Thermodynamic consistency test of VLE data with activity coefficients.

Method II: Without considering the activity coefficients, it has been possible to test the consistency of vapour–liquid equilibrium data of a binary system. The Gibbs–Duhem equation can be presented in the following form:

$$\sum x_i d \ln \bar{f}_i = 0 \qquad (10.104)$$

In the binary system at low to moderate pressure, the preceding equation can be written as

$$x_1 \frac{d \ln p_1}{dx_1} = x_2 \frac{d \ln p_2}{dx_1}$$

or

$$\frac{x_1}{p_1} \frac{dp_1}{dx_1} = \frac{x_2}{p_2} \frac{dp_2}{dx_2} \qquad (10.105)$$

This test is similar to Method I in which we require plotting partial pressure versus mole fraction of the liquid x_1 as shown in Fig. 10.17. The VLE data for chloroform–ethanol system at 45°C is represented in Fig. 10.17. Now, we determine the slopes and the magnitudes of $\dfrac{x_1}{p_1}$ and $\dfrac{x_2}{p_2}$ to substitute into Eq. (10.105) for checking the validity. If both sides of Eq. (10.105) are found to the same in magnitude, then the data are thermodynamically consistent.

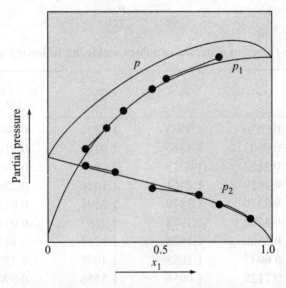

Fig. 10.17 Thermodynamic consistency test of VLE data with partial pressure.

EXAMPLE 10.19 For the system acetone (1)–carbon tetrachloride (2), the vapour–liquid equilibrium data at 45°C were reported by an experimenter after several observations. Test whether the following data are thermodynamically consistent or not:

P (torr)	315.32	339.70	397.77	422.46	448.88	463.92	472.84	485.16	498.07	513.20
x_1	0.0556	0.0903	0.2152	0.2929	0.3970	0.4769	0.5300	0.6047	0.7128	0.9636
y_1	0.2165	0.2910	0.4495	0.5137	0.5832	0.6309	0.6621	0.7081	0.7718	0.9636

Solution: The saturated vapour pressure of components 1 and 2 can be calculated using Antoine equation. For the acetone–carbon tetrachloride system, Antoine constants of the components obtained from the standard data are:

	A	B	C
Acetone	7.117	1210.595	229.664
CCl_4	6.840	177.910	220.576

Substituting all the values in Antoine equation, we get

$$P_1^{Sat} = 512.38 \text{ torr} \quad \text{and} \quad P_2^{Sat} = 254.41 \text{ torr}$$

Now, we know that if the vapour phase behaves like an ideal gas, then the activity coefficients of the components can be computed by using the following expression:

for component 1

$$\gamma_1 = \frac{y_1 P}{x_1 P_1^{Sat}}$$

for component 2

$$\gamma_2 = \frac{y_2 P}{x_2 P_2^{Sat}}$$

Substitution of the values of different variables yields the following values:

x_1	γ_1	γ_2	$\ln\left(\dfrac{\gamma_1}{\gamma_2}\right)$
0.0556	2.3963	1.023	0.8460
0.0903	2.1365	1.0407	0.7193
0.2152	1.6215	1.0967	0.3910
0.2929	1.4461	1.1420	0.2361
0.3970	1.2870	1.2196	0.0538
0.4769	1.1978	1.2867	−0.0716
0.5300	1.1528	1.3362	−0.1476
0.6047	1.1088	1.4082	−0.2390
0.7128	1.0055	1.5556	−0.3907
0.9636	1.0016	2.0172	−0.7001

The $\ln\left(\dfrac{\gamma_1}{\gamma_2}\right)$ values are plotted against x_1. The curve is shown in Fig. 10.18. We get from the plot that the area above X-axis = 1560 and area below X-axis = 2190. Therefore

$$\frac{\text{Area above X-axis} - \text{Area below X-axis}}{\text{Area above X-axis} + \text{Area below X-axis}} = \frac{1560 - 2190}{1560 + 2190} = 0.168 > 0.02$$

The given area test data does not satisfy the condition. Hence, the data is not consistent thermodynamically.

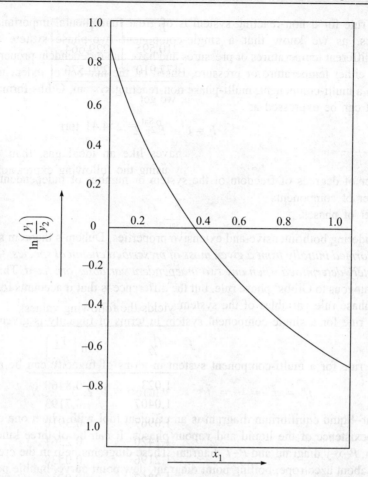

Fig. 10.18 Graphical representation for testing thermodynamic consistency.

SUMMARY

In order to achieve separation between the components of a mixture, chemical process engineers need to carry out several mass transfer operations such as distillation, absorption, drying, adsorption, leaching, evaporation and extraction. It is well known to us that mass can be transferred from one phase to another phase when the concentration gradient exists. This process continues till the concentration difference is equalized or phase equilibrium is attained. Therefore, one should have sufficient knowledge about the various forms of phase equilibrium such as vapour–liquid equilibrium and liquid–liquid equilibrium for proper design and analysis of mass transfer equipment such as distillation column, adsorber and drier. The criteria of phase equilibrium for a system undergoing a process occurring at constant temperature and pressure can be represented by

$$dG_{T,P} = 0$$

The phase rule for a non-reacting system is of great fundamental importance in solution thermodynamics, as we know that a single-component two-phase system may exist in equilibrium at different temperatures or pressures and have one independent property, which may be taken to be either temperature or pressure. Based on the number of independent variables associated with a multi-component, multi-phase non-reacting system, Gibbs formulated a phase rule in 1875. It can be expressed as

$$F = C - P + 2$$

where

F = Number of degrees of freedom of the system or number of independent variables
C = Number of components
P = Number of phases.

Now, considering both intensive and extensive properties, Duhem's theorem states: *For any closed system formed initially from a given mass of prescribed chemical species, the equilibrium state is completely determined when any two independent variables are fixed.* This theorem is a kind of rule analogous to Gibbs' phase rule, but the difference is that it accounts for the extensive and intensive phase rule variables of the system.

The phase rule for a single-component system in terms of fugacity is given by

$$f^v = f^l$$

The phase rule for a multi-component system in terms of fugacity can be represented by

$$\bar{f}_i^a = \bar{f}_i^b = \cdots = \bar{f}_i^\pi \qquad \text{where } i = 1, 2, 3, \ldots, N$$

The vapour–liquid equilibrium diagram is an efficient tool with which one can understand the state of coexistence of the liquid and vapour phases. It can be of three kinds, namely the T–x–y diagram, P–x–y diagram and P–T diagram. These diagrams help in the creation of better understanding about azeotrope, boiling point diagram, dew point curve, bubble point curve, etc. It is needless to mention that the design and development of mass transfer equipment needs the availability of vapour–liquid equilibrium data. In most of the cases, accurate data are not available and therefore assignments are completed with fragmentary data. Therefore, a chemical engineer often needs to use important correlations to check the correctness of the data. Such correlations are presented and derived in this chapter with the help of the principle of phase equilibrium. For the computation of activity coefficients at low to moderate pressure, Wohl's equation, van Laar equation, Margules equation and UNIFAC method are discussed. Some simple methods have been discussed to test the thermodynamic consistency of VLE data.

Wohl's three-suffix equation is given by

$$\frac{G^E}{RT(q_1 x_1 + q_2 x_2)} = 2a_{12}z_1z_2 + 3a_{112}z_1^2z_2 + 3a_{122}z_1z_2^2$$

where

x_1, x_2 = Mole fractions of components 1 and 2
q_1, q_2 = Effective molal volume
z_1, z_2 = Effective volume fraction
a_{12}, a_{112}, a_{122} = Empirical constants.

With the help of Wohl's equation, the activity coefficients can be expressed as

$$\ln \gamma_1 = z_2^2 \left[A + 2 \left(B \frac{q_1}{q_2} - A \right) z_1 \right]$$

$$\ln \gamma_2 = z_1^2 \left[B + 2 \left(A \frac{q_2}{q_1} - B \right) z_2 \right]$$

Margules two-parameter (three-suffix) equation is given by

$$\ln \gamma_1 = x_2^2[A + 2(B - A)x_1] = (2B - A)x_2^2 + 2(A - B)x_2^3$$
$$\ln \gamma_2 = x_1^2[B + 2(A - B)x_2] = (2A - B)x_1^2 + 2(B - A)x_1^3$$

Margules one-parameter (two-suffix) equation is given by

$$\ln \gamma_1 = Ax_2^2$$
$$\ln \gamma_2 = Ax_1^2$$

van Laar equations are given by

$$\ln \gamma_1 = Az_2^2 = \frac{Ax_2^2}{\left[\left(\frac{A}{B} \right) x_1 + x_2 \right]^2} = \frac{A}{\left(1 + \frac{Ax_1}{Bx_2} \right)^2} = A \left(1 + \frac{Ax_1}{Bx_2} \right)^{-2}$$

$$\ln \gamma_2 = Bz_1^2 = \frac{Bx_1^2}{\left[x_1 + \left(\frac{B}{A} \right) x_2 \right]^2} = \frac{B}{\left(1 + \frac{Bx_2}{Ax_1} \right)^2} = B \left(1 + \frac{Bx_2}{Ax_1} \right)^{-2}$$

UNIQUAC equations for estimation of activity are given by

$$\ln \gamma_i = \ln \gamma_i^C \text{(Combinatorial)} + \ln \gamma_i^R \text{(Residual)}$$

$$\ln \gamma_i^C = \ln \frac{\phi_i}{x_i} + \frac{z}{2} q_i \ln \frac{\theta_i}{\phi_i} + l_i - \frac{\phi_i}{x_i} \sum_j x_j l_j$$

$$\ln \gamma_i^R = q_i \left[1 - \ln \left(\sum_j \theta_j \tau_{ji} \right) - \sum_j \frac{\theta_j \psi_{ij}}{\sum_k \theta_k \psi_{kj}} \right]$$

Raoult's law is valid for the system at low to moderate pressure. But the deviation from Raoult's law measures the non-ideality of the solution. Finally, the stepwise procedure for the estimation of dew point, bubble point and flash point has been discussed.

KEY TERMS

Azeotrope   A mixture of two or more pure components in such a ratio that its composition cannot be changed by simple distillation

Boiling Point Diagram   A T–x–y diagram in which temperature along the ordinate is plotted against composition of liquid and vapour along the abscissa in order to represent the vapour–liquid equilibrium of a binary mixture at constant pressure.

Bubble Point Curve   The lower curve in the boiling point diagram, which is also called the saturated liquid curve.

Bubble Point Curve   The saturated liquid curve in the VLE diagram.

Bubble Point   The point at which the first bubble of vapour is formed upon heating a liquid consisting of two or more components at a given pressure.

Dew Point Curve   The upper curve in the boiling point diagram, which is also called the saturated vapour curve.

Dew Point   The point at which the first drop of liquid (condensate) is formed upon cooling the vapour at a given pressure.

Duhem's Theorem   *For any closed system formed initially from a given mass of prescribed chemical species, the equilibrium state is completely determined when any two independent variables are fixed.*

Frost Point   The point at which the dew point temperature falls below freezing.

Maxcondenbar   The maximum pressure on the phase boundary envelope of the pressure–temperature diagram of VLE binary or multicomponent mixture, represented by the point G in Fig. 10.6.

Maxcondentherm   The maximum temperature on the phase boundary envelope of the pressure–temperature diagram of VLE binary or multicomponent mixture, represented by point H in Fig. 10.6.

Retrograde Condensation   The condensation that takes place in the region bounded by the maxcondenbar, the maxcondentherm and the critical points.

Universal Functional Activity Coefficient (UNIFAC) Method   An efficient method by which the activity coefficients of components in a solution mixture can be computed with the assistance of the group contribution technique.

IMPORTANT EQUATIONS

1. Criteria for equilibrium are given by
   (i) $dS_{U,V} \leq 0$ (10.4)
   (ii) $dA_{T,V} \leq 0$ (10.8)
   (iii) $dG_{T,P} \leq 0$ (10.13)
2. Criterion for phase equilibrium is given by
$$dG_{T,P} = 0$$ (10.15)
3. Gibbs phase rule for non-reacting system is given by
$$F = C - P + 2$$ (10.17)

where

F = Number of degrees of freedom of the systems or number of independent variables

C = Number of components

P = Number of phases.

4. For binary vapour–liquid equilibrium mixture

$$f^{\mathrm{v}} = f^{\mathrm{l}} \qquad (10.26)$$

5. The chemical potential of a component in multiple phases must be the same under equilibrium conditions. This is shown by

$$\mu_i^a = \mu_i^b = \cdots = \mu_i^\pi \ (i = 1, 2, 3, \ldots, N) \qquad (10.34)$$

6. Wohl's three-suffix equation is given by

$$\frac{G^{\mathrm{E}}}{RT(q_1 x_1 + q_2 x_2)} = 2a_{12}z_1 z_2 + 3a_{112}z_1^2 z_2 + 3a_{122}z_1 z_2^2 \qquad (10.49)$$

where

x_1, x_2 = Mole fractions of components 1 and 2

q_1, q_2 = Effective molal volume

z_1, z_2 = Effective volume fraction

a_{12}, a_{112}, a_{122} = Empirical constants.

7. Margules three-suffix equations are given by

$$\ln \gamma_1 = x_2^2[A + 2(B - A)x_1] = (2B - A)x_2^2 + 2(A - B)x_2^3 \qquad (10.54)$$
$$\ln \gamma_2 = x_1^2[B + 2(A - B)x_2] = (2A - B)x_1^2 + 2(B - A)x_1^3 \qquad (10.55)$$

8. van Laar equations are given by

$$\ln \gamma_1 = A z_2^2 = \frac{Ax_2^2}{\left[\left(\dfrac{A}{B}\right)x_1 + x_2\right]^2} = \frac{A}{\left(1 + \dfrac{Ax_1}{Bx_2}\right)^2} = A\left(1 + \frac{Ax_1}{Bx_2}\right)^{-2} \qquad (10.58)$$

$$\ln \gamma_2 = B z_1^2 = \frac{Bx_1^2}{\left[x_1 + \left(\dfrac{B}{A}\right)x_2\right]^2} = \frac{B}{\left(1 + \dfrac{Bx_2}{Ax_1}\right)^2} = B\left(1 + \frac{Bx_2}{Ax_1}\right)^{-2} \qquad (10.59)$$

where A and B are the adjustable parameters.

9. Wilson equation is given by

$$\frac{G^{\mathrm{E}}}{RT} = -x_1 \ln (x_1 + x_2 \Lambda_{12}) - x_2 \ln (x_2 + x_1 \Lambda_{21}) \qquad (10.64)$$

where Λ_{12} and Λ_{21} are the two Wilson parameters.

10. NRTL equation is given by

$$\frac{G^{\mathrm{E}}}{RT} = x_1 x_2 \left(\frac{\tau_{21} G_{21}}{x_1 + x_2 G_{21}} + \frac{\tau_{12} G_{12}}{x_2 + x_1 G_{12}} \right) \qquad (10.69)$$

11. UNIFAC equation is given by

$$\ln \gamma_i^C = \ln \frac{\phi_i}{x_i} + \frac{z}{2} q_i \ln \frac{\theta_i}{\phi_i} + l_i - \frac{\phi_i}{x_i} \sum x_i l_j \qquad (10.80)$$

$$\ln \gamma_i^R = \sum_k v_k^i (\ln \Gamma_k - \ln \Gamma_k^i) \qquad (10.81)$$

12. Relative volatility is given by

$$\alpha = \frac{K_i}{K_j} = \frac{\dfrac{y_i}{x_i}}{\dfrac{y_j}{x_j}} \qquad (10.93)$$

13. Material balance for the components in flash vaporizer is given by

$$Fz_i = Lx_i + Vy_i \qquad (10.96)$$

14. Basic equation for testing the thermodynamic consistency of VLE data is given by

$$x_1 \frac{d \ln \gamma_1}{dx_1} = x_2 \frac{d \ln \gamma_2}{dx_2} \qquad (10.102)$$

EXERCISES

A. Review Questions

1. What do you mean by criterion for equilibrium of a system?
2. Mention the different criteria for phase equilibrium. What is the necessity for a system to be in equilibrium?
3. How can equilibrium be classified? When does a system reach equilibrium?
4. Explain the phase rule for non-reacting system. State Duhem's theorem in substantiating the phase rule.
5. Enumerate the role of independent variables in vapour–liquid equilibrium.
6. Show that the chemical potentials of a component in different phases are equal.
7. What is the condition of phase equilibrium for a single-component system? How does it differ from that of a multicomponent system?
8. Justify the statement: *Fugacities of a pure component in vapour and liquid phases are identical.*
9. With the help of neat schematic representation, explain the importance of vapour–liquid equilibrium of a binary system.
10. What is the significance of Poynting pressure correction factor in vapour–liquid equilibrium?
11. What is boiling point diagram? Discuss it for benzene–toluene system.
12. With the help of temperature–composition diagram, discuss the distillation process. What are the roles of bubble point and curve dew point curve?

13. Define the term *tie line*. How does it play an important role in boiling point diagram?
14. Differentiate between P–x–y and T–x–y diagrams in explaining the VLE of a binary system.
15. Draw the T–x–y diagram at constant pressure for a minimum boiling azeotrope.
16. Draw the P–x–y diagram at constant temperature for a maximum boiling azeotrope.
17. What is retrograde condensation?
18. What do you mean by maxcondentherm and maxcondenbar? How do they differ from each other?
19. What is Raoult's law? How do you explain the deviation from Raoult's law? What are positive and negative deviation of a solution from ideality?
20. Write down the characteristics of an ideal solution.
21. What is an azeotrope? How do you categorize an azeotrope? Discuss the minimum boiling azeotrope with the help of phase diagram for a particular system.
22. Explain the azeotrope formation for nitric acid–water system with graphical representation.
23. Explain the vapour–liquid equilibrium for a binary system at low pressure.
24. Derive Margules two-suffix three-parameter equation for the calculation of activity coefficients.
25. Mention the usefulness of the van Laar equation to determine the activity coefficients of a binary solution.
26. Differentiate between UNIFAC and UNIQUAC methods.
27. What is K-factor? Illustrate the importance of K-factor.
28. Define bubble point and dew point. How do you calculate the dew point and bubble point of binary VLE mixture?
29. How does the K-factor relate to the bubble point and dew point calculation of a binary vapour–liquid equilibrium mixture?
30. In case of flash distillation, how do you calculate the flash?
31. How does the Gibbs–Duhem equation play a significant role for the thermodynamic consistency of VLE data?
32. Discuss the methods involved in checking the thermodynamic consistency of the vapour–liquid equilibrium data of a binary system.

B. Problems

1. A piston–cylinder assembly contains a binary liquid mixture which comprises 30% ethylene by mole and 70% propylene by mole at 1 MPa. Applying K-method, calculate the temperature at which vaporization begins and the composition of the first bubble formed.
2. A vapour mixture of 20 mol per cent methane, 35 mol per cent ethane and 45 mol per cent propane is available at 30°C. Using K-factors, determine the pressure at which condensation begins if the mixture is isothermally compressed and also calculate the composition of the first drop of liquid formed.

3. For a mixture of 15 mol per cent methane, 25 mol per cent ethane and 60 mol per cent propane at 10°C. Determine

   (a) Dew point pressure   (b) Bubble point pressure.

4. A liquid mixture of 25 mol percent ethylene and 75 mol percent propylene at −40°C is kept in a piston–cylinder assembly. The piston exerts a constant pressure of 1 MPa. Using the K-method, determine the dew point temperature.

5. A binary liquid mixture consists of 50 mol per cent n-hexane and 50 mol per cent n-butane. At 423 K, the vapour pressure of n-hexane and n-butane are 13.1 atm and 7.5 atm respectively. Calculate the total pressure and equilibrium composition of the vapour phase. Assume that the mixture behaves like an ideal solution.

6. An equimolar binary liquid mixture consists of two components—methane and propane. At 443 K, the vapour pressure of methane and propane are 16.5 atm and 9.2 atm respectively. Calculate the total pressure and equilibrium composition of the vapour phase. Assume the mixture follows Raoult's law.

7. The pure component vapour pressure of two organic liquids—acetone and acetonitrile—by Antoine equations are given by

$$\ln P_1^{Sat} = 14.54 - \frac{2940}{T - 36}$$

and

$$\ln P_2^{Sat} = 14.27 - \frac{2945}{T - 50}$$

where P_1^{Sat} and P_2^{Sat} are in KPa and T in K.

8. Calculate the composition of liquid and vapour in equilibrium at 327 K and 65 KPa. The azeotrope of the ethanol–benzene system has a composition of 44.8 mol per cent ethanol with a boiling point of 68.24°C at 760 mm Hg. At 68.24°C, the vapour pressure of pure benzene is 517 mm Hg. Calculate the van Laar constants for the system, and evaluate the activity coefficients for a solution containing 10 mol per cent ethanol.

9. For the system methanol (1)–methyl acetate (2), the activity coefficients for components 1 and 2 are represented by

$$\ln \gamma_1 = Ax_2^2 \quad \text{and} \quad \ln \gamma_2 = Ax_1^2$$

where

$$A = 2.771 - 0.00523\,T$$

The vapour pressures of the components are given by the Antoine equations

$$\ln P_1^{Sat} = 16.5915 - \frac{3,643.31}{T - 33.424} \quad \text{and} \quad \ln P_2^{Sat} = 14.2532 - \frac{2,665.54}{T - 53.424}$$

where P is in KPa and T in K. Calculate

   (a) P and y_i at $T = 318.15$ K and $x_1 = 0.25$
   (b) P and x_i at $T = 318.15$ K and $y_1 = 0.60$
   (c) T and y_i at $P = 101.33$ KPa and $x_1 = 0.85$
   (d) T and x_i at $P = 101.33$ KPa and $x_1 = 0.40$

10. For a binary system, the excess Gibbs free energy of components 1 and 2 at 30°C is given by $\dfrac{G^E}{RT} = 0.500x_1x_2$. The vapour pressures of components 1 and 2 are given by

$$\ln P_1^{Sat} = 11.92 - \frac{4050}{T} \quad \text{and} \quad \ln P_2^{Sat} = 12.12 - \frac{4050}{T}$$

where P_1^{Sat} and P_2^{Sat} are in bar and T is in K. Calculate the composition of the azeotrope.

11. An azeotrope consists of 40.0 mol per cent acetone (1) and 60.0 mol per cent methanol at 760 mm Hg and 310 K. At 310 K, the vapour pressure of acetone and methanol are 780 mm Hg and 542 mm Hg respectively. Calculate the van Laar constants.

12. The azeotrope of the benzene–toluene system has a composition of 53.2 mol per cent benzene with a boiling point of 350.6 K at 101.3 kPa. At this temperature, the vapour pressure of benzene (1) is 112.23 kPa and the vapour pressure of toluene (2) is 88.27 kPa. Determine the activity coefficients for the solution containing 10 mol per cent benzene.

13. Using the Wilson and NRTL equations, estimate the activity coefficients of the components of a binary system consisting of toluene (1) and cyclohexane (2) at 60°C and $x_1 = 0.4$. At this temperature, the molar volumes of the components are $V_1 = 75.2$ cm^3/mol and $V_2 = 20.34$ cm^3/mol.

    Data given for
    Wilson equation:

$$a_{12} = 312.33 \text{ cal/mol}$$
$$a_{21} = 1678.32 \text{ cal/mol}$$

    NRTL equation:

$$b_{12} = 682.21 \text{ cal/mol}$$
$$b_{21} = 1231.47 \text{ cal/mol}$$
$$\alpha = 0.568.$$

14. Wilson's parameters for the chloroform–methanol system at 35°C are given by $\lambda_{12} - \lambda_{11} = -1.522$ kJ/mol-K and $\lambda_{12} - \lambda_{22} = 7.559$ kJ/mol-K. Estimate the VLE data for the system at 35°C if $V_1^l = 80.67 \times 10^{-6}$ m^3/mol and $V_2^l = 40.73 \times 10^{-6}$ m^3/mol.

15. Carbon tetrachloride–ethanol forms an azeotrope at 760 torr, where $x_1 = 0.613$ and $T = 64.95$°C. Using the van Laar model, predict the P–x–y data at 64.95°C.

16. Construct the P–x–y diagram for the cyclohexane–benzene system at 313 K, given that at this temperature $P_1^{Sat} = 24.62$ kPa and $P_2^{Sat} = 24.4$ kPa. The liquid phase activity coefficients are given by $\ln \gamma_1 = 0.458x_2^2$ and $\ln \gamma_1 = 0.458x_1^2$.

17. The ternary system A–B–C at 80°C and 100 kPa has the overall composition $z_1 = 0.40$, $z_2 = 0.30$, and $z_3 = 0.30$. Assuming that Raoult's law is appropriate to this system, determine L, V, x_i and y_i. The vapour pressure of pure species are given by

$$P_1^{Sat} = 187.32 \text{ kPa}$$
$$P_2^{Sat} = 95.48 \text{ kPa}$$
$$P_3^{Sat} = 50.55 \text{ kPa}.$$

18. For the acetone–acetonitrile system, $z_1 = 0.75$ at 60°C and 70 kPa. The vapour pressures of the components are given by the Antoine equation

$$\ln P = A - \frac{B}{T + C}$$

where T is in °C and P is in kPa. Antoine constants for the components are as follows:

Compound	A	B	C
Acetone	7.117	1210.595	229.664
Acetonitrile	7.559	1482.290	250.523

Determine the moles and composition of the vapour and liquid streams, V, L and x, y respectively.

19. A mixture containing 15 mol per cent ethane, 35 mol per cent propane, and 50 mol per cent n-butane is brought to a condition of 40°C at pressure P. If the mole fraction of the liquid in the system is 0.40, what is pressure P and what are the compositions of the liquid and vapour phases?

20. For the binary system n-pentanol (1)–n-hexane (2), determine the activity coefficients at 31 K in an equimolar mixture. The Wilson parameters are as follows:

$$V_1 = 109.2 \times 10^{-6} \text{ m}^3/\text{mol}$$
$$V_2 = 132.5 \times 10^{-6} \text{ m}^3/\text{mol}$$
$$a_{12} = 7194.18 \text{ J/mol}$$
$$a_{21} = 697.52 \text{ J/mol.}$$

21. n-heptane and toluene form an ideal solution mixture. At 373 K, their vapour pressures are 106 kPa and 74 kPa respectively. Determine the composition of the liquid and vapour in equilibrium at 373 K and 101.3 kPa.

22. Prove that at the azeotropic point, the composition of the vapour and liquid phases are the same.

23. Ethyl alcohol and hexane form an azeotrope containing 33.2 mol per cent ethanol at 58.7°C and 1 atm pressure. Determine the van Laar parameters at this temperature. Assume the system to follow modified Raoult's law. The vapour pressures by Antoine equation are

$$\ln P_1^{Sat} = 16.6758 - \frac{3674.49}{T + 226.45}$$

and

$$\ln P_2^{Sat} = 13.8216 - \frac{2697.55}{T + 224.37}$$

where P is in kPa and T in °C.

24. For the ethyl ethanoate–n-heptane system at 343.15 K, assuming the system to follow modified Raoult's law, predict whether an azeotrope gets formed or not. If it is so, calculate the azeotrope composition and pressure at $T = 343.15$ K. We are given that

$$\ln \gamma_1 = 0.95 x_2^2$$

$$\ln \gamma_2 = 0.95 x_1^2$$
$$P_1^{Sat} = 79.80 \text{ kPa}$$
$$P_2^{Sat} = 40.50 \text{ kPa}.$$

25. For the acetone–cyclohexane system at 50°C and $x_1 = 0.35$, calculate the activity coefficients of the components using the NRTL equation. We are given that

$$b_{12} = 682.21 \text{ cal/mol}$$
$$b_{21} = 1231.47 \text{ cal/mol}$$
$$\alpha = 0.568.$$

26. A binary system of species 1 and 2 has vapour and liquid phases in equilibrium at temperature T, for which

$$\ln \gamma_1 = 1.8 x_2^2$$
$$\ln \gamma_2 = 1.8 x_1^2$$
$$P_1^{Sat} = 1.24 \text{ bar}$$
$$P_2^{Sat} = 0.89 \text{ bar}.$$

Determine the pressure and composition of the azeotrope at temperature T.

27. A mixture comprising 30 mol per cent methane, 10 mol per cent propane and 30 mol per cent n-butane is brought to a condition of -15°C at pressure P, where it exists as a vapour–liquid mixture in equilibrium. If the mole fraction of methane in the vapour phase is 0.80, what is the pressure P in bar?

28. Show that the van Laar and Margules equations are consistent with the Gibbs–Duhem equation.

29. For a mixture of 20 mol per cent methane, 30 mol per cent ethane and 50 mol per cent propane at 10°C, determine

    (a) Dew point pressure
    (b) Bubble point pressure.

    Get K-values from the standard nomograph.

30. The following data are reported for vapour–liquid equilibrium for the ethanol–water system at 298 K. Test whether the data are thermodynamically consistent or not.

x_1	0.122	0.163	0.226	0.320	0.337	0.437	0.440	0.579	0.830
y_1	0.474	0.531	0.562	0.582	0.589	0.620	0.619	0.685	0.849
P(kPa)	5.57	6.02	6.38	6.76	6.80	7.02	7.04	7.30	7.78

Additional Topics in Phase Equilibrium

11

LEARNING OBJECTIVES

After reading this chapter, you will be able to:
- Know about the importance of liquid–liquid equilibrium and its temperature dependence
- Discuss the ternary liquid–liquid equilibrium and its significance
- Know about the solid–liquid equilibrium
- Develop some important relationships based on the solid–vapour equilibrium
- Derive an expression to estimate the osmotic pressure at equilibrium
- Solve some important problems pertaining to the different equilibriums

Apart from the vapour–liquid equilibrium in a binary system, it is also essential to discuss the other phase equilibriums such as liquid–liquid, solid–liquid, solid–vapour, and osmotic equilibrium, because these kinds of equilibria are also of great theoretical and practical importance. Therefore, they have been introduced with suitable practical examples. Some important problems are solved in corroboration with the theoretical background. The development of the relationship between different variables like temperature, pressure, and composition is also substantiated.

11.1 LIQUID–LIQUID EQUILIBRIUM (LLE)

Let us consider liquid–liquid equilibrium of a binary system, say water and chloroform. At uniform temperature and pressure, the equilibrium criteria of liquid–liquid equilibrium in a system is expressed as

$$\bar{f}_i^\alpha = \bar{f}_i^\beta \tag{11.1}$$

where $i = 1, 2, ..., N$.

The superscripts α and β represent the two liquid phases.

Incorporating the activity coefficients to Eq. (11.1), we get

$$x_i^\alpha \gamma_i^\alpha f_i^\alpha = x_i^\beta \gamma_i^\beta f_i^\beta \tag{11.2}$$

Now, if the pure components are in equilibrium and existing as liquids at constant temperature and pressure, then the fugacity of the pure component in both the phases α and β is the same, i.e.

$$f_i^{\alpha} = f_i^{\beta} = f_i \tag{11.3}$$

Then, Eq. (11.2) reduces to

$$x_i^{\alpha}\gamma_i^{\alpha} = x_i^{\beta}\gamma_i^{\beta} \tag{11.4}$$

These activity coefficients, γ_i^{α} and γ_i^{β}, may be derived from excess Gibbs free energy $\dfrac{G^{E}}{RT}$.

Hence they are identical in function. For a binary liquid–liquid equilibrium system at constant pressure or if the reduced temperature is low enough, the influence of pressure on activity coefficients may be considered to be negligible. Therefore, for the system consisting of components 1 and 2, we may write

for component 1

$$x_1^{\alpha}\gamma_1^{\alpha} = x_1^{\beta}\gamma_1^{\beta} \tag{11.5}$$

and

for component 2

$$x_2^{\alpha}\gamma_2^{\alpha} = x_2^{\beta}\gamma_2^{\beta} \tag{11.6}$$

For a binary system, we know that $x_1 + x_2 = 1$. Then

$$x_1^{\alpha} + x_2^{\alpha} = 1 \tag{11.7}$$

$$x_1^{\beta} + x_2^{\beta} = 1 \tag{11.8}$$

Equation (11.6) can be expressed as

$$(1 - x_1^{\alpha})\gamma_2^{\alpha} = (1 - x_1^{\beta})\gamma_2^{\beta} \tag{11.9}$$

Equations (11.5) and (11.9) can be reproduced in the following form by taking logarithm of both sides:

$$\ln\frac{\gamma_1^{\alpha}}{\gamma_1^{\beta}} = \ln\frac{x_1^{\beta}}{x_1^{\alpha}} \tag{11.10}$$

$$\frac{1 - x_1^{\alpha}}{1 - x_1^{\beta}} = \frac{\gamma_2^{\beta}}{\gamma_2^{\alpha}} \tag{11.11}$$

The liquid–liquid equilibrium data are used to provide information for the calculation of vapour–liquid equilibrium. As Eqs. (11.10) and (11.11) are simple expressions, they can be utilized for the computation of activity coefficients.

11.1.1 Effect of Temperature on Liquid–Liquid Equilibrium

The effect of temperature on liquid–liquid equilibria is quite strong. The temperature of two liquid phases of partially miscible liquids increases with miscibility. The temperature at which the miscibility of two components takes place is known as *critical solution temperature*. When two liquid phases merge at critical solution temperature, the first and second derivatives of the activity coefficients are zero. Mathematically for component 1

$$\frac{\partial \ln \gamma_1 x_1}{\partial x_1} = 0$$

and

for component 2

$$\frac{\partial \ln \gamma_2 x_2}{\partial x_2} = 0$$

Consider the aniline–hexane system at 1 atm. At 25°C, if hexane is added to aniline, small amounts of hexane become miscible in aniline. Further addition of hexane will cause the separation of two phases P and Q. The P phase contains an appreciable amount of hexane dissolved in aniline and the Q phase contains small amounts of aniline dissolved in hexane. If such mixture is heated gradually, more hexane will dissolve and heterogeneous phases will become homogeneous. At critical solution, temperature components are completely miscible. When the pressure has no effect on the liquid–liquid equilibrium data, the influence of temperature on composition (T versus x) can be represented on a solubility diagram, as shown in Figures 11.1, 11.2 and 11.3.

The solubility diagram may be of three kinds. These are discussed in the following way:

I. *Binodal Curves*: These island-shaped curves represent the coexistence of the two liquid phases along with their compositions. In Fig. 11.1, the curve BAC represents A phase (B-rich phase) and BDC represents B phase (A-rich phase). T_U is the upper critical solution temperature and T_L is the lower critical solution temperature.

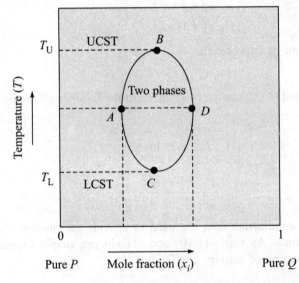

Fig. 11.1 Constant pressure liquid–liquid solubility diagram.

II. *System with UCST*: When the miscibility occurs on increasing temperature, the critical solution temperature is known as *upper critical solution temperature (UCST)*. The mutual solubility of two liquids are shown in Fig. 11.2 by the dome-shaped curve. The point R in the tie line represents the separation of two phases in equilibrium. Then the ratio of the moles of two phases is given by lever rule as

$$\frac{N_P}{N_Q} = \frac{RT}{SR}$$

At temperatures between T_U and T_L, the attainment of liquid–liquid equilibrium is possible. The upper critical solution temperature of other binary systems at atmospheric pressure are:

1. Methanol + n-hexane: 33°C
2. Methanol + cyclohexane: 49°C
3. Cyclohexane + aniline: 29.5°C

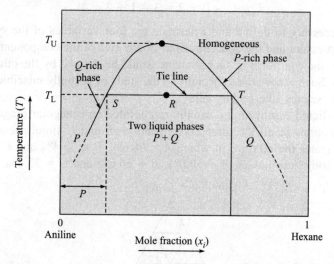

Fig. 11.2 Constant pressure liquid–liquid solubility diagram with upper critical solution temperature (UCST).

III. *System with LCST*: There are some systems in which solubility decreases with increase in temperature. These systems are exothermic in nature. The lower critical solution temperature (LCST) is shown in Fig. 11.3.

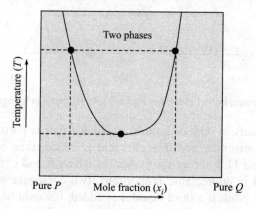

Fig. 11.3 Constant pressure liquid–liquid solubility diagram with lower critical solution temperature (LCST).

The lower critical solution temperatures of other binary systems at atmospheric pressure are:

1. Water + nicotine: 60°C
2. Water + β-picoline: 86°C.

11.2 TERNARY LIQUID–LIQUID EQUILIBRIUM

If all the three components in a ternary system form a homogeneous single phase, which is the minimum number of phases, the degrees of freedom will be maximum. According to Gibbs phase rule, it is given by

$$F = C - P + 2 = 3 - 1 + 2 = 4$$

Hence, it is necessary to define and determine the four variables of the system. These are temperature, total pressure and the concentrations of any two of the components. If the first two variables are fixed, the equilibria in such systems would be defined by the other variables, i.e., two concentrations. Such cases arise especially in the study of partially miscible liquids or in the case of condensed systems like ternary alloys.

Ternary liquid–liquid equilibria are usually represented on triangular diagrams. Figure 11.4 shows the concentrations of all the three components A, B and C simultaneously. The point y in the diagram indicates the mixture in which $A = 66.6\%$, $B = 33.4\%$ and $C = 0.0\%$, and the point x represents a mixture having $A = 50.0\%$, $B = 16.6\%$ and $C = 33.4\%$.

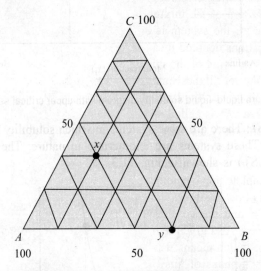

Fig. 11.4 Demonstration of ternary liquid–liquid equilibria in triangular diagram.

Let us consider a ternary system of liquids in which a pair is partially miscible. Water and chloroform are partially miscible, and if acetic acid is added, then a complete homogeneous mixture is possible. Figure 11.5 shows the system in which B and C form a partially miscible binary pair. When mixed in appreciable proportions, two conjugate solutions will be formed, separated into two layers. Now, if a third liquid A is added, it would be distributed between two layers. A is completely miscible with B and C. It is observed that when sufficient amounts of

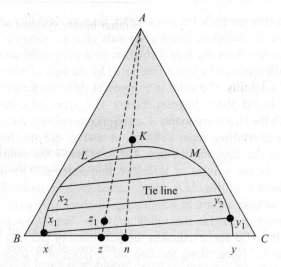

Fig. 11.5 Miscibility diagram for the ternary system.

A are added to the system, the two layers are converted into a single homogeneous mixture. The variation in the miscibility of B and C, when increasing amounts of A are added, is shown in Fig. 11.5.

Suppose we have a mixture of B and C in the proportion found at point z. This mixture will separate into two layers of conjugate solutions having compositions x and y. As A is gradually added, the overall composition of the mixtures of three components will change along Az. For a mixture of composition z_1, the system is still heterogeneous and the two ternary conjugate layers will have concentrations denoted by x_1 and y_1. The line x_1y_1 passing through z_1 is called a *tie line*. The tie line is a measure of the miscibility gap. As more and more A is added, the tie line becomes shorter and shorter till these merge at a point K. The curve xKy is obtained by joining the terminuses of the tie lines, i.e., the limiting values of the miscibilities of B in C and C in B in presence of A. The curve xKy is called a *binodal curve*. Complete miscibility by merging of two layers would occur only at K, i.e., the composition of the two layers at this point becomes identical and the point K is called the *plait point* of the system. In locating the plait point on this curve, it is necessary to start with a mixture of B and C in definite proportion. If we add A to a mixture of B and C, complete homogeneity is obtained on reaching L, which is not necessarily the plait point. It is the tie line terminus, showing disappearance of C-rich M-phase. Only if we start with compositions B and C, we shall reach the plait point K, when progressively the proportion of A is increased. At a given temperature and pressure, the plait point will therefore have a fixed composition, which means it is an invariable point at a given temperature, $F = 0$. Other well-known ternary systems of this type are water–alcohol–ether and lead–zinc–silver.

11.3 SOLID–LIQUID EQUILIBRIUM

In order to study the solid–liquid binary system, we can consider crystallization or melting process, as they belong to solid–liquid mass transfer operations. In the crystallization process, solid particles are formed from a liquid solution. The solution is concentrated and usually cooled until the solute in solute concentration becomes higher than its solubility at that temperature. The solute then comes out of the solution, forming crystals of approximately pure solute. Actually, the transfer of solute occurs from the liquid solution to a pure solid crystalline phase.

A solid–liquid equilibrium can be better illustrated by the help of a temperature–composition diagram. Basically the solubility of a solid in a liquid at different temperatures is conveniently represented on a solid–liquid phase diagram. Figure 11.6 represents the phase diagram of an eutectic system in which the two components *A* and *B* are completely miscible in liquid state and pure components only crystallize from solution. *A* and *B* are the freezing points of pure components *A* and *B*. In the diagram, the line *AEB* represents the saturation concentration or solubility of the solute in the solvent at different temperatures. The line *AEB* is known as *solubility curve*. The point *E* on the *AEB* line is called *eutectic point*. Thus T_E is the eutectic temperature or the lowest temperature at which a liquid phase may exist in the system and *L* is the eutectic composition. Any point in the region above the line *AEB* represents the unsaturated solution. The point on the *AEB* curve indicates the beginning of freezing and the corresponding curve is known as *liquidus*. Here, along the line *AE*, crystals of *A* separate from the solution. Likewise, the line *BE* is the other liquidus which represents the beginning of freezing of *B* from the solution. The end of freezing or beginning of melting is represented by line *CED*. This is known as *solidus*. Now consider a mixture is sufficiently heated to give a homogeneous solution. If the solution is cooled, then the temperature of the solution will come down to the point *K* on the liquidus and the crystals of *A* will start to separate. On further cooling, more crystals separate and the liquid composition changes along the curve *AE*. With progressive cooling, when the eutectic temperature is reached, the entire liquid will have sufficient tendency to solidify.

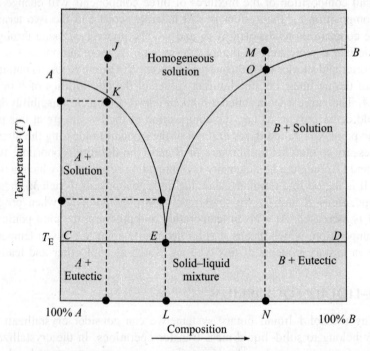

Fig. 11.6 Pressure–temperature composition diagram of binary VLE mixture.

At equilibrium between a saturated solid phase and a liquid phase, we may write for the preceding binary system

$$f_1^s = \bar{f}_1^l \tag{11.12}$$

where

f_1^s = Fugacity of pure solid

$\bar{f}_1^l$ = Fugacity of component i in solution.

Again, we know that

$$\bar{f}_1^l = x_1 \gamma_1 f_1^l \tag{11.13}$$

and

$$f_1^s = x_1 \gamma_1 f_1^l \tag{11.14}$$

Comparing Eqs. (11.13) and (11.14), we get

$$\frac{f_1^s}{f_1^l} = x_1 \gamma_1 \tag{11.15}$$

where x_1 is the mole fraction of component 1 at saturated temperature and pressure.

11.4 DEPRESSION OF FREEZING POINT OF SOLUTION

The difference between the freezing points of the pure solvent and the solution containing the involatile solute is called the *depression of freezing point*. The normal freezing point of a liquid is the temperature at which the liquid and the solid phases have the same vapour pressure. In Fig. 11.7, AB and AC are the vapour–pressure curves of liquid and solid phases of pure solvent. These intersect at A, which indicates the freezing point T_f of the pure solvent, the vapour pressure

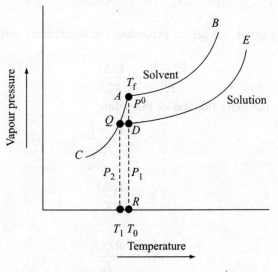

Fig. 11.7 Depression of freezing point of solution.

at A being P_1^0. At the temperature T_f, the vapour pressure of the solution is P_1 (say) at the point R; and this pressure is less than that of the solid, which is P_1^0. Hence if solute (solid) is added to the solution, the solid will melt to attain equilibrium. On further cooling of the solution, the vapour pressure curve will meet the curve AC at the point Q, where the pressure and temperature are supposed to be P_s and T_f'. Therefore, T_f' is the freezing point of the solution, and $T_f' < T_f$. The freezing point is lowered due to the presence of the solute.

Using the Clausius–Clapeyron equation, a quantitative relation between the concentration of the solution and the depression of the freezing point can be derived.

Case 1: The vapour pressures at A and Q are P_1^0 and P_s, and both of them lie on the vapour pressure curve AC of the solid phase.

$$\ln \frac{P_s}{P_1^0} = -\frac{L_s}{R}\left(\frac{1}{T_f'} - \frac{1}{T_f}\right) = -\frac{L_s}{R}\frac{\Delta T}{T_f^2} \quad \text{(putting } T_f - T_f' = \Delta T \text{ and assuming } T_f T_f' \approx T_f^2) \quad (11.16)$$

where L_s is the molar latent heat of sublimation.

Case 2: The vapour pressures at Q and R are P_s and P_1 and both of them lie on the vapour pressure curve DE of the solution.

$$\ln \frac{P_s}{P_1} = -\frac{L_v}{R}\left(\frac{1}{T_f'} - \frac{1}{T_f}\right) = -\frac{L_v}{R}\frac{\Delta T}{T_f^2} \quad (11.17)$$

Subtracting Eq. (11.17) from Eq. (11.16), we get

$$\ln \frac{P_1}{P_1^0} = -\frac{\Delta T}{RT_f^2}(L_s - L_v) = -\frac{L_f \Delta T}{RT_f^2} \quad \text{(since molar latent heat of fusion, } L_f = L_s - L_v) \quad (11.18)$$

Equation (11.18) can be rewritten as

$$\ln\left(1 - \frac{P_1^0 - P_1}{P_1^0}\right) = -\frac{L_f \Delta T}{RT_f^2} \quad (11.19)$$

As $\dfrac{P_1^0 - P_1}{P_1^0}$ is quite small, we get on expanding the logarithmic series

$$\frac{P_1^0 - P_1}{P_1^0} = \frac{L_f \Delta T}{RT_f^2} \quad (11.20)$$

From Raoult's law, the mole fraction of the solute

$$x_2 = \frac{P_1^0 - P_1}{P_1^0}$$

Hence

$$x_2 = \frac{L_f \Delta T}{RT_f^2}$$

or

$$\Delta T = \frac{RT_f^2 x_2}{L_f} \quad (11.21)$$

That is, the depression of freezing point is proportional to the mole fraction of the solute. Equation (11.21) is known as *van't Hoff's law of freezing point depression*.

In case of dilute solution, mole fraction

$$x_2 \approx \frac{n_2}{n_1} = \frac{\dfrac{w_2}{M_2}}{\dfrac{w_1}{M_1}}$$

where M_2 and M_1 are the molecular weights and w_1 and w_2 are the weights of solvent and solute. Thus

$$\Delta T = \frac{RT_f^2}{1000 \times L_f}\left(\frac{w_2 \times 1000}{w_1 M_2}\right) \tag{11.22}$$

or

$$\Delta T = \frac{RT_f^2}{1000 \times L_f}m \tag{11.23}$$

where

L_f = Latent heat of fusion per gram
m = Molality of the solution.

Hence, Eq. (11.23) reduces to

$$\Delta T = K_f m \tag{11.24}$$

where K_f is the molal depression constant.

Thus the depression of freezing point is proportional to the molality of the solution.

EXAMPLE 11.1   A solution containing 2.423 g of sulphur in 100 g of naphthalene (melting point 80.1°C) gave a freezing point depression of 0.64°C. The latent heat of fusion of naphthalene is 35.7 cal/g. What is the molecular formula of sulphur in the solution?

Solution: We are given that

$\Delta T_f = 0.64°C$
$T_f = 353.1°C$
$L_f = 35.7$ cal/g
$a = 2.423$ g
$b = 100$ g.

We know that

$$\Delta T_f = \frac{RT_f^2}{100 \times L_f}\frac{a \times 1000}{b \times M_2}$$

$$0.64 = \frac{2 \times (353.1)^2}{1000 \times 35.7} \times \frac{2.423 \times 1000}{100 \times M_2} \qquad \left(\begin{array}{l}\text{Since } R = 1.987 \text{ cal/mol-K} \\ \simeq 2 \text{ cal/mol-K}\end{array}\right)$$

$$M_2 = 264$$

But the atomic weight of sulphur is 32. So the molecular formula of sulphur here is S_8.

EXAMPLE 11.2 Calculate the molal freezing point depression constant for benzene if its latent heat of fusion at 5°C is 9.83 kJ/mol.

Solution: We are given that

$$T_f = 5 + 273 = 278 \text{ K}$$
$$L_f = 9.83 \text{ kJ/mol} = 9830 \text{ J/mol}$$
$$R = 8.314 \text{ J/mol-K}.$$

Substituting these values into the following equation, we get

$$K_f = \frac{RT_f^2 M_1}{1000 \times L_f} = \frac{8.314 \times (278)^2 \times 78}{1000 \times 9830} = 5.08 \text{ kmol}^{-1}\text{kg}$$

EXAMPLE 11.3 The melting point of phenol is 40°C. A solution containing 0.172 g acetanilide (C_8H_9ON) in 12.54 g phenol freezes at 39.25°C. Calculate the freezing point depression constant and the latent heat of fusion of phenol.

Solution: We are given that
Freezing point depression, $\Delta T = 40 - 39.25 = 0.75°C$
Molecular weight of acetanilide, $M_2 = 135$.
Substituting these values into the following equation, we get

$$K_f = \frac{\Delta T \cdot b \cdot M_2}{1000 \times a} = \frac{0.75 \times 12.54 \times 135}{0.172 \times 1000} = 7.38$$

Therefore, the freezing point depression constant is 7.38.
Now, the latent heat of fusion of phenol can be computed as

$$K_f = 7.38 = \frac{RT_f^2}{1000 \times L_v} = \frac{2 \times (313)^2}{1000 \times L_v}$$

or

$$L_v = \frac{2 \times (313)^2}{1000 \times 7.38} = 26.5 \text{ cal/g}$$

11.5 ELEVATION OF BOILING POINT OF SOLUTION

Like depression of freezing point of a solution, we can determine the elevation of boiling point, considering the equilibrium between the liquid and vapour phases. The vapour pressure of a pure solvent increases with rise in temperature. When a non-volatile solute is added to a solvent, the vapour pressure of the solution will be lower than that of the pure solvent at any given temperature. A liquid boils when its vapour pressure becomes equal to that of the super-incumbent pressure. Let T_b be the boiling temperature of the pure solvent. Its vapour pressure equals one atmosphere at this temperature. But at this temperature, the solution will have a boiling point lower than one atmosphere and it will not boil. On further increase in temperature, say to T_b', the vapour pressure rises to one atmosphere and at this point the solution starts boiling. This means that the solution has a higher boiling point than that of the pure solvent. The increase in boiling temperature is given by

$$PR = T_b' - T_b = \Delta T$$

and increase in pressure is given by

$$PQ = P_1^0 - P_1 = \Delta P$$

Since the solutions are dilute, ΔT and ΔP are small quantities.

In the solution vapour pressure curve CD of Fig. 11.8, at the point Q, the pressure is P_1 and temperature is T_b. At the point R, the pressure is P_1^0 and temperature is T_b'.

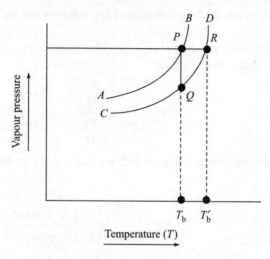

Fig. 11.8 Elevation of boiling point of solution.

Using the Clausius–Clapeyron equation, a quantitative relation between the concentration of the solution and the elevation of the boiling point can be derived.

$$\ln \frac{P_1}{P_1^0} = -\frac{L_v}{R}\left(\frac{1}{T_b} - \frac{1}{T_b'}\right) = -\frac{L_v}{R}\frac{T_b' - T_b}{T_b' T_b} \tag{11.25}$$

where L_v is the molar latent heat of vaporization
or

$$\ln\left(1 - \frac{P_1^0 - P_1}{P_1^0}\right) = -\frac{L_v \Delta T}{RT_b^2} \qquad \text{(putting } T_b - T_b' = \Delta T \text{ and assuming } T_b T_b' \approx T_b^2\text{)}$$

or

$$\ln\left(1 - \frac{\Delta T}{P_1^0}\right) = -\frac{L_v \Delta T}{RT_b^2} \tag{11.26}$$

As $\dfrac{\Delta P}{P_1^0}$ is quite small, we get on expanding and neglecting higher powers

$$-\frac{\Delta P}{P_1^0} = -\frac{L_v \Delta T}{RT_b^2} \tag{11.27}$$

But from Raoult's law, we get $\dfrac{\Delta P}{P_1^0} = x_2 =$ mole fraction of the solute.

Therefore, Eq. (11.27) becomes

$$x_2 = \frac{L_v \Delta T}{RT_b^2}$$

or

$$\Delta T = \frac{RT_b^2}{L_v} x_2 \tag{11.28}$$

This is the relation between the concentration of the solution and the elevation of the boiling point.

Equation (11.28) is known as *van't Hoff's law of boiling point elevation*.

In case of dilute solution, mole fraction

$$x_2 \approx \frac{n_2}{n_1} = \frac{\dfrac{w_2}{M_2}}{\dfrac{w_1}{M_1}}$$

where M_2 and M_1 are the molecular weights and w_1 and w_2 are grams of solute and solvent respectively.

Thus

$$\Delta T = \frac{RT_b^2}{1000 \times L_v}\left(\frac{w_2 \times 1000}{w_1 M_2}\right) \tag{11.29}$$

or

$$\Delta T = \frac{RT_b^2}{1000 \times L_v} m \tag{11.30}$$

where

$L_v =$ Latent heat of vaporization per gram
$m =$ Molality of the solution.

Hence, Eq. (11.30) reduces to

$$\Delta T = K_b m \tag{11.31}$$

where $K_b =$ Molal elevation constant $= \dfrac{RT_b^2}{1000 \times L_v}$.

Therefore, the elevation of the boiling point is proportional to the molality of the solution.

EXAMPLE 11.4   The boiling point of acetic acid is 118.1°C and its latent heat of vaporization is 121 cal/g. A solution containing 0.4344 g anthracene in 44.16 g acetic acid boils at 118.24°C. What is the molecular weight of anthracene?

Solution: Now

$$\Delta T_b = 118.24 - 118.1 = 0.14°C$$

If a g solute is dissolved in b g solvent, then the molecular weight of solute can be expressed as

$$M_2 = K_b \frac{a \times 1000}{b \times \Delta T_b} \tag{i}$$

Putting $K_b = \dfrac{RT_b^2}{1000 \times L_v}$ in Eq. (i), we get

$$M_2 = \frac{RT_b^2}{1000 \times L_v} \cdot \frac{a \times 1000}{b \times \Delta T_b}$$

$$= \frac{2 \times (391.1)^2}{1000 \times 121} \cdot \frac{0.4344 \times 1000}{44.16 \times 0.14} = 178$$

EXAMPLE 11.5   The boiling point elevation of a solvent is observed to be 2.3 K when 13.86 g of a solute is added to 100 g. Calculate molal elevation constant K_b and molar latent heat of vaporization L_v of the solvent. Given that
Molar mass of solvent = 78 g/mol
Boiling point = 353.1 K
Molar mass of solute = 154 g/mol.

Solution: We have

$$w_1 = 100 \text{ g} \qquad w_2 = 13.86 \text{ g}$$
$$M_1 = 78 \text{ g/mol} \qquad M_2 = 154 \text{ g/mol}$$
$$\Delta T_b = 2.3 \text{ K} \qquad T_b = 353.1 \text{ K}$$
$$R = 8.314 \text{ J/kmol.}$$

Molality

$$m = \frac{w_2 \times 1000}{w_1 M_2} = \frac{13.86 \times 1000}{154 \times 100} = 0.9 \text{ mol/kg}$$

We know that the molal elevation constant

$$K_b = \frac{\Delta T_b}{m} = \frac{2.3 \text{ K}}{0.9 \text{ mol/kg}} = 2.55 \text{ K-kg/mol}$$

Now, the molar latent heat of vaporization can be estimated as

$$L_v = \frac{RT_b^2}{1000 \times \Delta T} m = \frac{RT_b^2 M_1}{1000 \times K_b} = \frac{8.314 \times (353.1)^2 \times 78}{1000 \times 2.55} = 31701 \text{ J/mol}$$

11.6   SOLID–VAPOUR EQUILIBRIUM

Let us consider a binary system consisting of a pure solid (1) and its vapour (2). Here, vapour is assumed to be insoluble in solid phase. Here, the solid is solute component and vapour is solvent component. Now, when the solid is in equilibrium with its vapour, the mole fraction of the solute in the vapour phase, i.e. y_1, can be estimated by considering it as a function of temperature and pressure.

Since the solute is pure solid and one in which its vapour cannot be dissolved, then from the criterion for equilibrium between two phases, we may write

$$f_1^s = f_1^v \qquad (11.32)$$

The fugacity of pure solid at T and P is given by

$$f_1^s = P_1^{Sat}\phi_1^{Sat}\exp\left(\frac{V_1^s(P - P_1^{Sat})}{RT}\right) \qquad (11.33)$$

where

V_1^s = Molar volume of solid;

P_1^{Sat} = Solid–vapour saturation pressure at temperature T;

ϕ_1^{Sat} = Fugacity coefficient at solid–vapour saturation pressure P_1^{Sat}.

Again, we know that at phase equilibrium, the fugacity for the vapour phase at saturated condition is given by

$$\bar{f}_1^v = y_1\bar{\phi}_1 P \qquad (11.34)$$

Substituting Eq. (11.34) into Eq. (11.32), we have

$$y_1\bar{\phi}_1 P = P_1^{Sat}\phi_1^{Sat}\exp\left[\frac{V_1^s(P - P_1^{Sat})}{RT}\right]$$

$$y_1 = \frac{P_1^{Sat}}{P}\frac{\phi_1^{Sat}}{\bar{\phi}_1}\exp\left[\frac{V_1^s(P - P_1^{Sat})}{RT}\right] \qquad (11.35)$$

$$y_1 = \frac{P_1^{Sat}}{P}W \qquad (11.36)$$

where $W = \dfrac{\phi_1^{Sat}}{\bar{\phi}_1}\exp\left[\dfrac{V_1^s(P - P_1^{Sat})}{RT}\right]$ = Enhancement factor.

At sufficiently low pressure, the effect of pressure on the fugacity of the solid is considerably negligible. In that case, $W \approx 1$, so the expression in Eq. (11.36) becomes

$$y_1 = \frac{P_1^{Sat}}{P} \qquad (11.37)$$

At moderate and high pressures, as the enhancement factor is always greater than unity, i.e., $W > 1$, the Poynting pressure correction factor can not be avoided. It is also greater than one. Therefore, $W \rightarrow 1$ as $P \rightarrow P_1^{Sat}$.

11.7 OSMOTIC PRESSURE AND EQUILIBRIUM

Osmotic pressure of the solution is the pressure required to prevent osmosis when the solution is separated from pure solvent by a semi-permeable membrane due to a differential in the concentrations of solute. The magnitude of the osmotic pressure depends on various factors, such as temperature and concentration. But it is independent of the nature of the semi-permeable membrane.

Let us consider a vessel divided into two compartments by a semi-permeable membrane, as shown in Fig. 11.9. Compartment 1 contains a liquid mixture consisting of solvent (1) and solute (2). Compartment 2 contains pure solvent. The partition is permeable to solvent of liquid mixture only. Temperature is uniform and constant throughout the vessel. The piston moves around the vessel in such a way that the two compartments can adjust the pressure. Now, imagine that the pressures P_1 and P_2 inside both the compartments are the same, i.e., $P_1 = P_2$.

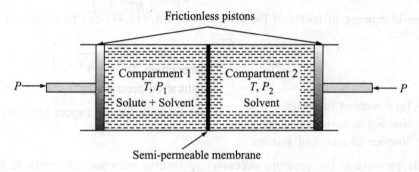

Fig. 11.9 Demonstration of osmotic pressure.

Since the pressures are equal, fugacity and concentration of the solvent belonging to compartment 1 are comparatively smaller than that of compartment 2. The difference in concentration as well as fugacity causes a driving force for mass transfer. If the pressure P_1 is increased to the level P^*, then the system will attain an equilibrium. The equilibrium is known as *osmotic equilibrium*.

Therefore, the difference in pressure known as osmotic pressure is given by

$$\Pi = P_2 - P^* \tag{11.38}$$

From the knowledge of criterion of equilibrium, we have

$$f_1^1 = \bar{f}_1^2 = \gamma_1 x_1 f_1^0 \tag{11.39}$$

Applying the criterion of equilibrium to Poynting pressure correction factor, we have

$$f_1^0 = f_1^1 \exp\left[\frac{V_1^1(P - P^*)}{RT}\right] \tag{11.40}$$

where

f_1^0 = Fugacity of pure solvent at standard state

f_1^1 = Fugacity of the pure solvent at pressure P_1.

Substituting Eq. (11.39) into Eq. (11.40), we obtain

$$f_1^0 = \gamma_1 x_1 f_1^0 \exp\left[\frac{V_1^1(P_2 - P^*)}{RT}\right] \tag{11.41}$$

or

$$\gamma_1 x_1 \exp\left[\frac{V_1^1(P_2 - P^*)}{RT}\right] = 1 \tag{11.42}$$

Since $\Pi = P_2 - P$, then Eq. (11.42) reduces to

$$\Pi = \frac{RT}{V_1^l} \ln(\gamma_1 x_1) \tag{11.43}$$

In case of dilute solution

$$\Pi = -\frac{RT}{V_1^l} \ln(\gamma_1 x_1) = \frac{RT x_2}{V_1^l} \tag{11.44}$$

In terms of number of moles of the components, Eq. (11.44) can be expressed as

$$\Pi = \frac{RT x_2}{V_1^l} = \frac{RT n_2}{V_1^l N} \qquad \left(\text{as } x_2 = \frac{n_2}{N} = \frac{n_2}{n_1 + n_2} \right) \tag{11.45}$$

where

N = Total number of moles;
n_2 = Number of moles of solute;
n_1 = Number of moles of solvent.

Osmotic pressure is the pressure necessary to reverse osmosis and return to the initial condition. It can normally be measured by an instrument known as *osmometer*, which measures the osmotic pressure in suitable pressure units. Osmotic pressure at a given temperature depends upon the molar concentration.

EXAMPLE 11.6 Calculate the osmotic pressure at 50°C of a glucose ($C_6H_{12}O_6$) solution that has 60 g of glucose dissolved in enough water to make 1500 g of solution.

Solution: Molecular weight of glucose = 180 g
Mole fractions of glucose

$$x_2 = \frac{\dfrac{60}{180}}{\dfrac{60}{180} + \dfrac{1500}{18}} = 0.004$$

Applying Eq. (11.45), we get

$$\Pi = \frac{RT x_2}{V_1^l} = \frac{8.314 \times 323 \times 0.004}{V_1^l}$$

Molar volume of water

$$= V_1^l = 18 \times 10^{-6} \text{ m}^3/\text{mol}$$

Therefore, the osmotic pressure

$$\Pi = \frac{8.314 \times 323 \times 0.004}{18 \times 10^{-6}} = 5.96 \times 10^5 \text{ N/m}^2 = 5.96 \text{ bar} = 596.76 \text{ kPa}$$

EXAMPLE 11.7 A solution contains 0.6 g urea and 1.8 g glucose in 100 cc of water at 27°C. Calculate the osmotic pressure of the solution.

Solution: Molecular weight of urea is 60 g and that of glucose is 180 g. We know that osmotic pressure

$$\Pi = \frac{RTx_2}{V_1^l} = CRT$$

where C is the concentration per litre.
  Here

$$C = \frac{6}{60} + \frac{18}{180} = 0.2$$

Therefore, the osmotic pressure is given by

$$\Pi = CRT = 0.2 \times 0.082 \times 300 = 4.92 \text{ atm}$$

KEY TERMS

Binodal Curve   The island-shaped curve in the constant-pressure liquid–liquid solubility diagram which represents the coexistence of the two liquid phases along with their compositions.

Critical Solution Temperature   The temperature at which the mixing of two components takes place.

Osmotic Pressure   The pressure required to prevent osmosis when the solution is separated from pure solvent by a semi-permeable membrane due to a differential in the concentrations of solute.

Triangular Diagram   The diagram indicating the variation of concentrations in equilibria of three-component systems.

Upper Critical Solution Temperature (UCST)   The critical solution temperature when miscibility occurs on increasing temperature.

IMPORTANT EQUATIONS

1. The equilibrium criterion of liquid–liquid equilibrium in a system is given by

$$\bar{f}_i^\alpha = \bar{f}_i^\beta \quad \text{where } i = 1, 2, ..., N \tag{11.1}$$

2. In terms of the activity coefficients

$$x_i^\alpha \gamma_i^\alpha f_i^\alpha = x_i^\beta \gamma_i^\beta f_i^\beta \tag{11.2}$$

3. For solid–liquid equilibrium

$$\frac{f_1^s}{f_1^l} = x_1 \gamma_1 \tag{11.15}$$

4. In case of solid–vapour equilibrium

$$y_1 = \frac{P_1^{Sat}}{P} \frac{\phi_1^{Sat}}{\bar{\phi}_1} \exp\left[\frac{V_1^s(P - P_1^{Sat})}{RT}\right] \tag{11.35}$$

$$= \frac{P_1^{Sat}}{P} W \tag{11.36}$$

where $W = \dfrac{\phi_1^{Sat}}{\phi_1} \exp\left[\dfrac{V_1^s(P - P_1^{Sat})}{RT}\right] =$ Enhancement factor.

5. The osmotic pressure is given by

$$\Pi = P_2 - P^* \tag{11.38}$$

6. For osmotic equilibrium

$$f_1^0 = f_1^1 \exp\left[\frac{V_1^1(P - P^*)}{RT}\right] \tag{11.40}$$

where

$f_1^0 =$ Fugacity of pure solvent at standard state

$f_1^1 =$ Fugacity of pure solvent at pressure P_1.

7. Another important form of osmotic pressure is given by

$$\Pi = \frac{RTx_2}{V_1^1} = \frac{RTn_2}{V_1^1 N} \tag{11.45}$$

EXERCISES

A. Review Questions

1. Discuss the significance of liquid–liquid equilibrium in extraction process.
2. Explain the effect of temperature on liquid–liquid equilibrium with the help of a neat graphical representation.
3. Define the term *critical solution temperature*.
4. What do you mean by depression of freezing point? Discuss van't Hoff's law of freezing point depression of a solution.
5. What is ternary liquid–liquid equilibrium? How does a triangular diagram play an important role in explaining the phase equilibrium in a ternary system?
6. Give an informatory note on miscibility diagram for the ternary system.
7. Discuss the solid–vapour equilibrium for sublimation process.
8. What is boiling point elevation? Derive an expression on the relationship between the concentration of the solute and the boiling point elevation.
9. Show that the elevation of the boiling point is proportional to the molality of the solution.
10. What is osmotic pressure? How does it play an important role in osmotic equilibrium?
11. Explain osmotic equilibrium with the help of a neat sketch.
12. Justify the statement: *Osmosis is of paramount importance in biological systems.*

B. Problems

1. The freezing point of pure benzene is 5.44°C and that of the solution containing 2.092 g of benzaldehyde in 100 g of benzene is 4.44°C. Calculate the molecular weight of benzaldehyde when K_f for benzene is 5.1.

2. The molality of dissolved gases in water at 0°C and 1 atm is 1.29×10^{-3}. The decrease in volume during melting of ice is 0.0907 cc/g. The latent heat of fusion is 1436.3 cal/mol. The vapour pressure at triple point is 4.58 mm. Calculate the triple point temperature.

3. Calculate the mass of methyl alcohol which, when dissolved in 100 g of water, would just prevent the formation of ice at −10°C, given that K_f is 1.86 K/mol-kg.

4. A hydrocarbon $H_2(CH_2)_n$ is dissolved in a solvent S which freezes at 9.0°C. A solution which contains 0.90 g of hydrocarbon per 180 g of the solvent freezes at 8.48°C. Calculate the molecular weight of the hydrocarbon and its formula. $K_f = 12$ K/mol-kg.

5. Estimate the depression in the freezing point if 58.5 g of NaCl is added to 1 L of water at atmospheric pressure.

6. Calculate the molal boiling point elevation constant of benzene if its heat of vaporization at 80.1°C is 30.67 kJ/mol.

7. Determine the boiling point elevation of water when 5 g sucrose is added to 100 g of pure water at 100°C. Molecular weight of sucrose is 342.30.

8. A sample of benzene which is contaminated with non-volatile substances boils at 3°C higher than that of a sample of pure benzene. The normal boiling point of benzene is 80.1°C and the latent heat of vaporization is 30.76 kJ/mol. Estimate the mol per cent of the contaminants in the sample of benzene.

9. A solution containing 0.50 g of a sample X in 50 g of a solvent yields a boiling point elevation of 0.40 K, while a solution of 0.60 g of sample Y in the same mass of solvent gives a boiling point elevation of 0.60 K. Find out the molar mass of sample Y. The molar mass of X is 128 g/mol.

10. Calculate the osmotic pressure of a sugar solution that contains 100 g of sucrose ($C_{12}H_{22}O_{11}$) dissolved in enough water to make 1 L of solution at 25°C.

11. Estimate the osmotic pressure of an aqueous solution containing 5 g of glucose ($C_6H_{12}O_6$) and 5 g of sucrose ($C_{12}H_{22}O_{11}$) in 1 L of water at 25°C.

12. The average osmotic pressure of human blood is 0.78 MPa at the body temperature of 37°C. Calculate the concentration of NaCl in water which can be used as an extender of blood in case of emergency.

13. Human blood is isotonic with 0.9% NaCl solution at 27°C. What is the osmotic pressure?

14. An evaporator operating at 150 kPa is fed with 5 mol per cent sodium hydroxide solution. Estimate the temperature at which the solution boils in the evaporator.

Chemical Reaction Equilibria

12

LEARNING OBJECTIVES

After reading this chapter, you will be able to:

- Know the definition and importance of reaction co-ordinates
- Develop the general criterion for chemical reaction equilibrium
- Define and evaluate the standard Gibbs free energy change and the equilibrium constant
- Establish the relationship between the standard Gibbs free energy change and the equilibrium constant
- Predict the effect of temperature on the equilibrium constant and deduce van't Hoff's equation
- Judge the pressure dependence of the equilibrium constant of a chemical reaction
- Discuss the influence of other factors like inert materials, products and excess reactants on the equilibrium constant
- Analyze the equilibrium conversion for homogeneous and heterogeneous reactions
- Represent the phase rule for reacting systems
- Evaluate the equilibrium composition for simultaneous reactions
- Understand the importance of fuel cell

The reactor is the heart of the chemical process plants, in which the raw materials are converted into useful products by means of chemical reaction. The foremost responsibility of a chemical engineer is to be familiar with the proper design of a chemical reactor and the operations which take place in reactors. This is because chemical engineers need to apply the principle of thermodynamics and chemical kinetics to the chemically reacting systems to determine the degree of conversion at equilibrium. The rate and equilibrium conversion of a chemical reaction depends upon different operating conditions like temperature, pressure, and composition of the reactants. Some of the reactions are greatly influenced by the presence of catalysts. For example, oxygen is produced from potassium chlorate in the presence of a positive catalyst—manganese dioxide. Sulphur dioxide is oxidized to form sulphur trioxide. The reaction takes place in the presence of vanadium pentaoxide. Again, the reaction leading to the formation of ammonia is affected strongly by temperature and pressure. An appreciable extent of yield can be achieved at optimum pressure and proper temperature. Therefore, the aim of this chapter is to predict the effect of temperature and pressure on the degree of conversion of a chemical reaction and also to estimate the equilibrium constant.

12.1   REACTION COORDINATE

The progress of a chemical reaction along a pathway can be better represented by introducing an important factor known as *reaction coordinate*. This is one-dimensional coordinate. It is a geometric parameter that changes during the conversion of one or more molecular entities. The reaction coordinate implies the extent to which a chemical reaction taken place.

Consider the following chemical reaction:

$$P + 3Q \rightarrow 2R$$

For this reaction, the stoichiometric numbers are

$$V_R = 2 \qquad V_P = -1 \qquad \text{and} \qquad V_Q = -3$$

The reaction can also be represented in the following form:

$$0 = 2R - P - 3Q$$

The stoichiometric numbers are as follows:

Products +ve

Reactants −ve

Inert 0

Now, the changes in the number of moles of the species present in the reaction are directly proportional to their respective stoichiometric numbers.

$$\frac{\Delta n_P}{V_P} = \frac{\Delta n_Q}{V_Q} = \frac{\Delta n_R}{V_R} \tag{12.1}$$

It can be stated in another way. For the formation of 1 mol of R, 0.5 mol of P and 1.5 mol of Q are consumed. In that case

$$\Delta n_P = -0.5 \qquad \Delta n_Q = -1.5 \qquad \Delta n_R = 1$$

Equation (12.1) can be represented in differential form as

$$\frac{dn_P}{V_P} = \frac{dn_Q}{V_Q} = \frac{dn_R}{V_R} = \cdots = d\varepsilon \tag{12.2}$$

Here, ε is the reaction coordinate, which is defined as the degree to which a reaction has advanced. Generally, the change in the extent of reaction is the same for each component, but the change in number of moles dn_i are different for species i. It can be expressed as

$$\frac{dn_i}{V_i} = d\varepsilon$$

or

$$dn_i = V_i d\varepsilon \qquad (12.3)$$

In the initial state at $\varepsilon = 0$, the number of moles of individual component may be specified as $n_i = n_{i0}$. Then, Eq. (12.3) can be integrated to give

$$\int_{n_{i0}}^{n_i} dn_i = v_i \int_0^\varepsilon d\varepsilon \qquad (12.4)$$

or

$$n_i - n_{i0} = v_i \varepsilon$$

or

$$n_i = n_{i0} + v_i \varepsilon \qquad (12.5)$$

where

n_i = Number of moles present after the reaction
n_{i0} = Number of moles of the species initially present in the system.

Then the total number of moles are given by

$$n = \sum_i n_i = \sum_i n_{i0} + \varepsilon_i \sum_i v_i$$
$$= n_0 + v\varepsilon \qquad (12.6)$$

where

$$\sum_i n_{i0} = n_0 \quad \text{and} \quad \sum_i v_{i0} = v_0$$

Therefore, the mole fraction of component i in the reaction mixture is given by

$$y_i = \frac{n_i}{n} = \frac{n_{i0} + v_i \varepsilon}{n_0 + v\varepsilon} \qquad (12.7)$$

EXAMPLE 12.1   The following reaction takes place in a system consisting of 3 mol CH_4, 1 mol H_2O, 1 mol CO and 4 mol H_2 initially:

$$CH_4 + H_2O \rightarrow CO + 3H_2$$

Express the composition of the mixture in terms of mole fraction as a function of extent of reaction.

Solution: For the given reaction, the total number of moles of various components initially present

$$n_0 = \sum_i n_{i0} = 3 + 1 + 1 + 4 = 9$$

The total stoichiometric numbers

$$v = \sum_i v_{i0} = 1 + 3 - 1 - 1 = 2$$

Let

ε = Extent of reaction

Moles of $CH_4 = 3 - \varepsilon$
Moles of $H_2O = 1 - \varepsilon$
Moles of $CO = 1 + \varepsilon$
Moles of $H_2 = 4 + 3\varepsilon$.

Total moles of the reaction mixture

$$n = \sum_i n_i = n_0 + v\varepsilon$$

$$= (3 - \varepsilon) + (1 - \varepsilon) + (1 + \varepsilon) + (4 + 3\varepsilon) = 9 + 2\varepsilon$$

Then the mole fractions of different components

$$y_{CH_4} = \frac{3 - \varepsilon}{9 + 2\varepsilon} \qquad y_{H_2O} = \frac{1 - \varepsilon}{9 + 2\varepsilon} \qquad y_{CO} = \frac{1 + \varepsilon}{9 + 2\varepsilon} \qquad y_{H_2} = \frac{4 + 3\varepsilon}{9 + 2\varepsilon}$$

12.2 MULTIREACTION STOICHIOMETRY

When two or more independent reactions take place simultaneously, then the change in number of moles of species n_i due to the several reactions is generally accounted for by introducing a separate reaction coordinate ε_j. Then Eq. (12.3) can be represented for the multireaction by

$$dn_i = \sum_i v_{i,j} \, d\varepsilon_j \qquad (i = 1, 2, ..., N) \tag{12.8}$$

where

$v_{i,j}$ = Stoichiometric number of species i of reaction j
ε_j = Extent or advancement of reaction.

On integration with the integral limit, $n_i = n_{i_0}$ and $\varepsilon_j = 0$. Equation (12.8) gives

$$\int_{n_{i_0}}^{n_i} dn_i = v_{i,j} \int_0^{\varepsilon_j} d\varepsilon_j \tag{12.9}$$

or

$$n_i = n_{i_0} + \sum_i v_{i,j} \varepsilon_j \tag{12.10}$$

where

n_i = Number of moles present after the reaction
n_{i_0} = Number of moles of species initially present in the system.

Then the total number of moles of species in the mixture are given by

$$n = \sum_i n_{i_0} + \sum_i \sum_{i,j} v_{i,j} \varepsilon_j$$

$$= n_0 + \sum_j \left(\sum_i v_{i,j} \right) \varepsilon_j \tag{12.11}$$

Now, if the stoichiometric number $v_j = \sum_i v_{i,j}$, then Eq. (12.11) becomes

$$n = n_0 + \sum_j v_j \varepsilon_j \tag{12.12}$$

Substituting the value of n in Eq. (12.12), the mole fraction y_i of species i after the reaction can be expressed as

$$y_i = \frac{n_i}{n} = \frac{n_{i_0} + \sum_j v_{i,j}\varepsilon_j}{n_0 + \sum_j v_j \varepsilon_j} \tag{12.13}$$

EXAMPLE 12.2   A system initially charged with 3 mol CH_4 and 4 mol H_2O is undergoing the following reaction:

$$CH_4 + H_2O \rightarrow CO + 3H_2 \tag{A}$$
$$CH_4 + 2H_2O \rightarrow CO_2 + 4H_2 \tag{B}$$

Derive the expression for the mole fractions of the components in terms of extent of the reactions.

Solution: For the two given reactions, the total number of moles of various components initially present

$$n_0 = \sum_i n_{i_0} = 3 + 4 = 7$$

For reaction (A), total stoichiometric numbers, $v_A = 1 + 3 - 1 - 1 = 2$.
For reaction (B), total stoichiometric numbers, $v_B = 1 + 4 - 1 - 2 = 2$.
Let

ε_A = Extent of reaction (A)

and

ε_B = Extent of reaction (B).

Then the required expression is as follows:

Components	$v_{i,j}$		n_{i_0}	$n_i = n_{i_0} + \sum_j v_{i,j}\varepsilon_j$	$y_i = \dfrac{n_i}{n}$
	v_A	v_B			
CH_4	1	1	3	$3 - \varepsilon_A - \varepsilon_B$	$\dfrac{3 - \varepsilon_A - \varepsilon_B}{7 + 2\varepsilon_A + 2\varepsilon_B}$
H_2O	1	2	4	$4 - \varepsilon_A - 2\varepsilon_B$	$\dfrac{4 - \varepsilon_A - 2\varepsilon_B}{7 + 2\varepsilon_A + 2\varepsilon_B}$
CO	1	–	0	ε_A	$\dfrac{\varepsilon_A}{7 + 2\varepsilon_A + 2\varepsilon_B}$
CO_2	–	1	0	ε_B	$\dfrac{\varepsilon_B}{7 + 2\varepsilon_A + 2\varepsilon_B}$
H_2	3	4	0	$3\varepsilon_A + 4\varepsilon_B$	$\dfrac{3\varepsilon_A + 4\varepsilon_B}{7 + 2\varepsilon_A + 2\varepsilon_B}$

$$n = n_0 + \sum_j v_j \varepsilon_j = 7 + 2\varepsilon_A + 2\varepsilon_B$$

12.3 EQUILIBRIUM CRITERION OF A CHEMICAL REACTION

We know that as the process proceeds the total Gibbs free energy of a closed system in an actual process must decrease at constant temperature and pressure (during an irreversible process), i.e.

$$(dG)_{T,P} \leq 0 \tag{12.14}$$

Accordingly, a chemical reaction must proceed in the direction of decreasing G. Hence the reaction will ultimately stop and the equilibrium will be reached when Gibbs free energy G attains the minimum value.

At the equilibrium state

$$(dG)_{T,P} = 0 \tag{12.15}$$

This is the equilibrium condition or criterion in a chemical reaction. So, the virtual change occurs in a system at constant temperature and produces no change in G of the system. We can prove it.

We know that Gibbs free energy

$$G = H - TS$$
$$dG = dH - TdS - SdT$$
$$dG = dU + PdV + VdP - TdS - SdT$$
$$(dG)_{T,P} = dU + PdV - TdS = dQ - TdS = 0$$

It can alternatively be shown as

$$dG = -SdT + VdP + \sum \mu_i dn_i \tag{12.16}$$

At constant temperature and pressure, Eq. (12.16) reduces to

$$(dG)_{T,P} = \sum \mu_i dn_i = 0 \tag{12.17}$$

For a reaction

$$aA + bB \rightarrow rR + sS$$

$$\frac{dn_A}{a} = \frac{dn_B}{b} = \frac{dn_R}{r} = \frac{dn_S}{s} = \cdots = d\varepsilon \tag{12.18}$$

or

$$\frac{dn_i}{V_i} = d\varepsilon$$

or

$$dn_i = V_i d\varepsilon$$

Substituting the value of dn_i into Eq. (12.17), we have

$$(dG)_{T,P} = \sum \mu_i dn_i = \sum \mu_i v_i d\varepsilon = 0$$

or

$$(dG)_{T,P} = \sum \mu_i v_i = 0 \tag{12.19}$$

This is the criterion of equilibrium of a chemical reaction. The change in free energy with extent of reaction can be graphically represented with the help of Fig. 12.1.

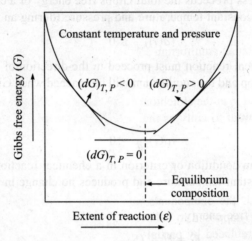

Fig. 12.1   Illustration of change in Gibbs free energy with extent of reaction.

12.4   EQUILIBRIUM CONSTANT

In order to quantify the chemical equilibria, equilibrium constants are determined. Consider the chemical reaction

$$\alpha A + \beta B \rightleftharpoons \sigma S + \tau T \tag{12.20}$$

According to the law of mass action, the rate of the forward reaction between A and B is given by

$$R_{AB} = k_1 C_A^\alpha C_B^\beta \tag{12.21}$$

Similarly, the rate of the opposite reaction between S and T is given by

$$R_{ST} = k_2 C_S^\sigma C_T^\tau \tag{12.22}$$

At equilibrium, $R_{AB} = R_{ST}$, i.e., rates of the reaction in the two opposite directions would be equal. This implies that

$$k_1 C_A^\sigma C_B^\beta = k_2 C_S^\sigma C_T^\tau \tag{12.23}$$

Then the equilibrium constant for the preceding reaction can be defined in terms of concentration as

$$K_c = \frac{k_1}{k_2} = \frac{C_S^\sigma C_T^\tau}{C_A^\alpha C_B^\beta} = \text{Constant} \tag{12.24}$$

Here, the equilibrium constant K_c is a concentration quotient and depends on temperature and pressure. The value of equilibrium constant can be determined if any one of the concentrations can be measured.

Now, the equilibrium constant for the given reaction can be expressed in terms of activity as

$$K_a = \frac{a_S^\sigma a_T^\tau}{a_A^\alpha a_B^\beta} = \Pi a_i^{v_i} \tag{12.25}$$

while K_a is the thermodynamic equilibrium constant. It is the activity quotient that expresses the activity of the chemical species, as it is conventional to put the activities of the products in the numerator and those of the reactant in the denominator. Here, a_i is the activity of component i in the reaction mixture and v_i is the stoichiometric number of component i. In ideal systems, the activity coefficients are equal to unity as the gas mixture or dilute solutions behave ideally.

Then

$$K_a = K_c = \frac{C_S^\sigma C_T^\tau}{C_A^\alpha C_B^\beta}$$

In a gaseous reaction, the components are generally expressed in terms of their partial pressures. If the gases are supposed to behave ideally, then the activities are equal to fugacity and the fugacity may be replaced by partial pressures.

The activity a_i is defined as the ratio of the fugacity of the component in the solution to that in the standard state.

It can mathematically be expressed as

$$a_i = \frac{\bar{f}_i}{f_i^0} \tag{12.26}$$

Putting the value of a_i into Eq. (12.26), we get

$$K_a = \Pi \left(\frac{\bar{f}_i}{f_i^0} \right)^{v_i} = K_f \tag{12.27}$$

where K_f denotes the equilibrium constant related to fugacity of the components.

For gaseous system, the value of f_i^0 approaches unity for a pure component. Hence, Eq. (12.27) becomes

$$K_f = \Pi (\bar{f}_i)^{v_i} \tag{12.28}$$

Again, when it is necessary to consider the gaseous reaction, the equilibrium constant is expressed in terms of partial pressure and given by

$$K_p = \Pi (\bar{p}_i)^{v_i} \tag{12.29}$$

Considering Eqs. (12.27), (12.28), and (12.29), we get

$$K_a = K_p = K_f \tag{12.30}$$

An example is cited here to understand clearly the relation between K_p and K_c. Consider the chemical reaction

$$N_2(g) + 3H_2(g) \rightleftharpoons 2NH_3(g)$$

$$K_p = \frac{(P_{NH_3})^2}{(P_{N_2}) \times (P_{H_2})} = \frac{\{[NH_3] \cdot RT\}^2}{\{[N_2] \cdot RT\} \times \{[H_2] \cdot RT\}^3}$$

$$\left(\text{as } P = \frac{n}{V} RT = CRT \text{ and } C = \text{molar concentration} \right)$$

$$= \frac{[NH_3]^2}{[N_2] \times [H_2]^3} \cdot \frac{(RT)^2}{(RT)^4} = K_C \times \frac{1}{(RT)^2} \qquad (12.31)$$

Hence, for this reaction, $K_p \neq K_c$. But it has been observed that, in a chemical reaction, if the number of moles of reactants are equal to the number of moles of the resultants, then $K_p = K_c$.

For instance

$$H_2(g) + I_2(g) \rightleftharpoons 2HI(g)$$

Hence, the characteristics of K_p for ideal gas can be summarized as follows:

- K_p of a chemical reaction depends on temperature only.
- It is independent of pressure of the reaction mixture.
- It is unaffected by the presence of inert gases.
- K_p of the reverse reaction is $1/K_p$.

- When the stoichiometric coefficients are doubled, the value of K_p is squared.

For example

$$H_2 + \frac{1}{2} O_2 \rightleftharpoons H_2O \qquad K_p = \frac{P_{H_2O}}{P_{H_2} P_{O_2}^{1/2}}$$

If the reaction is represented as

$$2H_2 + O_2 \rightleftharpoons 2H_2O$$

then

$$K_{P_2} = \frac{P_{H_2O}^2}{P_{H_2}^2 P_{O_2}} = (K_{P_1})^2$$

12.5   STANDARD GIBBS FREE ENERGY CHANGE

In thermodynamics, we are concerned with the changes in the energy of a system during a process, and not the energy values at the particular state. Therefore, we can choose any state as the reference state and assign a zero value at that state. When a process involves no change in chemical composition, the reference state chosen has no effect on the results. When the process involves chemical reaction, however, the composition of the system at the end of a process is no longer the same as that at the beginning of the process. In this case, it becomes necessary to have a common reference state for all substances. The chosen reference state is 25°C and 1 atm, which is known as the *standard reference state*. Property values at standard reference state are indicated by a superscript like h^0. Standard Gibbs free energy change is denoted by ΔG^0 at reference state and it is a state function.

Consider the reaction

$$aA + bB \rightleftharpoons rR + sS$$

The standard Gibbs free energy change for the reaction will be

$$\Delta G^0_{298} = r\Delta G^0_{f\,298}(R) + s\Delta G^0_{f\,298}(S) - a\Delta G^0_{f\,298}(A) - b\Delta G^0_{f\,298}(B) \qquad (12.32)$$

where ΔG_f^0 is the standard free energy of formation, which is the free energy change that occurs when one mole of a compound is formed from its elements, all in standard state.

Let us consider the chemical reaction represented by

$$aA + bB \rightarrow rR + sS$$

The free energy change for the reaction is given by

$$\Delta G^0 = (r\mu_R + s\mu_S) - (a\mu_A + b\mu_B) \qquad (12.33)$$

The most commonly used methods for the determination of free energy change of the process are

(i) $\Delta G^0 = \Delta H^0 - T\Delta S^0$

(ii) $\Delta G^0 = \sum \Delta G^0_{f(\text{Products})} - \sum \Delta G^0_{f(\text{Reactants})}.$

Applying these methods, some problems have been solved in this chapter.

12.6 FEASIBILITY OF CHEMICAL REACTION

The feasibility of a chemical reaction can be better judged by the change in Gibbs free energy and from the knowledge of the criterion for chemical equilibrium, i.e., $(dG)_{T,P} = 0$, because the Gibbs free energy of a system reaches a minimum value in the state of equilibrium. The equation $\Delta G^0 = -RT \ln K_p$ provides the following important information:

- If ΔG^0 has a negative value, the equilibrium constant has a value greater than one, and the reactants in their standard state spontaneously react to give the products in their standard state.
- If ΔG^0 is equal to zero, the equilibrium constant value is one, and the reactants in their standard states are in equilibrium with the products in their standard state.
- If ΔG^0 has a positive value, the equilibrium constant has a value less than one, and it does necessarily mean that the indicated moles of reactants in their standard states are not covered completely in the products.

Therefore, the conditions for the feasibility of a reaction can be represented as

(i) $\Delta G < 0$, then $K > 1$. The reaction is spontaneous and feasible.

(ii) $\Delta G = 0$, then $K = 1$. The system has attained a state of equilibrium.

(iii) $\Delta G > 0$. The reaction proceeds spontaneously in the reverse direction.

EXAMPLE 12.3 In a chemical laboratory, it is decided to carry out the reaction

$$C_2H_5OH(g) + \frac{1}{2}O_2(g) \rightarrow CH_3CHO(g) + H_2O(g)$$

at 1 bar and 298 K. Calculate the standard Gibbs free energy change at 298 K and predict whether it is feasible to carry out the given reaction or not. If possible, then calculate the equilibrium constant. We are given that

$$\Delta G^{\theta}_{f,CH_3CHO} = -133.978 \text{ kJ} \qquad \Delta G^{0}_{f,H_2O} = -228.60 \text{ kJ} \qquad \Delta G^{0}_{f,C_2H_5OH} = -174.883 \text{ kJ}$$

Solution: The reaction supposed to carry out is given by

$$C_2H_5OH(g) + \frac{1}{2}O_2(g) \rightarrow CH_3CHO(g) + H_2O(g)$$

The standard Gibbs free energy at 298 K for the reaction can be determined by

$$\Delta G^{0}_{298} = \Delta G^{0}_{f298}[CH_3CHO(g)] - \Delta G^{0}_{f298}[H_2O(g)] - \Delta G^{0}_{f298}[C_2H_5OH(g)]$$
$$= (-133.978 \text{ kJ}) - (-228.6 \text{ kJ}) - (-174.883 \text{ kJ}) = 269.505 \text{ kJ}$$

So, the standard Gibbs free energy, ΔG^{0}_{298} is $-$ve, i.e., $\Delta G^{0}_{298} < 0$, and the reaction is feasible to carry out.

Now the equilibrium constant can be evaluated by the following relation:

$$\Delta G^{0} = -RT \ln K$$

or

$$\ln K = -\frac{\Delta G^{0}}{RT} = -\frac{-187695 \text{ J}}{8.314 \times 298} = 75.757$$

or

$$K = e^{75.757} = 7.958 \times 10^{32}$$

EXAMPLE 12.4 At 1 atm and 27°C, will the vaporization of liquid water be spontaneous? Given that $\Delta H = 9710$ cal and $\Delta S = 26$ e.u.

Solution: The process can be represented as

$$H_2O(l) = H_2O(g) \qquad P = 1 \text{ atm}$$

For the given process, the standard Gibbs free energy can be determined by the following relation:

$$\Delta G = \Delta H - T\Delta S$$
$$= 9710 - 26 \times 300 = +1910 \text{ cal}$$

Since $\Delta G = +$ve, then vaporization is not possible. Rather, the reverse process of condensation will take place.

The temperature at which the liquid and vapour will be in equilibrium can be obtained by putting $\Delta G = 0$, i.e.

$$\Delta G = 9710 - 26 T = 0$$

or

$$T = 373.4 \text{ K}$$

This is the actual boiling point of water at 1 atm.

12.7 RELATION BETWEEN EQUILIBRIUM CONSTANT AND STANDARD GIBBS FREE ENERGY CHANGE

Consider the following reaction:

$$aA + bB \rightarrow rR + sS$$

From the knowledge of criterion of equilibrium for a chemically reacting system

$$(r\mu_R + s\mu_S) - (a\mu_A + b\mu_B) = 0$$

or

$$\sum_i v_i \mu_i = 0 \tag{12.34}$$

We also know that the chemical potential μ_i of a component i is defined as

$$\mu_i = RT \ln \bar{f}_i + \mathring{G}_i \tag{12.35}$$

where $\bar{f}_i$ is the fugacity of component i.

At standard state, the chemical potential of component i is given by

$$\mathring{\mu}_i = RT \ln \mathring{f}_i + \mathring{G}_i$$

where

$\mathring{G}_i$ = Constant that depends only on temperature;

$\mathring{f}_i$ = Standard state fugacity of component i.

$$\mu_i = \mathring{\mu}_i - RT \ln \mathring{f}_i + RT \ln \bar{f}_i$$

$$= \mathring{\mu}_i + RT \ln \frac{\bar{f}_i}{\mathring{f}_i}$$

$$= \mathring{\mu}_i + RT \ln a_i \tag{12.36}$$

where $a_i = \dfrac{\bar{f}_i}{\mathring{f}_i}$ is the activity coefficient of species i.

Substituting the value of μ_i in Eq. (12.35), we get

$$\sum_i v_i \mu_i = 0$$

or

$$\sum v_i (\mathring{\mu}_i + RT \ln a_i) = 0$$

or

$$\sum v_i \mathring{\mu}_i + \sum v_i \ln a_i = 0$$

or

$$\Delta \mathring{G} = -RT \sum v_i \ln a_i = -RT \sum \ln a_i^{v_i}$$

or

$$\Delta \overset{\circ}{G} = -RT \ \ln \ (\Pi a_i^{\nu_i}) = -RT \ \ln K$$

or

$$K = e^{\frac{-\Delta \overset{\circ}{G}}{RT}} \tag{12.37}$$

This is a useful relation between Gibbs free energy change and equilibrium constant, by which the equilibrium constant can be determined at the given Gibbs free energy change and the temperature.

EXAMPLE 12.5   Calculate the standard Gibbs free energy change and the equilibrium constant at 1 bar and 298 K for the ammonia synthesis reaction

$$N_2(g) + 3H_2(g) \rightarrow 2NH_3(g)$$

Given that the standard free energies of formation at 298 K for NH_3 are -16.750 kJ/mol.

Solution: For the reaction, the standard Gibbs free energy change at 298 K can be estimated as

$$\Delta G_{298}^0 = 2\Delta G_{f298}^0[NH_3(g)] - \Delta G_{f298}^0[N_2(g)] - 3\Delta G_{f298}^0[H_2(g)]$$
$$= 2 \times (-16.750 \ \text{kJ/mol}) - 0 - 0 = -33.5 \ \text{kJ}$$

Using the relation between standard Gibbs free energy change and equilibrium constant, we get

$$\Delta G^0 = -RT \ \ln K$$

or

$$\ln K = -\frac{\Delta G^0}{RT} - \frac{33500}{8.314 \times 298} = -13.521$$

or

$$K = e^{-13.521} = 1.342 \times 10^6$$

EXAMPLE 12.6   Calculate the equilibrium constant K_P at 25°C for the water–gas reaction

$$CO \ (g) + H_2O \ (g) \rightarrow CO_2 \ (g) + H_2 \ (g)$$

Given that

$$\Delta G_{f,CO}^0 = -32.8 \ \text{kcal} \qquad \Delta G_{f,H_2O}^0 = -54.64 \ \text{kcal} \qquad \Delta G_{f,CO_2}^0 = -94.26 \ \text{kcal}$$

Solution: For the reaction, the standard Gibbs free energy change at 25°C can be estimated as

$$\Delta G_{298}^0 = \Delta G_{f298}^0[CO_2(g)] - \Delta G_{f298}^0[H_2(g)] - \Delta G_{f298}^0[CO(g)] - \Delta G_{f298}^0[H_2O(g)]$$
$$= -94.26 + 0 - (-32.8) - (-54.64) = -6.82 \ \text{kcal}$$

Now the equilibrium constant can be evaluated by the following relation:

$$\Delta G^0 = -RT \ \ln K$$

or

$$\ln K = \frac{\Delta G^0}{RT} = \frac{6820}{1.987 \times 298} = 11.517$$

or

$$K = e^{11.517} = 1.06 \times 10^5$$

12.8 EFFECT OF TEMPERATURE ON EQUILIBRIUM CONSTANT: VAN'T HOFF EQUATION

The influence of temperature on equilibrium constant of a chemical reaction at given enthalpy change can be better explained by van't Hoff's equation. This is also known as *van't Hoff's isochore*.

We know that

$$dG = -SdT + VdP$$

or

$$\left(\frac{\partial G}{\partial T}\right)_P = -S \tag{12.38}$$

Then we can obtain at standard state

$$\left(\frac{\partial \Delta \overset{\circ}{G}}{\partial T}\right)_P = \left[\frac{\partial(\Sigma v_i G_i)}{\partial T}\right]$$

$$= \left[\frac{\partial}{\partial T}\left(\Sigma v_i \overset{\circ}{H_i} - T \Sigma v_i \overset{\circ}{S_i}\right)\right]$$

where ΔS is the standard enthalpy change of the reaction

$$= \left[\frac{\partial}{\partial T}\left(\sum v_i \overset{\circ}{H_i}\right)\right]_P - \left[\frac{\partial}{\partial T}\left(T \sum v_i \overset{\circ}{S_i}\right)\right]_P$$

$$= 0 - \sum v_i \overset{\circ}{S_i} = -\Delta \overset{\circ}{S} \tag{12.39}$$

Now

$$\left(\frac{\partial\left(\frac{\Delta \overset{\circ}{G}}{T}\right)}{\partial T}\right) = \frac{1}{T}\left(\frac{\partial \Delta \overset{\circ}{G}}{\partial T}\right)_P - \frac{\Delta \overset{\circ}{G}}{T^2}$$

$$= -\frac{\Delta \overset{\circ}{S}}{T} - \frac{\Delta \overset{\circ}{G}}{T^2} \tag{12.40}$$

(since $\Delta \overset{\circ}{G} = \Delta \overset{\circ}{H} - T\Delta \overset{\circ}{S}$)

Putting the value of $\Delta \overset{\circ}{G}$ into Eq. (12.40), we get

$$\left(\frac{\partial \left(\dfrac{\Delta \overset{\circ}{G}}{T} \right)}{\partial T} \right)_P = \frac{\Delta \overset{\circ}{S}}{T} - \frac{\Delta \overset{\circ}{G}}{T^2}$$

$$= -\left(\frac{T\Delta \overset{\circ}{S} + \Delta \overset{\circ}{G}}{T^2} \right) = -\frac{\Delta \overset{\circ}{H}}{T^2} \tag{12.41}$$

We know that

$$\Delta \overset{\circ}{G} = -RT \ln K$$

or

$$\frac{\Delta \overset{\circ}{G}}{T} = -R \ln K \tag{12.42}$$

Therefore

$$\left(\frac{\partial (\Delta \overset{\circ}{G}/T)}{\partial T} \right) = -R \left(\frac{\partial \ln K}{\partial T} \right) \tag{12.43}$$

Comparing Eq. (12.41) with Eq. (12.43), we get

$$\frac{\partial \ln K}{\partial T} = \frac{\Delta \overset{\circ}{H}}{T^2} \tag{12.44}$$

If the reaction is an exothermic one, then $\Delta \overset{\circ}{H} < 0$, and the equilibrium constant K decreases with increase in temperature. On the other hand, for an endothermic reaction, $\Delta \overset{\circ}{H} > 0$ and the equilibrium constant K increases with increasing temperature for an endothermic reaction. Van't Hoff's equation predicts the effect of temperature on the equilibrium constant.

If $\Delta \overset{\circ}{H}$ is assumed to be constant in the temperature range T_1 to T_2, van't Hoff's equation can be integrated to obtain

$$\ln \frac{K_2}{K_1} = \frac{\Delta \overset{\circ}{H}}{R} \left(\frac{1}{T_1} - \frac{1}{T_2} \right) \tag{12.45}$$

where

K_1 = Equilibrium constant at absolute temperature T_1
K_2 = Equilibrium constant at absolute temperature T_2.

Now, if the $\ln K$ is plotted against $1/T$, then the nature of the graph will be a straight line as shown Fig. 12.2.

The standard heat of reaction can appreciably vary with temperature. Here, it can be expressed by molal heat capacities in terms of temperature.

From Kirchhoff's equation, we get

$$\frac{\partial (\Delta H)}{\partial T} = \Delta C_P = C_{P_{\text{Products}}} - C_{P_{\text{Reactants}}} \tag{12.46}$$

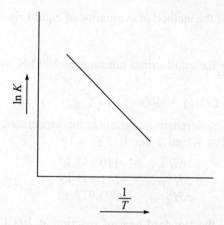

Fig. 12.2 Demonstration of van't Hoff's equation on influence of T on K.

We know that the standard heat of reaction at temperature T is given by

$$\Delta H_T^0 = \Delta H_{T_0}^0 + \int_{T_0}^{T} \Delta C_P dT \qquad (12.47)$$

where

$$\Delta C_P = \Delta\alpha + \Delta\beta T + \Delta\gamma T^2$$

Thus, (12.44) the equation becomes

$$\Delta H_T^0 = \Delta H_{T_0}^0 + \Delta\alpha T + \frac{\Delta\beta}{2}T^2 + \frac{\Delta\gamma}{3}T^3 + \cdots \qquad (12.48)$$

van't Hoff's Eq. (12.44) will then be written as

$$\frac{\partial \ln K}{\partial T} = \frac{\Delta H_T^0}{RT^2} = \frac{\Delta H_{T_0}^0 + \Delta\alpha T + \dfrac{\Delta\beta}{2}T^2 + \dfrac{\Delta\gamma}{3}T^3 + \cdots}{RT^2}$$

This equation can be rearranged as

$$\frac{\partial \ln K}{\partial T} = \frac{\Delta H_{T_0}^0}{RT^2} + \frac{\Delta\alpha}{RT} + \frac{\Delta\beta}{2R} + \frac{\Delta\gamma T}{3R} \qquad (12.49)$$

On integration

$$\ln K = -\frac{\Delta H_{T_0}^0}{RT} + \frac{\Delta\alpha \ln T}{R} + \frac{\Delta\beta T}{2R} + \frac{\Delta\gamma T^2}{6R} + I \qquad (12.50)$$

where I is the integration constant of the isochore.

We can put the value of $\ln K$ in the equation $\Delta G = -RT \ln K$ to evaluate the Gibbs free energy change. It can be determined by the following way:

$$\Delta G^0 = \Delta H_{T_0}^0 - \Delta\alpha T \ln T - \frac{\Delta\beta T^2}{2} - \frac{\Delta\gamma T^3}{6} - IRT \qquad (12.51)$$

The integration constant I can easily be determined from the value of the equilibrium constant and standard Gibbs free energy change at a single temperature. Some problems are being worked out for better understanding of the method of evaluation of equilibrium constant and standard free energy change.

EXAMPLE 12.7 Calculate the equilibrium constant at 1000 K and 1 bar for the following reaction:

$$CO(g) + H_2O(g) \rightarrow CO_2(g) + H_2(g)$$

Assuming that the heat of reaction remains constant in the temperature range involved. Given that Equilibrium constant K at 298 K and 1 bar 1.1582×10^5

$$\Delta H^0_{f,CO} = -110.532 \text{ kJ}$$
$$\Delta H^0_{f,H_2O} = -241.997 \text{ kJ}$$
$$\Delta H^0_{f,CO_2} = -393.978 \text{ kJ}.$$

Solution: For the reaction, the standard heat of reaction at 298 K can be estimated as

$$\Delta H^0_{298} = \Delta H^0_{f298}[CO_2(g)] - \Delta H^0_{f298}[CO(g)] - \Delta H^0_{f298}[H_2O(g)]$$
$$= (-393.978 \text{ kJ}) - (-110.532 \text{ kJ}) - (-241.997 \text{ kJ}) = -41.449 \text{ kJ}$$

From van't Hoff's equation, we get

$$\ln \frac{K_2}{K_1} = \frac{\Delta \overset{\circ}{H}}{R} \left(\frac{1}{T_1} - \frac{1}{T_2} \right)$$

At 298 K

$$K_1 = 1.1582 \times 10^5 \qquad K_2 = ? \qquad T_1 = 298 \text{ K} \qquad T_2 = 1000 \text{ K}$$

Substituting the values, we get

$$\ln \left(\frac{K_2}{1.1582 \times 10^5} \right) = \frac{-41449}{8.314} \left(\frac{1}{298} - \frac{1}{1000} \right)$$
$$K_2 = 0.9189$$

EXAMPLE 12.8 Assuming that the standard enthalpy changes of the reaction are constant in the temperature range of 298 K to 700 K, estimate the equilibrium constant at 700 K for the ammonia synthesis reaction

$$N_2(g) + 3H_2(g) \rightarrow 2NH_3(g)$$

Given that

$$\Delta H^0_{f,NH_3} = -46.1 \text{ kJ}$$
$$\Delta G^0_{f,NH_3} = -16.747 \text{ kJ}$$

Solution: For the reaction, the standard heat of reaction at 298 K can be estimated as

$$\Delta H^0_{298} = 2\Delta H^0_{f298}[NH_3(g)] - \Delta H^0_{f298}[N_2(g)] - 3\Delta H^0_{f298}[H_2(g)]$$
$$= 2 \times (-46.1 \text{ kJ}) - 0 - 0 = -92.2 \text{ kJ}$$

Now, the standard Gibbs free energy change for the reaction at 298 K is given by

$$\Delta G^0_{298} = 2\Delta G^0_{f298}[NH_3(g)] - \Delta G^0_{f298}[N_2(g)] - 3\Delta G^0_{f298}[H_2(g)]$$
$$= 2 \times (-16.747 \text{ kJ}) - 0 - 0 = -33.494 \text{ kJ}$$

The equilibrium constant K is given by

$$\ln K = -\frac{\Delta G^0}{RT} = -\frac{-33494}{8.314 \times 298} = 13.518$$

$$K = e^{13.518} = 7.426 \times 10^5$$

This is the value of equilibrium constant at 298 K, i.e., $K_{298} = 7.426 \times 10^5$.
Now, the equilibrium constant at 700 K, i.e., K_{700} can be determined as

$$\ln\left(\frac{K_{700}}{K_{298}}\right) = \frac{-92200 \text{ kJ}}{8.314} = -21.371$$

or

$$\frac{K_{700}}{K_{298}} = e^{-21.371}$$

or

$$K_{700} = 7.426 \times 10^5 \times e^{-21.371} = 7.426 \times 10^5 \times 5.23 \times 10^{-10}$$
$$K_{700} = 3.883 \times 10^{-4}$$

Hence, the equilibrium constant for ammonia synthesis reaction at 700 K is 3.883×10^{-4}.

EXAMPLE 12.9 Estimate the standard Gibbs free energy and the equilibrium constant at 1000 K for the reaction

$$CO(g) + H_2O(g) \rightarrow CO_2(g) + H_2(g)$$

Given that the standard heat of reaction ΔH^0_{298} and standard Gibbs free energy of the reaction $\Delta G^0_{f,298}$ for the formation of products are −41,450 J/mol and −28.888 J/mol respectively. The specific heat data in kJ as function of temperature are as follows:

Components	α	$\beta \times 10^3$
$CO_2(g)$	45.369	8.688
$H_2(g)$	27.012	3.509
$CO(g)$	28.068	4.631
$H_2O(g)$	28.850	12.055

Solution: We are given that

$$\Delta H^0_{298} = -41,450 \text{ J/mol} \quad \text{and} \quad \Delta G^0_{f,298} = -28,888 \text{ J/mol}$$

Now

$$\Delta\alpha = 45.369 + 27.012 - 28.068 - 28.068 - 28.850 = 15.463$$
$$\Delta\beta = (8.688 + 3.509 - 4.631 - 12.055) \times 10^{-3} = -4.489 \times 10^{-3}$$

To obtain the standard heat of reaction at the desired temperature, we can use the following equation:

$$\Delta H^0_T = \Delta H^0_{T_0} + \Delta\alpha T + \frac{\Delta\beta}{2}T^2 + \frac{\Delta\gamma}{3}T^3 + \cdots$$

Since the values of coefficient γ are not given, then it can be safely neglected. So the preceding equation becomes

$$\Delta H_T^0 = \Delta H_{T_0}^0 + \Delta\alpha T + \frac{\Delta\beta}{2}T^2$$

Substituting these values in the following equation, we get

$$-41,450 \text{ J/mol} = \Delta H_{T_0}^0 + 15.463 \times 298 - \frac{4.489\times10^5}{2}(298)^2$$

$$\Delta H_{T_0}^0 = -46.056 \text{ kJ}$$

We know

$$\Delta G_{f,298}^0 = -28,888 \text{ J/mol}$$

Putting the value of standard Gibbs free energy into the equation to get the value of integration constant I at 298 K, we get

$$\Delta G^0 = \Delta H_{T_0}^0 - \Delta\alpha T \ln T - \frac{\Delta\beta T^2}{2} - IRT \qquad (12.52)$$

or

$$-28,888 = -46056 - 15.463 \times 298 \ln 298 + \frac{4.489\times10^{-3}}{2}(298)^2 - IR \times 298$$

or

$$IR = 133.67$$

Now, substituting $T = 1000$ K and the value of IR in Eq. (12.51), we get

$$\Delta G^0 = -46056 - 15.463 \times 1000 \times \ln 1000 + \frac{4.489\times10^{-3}(1000)^2}{2} - 133.67 \times 1000$$

$$= -284.296 \text{ kJ}$$

The equilibrium constant K_{1000} can be determined as

$$\ln K = -\frac{\Delta G^0}{RT} = -\frac{-2842}{8.314\times1000} = 0.3418$$

or

$$K_{1000} = e^{0.3418} = 1.4$$

Therefore, the standard Gibbs free energy and equilibrium constant are 284.296 kJ and 1.4 respectively.

12.9 HOMOGENEOUS GAS-PHASE REACTION EQUILIBRIUM

Homogeneous reaction is defined as the reaction in which the reacting substances and the products are required to be in the single phase. In order to describe the equilibrium in homogeneous reaction, a standard state is necessary to be specified and chosen for the species associated with the chemical reactions.

For a gas, the standard state is the ideal gas state of the pure gas at the standard state of 1 bar or atm, that is, $P^0 = 1$. Now, for a homogeneous gas-phase reaction, if the reaction is assumed to behave like an ideal gas, then the fugacity of the gas is equal to its pressure, i.e., $f_i^0 = P^0 = 1$. Then the activity of the species i is given by

$$a_i = \frac{\bar{f}_i}{f_i^0} = \bar{f}_i = \overline{\Phi}_i y_i P \qquad (\text{as } f_i^0 = 1) \qquad (12.53)$$

where $\overline{\Phi}_i$ is the fugacity coefficient, a dimensionless ratio. For an ideal gas, the partial pressure is given by

$$p_i = y_i P$$

where P = Total pressure
$ p$ = Partial pressure.

Substituting the value of a_i in the expression of equilibrium constant applicable to all chemical reactions, we get

$$K_a = \Pi a_i^{v_i} = \Pi(\overline{\Phi}_i y_i P) = \Pi(\overline{\Phi}_i P_i)^{v_i} \qquad (12.54)$$

Equation (12.54) can be rearranged as

$$K_a = \Pi(\overline{\Phi}_i P)^{v_i} = (\Pi\overline{\Phi}_i^{v_i})(\Pi P_i^{v_i}) \qquad (12.55)$$

Now, if the equilibrium mixture is assumed to behave ideally, then $\overline{\Phi}_i = \Phi_i$. Therefore

$$K_a = (\Pi\overline{\Phi}_i^{v_i})(\Pi P_i^{v_i}) = (\Pi\Phi_i^{v_i})(\Pi P_i^{v_i}) = K_\Phi K_P = K_\Phi K_y P^{\Sigma v_i} \qquad (12.56)$$

where $K_y (= \Pi y_i^{v_i})$ is the equilibrium constant in terms of mole fractions.

Since at sufficiently low pressure and high temperature, the gas mixture behaves like an ideal gas, $\overline{\Phi}_i = 1$ and consequently $K_\Phi = 1$.

Thus, Eq. (12.56) becomes

$$K_a = K_P = K_y P^{\Sigma v_i} \qquad (12.57)$$

Consider the following chemical reaction:

$$aA + bB \rightarrow rR + sS$$

Then the three equilibrium constants can be represented as

$$K_a = \Pi a_i^{v_i} = \frac{a_R^r a_S^s}{a_A^a a_B^b}$$

$$K_p = \Pi p_i^{v_i} = \frac{p_R^r p_S^s}{p_A^a p_B^b}$$

$$K_y = \Pi y_i^{v_i} = \frac{y_R^r y_S^s}{y_A^a y_B^b}$$

EXAMPLE 12.10 An equimolar mixture of $CO(g)$ and $H_2O(g)$ enters a reactor which is maintained at 10 bar and 1000 K. The reaction involved is

$$CO(g) + H_2O(g) \rightarrow CO_2(g) + H_2(g)$$

Given that the equilibrium constant for this reaction at 1000 K is 1.5, calculate the degree of conversion and the composition of the gas mixture that leaves the reactor. The reaction mixture is assumed to behave like an ideal gas.

Solution: Let ε be the degree of conversion at equilibrium.

Component	$CO(g)$	$H_2O(g)$	$CO_2(g)$	$H_2(g)$
Moles in feed	1	1	0	0
Moles at equilibrium	$1 - \varepsilon_e$	$1 - \varepsilon_e$	ε_e	ε_e
Mole fraction	$\dfrac{1 - \varepsilon_e}{2}$	$\dfrac{1 - \varepsilon_e}{2}$	$\dfrac{\varepsilon_e}{2}$	$\dfrac{\varepsilon_e}{2}$
Total moles at equilibrium	$(1 - \varepsilon_e) + (1 - \varepsilon_e) + \varepsilon_e + \varepsilon_e = 2$			

Now, the sum of the stoichiometric coefficients

$$= \Sigma v_i = 1 + 1 - 1 - 1 = 0$$

$$K_y = \prod_i y_i^{v_i} = \frac{\left(\dfrac{\varepsilon_e}{2}\right)\left(\dfrac{\varepsilon_e}{2}\right)}{\left(\dfrac{1-\varepsilon_e}{2}\right)\left(\dfrac{1-\varepsilon_e}{2}\right)} = \left(\frac{\varepsilon_e}{1-\varepsilon_e}\right)^2$$

$$P = 10 \text{ bar}$$

We know that

$$K = K_y P^{\Sigma v_i} = K_y (10)^0 = K_y$$

Therefore

$$\left(\frac{\varepsilon_e}{1-\varepsilon_e}\right)^2 = 1.5$$

Solving for ε_e, we get $\varepsilon_e = 0.5505$. Hence the degree of conversion at equilibrium = 0.5505. The composition of the gas mixture leaving the reactor is given by

$$y_{CO} = y_{H_2O} = \frac{1-\varepsilon_e}{2} = 0.2247$$

$$y_{CO_2} = y_{H_2} = \frac{\varepsilon_e}{2} = 0.2753$$

EXAMPLE 12.11 Ammonia synthesis reaction is represented by

$$N_2(g) + 3H_2(g) \rightarrow 2NH_3(g)$$

Initially, 1 mol of nitrogen and 5 mol of hydrogen gas are fed into the reactor. The process is maintained at 800 K and 250 bar. Estimate the molar composition of the gases at equilibrium, assuming that the reaction mixture behaves like an ideal gas. Given that equilibrium constant, $K = 1.106 \times 10^{-5}$.

Solution: The reaction involved is

$$N_2(g) + 3H_2(g) \rightarrow 2NH_3(g)$$

Let ε be the degree of conversion of nitrogen into ammonia at equilibrium.

Component	$N_2(g)$	$H_2(g)$	$NH_3(g)$
Moles in feed	1	5	0
Moles at equilibrium	$1 - \varepsilon_e$	$5 - 3\varepsilon_e$	$2\varepsilon_e$
Mole fraction	$\dfrac{1 - \varepsilon_e}{2(3 - \varepsilon_e)}$	$\dfrac{5 - \varepsilon_e}{2(3 - \varepsilon_e)}$	$\dfrac{2\varepsilon_e}{2(3 - \varepsilon_e)}$
Total moles at equilibrium	$1 - \varepsilon_e + 5 - 3\varepsilon_e + 2\varepsilon_e = 6 - 2\varepsilon_e = 2(3 - \varepsilon_e)$		

Now, the sum of the stoichiometric coefficients = $\Sigma v_i = 2 - 3 - 1 = 2$.
Operating pressure, $P = 250$ bar.
We know that

$$K_y = \prod_i y_i^{v_i} = \frac{\left(\dfrac{2\varepsilon_e}{2(3 - \varepsilon_e)}\right)^2}{\left(\dfrac{1 - \varepsilon_e}{2(3 - \varepsilon_e)}\right)\left(\dfrac{5 - 3\varepsilon_e}{2(3 - \varepsilon_e)}\right)} = \frac{[4\varepsilon_e(3 - \varepsilon_e)]^2}{(1 - \varepsilon_e)(5 - 3\varepsilon_e)^3}$$

Since the equilibrium constant, $K = 1.106 \times 10^{-5}$, we have

$$1.107 \times 10^{-5} = K_y(250)^{-2} = \left[\frac{\{4\varepsilon_e(3 - \varepsilon_e)\}^2}{(1 - \varepsilon_e)(5 - 3\varepsilon_e)^3}\right] \times (250)^{-2}$$

Solving for ε_e, we get $\varepsilon_e = 0.4337$. Hence, the degree of conversion at equilibrium = 0.4337. Therefore, the molar composition of the gases is given by

$$y_{N_2} = \frac{1 - \varepsilon_e}{2(3 - \varepsilon_e)} = 0.1103 \quad y_{H_2} = \frac{5 - 3\varepsilon_e}{2(3 - \varepsilon_e)} = 0.7207 \quad \text{and} \quad y_{NH_3} = \frac{2\varepsilon_e}{2(3 - \varepsilon_e)} = 0.1690$$

EXAMPLE 12.12 The reaction for oxidation of sulphur is given by

$$SO_2(g) + \frac{1}{2}O_2(g) \rightarrow SO_3(g)$$

A mixture of sulphur dioxide and oxygen is fed to the reactor in the mole ratio of 5:4. The process is maintained at 750 K and 1 bar. Estimate the molar composition of the reactor-effluent gases at equilibrium, assuming that the reaction mixture behaves like an ideal gas. Given that equilibrium constant $K = 74$.

Solution: The reaction involved is

$$SO_2(g) + \frac{1}{2}O_2(g) \rightarrow SO_3(g)$$

Let ε be the degree of conversion of nitrogen into ammonia at equilibrium.

Component	$SO_2(g)$	$O_2(g)$	$SO_3(g)$
Moles in feed	1	0.5	0
Moles at equilibrium	$10 - \varepsilon_e$	$8 - 0.5\varepsilon_e$	ε_e
Mole fraction	$\dfrac{10 - \varepsilon_e}{18 - 0.5\varepsilon_e}$	$\dfrac{8 - 0.5\varepsilon_e}{18 - 0.5\varepsilon_e}$	$\dfrac{\varepsilon_e}{18 - 0.5\varepsilon_e}$

Total moles at equilibrium	$10 - \varepsilon_e + 8 - 0.5\varepsilon_e + \varepsilon_e = 18 - 0.5\varepsilon_e$

Now, the sum of the stoichiometric coefficients $= \Sigma v_i = 1 - 1 - 0.5 = -0.5$
Operating pressure, $P = 1$ bar.

We know that

$$K_y = \prod_i y_i^{v_i} = \frac{\left(\dfrac{\varepsilon_e}{18 - 0.5\varepsilon_e}\right)}{\left(\dfrac{10 - \varepsilon_e}{18 - 0.5\varepsilon_e}\right)\left(\dfrac{8 - 0.5\varepsilon_e}{18 - 0.5\varepsilon_e}\right)} = \frac{\varepsilon_e(18 - 0.5\varepsilon_e)}{(10 - \varepsilon_e)(8 - 0.5\varepsilon_e)}$$

Since the equilibrium constant $K = 74$, we have

$$74 = K_y(1)^{-0.5} = \left\{\frac{\varepsilon_e(18 - 0.5\varepsilon_e)}{(10 - \varepsilon_e)(8 - 0.5\varepsilon_e)}\right\} \times (1)^{-0.5}$$

Solving for ε_e, we get $\varepsilon_e = 9.73$. Hence the degree of conversion at equilibrium $= 9.73$. Therefore, the molar composition of the gases is given by

$$y_{SO_2} = \frac{10 - \varepsilon_e}{18 - 0.5\varepsilon_e} = 0.27 \quad y_{O_2} = \frac{8 - 0.5\varepsilon_e}{18 - 0.5\varepsilon_e} = 3.135 \quad \text{and} \quad y_{SO_3} = \frac{\varepsilon_e}{18 - 0.5\varepsilon_e} = 9.73$$

12.10 EFFECT OF PRESSURE ON CHEMICAL EQUILIBRIUM

The pressure of the reaction mixture affects the equilibrium composition in a gas phase reaction, but it has no effect on equilibrium constant. This can be substantiated in the following way.

Consider the chemical reaction

$$aA + bB \rightarrow rR + sS$$

The equilibrium constant is given by

$$K = K_a = \frac{y_R^r y_S^s}{y_A^a y_B^b} P^{\Delta v} = \Pi y_i^{v_i} P^{\Delta v} = K_y P^{\Delta v} \tag{12.58}$$

where

$$\Delta v = v_{\text{Products}} - v_{\text{Reactants}} = v_C + v_D - v_A - v_B$$

The influence of pressure on equilibrium of chemical reaction greatly depends on the sign of Δv or the difference between the number of moles of the products and the number of moles of the reactants in the stoichiometric reaction. If $\Delta v = +ve$, then an increase in pressure at a specified temperature increases the number of moles of the reactants, decreases the number of moles of the products, and reduces the degree of conversion at equilibrium. On the contrary, if $\Delta v = -ve$, there will be an opposite effect, i.e., an increase in pressure increases the degree of conversion. When $\Delta v = 0$, there is no change in mole number and $P^{\Delta v} = 1$. As a result, $K = K_y$. The pressure will have no effect on degree of conversion at equilibrium.

To illustrate, the ammonia synthesis reaction can be represented as

$$N_2(g) + 3H_2(g) \rightleftharpoons 2NH_3(g) \qquad \Delta v = 2 - 3 - 1 = -2$$

Since $\Delta v = -ve$, an increase in pressure at specified temperature decreases the number of moles of the reactants, increases the number of moles of the products, and increases the degree of conversion at equilibrium. Hence, the formation of ammonia will be favoured by the increase in pressure, because the system will respond in such a way that tends to minimize the effect of the increase in pressure.

Hence, it can be concluded that in the gas-phase reaction

- when $\Delta v > 0$, a rise in pressure will have an unfavourable effect, and the equilibrium of the reaction will be shifted to the right by a decrease in pressure; and
- when $\Delta v < 0$, it will be advantageous to raise the pressure, and the equilibrium of the reaction will be shifted towards the formation of products.

The equilibrium constant is independent of pressure. This is well-supported by the following relation:

$$K = \frac{a_C^c a_D^d}{a_A^a a_B^b} = \Pi a_i^{v_i} = K_a \qquad \text{for the reaction}$$

$$aA + bB \rightarrow cC + dD$$

This is a function of temperature only as it is observed from the relationship between the standard Gibbs free energy change and temperature, i.e., $\Delta G^0 = -RT \ln K$.

EXAMPLE 12.13 The dissociation of phosphorus pentachloride takes place at 250°C as

$$PCl_5 \rightleftharpoons PCl_3 + Cl_2 \qquad K_P = 1.8$$

Calculate the pressure (in atm) that is necessary to obtain 50% conversion of PCl_5.

Solution: Basis: 1 mol of PCl_5.
At equilibrium

$$PCl_5 = 0.5 \text{ mol} \qquad PCl_3 = 0.5 \text{ mol} \qquad \text{and} \qquad Cl_2 = 0.5 \text{ mol}$$

Total moles at equilibrium = 1.5 mol.
Therefore, the partial pressures

$$P_{PCl_5} = P_{PCl_3} = P_{Cl_2} = \frac{0.5}{1.5} P$$

where P is total pressure.

The equilibrium constant is given by

$$K_P = 1.8 = \frac{P_{PCl_3} \times P_{Cl_2}}{P_{PCl_5}} = \frac{\left(\dfrac{0.5}{1.5}P\right)\left(\dfrac{0.5}{1.5}P\right)}{\left(\dfrac{0.5}{1.5}P\right)} = \frac{P}{3}$$

or

$$P = 5.4 \text{ atm}$$

This implies that the total pressure required for 50% dissociation is numerically three times the K_P value.

12.11 EFFECT OF OTHER FACTORS ON EQUILIBRIUM CONVERSION

Apart from the effect of temperature and pressure, the other factors affecting the equilibrium of a chemical reaction are the presence of inert gases, presence of excess reactants, and presence of the products in the initial reaction mixture of the reacting system.

12.11.1 Effect of Presence of Inert Gases

The presence of an inert gas as shown in Fig. 12.3 influences the equilibrium composition, but it does not affect the equilibrium constant. If the reaction goes on with a decrease in the number of moles ($\Delta n < 0$), the dilution with an inert gas shifts the equilibrium of reaction to the left. On the contrary, an increase in the number of moles ($\Delta n > 0$) shifts the equilibrium to the right.

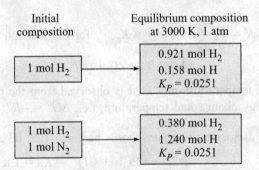

Fig. 12.3 Illustration on the effect of presence of inert gases on equilibrium constant.

We have obtained the relation

$$K_P = n_i^{\nu_i}\left(\frac{P}{N_{total}}\right)^{\Delta\nu} \tag{12.59}$$

where N_{total} is the total number of moles of the ideal gas mixture including inert gas.

Since the stoichiometric coefficient plays an important role in determining the equilibrium composition of the reacting substances, here it will indicate the presence of inert gases affects the equilibrium composition. It has been observed that at a specified temperature and pressure, if ΔV is +ve, an increase in number of moles of inert gases decreases the number of moles of the reactants and increases the number of moles of the products. The situation will be reverse if ΔV is −ve, i.e., the presence of inert gases will decrease the conversion if the reaction is accompanied by a decrease in the number of moles, and if ΔV is zero, then the presence of inert gases will have no effect. For example, if nitrogen–hydrogen–ammonia mixture taken at $P = 100$ MPa contains 10% of an inert gas, this would be equivalent to a reduction of 25 MPa in pressure.

EXAMPLE 12.14 A mixture of 1 mol of sulphur dioxide gas, 0.5 mol of oxygen gas and 2 mol of argon gas are fed into a reactor at 30 bar and 900 K to produce sulphur trioxide gas. The equilibrium constant for the reaction is 6. Calculate the degree of conversion and equilibrium composition of the reaction mixture, assuming that the mixture behaves like an ideal gas.

Solution: The reaction involved is

$$SO_2(g) + \frac{1}{2}O_2(g) \rightarrow SO_3(g)$$

Let ε_e be the degree of conversion at equilibrium.

Component	$SO_2(g)$	$O_2(g)$	$SO_3(g)$	$Ar(g)$
Moles in feed	1	$\dfrac{1}{2}$	0	2
Moles at equilibrium	$1 - \varepsilon_e$	$\dfrac{1}{2}(1 - \varepsilon_e)$	ε_e	2
Mole fraction	$\dfrac{2(1 - \varepsilon_e)}{7 - \varepsilon_e}$	$\dfrac{1 - \varepsilon_e}{7 - \varepsilon_e}$	$\dfrac{2\varepsilon_e}{7 - \varepsilon_e}$	$\dfrac{4}{7 - \varepsilon_e}$
Total moles at equilibrium		$(1 - \varepsilon_e) + \dfrac{1}{2}(1 - \varepsilon_e) + \varepsilon_e + 2 = \dfrac{7 - \varepsilon_e}{2}$		

Now, the sum of the stoichiometric coefficients $= \Sigma v_i = 1 - 1 - \dfrac{1}{2} = -\dfrac{1}{2}$

The pressure at which reaction maintained is $P = 30$ bar.
The equilibrium constant is given by

$$K = K_y P^{\Sigma v_i} = \frac{\dfrac{2\varepsilon_e}{7 - \varepsilon_e}}{\dfrac{2(1 - \varepsilon_e)}{7 - \varepsilon_e}\left(\dfrac{1 - \varepsilon_e}{7 - \varepsilon_e}\right)^{\frac{1}{2}}}(30)^{-\frac{1}{2}} = 6$$

Solving for ε_e, we get $\varepsilon_e = 0.850$. Hence, the degree of conversion at equilibrium $= 0.850$. The equilibrium composition of the reaction mixture is given by

$$y_{SO_2} = \frac{2(1 - \varepsilon_e)}{7 - \varepsilon_e} = 0.0487 \qquad y_{O_2} = \frac{1 - \varepsilon_e}{7 - \varepsilon_e} = 0.0243$$

$$y_{SO_3} = \frac{2\varepsilon_e}{7 - \varepsilon_e} = 0.2764 \qquad y_A = \frac{4}{7 - \varepsilon_e} = 0.6504$$

12.11.2   Effect of Presence of Excess Reactants

An increase in the number of the moles of the excess reactants increases the concentration of the reacting substances, which in turn increases the number of moles of the products as well as the equilibrium conversion of the limiting reactant.

Suppose the chemical reaction is given by

$$aA + bB \rightleftharpoons lL + mM$$

$$K_n = \frac{n_L^l n_M^m}{n_A^a n_B^b} = K \left(\frac{N_{total}}{P} \right)^{\Sigma(l+m-a-b)} \tag{12.60}$$

Here, if component A is considered to be the limiting reactant, then the increase in B increases the number of moles of L and M, and also the degree of conversion at equilibrium, to keep the equilibrium constant unchanged.

Therefore, the presence of excess of one reactant tends to increase the degree of conversion of the reactants into products.

12.11.3   Effect of Presence of Products in Initial Mixture of Reacting System

Consider the following reaction:

$$A \rightleftharpoons B + C$$

Equilibrium constant

$$K_p = \frac{P_B P_C}{P_A}$$

At equilibrium, if an excess amount of B is added to the system, then P_C should decrease and P_A should increase to maintain the constancy of K_p. Consequently, dissociation will be suppressed.

Again, if we consider the reaction

$$aA + bB \rightarrow lL + mM$$

The equilibrium constant of the reaction is given by

$$K = \frac{n_L^l n_M^m}{n_A^a n_B^b} \left(\frac{P}{N_{total}} \right)^{\Sigma v_i} = K_n \left(\frac{P}{N_{total}} \right)^{\Sigma v_i}$$

or

$$K_n = K \left(\frac{N_{total}}{P} \right)^{\Sigma v_i} \tag{12.61}$$

It is very much clear from the preceding expression that if the initial reaction composition largely consists of products, then in order to maintain the equilibrium the total number of moles of the product would decrease, and the partial pressure of the reactants would be larger than that of the product. The equilibrium constant will approach a small value. Therefore, the addition or presence of the products in the initial feed of the reacting system reduces the equilibrium conversion of the reactants.

12.12 HOMOGENEOUS LIQUID-PHASE REACTION EQUILIBRIUM

The equilibrium constant of a chemical reaction is given by

$$K_a = \Pi a_i^{v_i} \qquad (12.62)$$

where Π indicates the multiplication of various terms and a_i indicates the activity of species i.

We know that the activity of species i in the liquid mixture from low to moderate pressure is nothing but the ratio of the fugacity to the standard fugacity, and it can be mathematically expressed as

$$a_i = \frac{\bar{f}_i}{f_i^0} \qquad (12.63)$$

where

$\bar{f}_i$ = Fugacity of species i in the reaction mixture at the actual temperature and pressure
f_i^0 = Fugacity of pure liquid in the standard state of 1 atmospheric pressure (or 1 bar) and reaction temperature.

From the definition of activity coefficient, we have

$$\gamma_i = \frac{\bar{f}_i}{f_i x_i}$$

or

$$\bar{f}_i = \gamma_i x_i f_i \qquad (12.64)$$

where
γ_i = Activity coefficient
f_i = Fugacity of the pure liquid at the temperature and pressure of equilibrium mixture.

Substituting Eq. (12.64) in Eq. (12.63), we obtain

$$a_i = \frac{\bar{f}_i}{f_i^0} = \gamma_i x_i \left(\frac{f_i}{f_i^0} \right) \qquad (12.65)$$

Now, at this state the fugacity ratio $\dfrac{f_i}{f_i^0}$ of the pure liquid is considered to be unity. Then Eq. (12.65) becomes

$$a_i = \gamma_i x_i \qquad (12.66)$$

Putting the value of a_i into Eq. (12.62), we get the equilibrium constant as

$$K_a = \Pi a_i^{v_i} = \Pi(\gamma_i x_i)^{v_i} = (\Pi \gamma_i^{v_i})(\Pi x_i^{v_i}) = K_\gamma K_x \qquad (12.67)$$

where x_i is the mole fraction of species i in the equilibrium reaction mixture.

If the reaction mixture is an ideal solution, $\gamma_i = 1$. Then Eq. (12.67) will be

$$K_a = K_x = \Pi(x_i)^{v_i} \qquad (12.68)$$

But in case of chemical reaction occurring in a non-ideal liquid solution, $K_\gamma \neq 1$; therefore, in many of the liquid-phase reactions, the equilibrium constant is given by $K_c = \Pi(c_i)^{v_i}$, where K_c is the equilibrium constant in terms of concentration.

For the solute components at low concentration in the solution, Henry's law can be applied to represent the behaviour of the solute. If the solute follows Henry's law upto unit molality (i.e., mole number of the solute per kg of water), then a fictitious or hypothetical state would exist. The fictitious state is shown in Fig. 12.4 as consisting of a plot of fugacity of the solute versus molality. According to Henry's law, the relation between fugacity and molality is given by

$$\bar{f}_i = K_i m_i \tag{12.69}$$

where K_i is Henry's constant.

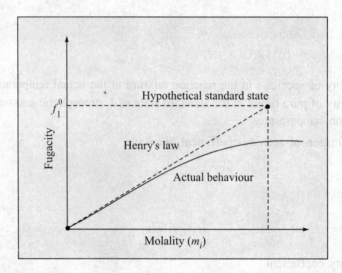

Fig. 12.4 Graphical representation of standard state for dilute aqueous solution.

At hypothetical standard state, $m_i = 1$. Then the fugacity is expressed as

$$f_1^0 = K_i m_i = K_i \tag{12.70}$$

So the activity a_i is given by

$$a_i = \frac{\bar{f}_i}{f_i^0} = \frac{K_i m_i}{K_i} = m_i \tag{12.71}$$

This expression gives a simple relation between fugacity and concentration for which Henry's law is applicable.

EXAMPLE 12.15 In a laboratory investigation, acetic acid is esterified in the liquid phase with ethanol at 373.15 K and 1 atm pressure to produce ethyl acetate and water according to the reaction

$$CH_3COOH(l) + C_2H_5OH(l) \rightarrow CH_3COOC_2H_5(l) + H_2O(l)$$

If initially there is 1 mol each of acetic acid and ethanol, estimate the mole fraction of ethyl acetate in the reacting mixture at equilibrium by using the following data:

Components	$\Delta H_{f,298}^{0}$ (kJ)	$\Delta G_{f,298}^{0}$ (kJ)
CH_3COOH (l)	-484.5	-389.9
C_2H_5OH(l)	-277.69	-174.78
$CH_3COOC_2H_5$(l)	-480.0	-332.2
H_2O(l)	-285.83	-237.13

Solution: The standard heat of reaction at 298 K can be evaluated as $\Delta H_{298}^{0} = \Delta H_{f298}^{0} = [CH_3COOC_2H_5(l)] + \Delta H_{f298}^{0}[H_2O(l)] - \Delta H_{f298}^{0}[CH_3COOH(l)] - \Delta H_{f298}^{0}[C_2H_5OH(l)] = -480 + (-285.83) - (-484.5) - (-277.69) = -3.64$ kJ

Now, the standard Gibbs free energy change for the reaction at 298 K is given by

$\Delta G_{298}^{0} = \Delta G_{f298}^{0} = [CH_3COOC_2H_5(l)] + \Delta G_{f298}^{0}[H_2O(l)] - \Delta G_{f298}^{0}[CH_3COOH(l)] - \Delta G_{f298}^{0}[C_2H_5OH(l)] = -332.2 + (-237.13) - (-389.9) - (-174.78) = -4.65$ kJ

The equilibrium constant K is given by

$$\ln K = -\frac{\Delta G^0}{RT} = -\frac{-4650}{8.314 \times 298.15} = 1.8759$$

or

$$K_{298} = e^{1.8759} = 6.5266$$

Now, the equilibrium constant at 373.15 K, i.e., K_{373} can be determined as

$$\ln\left(\frac{K_{373}}{K_{298}}\right) = \frac{-3640}{8.314}\left(\frac{1}{298.15} - \frac{1}{373.15}\right) = -0.2951$$

or

$$\frac{K_{373}}{K_{298}} = e^{-0.2951}$$

or

$$K_{373} = 6.5266 \times 0.7444$$
$$K_{373} = 4.8586$$

Hence the equilibrium constant for the reaction at 373 K is 4.8586.
Let ε_e be the degree of conversion at equilibrium.

Component	CH_3COOH (l)	C_2H_5OH (l)	$CH_3COOC_2H_5$ (l)	H_2O (l)
Moles in feed	1	1	0	0
Moles at equilibrium	$1 - \varepsilon_e$	$1 - \varepsilon_e$	ε_e	ε_e
Mole fraction	$\dfrac{1 - \varepsilon_e}{2}$	$\dfrac{1 - \varepsilon_e}{2}$	$\dfrac{\varepsilon_e}{2}$	$\dfrac{\varepsilon_e}{2}$
Total moles at equilibrium	$(1 - \varepsilon_e) + (1 - \varepsilon_e) + \varepsilon_e + \varepsilon_e = 2$			

Now, the sum of the stoichiometric coefficients

$$= \Sigma v_i = 1 + 1 - 1 - 1 = 0$$

$$K = K_x = \Pi(x_i)^{v_i} = \frac{\left(\dfrac{\varepsilon_e}{2}\right)\left(\dfrac{\varepsilon_e}{2}\right)}{\left(\dfrac{1-\varepsilon_e}{2}\right)\left(\dfrac{1-\varepsilon_e}{2}\right)} = \left(\frac{\varepsilon_e}{1-\varepsilon_e}\right)^2$$

Therefore, $\left(\dfrac{\varepsilon_e}{1-\varepsilon_e}\right)^2 = 4.8586$.

Solving for ε_e, we have $\varepsilon_e = 0.6879$. Hence the degree of conversion at equilibrium = 0.6879. The mole fraction of ethyl acetate is given by

$$y_{CH_3COOC_2H_5} = \frac{\varepsilon_e}{2} = \frac{0.6879}{2} = 0.344$$

12.13   HETEROGENEOUS REACTION EQUILIBRIA

The chemical reaction that we have discussed so far occurs in the single phase. Now our attention is to be concentrated on the analysis of the chemical reactions occurring in more than one phase, i.e., in heterogeneous phase. A standard state is necessary to be assigned to explain the heterogeneous reaction at equilibrium.

Heterogeneous reaction equilibrium can broadly be categorized into two classes. These are:

1. Solid–gas reaction equilibrium
2. Liquid–gas reaction equilibrium.

12.13.1   Solid–Gas Reaction Equilibrium

By definition, heterogeneous system is defined as the system that requires at least two phases to proceed. The standard state for the solid-phase reactions may be the pure solid at 1 bar and reaction temperature. In heterogeneous equilibrium, the equilibrium constant involves only the activity or partial pressures of the gaseous constituents and does not include any term for the concentration of either solids or liquids. The equilibrium constant is thus independent of the extent of pure solid or liquid phase.

At moderate pressure, the activity of the solid components is considered to be unity. Activity of the solid is given by

$$a_i = \frac{\bar{f}_i}{f_i^0} \tag{12.72}$$

where

$\bar{f}_i$ = Fugacity of species i in the reaction mixture at the actual temperature and pressure
f_i^0 = Fugacity of pure liquid in the standard state of 1 bar and reaction temperature.

For example, consider the dissociation of calcium carbonate.

$$CaCO_3(s) \rightleftharpoons CaO(s) + CO_2(g)$$

The equilibrium constant K_a is given by

$$K_a = \frac{a_{CaO} a_{CO_2}}{a_{CaCO_3}} \qquad (12.73)$$

Since the activities of solid components, CaO and $CaCO_3$, may be taken as unity, i.e.

$$a_{CaO} = a_{CaCO_3} = 1$$

Hence, Eq. (12.73) reduces to

$$K_a = a_{CO_2} \qquad (12.74)$$

Comparing Eq. (12.74) with Eq. (12.72), we get

$$K_a = a_{CO_2} = \frac{\bar{f}_{CO_2}}{f^0_{CO_2}} \qquad (12.75)$$

The standard state for gases is referred to as the *ideal gas state*, and at that state, $\bar{f}_{CO_2} = 1$. So, Eq. (12.75) becomes

$$K_a = a_{CO_2} = \bar{f}_{CO_2} \qquad (12.76)$$

Since CO_2 is assumed to behave like an ideal gas, then fugacity is exactly equal to idealized partial pressure p, i.e., $\bar{f}_{CO_2} = p_{CO_2}$.

$$K_P = p_{CO_2} \qquad (12.77)$$

Equation (12.77) can be expressed as

$$K_P = p_{CO_2} = e^{-\frac{\Delta G^0}{RT}} \qquad \text{(as } \Delta G^0 = -RT \ln K) \qquad (12.78)$$

Hence the dissociation pressure of calcium carbonate would be constant. This pressure exerted by gaseous CO_2 depends only on temperature.

Consider another instance—hydrogen from reduction of steam.

$$3Fe + 4H_2O \rightleftharpoons Fe_3O_4 + 4H_2$$

$$K_a = \frac{a_{Fe_3O_4} \times a_{H_2}^4}{a_{Fe}^3 \times a_{H_2O}^4} = \left(\frac{a_{H_2}}{a_{H_2O}}\right)^2 \approx \left(\frac{P_{H_2}}{P_{H_2O}}\right)^4$$

Hence

$$K_P \approx \left(\frac{P_{H_2}}{P_{H_2O}}\right)^4 \qquad P_{H_2} \propto P_{H_2O}$$

Thus the ratio of $\dfrac{P_{H_2}}{P_{H_2O}}$ is constant. It does necessarily imply that an increase in steam pressure yields a higher amount of hydrogen.

EXAMPLE 12.16 The thermal decomposition of limestone takes place as

$$CaCO_3(s) \rightleftharpoons CaO(s) + CO_2(g)$$

Calculate the decomposition pressure at 1000 K when standard enthalpy, ΔH^0_{1000}, and entropy, ΔS^0_{1000}, change of the reaction are 1.7553×10^5 J and 150.3 J/mol-K.

Solution: We know that the standard Gibbs free energy change is

$$\Delta G^0 = \Delta S^0 - T\Delta S^0 \tag{i}$$

Putting $\Delta H^0_{1000} = 1.7553 \times 10^5$ J, $\Delta S^0_{1000} = 150.3$ J/mol-K, and $T = 1000$ K into Eq. (i), we get

$$\Delta G^0 = \Delta H^0 - T\Delta S^0 = 1.7553 \times 10^5 - 1000 \times 150.3$$
$$= 25,230 \text{ J}$$

Again, the standard Gibbs free energy change of this reaction can be expressed as

$$\Delta G^0 = -RT \ln K$$

or

$$K = e^{-\frac{\Delta G^0}{RT}} = e^{-\frac{25,230}{8.314 \times 1000}} = e^{-3.034} = 0.0481 \text{ bar} = \overline{p}$$

Hence the decomposition pressure of limestone is 0.0481 bar at 1000 K.

EXAMPLE 12.17 The following thermal decomposition occurs at 400 K:

$$A(s) \rightarrow B(s) + C(g)$$

The standard Gibbs free energy of the reaction, $\Delta G^0 = 85,000 - 213.73T + 6.71T \ln T - 0.00028T^2$. Determine the decomposition pressure at 400 K and 700 K. Also determine the decomposition temperature at 1 bar.

Solution: At 400 K, the standard Gibbs free energy change of the reaction is given by

$$\Delta G^0_{400} = 85,000 - 213.73 \times 400 + 6.71 \times 400 \ln 400 - 0.00028(400)^2 = 15544.20 \text{ J}$$

Similarly, at 700 K

$$\Delta G^0_{700} = 85,000 - 213.73 \times 700 + 6.71 \times 700 \ln 700 - 0.00028(700)^2 = 33978.20 \text{ J}$$

At 400 K, the equilibrium constant is given by

$$K = e^{-\frac{\Delta G^0}{RT}} = e^{-\frac{15544.20}{8.314 \times 400}} = 0.0093 \text{ bar}$$

At 700 K, the equilibrium constant is given by

$$K = e^{-\frac{\Delta G^0}{RT}} = e^{-\frac{15544.20}{8.314 \times 400}} = 343.09 \text{ bar}$$

Now, the equilibrium constant for solid–gas reaction is given by

$$K = \frac{a_B a_C}{a_A} = a_C = \overline{f}_C = P_C$$

or

$$K = P_C = 1 = e^{-\frac{\Delta G^0}{RT}}$$

or

$$\ln 1 = -\frac{\Delta G^0}{RT}$$

or

$$\Delta G^0 = 0$$

Substituting the value of ΔG^0 in the given equation, we have

$$0 = 85{,}000 - 213.73T + 6.71T \ln T - 0.00028T^2$$

Solving for T, we get $T = 1020$ K.
Therefore, the decomposition temperature is 1020 K.

EXAMPLE 12.18   In a steel reactor, steam is passed over a bed of red-hot carbon at 875 K and 1 bar. At these conditions, the equilibrium constant for the reaction is 0.514. Calculate the equilibrium composition of the reactor effluent at 875 K. The reaction mixture is assumed to behave as an ideal gas.

Solution: The reaction involved is

$$Fe(s) + H_2O(g) \rightleftharpoons FeO(s) + H_2(g)$$

The equilibrium constant K is given by

$$K = \frac{a_{FeO} a_{H_2}}{a_{Fe} a_{H_2O}} \qquad \text{(i)}$$

Since the activities of the solid components Fe and FeO at equilibrium may be taken as unity, then $a_{Fe} = a_{FeO} = 1$. Hence Eq. (i) reduces to

$$K_a = \frac{a_{H_2}}{a_{H_2O}} \qquad \text{(ii)}$$

Let ε_e be the moles of H_2O decomposed per mole of H_2O supplied.

Component	Moles in feed	Moles at equilibrium	Mole fraction	Total moles at equilibrium
$H_2O(g)$	1	$1 - \varepsilon_e$	$1 - \varepsilon_e$	$1 - \varepsilon_e + \varepsilon_e = 1$
$H_2(g)$	0	ε_e	ε_e	

The reaction occurs at the pressure $P = 1$ bar.
For the reaction

$$K_y = \frac{\varepsilon_e}{1 - \varepsilon_e}$$

the equilibrium constant is given by

$$0.514 = K = K_y P^{\Sigma v_i} = \frac{\varepsilon_e}{1 - \varepsilon_e}$$

Solving for ε_e, we get

$$\varepsilon_e = 0.34$$

Mole fraction of H_2

$$y_{H_2} = \varepsilon_e = 0.34$$

Mole fraction of H_2O

$$y_{H_2O} = 1 - \varepsilon_e = 1 - 0.34 = 0.66$$

Hence, the ratio

$$\frac{a_{H_2}}{a_{H_2O}} = \frac{0.34}{0.66} = 0.515$$

EXAMPLE 12.19 The hydrate of sodium carbonate decomposes according to the following equation:

$$Na_2CO_3 \cdot H_2O(s) \rightarrow Na_2CO_3(s) + H_2O(g)$$

The equilibrium pressure of water vapour in this reaction is given by the following equation:

$$\log P = 7.944 - \frac{3000}{T}$$

where

P = Pressure in atm
T = Temperature in K.

Derive an expression for the standard free energy change ΔG^0 of this reaction.

Solution: The equilibrium constant K of the reaction is given by

$$K = \frac{a_{Na_2CO_3} a_{H_2O}}{a_{Na_2CO_3 \cdot H_2O}} \tag{i}$$

The activities of the solid components $Na_2CO_3 \cdot H_2O$ and Na_2CO_3 at equilibrium may be taken as unity. Then

$$a_{Na_2CO_3 \cdot H_2O} = a_{Na_2CO_3} = 1$$

Hence, Eq. (i) reduces to

$$K = a_{H_2O} = P_{H_2O} \tag{ii}$$

Therefore, the standard Gibbs free energy change can be estimated as

$$\Delta G^0 = -RT \ln K = -2.303 RT \log P$$

$$= -2.303 RT \left(7.944 - \frac{3000}{T} \right) = 57442 - 152.15 T \text{ J}$$

12.13.2 Liquid–Gas Reaction Equilibrium

Consider the following reaction occurring between a gas and a liquid:

$$aA(g) + bB(l) \rightarrow rR(aq)$$

The equilibrium in the preceding reaction can be understood using the following methodology:

- The reaction may be assumed to occur in the gas phase only. The mass transfer takes place between gas and liquid phases to maintain the phase equilibrium. In order to evaluate the equilibrium constant of the reaction, the conventional standard state for the gases is used, and it is ideal gas state at 1 bar and reaction temperature. The two equations based on standard state for the gases and the mass transfer between gas and liquid phases are coupled to get the desired result.
- The reaction may be assumed to take place in the liquid phase only to evaluate the equilibrium constant on the basis of conventional standard state for the liquids, and couple the resulting expression with the phase equilibrium relation.
- In this methodology, the mixed states shown in the reaction occurring are considered, and the ideal gas state is used as the conventional standard state for gas A, pure liquid state for liquid B, and hypothetical 1-molal aqueous state for aqueous solution C, to evaluate the equilibrium constant by using the following expression:

$$K = \frac{a_R^r}{a_A^a a_B^b} = \frac{m_R^r}{\left(\bar{f}_A\right)^a \left(\gamma_B x_B\right)^b} \qquad (12.79)$$

12.14 PHASE RULE FOR REACTING SYSTEMS

The number of independent variables associated with a multi-component, multi-phase non-reacting system is given by Gibbs phase rule and can be expressed as

$$F = C - P + 2$$

where

F = Number of degrees of freedom of the systems or number of independent variables
C = Number of components
P = Number of phases. 2 is included in the equation to account for the fact that the two intensive variables T and P must be specified to describe the state of equilibrium.

In order to apply to the reacting system, the phase rule must be modified. This is because for the occurrence of independent reaction, an additional constraint is required to be imposed on the reacting system by the relation $\sum v_i \mu_i = 0$. Since μ_i is the driving force of a chemically reacting system to be in equilibrium, it will be the function of temperature, pressure, and phase composition. One should keep in mind that for every occurrence of independent reaction, the number of degrees of freedom will be reduced by one. Thus, the phase rule for the reacting system is represented by

$$F = C - P + 2 - r \qquad (12.80)$$

where r is the number of independent reactions occurring.

It is necessary to remember that the number of degrees of freedom F is nothing but the difference between the number of variables required to specify the state of the system and the number of restricting relations due to their interdependence. Again, the number of independent components is defined as the total number of components minus the number of reactions which can occur between them.

The number of independent chemical reactions can be easily determined in the following way:

- Write down the chemical equation for the formation of each chemical compound present in the system from its constituent elements.
- Eliminate the elements which are not present in the system by the proper combination of the equations with others written in the former step.

On applying these methods, the set or number of equations available is found to be exactly equal to the number of independent chemical reactions. This methodology for the reduction of equation yields the following relation:

$r \geq$ Number of compounds present in the system
  − Number of constituent elements not present as elements

Now, under certain circumstances, if the special constraint is necessary to be taken into account, then the phase rule becomes

$$F = C - P + 2 - r - s \qquad (12.81)$$

where s stands for special constraints such as additional equations among variables.

Consider the reaction

$$NH_4Cl(s) \rightarrow NH_3(g) + HCl(g)$$

In this reaction, $C = 3$ but $P = 2$. Here the imposition of a special constraint is required because equal amounts of NH_3 and HCl are formed in the gas phase after thermal decomposition of NH_4Cl. Therefore, ammonia and hydrogen chloride are equimolar, i.e., $y_{NH_3} = y_{HCl}$. Hence the number of degrees of freedom of the system by the phase rule will be

$$F = C - P + 2 - r - s = 3 - 2 + 2 - 1 - 1 = 1$$

EXAMPLE 12.20 In a chemically reacting system at equilibrium, CO, CO_2, H_2, H_2O, and CH_4 are considered to be present in a single-gas phase. Determine the number of independent reactions that occur and the number of degrees of freedom.

Solution: There are no special constraints to be taken into consideration. Only the number of independent chemical reactions are to be determined.

According to the reduction procedure, the formation reactions of the different chemical compounds from its constituent elements are represented as

$$CO: C + \frac{1}{2}O_2 \rightarrow CO \qquad (i)$$

$$CO_2: C + O_2 \rightarrow CO_2 \qquad (ii)$$

$$H_2O: H_2 + \frac{1}{2}O_2 \rightarrow H_2O \qquad (iii)$$

$$CH_4: C + 2H_2 \rightarrow CH_4 \qquad (iv)$$

It is found from the preceding equations that C and O_2 are not present as the elements in the system. Therefore, these are to be eliminated systematically by combining the reactions properly.

Elimination of C:

By combining Eq. (ii) with (i), we get

$$CO + \frac{1}{2}O_2 \rightarrow CO_2 \tag{v}$$

By combining Eq. (ii) with (iv), we get

$$CH_4 + O_2 \rightarrow 2H_2 + CO_2 \tag{vi}$$

Elimination of O_2:

Equations (iii), (v), and (vi) form the new set from which O_2 needs to be eliminated.

By combining Eq. (iii) with (iv), we get

$$CO_2 + H_2 \rightarrow CO + H_2O \tag{vii}$$

By combining Eq. (iii) with (vi), we get

$$CH_4 + 2H_2O \rightarrow CO_2 + 4H_2 \tag{viii}$$

Equations (vii) and (viii) are an independent set, indicating $r = 2$. Hence the number of degrees of freedom of the reacting system

$$F = C - P + 2 - r - s = 5 - 1 + 2 - 1 - 1 = 4$$

EXAMPLE 12.21 Determine the number of degrees of freedom of a chemically reacting system prepared by partially decomposing $CaCO_3$ into an evacuated space. (WBUT '07)

Solution: The system undergoing the chemical reaction is

$$CaCO_3(s) \rightarrow CaO(s) + CO_2(g)$$

Here

Number of independent reactions, $r = 1$;

Number of components, $C = 3$;

Number of phases, $P = 3$ (solid CaO, solid $CaCO_3$ and gaseous CO_2).

No special constraint is required to be assumed for the system prepared for the decomposition of $CaCO_3$ into an evacuated space as the reaction zone. Hence $s = 0$. Therefore, the number of degrees of freedom of the reacting system is given by

$$F = C - P + 2 - r - s = 3 - 3 + 2 - 1 - 0 = 1$$

12.15 CHEMICAL EQUILIBRIUM FOR SIMULTANEOUS REACTIONS

The reacting system we have discussed so far involves a single chemical reaction. We have calculated for this system the equilibrium constant as well as the composition of the reaction mixture. But it is needless to mention that in many of the practical cases the chemical reactions involve two or more reactions that take place simultaneously, which makes them more difficult to dealt with. Therefore, it is necessary to apply the equilibrium criterion to all possible reactions that may occur in the reacting system. Here, we will treat and solve the problems of simultaneous reactions based on the number of independent reactions and their equilibrium constants.

For example, consider the occurrence of the following simultaneous reactions:

$$CO + \frac{1}{2}O_2 \rightleftharpoons CO_2$$

$$CO + H_2O \rightleftharpoons CO_2 + H_2$$

$$H_2 + \frac{1}{2}O_2 \rightleftharpoons H_2O$$

It is found that the preceding three reactions are not independent, because any one of these reactions can be obtained by the difference of the other two reactions. Therefore, the three reactions necessarily form one set in which the first reaction combines first with the second reaction to yield a new reaction and then with the third reaction to give a second new reaction. Thus, the the three initial reactions essentially represent only two independent reactions occurring simultaneously.

The independent reactions occurring simultaneously are expressed as

$$\sum_{i=1}^{c} v_{ji} A_{ji} = 0 \qquad (12.82)$$

Applying the equilibrium criterion in a chemical reaction, we get

$$(dG)_{T,P} = \sum_{j=1}^{n} \left(\sum_{i=1}^{c} \mu_i dn_{ji} \right) = 0 \qquad (12.83)$$

Now, putting the value of $dn_{ji} = v_{ji} d\varepsilon_j$ in the Eq. (12.83), we obtain

$$(dG)_{T,P} = \sum_{j=1}^{n} \left(\sum_{i=1}^{c} \mu_i v_{ji} d\varepsilon_j \right) = 0 \qquad (12.84)$$

The preceding equation can be simplified as

$$\sum_{i=1}^{c} \mu_i v_{ji} d\varepsilon_j = 0 \qquad (12.85)$$

It can be substituted in the equation $\mu_i = g_i^0 + RT \ln a_i$ to get

$$\sum v_{ji}(g_i^0 + RT \ln a_i) = 0$$

The standard Gibbs free energy is given by

$$\Delta G_j^0 = -RT \ln \Pi \, a_{ji}^{v_{ji}} = -RT \ln K_j$$

where

$$K_j = \Pi \, a_{ji}^{v_{ji}} \text{ for the } j\text{th reaction.}$$

Since the equilibrium constants of the chemical reactions depend on the mole fractions of the components, then K_j for r independent reactions can be expressed as

$$K_j = \Pi \, a_i^{v_{ji}} = P^{v_j} \Pi y_i^{v_{ji}} \qquad (12.86)$$

Now, assuming the reaction mixture behaves like an ideal gas, we can find out the value of K_j from the following relation:

$$K_j = \Pi \, a_i^{v_{ji}} = P^{v_j} \Pi \, y_i^{v_{ji}} = K_{y,j} P^{v_j}$$

where P is the operating pressure.

Considering the new set of equations that comprises two reactions (A) and (B) deduced from the first three equations, the preceding expressions will be

$$K_A = K_{yA} P^{\Sigma v_A} \quad \text{and} \quad K_B = K_{yB} P^{\Sigma v_B} \tag{12.87}$$

Again

$$K_A = e^{\frac{\Delta G^0}{RT}}$$

where K_A is the equilibrium constant of reaction A
and

$$K_B = e^{\frac{\Delta G_1^0}{RT}}$$

where K_B is the equilibrium constant of reaction B.

On substitution of the values of K_A and K_B into Eq. (12.87), the composition of the reaction mixture y_A and y_B can be estimated, as K_{yA} and K_{yB} are already expressed in terms of the extent of the reaction, ε.

EXAMPLE 12.22 A feed stock of pure n-butane is cracked at 750 K and 1.2 bar to produce olefins. The following two independent reactions occur favouring the equilibrium conversions at these conditions:

$$C_4H_{10} \rightarrow C_2H_4 + C_2H_6 \tag{A}$$
$$C_4H_{10} \rightarrow C_3H_6 + CH_4 \tag{B}$$

At 750 K and 1.2 bar, the equilibrium constants for the reactions (A) and (B) are $K_A = 3.856$ and $K_B = 268.4$. Determine the composition of the product gases at equilibrium.

Solution: Let

ε_A = Degree of conversion of C_4H_{10} in reaction (A)
ε_B = Degree of conversion of C_4H_{10} in reaction (B).

Component	C_4H_{10}	C_2H_4	C_2H_6	C_3H_6	CH_4
Moles in feed	1	1	0	0	0
Moles at equilibrium	$1 - \varepsilon_A - \varepsilon_B$	ε_A	ε_A	ε_B	ε_B
Mole fraction	$\dfrac{1-\varepsilon_A-\varepsilon_B}{1+\varepsilon_A+\varepsilon_B}$	$\dfrac{\varepsilon_A}{1+\varepsilon_A+\varepsilon_B}$	$\dfrac{\varepsilon_A}{1+\varepsilon_A+\varepsilon_B}$	$\dfrac{\varepsilon_B}{1+\varepsilon_A+\varepsilon_B}$	$\dfrac{\varepsilon_B}{1+\varepsilon_A+\varepsilon_B}$

Total moles at equilibrium	$1 - \varepsilon_A - \varepsilon_B + \varepsilon_A + \varepsilon_A + \varepsilon_B + \varepsilon_B = 1 + \varepsilon_A + \varepsilon_B$

Now, the sum of the stoichiometric coefficients for reaction (A)

$$= \sum v_A = 1 + 1 - 1 = 1$$

and
for reaction (B)

$$= \sum v_B = 1 + 1 - 1 = 1$$

We know that the equilibrium constant for reaction (A) and reaction (B) can be evaluated by $K_A = K_{yA} P^{\Sigma v_A}$ and $K_B = K_{yB} P^{\Sigma v_B}$ respectively.

The reactions occur at the pressure, $P = 1.2$ bar.
For reaction (A)

$$K_{yA} = \frac{\left(\dfrac{\varepsilon_A}{1 + \varepsilon_A + \varepsilon_B}\right)\left(\dfrac{\varepsilon_A}{1 + \varepsilon_A + \varepsilon_B}\right)}{\left(\dfrac{1 - \varepsilon_A - \varepsilon_B}{1 + \varepsilon_A + \varepsilon_B}\right)} = \frac{(\varepsilon_A)^2}{(1 - \varepsilon_A - \varepsilon_B)(1 + \varepsilon_A + \varepsilon_B)}$$

For reaction (B)

$$K_{yB} = \frac{\left(\dfrac{\varepsilon_B}{1 + \varepsilon_A + \varepsilon_B}\right)\left(\dfrac{\varepsilon_B}{1 + \varepsilon_A + \varepsilon_B}\right)}{\left(\dfrac{1 - \varepsilon_A - \varepsilon_B}{1 + \varepsilon_A + \varepsilon_B}\right)} = \frac{(\varepsilon_B)^2}{(1 - \varepsilon_A - \varepsilon_B)(1 + \varepsilon_A + \varepsilon_B)}$$

Substituting the value of K_{yA}, P and Σv_A in the equation $K_A = K_{yA} P^{\Sigma v_A}$

$$K_A = K_{yA} P^{\Sigma v_A} = \frac{(\varepsilon_A)^2}{(1 - \varepsilon_A - \varepsilon_B)(1 + \varepsilon_A + \varepsilon_B)}(1.2)^1 \qquad (i)$$

Putting $K_A = 3.856$ into Eq. (i) and simplifying, we get

$$\frac{1.2(\varepsilon_A)^2}{(1 + \varepsilon_A + \varepsilon_B)(1 - \varepsilon_A - \varepsilon_B)} = 3.856 \qquad (ii)$$

Similarly, substituting the value of K_{yB}, P and Σv_B in the equation $K_B = K_{yB} P^{\Sigma v_B}$

$$K_B = K_{yB} P^{\Sigma v_B} = \frac{(\varepsilon_B)^2}{(1 - \varepsilon_A - \varepsilon_B)(1 + \varepsilon_A + \varepsilon_B)} \cdot (1.2)^1 \qquad (iii)$$

Putting $K_B = 268.4$ into Eq. (iii), we get

$$\frac{1.2(\varepsilon_B)^2}{(1 - \varepsilon_A - \varepsilon_B)(1 + \varepsilon_A + \varepsilon_B)} = 268.4 \qquad (iv)$$

Dividing Eq. (iii) by (ii), we have

$$\left(\frac{\varepsilon_B}{\varepsilon_A}\right)^2 = \frac{268.4}{3.856} = 69.6$$

or

$$\varepsilon_B = 8.34\varepsilon_A \tag{v}$$

Putting $\varepsilon_B = 8.34\varepsilon_A$ into Eq. (iv), we get

$$\varepsilon_A = 0.1068$$

Hence

$$\varepsilon_B = 8.34\varepsilon_A = 0.8914$$

Therefore, the composition of the product is

$$y_{C_4H_{10}} = \frac{1 - \varepsilon_A - \varepsilon_B}{1 + \varepsilon_A + \varepsilon_B} = 0.0010 \qquad y_{C_2H_4} = y_{C_2H_6} = \frac{\varepsilon_A}{1 + \varepsilon_A + \varepsilon_B} = 0.0534$$

and

$$y_{C_3H_6} = y_{CH_4} = \frac{\varepsilon_B}{1 + \varepsilon_A + \varepsilon_B} = 0.4461$$

Note: The values of ε_A and ε_B can be estimated by another method. First a value for ε_A is to be assumed and then put into Eqs. (ii) and (iv) to get the two values of ε_B. Then these two values of ε_B are to be plotted against ε_A. The process is to be repeated for some other assumed values of ε_A. In this way, one can obtain the intersection of the two curves, which provides the values of ε_A and ε_B. Then the procedure to get the mole fraction of the different components of the reaction mixture is to be followed.

12.16 FUEL CELL

It is our common experience that the burning of fuel provides thermal energy. As a consequence, the work potential of the fuel is lost to a great extent. Fuel cell plays an important role in saving this energy.

A fuel cell is an electrochemical energy conversion device in which the chemical energy of a fuel such as methane and hydrogen is converted directly into electrical energy by their oxidation in suitable anode. A fuel cell comprises an anode and a cathode separated by an electrolyte and connected to each other through an electrical load in the external circuit. The fuel on the anode side and the oxidant (usually O_2 of air) on the cathode side reacts in the presence of an electrolyte. At the cathode, oxygen is dissolved to form OH^- ions. A fuel cell functions like a battery or electrolytic cell. But there is a difference between fuel cell and battery. In fuel cell the reactants are not stored, but they must be fed continuously and products must flow out. But the battery is useful only till the reactants initially put into it have been consumed, i.e., the reactants are stored in the battery.

The operation of a hydrogen–oxygen fuel cell is shown in Fig. 12.5, where H_2 and O_2 react to form water. Its operation is as follows:

The electron flows in the external circuit from anode to cathode. Oxygen enters the cathode side and hydrogen enters the anode side. At the ion-exchange membrane surface, hydrogen is ionized according to the reaction

$$2H_2 \rightarrow 4H^+ + 4e^-$$

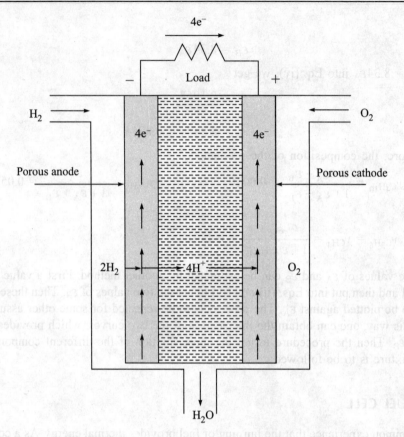

Fig. 12.5 Schematic representation of a H_2–O_2 fuel cell.

The electrons flow through the external circuit and the hydrogen ions flow through the membrane to the cathode, where the following reaction takes place:

$$4H^+ + 4e^- + O_2 \rightarrow 2H_2O$$

Thus electricity flows through the electrical load due to potential difference between cathode and anode, and work is produced. Also, heat transfer takes place between the fuel cell and the environment.

Some other fuel used are cells shown in Table 12.1. In case of $CH_4(g)$ as a fuel, the reactions at the anode and the cathode and the overall are reaction:

Anode: $CH_4 + 2H_2O \rightarrow CO_2 + 8H^+ + 8e^-$
Cathode: $2O_2 + 8e^- + 8H^+ \rightarrow 2H_2O + 2H_2O$
Overall: $CH_4 + 2O_2 \rightarrow CO_2 + 2H_2O$

In the fuel cell, electrons must be free from impurities and should be porous. The polarization due to adsorption and poisoning of the electrodes reduces the cell performance. The concentrations of electrolyte and temperature have great influence on the e.m.f. of cells.

Table 12.1 Examples of some fuel cells

Fuel cell	Electrode	Electrolyte	Reaction at anode
H_2-O_2	Pt/Ni	NaOH	$\frac{1}{2}H_2 + H_2O \rightarrow H_3O^+ + e^-$
CH_4-O_2	Pt/Pt	H_2SO_4	$CH_4 + 2H_2O \rightarrow 8H^+ + CO_2 + 8e^-$
$N_2H_4-O_2$	Ni/C	KOH	$N_2H_4 \rightarrow N_2 + 4H^+ + 4e^-$
$CO-O_2$	Cu/Ag	KOH	$CO + H_2O \rightarrow CO_2 + 2H^+ + 2e^-$

Advantages

The fuel cell

- produces power at low cost at considerable periods;
- requires easy maintenance;
- has harmless bye-products;
- is flexible in application;
- is compact, lightweight and has no major moving parts.

Areas of Application

The fuel cells are used

- as auxiliary power sources in remote locations such as spacecraft, submarines, remote weather stations, large parks, and in certain military applications;
- to replace gasoline and diesel-powered generators and automobile engines;
- to convert household sewage and other wastes to less noxious products.

SUMMARY

The chapter has been started with a discussion of the importance of reaction coordinate in a chemical reaction, by which one could know the extent to which a chemical reaction takes place. The extent of reaction is also important in the case of multireactions to compute the conversion of the components at equilibrium. A chemical engineer is very much concerned with the determination of equilibrium of a chemical reaction. The criterion of chemical equilibrium has been substantiated with the help of Gibbs free energy change and equilibrium constant by the relation

$$K = e^{\frac{-\Delta \overset{\circ}{G}}{RT}}$$

The factors influence the equilibrium factors such as temperature, pressure, presence of inert gases, presence of products in the initial reaction mixture, and presence of excess reactants. The effect of temperature on equilibrium constant is expressed by the van't Hoff equation as

$$\ln \frac{K_2}{K_1} = \frac{\Delta \overset{\circ}{H}}{R} \left(\frac{1}{T_1} - \frac{1}{T_2} \right)$$

where

K_1 = Equilibrium constant at absolute temperature T_1

K_2 = Equilibrium constant at absolute temperature T_2.

The effect of pressure on chemical equilibrium were explained on the basis of Le Chatelier's principle, which implies that the mixture pressure affects the equilibrium composition but not the equilibrium constant. For instance, the synthesis of ammonia can be favoured by increase in pressure. The methodology for the evaluation of equilibrium constant for homogeneous and heterogeneous reactions and the mole fractions of the reaction mixture at equilibrium have been discussed.

For homogeneous gas phase reaction, the equilibrium constant can be expressed as

$$K_a = (\Pi \, \overline{\Phi}_i^{y_i})(\Pi P_i^{y_i}) = (\Pi \Phi_i^{y_i})(\Pi P_i^{y_i}) = K_\Phi K_P = K_\Phi K_y \, P^{\Sigma v_i}$$

where $K_y = \Pi y_i^{y_i}$ is the equilibrium constant in terms of mole fractions.

For homogeneous liquid phase reaction, the equilibrium constant can be expressed as

$$K_a = \Pi \, a_i^{y_i} = \Pi (\gamma_i x_i)^{v_i} = (\Pi \gamma_i^{v_i})(\Pi x_i^{v_i}) = K_\gamma K_x$$

where x_i is the mole fraction of species i in the equilibrium reaction mixture.

If the reaction mixture is an ideal solution, $\gamma_i = 1$. Then the preceding equation will be

$$K_a = K_x = \Pi (x_i)^{v_i}$$

Similarly for heterogeneous reaction, the procedure for the determination of equilibrium constant and equilibrium conversion of the components have been discussed.

In case of chemical reaction occurring in non-ideal liquid solution, $K_\gamma \neq 1$. Therefore, in many of the liquid phase reactions, the equilibrium constant is given by

$$K_c = \Pi (c_i)^{v_i}$$

where K_c is the equilibrium constant in terms of concentration.

The phase rule can also be applied to the reacting system as

$$F = C - P + 2 - r$$

where r is the number of independent reactions occurring.

Considering the special constraint necessary to be taken into account, the phase rule can be represented as

$$F = C - P + 2 - r - s$$

where s stands for special constraints.

The equilibrium for more than one reaction simultaneously taking place is also of great theoretical and practical importance. Finally, the role of a fuel cell to produce electricity from chemical energy of a fuel has been explained.

KEY TERMS

Equilibrium Constant   A good qualitative indicator of the feasibility of a reaction, which is determined to quantify the chemical equilibria.

Fuel Cell   An electrochemical energy conversion device in which the chemical energy of a fuel is converted directly into electrical energy by its oxidation.

Heterogeneous Reaction   The reaction which requires at least two phases to proceed.

Homogeneous Reaction The reaction in which all the reactants and the products are found to be in a single phase, and which takes place in one phase alone.

Phase Rule for Reacting System A rule for the reacting system represented by

$$F = C - P + 2 - r$$

where r is the number of independent reactions occurring.

If special constraints are needed to be taken into consideration

$$F = C - P + 2 - r - s$$

where s is the number of special constraints.

Reaction Coordinate An important factor representing the progress of a chemical reaction along a pathway, which implies the extent to which a chemical reaction taken place.

Simultaneous Reactions Reactions that occur at the same time.

van't Hoff Equation Equation representing the effect of temperature on chemical equilibrium.

IMPORTANT EQUATIONS

1. The mole fraction of component i in the reaction mixture is given by

$$y_i = \frac{n_i}{n} = \frac{n_{i_0} + v_i \varepsilon}{n_0 + v\varepsilon} \tag{12.7}$$

2. For multireaction, the mole fraction y_i of species i after the reaction can be expressed as

$$y_i = \frac{n_i}{n} = \frac{n_{i_0} + \sum_j v_{i,j}\varepsilon_j}{n_0 + \sum_j v_j \varepsilon_j} \tag{12.13}$$

3. The criterion of equilibrium of a chemical reaction can be represented by

$$(dG)_{T,P} = \sum \mu_i v_i = 0 \tag{12.19}$$

4. For a gaseous system, the f^0_i value approaches unity for a pure component. The equilibrium constant is given by

$$K_f = \Pi(\bar{f}_i)^{v_i} \tag{12.28}$$

5. For a gaseous reaction, the equilibrium constant is expressed in terms of partial pressure and is given by

$$K_p = \Pi(\bar{p}_i)^{v_i} \tag{12.29}$$

6. For the reaction $aA + bB \rightarrow rR + sS$, the free energy change is expressed as

$$\Delta G^0 = (r\mu_R + s\mu_S) - (a\mu_A + b\mu_B) \tag{12.33}$$

7. The most commonly used equations for the determination of free energy change of the process are

(i) $\Delta G^0 = \Delta H^0 - T\Delta S^0$

(ii) $\Delta G^0 = \sum \Delta G^0_{f(\text{Products})} - \sum \Delta G^0_{f(\text{Reactants})}$.

8. The useful relation between Gibbs free energy change and equilibrium constant, by which the equilibrium constant can be determined from the given Gibbs free energy change and temperature, is given by

$$K = e^{\frac{-\Delta \mathring{G}}{RT}} \tag{12.37}$$

9. The effect of temperature on equilibrium constant by the van't Hoff equation is substantiated by

$$\ln\frac{K_2}{K_1} = \frac{\Delta \mathring{H}}{R}\left(\frac{1}{T_1} - \frac{1}{T_2}\right) \tag{12.45}$$

where

K_1 = Equilibrium constant at absolute temperature T_1

K_2 = Equilibrium constant at absolute temperature T_2.

10. The standard heat of reaction at temperature T is given by

$$\Delta H^0_T = \Delta H^0_{T_0} + \Delta\alpha T + \frac{\Delta\beta}{2}T^2 + \frac{\Delta\gamma}{3}T^3 + \cdots \tag{12.48}$$

11. The Gibbs free energy change in a reaction can be determined in the following way:

$$\Delta G^0 = \Delta H^0_{T_0} - \Delta\alpha T \ln T - \frac{\Delta\beta T^2}{2} - \frac{\Delta\gamma T^3}{6} - IRT \tag{12.51}$$

where I is the integration constant.

12. The equilibrium constant related to number of mole fractions and pressure is given by

$$K_n = K\left(\frac{N_{\text{total}}}{P}\right)^{\Sigma v_i} \tag{12.61}$$

13. The equilibrium constant for homogeneous gas-phase reaction can be evaluated by

$$K_a = K_P = K_y P^{\Sigma v_i}$$

14. The equilibrium constant for homogeneous gas-phase reaction can also be evaluated by

$$K_a = K_x = \Pi(x_i)^{v_i} \tag{12.68}$$

15. The equilibrium constant for heterogeneous solid–gas reaction can be determined by

$$K_P = p_{\text{gas}} = e^{\frac{-\Delta \mathring{G}}{RT}} \tag{12.78}$$

16. The equilibrium constant for heterogeneous liquid–gas reaction can also be determined by

$$K = \frac{a_R^r}{a_A^a a_B^b} = \frac{m_R^r}{(\bar{f}_A)^a (\gamma_B x_B)^b} \tag{12.79}$$

17. The phase rule for the reacting system is represented by

$$F = C - P + 2 - r \qquad (12.80)$$

where r is the number of independent reactions occurring.

18. If special constraints are needed to be taken into consideration

$$F = C - P + 2 - r - s \qquad (12.81)$$

where s is the number of special constraints.

EXERCISES

A. Review Questions

1. What is reaction coordinate? What is its significance in chemical reaction?
2. What do you mean by extent of reaction? Derive an expression of the relationship between mole fraction of the component and the extent of reaction.
3. What is the criterion of chemical equilibrium for a reacting system?
4. Give an informatory note on equilibrium constant of the chemical reaction.
5. Prove that $K_a = K_f = K_p$.
6. With an example, explain the importance of multireaction stoichiometry.
7. What is the standard Gibbs free energy change and how is it related to the equilibrium constant?
8. Define *standard state*. What is standard free energy of formation of a compound?
9. What is the influence of temperature on equilibrium constant? In light of this, derive van't Hoff's equation.
10. Discuss the condition of feasibility of a chemical reaction in terms of standard Gibbs free energy change.
11. What is the effect of pressure on equilibrium conversion of a gas-phase chemical reaction?
12. How does the addition of inert gases affect the degree of conversion of a chemical reaction?
13. Explain the effect of presence of product and excess of reactant in the initial mixture of the reacting system.
14. Discuss the procedure for the determination of equilibrium constant in a liquid-phase reaction.
15. Determine the equilibrium constant in a homogeneous gas-phase reaction.
16. Discuss the equilibria with simultaneous reaction and estimate the equilibrium constant.
17. Enumerate the importance of phase rule for chemically reacting system.
18. Mention the different steps for the prediction of the number of independent reactions in determining the degrees of freedom.
19. Explain in detail, with neat sketch, the working principle of a fuel cell in producing electrical energy from the chemical energy of a fuel.
20. Mention the areas of application of the fuel cell.

B. Problems

1. A system, consisting of 2 mol of N_2, 5 mol of H_2 and 2 mol of NH_3 initially, is undergoing the following reaction:

$$N_2 + 3H_2 \rightarrow 2NH_3$$

Derive the expression for the mole fractions of various constituents in the reaction mixture in terms of extent of reaction.

2. The following reaction takes place in a system consisting of 3 mol CH_4, 5 mol H_2O, 1 mol CO and 4 mol H_2 initially:

$$CH_4(g) + 2H_2O(g) \rightarrow CO(g) + 4H_2(g)$$

Express the composition of the mixture in terms of mole fraction as a function of extent of reaction.

3. Develop the expressions for the mole fractions of reacting species as functions of the reaction coordinate for a system initially containing 3 mol H_2S and 5 mol O_2 undergoing the following reaction:

$$2H_2S(g) + 3O_2(g) \rightarrow 2H_2O(g) + 2SO_2(g)$$

4. A system initially containing 3 mol CO_2, 5 mol H_2 and 1 mol H_2O is undergoing the following reactions:

$$CO_2(g) + 3H_2(g) \rightarrow CH_3OH(g) + H_2O(g)$$
$$CO_2(g) + H_2(g) \rightarrow CO(g) + H_2O(g)$$

Derive the expression for the mole fractions of various components in the reaction mixture in terms of extent of reaction.

5. A system is initially charged with 4 mol ethylene and 3 mol oxygen, and the following reaction occurs between the species:

$$C_2H_4(g) + O_2(g) \rightarrow 2[(CH_2)O](g)$$
$$C_2H_4(g) + 3O_2(g) \rightarrow 2CO_2(g) + 2H_2O(g)$$

Express the composition of the mixture in terms of mole fraction as a function of reaction coordinate.

6. Calculate the standard Gibbs free energy change and the equilibrium constant at 1 bar and 298.15 K for the ammonia synthesis reaction

$$C_2H_5OH(g) + \frac{1}{2} O_2(g) \rightarrow CH_3CHO(g) + H_2O(g)$$

Given that the standard free energies of formation at 298 K for CH_3CHO are -133.978 kJ/mol and for C_2H_5OH are -174.883 kJ/mol.

7. Calculate the equilibrium constant at 25°C for the following reaction:

$$N_2O_4(g) \rightarrow 2NO_2(g)$$

Given that $\Delta G^0_{f,NO_2} = 51.31$ kJ/mol and $\Delta G^0_{f,N_2O_4} = 97.54$ kJ/mol.

8. Calculate ΔG_{298}^0 at 1 bar and 298.15 K for the following reaction:

$$SO_2(g) + \frac{1}{2}O_2(g) \rightarrow SO_3(g)$$

Given that $\Delta G_{f,SO_3}^0 = -370.532$ kJ/mol and $\Delta G_{f,SO_2}^0 = -300.612$ kJ/mol.

9. At 2000 K, the free energy change in calories for the reaction

$$N_2 + O_2 \rightleftharpoons 2NO$$

is given by $\Delta G^0 = 22000 - 2.5T$. Calculate K_p at this temperature.

10. In a chemical laboratory, it is proposed to carry out the reaction

$$C_2H_4(g) + H_2O(g) \rightarrow C_2H_5OH(g)$$

at 1 bar and 298 K. Calculate the standard Gibbs free energy change at 298 K and predict whether it is feasible to carry out the given reaction or not. If possible, calculate the equilibrium constant.

11. For the chemical reaction

$$N_2O \rightarrow N_2 + \frac{1}{2}O_2$$

$\Delta H = -20000$ cal and $\Delta S = 18$ e.u. Will the reaction be spontaneous at 25°C?

12. For the endothermic reaction,

$$C + H_2O(g) \rightarrow CO(g) + H_2(g)$$

where $\Delta H = =+31400$ cal and $\Delta S = +32$ e.u. Discuss the feasibility of the reaction at 100°C and 1000°C.

13. Assuming that the standard enthalpy changes of the reaction are constant in the temperature range 298 K to 800 K, estimate the equilibrium constant at 800 K for the ammonia synthesis reaction

$$N_2(g) + 3H_2(g) \rightarrow 2NH_3(g)$$

Given that $\Delta H_{f,NH_3}^0 = -46.1$ kJ and $\Delta G_{f,NH_3}^0 = -16.747$ kJ.

14. The equilibrium constant for the reaction $A \rightarrow B$ is doubled when temperature is changed from 25°C to 35°C. Calculate the enthalpy of reaction.

15. Calculate the equilibrium constant at 750 K and 1 bar for the following reaction:

$$C_2H_5OH(g) + \frac{1}{2}O_2(g) \rightarrow CH_3CHO(g) + H_2O(g)$$

Assume that the heat of reaction remains constant in the temperature range involved, given that the equilibrium constant K at 298 K and 1 bar is 1.1582×10^5. Use data table for standard heat of reaction.

16. Evaluate the equilibrium constant at 600 K for the reaction

$$CO(g) + 2H_2(g) \rightarrow CH_3OH(g)$$

Data Table:

Components	$\Delta H_{f,298}^0$ (kJ)	$\Delta G_{f,298}^0$ (kJ)
CO(g)	-110.5	-203.9
H_2(g)	$-$	-136.78
CH_3OH(g)	-200.7	-249.83

17. Calculate ΔG^0 for the hydrogenation of ethylene at 298 K using the following data:

$\Delta G_{f,298}^0$ for $C_2H_4 = 68.12$ kJ/mol
$\Delta G_{f,298}^0$ for $C_2H_6 = -32.89$ kJ/mol.

The reaction involved is

$$C_2H_4(g) + H_2(g) \rightarrow C_2H_6(g)$$

18. Calculate the values of K_p at 25°C and 800°C for the water–gas reaction

$$CO(g) + H_2O(g) \rightarrow H_2(g) + CO_2(g)$$

Using the following data at 298 K and 1 atm:

Substance	H_2	CO	H_2O	CO_2
$\Delta G_{f,298}^0$, kJ/mol	0	-137.27	-228.59	-394.38
$\Delta H_{f,298}^0$, kJ/mol	0	-110.52	-241.83	-392.51

19. Hydrogen bromide is produced by the following reaction:

$$\frac{1}{2} H_2(g) + \frac{1}{2} Br_2(g) \rightarrow HBr(g)$$

The equilibrium constant at temperature 420 K is given as log $K_p = 2.44$, and $\Delta H_0 = -11970$ cal at 600 K.
Given that $C_{P_{H_2}} = 6.5 + 0.0009T$, $C_{P_{Br_2}} = 7.4 + 0.001T$ and $C_{P_{HBr}} = 6.5 + 0.001T$.
Develop general expressions for the standard free energy change as function of temperature.

20. Estimate the standard Gibbs free energy and the equilibrium constant at 1 bar and 800 K for the reaction

$$N_2(g) + 3H_2(g) \rightleftharpoons 2NH_3(g)$$

Given that the standard heat of reaction, ΔH_{298}^0, and standard Gibbs free energy of the reaction, ΔG_{298}^0, for the formation of products are -92.11 kJ/mol and -33.494 kJ/mol respectively. The specific heat data in kJ as function of temperature are as follows:

Components	α	$\beta \times 10^3$
NH_3(g)	29.747	25.108
N_2(g)	27.270	4.930
H_2(g)	27.012	3.509

21. For the vapour-phase hydration of ethylene to ethanol

$$C_2H_4 + H_2O \rightarrow C_2H_5OH$$

The equilibrium constants were measured at temperature 420 K and 600 K. They are 6.8×10^{-2} and 1.9×10^{-3} respectively. The specific heat (J/mol-K) data are

Components	C_P (J/mol-K)
Ethylene	$11.886 + 120.12 \times 10^{-3}T - 36.649 \times 10^{-6}T^2$
Water	$30.475 + 9.652 \times 10^{-3}T + 1.189 \times 10^{-6}T^2$
Ethanol	$29.358 + 166.9 \times 10^{-3}T - 50.09 \times 10^{-6}T^2$

Develop general expressions for the equilibrium constant and standard free energy change as function of temperature.

22. Methanol is produced by the following reaction:

$$CO(g) + 2H_2(g) \rightarrow CH_3OH(g)$$

The standard heat of formation of $CO(g)$ and $CH_3OH(g)$ at 298 K are $-110,500$ J/mol and $-200,700$ J/mol respectively. The standard free energies of formation are $-137,200$ J/mol and $-162,000$ J/mol respectively.

(a) Estimate the standard free energy and determine whether the reaction is feasible at 298 K.

(b) Determine the equilibrium constant at 400 K, assuming that the heat of reaction is constant.

23. Sulphur trioxide is produced according to the reaction

$$SO_2(g) + \frac{1}{2}O_2(g) \rightarrow SO_3$$

Given that

$\Delta H^0_{f,298}$ for SO_2 = -70.96 kcal/mol
$\Delta H^0_{f,298}$ for SO_3 = -94.45 kcal/mol
$\Delta G^0_{f,298}$ for SO_2 = -71.79 kcal/mol
$\Delta G^0_{f,298}$ for SO_3 = -88.52 kcal/mol.

Constants of isobaric molar heat capacities are tabulated as follows:

Components	α	$\beta \times 10^2$	$\gamma \times 10^5$
$SO_3(g)$	3.918	3.483	-2.675
$SO_2(g)$	6.157	1.384	-0.9103
$O_2(g)$	6.085	0.3631	-0.1709

(a) Determine the standard heat of reaction as a function of temperature

(b) Determine the equilibrium constant as a function of temperature at 500 K

(c) Calculate the standard free energy change at 500 K.

24. In the gaseous reaction

$$2A + B \rightleftharpoons A_2B$$

$\Delta G^0 = -1200$ cal at 227°C. What total pressure would be necessary to produce 60% conversion of B into A_2B when 2:1 mixture is used?

25. Phosphorus pentachloride dissociates according to the reaction:

$$PCl_5(g) \rightarrow PCl_3(g) + Cl_2(g)$$

Show that $K_P = \dfrac{\alpha^2 \left(\dfrac{P}{P^0}\right)}{1-\alpha^2}$, where P is the total pressure and α is the degree of dissociation.

26. If a mixture of 1 mol of CO(g), 1 mol of H_2O(g), and 2 mol of He(g) are fed into a reactor at 10 bar and 1000 K, the following reaction occurs to produce carbon dioxide and hydrogen gas:

$$CO(g) + H_2O(g) \rightarrow CO_2(g) + H_2(g)$$

The equilibrium constant for the reaction is 1.5. Calculate the degree of conversion and equilibrium composition of the reaction mixture, assuming that the mixture behaves like an ideal gas.

27. Ammonia synthesis reaction is represented by

$$N_2(g) + 3H_2(g) \rightarrow 2NH_3(g)$$

The mixture of nitrogen, hydrogen and helium gas is fed to the reactor in the mole ratio of 1:3:2. The process is maintained at 800 K and 250 bar. Estimate the molar composition of the gases at equilibrium, assuming that the reaction mixture behaves like an ideal gas, given that equilibrium constant, $K = 1.106 \times 10^{-5}$.

28. The esterification of ethanol with acetic acid takes place at 373 K as

$$CH_3COOH(l) + C_2H_5OH(l) \rightarrow CH_3COOC_2H_5(l) + H_2O(l)$$

The equilibrium constant is given by $K_x = 2.92$. Initially the liquid mixture contains 5.0 kmol CH_3COOH, 10.0 kmol C_2H_5OH and 10.0 kmol H_2O. Estimate the extent of the reaction at equilibrium concentration of the composition of the liquid mixture.

29. The thermal decomposition of limestone takes place as

$$CaCO_3(s) \rightleftharpoons CaO(s) + CO_2(g)$$

Calculate the decomposition pressure at 600 K when standard enthalpy (ΔH_{600}^0) and entropy (ΔS_{600}^0) change of the reaction are 1.123×10^5 J and 124.1 J/mol-K.

30. Determine the decomposition pressure of calcium carbonate at 600 K and 1000 K. Also determine the decomposition temperature at 1 bar.

31. Consider the following decomposition process:

$$A(s) \rightleftharpoons B(s) + C(g)$$

The pressure at equilibrium is $\ln P = 12 - \dfrac{3790}{T}$. Deduce the expression for ΔG^0 as a function of temperature for the given reaction.

32. Calculate ΔG^0 at 600°C and 850°C for the reaction

$$CaCO_3(s) \rightleftharpoons CaO(s) + CO_2(g)$$

Given that the decomposition pressure of carbon dioxide at 850°C and 600°C are 1 atm and 0.032 atm respectively.

33. In a steel reactor, steam is passed over a bed of white-hot carbon at 1000 K and 1 bar. At these conditions, the equilibrium constant for the reaction is 0.270. Calculate the equilibrium composition of the reactor effluent at 1000 K. The reaction mixture is assumed to behave as an ideal gas.

34. Acetic acid is esterified in the liquid phase with ethanol at 373.15 K and 1 atm pressure to produce ethyl acetate and water according to the reaction

$$CH_3COOH(l) + C_2H_5OH(l) \rightarrow CH_3COOC_2H_5(l) + H_2O(l)$$

If initially there is one mol each of acetic acid and ethanol, estimate the mole fraction of ethyl acetate in the reacting mixture at equilibrium by using the following data:

Components	$\Delta H^0_{f,298}$ (kJ)	$\Delta G^0_{f,298}$ (kJ)
$CH_3COOH(l)$	−116.2	−93.56
$C_2H_5OH(l)$	−66.35	−41.76
$CH_3COOC_2H_5(l)$	−110.72	−76.11
$H_2O(l)$	−68.31	−56.68

35. In a chemically reacting system at equilibrium, CO, C, CO_2, H_2, H_2O, N_2 and CH_4 are considered to be present in a single-gas phase. Determine the number of independent reactions occurring and the number of degrees of freedom.

36. A gaseous system comprises NO_2, NH_3, NO, O_2, H_2O, and N_2 at chemical equilibrium. Determine the number of degrees of freedom of the system.

37. In the steam cracking of methane, the following two independent reactions occur at 1000 K and 1 bar:

$$CH_4 + H_2O \rightarrow CO + 3H_2 \tag{A}$$
$$CO + H_2O \rightarrow CO_2 + H_2 \tag{B}$$

At 1000 K and 1 bar, the equilibrium constants for the reactions (A) and (B) are $K_A = 30$ and $K_B = 1.5$. Determine the composition of the product gases leaving the reactor. Assume that the reaction mixture behaves like an ideal gas.

38. The catalytic dehydrogenation of 1-butene

$$C_4H_8(g) \rightarrow C_4H_6(g) + H_2(g)$$

is carried out at 900 K and 1 atm and with a ratio of 10 mol of steam per mole of butane. Determine the extent of reaction at equilibrium. Also determine the extent of reaction in the absence of steam.

39. Propane is pyrolyzed by the following two independent gaseous reactions:

$$C_3H_8 \rightarrow C_3H_6 + H_2$$
$$C_3H_8 \rightarrow C_2H_4 + CH_4$$

At 1020 K and 1 bar, the equilibrium constants for the reactions (A) and (B) are $K_A =$ 28.2 and $K_B = 1.7$. Determine the composition of the equilibrium mixture. The reaction mixture is assumed to behave as an ideal gas.

40. The following reactions take place at 1500 K and 10 bar:

$$A + B \rightarrow C + D \qquad\qquad\qquad (A)$$
$$A + C \rightarrow 2E \qquad\qquad\qquad (B)$$

where $K_A = 2.67$ and $K_B = 3.20$.

Determine the composition of the equilibrium mixture, assuming that the reaction mixture behaves like an ideal gas.

41. In the catalytic dehydrogenation of acetylene, the following two reactions are expected to occur simultaneously in the reactor at 1500 K and 1 bar:

$$C_2H_2 \rightarrow 2C + H_2 \qquad\qquad\qquad (A)$$
$$2C + 2H_2 \rightarrow C_2H_4 \qquad\qquad\qquad (B)$$

At these conditions, the equilibrium constants for the reactions (A) and (B) are $K_A = 5.2$ and $K_B = 0.1923$. Assuming that the reaction mixture behaves like an ideal gas, determine the composition of the reactor effluent.

42. The following reactions take place at 1 bar and 1100 K:

$$A + B \rightarrow C + D \qquad\qquad\qquad (A)$$
$$A + C \rightarrow 2E \qquad\qquad\qquad (B)$$

where $K_A = 0.1429$ and $K_B = 2.0$.

Determine the composition of the equilibrium mixture. Assume that the reaction mixture behaves like an ideal gas.

Property Tables

TABLE A.1 Conversion Factors

Dimension	Metric	Metric/English
Pressure	1 Pa = 1 N/m^2 1 kPa = 10^3 Pa = 10^{-3} MPa 1 atm = 101.325 kPa = 1.01325 bar = 760 mm Hg at 0°C = 1.03323 kgf/cm^2 1 mm Hg = 0.1333 kPa 1 bar = 10^5 N/m^2	1 Pa = 1.4504 × 10^{-4} psia = 0.020886 lbf/ft^2 1 psi = 144 lbf/ft^2 = 6.894 kPa 1 atm = 14.696 psia = 29.92 in Hg at 30°F
Temperature	T(K) = T(°C) + 273.15 ΔT(K) = ΔT(°C)	T(R) = T(°F) + 459.67 = 1.8T(K) T(°F) = 1.8 T(°C) + 32 ΔT(°F) = ΔT(RC) = 1.8 T(K)
Volume	1 m^3 = 1000 L = 10^6 cm^3 (cc)	1 m^3 = 6.1024 × 10^4 in^3 = 35.315 ft^3 = 264.17 gal (US) 1 gal = 231 in^3 = 3.7854 L 1 fl ounce = 29.5735 cm^3 = 0.0295 L 1 gal = 128 fl ounces
Velocity	1 m/s = 3.60 km/hr	1m/s = 3.2808 ft/s = 2.237 mi/h 1 mi/h = 1.46667 ft/s 1 mi/h = 1.6093 km/h
Specific volume	1m^3/kg = 1000 L/g = 1000 cm^3/g	1 m^3/kg = 16.02 ft^3/lbm 1 ft^3/lbm = 0.06242 m^3/kg
Thermal conductivity	1 W/m-°C = 1 W/m-K	1 W/m°C = 0.5778 Btu/h-ft-°F
Specific heat	1 kJ/kg.°C = 1 kJ/kg-K = 1J/g°C	1 Btu/lbm-°F = 4.1868 kJ/kg-°C 1 kJ/kg-°C = 0.23885 Btu/lbm-°F
Power	1 W = 1 J/s 1 kW = 1000 W = 1.341 hp 1 hp = 745.7 W	1 kW = 3412.14 Btu/h 1 Btu/h = 1.055056 kJ/h 1 hp = 42.41 Btu/min = 0.7457 kW

Contd.

TABLE A.1 Conversion Factors (*Contd.*)

Dimension	Metric	Metric/English
	1 ton of refrigeration = 3.517 kW	1 ton of refrigeration = 200 Btu/min
Mass	1 kg = 1000 g	1 kg = 2.204 lbm
	1 metric ton = 1000 kg	1 lbm = 0.4535 kg
		1 ounce = 28.3495 g
		1 short ton = 2000 lbm = 907.184 kg
Length	1 m = 100 cm = 1000 mm = 10^6 mm	1 m = 39.37 in = 3.2808 ft
		1 ft = 12 in = 0.3048 m
		1 mile = 5280 ft = 1.6093 km
		1 in = 2.54 cm
Force	1 N = 1 kg-m/s^2 = 10^5 dyne	1 N = 0.22481 lbf
	1 kgf = 9.806 N	1 lbf = 32.174 lbm-ft/s^2 = 4.448 N
Energy	1 kJ = 1000 J = 1000 N-m = 1 kPa-m^3	1 kJ = 0.94782 Btu
	1 kJ/kg = 1000 m^2/s^2	1 Btu = 1.0550 kJ
	1 kWh = 3600 kJ	1 Btu/lbm = 25,037 ft^2/s^2 = 2.326 kJ/kg
	1 cal = 4.184 J	1 kJ/kg = 0.430 Btu/lbm
		1 kWh = 3412.14 Btu
Area	1m^2 = 10^4 cm^2 = 10^6 mm^2 = 10^{-6} km^2	1 m^2 = 1550 in^2 = 10.764 ft^2
		1 ft^2 = 144 in^2 = 0.0929 m^2
Density	1 g/cm^3 = 1 kg/L = 1000 kg/m^3	1 g/cm^3 = 62.428 lbm/ft^3 = 0.0361 lbm/in^3
Acceleration	1 m/s^2 = 100 cm/s^2	1 m/s^2 = 3.2808 ft/s^2
		1 ft/s^2 = 0.3048 m/s^2
Universal gas	R = 8.314 kJ/kmol-K	R = 8.314 kJ/kmol-K
constant	= 8.314 kPa-m^3/kmol-K	= 8.314 kPa-m^3/kmol-K
	= 1.987 cal/mol-K	= 82.05 L-atm/kmol-K

TABLE A.2 Properties of Pure Species

Substance	Molar mass	ω	T_C (K)	P_C (bar)	Z_C	V_C (cm^3 mol^{-1} or 10^{-3} m^3 kmol^{-1})	T_r (K)
Methane	16.043	0.012	190.6	45.99	0.286	98.6	111.4
Ethane	30.070	0.100	305.3	48.72	0.279	145.5	184.6
Propane	44.097	0.154	369.8	42.48	0.276	200.0	231.1
n-Butane	58.123	0.200	425.1	37.96	0.274	255.0	272.7
n-Pentane	72.150	0.252	469.7	33.70	0.270	313.0	309.2
n-Hexane	86.177	0.301	507.6	30.25	0.266	371.0	341.9
n-Heptane	100.204	0.350	540.2	27.40	0.261	428.0	371.6
n-Octane	114.231	0.400	568.7	24.90	0.256	486.0	398.8
n-Nonane	128.258	0.444	594.6	22.90	0.252	544.0	424.0
n-Decane	142.285	0.492	617.7	21.10	0.247	600.0	447.3
Isobutane	58.123	0.181	408.1	36.48	0.282	262.7	261.4

Contd.

TABLE A.2   Properties of Pure Species (*Contd.*)

Substance	Molar mass	ω	T_C (K)	P_C (bar)	Z_C	V_C (cm^3 mol^{-1} or 10^{-3} m^3 kmol^{-1})	T_r (K)
Isooctane	114.231	0.302	544.0	25.68	0.266	468.0	372.4
Cyclopentane	70.134	0.196	511.8	45.02	0.273	258.0	322.4
Cyclohexane	84.161	0.210	553.6	40.73	0.273	308.0	353.9
Methylcyclopentane	84.161	0.230	532.8	37.85	0.272	319.0	345.0
Methylcyclohexane	98.188	0.235	572.2	34.71	0.269	368.0	374.1
Ethylene	28.054	0.087	282.3	50.40	0.281	131.0	169.4
Propylene	42.081	0.140	365.6	46.65	0.289	188.4	225.5
1-Butene	56.108	0.191	420.0	40.43	0.277	239.3	266.9
cis-2-Butene	56.108	0.205	435.6	42.43	0.273	233.8	276.9
trans-2-Butene	56.108	0.218	428.6	41.00	0.275	237.7	274.0
1-Hexene	84.161	0.280	504.0	31.40	0.265	354.0	336.3
Isobutylene	56.108	0.194	417.9	40.00	0.275	238.9	266.3
1,3-Butadiene	54.092	0.190	425.2	42.77	0.267	220.4	268.7
Cyclohexane	82.145	0.212	560.4	43.50	0.272	291.0	356.1
Acetylene	26.038	0.187	308.3	61.39	0.271	113.0	189.4
Benzene	78.114	0.210	562.2	48.98	0.271	259.0	353.2
Toluene	92.141	0.262	591.8	41.06	0.264	316.0	383.8
Ethylbenzene	106.167	0.303	617.2	36.06	0.263	374.0	409.4
Cumene	120.194	0.326	631.1	32.09	0.261	427.0	425.6
o-Xylene	106.167	0.310	630.3	37.34	0.263	369.0	417.6
m-Xylene	106.167	0.326	617.1	35.36	0.259	376.0	412.3
p-Xylene	106.167	0.322	616.2	35.11	0.260	379.0	411.5
Styrene	104.152	0.297	636.0	38.40	0.256	352.0	418.3
Naphthalene	128.174	0.302	748.4	40.51	0.269	413.0	...
Biphenyl	154.211	0.365	789.3	38.50	0.295	502.0	528.2
Formaldehyde	30.026	0.282	408.0	65.90	0.223	115.0	254.1
Acetaldehyde	44.053	0.291	466.0	55.50	0.221	154.0	294.0
Methyl acetate	74.079	0.331	506.6	47.50	0.257	228.0	330.1
Ethyl acetate	88.106	0.366	523.3	38.80	0.255	286.0	350.2
Acetone	58.080	0.307	508.2	47.01	0.233	209.0	329.4
Methyl ethyl ketone	72.107	0.323	535.5	41.50	0.249	267.0	352.8
Diethyl ether	74.123	0.281	466.7	36.40	0.263	280.0	307.6
Methyl t-butyl ether	88.150	0.266	497.1	34.30	0.273	329.0	328.4
Methanol	32.042	0.564	512.6	80.97	0.224	118.0	337.9
Ethanol	46.069	0.645	513.9	61.48	0.240	167.0	351.4
1-Propanol	60.096	0.622	536.8	51.75	0.254	219.0	370.4
1-Butanol	74.123	0.594	563.1	44.23	0.260	275.0	390.8

Taken from Smith, J.M., H.C. Van Ness and M.M. Abbott, *Introduction to Chemical Engineering Thermodynamics*, 6th ed., McGraw-Hill, New York, 2001, pp. 632–633. Reprinted with permission of The McGraw-Hill Companies, Inc.

TABLE A.2   Properties of Pure Species (*Contd.*)

Substance	Molar mass	ω	T_C (K)	P_C (bar)	Z_C	V_C (cm^3 mol^{-1} or 10^{-3} m^3 kmol^{-1})	T_r (K)
1-Hexanol	102.177	0.579	611.4	35.10	0.263	381.0	430.6
2-Propanol	60.096	0.668	508.3	47.62	0.248	220.0	355.4
Phenol	94.113	0.444	694.3	61.30	0.243	229.0	455.0
Ethylene glycol	62.068	0.487	719.7	77.00	0.246	191.0	470.5
Acetic acid	60.053	0.467	592.0	57.86	0.211	179.7	391.1
n-Butyric acid	88.106	0.681	615.7	40.64	0.232	291.7	436.4
Benzoic acid	122.123	0.603	751.0	44.70	0.246	344.0	522.4
Acetonitrile	41.053	0.338	545.5	48.30	0.184	173.0	354.8
Methylamine	31.057	0.281	430.1	74.60	0.321	154.0	266.8
Ethylamine	45.084	0.285	456.2	56.20	0.307	207.0	289.7
Nitromethane	61.040	0.348	588.2	63.10	0.223	173.0	374.4
Carbon tetrachloride	153.822	0.193	556.4	45.60	0.272	276.0	349.8
Chloroform	119.377	0.222	536.4	54.72	0.293	239.0	334.3
Dichloromethane	84.932	0.199	510.0	60.80	0.265	185.0	312.9
Methyl chloride	50.488	0.153	416.3	66.80	0.276	143.0	349.1
Ethyl chloride	64.514	0.190	460.4	52.70	0.275	200.0	285.4
Chlorobenzene	112.558	0.250	632.4	45.20	0.265	308.0	404.9
Tetrafluoroethane	102.030	0.327	374.2	40.60	0.258	198.0	247.1
Argon	39.948	0.000	150.9	48.98	0.291	74.6	87.3
Krypton	83.800	0.000	209.4	55.02	0.288	91.2	119.8
Xenon	131.30	0.000	289.7	58.40	0.286	118.0	165.0
Helium 4	4.003	0.390	5.2	2.28	0.302	57.3	4.2
Hydrogen	2.016	0.216	33.19	13.13	0.305	64.1	20.4
Oxygen	31.999	0.022	154.6	50.43	0.288	73.4	90.2
Nitrogen	28.014	0.038	126.2	34.00	0.289	89.2	77.3
Air	28.851	0.035	132.2	37.45	0.289	84.8	
Chlorine	70.905	0.069	417.2	77.10	0.265	124.0	239.1
Carbon monoxide	28.010	0.048	132.9	34.99	0.299	93.4	81.7
Carbon dioxide	44.010	0.224	304.2	73.83	0.274	94.0	...
Carbon disulphide	76.143	0.111	552.0	79.00	0.275	160.0	319.4
Hydrogen sulphide	34.082	0.094	373.5	89.63	0.284	98.5	212.8
Sulphur dioxide	64.065	0.245	430.8	78.84	0.269	122.0	263.1
Sulphur trioxide	80.064	0.424	490.9	82.10	0.255	127.0	317.9
Nitric oxide (NO)	30.006	0.583	180.2	64.80	0.251	58.0	121.4
Nitrous oxide (N$_2$O)	44.013	0.141	309.6	72.45	0.274	97.4	184.7
Hydrogen chloride	36.461	0.132	324.7	83.10	0.249	81.0	188.2
Hydrogen cyanide	27.026	0.410	456.7	53.90	0.197	139.0	298.9
Water	18.015	0.345	647.1	220.55	0.229	55.9	373.2
Ammonia	17.031	0.253	405.7	112.80	0.242	72.5	239.7
Nitric acid	63.013	0.714	520.0	68.90	0.231	145.0	356.2
Sulphuric acid	98.080	...	924.0	64.00	0.147	177.0	610.0

TABLE A.3 Isobaric Molar Heat Capacities of Common Gases in the Ideal Gas State
$$C_P^0 = a + bT + cT^2 + dT^3 \ (T \text{ in K and } C_P^0 \text{ in kJ/kmol-K})$$

Substance	Formula	a	$b \times 10^2$	$c \times 10^5$	$d \times 10^9$	Temperature Range (K)
Common Gases						
Nitrogen	N_2	28.900	−0.157	0.808	−2.873	273–1800
Oxygen	O_2	25.480	1.520	−0.715	1.312	273–1800
Air	...	28.110	0.196	0.480	−1.966	273–1800
Hydrogen	H_2	29.110	−0.191	0.400	−0.870	273–1800
Carbon monoxide	CO	28.160	0.167	0.537	−2.222	273–1800
Carbon dioxide	CO_2	22.260	5.981	−3.501	7.469	273–1800
Water vapour	H_2O	32.240	0.192	1.055	−3.595	273–1800
Oxides of Nitrogen						
Nitric oxide	NO	29.340	0.990	−0.3234	0.366	273–3800
Nitrous oxide	N_2O	24.11	5.863	−3.562	10.580	273–1500
Nitrogen dioxide	NO_2	22.900	5.715	−3.520	7.870	273–1500
Paraffins						
Methane	CH_4	19.89	5.024	1.269	−11.01	273–1500
Ethane	C_2H_6	6.900	17.27	−6.406	7.285	273–1500
Propane	C_3H_8	−4.04	30.48	−15.72	31.74	273–1500
n-Butane	C_4H_{10}	3.96	37.15	−18.34	35.00	273–1500
iso-Butane	C_4H_{10}	−7.913	41.60	−23.01	49.91	273–1500
n-Hexane	C_6H_{14}	6.938	55.22	−28.65	57.69	273–1500
n-Pentane	C_5H_{12}	6.774	45.43	−22.46	42.29	273–1500
Alkenes						
Ethylene	C_2H_4	3.95	15.64	−8.344	17.67	273–1500
Propylene	C_3H_6	3.15	23.83	−12.18	24.62	273–1500
1-Butene	C_4H_8	−1.004	36.33	−21.46	50.69	273–1500
iso-Butene	C_4H_8	6.93	32.34	−16.72	33.68	273–1500
Aromatic Compounds						
Benzene	C_6H_6	−36.33	48.62	−31.66	77.86	273–1500
Toluene	C_7H_8	−34.49	56.07	−34.56	80.64	273–1500
Ethylbenzene	C_8H_{10}	−35.28	66.92	−42.00	100.59	273–1500
Styrene	C_8H_8	−25.06	60.28	−38.43	92.52	273–1500
Cumene	C_9H_{12}	−39.69	78.48	−49.84	120.96	273–1500

Taken from Hougen, O.A., K.M. Watson and R.A. Ragatz, *Chemical Process Principles*, Part 1, John Wiley and Sons, New York, 1954. Reprinted with permission.

TABLE A.3 Isobaric Molar Heat Capacities of Common Gases in the Ideal Gas State

$C_P^0 = a + bT + cT^2 + dT^3$ (T in K and C_P^0 in kJ/kmol-K) (Contd.)

Oxygenated Hydrocarbons

Formaldehyde	HCHO	22.87	4.090	0.715	−8.727	273–1500
Acetaldehyde	CH_3CHO	17.64	13.288	−2.163	−15.96	273–1000
Methanol	CH_3OH	19.15	9.181	−1.222	−8.064	273–1000
Ethanol	C_2H_5OH	19.95	21.025	−10.411	20.11	273–1500
Ethylene Oxide	C_2H_4O	−4.704	20.685	−10.033	13.225	273–1000

Miscellaneous Hydrocarbons

Acetone	C_3H_6O	6.825	27.976	−15.695	34.889	273–1500
Cyclopropane	C_3H_6	−27.220	34.465	−23.423	65.562	273–1000
Iso-pentane	C_5H_{12}	−9.546	52.222	−29.807	66.612	273–1500
Neo-pentane	C_5H_{12}	−16.233	55.881	−33.675	79.086	273–1500

Nitrogen Compounds

Ammonia	NH_3	27.586	2.563	0.990	−6.690	273–1500
Hydrazine	N_2H_4	16.338	14.926	−9.676	25.158	273–1500
Methylamine	CH_5N	12.579	15.162	−6.904	12.390	273–1500
Dimethylamine	C_2H_7N	−1.155	27.783	−14.624	30.034	273–1500
Trimethylamine	C_3H_9N	−8.811	40.395	−23.301	52.214	273–1500

Halogens and Halogenated Compounds

Fluorine	F_2	25.683	2.461	−1.755	4.111	273–2000
Chlorine	Cl_2	28.648	2.397	−2.144	6.497	273–1500
Bromine	Br_2	33.814	1.033	−0.893	2.688	273–1500
Iodine	I_2	35.716	0.551	−0.448	1.312	273–1800
Hydrogen fluoride	HF	30.244	−0.494	0.661	−1.579	273–2000
Hydrogen chloride	HCl	30.424	−0.764	1.331	−4.351	273–1500
Hydrogen bromide	HBr	30.109	−0.673	1.391	−4.876	273–1500
Hydrogen iodide	HI	28.148	0.1906	0.510	−2.021	273–1900

Chloromethanes

Methyl chloride	CH_3Cl	12.810	10.903	−5.224	9.660	273–1500
Methylene chloride	CH_2Cl_2	17.640	14.359	−9.870	25.485	273–1500
Chloroform	$CHCl_3$	31.962	14.536	−11.205	30.844	273–1500
Carbon tetrachloride	CCl_4	51.408	14.28	−12.579	37.077	273–1500

Sulphur Compounds

Sulphur	S_2	27.210	2.218	−1.628	3.986	273–1800
Sulphur dioxide	SO_2	25.780	5.795	−3.812	8.612	273–1800
Sulphur trioxide	SO_3	16.400	14.580	−11.200	32.420	273–1800
Hydrogen sulphide	H_2S	29.694	1.310	0.571	−3.301	273–1800
Carbon disulphide	CS_2	31.038	6.253	−4.603	11.592	273–1800

TABLE A.4 Van der Waals Constants of Some Selected Gases

Gases	$a \times 10^3$ [MPa-$(m^3/mol)^2$]	$b \times 10^3$ $(m^3/kmol)$
Acetylene	0.4520	0.0522
Air	0.1352	0.0365
Argon	0.1351	0.0320
Ammonia	0.4245	0.0372
Carbon dioxide	0.3658	0.0430
Carbon monoxide	0.1470	0.0395
Carbon tetrachloride	2.0667	0.1382
Chlorine	0.6578	0.0561
Chloroform	1.5371	0.1022
Ethane	0.5572	0.0650
Ethylene	0.4542	0.0581
Ethyl ether	1.7580	0.1362
n-Heptane	3.2001	0.2662
Hydrogen	0.0248	0.2661
Hydrogen chloride	0.3735	0.0426
Mercury	0.8241	0.0182
Methane	0.2302	0.0472
Neon	0.0213	0.019
Nitrogen	0.1366	0.0412
n-Octane	3.7895	0.2389
Oxygen	0.1379	0.0324
Propane	0.9355	0.0901
Sulphur dioxide	0.6925	0.0592
Sulphur trioxide	0.8354	0.0623
Water	0.5556	0.3352

TABLE A.5 Antoine Constants of Some Selected Compounds

$$\ln P^{sat} = A - \frac{B}{T + C} \quad (T \text{ in } °C \text{ and } P^{sat} \text{ in kPa})$$

Compounds	A	B	C
Acetaldehyde	14.416	2199.62	234.20
Acetic acid	16.042	3872.04	259.38
Acetone	14.2342	2691.46	230.00
Acetonitrile	14.6797	3125.82	240.76
Ammonia	15.1650	2212.43	243.92
Benzene	13.8575	2772.96	220.54
n-Butane	13.6836	2149.38	237.22
iso-Butane	13.4312	1990.78	236.42
Carbon tetrachloride	13.6816	2355.82	220.58

Contd.

TABLE A.5 Antoine Constants of Some Selected Compounds

$$\ln P^{sat} = A - \frac{B}{T+C} \quad (T \text{ in } °C \text{ and } P^{sat} \text{ in kPa}) \quad (Contd.)$$

Compounds	A	B	C
Chlorobenzene	13.9942	3294.95	217.68
Chloroform	13.9092	2341.93	226.30
1-Chlorobutane	13.9657	2825.98	223.75
1, 4-Dioxane	14.1210	2966.12	209.58
Ethane	13.4252	1520.55	256.40
Ethanol	16.6772	3674.55	226.82
Ethylbenzene	14.0236	3278.95	213.11
Formaldehyde	14.392	1941.20	244.20
n-Heptane	13.8582	2911.25	216.75
Methanol	16.5952	3644.40	239.68
Methyl acetate	14.3982	2739.60	223.20
n-Octane	14.1210	3219.95	212.80
Propane	13.7680	1892.65	248.45
Propylene	13.7134	1820.88	247.96
n-Pentane	13.8210	2477.25	233.56
1-Propanol	16.0685	3448.42	204.11
Toluene	14.0102	3102.92	219.75
p-Xylene	13.981	2906.86	216.32
Water	16.2634	3800.22	226.76

TABLE A.6 Properties of Saturated Water: Temperature Table

Temp. (°C) T	Sat. Press. (kPa) P^{sat}	Specific Volume (m³/kg) Sat. Liquid v_f	Sat. Vapour v_g	Internal Energy (kJ/kg) Sat. Liquid u_f	Evap. u_{fg}	Sat. Vapour u_g	Enthalpy (kJ/kg) Sat. Liquid h_f	Evap. h_{fg}	Sat. Vapour h_g	Entropy (kJ/kg-K) Sat. Liquid s_f	Evap. s_{fg}	Sat. Vapour s_g
0.01	0.6113	0.001000	206.14	0.00	2375.3	2375.3	0.01	2501.3	2501.4	0.0000	9.1562	9.1562
5	0.8721	0.001000	147.12	20.97	2361.3	2382.3	20.98	2489.6	2510.6	0.0761	8.9496	9.0257
10	1.2276	0.001000	106.38	42.00	2347.2	2389.2	42.01	2477.7	2519.8	0.1510	8.7498	8.9008
15	1.7051	0.001001	77.93	62.99	2333.1	2396.1	62.99	2465.9	2528.9	0.2245	8.5569	8.7814
20	2.339	0.001002	57.79	83.95	2319.0	2402.9	83.96	2454.1	2538.1	0.2966	8.3706	8.6672
25	3.169	0.001003	43.36	104.88	2304.9	2409.8	104.89	2442.3	2547.2	0.3674	8.1905	8.5580
30	4.246	0.001004	32.89	125.78	2290.8	2416.6	125.79	2430.5	2556.3	0.4369	8.0164	8.4533
35	5.628	0.001006	25.22	146.67	2276.7	2423.4	146.68	2418.6	2565.3	0.5053	7.8478	8.3531
40	7.384	0.001008	19.52	167.55	2262.6	2430.1	167.57	2406.7	2574.3	0.5725	7.6845	8.2570
45	9.593	0.001010	15.26	188.44	2248.4	2436.8	188.45	2394.8	2583.2	0.6387	7.5261	8.1648
50	12.349	0.001012	12.03	209.32	2234.2	2443.5	209.33	2382.7	2592.1	0.7038	7.3725	8.0763
55	15.758	0.001015	9.568	230.21	2219.9	2450.1	230.23	2370.7	2600.9	0.7679	7.2234	7.9913
60	19.940	0.001017	7.671	251.11	2205.5	2456.6	251.13	2358.5	2609.6	0.8312	7.0784	7.9096
65	25.03	0.001020	6.197	272.02	2191.1	2463.1	272.06	2346.2	2618.3	0.8935	6.9375	7.8310
70	31.19	0.001023	5.042	292.95	2176.6	2469.6	292.98	2333.8	2626.8	0.9549	6.8004	7.7553
75	38.58	0.001026	4.131	313.90	2162.0	2475.9	313.93	2321.4	2635.3	1.0155	6.6669	7.6824
80	47.39	0.001029	3.407	334.86	2147.4	2482.2	334.91	2308.8	2643.7	1.0753	6.5369	7.6122
85	57.83	0.001033	2.828	355.84	2132.6	2488.4	355.90	2296.0	2651.9	1.1343	6.4102	7.5445
90	70.14	0.001036	2.361	376.85	2117.7	2494.5	376.92	2283.2	2660.1	1.1925	6.2866	7.4791
95	84.55	0.001040	1.982	397.88	2102.7	2500.6	397.96	2270.2	2668.1	1.2500	6.1659	7.4159
100	101.35	0.001044	1.6729	418.94	2087.6	2506.5	419.04	2257.0	2676.1	1.3069	6.0480	7.3549
105	120.82	0.001048	1.4194	440.02	2072.3	2512.4	440.15	2243.7	2683.8	1.3630	5.9328	7.2958
110	143.27	0.001052	1.2102	461.14	2057.0	2518.1	461.30	2230.2	2691.5	1.4185	5.8202	7.2387

(Contd.)

Tables A-6 to A-8 taken from Van Wylen, G.J., and R.E. Sonntag, *Fundamentals of Classical Thermodynamics*, S.I. Version, 3rd ed., John Wiley and Sons, New York, 1985, pp. 635–640. Reprinted with permission of John Wiley and Sons, Inc.

TABLE A.6 Properties of Saturated Water: Temperature Table (Contd.)

Temp. (°C) T	Sat. Press. (kPa) P^{sat}	Specific Volume (m³/kg) Sat. Liquid v_f	Sat. Vapour v_g	Internal Energy (kJ/kg) Sat. Liquid u_f	Evap. u_{fg}	Sat. Vapour u_g	Enthalpy (kJ/kg) Sat. Liquid h_f	Evap. h_{fg}	Sat. Vapour h_g	Entropy (kJ/kg-K) Sat. Liquid s_f	Evap. s_{fg}	Sat. Vapour s_g
115	169.06	0.001056	1.0366	482.30	2041.4	2523.7	482.48	2216.5	2699.0	1.4734	5.7100	7.1833
120	198.53	0.001060	0.8919	503.50	2025.8	2529.3	503.71	2202.6	2706.3	1.5276	5.6020	7.1296
125	232.10	0.001065	0.7706	524.74	2009.9	2534.6	524.99	2188.5	2713.5	1.5813	5.4962	7.0775
130	270.10	0.001070	0.6685	546.02	1993.9	2539.9	546.31	2174.2	2720.5	1.6344	5.3925	7.0269
135	313.00	0.001075	0.5822	567.35	1977.7	2545.0	567.69	2159.6	2727.3	1.6870	5.2907	6.9777
140	361.30	0.001080	0.5089	588.74	1961.3	2550.0	589.13	2144.7	2733.9	1.7391	5.1908	6.9299
145	415.40	0.001085	0.4463	610.18	1944.7	2554.9	610.63	2129.6	2740.3	1.7907	5.0926	6.8833
150	475.80	0.001091	0.3928	631.68	1927.9	2559.5	632.20	2114.3	2746.5	1.8418	4.9960	6.8379
155	543.10	0.001096	0.3468	653.24	1910.8	2564.1	653.84	2098.6	2752.4	1.8925	4.9010	6.7935
160	617.80	0.001102	0.3071	674.87	1893.5	2568.4	675.55	2082.6	2758.1	1.9427	4.8075	6.7502
165	700.50	0.001108	0.2727	696.56	1876.0	2572.5	697.34	2066.2	2763.5	1.9925	4.7153	6.7078
170	791.70	0.001114	0.2428	718.33	1858.1	2576.5	719.21	2049.5	2768.7	2.0419	4.6244	6.6663
175	892.00	0.001121	0.2168	740.17	1840.0	2580.2	741.17	2032.4	2773.6	2.0909	4.5347	6.6256
180	1002.1	0.001127	0.19405	762.09	1821.6	2583.7	763.22	2015.0	2778.2	2.1396	4.4461	6.5857
185	1122.7	0.001134	0.17409	784.10	1802.9	2587.0	785.37	1997.1	2782.4	2.1879	4.3586	6.5465
190	1254.4	0.001141	0.15654	806.19	1783.8	2590.0	807.62	1978.8	2786.4	2.2359	4.2720	6.5079
195	1397.8	0.001149	0.14105	828.37	1764.4	2592.8	829.98	1960.0	2790.0	2.2835	5.1863	6.4698
200	1553.8	0.001157	0.12736	850.65	1744.7	2595.3	852.45	1940.7	2793.2	2.3309	4.1014	6.4323
205	1723.0	0.001164	0.11521	873.04	1724.5	2597.5	875.04	1921.0	2796.0	2.3780	4.0172	6.3952
210	1906.2	0.001173	0.10441	895.53	1703.9	2599.5	897.76	1900.7	2798.5	2.4248	3.9373	6.3585
215	2104.0	0.001181	0.09479	918.14	1682.9	2601.1	920.62	1879.9	2800.5	2.4714	3.8507	6.3221
220	2318.0	0.001190	0.08619	940.87	1661.5	2602.4	943.62	1858.5	2802.1	2.5178	3.7683	6.2861
225	2548.0	0.001199	0.07849	963.73	1639.6	2603.3	966.78	1836.5	2803.3	2.5639	3.6863	6.2503
230	2795.0	0.001209	0.07158	986.74	1617.2	2603.9	990.12	1813.8	2804.0	2.6099	3.6047	6.2146

(Contd.)

TABLE A.6   Properties of Saturated Water: Temperature Table (Contd.)

Temp. (°C) T	Sat. Press. (kPa) P^{sat}	Specific Volume (m³/Kg)		Internal Energy (kJ/kg)			Enthalpy (kJ/kg)			Entropy (kJ/kg-K)		
		Sat. Liquid v_f	Sat. Vapour v_g	Sat. Liquid u_f	Evap. u_{fg}	Sat. Vapour u_g	Sat. Liquid h_f	Evap. h_{fg}	Sat. Vapour h_g	Sat. Liquid s_f	Evap. s_{fg}	Sat. Vapour s_g
235	3060.0	0.001219	0.06537	1009.89	1594.2	2604.1	1013.62	1790.5	2804.2	2.6558	3.5233	6.1791
240	3344.0	0.001229	0.05976	1033.21	1570.8	2604.0	1037.32	1766.5	2803.8	2.7015	3.4422	6.1437
245	3648.0	0.001240	0.05471	1056.71	1546.7	2603.4	1061.23	1741.7	2803.0	2.7472	3.3612	6.1083
250	3973.0	0.001251	0.05013	1080.39	1522.0	2602.4	1085.36	1716.2	2801.5	2.7927	3.2802	6.0730
255	4319.0	0.001263	0.04598	1104.28	1496.7	2600.9	1109.73	1689.8	2799.5	2.8383	3.1992	6.0375
260	4688.0	0.001276	0.04221	1128.39	1470.6	2599.0	1134.37	1662.5	2796.9	2.8838	3.1181	6.0019
265	5081.0	0.001289	0.03877	1152.74	1443.9	2596.6	1159.28	1634.4	2793.6	2.9294	3.0368	5.9662
270	5499.0	0.001302	0.03564	1177.36	1416.3	2593.7	1184.51	1605.2	2789.7	2.9751	2.9551	5.9301
275	5942.0	0.001317	0.03279	1202.35	1387.9	2590.2	1210.07	1574.9	2785.0	3.0208	2.8730	5.8938
280	6412.0	0.001332	0.03017	1227.46	1358.7	2586.1	1235.99	1543.6	2779.6	3.0668	2.7903	5.8571
285	6909.0	0.001348	0.02777	1253.00	1328.4	2581.4	1262.31	1511.0	2773.3	3.1130	2.7070	5.8199
290	7436.0	0.001366	0.02557	1278.92	1297.1	2576.0	1289.07	1477.1	2766.2	3.1594	2.6227	5.7821
295	7993.0	0.001384	0.02354	1305.2	1264.7	2569.9	1316.3	1441.8	2758.1	3.2062	2.5375	5.7437
300	8581.0	0.001404	0.02167	1332.0	1231.0	2563.0	1344.0	1404.9	2749.0	3.2534	2.4511	5.7045
305	9202.0	0.001425	0.019948	1359.3	1195.9	2555.2	1372.4	1366.4	2738.7	3.3010	2.3633	5.6643
310	9856.0	0.001447	0.018350	1387.1	1159.4	2546.4	1401.3	1326.0	2727.3	3.3493	2.2737	5.6230
315	10547	0.001472	0.016867	1415.5	1121.1	2536.6	1431.0	1283.5	2714.5	3.3982	2.1821	5.5804
320	11274	0.001499	0.015488	1444.6	1080.9	2525.5	1461.5	1238.6	2700.1	3.4480	2.0882	5.5362
330	12845	0.001561	0.012996	1505.3	993.7	2498.9	1525.3	1140.6	2665.9	3.5507	1.8909	5.4417
340	14586	0.001638	0.010797	1570.3	894.3	2464.6	1594.2	1027.9	2622.0	3.6594	1.6763	5.3357
350	16513	0.001740	0.008813	1641.9	776.6	2418.4	1670.6	893.4	2563.9	3.7777	1.4335	5.2112
360	18651	0.001893	0.006945	1725.2	626.3	2351.5	1760.5	720.5	2481.0	3.9147	1.1379	5.0526
370	21030	0.001213	0.004925	1844.0	384.5	2228.5	1890.5	441.6	2332.1	4.1106	0.6865	4.7971
374.14	22090	0.003155	0.003155	2029.6	0	2029.6	2099.3	0	2099.3	4.4298	0	4.4298

TABLE A.7 Properties of Saturated Steam: Pressure Table

Press. (kPa) P	Sat. Temp. (°C) T^sat	Specific Volume (m³/kg)		Internal Energy (kJ/kg)			Enthalpy (kJ/kg)			Entropy (kJ/kg-K)		
		Sat. Liquid v_f	Sat. Vapour v_g	Sat. Liquid u_f	Evap. u_fg	Sat. Vapour u_g	Sat. Liquid h_f	Evap. h_fg	Sat. Vapour h_g	Sat. Liquid s_f	Evap. s_fg	Sat. Vapour s_g
0.6113	0.01	0.001000	206.14	0.00	2375.3	2375.3	0.01	2501.3	2501.4	0.0000	9.1562	9.1562
1.0	6.98	0.001000	129.21	29.30	2355.7	2385.0	29.30	2484.9	2514.2	0.1059	8.8697	8.9756
1.5	13.03	0.001001	87.98	54.71	2338.6	2393.3	54.71	2470.6	2525.3	0.1957	8.6322	8.8279
2.0	17.50	0.001001	67.00	73.48	2326.0	2399.5	73.48	2460.0	2533.5	0.2607	8.4629	8.7237
2.5	21.08	0.001002	54.25	88.48	2315.9	2404.4	88.49	2451.6	2540.0	0.3120	8.3311	8.6432
3.0	24.08	0.001003	45.67	101.04	2307.5	2408.5	101.05	2444.5	2545.5	0.3545	8.2231	8.5776
4	28.96	0.001004	34.80	121.45	2293.7	2415.2	121.46	2432.9	2554.4	0.4226	8.0520	8.4746
5	32.88	0.001005	28.19	137.81	2282.7	2420.5	137.82	2423.7	2561.5	0.4764	7.9187	8.3951
7.5	40.29	0.001008	19.24	168.78	2261.7	2430.5	168.79	2406.0	2574.8	0.5764	7.6750	8.2515
10	45.81	0.001010	14.67	191.82	2246.1	2437.9	191.83	2392.8	2584.7	0.6493	7.5009	8.1502
15	53.97	0.001014	10.02	225.92	2222.8	2448.7	225.94	2373.1	2599.1	0.7549	7.2536	8.0085
20	60.06	0.001017	7.649	251.38	2205.4	2456.7	251.40	2358.3	2609.7	0.8320	7.0766	7.9085
25	64.97	0.001020	6.204	271.90	2191.2	2463.1	271.93	2346.3	2618.2	0.8931	6.9383	7.8314
30	69.10	0.001022	5.229	289.20	2179.2	2468.4	289.23	2336.1	2625.3	0.9439	6.8247	7.7686
40	75.87	0.001027	3.993	317.53	2159.5	2477.0	317.58	2319.2	2636.8	1.0259	6.6441	7.6700
50	81.33	0.001030	3.240	340.44	2143.4	2483.9	340.49	2305.4	2645.9	1.0910	6.5029	7.5939
75	91.78	0.001037	2.217	384.31	2112.4	2496.7	384.39	2278.6	2663.0	1.2130	6.2434	7.4564
MPa												
0.100	99.63	0.001043	1.6940	417.36	2088.7	2506.1	417.46	2258.0	2675.5	1.3026	6.0568	7.3594
0.125	105.99	0.001048	1.3749	444.19	2069.3	2513.5	444.32	2241.0	2685.4	1.3740	5.0194	7.2844
0.150	111.37	0.001053	1.1593	466.94	2052.7	2519.7	467.11	2226.5	2693.6	1.4336	5.7897	7.2233
0.175	116.06	0.001057	1.0036	486.80	2038.1	2524.9	486.99	2213.6	2700.6	1.4849	5.6868	7.1717
0.200	120.23	0.001061	0.8857	504.49	2025.0	2529.5	504.70	2201.9	2706.7	1.5301	5.5970	7.1271
0.225	124.00	0.001064	0.7793	520.47	2013.1	2533.6	520.72	2191.3	2712.1	1.5706	5.5173	7.0878
0.250	127.44	0.001067	0.7187	535.10	2002.1	2537.2	535.37	2181.5	2716.9	1.6072	5.4455	7.0527

(Contd.)

TABLE A.7 Properties of Saturated Steam: Pressure Table *(Contd.)*

Press. (kPa) P	Sat. Temp. (°C) T^{sat}	Specific Volume (m³/kg)		Internal Energy (kJ/kg)			Enthalpy (kJ/kg)			Entropy (kJ/kg-K)		
		Sat. Liquid v_f	Sat. Vapour v_g	Sat. Liquid u_f	Evap. u_{fg}	Sat. Vapour u_g	Sat. Liquid h_f	Evap. h_{fg}	Sat. Vapour h_g	Sat. Liquid s_f	Evap. s_{fg}	Sat. Vapour s_g
0.275	130.60	0.001070	0.6573	548.19	1991.9	2540.5	548.19	2172.4	2721.3	1.6408	5.3801	7.0209
0.300	133.55	0.001073	0.6058	561.15	1982.4	2543.6	561.47	2163.8	2725.5	1.6718	5.3201	6.9919
0.325	136.30	0.001076	0.5620	572.90	1973.5	2546.4	573.25	2155.8	2729.0	1.7006	5.2646	6.9652
0.350	138.88	0.001079	0.5243	583.95	1965.0	2548.9	584.33	2148.1	2732.4	1.7275	5.2130	6.9405
0.375	141.32	0.001081	0.4914	594.40	1956.9	2551.3	594.81	2140.8	2735.6	1.7528	5.1647	6.9175
0.40	143.63	0.001084	0.4625	604.31	1949.3	2553.6	604.74	2133.8	2738.6	1.7766	5.1193	6.8959
0.45	147.93	0.001088	0.4140	622.77	1934.9	2557.6	623.25	2120.7	2743.9	1.8207	5.0359	6.8565
0.50	151.86	0.001093	0.3749	639.68	1921.6	2561.2	640.23	2108.5	2748.7	1.8607	4.9606	6.8213
0.55	155.48	0.001097	0.3427	655.32	1909.2	2564.5	655.93	2097.0	2753.0	1.8973	4.8920	6.7893
0.60	158.85	0.001101	0.3157	669.90	1897.5	2567.4	670.56	2086.3	2756.8	1.9312	4.8288	6.7600
0.65	162.01	0.001104	0.2927	683.56	1886.5	2570.1	684.28	2076.0	2760.3	1.9627	4.7703	6.7331
0.70	164.97	0.001108	0.2729	696.44	1876.1	2572.5	697.22	2066.3	2763.5	1.9922	4.7158	6.7080
0.75	167.78	0.001112	0.2556	708.64	1866.1	2574.7	709.47	2057.0	2766.4	2.0200	4.6647	6.6847
0.80	170.43	0.001115	0.2404	720.22	1856.6	2576.8	721.11	2048.0	2769.1	2.0462	4.6166	6.6628
0.85	172.96	0.001118	0.2270	731.27	1847.4	2578.7	732.22	2039.4	2771.6	2.0710	4.5711	6.6421
0.90	175.38	0.001121	0.2150	741.83	1838.6	2580.5	742.83	2031.1	2773.9	2.0946	4.5280	6.6226
0.95	177.69	0.001124	0.2042	751.95	1830.2	2582.1	753.02	2023.1	2776.1	2.1172	4.4869	6.6041
1.00	179.91	0.001127	0.19444	761.68	1822.0	2583.6	762.81	2015.3	2778.1	2.1387	4.4478	6.5865
1.10	184.09	0.001113	0.17753	780.09	1806.3	2586.4	781.34	2000.4	2781.7	2.1792	4.3744	6.5536
1.20	187.99	0.001139	0.16333	797.29	1791.5	2588.8	798.65	1986.2	2784.8	2.2166	4.3067	6.5233
1.30	191.64	0.001144	0.15125	813.44	1777.5	2591.0	814.93	1972.7	2787.6	2.2515	4.2438	6.4953
1.40	195.07	0.001149	0.14084	828.70	1764.1	2592.8	830.30	1959.7	2790.0	2.2842	4.1850	6.4693
1.50	198.32	0.001154	0.13177	843.16	1751.3	2594.5	844.89	1947.3	2792.2	2.3150	4.1298	6.4448
1.75	205.76	0.001166	0.11349	876.46	1721.4	2597.8	878.50	1917.9	2796.4	2.3851	4.0044	6.3896
2.00	212.42	0.001177	0.09963	906.44	1693.8	2600.3	908.79	1890.7	2799.5	2.4474	3.8935	6.3409

(Contd.)

TABLE A.7 Properties of Saturated Steam: Pressure Table (*Contd.*)

Press. (kPa) P	Sat. Temp. (°C) T^{sat}	Specific Volume (m³/kg)		Internal Energy (kJ/kg)			Enthalpy (kJ/kg)			Entropy (kJ/kg-K)		
		Sat. Liquid v_f	Sat. Vapour v_g	Sat. Liquid u_f	Evap. u_{fg}	Sat. Vapour u_g	Sat. Liquid h_f	Evap. h_{fg}	Sat. Vapour h_g	Sat. Liquid s_f	Evap. s_{fg}	Sat. Vapour s_g
2.25	218.45	0.001187	0.08875	933.83	1668.2	2602.0	936.49	1865.2	2801.7	2.5035	3.7937	6.2972
2.5	223.99	0.001197	0.07998	959.11	1644.0	2603.1	962.11	1841.0	2803.1	2.5547	3.7028	6.2575
3.0	233.90	0.001217	0.06668	1004.78	1599.3	2604.1	1008.42	1795.7	2804.2	2.6457	3.5412	6.1869
3.5	242.60	0.001235	0.05707	1045.43	1558.3	2603.7	1049.75	1753.7	2803.4	2.7253	3.4000	6.1253
4	250.40	0.001252	0.04978	1082.31	1520.0	2602.3	1087.31	1714.1	2801.4	2.7964	3.2737	6.0701
5	263.99	0.001286	0.03944	1147.81	1449.3	2597.1	1154.23	1640.1	2794.3	2.9202	3.0532	5.9734
6	275.64	0.001319	0.03244	1205.44	1384.3	2589.7	1213.35	1571.0	2784.3	3.0267	2.8625	5.8892
7	285.88	0.001351	0.02737	1257.55	1323.0	2580.5	1267.00	1505.1	2772.1	3.1211	2.6922	5.8133
8	295.06	0.001384	0.02352	1305.57	1264.2	2569.8	1316.64	1441.3	2758.0	3.2068	2.5364	5.7432
9	303.40	0.001418	0.02048	1350.51	1207.3	2557.8	1363.26	1378.9	2742.1	3.2858	2.3915	5.6772
10	311.06	0.001452	0.018026	1393.04	1151.4	2544.4	1407.56	1317.1	2724.7	3.3596	2.2544	5.6141
11	318.85	0.001489	0.015987	1433.7	1096.0	2529.8	1450.1	1255.5	2705.6	3.4295	2.1233	5.5527
12	324.75	0.001527	0.014263	1473.0	1040.7	2513.7	1491.3	1193.6	2684.9	3.4962	1.9962	5.4924
13	330.93	0.001567	0.012780	1511.1	985.0	2496.1	1531.5	1130.7	2662.2	3.5606	1.8718	5.4323
14	336.75	0.001611	0.011485	1548.6	928.2	2476.8	1571.1	1066.5	2636.6	3.6232	1.7485	5.3717
15	342.24	0.001658	0.010337	1585.6	869.8	2455.5	1610.5	1000.0	2610.5	3.6848	1.6249	5.3098
16	347.44	0.001711	0.009306	1622.7	809.0	2431.7	1650.1	930.6	2680.6	3.7461	1.4994	5.2455
17	352.37	0.001770	0.008364	1660.2	744.8	2405.0	1690.3	856.9	2547.2	3.8079	1.3698	5.1777
18	357.06	0.001840	0.007489	1698.9	675.4	2374.3	1732.0	777.1	2509.1	3.8715	1.2329	5.1044
19	361.54	0.001924	0.006657	1739.9	598.1	2338.1	1776.5	688.0	2464.5	3.9388	1.0839	5.0228
20	365.81	0.002036	0.005834	1785.6	507.5	2293.0	1826.3	583.4	2409.7	4.0139	0.9130	4.9269
21	369.89	0.002207	0.004952	1842.1	388.5	2230.6	1888.4	446.2	2334.6	4.0175	0.6938	4.8013
22	373.80	0.002742	0.003568	1961.9	125.2	2087.1	2022.2	143.4	2165.6	4.3110	0.2216	4.5327
22.09	374.14	0.003155	0.003155	2029.6	0	2029.6	2099.3	0	2099.3	4.4298	0	4.4298

TABLE A.8   Properties of Superheated Vapour

T (°C)	v (m³/kg)	u (kJ/kg)	h (kJ/kg)	s (kJ/kg-K)	v (m³/kg)	u (kJ/kg)	h (kJ/kg)	s (kJ/kg-K)
	P = 0.01 MPa = 0.1 bar				*P* = 0.05 MPa = 0.5 bar			
Sat.	14.674	2437.9	2584.7	8.1502	3.240	2483.9	2645.9	7.5939
50	14.869	2443.9	2592.6	8.1749				
100	17.196	2515.5	2687.1	8.4479	3.418	2511.6	2681.5	7.6947
150	19.512	2587.9	2783.0	8.6882	3.889	2585.6	2780.1	7.9401
200	21.825	2661.3	2879.5	8.9038	4.356	2659.9	2877.7	8.1580
250	24.136	2736.0	2977.3	9.1002	4.820	2735.0	2976.0	8.3556
300	26.445	2812.1	3076.5	9.2813	5.284	2811.3	3075.5	8.5373
400	31.063	2968.9	3279.6	9.6077	6.209	2968.5	3278.9	8.8642
500	35.679	3132.3	3489.1	9.8978	7.134	3132.0	3488.7	9.1546
600	40.295	3302.5	3705.4	10.1608	8.057	3302.2	3705.1	9.4178
700	44.911	3479.6	3928.7	10.4028	8.981	3479.4	3928.5	9.6599
800	49.526	3663.8	4159.0	10.6281	9.904	3663.6	4158.9	9.8852
900	54.141	3855.0	4396.4	10.8396	10.828	3854.9	4396.3	10.0967
1000	58.758	4055.3	4642.8	11.0429	11.751	4055.2	4642.7	10.3000
	P = 0.10 MPa = 1.0 bar				*P* = 0.20 MPa = 2.0 bar			
Sat.	1.694	2506.1	2675.5	7.3594	0.885	2529.5	2706.7	7.1272
50	1.695	2506.7	2676.2	7.3614				
100	1.936	2582.8	2776.4	7.6134	0.959	2576.9	2768.8	7.2795
200	2.172	2658.1	2875.3	7.8343	1.080	2654.4	2870.5	7.5066
250	2.406	2733.7	2974.3	8.0333	1.198	2731.2	2971.0	7.7086
300	2.639	2810.4	3074.3	8.2158	1.312	2808.6	3071.8	7.8926
400	3.103	2967.9	3278.2	8.5435	1.549	2966.7	3276.6	8.2218
500	3.565	3131.6	3488.1	8.8342	1.781	3130.8	3487.1	8.5133
600	4.028	3301.9	3704.4	9.0976	2.013	3301.4	3704.0	8.7770
700	4.490	3479.2	3928.2	9.3398	2.244	3478.8	3927.6	9.0194
800	4.952	3363.5	4158.6	9.5652	2.475	3663.1	4158.2	9.2449
900	5.414	3854.8	4396.1	9.7767	2.705	3854.5	4395.8	9.4566
1000	5.875	4055.0	4642.6	9.9800	2.937	4054.8	4642.3	9.6599
	P = 0.30 MPa = 3.0 bar				*P* = 0.40 MPa = 4.0 bar			
Sat.	0.6058	2543.6	2725.3	6.9919	0.4625	2553.6	2738.6	6.8959
150	0.6339	2570.8	2761.0	7.0778	0.4708	2564.5	2752.8	6.9299
200	0.7163	2650.7	2865.6	7.3115	0.5342	2646.8	2860.5	7.1706
250	0.7964	2728.7	2967.6	7.5166	0.5951	2726.1	2964.2	7.3789
300	0.8753	2806.7	3069.3	7.7022	0.6548	2804.8	3066.8	7.5662
400	1.0315	2965.6	3275.0	8.0330	0.7726	2964.4	3273.4	7.8985
500	1.1867	3130.0	3486.0	8.3251	0.8893	3129.2	3484.9	8.1913
600	1.3414	3300.8	3703.2	8.5892	1.0055	3300.2	3702.4	8.4558

Contd.

TABLE A.8 Properties of Superheated Vapour (*Contd.*)

T (°C)	v (m³/kg)	u (kJ/kg)	h (kJ/kg)	s (kJ/kg-K)	v (m³/kg)	u (kJ/kg)	h (kJ/kg)	s (kJ/kg-K)
700	1.4957	3478.4	3927.1	8.8319	1.1215	3477.9	3926.5	8.6987
800	1.6499	3662.9	4157.8	9.0576	1.2372	3662.4	4157.3	8.9244
900	1.8041	3854.2	4395.4	9.2692	1.3529	3853.9	4395.1	9.1362
1000	1.9582	4054.5	4642.0	9.4726	1.4685	4054.3	4641.7	9.3396
	P = 0.50 MPa = 5.0 bar				**P = 0.60 MPa = 6.0 bar**			
Sat.	0.3749	2561.2	2748.7	6.8213	0.3157	2567.4	2756.8	6.7600
200	0.4249	2642.9	2855.4	7.0592	0.3520	2638.9	2850.1	6.9665
250	0.4744	2723.5	2960.7	7.2709	0.3938	2720.9	2957.2	7.1816
300	0.5226	2802.9	3064.2	7.4599	0.4344	2801.0	3061.6	7.3724
350	0.5701	2882.6	3167.7	7.6329	0.4742	2881.2	3165.7	7.5464
400	0.6173	2963.2	3271.9	7.7938	0.5137	2962.1	3270.3	7.7079
500	0.7109	3128.4	3483.9	8.0873	0.5920	3127.6	3482.8	8.0021
600	0.8041	3299.6	3701.7	8.3522	0.6697	3299.1	3700.9	8.2674
700	0.8969	3477.5	3925.9	8.5952	0.7472	3477.0	3925.3	8.5107
800	0.9896	3662.1	4156.9	8.8211	0.8245	3661.8	4156.5	8.7367
900	1.0822	3853.6	4394.7	9.0329	0.9017	3853.4	4394.4	8.9486
1000	1.1748	4054.0	4641.4	4892.6	0.97893	4053.8	4641.1	9.1521
	P = 0.80 MPa = 8.0 bar				**P = 1.00 MPa = 10.0 bar**			
Sat.	0.2404	2576.8	2769.1	6.6628	0.1944	2583.6	2778.1	6.5865
200	0.2608	2630.6	2839.3	6.8158	0.2060	2621.9	2827.9	6.6940
250	0.2931	2715.5	2950.0	7.0384	0.2327	2709.9	2942.6	6.9247
300	0.3241	2797.2	3056.5	7.2328	0.2579	2793.2	3051.2	7.1229
350	0.3544	2878.2	3161.7	7.4089	0.2825	2875.2	3157.7	7.3011
400	0.3843	2959.7	3267.1	7.5716	0.3066	2957.3	3263.9	7.4651
500	0.4433	316.0	3480.6	7.8673	0.3541	3124.4	3478.5	7.7622
600	0.5018	3297.9	3699.4	8.1333	0.4011	3296.8	3697.9	8.0290
700	0.5601	3476.2	3924.2	8.3770	0.4478	3475.3	3923.1	8.2731
800	0.6181	3661.1	4155.6	8.6033	0.4943	3660.4	4154.7	8.4996
900	0.6761	3852.8	4393.7	8.8153	0.5407	3852.2	4392.9	8.7118
1000	0.7341	4053.3	4640.5	9.0189	0.5872	4052.7	4640.0	8.9155
	P = 1.20 MPa = 12.0 bar				**P = 1.40 MPa = 14.0 bar**			
Sat.	0.1633	2588.8	2784.8	6.5233	0.1408	2592.8	2790.0	6.4693
200	0.1693	2612.8	2815.9	6.5898	0.1430	2603.1	2803.3	6.4975
250	0.1923	2704.2	2935.0	6.8294	0.1635	2698.3	2927.2	6.7467
300	0.2138	2789.2	3045.8	7.0317	0.1822	2785.2	3040.4	6.9534
350	0.2345	2872.2	3153.6	7.2121	0.2003	2869.2	3149.5	7.1360
400	0.2548	2954.9	3260.7	7.3774	0.2178	2952.5	3257.5	7.3026

Contd.

TABLE A.8   Properties of Superheated Vapour   (*Contd.*)

T (°C)	v (m³/kg)	u (kJ/kg)	h (kJ/kg)	s (kJ/kg-K)	v (m³/kg)	u (kJ/kg)	h (kJ/kg)	s (kJ/kg-K)
500	0.2946	3122.8	3476.3	7.6759	0.2521	3121.1	3474.1	7.6027
600	0.3339	3295.8	3696.3	7.9435	0.2860	3294.4	3694.8	7.8710
700	0.3729	3474.4	3922.0	8.1881	0.3195	3473.6	3920.8	8.1160
800	0.4118	3659.7	4153.8	8.4148	0.3528	3659.0	4153.0	8.3431
900	0.4505	3851.6	4392.2	8.6272	0.3861	3851.1	4391.5	8.5556
1000	0.4892	4052.2	4639.4	8.8310	0.4193	4051.7	4638.8	8.7595
	P = 1.60 MPa = 16.0 bar				*P* = 1.80 MPa = 18.0 bar			
Sat.	0.1238	2596.0	2794.0	6.4218	0.1104	2598.4	2797.1	6.3794
225	0.1328	2644.7	2857.3	6.5518	0.1167	2636.6	2846.7	6.4808
250	0.1418	2692.3	2919.2	6.6732	0.1249	2686.0	2911.0	6.6066
300	0.1586	2781.1	3034.8	6.8844	0.1402	2776.9	3029.2	6.8226
350	0.1745	2866.1	3145.4	7.0694	0.1545	2863.0	3141.2	7.0100
400	0.1900	2950.1	3254.2	7.2374	0.1684	2947.7	3250.9	7.1794
500	0.2203	3119.5	3472.0	7.5390	0.1955	3117.9	3469.8	7.4825
600	0.2500	3293.3	3693.2	7.8080	0.2220	3292.1	3691.7	7.7523
700	0.2794	3472.7	3919.7	8.0535	0.2482	3471.8	3918.5	7.9983
800	0.3086	3658.3	4152.1	8.2808	0.2742	3657.6	4151.2	8.2258
900	0.3377	3850.5	4390.8	8.4935	0.3001	3849.9	4390.1	8.4386
1000	0.3668	4051.2	4638.2	8.6974	0.3260	4050.7	4637.6	8.6427
	P = 2.0 MPa = 20.0 bar				*P* = 2.50 MPa = 25.0 bar			
Sat.	0.0996	2600.3	2799.5	6.3409	0.0799	2603.1	2803.1	6.2575
225	0.1037	2628.3	2835.8	6.4147	0.0802	2605.6	2806.3	6.2639
250	0.1114	2679.6	2902.5	6.5453	0.0870	2662.6	2880.1	6.4085
300	0.1254	2772.6	3023.5	6.7664	0.0989	2761.6	3008.8	6.6438
350	0.1385	2839.8	3137.0	6.9563	0.1097	2851.9	3126.3	6.8403
400	0.1512	2945.2	3247.6	7.1271	0.1201	2939.1	3239.3	7.0148
500	0.1756	3116.2	3467.6	7.4317	0.1399	3112.1	3462.1	7.3234
600	0.1996	3290.9	3690.1	7.7024	0.1593	3288.0	3686.3	7.5960
700	0.2232	3470.9	3917.4	7.9487	0.1783	3468.7	3914.5	7.8435
800	0.2467	3657.0	4150.3	8.1765	0.1971	3655.3	4148.2	8.0720
900	0.2700	3849.3	4389.4	8.3895	0.2159	3847.9	4387.6	8.2853
1000	0.2934	4050.2	4637.1	8.5936	0.2346	4049.0	4635.6	8.4897
	P = 3.0 MPa = 30.0 bar				*P* = 3.5 MPa = 35.0 bar			
Sat.	0.0666	2604.1	2804.2	6.1869	0.0570	2603.7	2803.4	6.1253
250	0.0705	2644.0	2855.8	6.2872	0.0587	2623.7	2829.2	6.1749
300	0.0811	2750.1	2993.5	6.5390	0.0684	2738.0	2977.6	6.4461
350	0.0905	2843.7	3115.3	6.7428	0.0767	2835.3	3104.0	6.6579

Contd.

TABLE A.8 Properties of Superheated Vapour (*Contd.*)

T (°C)	v (m³/kg)	u (kJ/kg)	h (kJ/kg)	s (kJ/kg-K)	v (m³/kg)	u (kJ/kg)	h (kJ/kg)	s (kJ/kg-K)
400	0.0993	2932.8	3230.9	6.9212	0.0845	2926.4	3222.3	6.8405
450	0.1078	3020.4	3344.0	7.0834	0.0919	3015.3	3337.2	7.0052
500	0.1161	3108.0	3456.5	7.2338	0.0991	3103.0	3450.9	7.1572
600	0.1324	3285.0	3682.3	7.5085	0.1132	3282.1	3678.4	7.4339
700	0.1483	3466.5	3911.7	7.7571	0.1269	3464.3	3908.8	7.6837
800	0.1641	3653.5	4145.9	7.9862	0.1405	3651.8	4143.7	7.9134
900	0.1798	3846.5	4385.9	8.1999	0.1540	3845.0	4384.1	8.1276
1000	0.1954	4047.7	4634.2	8.4045	0.1675	4046.4	4632.7	8.3324
	P = 4.0 MPa = 40.0 bar				*P* = 4.5 MPa = 45.0 bar			
Sat.	0.0497	2602.3	2801.4	6.0701	0.0440	2600.1	2798.3	6.0198
275	0.0545	2667.9	2886.2	6.2285	0.0473	2650.3	2863.2	6.1401
300	0.0588	2725.3	2960.7	6.3615	0.0513	2712.0	2943.1	6.2828
350	0.0664	2826.7	3092.5	6.5821	0.0584	2817.8	3080.6	6.5131
400	0.0734	2919.9	3213.6	6.7690	0.0647	2913.3	3204.7	6.7047
450	0.0800	3010.2	3330.3	6.9363	0.0707	3005.0	3323.3	6.8746
500	0.0864	3099.5	3445.3	7.0901	0.0765	3095.3	3439.6	7.0301
600	0.0988	3279.1	3674.4	7.3688	0.0876	3276.0	3670.5	7.3110
700	0.1109	3462.1	3905.9	7.6198	0.0984	3459.9	3903.0	7.5631
800	0.1228	3650.0	4141.5	7.8502	0.1091	3648.3	4139.3	7.7942
900	0.1346	3843.6	4382.3	8.0647	0.1196	3842.2	4380.6	8.0091
1000	0.1465	4045.1	4631.2	8.2698	0.1302	4043.9	4629.8	8.2144
	P = 5.0 MPa = 50.0 bar				*P* = 6.0 MPa = 60.0 bar			
Sat.	0.0394	2597.1	2794.3	5.9734	0.0324	2589.7	2784.3	5.8892
275	0.0414	2631.3	2838.3	6.0544				
300	0.0453	2698.0	2924.5	6.2084	0.0361	2667.2	2884.2	6.0674
350	0.0519	2808.7	3068.4	6.4493	0.0422	2789.6	3043.0	6.3335
400	0.0578	2906.6	3195.7	6.6459	0.0473	2892.9	3177.2	6.5408
450	0.0633	2999.7	3316.2	6.8186	0.0521	2988.9	3301.8	6.7193
500	0.0685	3091.0	3433.8	6.9759	0.0566	3082.2	3422.2	6.8803
600	0.0786	3273.0	3666.5	7.2589	0.0652	3266.9	3658.4	7.1677
700	0.0884	3457.6	3900.1	7.5122	0.0735	3453.1	3894.2	7.4234
800	0.0981	3646.6	4137.1	7.7440	0.0816	3643.1	4132.7	7.6566
900	0.1076	3840.7	4378.8	7.9593	0.0895	3837.8	4375.3	7.8727
1000	0.1171	4042.6	4628.3	8.1648	0.0975	4040.1	4625.4	8.0786
	P = 7.0 MPa = 70.0 bar				*P* = 8.0 MPa = 80.0 bar			
Sat.	0.0273	2580.5	2772.1	5.8133	0.0235	2569.8	2758.0	5.7432
300	0.0294	2632.2	2838.4	5.9305	0.0242	2590.9	2785.0	5.7906

Contd.

TABLE A.8   Properties of Superheated Vapour   (*Contd.*)

T (°C)	v (m³/kg)	u (kJ/kg)	h (kJ/kg)	s (kJ/kg-K)	v (m³/kg)	u (kJ/kg)	h (kJ/kg)	s (kJ/kg-K)
350	0.0352	2769.4	3016.0	6.2283	0.0299	2747.7	2987.3	6.1301
400	0.0399	2878.6	3158.1	6.4478	0.0343	2863.8	3138.3	6.3634
450	0.0441	2978.0	3287.1	6.6327	0.0381	2966.7	3272.0	6.5551
500	0.0481	3073.4	3410.3	6.7975	0.0417	3064.3	3398.3	6.7240
550	0.0519	3167.2	3530.9	6.9486	0.0451	3159.8	3521.0	6.8778
600	0.0556	3260.7	3650.3	7.0894	0.0484	3254.4	3642.0	7.0206
700	0.0628	3448.5	3888.3	7.3476	0.0548	3443.9	3882.4	7.2812
800	0.0698	3639.5	4128.2	7.5822	0.0609	3636.0	4123.8	7.5173
900	0.0766	3835.0	4371.8	7.7991	0.0670	3832.1	4368.3	7.7351
1000	0.0835	4037.5	4622.5	8.0055	0.0730	4035.0	4619.6	7.9419

	P = 9.0 MPa = 90.0 bar				*P* = 10.0 MPa = 100.0 bar			
Sat.	0.02048	2557.8	2742.1	5.6772	0.01802	2544.4	2724.7	5.6141
325	0.02327	2646.6	2856.0	5.8712	0.01986	2610.4	2809.1	5.7568
350	0.02580	2724.4	2956.6	6.0361	0.02242	2699.2	2923.4	5.9443
400	0.02993	2848.4	3117.8	6.2854	0.02641	2832.4	3096.5	6.2120
450	0.03350	2955.2	3256.6	6.4844	0.02975	2943.4	3240.9	6.4190
500	0.03677	3055.2	3386.1	6.6576	0.03279	3045.8	3373.7	6.5966
550	0.03987	3152.2	3511.0	6.8142	0.03564	3144.6	3500.9	6.7561
600	0.04285	3248.1	3633.7	6.9589	0.03837	3241.7	3625.3	6.9029
650	0.04574	3343.6	3755.3	7.0943	0.04101	3338.2	3748.2	7.0938
700	0.04857	3439.3	3876.5	7.2221	0.04358	3434.7	3870.5	7.1687
800	0.05409	3632.5	4119.3	7.4596	0.04859	3628.9	4114.8	7.4077
900	0.05950	3829.2	4364.8	7.6783	0.05349	3826.3	4361.2	7.6272
1000	0.06491	4032.4	4616.7	7.8855	0.05839	4029.9	4613.8	7.8349

	P = 12.5 MPa = 125.0 bar				*P* = 15.0 MPa = 150.0 bar			
Sat.	0.01349	2505.1	2673.8	5.4624	0.01033	2455.5	2610.5	5.3098
350	0.01612	2624.6	2826.2	5.7118	0.01147	2520.4	2692.4	5.4421
400	0.02000	2789.3	3039.3	6.0417	0.01564	2740.7	2975.5	5.8811
450	0.02299	2912.5	3199.8	6.2719	0.01844	2879.5	3156.2	6.1404
500	0.02560	3021.7	3341.8	6.4618	0.02080	2996.6	3308.6	6.3443
550	0.02801	3125.0	3475.2	6.6290	0.02293	3104.7	3448.6	6.5199
600	0.03029	3225.4	3604.0	6.7810	0.02491	3208.6	3582.3	6.6776
650	0.03248	3324.4	3730.4	6.9218	0.02680	3310.3	3712.3	6.8224
700	0.03460	3422.9	3855.3	7.0536	0.02861	3410.9	3840.1	6.9572
800	0.03869	3620.0	4103.6	7.2965	0.03210	3610.9	4092.4	7.2040
900	0.04267	3819.1	4352.5	7.5182	0.03546	3811.9	4343.8	7.4279
1000	0.04664	4023.5	4606.5	7.7269	0.03880	4017.1	4599.2	7.6378

Contd.

TABLE A.8 Properties of Superheated Vapour *(Contd.)*

T (°C)	v (m³/kg)	u (kJ/kg)	h (kJ/kg)	s (kJ/kg-K)	v (m³/kg)	u (kJ/kg)	h (kJ/kg)	s (kJ/kg-K)
	P = 17.5 MPa = 175.0 bar				P = 20.0 MPa = 200.0 bar			
Sat.	0.0079	2390.2	2528.8	5.1419	0.00583	2293.0	2409.7	4.9269
400	0.0124	2685.0	2902.9	5.7213	0.00994	2619.3	2818.1	5.5540
450	0.0151	2844.2	3109.7	6.0184	0.01269	2806.2	3060.1	5.9017
500	0.0173	2970.3	3274.1	6.2383	0.01476	2942.9	3238.2	6.1401
550	0.0192	3083.9	3421.4	6.4230	0.01655	3062.4	3393.5	6.3348
600	0.0210	3191.5	3560.1	6.5866	0.01817	3174.0	3537.6	6.5048
650	0.0227	3296.0	3693.9	6.7357	0.01969	3281.4	3675.3	6.6582
700	0.0243	3398.7	3824.6	6.8736	0.02113	3386.4	3809.0	6.7993
800	0.0273	3601.8	4081.1	7.1244	0.02385	3592.7	4069.7	7.0544
900	0.0303	3804.7	4335.1	7.3507	0.02645	3797.5	4326.4	7.2830
1000	0.0332	4010.7	4592.0	7.5616	0.02902	4004.3	4584.7	7.4950
	P = 25.0 MPa = 250.0 bar				P = 30.0 MPa = 300.0 bar			
375	0.0019731	1798.7	1848.0	4.0320	0.0017892	1737.8	1791.5	3.9305
400	0.006004	2430.1	2580.2	5.1418	0.002790	2067.4	2151.1	4.4728
425	0.007881	2609.2	2806.3	5.4723	0.005303	2455.1	2614.2	5.1504
450	0.009162	2720.7	2949.7	5.6744	0.006735	2619.3	2821.4	5.4424
500	0.011123	2884.3	3162.4	5.9592	0.008678	2820.7	3081.1	5.7905
550	0.012724	3017.5	3335.6	6.1765	0.010168	2970.3	3275.4	6.0342
600	0.014137	3137.9	3491.4	6.3602	0.011446	3100.5	3443.9	6.2331
650	0.015433	3251.6	3637.4	6.5229	0.012596	3221.0	3598.9	6.4058
700	0.016646	3361.3	3777.5	6.6707	0.013661	3335.8	3745.6	6.5606
800	0.018912	3574.3	4047.1	6.9345	0.015623	3555.5	4024.2	6.8332
900	0.021045	3783.0	4309.1	7.1680	0.017448	3768.5	4291.9	7.0718
1000	0.02310	3990.9	4568.5	7.3802	0.019196	3978.8	4554.7	7.2867
	P = 35.0 MPa = 350.0 bar				P = 40.0 MPa = 400.0 bar			
375	0.0017003	1702.9	1762.4	3.8722	0.0016407	1677.1	1742.8	3.8290
400	0.002100	1914.1	1987.6	4.2126	0.0019077	1854.6	1930.9	4.1135
425	0.003428	2253.4	2373.4	4.7747	0.002532	2096.9	2198.1	4.5029
450	0.004961	2498.7	2672.4	5.1962	0.003693	2365.1	2512.8	4.9459
500	0.006927	2751.9	2994.4	5.6282	0.005622	2678.4	2903.3	5.4700
550	0.008345	2921.0	3213.0	5.9026	0.006984	2869.7	3149.1	5.7785
600	0.009527	3062.0	3395.5	6.1179	0.008094	3022.6	3346.4	6.0114
650	0.010575	3189.8	3559.9	6.3010	0.009063	3158.0	3520.6	6.2054
700	0.011533	3309.8	3713.5	6.4631	0.009941	3283.6	3681.2	6.3750
800	0.013278	3536.7	4001.5	6.7450	0.011523	3517.8	3978.7	6.6662
900	0.014883	3754.0	4274.9	6.9886	0.012962	3739.4	4257.9	6.9150
1000	0.016410	3966.7	4541.1	7.2064	0.014324	3954.6	4527.6	7.1356

Contd.

TABLE A.8 Properties of Superheated Vapour *(Contd.)*

T (°C)	v (m³/kg)	u (kJ/kg)	h (kJ/kg)	s (kJ/kg-K)	v (m³/kg)	u (kJ/kg)	h (kJ/kg)	s (kJ/kg-K)
	P = 50.0 MPa = 500.0 bar				P = 60.0 MPa = 600.0 bar			
375	0.0015594	1638.6	1716.6	3.7639	0.0015028	1609.4	1699.5	3.7141
400	0.0017309	1788.1	1874.6	4.0031	0.0016335	1745.4	1843.4	3.9318
425	0.002007	1959.7	2060.0	4.2734	0.0018165	1892.7	2001.7	4.1626
450	0.002486	2159.6	2284.0	4.5884	0.002085	2053.9	2179.0	4.4121
500	0.003892	2525.5	2720.1	5.1726	0.002956	2390.6	2567.9	4.9321
550	0.005118	2763.6	3019.5	5.5485	0.003956	2658.8	2896.2	5.3441
600	0.006112	2942.0	3247.6	5.8178	0.004834	2861.1	3151.2	5.6452
650	0.006966	3093.5	3441.8	6.0342	0.005595	3028.8	3364.5	5.8829
700	0.007727	3230.5	3616.8	6.2189	0.006272	3177.2	3553.5	6.0824
800	0.009076	3479.8	3933.6	6.5290	0.007459	3441.5	3889.1	6.4109
900	0.010283	3710.3	4224.4	6.7882	0.008508	3681.0	4191.5	6.6805
1000	0.011411	3930.5	4501.1	7.0146	0.009480	3906.4	4475.2	6.9127

TABLE A.9 Properties of Saturated Refrigerant 22 (Liquid-Vapour): Temperature Table

Temp. (°C)	Press. (bar)	Specific Volume (m³/kg) Sat. Liquid $v_f \times 10^3$	Specific Volume (m³/kg) Sat. Vapour v_g	Internal Energy (kJ/kg) Sat. Liquid u_f	Internal Energy (kJ/kg) Sat. Vapour u_g	Enthalpy (kJ/kg) Sat. Liquid h_f	Enthalpy (kJ/kg) Evap. h_{fg}	Enthalpy (kJ/kg) Sat. Vapour h_g	Entropy (kJ/kg·K) Sat. Liquid s_f	Entropy (kJ/kg·K) Sat. Vapour s_g	Temp. (°C)
-60	0.3749	0.6833	0.5370	-21.57	203.67	-21.55	245.35	223.81	-0.0964	1.0547	-60
-50	0.6451	0.6966	0.3239	-10.89	207.70	-10.85	239.44	228.60	-0.0474	1.0256	-50
-45	0.8290	0.7037	0.2564	-5.50	209.70	-5.44	236.39	230.95	-0.0235	1.0126	-45
-40	1.0522	0.7109	0.2052	-0.07	211.68	0.00	233.27	233.27	0.0000	1.0005	-40
-36	1.2627	0.7169	0.1730	4.29	213.25	4.38	230.71	235.09	0.0186	0.9914	-36
-32	1.5049	0.7231	0.1468	8.68	214.80	8.79	228.10	236.89	0.0369	0.9828	-32
-30	1.6389	0.7262	0.1355	10.88	215.58	11.00	226.77	237.78	0.0460	0.9787	-30
-28	1.7819	0.7294	0.1252	13.09	216.34	13.22	225.43	238.66	0.0551	0.9746	-28
-26	1.9345	0.7327	0.1159	15.31	217.11	15.45	224.08	239.53	0.0641	0.9707	-26
-22	2.2698	0.7393	0.0997	19.76	218.62	19.92	221.32	241.24	0.0819	0.9631	-22
-20	2.4534	0.7427	0.0926	21.99	219.37	22.17	219.91	242.09	0.0908	0.9595	-20
-18	2.6482	0.7462	0.0861	24.23	220.11	24.43	218.49	242.92	0.0996	0.9559	-18
-16	2.8547	0.7497	0.0802	26.48	220.85	26.69	217.05	243.74	0.1084	0.9525	-16
-14	3.0733	0.7533	0.0748	28.73	221.58	28.97	215.59	244.56	0.1171	0.9490	-14
-12	3.3044	0.7569	0.0698	31.00	222.30	31.25	214.11	245.36	0.1258	0.9457	-12
-10	3.5485	0.7606	0.0652	33.27	223.02	33.54	212.62	246.15	0.1345	0.9424	-10
-8	3.8062	0.7644	0.0610	35.54	223.73	35.83	211.10	246.93	0.1431	0.9392	-8
-6	4.0777	0.7683	0.0571	37.83	224.43	38.14	209.56	247.70	0.1517	0.9361	-6
-4	4.3638	0.7722	0.0535	40.12	225.13	40.46	208.00	248.45	0.1602	0.9330	-4
-2	4.6647	0.7762	0.0501	42.42	225.82	42.78	206.41	249.20	0.1688	0.9300	-2
0	4.9811	0.7803	0.0470	44.73	226.50	45.12	204.81	249.92	0.1773	0.9271	0

(Contd.)

Tables A-9 to A-11 calculated on the basis of equations from A. Kamei, S.W. Beyerlein, and E.W. Lemmon, "A Fundamental Equation for Chlorodifluoronethene (R-22)," *Fluid Phase Equilibria*, Vol. 80, No. 11, 1992, pp. 71–86. Taken from Moran, M.J., and H.N. Shapiro, *Fundamentals of Engineering Thermodynamics*, 2nd ed., Wiley, New York, 1992, pp. 730–735. Reprinted with permission of John Wiley and Sons, Inc. and Du Pont.

TABLE A.9   Properties of Saturated Refrigerant 22 (Liquid-Vapour): Temperature Table   (Contd.)

| Temp. (°C) | Press. (bar) | Specific Volume (m³/kg) | | Internal Energy (kJ/kg) | | Enthalpy (kJ/kg) | | | Entropy (kJ/kg-K) | | Temp. (°C) |
		Sat. Liquid $v_f \times 10^3$	Sat. Vapour v_g	Sat. Liquid u_f	Sat. Vapour u_g	Sat. Liquid h_f	Evap. h_{fg}	Sat. Vapour h_g	Sat. Liquid s_f	Sat. Vapour s_g	
2	5.3133	0.7844	0.0442	47.04	227.17	47.46	203.18	250.64	0.1857	0.9241	2
4	5.6619	0.7887	0.0415	49.37	227.83	49.82	201.52	251.34	0.1941	0.9213	4
6	6.0275	0.7930	0.0391	51.71	228.48	52.18	199.84	252.03	0.2025	0.9184	6
8	6.4105	0.7974	0.0368	54.05	229.13	54.56	198.14	252.70	0.2109	0.9157	8
10	6.8113	0.8020	0.0346	56.40	229.76	56.95	196.40	253.35	0.2193	0.9129	10
12	7.2307	0.8066	0.0326	58.77	230.38	59.35	194.64	253.99	0.2276	0.9102	12
16	8.1268	0.8162	0.0291	63.53	231.59	64.19	191.02	255.21	0.2442	0.9048	16
20	9.1030	0.8263	0.0259	68.33	232.76	69.09	187.28	256.37	0.2607	0.8996	20
24	10.164	0.8369	0.0232	73.19	233.87	74.04	183.40	257.44	0.2772	0.8944	24
28	11.313	0.8480	0.0208	78.09	234.92	79.05	179.37	258.43	0.2936	0.8893	28
32	12.556	0.8599	0.0186	83.06	235.91	84.14	175.18	259.32	0.3101	0.8842	32
36	13.897	0.8724	0.0168	88.08	236.83	89.29	170.82	260.11	0.3265	0.8790	36
40	15.341	0.8858	0.0151	93.18	237.66	94.53	166.25	260.79	0.3429	0.8738	40
45	17.298	0.9039	0.0132	99.65	238.59	101.21	160.24	261.46	0.3635	0.8672	45
50	19.433	0.9238	0.0116	106.26	239.34	108.06	153.84	261.90	0.3842	0.8603	50
60	24.281	0.9705	0.0089	120.00	240.24	122.35	139.61	261.96	0.4264	0.8455	60

TABLE A.10 Properties of Saturated Refrigerant 22 (Liquid-Vapour): Pressure Table

Press. (bar)	Temp. (°C)	Specific Volume (m³/kg) Sat. Liquid $v_f \times 10^3$	Sat. Vapour v_g	Internal Energy (kJ/kg) Sat. Liquid u_f	Sat. Vapour u_g	Enthalpy (kJ/kg) Sat. Liquid h_f	Evap. h_{fg}	Sat. Vapour h_g	Entropy (kJ/kg-K) Sat. Liquid s_f	Sat. Vapour s_g	Press. (bar)
0.40	-58.86	0.6847	0.5056	-20.36	204.13	-20.34	244.69	224.36	-0.0907	1.0512	0.40
0.50	-54.83	0.6901	0.4107	-16.07	205.76	-16.03	242.33	226.30	-0.0709	1.0391	0.50
0.60	-51.40	0.6947	0.3466	-12.39	207.14	-12.35	240.28	227.93	-0.0542	1.0294	0.60
0.70	-48.40	0.6989	0.3002	-9.17	208.34	-9.12	238.47	229.35	-0.0397	1.0213	0.70
0.80	-45.73	0.7026	0.2650	-6.28	209.41	-6.23	236.84	230.61	-0.0270	1.0144	0.80
0.90	-43.30	0.7061	0.2374	-3.66	210.37	-3.60	235.34	231.74	-0.0155	1.0084	0.90
1.00	-41.09	0.7093	0.2152	-1.26	211.25	-1.19	233.95	232.77	-0.0051	1.0031	1.00
1.25	-36.23	0.7166	0.1746	4.04	213.16	4.13	230.86	234.99	0.0175	0.9919	1.25
1.50	-32.08	0.7230	0.1472	8.60	214.77	8.70	228.15	236.86	0.0366	0.9830	1.50
1.75	-28.44	0.7287	0.1274	12.61	216.18	12.74	225.73	238.47	0.0531	0.9755	1.75
2.00	-25.18	0.7340	0.1123	16.22	217.42	16.37	223.52	239.88	0.0678	0.9691	2.00
2.25	-22.22	0.7389	0.1005	19.51	218.53	19.67	221.47	241.15	0.0809	0.9636	2.25
2.50	-19.51	0.7436	0.0910	22.54	219.55	22.72	219.57	242.29	0.0930	0.9586	2.50
2.75	-17.00	0.7479	0.0831	25.36	220.48	25.56	217.77	243.33	0.1040	0.9542	2.75
3.00	-14.66	0.7521	0.0765	27.99	221.34	28.22	216.07	244.29	0.1143	0.9502	3.00
3.25	-12.46	0.7561	0.0709	30.47	222.13	30.72	214.46	245.18	0.1238	0.9465	3.25
3.50	-10.39	0.7599	0.0661	32.82	222.88	33.09	212.91	246.00	0.1328	0.9431	3.50
3.75	-8.43	0.7636	0.0618	35.06	223.58	35.34	211.42	246.77	0.1413	0.9399	3.75
4.00	-6.56	0.7672	0.0581	37.18	224.24	37.49	209.99	247.48	0.1493	0.9370	4.00
4.25	-4.78	0.7706	0.0548	39.22	224.86	39.55	208.61	248.16	0.1569	0.9342	4.25
4.50	-3.08	0.7740	0.0519	41.17	225.45	41.52	207.27	248.80	0.1642	0.9316	4.50
4.75	-1.45	0.7773	0.0492	43.05	226.00	43.42	205.98	249.40	0.1711	0.9292	4.75
5.00	0.12	0.7805	0.0469	44.86	226.54	45.25	204.71	249.97	0.1777	0.9269	5.00
5.25	1.63	0.7836	0.0447	46.61	227.04	47.02	203.48	250.51	0.1841	0.9247	5.25
5.50	3.08	0.7867	0.0427	48.30	227.53	48.74	202.28	251.02	0.1903	0.9226	5.50
5.75	4.49	0.7897	0.0409	49.94	227.99	50.40	201.11	251.51	0.1962	0.9206	5.75

(Contd.)

TABLE A.10   Properties of Saturated Refrigerant 22 (Liquid-Vapour): Pressure Table   (Contd.)

| Press. (bar) | Temp. (°C) | Specific Volume (m³/kg) | | Internal Energy (kJ/kg) | | Enthalpy (kJ/kg) | | | Entropy (kJ/kg-K) | | Press. (bar) |
		Sat. Liquid $v_f \times 10^3$	Sat. Vapour v_g	Sat. Liquid u_f	Sat. Vapour u_g	Sat. Liquid h_f	Evap. h_{fg}	Sat. Vapour h_g	Sat. Liquid s_f	Sat. Vapour s_g	
6.00	5.85	0.7927	0.0392	51.53	228.44	52.01	199.97	251.98	0.2019	0.9186	6.00
7.00	10.91	0.8041	0.0337	57.48	230.04	58.04	195.60	253.64	0.2231	0.9117	7.00
8.00	15.45	0.8149	0.0295	62.88	231.43	63.53	191.52	255.05	0.2419	0.9056	8.00
9.00	19.59	0.8252	0.0262	67.84	232.64	68.59	187.67	256.25	0.2591	0.9001	9.00
10.00	21.40	0.8352	0.0236	72.46	233.71	73.30	183.99	257.28	0.2748	0.8952	10.00
12.00	30.25	0.8546	0.0195	80.87	235.48	81.90	177.04	258.94	0.3029	0.8864	12.00
14.00	36.29	0.8734	0.0166	88.45	236.89	89.68	170.49	260.16	0.3277	0.8786	14.00
16.00	41.73	0.8919	0.0144	95.41	238.00	96.83	164.21	261.04	0.3500	0.8715	16.00
18.00	46.69	0.9104	0.0127	101.87	238.86	103.51	158.13	261.64	0.3705	0.8649	18.00
20.00	51.26	0.9291	0.0112	107.95	239.51	109.81	152.17	261.98	0.3895	0.8586	20.00
24.00	59.46	0.9677	0.0091	119.24	240.22	121.56	140.43	261.99	0.4241	0.8463	24.00

TABLE A.11 Properties of Superheated Refrigerant 22 Vapour

T (°C)	v (m³/kg)	u (kJ/kg)	h (kJ/kg)	s (kJ/kg-K)	v (m³/kg)	u (kJ/kg)	h (kJ/kg)	s (kJ/kg-K)
	p = 0.4 bar = 0.04 MFa (T_{sat} = −58.86°C)				*p* = 0.6 bar = 0.06 MPa (T_{sat} = −51.40°C)			
Sat.	0.50559	204.13	224.36	1.0512	0.34656	207.14	227.93	1.0294
−55	0.51532	205.92	226.53	1.0612				
−50	0.52787	208.26	229.38	1.0741	0.34895	207.80	228.74	1.0330
−45	0.54037	210.63	232.24	1.0868	0.35747	210.20	231.65	1.0459
−40	0.55284	211.02	235.13	1.0993	0.36594	212.62	234.58	1.0586
−35	0.56526	215.43	238.05	1.1117	0.37437	215.06	237.52	1.0711
−30	0.57766	217.88	240.99	1.1239	0.38277	217.53	240.49	1.0835
−25	0.59002	220.35	243.95	1.1360	0.39114	220.02	243.49	1.0956
−20	0.60236	222.85	246.95	1.1479	0.39948	222.54	246.51	1.1077
−15	0.61468	225.38	249.97	1.1597	0.40779	225.08	249.55	1.1196
−10	0.62697	227.93	253.01	1.1714	0.41608	227.65	252.62	1.1314
−5	0.63925	230.52	256.09	1.1830	0.42436	230.25	255.71	1.1430
0	0.65151	233.13	259.19	1.1944	0.43261	232.88	258.83	1.1545
	p = 0.8 bar = 0.08 MPa (T_{sat} = −45.73°C)				*p* = 1.0 bar = 0.10 MPa (T_{sat} = −41.09°C)			
Sat.	0.26503	209.41	230.61	1.0144	0.21518	211.25	232.77	1.0031
−45	0.26597	209.76	231.04	1.0163				
−40	0.27245	212.21	234.01	1.0292	0.21633	211.79	233.42	1.0059
−35	0.27890	214.68	236.99	1.0418	0.22158	214.29	236.44	1.0187
−30	0.28530	217.17	239.99	1.0543	0.22679	216.80	239.48	1.0313
−25	0.29167	219.68	243.02	1.0666	0.23197	219.34	242.54	1.0438
−20	0.29801	222.22	246.06	1.0788	0.23712	221.90	245.61	1.0560
−15	0.30433	224.78	249.13	1.0908	0.24224	224.48	248.70	1.0681
−10	0.31062	227.37	252.22	1.1026	0.24734	227.08	251.82	1.0801
−5	0.31690	229.98	255.34	1.1143	0.25241	229.71	254.95	1.0919
0	0.32315	232.62	258.47	1.1259	0.25747	232.36	258.11	1.1035
5	0.32939	235.29	261.64	1.1374	0.26251	235.04	261.29	1.1151
10	0.33561	237.98	264.83	1.1488	0.26753	237.74	264.50	1.1265
	p = 1.5 bar = 0.15 MPa (T_{sat} = −32.08°C)				*p* = 2.0 bar = 0.20 MPa (T_{sat} = −25.18°C)			
Sat.	0.14721	214.77	236.86	0.9830	0.11232	217.42	239.88	0.9691
−30	0.14872	215.85	238.16	0.9883				
−25	0.15232	218.45	241.30	1.0011	0.11242	217.51	240.00	0.9696
−20	0.15588	221.07	244.45	1.0137	0.11520	220.19	243.23	0.9825
−15	0.15941	223.70	247.61	1.0260	0.11795	222.88	246.47	0.9952
−10	0.16292	226.35	250.78	1.0382	0.12067	225.58	249.72	1.0076

Contd.

TABLE A.11   Properties of Superheated Refrigerant 22 Vapour   (*Contd.*)

T (°C)	v (m³/kg)	u (kJ/kg)	h (kJ/kg)	s (kJ/kg-K)	v (m³/kg)	u (kJ/kg)	h (kJ/kg)	s (kJ/kg-K)
−5	0.16640	229.02	253.98	1.0502	0.12336	228.30	252.97	1.0199
0	0.16987	231.70	257.18	1.0621	0.12603	231.03	256.23	1.0310
5	0.17331	234.42	260.41	1.0738	0.12868	233.78	259.51	1.0438
10	0.17674	237.15	263.66	1.0854	0.13132	236.54	262.81	1.0555
15	0.18015	239.91	266.93	1.0968	0.13393	239.33	266.12	1.0671
20	0.18355	242.69	270.22	1.1081	0.13653	242.14	269.44	1.0786
25	0.18693	245.49	273.53	1.1193	0.13912	244.97	272.79	1.0899

		p = 2.5 bar = 0.25 MPa (T_{sat} = −19.51°C)					p = 3.0 bar = 0.30 MPa (T_{sat} = −14.66°C)		
Sat.	0.09097	219.55	242.29	0.9586	0.07651	221.34	244.29	0.9502
−15	0.09303	222.03	245.29	0.9703				
−10	0.09528	224.79	248.61	0.9831	0.07833	223.96	247.46	0.9623
−5	0.09751	227.55	251.93	0.9956	0.08025	226.78	250.86	0.9751
0	0.09971	230.33	255.26	1.0078	0.08214	229.61	254.25	0.9876
5	0.10189	233.12	258.59	1.0199	0.08400	232.44	257.64	0.9999
10	0.10405	235.92	261.93	1.0318	0.08585	235.28	261.04	1.0120
15	0.10619	238.74	265.29	1.0436	0.08767	238.14	264.44	1.0239
20	0.10831	241.58	268.66	1.0552	0.08949	241.01	267.85	1.0357
25	0.11043	244.44	272.04	1.0666	0.09128	243.89	271.28	1.0472
30	0.11253	247.31	275.44	1.0779	0.09307	246.80	274.72	1.0587
35	0.11461	250.21	278.86	1.0891	0.09484	249.72	278.17	1.0700
40	0.11669	253.13	282.30	1.1002	0.09660	252.66	281.64	1.0811

		p = 3.5 bar = 0.35 MPa (T_{sat} = −10.39°C)					p = 4.0 bar = 0.40 MPa (T_{sat} = −6.56°C)		
Sat.	0.06605	222.88	246.00	0.9431	0.05812	224.24	247.48	0.9370
−10	0.06619	223.10	246.27	0.9441				
−5	0.06789	225.99	249.75	0.9572	0.05860	225.16	248.60	0.9411
0	0.06956	22&.86	253.21	0.9700	0.06011	228.09	252.14	0.9542
5	0.07121	231.74	256.67	0.9825	0.06160	231.02	225.66	0.9670
10	0.07284	234.63	260.12	0.9948	0.06306	233.95	259.18	0.9795
15	0.07444	237.52	263.57	1.0069	0.06450	236.89	262.69	0.9918
20	0.07603	240.42	267.03	1.0188	0.06592	239.83	266.19	1.0039
25	0.07760	243.34	270.50	1.0305	0.06733	242.77	269.71	1.0158
30	0.07916	246.27	273.97	1.0421	0.06872	245.73	273.22	1.0274
35	0.08070	249.22	227.46	1.0535	0.07010	248.71	276.75	1.0390
40	0.08224	252.18	280.97	1.0648	0.07146	251.70	280.28	1.0504
45	0.08376	255.17	284.48	1.0759	0.07282	254.70	283.83	1.0616

Contd.

TABLE A.11   Properties of Superheated Refrigerant 22 Vapour   (*Contd.*)

T (°C)	v (m³/kg)	u (kJ/kg)	h (kJ/kg)	s (kJ/kg-K)	v (m³/kg)	u (kJ/kg)	h (kJ/kg)	s (kJ/kg-K)
	p=4.5				p=5.0			

<table>
<tr><td>T
(°C)</td><td>v
(m³/kg)</td><td>u
(kJ/kg)</td><td>h
(kJ/kg)</td><td>s
(kJ/kg-K)</td><td>v
(m³/kg)</td><td>u
(kJ/kg)</td><td>h
(kJ/kg)</td><td>s
(kJ/kg-K)</td></tr>
<tr><td colspan="5" align="center">p = 4.5 bar = 0.45 MPa
(T_sat = −3.08°C)</td><td colspan="4" align="center">p = 5.0 bar = 0.50 MPa
(T_sat = 0.12°C)</td></tr>
<tr><td>Sat.</td><td>0.05189</td><td>225.45</td><td>248.80</td><td>0.9316</td><td>0.04686</td><td>226.54</td><td>249.97</td><td>0.9269</td></tr>
<tr><td>0</td><td>0.05275</td><td>227.29</td><td>251.03</td><td>0.9399</td><td></td><td></td><td></td><td></td></tr>
<tr><td>5</td><td>0.05411</td><td>230.28</td><td>254.63</td><td>0.9529</td><td>0.04810</td><td>229.52</td><td>253.57</td><td>0.9399</td></tr>
<tr><td>10</td><td>0.05545</td><td>233.26</td><td>258.21</td><td>0.9657</td><td>0.04934</td><td>232.55</td><td>257.22</td><td>0.9530</td></tr>
<tr><td>15</td><td>0.05676</td><td>236.24</td><td>261.78</td><td>0.9782</td><td>0.05056</td><td>235.57</td><td>260.85</td><td>0.9657</td></tr>
<tr><td>20</td><td>0.05805</td><td>239.22</td><td>265.34</td><td>0.9904</td><td>0.05175</td><td>238.59</td><td>264.47</td><td>0.9781</td></tr>
<tr><td>25</td><td>0.05933</td><td>242.20</td><td>268.90</td><td>1.0025</td><td>0.05293</td><td>241.61</td><td>268.07</td><td>0.9903</td></tr>
<tr><td>30</td><td>0.06059</td><td>245.19</td><td>272.46</td><td>1.0143</td><td>0.05409</td><td>244.63</td><td>271.68</td><td>1.0023</td></tr>
<tr><td>35</td><td>0.06184</td><td>24&.19</td><td>276.02</td><td>1.0259</td><td>0.05523</td><td>247.66</td><td>275.28</td><td>1.0141</td></tr>
<tr><td>40</td><td>0.06308</td><td>251.20</td><td>279.59</td><td>1.0374</td><td>0.05636</td><td>250.70</td><td>278.89</td><td>1.0257</td></tr>
<tr><td>45</td><td>0.06430</td><td>254.23</td><td>283.17</td><td>1.0488</td><td>0.05748</td><td>253.76</td><td>282.50</td><td>1.0371</td></tr>
<tr><td>50</td><td>0.06552</td><td>257.28</td><td>286.76</td><td>1.0600</td><td>0.05859</td><td>256.82</td><td>286.12</td><td>1.0484</td></tr>
<tr><td>55</td><td>0.06672</td><td>260.34</td><td>290.36</td><td>1.0710</td><td>0.05969</td><td>259.90</td><td>289.75</td><td>1.0595</td></tr>
<tr><td colspan="5" align="center">p = 5.5 bar = 0.55 MPa
(T_sat = 3.08°C)</td><td colspan="4" align="center">p = 6.0 bar = 0.60 MPa
(T_sat = 5.85°C)</td></tr>
<tr><td>Sat.</td><td>0.04271</td><td>227.53</td><td>251.02</td><td>0.9226</td><td>0.03923</td><td>228.44</td><td>251.98</td><td>0.9186</td></tr>
<tr><td>5</td><td>0.04317</td><td>228.72</td><td>252.46</td><td>0.9278</td><td></td><td></td><td></td><td></td></tr>
<tr><td>10</td><td>0.04433</td><td>231.81</td><td>256.20</td><td>0.9411</td><td>0.04015</td><td>231.05</td><td>255.14</td><td>0.9299</td></tr>
<tr><td>15</td><td>0.04547</td><td>234.89</td><td>259.90</td><td>0.9540</td><td>0.04122</td><td>234.18</td><td>258.91</td><td>0.9431</td></tr>
<tr><td>20</td><td>0.04658</td><td>237.95</td><td>263.57</td><td>0.9667</td><td>0.04227</td><td>237.29</td><td>262.65</td><td>0.9560</td></tr>
<tr><td>25</td><td>0.04768</td><td>241.01</td><td>267.23</td><td>0.9790</td><td>0.04330</td><td>240.39</td><td>266.37</td><td>0.9685</td></tr>
<tr><td>30</td><td>0.04875</td><td>244.07</td><td>270.88</td><td>0.9912</td><td>0.04431</td><td>243.49</td><td>270.07</td><td>0.9808</td></tr>
<tr><td>35</td><td>0.04982</td><td>247.13</td><td>274.53</td><td>1.0031</td><td>0.04530</td><td>246.58</td><td>273.76</td><td>0.9929</td></tr>
<tr><td>40</td><td>0.05086</td><td>250.20</td><td>278.17</td><td>1.0148</td><td>0.04628</td><td>249.68</td><td>277.45</td><td>1.0048</td></tr>
<tr><td>45</td><td>0.05190</td><td>253.27</td><td>281.82</td><td>1.0264</td><td>0.04724</td><td>252.78</td><td>281.13</td><td>1.0164</td></tr>
<tr><td>50</td><td>0.05293</td><td>256.36</td><td>285.47</td><td>1.0378</td><td>0.04820</td><td>255.90</td><td>284.82</td><td>1.0279</td></tr>
<tr><td>55</td><td>0.05394</td><td>259.46</td><td>289.13</td><td>1.0490</td><td>0.04914</td><td>259.02</td><td>288.51</td><td>1.0393</td></tr>
<tr><td>60</td><td>0.05495</td><td>262.58</td><td>292.80</td><td>1.0601</td><td>0.05008</td><td>262.15</td><td>292.20</td><td>1.0504</td></tr>
<tr><td colspan="5" align="center">p = 7.0 bar = 0.70 MPa
(T_sat = 10.91°C)</td><td colspan="4" align="center">p = 8.0 bar = 0.80 MPa
(T_sat = 15.45°C)</td></tr>
<tr><td>Sat.</td><td>0.03371</td><td>230.04</td><td>253.64</td><td>0.9117</td><td>0.02953</td><td>231.43</td><td>255.05</td><td>0.9056</td></tr>
<tr><td>15</td><td>0.03451</td><td>232.70</td><td>256.86</td><td>0.9229</td><td></td><td></td><td></td><td></td></tr>
<tr><td>20</td><td>0.03547</td><td>235.92</td><td>260.75</td><td>0.9363</td><td>0.03033</td><td>234.47</td><td>258.74</td><td>0.9182</td></tr>
<tr><td>25</td><td>0.03639</td><td>239.12</td><td>264.59</td><td>0.9493</td><td>0.03118</td><td>237.76</td><td>262.70</td><td>0.9315</td></tr>
<tr><td>30</td><td>0.03730</td><td>242.29</td><td>268.40</td><td>0.9619</td><td>0.03202</td><td>241.04</td><td>266.66</td><td>0.9448</td></tr>
</table>

Contd.

TABLE A.11   Properties of Superheated Refrigerant 22 Vapour   (*Contd.*)

T (°C)	v (m³/kg)	u (kJ/kg)	h (kJ/kg)	s (kJ/kg-K)	v (m³/kg)	u (kJ/kg)	h (kJ/kg)	s (kJ/kg-K)
35	0.03819	245.46	272.19	0.9743	0.03283	244.28	270.54	0.9574
40	0.03906	248.62	275.96	0.9865	0.03363	247.52	274.42	0.9700
45	0.03992	251.78	279.72	0.9984	0.03440	250.74	278.26	0.9821
50	0.04076	254.94	283.48	1.0101	0.03517	253.96	282.10	0.9941
55	0.04160	258.11	287.23	1.0216	0.03592	257.18	285.92	1.0058
60	0.04242	261.29	290.99	1.0330	0.03667	260.40	289.74	1.0174
65	0.04324	264.48	294.75	1.0442	0.03741	263.64	293.56	1.0287
70	0.04405	267.68	298.51	1.0552	0.03814	266.87	297.38	1.0400

	p = 9.0 bar = 0.90 MPa (T_{sat} = 19.59°C)				p = 10.0 bar = 1.00 MPa (T_{sat} = 23.40°C)			
Sat.	0.02623	232.64	256.25	0.9001	0.02358	233.71	257.28	0.8952
20	0.02630	232.92	256.59	0.9013				
30	0.02789	239.73	264.83	0.9289	0.02457	238.34	262.91	0.9139
40	0.02939	246.37	272.82	0.9549	0.02598	245.18	271.17	0.9407
50	0.03082	252.95	280.68	0.9795	0.02732	251.90	279.22	0.9660
60	0.03219	259.49	288.46	1.0033	0.02860	258.56	287.15	0.9902
70	0.03353	266.04	296.21	1.0262	0.02984	265.19	295.03	1.0135
80	0.03483	272.62	303.96	1.0484	0.03104	271.84	302.88	1.0361
90	0.03611	279.23	311.73	1.0701	0.03221	278.52	310.74	1.0580
100	0.03736	285.90	319.53	1.0913	0.03337	285.24	318.61	1.0794
110	0.03860	292.63	327.37	1.1120	0.03450	292.02	326.52	1.1003
120	0.03982	299.42	335.26	1.1323	0.03562	298.85	334.46	1.1207
130	0.04103	306.28	343.21	1.1523	0.03672	305.74	342.46	1.1408
140	0.04223	313.21	351.22	1.1719	0.03781	312.70	350.51	1.1605
150	0.04342	320.21	359.29	1.1912	0.03889	319.74	358.63	1.1790

	p = 12.0 bar = 1.20 MPa (T_{sat} = 30.25°C)				p = 14.0 bar = 1.40 MPa (T_{sat} = 36.29°C)			
Sat.	0.01955	235.48	258.94	0.8864	0.01662	236.89	260.16	0.8786
40	0.02083	242.63	267.62	0.9146	0.01708	239.78	263.70	0.8900
50	0.02204	249.69	276.14	0.9413	0.01823	247.29	272.81	0.9186
60	0.02319	256.60	284.43	0.9666	0.01929	254.52	281.53	0.9452
70	0.02428	263.44	292.58	0.9907	0.02029	261.60	290.01	0.9703
80	0.02534	270.25	300.66	1.0139	0.02125	268.60	298.34	0.9942
90	0.02636	277.07	308.70	1.0363	0.02217	275.56	306.60	1.0172
100	0.02736	283.90	316.73	1.0582	0.02306	282.52	314.80	1.0395
110	0.02834	290.77	324.78	1.0794	0.02393	289.49	323.00	1.0612
120	0.02930	297.69	332.85	1.1002	0.02478	296.50	331.19	1.0823
130	0.03024	304.65	340.95	1.1205	0.02562	303.55	339.41	1.1029

Contd.

TABLE A.11   Properties of Superheated Refrigerant 22 Vapour   (Contd.)

T (°C)	v (m³/kg)	u (kJ/kg)	h (kJ/kg)	s (kJ/kg-K)	v (m³/kg)	u (kJ/kg)	h (kJ/kg)	s (kJ/kg-K)
140	0.03118	311.68	349.09	1.1405	0.02644	310.64	347.65	1.1231
150	0.03210	318.77	357.29	1.1601	0.02725	317.79	355.94	1.1429
160	0.03301	325.92	365.54	1.1793	0.02805	324.99	364.26	1.1624
170	0.03392	333.14	373.84	1.1983	0.02884	332.26	372.64	1.1815

	p = 16.0 bar = 1.60 MPa (T_{sat} = 41.73°C)					p = 18.0 bar = 1.80 MPa (T_{sat} = 46.69°C)			
Sat.	0.01440	238.00	261.04	0.8715	0.01265	238.86	261.64	0.8649
50	0.01533	244.66	269.18	0.8971	0.01301	241.72	265.14	0.8758
60	0.01634	252.29	278.43	0.9252	0.01401	249.86	275.09	0.9061
70	0.01728	259.65	287.30	0.9515	0.01492	257.57	284.43	0.9337
80	0.01817	266.86	295.93	0.9762	0.01576	265.04	293.40	0.9595
90	0.01901	274.00	304.42	0.9999	0.01655	272.37	302.16	0.9839
100	0.01983	281.09	312.82	1.0228	0.01731	279.62	310.77	1.0073
110	0.02062	288.18	321.17	1.0448	0.01804	286.83	319.30	1.0299
120	0.02139	295.28	329.51	1.0663	0.01874	294.04	327.78	1.0517
130	0.02214	302.41	337.84	1.0872	0.01943	301.26	336.24	1.0730
140	0.02288	309.58	346.19	1.1077	0.02011	308.50	344.70	1.0937
150	0.02361	316.79	354.56	1.1277	0.02077	315.78	353.17	1.1139
160	0.02432	324.05	362.97	1.1473	0.02142	323.10	361.66	1.1338
170	0.02503	331.37	371.42	1.1666	0.02207	330.47	370.19	1.1532

	p = 20.0 bar = 2.00 MPa (T_{sat} = 51.26°C)					p = 24.0 bar = 2.4 MPa (T_{sat} = 59.46°C)			
Sat.	0.01124	239.51	261.98	0.8586	0.00907	240.22	261 99	0.8463
60	0.01212	247.20	271.43	0.8873	0.00913	240.78	262.68	0.8484
70	0.01300	255.35	281.36	0.9167	0.01006	250.30	274.43	0.8831
80	0.01381	263.12	290.74	0.9436	0.01085	258.89	284.93	0.9133
90	0.01457	270.67	299.80	0.9689	0.01156	267.01	294.75	0.9407
100	0.01528	278.09	308.65	0.9929	0.01222	274.85	304.18	0.9663
110	0.01596	285.44	317.37	1.0160	0.01284	282.53	313.35	0.9906
120	0.01663	292.76	326.01	1.0383	0.01343	290.11	322.35	1.0137
130	0.01727	300.08	334.61	1.0598	0.01400	297.64	331.25	1.0361
140	0.01789	307.40	343.19	1.0808	0.01456	305.14	340.08	1.0577
150	0.01850	314.75	351.76	1.1013	0.01509	312.64	348.87	1.0787
160	0.01910	322.14	360.34	1.1214	0.01562	320.16	357.64	1.0992
170	0.01969	329.56	368.95	1.1410	0.01613	327.70	366.41	1.1192
180	0.02027	337.03	377.58	1.1603	0.01663	335.27	375.20	1.1388

Solved Model Question Papers

Paper I

1. Show that for steady state flow process, $\Delta H = Q - W_S$.

   Solution: See Section 2.18 (First Law of Thermodynamics to Flow Processes).

2. An ideal gas ($C_P = 5$, $C_V = 3$) is changed from 1 atm and 22.4 m³ to 10 atm and 2.24 m³ by the reversible process of heating at constant volume followed by cooling at constant pressure. Draw the path in a P–V diagram and calculate Q, W, ΔU and ΔH of the overall process.

   Solution: Basis: 1 kmol of ideal gas

   At initial state

   $$V_1 = 22.4 \text{ m}^3 \qquad P_1 = 1 \text{ atm}$$

   At final state

   $$V_2 = 2.24 \text{ m}^3 \qquad P_2 = 10 \text{ atm}$$
   $$C_P = 5 \text{ kcal/kmol°C} = 20.914 \text{ kJ/kmol°C}$$
   $$C_V = 3 \text{ kcal/kmol°C} = 12.6914 \text{ kJ/kmol°C}$$

   For ideal gas

   $$T_1 = \frac{P_1 V_1}{nR} = \frac{22.4 \times 1}{1 \times 0.082} = 273 \text{ K}$$

   Process I: Heating at constant volume followed by cooling at constant pressure, which comprises two steps:

   Step 1: Heating at constant volume
   Since the process is a constant-volume one with ideal gas, $V_1 = V_2 = V$. We may write

   $$T_2 = T_1 \frac{P_2}{P_1} = 273 \times \left(\frac{10}{1}\right) = 2730 \text{ K}$$

For constant-volume process, $dV = 0$. Then the work done $\Delta W = P\Delta V = 0$.

$$\Delta Q = \Delta U = C_V(T_2 - T_1) = 12.6914(2730 - 273) = 31182.769 \text{ kJ}$$

Step 2: Cooling at constant pressure

Here, the temperature of the gas goes down while the pressure reduces from 10 atm to 1 atm, i.e., $P_1 = 10$ atm and $P_2 = 1$ atm and $T_1 = 2730$ K and $T_2 = 273$ K, as the ideal gas is supposed to be cooled by cooling process from 2730 K to 273 K.

Since the process is a constant-pressure one with ideal gas, then P = constant.

Heat interaction = $\Delta Q_P = \Delta H_P = C_P(T_2 - T_1) = 20.914(273 - 2730) = -51385.698$ kJ

Internal energy

$$\begin{aligned} \Delta U &= \Delta H - P\Delta V = \Delta H - P(V_2 - V_1) \\ &= -51385.698 - [10 \times 101.325(2.24 - 22.4)] = 31091.232 \text{ kJ} \end{aligned}$$

Now, for the two steps combined

$$\Delta Q = 31182.769 \text{ kJ} - 51385.698 \text{ kJ} = -20202.929 \text{ kJ}$$
$$\Delta U = 31182.769 \text{ kJ} - 31091.232 \text{ kJ} \approx 0$$

Total work done

$$= \Delta W = \Delta Q - \Delta U = -20202.929 \text{ kJ} - 0 = -20202.929 \text{ kJ}$$

3. The following experimental data are available for CO_2 gas at 92°C:

P_R	1	2	3	4	6	8	10
Z	0.856	0.583	0.535	0.620	0.800	0.975	1.160

Find the fugacity of CO_2 at 100 atm.

Data: For CO_2, $T_c = 304.1$ K and $P_c = 72.9$ atm.

Solution: We are given that temperature, $T = (92 + 273)$ K = 365 K

Now, on the basis of given value, some important data can be represented as follows:

P_R	1	2	3	4	6	6	10
Z	0.856	0.583	0.535	0.620	0.800	0.975	1.160
$P = P_R \times P_c$ $P_c = 72.9$ atm	72.9	145.8	218.7	291.6	437.4	583.2	729
$B = (Z-1) \times \dfrac{RT}{P}$	-5.99	-8.68	-6.45	-3.96	-1.39	-0.13	0.67
$\gamma = \exp\left(\dfrac{BP}{RT}\right)$	0.87	0.66	0.63	0.68	0.82	0.98	1.18
$f = \gamma \times P$	63.42	96.23	137.78	198.29	358.67	571.54	860.22

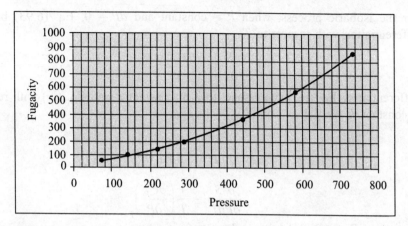

From the pressure versus fugacity graph, we get that the fugacity corresponding to 100 atm pressure is 80 (approximately).

4. Define Joule–Thomson coefficient μ_{JT}. Express μ_{JT} in terms of C_P and compressibility factor Z in the following form:

$$\mu_{JT} = \frac{RT^2}{PC_P}\left(\frac{\partial Z}{\partial T}\right)_P \qquad \text{where } PV = ZRT$$

Solution: The relation is derived under Section 6.4.11 (Joule–Thomson Coefficient).

5. Derive the following expression:

$$\left(\frac{\partial C_P}{\partial P}\right)_T = -T\left(\frac{\partial^2 V}{\partial T^2}\right)_P$$

Solution: If entropy S is considered to be a function of T and P, i.e., $S = f(T, P)$, then its total differential is given by

$$dS = \left(\frac{\partial S}{\partial T}\right)_P dT + \left(\frac{\partial S}{\partial P}\right)_T dP \tag{i}$$

Using the relation $\left(\frac{\partial S}{\partial T}\right)_P = \frac{C_P}{T}$ and the fourth Maxwell's relation $\left(\frac{\partial V}{\partial T}\right)_P = -\left(\frac{\partial S}{\partial P}\right)_T$, we have

$$dS = \frac{C_P}{T} dT - \left(\frac{\partial V}{\partial T}\right)_P dP \tag{ii}$$

For an isothermal change of the process, when $T = $ constant and $dT = 0$, Eq. (ii) becomes on differentiation with respect to P

$$\left(\frac{\partial S}{\partial P}\right)_T = -\left(\frac{\partial V}{\partial T}\right)_P \tag{iii}$$

For the isobaric process, when P = constant and dP = 0, Eq. (6.93) becomes on differentiation with respect to T

$$\left(\frac{\partial S}{\partial T}\right)_P = \frac{C_P}{T} \tag{iv}$$

Differentiating Eq. (iii) with respect to T at constant P and Eq. (iv) with respect to P at constant T, we get

$$\frac{\partial^2 S}{\partial T \partial P} = -\left(\frac{\partial^2 V}{\partial T^2}\right)_P \tag{v}$$

$$\frac{\partial^2 S}{\partial P \partial T} = \frac{1}{T}\left(\frac{\partial C_P}{\partial P}\right)_T \tag{vi}$$

Comparing Eqs. (v) and (vi), we have

$$\left(\frac{\partial C_P}{\partial P}\right)_T = -T\left(\frac{\partial^2 V}{\partial T^2}\right)_P$$

Hence proved.

6. What is Clausius–Clapeyron equation?

   Solution: See Section 6.4.3 (Clapeyron equation).

7. With the help of a T–S diagram describe the vapour compression refrigeration cycle and derive its COP.

   Solution: See Section 8.1.5 (Vapour–Compression Refrigeration Cycle).

8. For a binary liquid mixture at constant temperature and pressure, excess Gibbs energy is given by

$$\frac{G^E}{RT} = (-2.6x_1 - 1.8x_2)x_1 x_2$$

   Show that these expressions satisfy the Gibbs–Duhem equation.

   Solution: See Example 9.20.

9. Under atmospheric conditions the acetone–chloroform azeotrope boils at 64.6°C and contains 33.5 mole percent acetone. The vapour pressures of acetone and chloroform at this temperature are 995 and 855 mm Hg respectively. Calculate the composition of the vapour at this temperature in equilibrium with a liquid analyzing 11.1 mole percent acetone. We are given that van Laar equation

$$\ln \gamma_1 = \frac{Ax_2^2}{\left(\dfrac{A}{B}x_1 + x_2\right)^2} \quad \text{and} \quad \ln \gamma_2 = \frac{Bx_1^2}{\left(x_1 + \dfrac{B}{A}x_2\right)^2}$$

   Solution: See Example 10.11.

10. For the reaction

$$CO_2(g) + H_2(g) \rightarrow CO(g) + H_2O(g)$$

at 298 K is 8.685×10^{-6}. Estimate the value of K_a at 1000 K, assuming that ΔH^0 is constant in the temperature range of 298 K to 1000 K. The data given is:

Component	ΔH_f^0 (kJ/mole)
CO(g)	110.532
H$_2$O(g)	−241.997
CO$_2$(g)	−393.978
H$_2$(g)	0

We are given that van't Hoff's equation

$$\left(\frac{\partial \ln K_a}{\partial T}\right) = \frac{\Delta H^0}{RT^2}$$

where ΔH^0 is the standard heat of reaction.

Solution: We know that

$$\Delta H^0 = \sum \Delta H_{f(\text{Products})}^0 - \sum \Delta H_{f(\text{Reactants})}^0$$
$$= -110.532 + (-241.997) - (-393.978) - 0 = 262.513 \text{ kJ}$$

We have

$$\ln \frac{K_{a2}}{K_{a1}} = \frac{\Delta H^0}{R}\left(\frac{1}{T_1} - \frac{1}{T_2}\right)$$

We are given that the equilibrium constant at 298.15 K is $K_{a1} = 8.685 \times 10^{-6}$, $\Delta H^0 = -262.513$ kJ and $R = 8.314$.

Substituting the values, we get

$$\ln \frac{K_{a2}}{8.685 \times 10^{-6}} = \frac{262.513}{8.314}\left(\frac{1}{298} - \frac{1}{1000}\right)$$
$$K_{a2} = 1.746 \times 10^{27}$$

Therefore, the equilibrium constant at 800 K is $K_{a2} = 1.746 \times 10^{27}$.

11. For methanol synthesis reaction

$$CO(g) + 2H_2(g) \rightleftharpoons CH_3OH(g)$$

Using a feed mixture of carbon monoxide and hydrogen in the stoichiometric proportion, state what equilibrium mole fraction of methanol at 50 bar pressure and 450 K can be found from the following data:

Components	ΔG_{f300}^0	ΔH_{f300}^0
Methanol (g)	−166.270	−238.660
Carbon monoxide (g)	−137.169	−110.525

Solution: The standard heat of reaction and standard free energy of reaction at 298 K are estimated as

$$\Delta H^0 = -128,135 \text{ J/mol}$$
$$\Delta G^0 = -29,101 \text{ J/mol}.$$

For the equation

$$\Delta H_T^0 = \Delta H' + \Delta aT + \frac{\Delta b}{2}T^2 + \frac{\Delta c}{3}T^3 + \frac{\Delta d}{4}T^4 \qquad \text{(i)}$$

$$\Delta a = 18.382 - 28.068 - 2 \times 27.012 = -63.710$$
$$\Delta b = (101.564 - 4.631 - 2 \times 3.509) \times 10^{-3} = 89.915 \times 10^{-3}$$
$$\Delta c = -28.683 \times 10^{-6}$$
$$\Delta d = 0$$

On putting $\Delta H_T^0 = \Delta H_{300}^0 = -128.135$ kJ/mol, Eq. (i) gives

$$-128,135 = \Delta H^0 - 63.710 \times 300 + \frac{89.91 \times 10^{-3}}{2} \times (300)^2 - \frac{28.683 \times 10^{-6}}{3} \times (300)^3$$
$$+ 0 - 63.710 \times 300 \quad \Delta H^0 = -113.184 \text{ kJ}$$

Again, we know that

$$\Delta G_{300}^0 = \Delta H - \Delta aT \ln T - \frac{\Delta b}{2}T^2 - \frac{\Delta c}{6}T^3 - \frac{\Delta d}{12}T^4 - ART \qquad \text{(ii)}$$

where A is the constant of integration.

$$-29101 = -113184 + 63.710 \times 300 \ln 300 - \frac{89.915 \times 10^{-3}}{2}(300)^2$$

$$+ \frac{28.683 \times 10^{-6}}{6}(300)^3 + \frac{1.122 \times 10^5}{2 \times 300} - A \times 8.314 \times 300 \qquad \text{(iii)}$$

On putting $T = 300$ K in the preceding equations, Eq. (iii) gives $A = 8.501$.

Now, let us evaluate ΔG_{450}^0.

$$\Delta G_{450}^0 = -113184 + 63.710 \times 450 \ln 450 - \frac{89.915 \times 10^{-3}}{2}(450)^2$$

$$+ \frac{28.683 \times 10^{-6}}{6}(450)^3 + \frac{1.122 \times 10^5}{2 \times 450} \qquad \text{(iv)}$$

$$- 8.501 \times 8.314 \times 450 = 21.617 \text{ kJ}$$

Using the expression

$$K_a = \exp\left(\frac{-\Delta G_{450}^0}{RT}\right) = \exp\left(\frac{-2167}{8.314 \times 450}\right) = 3.095 \times 10^{-3}$$

Let ε be the degree of conversion at equilibrium.

Component	CO	H_2	CH_3OH
Moles in feed	1	2	0
Moles at equilibrium	$1 - \varepsilon_e$	$2(1 - \varepsilon_e)$	ε_e
Mole fraction	$\dfrac{1 - \varepsilon_e}{3 - 2\varepsilon_e}$	$\dfrac{2(1 - \varepsilon_e)}{3 - 2\varepsilon_e}$	$\dfrac{\varepsilon_e}{3 - 2\varepsilon_e}$
Total moles at equilibrium	$1 - \varepsilon_e + 2(1 - \varepsilon_e) + \varepsilon_e = 3 - 2\varepsilon_e$		

Now, the sum of the stoichiometric coefficients $= \Sigma v_i = 1 - 1 - 2 = -2$.

$$K = K_x = \Pi(x_i)^{v_i} = \frac{\dfrac{\varepsilon_e}{3 - 2\varepsilon_e}}{\left(\dfrac{1 - \varepsilon_e}{3 - 2\varepsilon_e}\right) \times \left[\dfrac{2(1 - \varepsilon_e)}{3 - 2\varepsilon_e}\right]^2} = 3.095 \times 10^{-3}(50)^{-2}$$

Therefore, find out the value of ε_e

Hence, the degree of conversion at equilibrium is equal to the calculated value.

The mole fraction of methanol is given by

$$y_{CH_3OH} = \frac{\varepsilon_e}{3 - 2\varepsilon_e} \qquad \text{(on putting the value of } \varepsilon_e\text{)}$$

Paper II

1. Calculate the second and third virial coefficients of a gas which obeys van der Waal's equation of state.

   Solution: See Example 3.7.

2. The following experimental data are available for CO_2 gas at 92°C:

P_R	1	2	3	4	6	8	10
Z	0.856	0.583	0.535	0.620	0.800	0.975	1.160

   The critical temperature for CO_2 is 304.1 K and critical pressure is 72.9 atm. Calculate the pressure of gas if the molar volume is 0.13 L/mol. Take $R = 0.082$ L-atm/mol-K.

   Solution: We are given that temperature, $T = (92 + 273)$ K = 365 K.
   Now, on the basis of given values, some important data can be represented as follows:

	P_R	1	2	3	4	6	8	10
	Z	0.856	0.583	0.535	0.620	0.800	0.975	1.160
$P = P_R \times P_c$								
$P_c = 72.9$ atm		72.9	145.8	218.7	291.6	437.4	583.2	729
$V = \dfrac{ZRT}{P}$		0.35	0.12	0.073	0.064	0.055	0.05	0.048

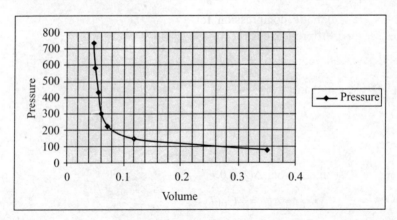

   From the pressure versus volume graph, we get that the pressure corresponding to 0.13 L/mol molar volume is 148 atm (approx).

3. An ideal gas ($C_P = 5$ kcal/kmol°C, $C_V = 3$ kcal/kmol°C) is changed from 1 atm and 22.4 m$_3$ to 10 atm and 2.24 m^3 by the following reversible process:

   (a) Isothermal compression;

   (b) Adiabatic compression followed by cooling at constant volume;

(c) Heating at constant volume followed by cooling at constant pressure.
Calculate Q, W, ΔU and ΔH of the overall process in each case.

Solution: Basis: 1 kmol of ideal gas
We are given that

$V_1 = 22.4$ m^3

$P_1 = 1$ atm

$V_2 = 2.24$ m^3

$P_2 = 10$ atm

$C_P = 5$ kcal/kmol°C $= 20.914$ kJ/kmol°C

$C_V = 3$ kcal/kmol°C $= 12.6914$ kJ/kmol°C.

For ideal gas

$$T_1 = \frac{P_1 V_1}{nR} = \frac{22.4 \times 101.325}{1 \times 8.314} = 273 \text{ K}$$

The ideal gas undergoes the reversible processes.

Process I: Isothermal compression (T = constant)

Change in internal energy, $\Delta U = 0$. Then the first law $\Delta Q = \Delta U + \Delta W$ gives

$$\Delta Q = \Delta W = nRT \ln \frac{P_1}{P_2} = 1 \times 8.314 \times 273 \times \ln \frac{1}{10} = -5226.22 \text{ KJ}$$

So, $\Delta U = 0$ and $\Delta H = 0$.

Process II: Adiabatic compression followed by cooling at constant volume, which comprises two different steps:

Step 1: Adiabatic compression ($dQ = 0$)

The heat capacity ratio $= \gamma = \dfrac{C_P}{C_V} = \dfrac{5}{3} = 1.67$

$$T_2 = T_1 \times \left(\frac{V_1}{V_2}\right)^{\gamma - 1} = 273 \times \left(\frac{22.4}{2.24}\right)^{0.67} = 1276.92 \text{ K}$$

Change in heat interaction, $\Delta Q = 0$. Then first law $\Delta Q = \Delta U + \Delta W$ gives

$$\Delta W = -\Delta U = -C_V(T_2 - T_1) = -12.6(1276.92 - 273) = -12649.39 \text{ kJ}$$

Then $\Delta U = 12649.39$ kJ

Step 2: Cooling at constant volume (V = constant and $\Delta V = 0$)

Hence, work done

$$\Delta W = P\Delta V = 0$$

Here, $T_1 = 1276.92$ K and $T_2 = 273$ K as the ideal gas is supposed to be cooled by cooling process from 1276.92 K to 273 K. Therefore

$$\Delta Q = \Delta U = C_V(T_2 - T_1) = 12.6(273 - 1276.92) = -12649.39 \text{ kJ}$$

Total internal energy

$$\Delta U = 12649.39 \text{ kJ} - 12649.39 \text{ kJ} = 0$$
$$\Delta H = 0$$

Process III: Heating at constant volume followed by cooling at constant pressure, which comprises two steps:

Step 1: Heating at constant volume (V = constant and $\Delta V = 0$)
Here, $T_1 = 273$ K and T_2 is unknown as the ideal gas is supposed to be heated by heating process, but at what temperature remains to be determined.
Then, we may write

$$T_2 = T_1 \frac{P_2}{P_1} = 273 \times \left(\frac{10}{1}\right) = 2730 \text{ K}$$

For constant-volume process, $dV = 0$. Then, work done

$$dW = PdV = 0$$

Heat interaction

$$\Delta Q = \Delta U = C_V(T_2 - T_1) = 12.6(2730 - 273) = 30958.2 \text{ kJ}$$

Step 2: Cooling at constant pressure
Here, $T_1 = 2730$ K and $T_2 = 273$ K as the ideal gas is supposed to be cooled by cooling process from 2730 K to 273 K. Therefore

$$\Delta Q = \Delta U = C_P(T_2 - T_1) = 20.914(273 - 2730) = -51385.698 \text{ kJ}$$

Again, internal energy

$$\Delta U = \Delta H - P\Delta V = \Delta H - P(V_2 - V_1) = -51385.698 - [10 \times 101.325(2.24 - 22.4)]$$
$$= -30958.578 \text{ kJ}$$

Now, for the two steps combined

$$\Delta Q = 30958.2 \text{ kJ} - 51385.698 \text{ kJ} = -20427.498 \text{ kJ}$$
$$\Delta U = 30958.2 \text{ kJ} - 30958.578 \text{ kJ} \approx 0$$

Total work done

$$\Delta W = \Delta Q - \Delta U = -20427.498 \text{ kJ} - 0 = -20427.498 \text{ kJ}$$

4. A steel casting at a temperature of 725 K and weighing 35 kg is quenched in 150 kg oil at 275 K. If there is no heat losses, determine the change in entropy. The specific heat (C_P) of steel is 0.88 kJ/kg-K and that of oil is 2.5 kJ/kg-K.

Solution: Let the final temperature of the system be T.
Then, from the energy balance over the process, we have

$$35(0.88)(725 - T) = 150(2.5)(T - 275)$$

Solving the equation yields $T = 309.16$ K.

Suppose that the change in entropy of the steel casting $= \Delta S_1$, and the change in entropy of oil $= \Delta S_2$. Then

(a) The change in entropy of the steel casting can be estimated as

$$\Delta S_1 = mC_P \int_{725}^{309.16} \frac{dT}{T} = 35 \times 0.88 \times \ln \frac{309.16}{725}$$

$$= -26.28 \text{ kJ/K}$$

(b) The change in entropy of water is given by

$$\Delta S_2 = mC_P \int_{275}^{309.16} \frac{dT}{T} = 150 \times 2.5 \times \ln \frac{309.16}{275}$$

$$= 43.83 \text{ kJ/K}$$

(c) The total entropy change of the process

$$\Delta S_{total} = \Delta S_1 + \Delta S_2$$

$$= -26.28 \text{ kJ/K} + 43.83 \text{ kJ/K}$$

$$= 17.55 \text{ kJ/K}$$

5. State and explain the third law of thermodynamics.

   Solution: See the Section 5.11 (The Third Law of Thermodynamics)

6. (a) With the help of T–S and P–H diagrams, explain in detail the working principle of vapour compression cycle of refrigeration.

   (b) Define

   (i) Ton of refrigeration

   (ii) COP.

   Solution: See Section 8.1.5 (Vapour–Compression Refrigeration Cycle).

7. (a) Deduce Gibbs–Duhem equation.

   Solution: See Section 9.4 (Gibbs–Duhem Equation).

   (b) The excess Gibbs energy of a binary liquid mixture at constant temperature and pressure is given by

$$\frac{G^E}{RT} = (-2.6x_1 - 1.8x_2)x_1x_2$$

   Show that the expression satisfies the Gibbs–Duhem equation.

   Solution: See Example 9.20.

8. Under atmospheric conditions the acetone–chloroform azeotrope boils at 64.6°C and contains 33.5 mole per cent acetone. The vapour pressures of acetone and chloroform at this temperature are 995 and 855 mm Hg respectively. Calculate the composition of the vapour at this temperature in equilibrium with a liquid analyzing 11.1 mol per cent acetone. Use van Laar equation to solve this problem.

$$\ln \gamma_1 = \frac{Ax_2^2}{\left(\dfrac{A}{B}x_1 + x_2\right)^2} \quad \text{and} \quad \ln \gamma_2 = \frac{Bx_1^2}{\left(x_1 + \dfrac{B}{A}x_2\right)^2}$$

Solution: See Example 10.11.

9. (a) Describe the effect of temperature on the equilibrium constant.
   Solution: See Section 12.8 (Effect of Temperature on Equilibrium Constant—van't Hoff Equation).

   (b) Industrial CH_3OH is prepared according to the reaction

   $$CO(g) + 2H_2(g) = CH_3OH(g)$$

   Assuming the reaction mixture attains a state of equilibrium, estimate the degree of conversion at 500 K and 5 bar. If the reaction is satisfied with a stoichiometric mixture of CO and H_2, determine the composition of equilibrium mixture at 500 K and 5 bar. Data: $\Delta G^0 = 22.048$ kJ/mol.

   Solution: The reaction involved is

   $$CO(g) + 2H_2(g) = CH_3OH(g)$$

We are given that

$$\Delta G^0 = 22.048 \text{ kJ/mol} = 22048 \text{ J/mol}$$
$$T = 500 \text{ K}$$
$$P = 5 \text{ bar.}$$

Using the relation $\ln K_a = \left(-\dfrac{\Delta G^0}{RT}\right)$, we get

$$\ln K_a = \left(-\frac{22048}{8.314 \times 500}\right)$$

or

$$K_a = 4.973 \times 10^{-3}$$

Let ε_e be the degree of conversion at equilibrium.

Component	CO(g)	H$_2$(g)	CH$_3$OH(g)
Moles in feed	1	1	0
Moles at equilibrium	$1 - \varepsilon_e$	$2(1 - \varepsilon_e)$	ε_e
Mole fraction	$\dfrac{1 - \varepsilon_e}{3 - 2\varepsilon_e}$	$\dfrac{2(1 - \varepsilon_e)}{3 - 2\varepsilon_e}$	$\dfrac{\varepsilon_e}{3 - 2\varepsilon_e}$
Total moles at equilibrium	$1 - \varepsilon_e + 2(1 - \varepsilon_e) + \varepsilon_e = 3 - 2\varepsilon_e$		

Now, the sum of the stoichiometric coefficients $= \Sigma \nu_i = 1 - 1 - 2 = -2$
The pressure at which reaction maintained is $P = 5$ bar.

The equilibrium constant is given by

$$K_a = K_y P^{\Sigma \nu_i} = \dfrac{\dfrac{\varepsilon_e}{3 - 2\varepsilon_e}}{\dfrac{1 - \varepsilon_e}{3 - 2\varepsilon_e}\left(\dfrac{2(1 - \varepsilon_e)}{3 - 2\varepsilon_e}\right)^2}(5)^{-2} = 4.973 \times 10^{-3}$$

Solving for ε_e, we get $\varepsilon_e = 0.0506$. Hence, the degree of conversion at equilibrium
= 0.0506.

The equilibrium composition of the reaction mixture is given by

$$y_{CO} = \frac{1 - \varepsilon_e}{3 - 2\varepsilon_e} = 0.3275 \qquad y_{H_2} = \frac{2(1 - \varepsilon_e)}{3 - 2\varepsilon_e} = 0.6550$$

$$y_{CH_3OH} = \frac{\varepsilon_e}{3 - 2\varepsilon_e} = \frac{0.0506}{3 - 2 \times 0.0506} = 0.0175$$

Therefore, the required composition of the reaction mixture is $CH_3OH = 1.75\%$,
$CO = 32.75\%$, and $H_2 = 65.50\%$.

10. Estimate the standard free energy change and equilibrium constant at 700 K for the reaction

$$N_2(g) + 3H_2(g) = 2NH_3(g)$$

given that the standard heat of formation and standard free energy of formation of NH_3 at 298 K is $-46,100$ J/mol and $-16,500$ J/mol respectively. The specific heats (in J/mol-K) as functions of temperature (K) are as follows:

$$\begin{aligned}
C_P &= 27.27 + 4.93 \times 10^{-3}T && \text{for } N_2 \\
C_P &= 27.01 + 3.51 \times 10^{-3}T && \text{for } H_2 \\
C_P &= 29.75 + 25.11 \times 10^{-3}T && \text{for } NH_3
\end{aligned}$$

Solution: The standard heat of reaction and standard free energy of reaction at 298 K are estimated as

$$\Delta H^0 = -92,200 \text{ J/mol}$$
$$\Delta G^0 = -33,000 \text{ J/mol}.$$

For equation

$$\Delta H_T^0 = \Delta H' + \Delta\alpha T + \frac{\Delta\beta}{2}T^2 + \frac{\Delta\gamma}{3}T^3 \qquad \text{(i)}$$

$$\Delta\alpha = 2 \times 29.75 - 27.27 - 3 \times 27.01 = -48.8$$
$$\Delta\beta = (2 \times 25.11 - 4.93 - 3 \times 3.51) \times 10^{-3} = 34.76 \times 10^{-3}$$

Equation (i) gives

$$\begin{aligned}
-92,200 &= \Delta H' - 48.8T + 17.38 \times 10^{-3}T^2 \\
&= \Delta H' - 48.8 \times 298 + 17.38 \times 10^{-3} \times (298)^2 \\
&= \Delta H' - 1.3 \times 10^4
\end{aligned}$$

Therefore, $\Delta H' = -7.9201 \times 10^4$.

Again, we know that

$$\Delta G^0 = \Delta H' - \Delta\alpha T \ln T - \frac{\Delta\beta}{2}T^2 - \frac{\Delta\gamma}{2}T^3 - ART \qquad \text{(ii)}$$

where A is constant of integration.

Substituting the preceding values, we get

$$-33,000 = \Delta H' - \Delta\alpha T \ln T - \frac{\Delta\beta}{2}T^2 - ART$$

$$= -7.9201 \times 10^4 + 48.8 \times 298 \times \ln 298 - 17.38 \times 10^{-3} \times 298^2 - A \times 8.314 \times 298$$

$$= 2105 - 2477.57\, A$$

or

$$A = 14.169$$

Now, substituting $\Delta H'$ and A into Eq. (i) and Eq. (ii), we get

$$\ln K = \frac{79,201}{RT} - \frac{48.8}{R}\ln T + \frac{17.38 \times 10^{-3}}{R}T + 14.169 \qquad \text{(iii)}$$

$$\Delta G^0 = -79,201 + 48.8 \ln T - 17.38 \times 10^{-3}T^2 - 14.169RT \qquad \text{(iv)}$$

On putting $T = 70$ K in the preceding equations, Eq. (iii) gives $K = 1 \times 10^{-4}$ and Eq. (iii) gives $\Delta G^0 = 53,607$ J/mol.

Paper III

1. Define and explain triple point with the help of a P–T diagram.

   Solution: See Section 3.4.3 (*P–T Diagram*).

2. Derive the following relation:

$$\left(\frac{\partial C_V}{\partial V}\right)_T = T\left(\frac{\partial^2 P}{\partial T^2}\right)_V$$

   Solution: See Example 6.18.

3. What is an azeotropic solution? Explain the maximum boiling azeotrope with the help of a T–x–y diagram.

   Solution: See Section 10.10 (*Azeotrope Formation*).

4. Prove that chemical potentials of two phases in equilibrium are equal.

   Solution: See Section 9.2 (*Chemical Potential*).

5. A system was prepared by partially decomposing $CaCO_3$ into an evacuated space. What is the number of degrees of freedom (f) for the system?

   Solution: See Example 12.21.

6. (a) What are meant by reduced temperature and reduced pressure?

   Solution: See Chapter 6.

   (b) Prove that van der Waals constants (a, b) can be expressed in terms of critical temperature and pressure as follows:

$$a = \frac{27R^2T_c^2}{64P_c} \qquad b = \frac{RT_c}{8P_c}$$

   Solution: See Chapter 3.

   (c) 1 mol of an ideal gas is reversibly and adiabatically compressed from an initial volume of 30×10^{-3} m^3 to a final volume of 10×10^{-3} m^3. The initial temperature of the gas is 300 K. What are the final pressure and temperature? Calculate the work done. C_V of the gas is $\frac{5}{2}R$.

   Solution: We are given that

$$V_1 = 30 \times 10^{-3} \text{ m}^3$$
$$T_1 = 300 \text{ K}$$
$$V_2 = 10 \times 10^{-3} \text{ m}^3$$
$$T_2 = ?$$
$$C_V = \frac{5}{2}R$$

We know that

$$C_P - C_V = R$$

or

$$C_P - C_V = R = \frac{5}{2}R + R = \frac{7}{2}R$$

Therefore, the heat capacity ratio

$$\gamma = \frac{C_P}{C_V} = \frac{\frac{7}{2}R}{\frac{5}{2}R} = \frac{7}{5}$$

Since the gas is an ideal one, the ideal gas law $P_1 V_1 = nRT_1$ can be applied to calculate P_1.

$$P_1 = \frac{nRT_1}{V_1} = \frac{1 \times 8.314 \times 300}{30 \times 10^{-3}} = 83140 \text{ Pa}$$

Final pressure

$$P_2 = P_1 \left(\frac{V_1}{V_2} \right)^{\gamma} = 83140 \text{ Pa} \times \left(\frac{30}{10} \right)^{1.4} = 382061.323 \text{ Pa} = 387.06 \text{ kPa}$$

For adiabatic process

$$TV^{\gamma-1} = \text{Constant}$$

or

$$T_1 V_1^{\gamma-1} = T_2 V_2^{\gamma-1} = \text{Constant}$$

or

$$T_2 = T_1 \frac{V_1^{\gamma-1}}{V_2^{\gamma-1}} = T_1 \left(\frac{V_1}{V_2} \right)^{\gamma-1} = 300 \times \left(\frac{30}{10} \right)^{1.4-1} = 465.55 \text{ K}$$

Since the gas is compressed adiabatically, $dQ = 0$.
From first law, we get

$$dQ = dU + dW$$

or

$$dW = -dU + -C_V(T_2 - T_1) = C_V(T_1 - T_2)$$

$$= \frac{5}{2}R (300 - 465.55) = -3440.96 \text{ J}$$

Hence, work done = −3440.956 J (due to compression).

7. (a) Prove that the entropy of an isolated system either increases or remains constant, but can never decrease.

Solution: See Section 5.9 (Mathematical Statement of Second Law of Thermodynamics). The student may also go through the illustration in it.

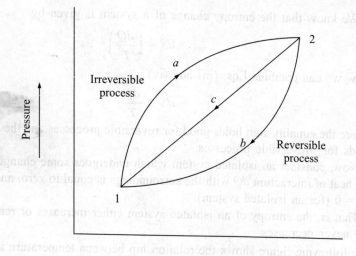

A reversible cycle and an irreversible cycle undergone by a system; (1–a–2–b–1) is reversible and (1–a–2–c–1) is irreversible.

Suppose a system is taken from the initial state 1 to the final system 2 along a reversible path a and then restored to the initial state through another reversible path b as shown in the figure.

Then 1–a–2–b–1 constitutes a reversible cycle. If the system is restored through the Clausius inequality to reversible cycle 1–a–2–b–1, and the irreversible cycle 1–a–2–c–1, gives

$$\oint_{1a2b1} \frac{dQ}{T} = \int_{1a2} \left(\frac{dQ}{T}\right)_R + \int_{2b1} \left(\frac{dQ}{T}\right)_R = 0 \tag{i}$$

$$\oint_{1a2c1} \frac{dQ}{T} = \int_{1a2} \left(\frac{dQ}{T}\right)_R + \int_{2c1} \left(\frac{dQ}{T}\right)_R < 0 \tag{ii}$$

Substituting Eq. (ii) into Eq. (i), we get

$$\int_{2b1} \left(\frac{dQ}{T}\right)_R - \int_{2c1} \frac{dQ}{T} > 0$$

or

$$\int_{2b1} \left(\frac{dQ}{T}\right)_R = \int_{2c1} \frac{dQ}{T}$$

or

$$dS > \frac{dQ}{T} \tag{iii}$$

Here, the subscript R represents the processes 1–a–2 and 2–b–1. Equation (iii) shows that the change in the entropy of a system is greater than dQ/T, where dQ denotes the heat interaction along the irreversible path.

We know that the entropy change of a system is given by

$$dS = \left(\frac{dQ}{T}\right) \tag{iv}$$

Now we can combine Eqs. (iii) and (iv)

$$dS \geq \frac{dQ}{T} \tag{v}$$

where the equality sign holds good for reversible processes and the inequality sign holds for irreversible processes.

Now, consider an isolated system which undergoes some change of state, then the heat of interaction dQ with the surroundings is equal to zero, and Eq. (v) gives $dS = 0$ (for an isolated system).

That is, the entropy of an isolated system either increases or remains constant, but never decreases.

(b) The following figure shows the relationship between temperature and entropy for a closed *PVT* system during a reversible process. Calculate the heat added to the system for each of the three steps 1–2, 2–3 and 3–1 and for the entire process.

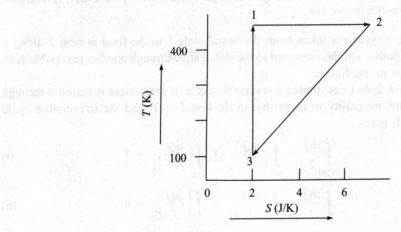

Solution: For process 3–1

$$dQ_{3-1} = T(S_2 - S_1) = 0$$

For process 1–2

$$dQ_{1-2} = \oint T dS = \text{Area of } \Delta 123$$

$$= \frac{1}{2} \times (6 - 2) \times (400 - 100) = 600 \text{ J}$$

For a reversible process, $\Delta Q = 0$, since

$$Q_{3-1} + Q_{1-2} + Q_{2-3} = 0$$

or

$$Q_{2-3} = -600 \text{ J}$$

(c) A vessel is divided into two parts and contains 2 mol N_2 gas at 80°C and 40 bar on one side and 3 mol of argon gas at 150°C and 15 bar on the other side. Calculate the change in entropy when the partition is removed and the gases are mixed adiabatically and completely. Assume both gases are ideal.

Solution: The change in entropy due to mixing is given by

$$\Delta S_{mixture} = -R \left(n_A \ln X_A + n_B \ln X_B \right)$$

$$= -R \left(n_A \ln \frac{p_A}{P} + n_B \ln \frac{p_B}{P} \right)$$

$$= -8.314 \left(\frac{2}{5} \ln \frac{40}{55} + \frac{3}{5} \ln \frac{15}{55} \right)$$

$$= 7.578 \text{ J/mol-K}$$

8. (a) With the help of a schematic diagram, describe the Linde liquefaction process.

Solution: See Section 8.2.2 (Joule–Thomson Expansion).

(b) With the help of a *T–S* diagram, describe the vapour compression refrigeration cycle. Derive the expression for COP.

Solution: See Section 8.1.5 (Vapour–Compression Refrigeration Cycle).

(c) Calculate the efficiency of an absorption refrigeration cycle if the temperature of the space to be refrigerated is −10°C, the temperature of the heat source is 100°C and the ambient temperature is 30°C.

Solution: We are given that
Source temperature = $T_1 = 100°C = 373$ K
Ambient temperature = $T_2 = 30°C = 303$ K
Sink temperature = $T_R = -10°C = 263$ K.
We know that

$$\text{COP}_{max} = \frac{T_R}{T_2 - T_R} \times \frac{T_1 - T_2}{T_1} = \frac{263}{303 - 263} \times \frac{373 - 303}{373}$$

$$= 1.234$$

9. (a) Show that in a binary solution if the solute obeys Henry's law, the solvent obeys the Lewis–Randall rule.

Solution: See Example 9.21.

(b) Estimate the fugacity of liquid acetone at 110°C and 275 bar. At 110°C the vapour pressure of acetone is 4.360 bar and the molar volume of saturated liquid acetone is 73 cm^3/mol.

Solution: See Example 9.15.

(c) The following experimental data are available for CO_2 gas at 92°C:

P_R	1	2	3	4	6	8	10
Z	0.856	0.583	0.535	0.620	0.800	0.975	1.160

Find the fugacity of CO_2 at 100 atm.

Data: For CO_2

$$T_c = 304.1 \text{ K}$$
$$P_c = 72.9 \text{ atm.}$$

Solution: See the solution of Question 3 of Solved Model Question Paper I.

10. (a) What is Poynting pressure correction factor? Discuss its application.

Solution: See Chapter 10.

(b) It is known that benzene (1) and toluene (2) form an ideal liquid solution. If a liquid mixture of benzene and toluene having $x_1 = 0.6$ is heated in a closed vessel at 760 torr, determine the temperature at which vaporization starts and composition of the vapour that forms. Calculate upto 3 iterations.

The Antoine constants for benzene and toluene are as follows:

	A	B	C
Benzene (1)	6.87987	1196.760	219.161
Toluene (2)	6.95087	1342.310	219.187

Solution: The vapour pressure of benzene and toluene are adequately represented by the Antoine equation

$$\log_{10} P^S = A - \frac{B}{T + C}$$

where P is in torr and T is in °C.

For ideal solution, the phase equilibrium relation is given by

$$y_i P = x_i P_i^S$$

For two components

$$(y_1 + y_2)P = x_1 P_1^S + x_2 P_2^S = x_1 P_1^S + (1 - x_1)P_2^S$$

or

$$P = x_1 P_1^S + x_2 P_2^S = P_2^S + (P_1^S - P_2^S)x_1$$

In this problem, the temperature is not given. So we find out the temperature by trial and error method.

At 95°C

$$P_1^S = 6.87987 - \frac{1196.760}{(95 + 219.161)} = 1176.21 \text{ torr}$$

$$P_2^S = 477.03 \text{ torr}$$

$$P = P_2^S + (P_1^S + P_2^S) \times X_1 \quad \text{(given that } X_1 = 0.6\text{)}$$

$$P = 477.03 + (1176.21 - 477.03) \times 0.6 = 896.538 \text{ torr}$$

which is greater than 760 torr.

So, we take temperature less than 95°C, i.e., 89.33°C.

Then

$$P_1^S = 1001.0804 \text{ torr}$$

$$P_2^S = 398.129 \text{ torr}$$
$$P = 398.129 + (1001.0804 - 398.129) \times 0.6$$
$$= 759.899 \approx 760 \text{ torr.}$$

So the temperature at which the vaporization starts is 89.33°C.
So

$$y_1 = \frac{x_1 P_1^S}{P} = \frac{0.6 \times 1001.0804}{760} = 0.7903$$

(c) Assuming ΔH^0 is constant in the temperature range 298.15 K to 800 K, estimate the equilibrium constant at 800 K for the following reaction:

$$SO_2(g) + \frac{1}{2} O_2(g) \rightarrow SO_3(g)$$

At 298.15 K, equilibrium constant K is 1.779×10^{12}.

Components	ΔH_f^0 (kJ/mol)
$SO_2(g)$	-297.263
$SO_3(g)$	-395.234

Solution: We know that

$$\Delta H^0 = \sum \Delta H_{f(\text{Products})}^0 - \sum \Delta H_{f(\text{Reactants})}^0$$
$$= -395.234 - (-297.263) = -97.971 \text{ kJ/mol}$$

We have

$$\ln \frac{K_{a2}}{K_{a1}} = \frac{\Delta H^0}{R} \left(\frac{1}{T_1} - \frac{1}{T_2} \right)$$

We are given that
Equilibrium constant at 298.15 K $(K_{a1}) = 1.779 \times 10^{12}$

$$\Delta H^0 = -97.971 \text{ kg/mol}$$
$$R = 8.314 \text{ k}^3/\text{kmol-K}$$

Substituting the values, we get

$$\ln \frac{K_{a2}}{1.779 \times 10^{12}} = \frac{-97.971}{8.314} \left(\frac{1}{298.15} - \frac{1}{800} \right)$$

or

$$\frac{K_{a2}}{1.779 \times 10^{12}} = 10^{-0.0247}$$

or

$$\frac{K_{a2}}{1.779 \times 10^{12}} = 0.9447$$
$$K_{a2} = 1.737 \times 10^{12}$$

Therefore, the equilibrium constant at 800 K is $K_{a2} = 1.737 \times 10^{12}$.

Paper IV

1. Discuss the procedure to find out fugacity using generalized virial coefficient correlation.

   Solution: See Section 6.9.3 (Fugacity Coefficient from Virial Equation of State).

2. What is the change in Gibbs free energy if N_A moles of an ideal gas A at temperature T and pressure P is mixed with N_B moles of an ideal gas B at the same temperature and pressure?

   Solution: See Example 9.17.

3. Show that Joule–Thomson coefficient (μ_{JT}) of an ideal gas is zero.

   Solution: See Example 6.20.

4. Derive the following expression:

$$C_P - C_V = TV\frac{\beta^2}{\alpha}$$

   where
   α = Isothermal compressibility
   β = Volume expansivity.

   Solution: See Example 6.10.

5. The vapour pressure of a substance can be estimated from the relation

$$\log_{10} P^{\text{Sat}} = A - \frac{B}{T + C}$$

   where P is in torr and T is in °C. The values of the constants A, B, and C for the substance are given in the following lines. Calculate the acentric factor.

	A	B	C	T_c	P_c
Benzene	6.87987	1196.760	219.161	562.1 K	49.24 bar

   Solution: We know that

$$T_R = \frac{T}{T_c}$$

   Using this relation, we get

$$T = 0.7 \times 562.1 = 393.47 \text{ K} = 120.47°C$$

   Substituting the value of T in the following expression

$$\log_{10} P^{\text{Sat}} = A - \frac{B}{T + C} = 6.87987 - \frac{1196.76}{120.47 + 219.161} = 3.3562$$

   we get

$$P^{\text{Sat}} = 2270.91 \text{ torr}$$

Again, we know that

$$P_R^{Sat} = \frac{P^{Sat}}{P_c} = \frac{2270.91}{750.061 \times 49.24} = 0.06149$$

$$49.24 \text{ bar} = 49.24 \times 750.061 \text{ torr}$$

Now using the expression for acentric factor

$$\omega = -1.0 - \log(P^{Sat})$$

or

$$\omega = -1.0 - \log(0.0614)$$

or

$$\omega = 0.2112 \approx 0.2$$

Thus, the required acentric factor = 0.2.

6. (a) Show that for steady flow process, $\Delta H = Q - W_S$.

   Solution: See Section 2.18 (First Law of Thermodynamics to Flow Processes).

   (b) One mole of air, initially at 423.15 K (150°C) and 8 bar, undergoes the following mechanically reversible changes: It expands isothermally to a pressure such that when it is cooled at constant volume to 323.15 K (50°C), its final pressure is 3 bar. Assuming air to be an ideal gas for which $C_P = 7/2R$ and $C_V = 5/2R$, calculate W, Q, ΔU and ΔH.

   Solution: We are given that

   $$T_1 = 423.15 \text{ K}$$
   $$P_1 = 8 \text{ bar}$$
   $$P_2 = 3 \text{ bar.}$$

Step 1: Isothermal Expansion ($\Delta U = 0$, $\Delta H = 0$)

The first law of thermodynamics gives

$$\Delta Q = \Delta W = RT \ln\frac{P_1}{P_2} = 8.314 \times 423.15 \times \ln\frac{8}{3} = 3449.74 \text{ J}$$

Step 2: Constant-Volume Process

$$\Delta Q = \Delta U = C_V(T_2 - T_1) = \frac{5}{2}R(423.15 - 323.15) = 2078.5 \text{ J}$$

$$\Delta H = C_P(T_2 - T_1) = \frac{7}{2}R(423.15 - 323.15) = 2909.9 \text{ J}$$

For the overall process

$$\Delta U = 2078.5 \text{ J} \qquad \Delta H = 2909.9 \text{ J} \qquad \Delta W = 3449.74 \text{ J}$$
$$\Delta Q = 3449.74 \text{ J} + 2078.5 \text{ J} = 5528.24 \text{ J}$$

7. (a) Derive Clausius–Clapeyron equation in the study of phase change of a pure substance.

   Solution: See Section 6.4.3 (Clapeyron Equation).

   (b) With the help of a neat sketch, explain the working principle of vapour absorption refrigeration system and derive the expression for COP of this system.

   Solution: See Section 8.1.6 (Absorption Refrigeration Cycle).

   (c) What is ton of refrigeration? Define and mention its different units.

   Solution: See Section 8.1.3 (Capacity of Refrigerator).

8. (a) A gas obeys the equation of state $P(V - b) = RT$. For this gas, $b = 0.0391$ L/ mol. Calculate the fugacity and fugacity coefficient for the gas at 1000°C and 1000 atm.

   Solution: We know that

   $$\ln\left(\frac{f}{P}\right) = \int_0^P (V_{real} - V_{id})dP \tag{i}$$

   Here
   for real gas

   $$V_{real} = \frac{RT}{P} + b = \frac{RT}{P} + 0.0391$$

   and
   for ideal gas

   $$V_{id} = \frac{RT}{P}$$

   Now, substituting $V_{id} = \frac{RT}{P}$ into Eq. (i), we obtain

   $$RT \ln\left(\frac{f}{P}\right) = \int_0^P (V_{real} - V_{id})dP = \left[\int_0^P \left(\frac{RT}{P} + 0.0391\right) - \left(\frac{RT}{P}\right)\right]dP$$

   or

   $$RT \ln\left(\frac{f}{P}\right) = 0.0391 \int_0^P dP = 0.0391 [P]_0^{1000} = 39.1$$

   or

   $$\ln\left(\frac{f}{P}\right) = \frac{39.1}{0.082 \times 1273} = 0.374$$

   or

   $$\phi = \frac{f}{P} = e^{0.374} = 1.453 \text{ atm}$$

   or

   $$f = P \times 1.453 = 1000 \times 1.453 = 1.453 \text{ atm}$$

   Therefore, the fugacity and fugacity coefficient of the gas are 1453 atm and 1.453 atm respectively.

(b) For a binary liquid mixture at constant temperature and pressure, the excess Gibbs energy is given by

$$\frac{G^E}{RT} = (-2.6x_1 - 1.8x_2)x_1x_2$$

Show that this expression satisfies the Gibbs–Duhem equation.

Solution: See Example 9.20.

9. (a) What is the basis for the group contribution methods in the prediction of activity coefficients? What are UNIQUAC and UNIFAC methods?

Solution: See Section 10.11.6 (UNIFAC Method).

(b) Under atmospheric condition the acetone–chloroform azeotrope boils at 64.6°C and contains 33.5 mol percent acetone. The vapour pressures of acetone and chloroform at this temperature are 995 and 855 mm Hg respectively. Calculate the composition of the vapour at this temperature in equilibrium with a liquid analyzing 11.1 mole percent acetone, given the van Laar equation

$$\ln \gamma_1 = \frac{Ax_2^2}{\left(\dfrac{A}{B}x_1 + x_2\right)^2} \qquad \text{and} \qquad \ln \gamma_2 = \frac{Bx_1^2}{\left(x_1 + \dfrac{B}{A}x_2\right)^2}$$

Solution: See Example 10.11.

10. (a) Define the fuel cell. Give the basic construction of the cell.

Solution: See Section 12.16 (Fuel Cell).

(b) In a laboratory investigation, ethanol is esterified to produce ethyl acetate and water at 100°C and 1 atm pressure according to the following equation:

$$CH_3COOH(l) + C_2H_5OH(l) = CH_3COOC_2H_5(l) + H_2O(l).$$

What is the equilibrium constant for the reaction at 100°C? What is the composition of the reaction mixture if initially one mol each of acetic acid and ethanol are present? The data given for the reaction are $\Delta G_{298}^0 = 1160$ cal and $\Delta H_{298}^0 = 1713$ cal.

Solution: We are given that
    standard heat of reaction at 298 K $= \Delta H_{298}^0 = 1160$ cal $= 4853$ J
and
    standard Gibbs free energy change for the reaction at 298 K

$$\Delta G_{298}^0 = 1713 \text{ cal} = 7167.19 \text{ J}$$

The equilibrium constant K at 298 K is given by

$$\ln K_{298} = -\frac{\Delta G^0}{RT} = -\frac{-4853.44}{8.314 \times 298} = -1.959$$

or

$$K_{298} = e^{-1.959} = 0.141$$

Now, the equilibrium constant at 373 K, i.e., K_{373} can be determined as

$$\ln\left(\frac{K_{373}}{K_{298}}\right) = \frac{-7167.19}{8.314}\left(\frac{1}{298} - \frac{1}{373}\right) = -0.581$$

or

$$\frac{K_{373}}{K_{298}} = e^{-0.581} = 0.559$$

or

$$K_{373} = 0.141 \times 0.559$$

or

$$K_{373} = 0.0788$$

Hence, the equilibrium constant for the reaction at 373 K is 0.0788.
Let ε_e be the degree of conversion at equilibrium.

Component	$CH_3COOH(l)$	$C_2H_5OH(l)$	$CH_3COOC_2H_5(l)$	$H_2O(l)$
Moles in feed	1	1	0	0
Moles at equilibrium	$1 - \varepsilon_e$	$1 - \varepsilon_e$	ε_e	ε_e
Mole fraction	$\dfrac{1 - \varepsilon_e}{2}$	$\dfrac{1 - \varepsilon_e}{2}$	$\dfrac{\varepsilon_e}{2}$	$\dfrac{\varepsilon_e}{2}$
Total moles at equilibrium	$(1 - \varepsilon_e) + (1 - \varepsilon_e) + \varepsilon_e + \varepsilon_e = 2$			

Now, the sum of the stoichiometric coefficients

$$\Sigma \nu_i = 1 + 1 - 1 - 1 = 0$$

$$K = K_x = \Pi(x_i)^{\nu_i} = \frac{\left(\dfrac{\varepsilon_e}{2}\right)\left(\dfrac{\varepsilon_e}{2}\right)}{\left(\dfrac{1-\varepsilon_e}{2}\right)\left(\dfrac{1-\varepsilon_e}{2}\right)} = \left(\frac{\varepsilon_e}{1-\varepsilon_e}\right)^2$$

Therefore, $\left(\dfrac{\varepsilon_e}{1-\varepsilon_e}\right)^2 = 0.0788$.

Solving for ε_e, we have $\varepsilon_e = 0.2186$. Hence, the degree of conversion at equilibrium = 0.2186.

The composition of the reaction mixture can be computed as

$$y_{CH_3COOH} = y_{C_2H_5OH} = \frac{1-\varepsilon_e}{2} = \frac{1-0.2186}{2} = 0.3907$$

$$y_{CH_3COOC_2H_5} = y_{H_2O} = \frac{\varepsilon_e}{2} = \frac{0.2186}{2} = 0.1093$$

(c)   A system consisting of 2 mol methane and 3 mol water is undergoing the following reaction:

$$CH_4 + H_2O \rightarrow CO + 3H_2 \tag{A}$$
$$CH_4 + 2H_2O \rightarrow CO_2 + 4H_2 \tag{B}$$

Derive expressions for mole fractions in terms of the extent of reaction.

Solution: Let

ε_1 = Degree of conversion of H_2O in reaction (A)

ε_2 = Degree of conversion of H_2O in reaction (B).

Component	CH_4	H_2O	CO	CO_2	H_2
Moles in feed	2	3	0	0	0
Moles at equilibrium	$2 - \varepsilon_1 - \varepsilon_2$	$3 - \varepsilon_1 - 2\varepsilon_2$	ε_1	ε_2	$3\varepsilon_1 + 4\varepsilon_2$
Mole fraction	$\dfrac{2 - \varepsilon_1 - \varepsilon_2}{5 + 2\varepsilon_1 + 2\varepsilon_2}$	$\dfrac{3 - \varepsilon_1 - 2\varepsilon_2}{5 + 2\varepsilon_1 + 2\varepsilon_2}$	$\dfrac{\varepsilon_1}{5 + 2\varepsilon_1 + 2\varepsilon_2}$	$\dfrac{\varepsilon_2}{5 + 2\varepsilon_1 + 2\varepsilon_2}$	$\dfrac{3\varepsilon_1 + 4\varepsilon_2}{5 + 2\varepsilon_1 + 2\varepsilon_2}$

Total moles at $2 - \varepsilon_1 - \varepsilon_2 + 3 - \varepsilon_1 - 2\varepsilon_2 + \varepsilon_1 + \varepsilon_2 + 3\varepsilon_1 + 4\varepsilon_2 = 5 + 2\varepsilon_1 + 2\varepsilon_2$
equilibrium

Paper V

1. Prove that van der Waals constants (a, b) can be expressed in terms of critical temperature and pressure as follows:

$$a = \frac{27R^2T_c^2}{64P_c} \qquad b = \frac{RT_c}{8P_c}$$

   Solution: See Chapter 3.

2. Derive Maxwell's relations.

   Solution: See Chapter 6.

3. Justify the following statement with illustration:
   'Violation of Kelvin–Planck statement leads to the violation of Clausius statement'.

   Solution: See 'Equivalence of Kelvin–Planck and Clausius Statements' in Chapter 5.

4. Show that

$$\left(\frac{\partial U}{\partial V}\right)_T = \frac{T\beta}{\kappa} - P$$

   where
   β = Coefficient of volume expansion
   κ = Isothermal compressibility.

   Solution: The first thermodynamic relation is given by

$$dU = TdS - PdV$$

   Differentiating it with respect to V at constant T, we get

$$\left(\frac{\partial U}{\partial V}\right)_T = T\left(\frac{\partial S}{\partial V}\right)_T - P = T\left(\frac{\partial P}{\partial T}\right)_V - P \qquad \text{[using Maxwell's relation)}$$

   From the cyclic relation among the three variables P, V and T, we have

$$\left(\frac{\partial P}{\partial V}\right)_T \left(\frac{\partial V}{\partial T}\right)_P \left(\frac{\partial T}{\partial P}\right)_V = -1$$

   or

$$\left(\frac{\partial P}{\partial T}\right)_V = -\left(\frac{\partial V}{\partial T}\right)_P \left(\frac{\partial P}{\partial V}\right)_T$$

   Substituting this relation, we have

$$\left(\frac{\partial U}{\partial V}\right)_T = \left(\frac{\partial P}{\partial T}\right)_V - P = -T\left(\frac{\partial V}{\partial T}\right)_P \left(\frac{\partial P}{\partial V}\right)_T - P = \frac{T\beta}{\alpha} - P$$

$$\left(\text{since } \alpha = \frac{1}{V}\left(\frac{\partial V}{\partial P}\right)_T \text{ and } \beta = \frac{1}{V}\left(\frac{\partial V}{\partial T}\right)_P\right).$$

5. A spherical balloon of 1 m diameter contains a gas at 120 kPa. The gas inside the balloon is heated until the pressure reaches 360 kPa. During heating the pressure of the gas inside the balloon is proportional to the cube of the diameter of the balloon. Determine the work done by the gas inside the balloon.

Solution: The pressure (P) of the gas inside the balloon is proportional to the cube of the diameter of the balloon, i.e.,

$$P = kD^3$$

where k is the proportionality constant.
Now, work done by the gas inside the balloon can be estimated as

$$W = \int PdV = \int kD^3 \cdot d\left(\frac{4}{3}\pi r^3\right) = \int kD^3 \cdot d\left(\frac{4}{3}\pi \frac{D^3}{8}\right)$$

$$= \frac{\pi k}{2} \int_{D_1}^{D_2} D^5 dD = \frac{\pi k}{12}(D_2^6 - D_1^6)$$

We are given that

$$P_1 = kD_1^3$$

or

$$k = \frac{P_1}{D_1^3} = \frac{120 \times 10^3}{1} = 120 \times 10^3$$

The final pressure

$$P_2 = kD_2^3$$

or

$$360 \times 10^3 = 120 \times 10^3 \times D_2^3$$

where $D_2 = 1.442$ m = Diameter of the balloon at the final pressure.
Hence, work done by the gas would be

$$W = \frac{120 \times 10^3}{6} \times \frac{\pi}{2}(1.4426 - 1) = 251013.25 \text{ J} = 251.01 \text{ kJ}$$

6. (a)   The heat capacity at 1 atm pressure of solid magnesium in the temperature range of 0 to 560°C is given by the expression

$$C_P = 6.2 + 1.33 \times 10^{-3}T + 6.78 \times 10^4 T^{-2} \text{ cal/deg-g-atom}$$

Determine the increase of entropy, per g-atom, for an increase of temperature from 300 K to 800 K at 1 atm pressure.

Solution: Given that the expression for heat capacity (C_P) for the temperature range 0 to 560°C (273 to 833 K) is $C_P = 6.2 + 1.33 \times 10^{-3}T + 6.78 \times 10^{-4}T^2$ cal/ deg-g atom.
This expression is also applicable for temperature range $T_1 = 300$ K to $T_2 = 800$ K.

This is a constant-pressure system ($P = 1$ atm)

Thus, using $\Delta S = \int\limits_{T_1}^{T_2} \dfrac{C_P dT}{T}$, we get

$$\Delta S = \int\limits_{300}^{800} \dfrac{6.2 + 1.33 \times 10^{-3} T + 6.78 \times 10^{-4} T^2}{T} dT$$

or

$$\Delta S = \int\limits_{300}^{800} \dfrac{6.2}{T} dT + \int\limits_{300}^{800} 1.33 \times 10^{-3} dT + \int\limits_{300}^{800} 6.78 \times 10^{-4} T \, dT$$

or

$$\Delta S = 6.2 \left[\ln T\right]_{300}^{800} + 1.33 \times 10^{-3} [T]_{300}^{800} + \dfrac{6.78 \times 10^{-4}}{2} [T^2]_{300}^{800}$$

or

$$\Delta S = 6.2 \ln\left(\dfrac{800}{300}\right) + 1.33 \times 10^{-3} (800 - 300) + \dfrac{6.78 \times 10^{-4}}{2} (640000 - 90000)$$

or

$$\Delta S = 6.0811 + 0.665 + 186.45$$

or

$$\Delta S = 193.196$$

Hence the required increase in entropy is equal to 193.196 cal/k per g-atom for the given system.

(b) A steam turbine developing 34 kW receives steam at 15 bar with an internal energy of 2720 kJ/kg, specific volume of 0.17 m³/kg and velocity of 110 m/s. Steam is exhausted from turbine at 0.1 bar with internal energy 2177 kJ/kg, specific volume 15 m³/kg and velocity 320 m/s. The heat loss over the surface of the turbine is 20 kJ/kg. Neglecting change in potential energy, determine

(i) Work done per kg of steam
(ii) Steam flow through the turbine in kg/h.

Solution: The steady flow energy equation for the control volume is

$$u_1 + P_1 v_1 + \dfrac{V_1^2}{2} + gZ_1 + \dfrac{\partial q}{\partial m} = u_2 + P_2 v_2 + \dfrac{V_2^2}{2} + gZ_2 + \dfrac{\partial W_x}{\partial m}$$

$$\dfrac{\partial W_x}{\partial m} = \dfrac{\partial q}{\partial m} + (P_1 v_1 - P_2 v_2) + (u_1 - u_2) + \dfrac{1}{2}(V_1^2 - V_2^2) + g(Z_1 - Z_2)$$

According to the problem

$$g(Z_1 - Z_2) = 0$$

We are given that

$u_1 = 2720$ kJ/kg (heat loss)

$P_1 = 15$ bar $= 15 \times 10^2$ kPa

$v_1 = 0.17$ m^3/kg

$P_2 = 0.1$ bar $= 0.1 \times 10^2$ kPa

$V_1 = 110$ m/s

$u_2 = 2177$ kJ/kg

$v_2 = 15$ m^3/kg

$V_2 = 320$ m/s

$\dfrac{\partial q}{\partial m} = -20$ kJ/kg.

$$\frac{\partial W_x}{\partial m} = \frac{\partial q}{\partial m} + (P_1 v_1 - P_2 v_2) + (u_1 - u_2) + \frac{1}{2}(V_1^2 - V_2^2)$$

Putting the value in the equation, we get

$$\frac{\partial W_x}{\partial m} = -20 + (2720 - 2177) + (15 \times 10^2 \times 0.17 - 0.1 \times 10^2 \times 15)$$

$$+ \frac{1}{2}(110^2 - 320^2) \times 10^{-3}$$

$$= 582.85 \text{ kJ/kg}$$

Power output

$$= \frac{\partial W_x}{\partial m} \times \text{Steam rate}$$

34 kW $= 582.85$ kJ/kg $\times$ Steam rate

Steam rate $= 0.05833$ kg/s $= 209.988$ kg/hr

7. (a) State and prove Clausius inequality theorem.

   Solution: See Section 5.14 (Clausius Inequality).

   (b) Nitrogen is compressed from an initial state of 1 bar and 25°C to a final state of 5 bar and 25°C by three different reversible processes in a closed system.

   (i) Heating at constant volume followed by cooling at constant pressure;
   (ii) Isothermal expansion;
   (iii) Adiabatic compression followed by cooling at constant volume.

   Nitrogen is assumed to behave an ideal gas with $C_V = \dfrac{5}{2}R$, $C_P = \dfrac{7}{2}R$, and the molar volume of nitrogen $= 0.02479$ m^3/mol. Estimate the heat, work requirement, ΔU and ΔH of nitrogen for the each process.

   Solution: Basis: 1 mol of nitrogen gas
   Initial state:

   $$P_1 = 1 \text{ bar} \qquad T_1 = 298 \text{ K} \qquad V_1 = 0.02479 \text{ m}^3/\text{mol}$$

Final state:

$$P_2 = 5 \text{ bar} \qquad T_2 = 298 \text{ K} \qquad V_2 = ?$$

$$C_P = \frac{7}{2}R = \frac{7}{2} \times 8.314 = 29.099 \text{ J/mol-K}$$

and

$$C_V = \frac{5}{2} \times 8.314 = 20.785 \text{ J/mol-K}$$

The final volume of the gas can be calculated as

$$V_2 = \frac{P_1 V_1}{P_2} = \frac{1 \times 0.02479}{5} = 0.004958 \text{ m}^3/\text{mol}$$

(i) In the first step of the first process, the air is heated at a constant volume equal to its initial volume until the final pressure of 5 bar is reached. Since it is a constant-volume process, i.e., $V_1 = V_2 = V$, then

$$T_2 = \frac{T_1 P_2}{P_1} = \frac{298 \times 5}{1} = 1490.75 \text{ K}$$

$$Q_V = \Delta U_V = C_V \Delta T = C_V(T_2 - T_1) = 20.781(1490.75 - 298) = 24,788 \text{ J/mol}$$

In the second step, the air is cooled at $P = 5$ bar to its final state. Since the process is a constant-pressure process, then

$$Q_P = \Delta H_P = C_P \Delta T = C_P(T_2 - T_1) = 29.10(298 - 1490.75) \text{ K} = -34,703 \text{ J}$$

ΔU_P Change in internal energy, is given by

$$\begin{aligned}
\Delta U &= \Delta H - \Delta(PV) \\
&= \Delta H - P\Delta V \\
&= \Delta H - P(V_2 - V_1) \\
&= -34703 \text{ J} - 5 \times 10^5 \, (0.004958 - 0.02479) \text{ N/m}^2 \text{xm}^3 \\
&= -24788 \text{ J}
\end{aligned}$$

Since the process comprises two steps, therefore the actual Q and ΔH can be obtained by summing up of respective values for two different steps.

$$Q = Q_P + Q_V = 24788 \text{ J} - 34703 \text{ J} = -9915 \text{ J}$$
$$\Delta U = 24788 \text{ J} - 24788 \text{ J} = 0$$

Hence

$$W = Q - \Delta U = -9915 \text{ J}$$

(ii) For isothermal compression of an ideal gas

$$Q = W = nRT \ln \frac{P_1}{P_2} = 1 \times 8.314 \times 298 \times \ln \frac{1}{5} = 3990 \text{ J}$$

(iii) This step comprises two processes—adiabatic compression and cooling at constant volume.

For the first process, $Q = 0$.

Work of compression can be calculated by the equation

$$W = C_V(T_2 - T_1)$$

Here
$$T_1 = 298 \text{ K}$$
$$T_2 = ?$$
$$V_1 = 0.02479 \text{ m}^3/\text{mol}$$
$$V_2 = 0.004958 \text{ m}^3/\text{mol}.$$

$$T_2 = T_1 \left(\frac{V_1}{V_2} \right)^{\nu-1} = 298 \left(\frac{0.02479}{0.004958} \right)^{1.4-1} = 567.67 \text{ K}$$

So

$$W = C_V(T_2 - T_1) = 20.785(298 - 567.67) = -5600 \text{ J}$$

Therefore

$$\Delta U = -W = -(-5600 \text{ J}) = 5600 \text{ J}$$

For the second process (constant-volume process)
(a) Q_V (Heat requirement) is given by

$$Q_V = \Delta U_V = C_V \Delta T = C_V(T_2 - T_1) = 20.785 (298 - 567.67) = -5600 \text{ J}$$
$$W = 5600 \text{ J}$$

8. (a) Two iron blocks of same size and at distinct temperatures T_1 and T_2 are brought in thermal contact with each other. The transfer process is allowed to take place until the thermal equilibrium is attained. Suppose, after the attainment of equilibrium, that the blocks are at final temperature T. Show that the change in entropy of the process is given by

$$\Delta S = C_P \ln \left[\frac{(T_2 - T_1)^2}{4T_1 T_2} + 1 \right]$$

Solution: See Example 5.23.

(b) Find the second and third virial coefficients of the van der Waals equation of state.

Solution: See Example 3.7.

(c) The total energy of a typical closed system is given by $E = 50 + 25T + 0.05T^2$ in Joules. The amount of heat absorbed by the system can be expressed as $Q = 4000 + 10T$

in Joules. Estimate the work done during the processes in which temperature rises from 400 K to 800 K.

Solution: See Example 2.1.

9. (a) Derive the relation between standard Gibbs energy change ΔG^0 and equilibrium constant K.
Solution: See Section 12.7 (Relation between Equilibrium Constant and Standard Gibbs Free Energy Change).

(b) The water gas shift reaction

$$CO(g) + H_2O(g) \rightarrow CO_2 + H_2$$

is carried out under the following conditions:

The reactants are 2 mol of H_2O and 1 mol of CO. The temperature is 1100 K and pressure is 1 bar. Data given: ln $K = 0$. Find the equilibrium composition of the components.

Solution: Given the reacting system

$$CO(g) + H_2O(g) \rightarrow CO_2 + H_2$$

Number of moles of H_2O = 2
Number of moles of CO = 1
Temperature (T) = 1100 K
Pressure (P) = 1 bar
We are given that ln $K = 0$, which implies that $K = 1$.
Let the degree of conversion be ε.
For this reaction

$$\Sigma V_i = 1 + 1 - 1 - 1 = 0$$

and mole fractions at equilibrium are

$$y_{CO} = \frac{1 - \varepsilon}{3} \qquad y_{H_2O} = \frac{2 - \varepsilon}{3} \qquad y_{CO_2} = \frac{\varepsilon}{3} \qquad y_{H_2} = \frac{\varepsilon}{3}$$

We know for this system

$$\frac{y_{H_2} \cdot y_{CO_2}}{y_{CO} \cdot y_{H_2O}} = 1$$

or

$$\frac{\left(\dfrac{\varepsilon}{3}\right)\left(\dfrac{\varepsilon}{3}\right)}{\left(\dfrac{1 - \varepsilon}{3}\right)\left(\dfrac{2 - \varepsilon}{3}\right)} = 1$$

or

$$\frac{\varepsilon^2}{(1 - \varepsilon)(2 - \varepsilon)} = 1$$

or

$$\varepsilon = 0.667$$

Thus equilibrium composition is

$$y_{CO} = \frac{1 - \varepsilon}{3} = \frac{1 - 0.667}{3} = 0.111 \qquad y_{H_2O} = \frac{2 - \varepsilon}{3} = \frac{2 - 0.667}{3} = 0.444$$

$$y_{CO_2} = \frac{\varepsilon}{3} = \frac{0.667}{3} = 0.222 \qquad y_{H_2} = \frac{\varepsilon}{3} = \frac{0.667}{3} = 0.222$$

Thus the required composition is CO = 11.1%, H_2O = 44.4%, CO_2 = 22.2%, and H_2 = 22.2%.

10. **(a)** The following equations have been proposed to represent activity coefficient data for a system at fixed T and P.

$$\ln \gamma_1 = x_2^2(0.5 + 2x_1)$$
$$\ln \gamma_2 = x_1^2(1.5 - 2x_2)$$

Do these equations satisfy Gibbs–Duhem equation? Determine an expression for $\dfrac{G^E}{RT}$ for the system.

Solution: Checking can be done by the following ways:

Method 1

$$\ln \gamma_1 = x_2^2(0.5 + 2x_1)$$

or

$$\ln \gamma_1 = 0.5x_2^2 + 2x_1x_2^2$$

or

$$\ln \gamma_1 = 2.5x_2^2 - 2x_2^3 \qquad \text{(since } x_1 + x_2 = 1 \text{ or } x_1 = 1 - x_2)$$

or

$$d \ln \gamma_1 = (5x_2 - 6x_2^2)dx_2 \qquad\qquad\qquad\qquad \text{(i)}$$

Similarly

$$\ln \gamma_2 = x_1^2(1.5 - 2x_2)$$

or

$$\ln \gamma_2 = -0.5x_1^2 + 2x_1^3$$

or

$$d \ln \gamma_2 = (-x_1 + 6x_1^2)dx_1 \qquad\qquad\qquad\qquad \text{(ii)}$$

From Gibbs–Duhem equation, we get

$$x_1 d \ln \gamma_1 + x_2 d \ln \gamma_2 = 0$$

Thus, by Eqs. (i) and (ii), we have

$$x_1 d \ln \gamma_1 + x_2 d \ln \gamma_2 = x_1(5x_2 - 6x_2^2)dx_2 + x_2(-x_1 + 6x_1^2)dx_1$$

or

$$x_1 d \ln \gamma_1 + x_2 d \ln \gamma_2 = 5x_1x_2dx_2 - 6x_1x_2^2dx_2 + x_2x_1dx_1 + 6x_1^2x_2dx_1$$
$$= 5x_1x_2dx_2 - x_2x_1dx_1 + 6x_1x_2^2dx_1 + 6x_1^2x_2dx_1$$

$$= (5x_1x_2dx_2 - x_2x_1dx_1) + 6x_1x_2dx_1 \, (x_1 + x_2) \qquad \text{(as } x_1 + x_2 = 1)$$
$$= (5x_1x_2dx_2 - x_2x_1dx_1) + 6x_1x_2dx_1$$
$$= (5x_1x_2dx_2 - 5x_2x_1dx_1) = (5x_1x_2dx_2 - 5x_2x_1dx_1) \qquad \text{(as } dx_1 = -dx_2)$$
$$= 0$$

Hence the Gibbs–Duhem equation is satisfied.

Method 2

The Gibbs–Duhem equation can also be satisfied if

$$\int_0^1 \ln \frac{\gamma_2}{\gamma_1} dx_1 = 0$$

We get

$$\ln \gamma_1 = x_2^2(0.5 + 2x_1) \quad \text{and} \quad \ln \gamma_2 = x_1^2(1.5 - 2x_2)$$

In terms of x_1

$$\ln \gamma_2 - \ln \gamma_1 = -0.5 - x_1 + 3x_1^2$$

Substituting the values, we obtain

$$\int_0^1 (-0.5 - x_1 + 3x_1^2) dx_1 = -0.5(1 - 0) - \frac{1}{2}(1 - 0) + \frac{3}{3}(1 - 0) = 0$$

Therefore, the Gibbs–Duhem equation is satisfied by this method too.

Expression for $\dfrac{G^E}{RT}$

$$\frac{G^E}{RT} = x_1 \ln \gamma_1 + x_2 \ln \gamma_2$$
$$= x_1(0.5x_2^2 + 2x_1x_2^2) + x_2(1.5x_1^2 - 2x_1^2x_2)$$
$$= (0.5x_1x_2^2 + 2x_1^2x_2^2) + (1.5x_1^2x_2 - 2x_1^2x_2^2)$$
$$= x_1x_2(0.5x_2 + 1.5x_1)$$

(b) Express volumes (cm^{-3} mol^{-1}) for the ethanol (1)–methyl butyl ether (2) system at 25°C are given by the equation

$$V^E = x_1x_2[-1.026 - 0.220(x_1 - x_2)]$$

If $V_1 = 58.63$ and $V_2 = 118.46$ cm^{-3} mol^{-1}, what volume of mixture is formed when 1000 m^3 each of pure species 1 and 2 are mixed at 25°C?

Solution: We are given that

$$V_1 = 58.63 \quad \text{and} \quad V_2 = 118.46 \text{ cm}^{-3} \text{ mol}^{-1}$$

$$1000 \text{ cm}^3 \text{ of species } 1 = \frac{1000}{58.63} = 17.05$$

and

$$1000 \text{ cm}^3 \text{ of species } 2 = \frac{1000}{118.46} = 8.441$$

The mole fraction of species 1 is

$$x_1 = \frac{17.05}{17.05 + 8.44} = 0.668$$

Therefore

$$x_2 = 1 - 0.668 = 0.331$$

The volume change of mixing can be expressed as

$$V^E = x_1 x_2 [-1.026 - 0.220(x_1 - x_2)] + x_1(1 - x_1)[-1.026 - 0.220(2x_1 - 1)]$$

Differentiating the equation with respect to x_1, we get

$$\frac{dV^E}{dx_1} = (1 - 2x_1)[-1.026 - 0.220(2x_1 - 1)] + (x_1 - x_1^2)[-0.220 \times 2]$$

When $x_1 = 0.668$

$$V^E = x_1(1 - x_1)[-1.026 - 0.220(2x_1 - 1)] = -0.243 \ \text{cm}^3/\text{mol}$$

and

$$\frac{dV^E}{dx_1} = (1 - 2x_1)[-1.026 - 0.220(2x_1 - 1)] + (x_1 - x_1^2)(-0.220 \times 2)$$

$$= 0.272 \ \text{cm}^3/\text{mol}$$

Now

$$\overline{V}_1 - V_1 = V^E + x_2 \frac{dV^E}{dx_1} = -0.243 + (0.331 \times 0.272) = -0.152 \ \text{cm}^3/\text{mol}$$

$$\overline{V}_2 - V_2 = V^E - x_1 \frac{dV^E}{dx_1} = -0.243 - (0.668 \times 0.272) = 0.424 \ \text{cm}^3/\text{mol}$$

Thus, at $x_1 = 0.668$

$$\overline{V}_1 = V_1 - 0.152 = 58.63 - 0.152 = 58.478 \ \text{cm}^3/\text{mol}$$

and

$$\overline{V}_2 = V_2 - 0.424 = 118.46 - 0.424 = 118.036 \ \text{cm}^3/\text{mol}$$

The total volume of the solution is

$$nV = n_1 \overline{V}_1 + n_2 \overline{V}_2 = (17.05 \times 58.478) + (8.441 \times 118.036) = 1993.38 \ \text{cm}^3$$

Hence, the volume of the mixture formed is 1993.38 cm^3.

Paper VI

1. The pressure of a gas is given by $P = \dfrac{10}{V}$, where P is in atm and V is in L. If the gas expands from 10 to 50 L and undergoes an increase in internal energy of 200 cal, how much heat will be absorbed during the process?

   Solution: The mathematical expression of first law of thermodynamics is given by

   $$dQ = dU + PdV$$

   Putting the value of P, we get

   $$dQ = dU + \frac{10}{V} \int_{V_1}^{V_2} dV$$

   On integration, it yields

   $$Q_2 - Q_1 = U_2 - U_1 + 10 \ln \frac{V_2}{V_1}$$

   $$
   \begin{aligned}
   \Delta Q &= \Delta U + 10 \ln \frac{50}{10} \\
   &= \Delta U + 10 \ln 5 \\
   &= 200 \text{ cal} + 6.99 \text{ L-atm} \\
   &= 200 \text{ cal} + 169.2 \text{ cal} \qquad \text{(since 1 L-atm = 24.2 cal)} \\
   &= 369.2 \text{ cal}
   \end{aligned}
   $$

2. Show that the Joule–Thomson coefficient of a gas

   $$\mu = \frac{V}{C_P}(T\beta - 1)$$

   where β is the coefficient of volume expansion and others have their usual meanings.

   Solution: If $f(H, T, P) = 0$, then from the cyclic relation we get

   $$\left(\frac{\partial H}{\partial T}\right)_P \left(\frac{\partial T}{\partial P}\right)_H \left(\frac{\partial P}{\partial H}\right)_T = -1$$

   It can be rearranged as

   $$\left(\frac{\partial T}{\partial P}\right)_H = -\frac{\left(\dfrac{\partial H}{\partial P}\right)_T}{\left(\dfrac{\partial H}{\partial T}\right)_P} \tag{i}$$

We know that $dH = TdS + VdP$. Differentiating it with respect to P at constant T, we have

$$\left(\frac{\partial H}{\partial P}\right)_T = \left(\frac{\partial S}{\partial P}\right)_T + V$$

Using the fourth Maxwell's relation $\left(\frac{\partial V}{\partial T}\right)_P = -\left(\frac{\partial S}{\partial P}\right)_T$, we get

$$\left(\frac{\partial H}{\partial P}\right)_T = T\left(\frac{\partial S}{\partial P}\right)_T + V = -T\left(\frac{\partial V}{\partial T}\right)_P + V = -\beta VT + V = -V(\beta T - 1)$$

Substitution of this into Eq. (i) yields

$$\mu_{JT} = \left(\frac{\partial T}{\partial P}\right)_H = -\frac{\left(\frac{\partial H}{\partial P}\right)_T}{\left(\frac{\partial H}{\partial T}\right)_P} = \frac{V(T\beta - 1)}{C_P}$$

3. Deduce the Gibbs–Duhem equation in light of chemical potential to a multi-component system.

    Solution: See Section 9.4.

4. Prove that chemical potentials of two phases in equilibrium are equal.

    Solution: See Section 9.2.

5. Calculate the standard Gibbs free energy change and the equilibrium constant at 1 bar and 298 K for the ammonia synthesis reaction

    $$N_2\,(g) + 3H_2\,(g) \rightarrow 2NH_3\,(g)$$

    given that the standard free energies of formation at 298 K for NH_3 are -16.750 kJ/mol.

    Solution: For the reaction, the standard Gibbs free energy change at 298 K can be estimated as

    $$\Delta G^0_{298} = 2\Delta G^0_{f298}[NH_3\,(g)] - \Delta G^0_{f298}[N_2\,(g)] - 3\Delta G^0_{f298}[H_2\,(g)]$$
    $$= 2 \times (-16.750 \text{ kJ/mol}) - 0 - 0 = -33.5 \text{ kJ} = -33500 \text{ J}$$

    Using the relation between standard Gibbs free energy change and equilibrium constant, we get

    $$\Delta G^\circ = -RT \ln K$$

    or

    $$\ln K = -\frac{\Delta G^0}{RT} - \frac{33500}{8.314 \times 298} = -13.521$$

    or

    $$K = e^{-13.521} = 1.342 \times 10^6$$

6. (a) Derive the Clausius–Clapeyron equation in the study of the vaporization process of a pure substance.

Solution: See Section 6.4.3.

(b) In a frictionless piston–cylinder assembly, an ideal gas undergoes a compression process from an initial state of 1 bar at 300 K to 10 bar at 300 K. The entire process comprises the following two mechanically reversible processes:

    (a) Cooling at constant pressure followed by heating at constant volume;
    (b) Heating at constant volume followed by cooling at constant pressure.

Calculate Q, ΔU and ΔH for the two processes, given that $C_P = 29.10$ kJ/kmol-K, $C_V = 20.78$ kJ/kmol-K and Volume of the gas at initial state-24.942 m³/kmol.

Solution: Basis: 1 kmol of ideal gas

Initial state: $P_1 = 1$ bar, $T_1 = 300$ K, $V_1 = 24.942$ m³/kmol
Final state: $P_2 = 10$ bar, $T_2 = 300$ K, $V_2 = ?$

The final volume of the gas can be calculated as

$$V_2 = \frac{P_1 V_1}{P_2} = \frac{1 \times 24.942}{10} = 2.494 \text{ m}^3$$

(a) In the first step of the first process, the cooling of gas takes place at constant pressure process in which the volume is reduced appreciably and consequently the temperature decreases as, say T'.

$$T' = \frac{T_1 V_2}{V_1} = \frac{300 \times 2.494}{24.942} = 30 \text{ K}$$

Heat requirement (Q_P):
Since the process is a constant pressure process

$$Q_P = \Delta H_P = C_P \Delta T = C_P(T_2 - T_1) = 29.10(30 - 300) = -7857 \text{ kJ}$$

Change in internal energy (ΔU_P):

$$\begin{aligned} \Delta U &= \Delta H - \Delta(PV) = \Delta H - P\Delta V \\ &= \Delta H - P(V_2 - V_1) \\ &= -7857 - 100 (2.494 - 24.942) = -5612 \text{ kJ} \end{aligned}$$

In the second step, the gas is heated at constant volume, i.e., V = constant or $dV = 0$. Then from first law of thermodynamics, we have

$$dQ = dU$$

Heat requirement (Q_V):

$$Q_V = \Delta U_V = C_V \Delta T = C_V(T_2 - T_1) = 20.78 (300 - 30) = 5612 \text{ kJ}$$

Change in internal energy (ΔU_V):

$$\Delta U_V = 5612 \text{ kJ}$$

Since the process comprises two steps, actual Q and ΔH can be obtained by summing up the respective values for 2 different steps.

$$Q = Q_P + Q_V = -7857 + 5612 = -2245 \text{ kJ}$$
$$\Delta U = -5612 \text{ kJ} + 5612 \text{ kJ} = 0$$

(b) In the first step of the second process, the gas is heated at constant volume, i.e., V = constant or dV = 0.

The temperature of the air, say T', at the end of the first step can be calculated as

$$T' = \frac{T_1 P_2}{P_1} = \frac{300 \times 10}{1} = 3000 \text{ K}$$

Heat requirement (Q_V):

$$Q_V = \Delta U_V = C_V \Delta T = C_V(T_2 - T_1) = 20.78 \ (3000 - 300) = 56108 \text{ kJ}$$

Change in internal energy (ΔU_V):

$$\Delta U_V = 56108 \text{ kJ}$$

In the second step, the gas is cooled at a constant pressure of 10 bar.
Heat requirement (Q_P):

$$Q_P = \Delta H_P = C_P \Delta T = C_P(T_2 - T_1) = 29.10 \ (300 - 3000) = -78570 \text{ kJ}$$

Change in internal energy (ΔU_P):

$$\Delta U = \Delta H - \Delta(PV) = \Delta H - P\Delta V$$
$$= \Delta H - P(V_2 - V_1)$$
$$= -7857 - 100 \ (2.494 - 24.942) = -5612 \text{ kJ}$$

For the two steps, the above quantities are summed up as

$$Q = Q_P + Q_V = 56108 - 78570 = -22462 \text{ kJ}$$
$$\Delta U = 56108 \text{ kJ} - 56122 \text{ kJ} = -14 \text{ kJ}$$

7. (a) With the help of the T–S diagram, explain the working principle of a vapour-compression refrigeration system. Derive the expression for its C.O.P.

Solution: See Section 8.1.5.

(b) An inventor claims to have developed an absorption refrigerating machine that receives heat from a source at 125°C, maintains the refrigerated space at −5°C and has a COP of 2. If the environment temperature is 28°C, can this claim be valid?

Solution: The mathematical expression for the COP of an absorption refrigerating machine is given by

$$\text{COP}_{\text{absorption cycle}} = \frac{T_L}{T_S - T_L} \cdot \frac{T_H - T_S}{T_H}$$

where

T_H = Temperature at which heat received = 125°C = 398 K
T_L = Temperature of the refrigerated space = −5°C = 268 K
T_S = Temperature of the environment or surroundings = 28°C = 301 K.

On substitution of these values in the above equation, we get

$$\text{COP}_{\text{absorption cycle}} = \frac{T_L}{T_S - T_L} \cdot \frac{T_H - T_S}{T_H} = \frac{268}{301 - 268} \cdot \frac{398 - 301}{398}$$

$$= 8.121 \times 0.243$$

$$= 1.97 \approx 2$$

Hence, the claim of the inventor is valid and reasonable.

(c) "Natural refrigerant is the best choice towards replacement of CFCs." Justify the statement by explaining the desirable properties of a refrigerant.

Solution: Chlorofluorocarbons (CFCs) have been the subject of worldwide attention due to the stratospheric ozone depletion issue, leading to the historic Montreal Protocol, an international agreement on control measures for ozone protection. The ozone layer in the stratosphere, which protects life on earth from the sun's harmful ultraviolet radiation, is getting progressively depleted due to the use of CFCs and other harmful substances in most of the conventional refrigeration systems. The problem concerns not only human beings, but also the plant and animal kingdoms on the earth's surface. In view of this alarming problem, replacements of CFCs by eco-friendly refrigerants becomes an urgent necessity.

In this alarming situation, natural refrigerants could be the promising solution towards replacement of CFCs for the application in a conventional refrigeration system due to its negligible global warming impact, zero ozone depletion potential, energy efficiency and cost-effectiveness.

Natural refrigerants are those compounds which are present in substantial quantities in the environment and have been found to be harmless. In fact, there are large numbers of natural refrigerants present in the biosphere (environment) with different physical properties, such as carbon dioxide, water, air, ammonia, hydrocarbon, helium and nitrogen.

The properties that a promising refrigerant should possess can be described in the following way:

Boiling Point: The boiling point of a refrigerant should be appreciably lower than the temperature levels at which the refrigerator works; otherwise there is a chance of moisture and air leakage into the system.

Critical Temperature and Pressure: A refrigerant should have the critical temperature and pressure well above the operating temperature and pressure of the system. If the system operates above the critical temperature, then condensation of the gas becomes impossible after compression at high pressure.

Latent Heat: Evaporation of the liquid is an important step in the refrigeration cycle which produces cooling; therefore the latent heat of the refrigerant should be as large as possible. Also, the weight of the refrigerant to be circulated in the system will be less if its latent heat is high. This reduces the initial cost of the refrigerant. The size of the system will also be small, which further reduces the initial cost.

Flammability and Explosiveness: A good refrigerant must be non-flammable and non-explosive. In the refrigeration system, to avoid the danger of explosion or fire hazard during high compression, the refrigerants should not have flammability and explosiveness even when mixed with air or oil. Hydrocarbons such as propane, ethane and butane are highly flammable and explosive. Ammonia is explosive when mixed with air in concentrations of 16 to 25 per cent ammonia by volume. The halogenated hydrocarbons are considered to be non-flammable.

Toxicity: Toxicity is one of the major factors in opting for a refrigerant. A toxic refrigerant is always injurious to human beings when mixed in a small percentage with air. For instance, ammonia causes toxicity.

Cost and Availability: If a refrigerant is cheap, then it would always be a refrigerant of choice. This factor plays an important role in the case of large refrigeration units. For instance, natural refrigerants (which are abundant in the biosphere or natural environment) such as air, CO_2, NH_3 and water are cheap. But the halogenated refrigerants such as CFC-12, HCFC-22, HFC-134a are costly and do not have easy availability.

Ozone Depletion Potential (ODP): Now-a-days, ODP is an environmental issue and an index by which the capability of a man-made synthetic chemical, used as a refrigerant, to deplete the ozone layer is measured. The halogenated refrigerants have ozone depletion potential while the natural refrigerants (hydrocarbon, air, CO_2, NH_3, water, etc.) do not. Hence, a good refrigerant should have no ozone depletion potential.

Global Warming Potential (GWP): GWP is also an index. It indicates the ability of a gas to absorb infra-red rays. Greenhouse gases such as CO_2, SO_2, CH_4, chlorofluoro-carbons, hydrofluorocabons, hydrochlorofluorocarbons, etc. have high GWP and are therefore responsible for global warming. Therefore a good refrigerant should have zero or negligible GWP.

Refrigerant	R-12	R-22	R-134a	R-290 (Propane)	NH_3	CO_2
Natural Fluid	No	No	No	Yes	Yes	Yes
Ozone Depletion Potential	0.82	0.005	0	0	0	0
Global Warming Potential (100 yr) IPCC[1] values	8100	1500	1300	20	<1	1
Global Warming Potential (100 yr) WMO[2] values	10600	1900	1600	20	<1	1
Critical Temperature (°C)	112.0	96.2	101.2	96.7	132.3	31.1
Critical Pressure (MPa)	4.14	4.99	4.06	4.25	11.27	7.38
Flammable	No	No	No	Yes	Yes	No
Toxic	No	No	No	No	Yes	No
Relative Price	—	1.0	4.0	0.3	0.2	0.1
Volumetric Capacity	1.0	1.6	1.0	1.4	1.6	8.4

[1] Intergovernmental Panel on Climate Change
[2] World Meteorological Organization

From the above table, it is very much clear that the natural refrigerants (CO_2 and NH_3) have zero ozone depletion potential and very low global warming potential, and so have a clear edge over conventional refrigerants.

8. Calculate the equilibrium constant for the vapour phase hydration of ethylene at 145°C. The following data are given:

$$\frac{C_P}{R} = A + BT + CT^2 + \frac{D}{T^2}$$

	A	$B \times 10^3$	$C \times 10^6$	$D \times 10^{-5}$
C_2H_4 (g)	1.424	14.394	−4.392	—
H_2O (g)	3.47	1.45	—	0.121
C_2H_5OH (g)	3.518	20.001	−6.002	—

$$\Delta H_{298K} = -45792 \text{ J/mol}, \ \Delta G_{298K} = -8378 \text{ J/mol}.$$

We are given that

$$\frac{\Delta H}{R} = \frac{C_1}{R} + \Delta AT + \frac{\Delta B}{2}T^2 + \frac{\Delta C}{3}T^3 - \frac{\Delta D}{T}$$

$$\ln K_a = -\frac{C_1}{RT} + \Delta A \ln T + \frac{\Delta B}{2}T + \frac{\Delta C}{6}T^2 + \frac{\Delta D}{2T^2} + C_2$$

C_1 and C_2 are integration constants.

Solution: The chemical reaction involved is

$$C_2H_4 (g) + H_2O (g) \rightarrow C_2H_5OH (g)$$

First of all, we need to determine the values for ΔA, ΔB, ΔC and ΔD for the above reaction. Here $\Delta = (C_2H_5OH) - (C_2H_4) - (H_2O)$.

$$\Delta A = 3.518 - 1.424 - 3.470 = -1.376$$
$$\Delta B = (20.001 - 14.394 - 1.450) \times 10^{-3} = 4.157 \times 10^{-3}$$
$$\Delta C = (-6.002 + 4.392 - 0.000) \times 10^{-6} = -1.610 \times 10^{-6}$$
$$\Delta D = (-0.000 - 0.000 - 0.121) \times 10^5 = -0.121 \times 10^5$$

We are given that the standard heat of reaction and Gibb's free energy at 298 K are

$$\Delta H_{298K} = -45792 \text{ J/mol} \qquad \Delta G_{298K} = -8378 \text{ J/mol}$$

We know that the standard heat of reaction at temperature T is given by

$$\Delta H_T^0 = \Delta H_{T_0}^0 + \int_{T_0}^{T} \Delta C_P dT$$

or

$$\frac{\Delta H_T^0}{R} = \frac{\Delta H_{T_0}^0}{R} + \int_{T_0}^{T} \frac{\Delta C_P}{R} dT \qquad \text{(i)}$$

Now

$$\int_{T_0}^{T} \frac{\Delta C_P}{R} \, dT = \left[\Delta A (T - T_0) + \frac{\Delta B}{2} (T^2 - T_0^2) + \frac{\Delta C}{3} (T^3 - T_0^3) - \Delta D \left(\frac{1}{T} - \frac{1}{T_0} \right) \right] \quad \text{(ii)}$$

Substitution of values of ΔA, ΔB, ΔC, ΔD, T (=145 + 273.15 = 418.15 K) and T_0 (=298.15 K) in the above Eq. (ii) yields

$$\int_{T_0}^{T} \frac{\Delta C_P}{R} \, dT = -23.121$$

Putting $\quad \Delta H_{298K} = -45792$ J/mol and $\displaystyle\int_{T_0}^{T} \frac{\Delta C_P}{R} \, dT = -23.121$ in Eq. (i), we get

$$\frac{\Delta H_{T_0}^0}{R} = -5484.68 \text{ J/mol} \quad \text{or} \quad \Delta H_{T_0}^0 = 45599.70 \text{ J/mol}$$

Now, substituting the value of $\Delta G_{298K} = -8378$ J/mol while $T = 298.15$ K in Eq. (iii)

$$\Delta G^0 = \Delta H_{T_0}^0 - \Delta \alpha T \ln T - \frac{\Delta \beta T^2}{2} - \frac{\Delta \gamma T^3}{6} - IRT \quad \text{(iii)}$$

we get $I = 0.0692$.

For $T = 418.15$ K, the equation

$$\ln K = -\frac{\Delta H_{T_0}^0}{RT} + \frac{\Delta \alpha \ln T}{R} + \frac{\Delta \beta T}{2R} + \frac{\Delta \gamma T^2}{6R} + I$$

gives

$$\ln K = -1.9536$$
$$K = 1.443 \times 10^{-1}$$

Therefore, the equilibrium constant K at 145°C is estimated to be 1.443×10^{-1}.

9. The following equations have been proposed to represent activity coefficient data for a system at fixed T and P:

$$\ln \gamma_1 = x_2^2 (0.5 + 2x_1)$$
$$\ln \gamma_2 = x_1^2 (0.5 - 2x_2)$$

Do these equations satisfy the Gibbs–Duhem equation? Determine an expression for G^E/RT for the system.

Solution: Checking can be done in the following ways:

Method 1: $\ln \gamma_1 = x_2^2 (0.5 + 2x_1)$

or

$$\ln \gamma_1 = 0.5x_2^2 + 2x_1 x_2^2$$

or

$$\ln \gamma_1 = 2.5x_2^2 - 2x_2^3 \quad \text{(since } x_1 + x_2 = 1, \text{ or } x_1 = 1 - x_2\text{)}$$

or
$$d \ln \gamma_1 = (5x_2 - 6x_2^2)dx_2 \tag{i}$$

Similarly
$$\ln \gamma_2 = x_1^2(1.5 - 2x_2)$$

or
$$\ln \gamma_2 = -0.5x_1^2 + 2x_1^3$$

or
$$d \ln \gamma_2 = (-x_1 + 6x_1^2)dx_1 \tag{ii}$$

From the Gibbs–Duhem equation, we get
$$x_1 d \ln \gamma_1 + x_2 d \ln \gamma_2 = 0$$

Thus by Eqs. (i) and (ii), we have
$$x_1 d \ln \gamma_1 + x_2 d \ln \gamma_2 = x_1(5x_2 - 6x_2^2)dx_2 + x_2(-x_1 + 6x_1^2)dx_1$$

or
$$
\begin{aligned}
x_1 d \ln \gamma_1 + x_2 d \ln \gamma_2 &= 5x_1x_2dx_2 - 6x_1x_2^2dx_2 - x_2x_1dx_1 + 6x_1^2x_2dx_1 \\
&= 5x_1x_2dx_2 - x_2x_1dx_1 + 6x_1x_2^2dx_1 + 6x_1^2x_2dx_1 \\
&= (5x_1x_2dx_2 - x_2x_1dx_1) + 6x_1x_2dx_1\,(x_1 + x_2) \quad \text{(as } x_1 + x_2 = 1) \\
&= (5x_1x_2dx_2 - x_2x_1dx_1) + 6x_1x_2dx_1 \\
&= (5x_1x_2dx_2 + 5x_2x_1dx_1) = (5x_1x_2dx_2 - 5x_2x_1dx_2) \\
&\qquad\qquad\qquad\qquad\qquad\qquad\quad \text{(as } dx_1 = -dx_2) \\
&= 0
\end{aligned}
$$

Hence, the Gibbs–Duhem equation is satisfied.

Method 2:   The Gibbs–Duhem equation can also be satisfied if
$$\int_0^1 \ln \frac{\gamma_2}{\gamma_1}\, dx_1 = 0$$

We get
$$\ln \gamma_1 = x_2^2(0.5 + 2x_1) \text{ and } \ln \gamma_2 = x_1^2(1.5 - 2x_2)$$

In terms of x_1
$$\ln \gamma_2 - \ln \gamma_1 = -0.5 - x_1 + 3x_1^2$$

Substituting the value, we obtain
$$\int_0^1 (-0.5 - x_1 + 3x_1^2)dx_1 = -0.5(1 - 0) - \frac{1}{2}(1 - 0) + \frac{3}{3}(1 - 0) = 0$$

Therefore, the Gibbs–Duhem equation is satisfied by this method too

The expression for $\dfrac{G^E}{RT}$ is
$$
\begin{aligned}
\frac{G^E}{RT} &= x_1 \ln \gamma_1 + x_2 \ln \gamma_2 \\
&= x_1(0.5x_2^2 + 2x_1x_2^2) + x_2(1.5x_1^2 - 2x_1^2x_2)
\end{aligned}
$$

$$= (0.5x_1x_2^2 + 2x_1^2x_2^2) + (1.5x_1^2x_2 - 2x_1^2x_2^2)$$
$$= x_1x_2(0.5x_2 + 1.5x_1)$$

10. (a) In a binary liquid system, the enthalpy of species 1 and 2 at constant temperature and pressure is represented by the following equation:

$$H = 400x_1 + 600x_2 + x_1x_2(40x_1 + 20x_2) \qquad \text{where } H \text{ is in J/mol}$$

Determine the expressions for $\overline{H}_1$ and $\overline{H}_2$ as functions of x_1 and the numerical values for the pure-species enthalpies H_1 and H_2.

(b) The activity coefficient of component 1 in a binary solution is represented by

$$\ln \gamma_1 = ax_2^2 + bx_2^3 + cx_2^4$$

where a, b and c are constants independent of concentrations. Obtain an expression for γ_2 in terms of x_1.

Solution: (a) The enthalpy equation is given by

$$H = 400x_1 + 600x_2 + x_1x_2(40x_1 + 20x_2) \qquad \text{(i)}$$

Since $x_1 + x_2 = 1$, then replace x_2 of Eq. (i) by $1 - x_1$ to get

$$H = 400x_1 + 600\,(1 - x_1) + x_1(1 - x_1)\,[40x_1 + 20\,(1 - x_1)]$$
$$= 600 - 180x_1 - 20x_1^3$$

Differentiating it with respect to x_1

$$\frac{dH}{dx_1} = -180 - 60x_1^2 \qquad \text{(ii)}$$

Following the Eq. (9.13) after substitution of $dy_2 = -dy_1$, we have in terms of x

$$\overline{H}_1 = H + x_2\frac{dH}{dx_1} \qquad \text{(iii)}$$

Substituting Eqs. (i) and (ii) into (iii), we have

$$\overline{H}_1 = 600 - 180x_1 - 20x_1^3 - 180x_2 - 60x_1^2x_2 \qquad \text{(iv)}$$

Replacing x_2 by $1 - x_1$ and simplifying, we have

$$\overline{H}_1 = 420 - 60x_1^2 + 40x_1^3 \qquad \text{(v)}$$

Similarly, following Eq. (9.13) after substitution of $x_1 = 1 - x_2$ and $dy_2 = -dy_1$, we have in terms of x

$$\overline{H}_2 = H - x_1\frac{dH}{dx_1} \qquad \text{(vi)}$$

On substitution of H and Eq. (ii) into Eq. (vi)

$$\overline{H}_2 = H - x_1\frac{dH}{dx_1} = 600 - 180x_1 - 20x_1^3 + 180x_1 + 60x_1^3$$

$$= 600 + 40x_1^3 \qquad \text{(vii)}$$

Numerical values for the pure-species enthalpies H_1:

We put $H_1 = \overline{H}_1$ when $x_1 = 1$, $x_2 = 0$ in Eq. (v) to obtain

$$H_1 = 400 \text{ J/mol}$$

The same value can be obtained from Eq. (iv).

Numerical values for the pure-species enthalpies H_2:

Putting $H_2 = \overline{H}_2$ when $x_1 = 0$ and $x_2 = 1$ in Eq. (vii), we obtain

$$H_2 = 600 \text{ J/mol}$$

(b) We are given that

$$\ln \gamma_1 = ax_2^2 + bx_2^3 + cx_2^4$$

Differentiating it we get

$$d \ln \gamma_1 = 2ax_2 dx_2 + 3bx_2^2 dx_2 + 4cx_2^3 dx_2 \qquad \text{(i)}$$

Multiplying both sides by x_1, we have

$$x_1 d \ln \gamma_1 = 2ax_1 x_2 dx_2 + 3bx_1 x_2^2 dx_2 + 4cx_1 x_2^3 dx_2 \qquad \text{(ii)}$$

From the Gibbs–Duhem equation (9.63), we get

$$x_2 d \ln \gamma_2 = -x_1 d \ln \gamma_1 \qquad \text{(iii)}$$

Thus, Eq. (iii) can be expressed as

$$x_2 d \ln \gamma_2 = -2ax_1 x_2 dx_2 - 3bx_1 x_2^2 dx_2 - 4cx_1 x_2^3 dx_2$$

or

$$d \ln \gamma_2 = -2ax_1 dx_2 - 3bx_1 x_2 dx_2 - 4cx_1 x_2^2 dx_2 \qquad \text{(since } x_1 + x_2 = 1\text{)}$$

or

$$d \ln \gamma_2 = 2ax_1 dx_2 + 3bx_1(1 - x_1) dx_1 + 4cx_1(1 - x_1)^2 dx_1$$

Integrating it from $x_1 = 0$ (at which $x_2 = 1$, $\gamma_2 = 1$, $\ln \gamma_2 = 0$) to $x_1 = x_1$ (at which $\ln \gamma_2 = \ln \gamma_2$), we get

$$\int_0^{\ln \gamma_2} d \ln \gamma_2 = (2a + 3b + 4c) \int_0^{x_1} x_1 dx_1 + 3bx_1 - (3b + 8c) \int_0^{x_1} x_1^2 dx_1 + 4c \int_0^{x_1} x_1^3 dx_1$$

or

$$\ln \gamma_2 = (2a + 3b + 4c) \left(\frac{x_1^2}{2} \right) - (3b + 8c) \left(\frac{x_1^3}{3} \right) + 4c \left(\frac{x_1^4}{4} \right)$$

or

$$\ln \gamma_2 = \left(a + \frac{3b}{2} + 2c \right) x_1^2 - \left(b + \frac{8c}{3} \right) x_1^3 + cx_1^4$$

Paper VII

1. A rigid and insulated tank of volume of $2m^3$ is divided into two equal compartments by a partition. One compartment contains an ideal gas at 400 K and 3 MPa, while the second compartment contains the same gas at 600 K and 1 MPa. This partition is punctured and the gases are allowed to mix. Determine the entropy change of the gas. The isobaric molar heat capacity of the gas is equal to $(5/2)R$.

   Solution: Let us assume two equal compartments named as compartment 1 and compartment 2, where each compartment has volume equal to 1 m^3. Thus,

   For compartment 1, we have $T_1 = 400$ K, $P_1 = 3$ MPa, $V_1 = 1$ m^3 and for compartment 2, we have $T_2 = 600$ K, $P_2 = 1$ MPa, $V_2 = 1$ m^3. Thus, $N_1 = (P_1 V_1)/(RT_1) = 902.092$ mol and similarly, $N_2 = 200.465$ mol.

   On mixing, let the final temperature of mixture be T. Then, internal energy of the system according to Eq. (2.27) can be written as:

   $$N_1 C_V(T - T_1) + N_2 C_V(T - T_2) = 0 \quad \text{or} \quad N_1(T - T_1) + N_2(T - T_2) = 0$$

   Solving it with known values, we get, $T = 436.36$ K

   Now total number of moles of mixture, (say N)

   $$= N_1 + N_2 = 902.093 + 200.465 = 1102.558 \text{ mol.}$$

   Thus, final pressure of mixture, (say P) $= \dfrac{NRT}{V} = 2$ MPa

   Now using Eq. (5.48), the entropy change per mole of the gas can be written as

   $$\Delta S = N_1 \left(C_P \ln \frac{T}{T_1} + R \ln \frac{P_1}{P} \right) + N_2 \left(C_P \ln \frac{T}{T_2} + R \ln \frac{P_2}{P} \right)$$

   Putting $C_P = \left(\dfrac{5}{2} \right) R$ and other known values of T, T_1, T_2, R, P, P_1, P_2, N_1, N_2, we get,

   $$\Delta S = 2.19 \text{ kJ/K}$$

2. What is the standard Gibbs' free energy change of a chemical reaction and how is it related to equilibrium constant?

   Solution: See Section 12.5 and 12.7.

3. Derive the relation between the second virial coefficient of van der Waals equation of state with 'a' and 'b'.

   Solution: See Example 3.7.

4. Find the change in entropy of an equimolar mixture of oxygen and nitrogen.

   Solution: Let there be N' moles of oxygen and N' moles of nitrogen in the mixture. Therefore, total number of moles of mixture $= 2N'$

Hence, mole fraction of Oxygen $(x_A) = \dfrac{1}{2}$

Similarly, mole fraction of Nitrogen $(x_B) = \dfrac{1}{2}$

Using Eq. (5.54), we can write

$$\Delta S_M = -NR(x_A \ln x_A + x_B \ln x_B) = -2N'R\left(\frac{1}{2}\ln\frac{1}{2} + \frac{1}{2}\ln\frac{1}{2}\right) = -2N'\ln\frac{1}{2}$$

For $N' = 1$ kmol, we get

$$\Delta S_M = -2R\,\ln\frac{1}{2} = 11.52 \text{ kJ/K}$$

5. Prove the following relation between fugacity and pressure:

$$\ln\left(\frac{f}{P}\right) = \int_0^P \frac{z-1}{P}\,dp$$

where, z = two parameter compressibility factor.

Solution: See Section 6.9.2 (Fugacity coefficient from compressibility factor).

6. Consider the dissociation of 1 mole of nitrogen tetroxide according to the following reaction:

$$N_2O_4 \rightarrow 2NO_2$$

Prove that the equilibrium constant in terms of reaction co-ordinate Y at equilibrium pressure is

$$K = \frac{4Y^2 P}{1 - Y^2}$$

The gas phase reaction is assumed to behave ideally.

Solution: The chemical reaction to be considered is

$$N_2O_4 \rightarrow 2NO_2$$

Basis: 1 mole of N_2O_4:
Let Y be the degree of conversion at equilibrium. Therefore, moles of N_2O_4 at equilibrium are $1-Y$ while the number of moles of NO_2 are $2Y$.
Hence the total number of moles $= 1 - Y + 2Y = 1 + Y$
At equilibrium, the molar composition of the gases is given by

$$Y_{NO_2} = \frac{2Y}{1+Y}, \quad Y_{N_2O_4} = \frac{1-Y}{1+Y}$$

Then the equilibrium constant can be expressed as

$$K_y = \frac{y_{NO_2}^2}{y_{N_2O_4}} = \frac{\left(\dfrac{2Y}{1+Y}\right)^2}{\left(\dfrac{1-Y}{1+Y}\right)} = \frac{4Y^2}{1-Y^2}$$

Here,

$$K_P = P^{\sum v_i} = P^{2-1} = P$$

On substitution of the values of K_y and K_p, we get

$$K = K_y K_p = \frac{4Y^2 P}{1-Y^2}$$

Hence proved.

7. (a) Distinguish between (i) Intensive and extensive properties (ii) Reversible and irreversible processes.

   Solution: (i) See Section 1.4.6; (ii) See Section 1.9 and 1.10.

   (b) An ideal gas (C_P = 5 kcal/kmol and C_V = 3 kcal/kmol) is changed from 1 atm and 22.4 m^3 to 10 atm and 2.24 m^3 by the following reversible process:
   (i) Isothermal compression
   (ii) Adiabatic compression followed by cooling at constant volume.
   (iii) Heating at constant volume followed by cooling at constant pressure.

   Calculate the heat, work requirement, ΔU and ΔH for each process.

   Solution: See model solved paper II, Q.No. 3.

8. (a) What are the conditions when maximum and minimum boiling azeotropes are formed? Prove that for a closed multi-component system, in equilibrium, chemical potential of a component in all phases is the same.

   Solution: See Section 10.10.1 (Types of Azeotropes) and
   See Section 10.5 (Phase equilibrium for multi-component system).

   (b) The molar volume of a binary liquid mixture at T and P is given by

   $$V = 120x_1 + 70x_2 + (15x_1 + 8x_2)x_1x_2$$

   Find the expression for the partial molar volume of species 1 and 2. Check whether the expression satisfies Gibbs–Duhem equation.

   Solution: Similar to Example 9.5.

   We are given that $\quad V = 120x_1 + 70x_2 + (15x_1 + 8x_2)x_1x_2$ \hfill (A)

   $$\frac{dV}{dx_2} = -35 - 44x_2 + 21x_2^2$$ \hfill (B)

Now, the mathematical expression obtained from the graphical method for the evaluation of partial molar volume is

$$\bar{V}_1 = V - x_2 \frac{dV}{dx_2} \tag{C}$$

Putting the value of V and $\dfrac{dV}{dx_2}$ in Eq. (C), we get

$$\bar{V}_1 = 120x_1 + 105x_2 + 15x_1^2 x_2 + 8x_1 x_2^2 + 44x_2^2 - 21x_2^3 \tag{D}$$

Similarly, we obtain the expression for $\bar{V}_2$ as

$$\bar{V}_2 = \bar{V}_1 + \frac{dV}{dx_2} = 120x_1 + 61x_2 + 15x_1^2 x_2 + 8x_1 x_2^2 + 65x_2^2 - 21x_2^3 - 35 \tag{E}$$

Now, we know from the Gibbs–Duhem relation that

$$x_1 d\bar{V}_1 + x_2 d\bar{V}_2 = 0 \tag{F}$$

We, substitute the values into Eq. (F) to satisfy Gibbs–Duhem relation.

9. (a) Deduce Gibbs–Duhem relation.

    Solution: See Section 9.4 (Gibbs–Duhem Equation).

    (b) A binary liquid mixture of (i) benzene and (ii) toluene is at 760 torr. Vapour pressures of benzene and toluene are represented by the Antoine equation.

$$\log_{10} P_1^{\text{sat}} (\text{torr}) = 6.879 - \frac{1196.76}{t(^\circ\text{C}) + 219.61}$$

$$\log_{10} P_2^{\text{sat}} (\text{torr}) = 6.950 - \frac{1342.310}{t(^\circ\text{C}) + 219.187}$$

Assume bubble and dew temperature at 95°C. Calculate bubble and dew composition at that temperature. Total pressure is taken as 760 torr. Mole fractions of benzene in liquid and vapour are 0.4047 and 0.6263, respectively.

Solution: Putting $t = 95°C$ in the given expressions, we get

$$P_1^{\text{sat}} = 1174.90 \text{ torr}$$

$$P_2^{\text{sat}} = 478.63 \text{ torr}$$

The total pressure given is, $P = 760$ torr.
Substituting the values into the expressions mentioned in the problem, we get

$$x_1 = \frac{y_1 P}{P_1^{\text{sat}}} = 0.405 \text{ and}$$

$$x_2 = \frac{y_2 P}{P_2^{\text{sat}}} = 0.593$$

Hence, bubble composition is $y_1 = 0.6263$ and $y_2 = 0.3737$ and dew composition is $x_1 = 0.4047$ and $x_2 = 0.593$.

10. (a) What are the criteria based on which refrigerant is chosen for refrigeration? Explain the absorption refrigeration with a flow diagram. Determine the expression of performance efficiency of an ideal absorption refrigeration cycle.

Solution: See Section 8.1.9 (Selection of right refrigerant) and see Section 8.1.6 (Absorption refrigeration cycle).

Performance efficiency of absorption refrigeration cycle means COP of absorption refrigeration cycle.

(b) What is the difference between Linde and Claude liquefaction processes?

Solution: See Section 8.2 (Comparison between Joule–Thomson and isentropic cooling).

Linde liquefaction process refers to Joule–Thomson expansion (isenthalpic expansion) and Claude liquefaction process refers to isentropic cooling.

(c) A rigid insulated tank of volume 3 m³ is divided into two compartments by a removable partition of negligible volume. One compartment of volume 1 m³ contains oxygen at 500 K and 10 bar while second one contains nitrogen at 800 K and 20 bar. The partition is removed and the gas is allowed to mix. After mixing, calculate the total change of entropy of the process. Assume both gases to be ideal with $\gamma = 1.4$.

Solution: Here, we have to find out the value of C_P.

We know that for ideal gas, $C_P - C_V = R$ and also $\gamma = \dfrac{C_P}{C_V}$

Thus, $\dfrac{C_P}{C_V} = 1.4$ and putting this in $C_P - C_V = R$, we get $C_P = \dfrac{7}{2} R$. Rest is similar to problem 1.

11. (a) What is van't Hoff equation and what does it predict?

Solution: See Section 12.8 (Effect of temperature on equilibrium constant: Van't Hoff equation).

(b) What is reaction co-ordinate? Discuss the Duhem's theorem for reacting system.

Solution: See Section 12.1 (Reaction co-ordinate) and Sections 10.3 and 12.14 (Duhem's theorem for reacting system).

(c) Assuming that the standard enthalpy changes of the reaction are constant in the temperature range of 298 K to 700 K, estimate the equilibrium constant at 700 K for the ammonia synthesis reaction:

$$N_2(g) + 3H_2(g) \rightarrow 2NH_3(g)$$

Given that $\Delta H^\circ f_{NH_3} = -46.1$ kJ at 298 K and $\Delta G^\circ f_{NH_3} = -16.747$ kJ at 298 K.

Solution: See Example 12.8.

Multiple Choice Questions

1. The study of thermodynamics enables us to understand
   (a) Whether the transformation of energy is feasible or not
   (b) To what extent the transformation will take place
   (c) In which direction the transformation will take place
   (d) All of these.

2. A thermodynamic system
   (a) Is a definite quantity of matter
   (b) Is surrounded by a boundary
   (c) Can exchange energy with its surroundings
   (d) All of these.

3. A system is said to be in thermodynamic equilibrium if its
   (a) Temperature remains unchanged
   (b) Pressure remains unchanged
   (c) Chemical potential remains unchanged
   (d) Temperature, pressure and chemical potential remain unchanged.

4. A process in which pressure remains constant is called
   (a) Isochoric process
   (b) Isobaric process
   (c) Isothermal process
   (d) Adiabatic process.

5. The variables of a system which are proportional to the size of the system are called
   (a) Extensive variables
   (b) Mass variables
   (c) Intensive variables
   (d) Thermodynamic variables.

6. A system in which mass as well as energy can not be exchanged with the surroundings is called
   (a) Open system
   (b) Closed system
   (c) Isolated system
   (d) None of these.

7. A centrifugal pump is an example of a/an
   (a) Closed system
   (b) Open system
   (c) Isolated system
   (d) None of these.

8. A system in which no mass transfer occurs across the boundary is called a/an
   (a) Open system
   (b) Closed system
   (c) Isolated system
   (d) Macroscopic system.

9. A pressure cooker is a good example of
   (a) Closed system
   (b) Open system
   (c) Isolated system
   (d) None of these.

10. The diathermal wall
    (a) Is incapable of exchanging heat with the surroundings
    (b) Permits the full flow of heat from the system to the surroundings and vice versa
    (c) Both (a) and (b)
    (d) None of these.

11. A quasi-static process
    (a) Proceeds very fast
    (b) Is a reversible process
    (c) Occurs when the system remains infinitesimally closed to an equilibrium state
    (d) None of these.

12. A process is called isochoric if the
    (a) Pressure is constant
    (b) Temperature is constant
    (c) Volume is constant
    (d) Mass is constant.

13. A change in which the temperature of the system remains unchanged is called a/an
    (a) Irreversible change
    (b) Adiabatic change
    (c) Reversible change
    (d) Isothermal change.

14. When no heat energy is allowed to enter or leave a system, the system is called a/an
    (a) Isothermal system
    (b) Reversible system
    (c) Adiabatic system
    (d) Irreversible system.

15. All the spontaneous processes are
    (a) Reversible
    (b) Irreversible
    (c) Quasi-static
    (d) None of these.

16. Which of the following is not an intensive property?
    (a) Temperature
    (b) Pressure
    (c) Density
    (d) Enthalpy.

17. Which one is not a state property?
    (a) Internal energy
    (b) Volume
    (c) Heat
    (d) Enthalpy.

18. Which of the following is not a property of the system?
    (a) Pressure
    (b) Volume
    (c) Work
    (d) Temperature.

19. A system is said to be in mechanical equilibrium if the
    (a) Composition of the system remains constant
    (b) Temperature of the system remains constant

(c) Pressure of the system remains constant

(d) None of these.

20. For an adiabatic process, which of the following is correct?

(a) $E = 0$ (b) $Q = 0$

(c) $Q = W$ (d) $PV = 0$.

21. One calorie is equivalent to

(a) 418.4 J   (b) 41.84 J

(c) 4.184 J   (d) 0.4184 J.

22. The first law of thermodynamics basically represents the law of conservation of

(a) Energy   (b) Mass

(c) Momentum   (d) None of these.

23. An exothermic reaction is one in which the reacting substances

(a) Have more energy than the product

(b) Have less energy than the product

(c) Have as much energy as the product

(d) Are at higher temperature than the product.

24. A minus sign of the free energy change indicates that the

(a) Reaction is spontaneous   (b) Reaction is non-spontaneous

(c) System is in equilibrium   (d) Reaction is very much unlikely.

25. A spontaneous reaction is defined as one which

(a) Proceeds by itself

(b) Proceeds only when heated

(c) Proceeds on heating at high pressure and in presence of catalyst

(d) Proceeds by itself or when properly initiated.

26. Which one of the following is a spontaneous process?

(a) Dissolution of $CuSO_4$ in H_2O

(b) Reaction between H_2 and O_2 to form H_2O

(c) Water falling downhill

(d) Flow of current from low potential to high potential.

27. In a reaction, if H and S are positive, the rate of reaction

(a) Increases with increase in temperature

(b) Decreases with increase in temperature

(c) Has no effect on temperature

(d) Decreases with increase in pressure.

28. All naturally occurring processes proceed spontaneously in a direction which leads to

(a) Decrease in entropy   (b) Increase in enthalpy

(c) Increase in free energy   (d) Decrease in free energy.

29. When a solid melts, there is

(a) An increase in enthalpy   (b) No change in enthalpy

(c) A decrease in enthalpy   (d) A decrease in internal energy.

30. For an endothermic reaction, H is

   (a) − ve

   (b) 0

   (c) + ve

   (d) Unpredictable.

31. Which of the following has/have the least entropy?

   (a) Ice

   (b) Liquid water

   (c) Steam

   (d) Water at triple point.

32. For a reversible process at equilibrium, the change in entropy is given by

   (a) $\Delta S = \dfrac{q_{rev}}{T}$

   (b) $\Delta S = T q_{rev}$

   (c) $\Delta S = \dfrac{\Delta H}{T}$

   (d) $\Delta S = T \Delta H$.

33. PV^γ = Constant (where $\gamma = C_P/C_V$) is valid for a/an

   (a) Isothermal process

   (b) Isentropic process

   (c) Isobaric process

   (d) Adiabatic process.

34. Which one of the following expressions represents the first law of thermodynamics?

   (a) $q = \Delta E + W$

   (b) $\Delta H = q - W$

   (c) $\Delta H = q + W$

   (d) $\Delta E = H - P \, \Delta V$.

35. The value of the Joule–Thomson coefficient for an ideal gas is

   (a) +ve

   (b) −ve

   (c) 0

   (d) α.

36. For the process $CO_2(s) \rightarrow CO_2(g)$

   (a) Both H and S are +ve

   (b) Both H and S are −ve

   (c) H is +ve and S is −ve

   (d) H is −ve and S is +ve.

37. Chemical potential is a/an

   (a) Extensive property

   (b) Intensive property

   (c) Reference property

   (d) None of these.

38. To obtain the integrated form of the Clausius–Clapeyron equation $\ln \dfrac{P_2}{P_1} = \dfrac{\Delta H_V}{R}\left(\dfrac{1}{T_1} - \dfrac{1}{T_2}\right)$ from the exact Clapeyron equation, it is assumed that the

   (a) Volume of the liquid phase is negligible compared to that of the vapour phase

   (b) Vapour phase behaves as an ideal gas

   (c) Heat of vaporization is independent of the temperature

   (d) All of these.

39. A gas shows deviation from ideal gas behaviour at

   (a) Low pressure and high temperature

   (b) Low pressure and low temperature

   (c) Low temperature and high pressure

   (d) High temperature and high pressure.

40. Which of the following statements is correct?

   (a) C_P can never be less than C_V

   (b) C_P is always equal to C_V

   (c) C_P may be less than C_V

   (d) C_P may be equal to or less than C_V.

41. A gas obeys the ideal gas law at
    (a) Low pressure and high temperature
    (b) Low pressure and low temperature
    (c) Low temperature and high pressure
    (d) High temperature and high pressure.

42. The processes in the Carnot cycle are carried out in a/an
    (a) Reversible fashion
    (b) Irreversible fashion
    (c) Neither (a) nor (b)
    (d) Both (a) and (b).

43. The compressibility factor Z of an ideal gas is always
    (a) 0
    (b) 1
    (c) >1
    (d) <1.

44. The Carnot cycle comprises four operational steps. These are
    (a) Two isothermals and two isochorics
    (b) Two isobarics and two isenthalpics
    (c) Two isothermals and two adiabatics
    (d) Two isothermals and two isentropics.

45. The kinetic energy of a gas molecule is zero at
    (a) 0°C
    (b) 273°C
    (c) 100°C
    (d) −273°C.

46. The refrigerator works on the principle of the
    (a) Zeroth law of thermodynamics
    (b) First law of thermodynamics
    (c) Second law of thermodynamics
    (d) Third law of thermodynamics.

47. Gibbs' phase rule finds application when heat transfer occurs by
    (a) Conduction
    (b) Convection
    (c) Radiation
    (d) Condensation.

48. The minimum number of degrees of freedom for any system is
    (a) 0
    (b) 1
    (c) 2
    (d) 3.

49. During Joule–Thomson expansion of gases
    (a) Enthalpy remains constant
    (b) Entropy remains constant
    (c) Temperature remains constant
    (d) None of these.

50. $C_P - C_V = R$ is valid for
    (a) Real gases
    (b) All gases
    (c) Ideal gases
    (d) None of these.

51. The necessary and sufficient condition for equilibrium between two phases is that
    (a) The concentration of each component in the two phases should be the same
    (b) The temperature of each phase should be the same
    (c) The pressure should be the same in the two phases
    (d) The chemical potential of each component in the two phases should be the same.

52. The change in entropy of an ideal gas during a reversible isothermal expansion is
    (a) Zero
    (b) Positive
    (c) Negative
    (d) Infinite.

53. Which of the following is true for the virial equation of state?
    (a) Virial co-efficients are universal constants
    (b) Virial co-efficient B represents three-body inter actions

(c) Virial co-efficients are functions of temperature only

(d) For gases, virial equations and ideal gas equations are the same.

54. The partial molal free energy of an element A in solution is the same as its

(a) Chemical potential   (b) Activity

(c) Fugacity   (d) Activity co-efficient.

55. The throttling process (Joule–Thompson effect) is a constant

(a) Entropy process   (b) Enthalpy process

(c) Pressure process   (d) None of these.

56. The second law of thermodynamics states that

(a) The energy change of a system undergoing any reversible process is zero

(b) It is not possible to transfer heat from a lower temperature to a higher temperature body

(c) The total energy of the system and the surroundings remains constant

(d) None of these.

57. A heat pump works on the principle of the

(a) First law of thermodynamics   (b) Second law of thermodynamics

(c) Zeroth law of thermodynamics   (d) Third law of thermodynamics.

58. The shape of the temperature–entropy diagram for the Carnot cycle is a

(a) Rectangle   (b) Rhombus

(c) Trapezoid   (d) Circle.

59. The absolute entropy for all crystalline substances at absolute zero temperature is

(a) Zero   (b) Negative

(c) More than zero   (d) None of these.

60. A closed system is cooled reversibly from 100°C to 50°C. If no work is done on the system, its

(a) U decreases and its S increases   (b) U and S both decrease

(c) U decreases but S remains constant (d) U remains constant but S decreases.

61. At the inversion point, the Joule–Thomson coefficient is

(a) +ve   (b) −ve

(c) Zero   (d) None of these.

62. The equation $dU = TdS - PdV$ is applicable to infinitesimal changes occurring in a/an

(a) Open system of constant composition

(b) Closed system of constant composition

(c) Open system with changes in composition

(d) Closed system with changes in composition.

63. A good refrigerant should have

(a) High viscosity and low thermal conductivity

(b) Low viscosity and high thermal conductivity

(c) High viscosity and high thermal conductivity

(d) Low viscosity and low thermal conductivity.

64. The theoretical minimum work required to separate 1 mol of a liquid mixture at 1 atm containing 50 mol % each of *n*-heptane and *n*-octane into pure compounds, each at 1 atm, is
    - (a) $-2\,RT \ln 0.5$
    - (b) $-RT \ln 0.5$
    - (c) $0.5\,RT$
    - (d) $2\,RT$.

65. The Mollier diagram is a plot of
    - (a) Temperature versus enthalpy
    - (b) Temperature versus entropy
    - (c) Enthalpy versus entropy
    - (d) Temperature versus internal energy.

66. The change in Gibbs' free energy for vaporization of a pure substance is
    - (a) +ve
    - (b) −ve
    - (c) 0
    - (d) Either +ve or −ve.

67. The activity coefficient of an ideal gas is
    - (a) Directly proportional to pressure
    - (b) Inversely proportional to pressure
    - (c) Unity at all pressures
    - (d) None of these.

68. Which of the following is not a refrigerant?
    - (a) SO_2
    - (b) NH_3
    - (c) CCl_2F_2
    - (d) $C_2H_4Cl_2$.

69. The number of phases in a colloidal system are
    - (a) 1
    - (b) 2
    - (c) 3
    - (d) 4.

70. The fugacity co-efficient of a substance is the ratio of its fugacity to the
    - (a) Mole fraction
    - (b) Activity
    - (c) Pressure
    - (d) Activity co-efficient.

71. The degrees of freedom at the triple point will be
    - (a) 0
    - (b) 1
    - (c) 2
    - (d) 3.

72. The free energy change for a chemical reaction is given by
    - (a) $RT \ln K$
    - (b) $-RT \ln K$
    - (c) $-R \ln K$
    - (d) $T \ln K$.

73. The ratio of the adiabatic compressibility to the isothermal compressibility is
    - (a) 1
    - (b) >1
    - (c) >>1
    - (d) <1.

74. The efficiency of a Carnot engine working between the temperatures T_1 and T_2 ($T_1 < T_2$) is
    - (a) $\dfrac{T_2 - T_1}{T_2}$
    - (b) $\dfrac{T_2 - T_1}{T_1}$
    - (c) $\dfrac{T_1 - T_2}{T_2}$
    - (d) $\dfrac{T_1 - T_2}{T_1}$.

75. 1 ton of refrigeration capacity is equivalent to the heat removal rate of
(a) 50 kcal/hr
(b) 200 kcal/hr
(c) 200 BTU/min
(d) 200 BTU/d.

76. In an ideal solution, the activity of a component is equal to its
(a) Fugacity
(b) Mole fraction
(c) Vapour pressure
(d) Partial pressure.

77. The efficiency of an Otto engine compared to that of a diesel engine for the same compression ratio will be
(a) More
(b) Less
(c) Equal
(d) Data insufficient.

78. Gibbs' free energy of a pure fluid approaches _____ as the pressure tends to zero at constant temperature
(a) Infinity
(b) Minus infinity
(c) Zero
(d) None of these.

79. The value of the activity coefficient for an ideal solution is
(a) 1
(b) 0
(c) Equal to Henry's law constant
(d) Equal to vapour pressure.

80. The phase rule for a system can be written as
(a) $F = C - P - 1$
(b) $F = C - P + 1$
(c) $F = C + P + 2$
(d) $F = C - P + 2$

where C = Number of components
P = Number of phases
F = Degrees of freedom.

81. Fugacity and pressure are numerically not equal for a gas
(a) At low temperature and high pressure
(b) At standard state
(c) Both (a) and (b)
(d) In an ideal state.

82. Which of the following is not affected by temperature?
(a) Fugacity
(b) Activity co-efficient
(c) Free energy
(d) All of these.

83. "The rate at which a substance reacts is proportional to its active mass and the rate of a chemical reaction is proportional to the product of the active masses of the reacting substances." This is the
(a) Lewis–Randall rule
(b) Statement of the Van't Hoff equation
(c) Le Chatelier's principle
(d) None of these.

84. For an ideal gas, slope of the pressure–volume curve at a given point will be
(a) Steeper for an isothermal process than for an adiabatic process
(b) Steeper for an adiabatic process than for an isothermal process
(c) Identical for both the processes
(d) Of opposite sign.

85. The point at which solid, liquid, and gas phases co-exist is known as the

    (a) Freezing point   (b) Triple point

    (c) Boiling point   (d) None of these.

86. The first law of thermodynamics is concerned with the

    (a) Direction of energy transfer   (b) Reversible processes only

    (c) Irreversible processes only   (d) None of these.

87. For an ideal solution, the value of the activity coefficient is

    (a) 0   (b) 1

    (c) <1   (d) >1.

88. A good halogenated refrigerant should have

    (a) High thermal conductivity   (b) Low freezing point

    (c) Zero ozone depletion potential   (d) All of these.

89. Gibbs' free energy of a pure fluid approaches—as the pressure tends to zero at constant temperature

    (a) Infinity   (b) Minus infinity

    (c) Zero   (d) None of these.

90. In the reaction represented by $2SO_2 + O_2 = 2SO_3$, $\Delta H = -42$ kcal, the forward reaction will be favoured by

    (a) Low temperature   (b) High pressure

    (c) Both (a) and (b)   (d) Neither (a) nor (b).

91. In a homogeneous solution, the fugacity of a component depends upon the

    (a) Pressure   (b) Composition

    (c) Temperature   (d) All of these.

92. Clausius–Clapeyron equation is applicable to _____ equilibrium processes.

    (a) Solid–vapour   (b) Solid–liquid

    (c) Liquid–vapour   (d) All of these.

93. For a gas whose P–V–T behaviour is given by $Z = 1 + \dfrac{BP}{RT}$, where B is a constant, the fugacity coefficient is

    (a) $\dfrac{BP}{RT}$ (b) $\exp \dfrac{BP}{RT}$

    (c) $\ln \dfrac{BP}{RT}$ (d) $1 + \dfrac{BP}{RT}$.

94. The reaction $A(l) \rightarrow R(g)$ is allowed to reach equilibrium condition in an autoclave. At equilibrium there are two phases—one a pure liquid phase of A and the other a vapour phase of A, R and S. Initially A alone is present. The number of degrees of freedom is

    (a) 1   (b) 2

    (c) 3   (d) 0.

95. The equilibrium constant for the reaction N_2 (g) + $3H_2$ (g) → $2NH_3$ is 0.1084. Under the same conditions, the equilibrium constant for the reaction (1/2) N_2 (g) + (3/2) H_2 (g) → NH_3 is
    (a) 0.1084
    (b) 0.3292
    (c) 0.0118
    (d) 0.0542.

96. In the ammonia synthesis reaction $N_2 + 3H_2 = 2NH_3 + 22.4$ kcal, the formation of NH_3 will be favoured by
    (a) High temperature
    (b) Low pressure
    (c) Low temperature only
    (d) Both low temperature and high pressure.

97. The internal energy of an ideal gas depends on
    (a) Temperature, specific heat and volume
    (b) Temperature and specific heat
    (c) Temperature, specific heat and pressure
    (d) Pressure, volume and temperature.

98. For steady flow, the first law of thermodynamics
    (a) Is concerned with heat interaction
    (b) Is an energy balance for a specified mass of fluid
    (c) Accounts for all energy entering and leaving a control volume
    (d) None of these.

99. A control volume refers to
    (a) A selected region in space for the analysis of a problem
    (b) An isolated system
    (c) A homogeneous system
    (d) A fixed mass.

100. At constant temperature, the volume of a given mass of a gas is inversely proportional to the pressure. This is known as
    (a) Avogadro's law
    (b) Charles' law
    (c) Boyle's law
    (d) none of these.

101. The major limitation of the first law of thermodynamics is that it does not consider
    (a) The direction of change
    (b) The extent of change
    (c) Both (a) and (b)
    (d) Neither (a) nor (b).

102. $C_P = C_V$ for a fluid which is
    (a) Incompressible
    (b) Compressible
    (c) Of very low volume expansivity
    (d) None of these.

103. The entropy change of a system is zero in a/an
    (a) Reversible process
    (b) Isochoric process
    (c) Isobaric process
    (d) Reversible adiabatic process.

104. For n moles of an ideal gas, $C_P - C_V$ is
    (a) nR
    (b) R
    (c) R^n
    (d) n^R.

105. Fugacity has the same dimension as
    (a) Temperature
    (b) Pressure
    (c) Activity coefficient
    (d) Work done.

106. At the triple point of water, the number of degrees of freedom is
    (a) One
    (b) Two
    (c) Zero
    (d) Three.

107. $C_P = C_V$ when

    (a) $\left(\dfrac{\partial V}{\partial T}\right)_P = 0$
    (b) $\left(\dfrac{\partial V}{\partial P}\right)_T = 0$

    (c) $\left(\dfrac{\partial P}{\partial T}\right)_V = 0$
    (d) None of these.

108. "If the two bodies are in thermal equilibrium with a third body, they are also in thermal equilibrium with each other." This is known as the
    (a) First law of thermodynamics
    (b) Second law of thermodynamics
    (c) Third law of thermodynamics
    (d) Zeroth law of thermodynamics.

109. The three-parameter corresponding state is concerned with
    (a) Temperature, pressure and volume
    (b) Reduced temperature, reduced pressure and reduced volume
    (c) Critical temperature, critical pressure and critical volume
    (d) Critical temperature, critical pressure and acentric factor.

110. The inversion point of a gas is defined as the temperature at which the Joule–Thomson coefficient (μ) is equal to
    (a) 0
    (b) 1
    (c) -1
    (d) ∞.

111. State which one of the following is correct:
    (a) $dG = -SdT + VdP$
    (b) $dU = TdS + PdV$
    (c) $dH = TdS - VdP$
    (d) $dA = -SdT + PdV$.

112. Enthalpy can be expressed as
    (a) $H = U - PV$
    (b) $H - U = PV$
    (c) $H = A - TS$
    (d) $H - A = TS$.

113. The mathematical expression of the first law of thermodynamics is
    (a) $dQ = dU + PdV$
    (b) $dQ = dU - dW$
    (c) $dU = dQ + PdV$
    (d) $dW = dQ + dU$.

114. Isothermal compressibility of a substance is given by

    (a) $\alpha = \dfrac{1}{V}\left(\dfrac{\partial V}{\partial P}\right)_T$
    (b) $\alpha = -\dfrac{1}{V}\left(\dfrac{\partial V}{\partial P}\right)_T$

    (c) $\alpha = -\dfrac{1}{V}\left(\dfrac{\partial P}{\partial V}\right)_T$
    (d) None of these.

115. Volume expansivity of a pure substance is given by

    (a) $\beta = -\dfrac{1}{V}\left(\dfrac{\partial V}{\partial T}\right)_P$

    (b) $\beta = \dfrac{1}{V}\left(\dfrac{\partial V}{\partial T}\right)_P$

    (c) $\beta = V\left(\dfrac{\partial V}{\partial T}\right)_P$

    (d) $\beta = \dfrac{1}{V}\left(\dfrac{\partial T}{\partial V}\right)_P$.

116. In the vapour absorption refrigeration system, the compression section is replaced by an assembly of
    (a) Absorber, heat exchanger and generator
    (b) Absorber, liquid pump, heat exchanger and generator
    (c) Absorber, liquid pump and heat exchanger
    (d) Absorber, liquid pump and generator.

117. The combined law of thermodynamics is represented by
    (a) $dU = TdS - PdV$
    (b) $dU = TdS + PdV$
    (c) $dU = dQ + PdV$
    (d) None of these.

118. The Clausius inequality is expressed mathematically as

    (a) $\oint \dfrac{dQ}{T} < 0$

    (b) $\oint \dfrac{dQ}{T} = 0$

    (c) $\oint \dfrac{dQ}{T} \le 0$

    (d) $\oint \dfrac{dQ}{T} \ge 0$.

119. The efficiency of a refrigerating machine operating between two thermal reservoirs depends upon the
    (a) Nature of the working fluid only
    (b) Two reservoir temperatures only
    (c) Mass of the working fluid
    (d) Both pressure and temperature of the working fluid.

120. Joule–Thomson expansion is an
    (a) Isothermal process
    (b) Isochoric process
    (c) Isenthalpic process
    (d) Isentropic process.

121. The work done due to isothermal expansion is

    (a) $W = nRT \ln \dfrac{P_1}{P_2}$

    (b) $W = nRT \ln \dfrac{V_2}{V_1}$

    (c) $W = nRT \ln \dfrac{V_1}{V_2}$

    (d) Both (a) and (b).

122. C_V for an ideal gas
    (a) Is independent of temperature only
    (b) Is independent of pressure only
    (c) Is independent of volume only
    (d) Is independent of both pressure and volume.

123. The highest temperature and pressure above which a liquid can not exist is known as the
   (a) Triple point   (b) Critical point
   (c) Freezing point   (d) Boiling point.

124. The equation which relates pressure, volume and temperature is known as the
   (a) Clausius–Clapeyron equation   (b) Kammerlingh–Onnes equation
   (c) Equation of state   (d) Maxwell's equation.

125. An irreversible process always leads to a/an
   (a) Decrease in entropy   (b) Increase in entropy
   (c) Constancy in entropy   (d) None of these.

126. Reduced temperature of a pure substance is the ratio of its
   (a) Critical temperature to temperature
   (b) Critical temperature to saturated temperature
   (c) Saturated temperature to critical temperature
   (d) Temperature to critical temperature.

127. The deviation from ideal gas behaviour can be accounted for by a correction factor called the
   (a) Acentric factor   (b) Solubility factor
   (c) Compressibility factor   (d) None of these.

128. For an ideal gas
   (a) $\left(\dfrac{\partial U}{\partial V}\right)_T = 1$ (b) $\left(\dfrac{\partial U}{\partial V}\right)_T = 0$

   (c) $\left(\dfrac{\partial U}{\partial V}\right)_T = -1$ (d) None of these.

129. The expansion of a gas takes place against zero pressure. This phenomenon is called
   (a) Isothermal expansion   (b) Joule–Thomson expansion
   (c) Isentropic expansion   (d) Free expansion.

130. The Joule–Thomson coefficient μ of a gas can be mathematically represented by
   (a) $\mu = \left(\dfrac{\partial T}{\partial P}\right)_H$ (b) $\mu = \left(\dfrac{\partial V}{\partial P}\right)_H$

   (c) $\mu = \left(\dfrac{\partial P}{\partial T}\right)_H$ (d) $\mu = \left(\dfrac{\partial P}{\partial V}\right)_H$.

131. The standard state assigned for a thermodynamic property is
   (a) 1 atm and 273 K   (b) 1 atm and 298 K
   (c) 1 atm and 300 K   (d) none of these.

132. Entropy is a measure of the
   (a) Randomness of a system   (b) Orderly configuration of molecules
   (c) Non-randomness of a system   (d) None of these.

133. For a constant pressure process

(a) $dH = C_V dT$

(b) $dH = \int C_V dT$

(c) $dH = C_P dT$

(d) $dH = dQ_P$.

134. For a constant volume process

(a) $dU = C_V dT$

(b) $\Delta U = \int C_V dT$

(c) $dU = TdS - PdV$

(d) $dU = C_P dT$.

135. According to Charles' law

(a) $V \propto P$ while $T = $ constant

(b) $V \propto 1/P$ while $T = $ constant

(c) $V \propto 1/T$ while $P = $ constant

(d) $V \propto T$ while $P = $ constant.

136. According to Boyle's law for gases

(a) $V \propto T$ while $P = $ constant

(b) $V \propto 1/P$ while $T = $ constant

(c) $V \propto 1/T$ while $P = $ constant

(d) $V \propto P$ while $T = $ constant.

137. The heat capacity at constant pressure (C_P) is defined as

(a) $C_P = \left(\dfrac{\partial U}{\partial T} \right)_P$

(b) $C_P = \left(\dfrac{\partial Q}{\partial T} \right)_P$

(c) $C_P = \left(\dfrac{\partial H}{\partial T} \right)_P$

(d) $C_P = \left(\dfrac{\partial W}{\partial T} \right)_P$.

138. The heat capacity at constant volume (C_V) is defined as

(a) $C_V = \left(\dfrac{\partial S}{\partial T} \right)_V$

(b) $C_V = \left(\dfrac{\partial Q}{\partial T} \right)_V$

(c) $C_V = \left(\dfrac{\partial H}{\partial T} \right)_V$

(d) $C_V = \left(\dfrac{\partial U}{\partial T} \right)_V$.

139. For an exothermic process, heat is

(a) Absorbed by the system

(b) Liberated from the system

(c) Accumulated in the system

(d) None of these.

140. For an endothermic reaction

(a) $\Delta H = +$ve

(b) $\Delta H = -$ve

(c) $\Delta H = \infty$

(d) $\Delta H = 0$.

141. "For a given chemical process, the net heat change will be the same whether the process occurs in one or in several stages." This is known as

(a) Kirchoff's law

(b) Hess's law

(c) Laplace's law

(d) Lavoisier's law.

142. The standard heat of reaction is expressed as

(a) $\Delta H^0_{\text{Reaction}} = \Sigma \Delta H^0_{\text{Reactants}} - \Sigma \Delta H^0_{\text{Products}}$

(b) $\Delta H^0_{\text{Reaction}} = \Sigma \Delta H^0_{\text{Products}} - \Sigma \Delta H^0_{\text{Reactants}}$

(c) $\Delta H^0_{\text{Reaction}} - \Sigma \Delta H^0_{\text{Products}} = \Sigma \Delta H^0_{\text{Reactants}}$

(d) $\Delta H^0_{\text{Reaction}} - \Sigma \Delta H^0_{\text{Reactants}} = \Sigma \Delta H^0_{\text{Products}}$.

143. The heat of combustion of a gas is generally determined with the help of the
  (a) Bomb calorimeter   (b) Dean and Stark apparatus
  (c) Optical pyrometer   (d) Orsat analyser.

144. If a reaction proceeds without loss or gain of heat and if all the products remain together in a single mass or stream of materials, these products will assume a definite temperature known as the
  (a) Adiabatic flame temperature   (b) Adiabatic reaction temperature
  (c) Flash point   (d) Both (a) and (b).

145. The work function or Helmholtz free energy is defined as
  (a) $H = U - TS$ (b) $H = U - TS$
  (c) $A = U - TS$ (d) $W = U - TS$.

146. Gibbs' free energy is defined as
  (a) $G = U - TS$ (b) $G = H - TS$
  (c) $G = H - TS$ (d) None of these.

147. The internal energy of an ideal gas is a function of
  (a) Temperature only   (b) Pressure only
  (c) Volume only   (d) All of these.

148. The commercial refrigerating machine follows the principle of the
  (a) Carnot cycle   (b) Reversed Carnot cycle
  (c) Stirling cycle   (d) None of these.

149. The condition for spontaneity of a chemical reaction is
  (a) $\Delta G_{T,P} < 0$ (b) $\Delta G_{T,P} > 0$
  (c) $\Delta G_{T,P} = 0$ (d) None of these.

150. The equilibrium constant of a chemical reaction is influenced by the
  (a) Pressure
  (b) Temperature
  (c) Initial concentration of the reacting substance
  (d) None of these.

151. In refrigeration cycle, heat is
  (a) Abstracted from the lower temperature region and discarded to the higher one
  (b) Absorbed from the higher temperature region and discarded to the lower one
  (c) Both (a) and (b)
  (d) Neither (a) nor (b).

152. The chemical potential of a component is given by

  (a) $\mu_i = \left(\dfrac{\partial G}{\partial n_i} \right)_{T,P}$ (b) $\mu_i = \left(\dfrac{\partial A}{\partial n_i} \right)_{T,P,n_i}$

(c) $\mu_i = \left(\dfrac{\partial G}{\partial n_i} \right)_{T, P, n_i}$ (d) $\mu_i = \left(\dfrac{\partial G}{\partial P} \right)_{T, n_i}$

153. The van Laar equation is useful in determining the activity co-efficient of a/an
    (a) Azeotropic mixture   (b) Ternary solution
    (c) Binary solution   (d) None of these.

154. The thermodynamic property relations are helpful in determining the
    (a) Measurable thermodynamic properties
    (b) Immeasurable thermodynamic properties
    (c) Change in free energy of the process
    (d) Reference properties only.

155. The influence of temperature on chemical equilibrium is substantiated by the
    (a) Arrhenius equation   (b) Le Chatelier's principle
    (c) Van't Hoff equation   (d) None of these.

156. The entropy change for a reversible process is always
    (a) 0   (b) > 0
    (c) < 0 (d) None of these.

157. The Claude gas liquefaction process employs the _____ for producing cooling effect
    (a) Isenthalpic expansion of gas   (b) Isentropic expansion of gas
    (c) Isochoric expansion of gas   (d) Isobaric expansion of gas.

158. At constant pressure and temperature, Gibbs' free energy due to mixing is always
    (a) 0   (b) ∞
    (c) +ve   (d) −ve.

159. The phase change process at varied equilibrium pressure and temperature is concerned with
    (a) Maxwell's equation   (b) Gibbs–Helmholtz equation
    (c) Clapeyron equation   (d) Gibbs–Duhem equation.

160. The free energy for a spontaneous process
    (a) Decreases   (b) Increases
    (c) Is zero   (d) None of these.

161. Free energy change at equilibrium is
    (a) Zero   (b) Positive
    (c) Negative   (d) Indeterminate.

162. For an irreversible process, the free energy is
    (a) −ve   (b) +ve
    (c) 0   (d) None of these.

163. The equilibrium constant of a chemical reaction does not depend upon
    (a) The temperature at equilibrium   (b) The pressure at equilibrium
    (c) Neither (a) nor (b)   (d) Both (a) and (b).

164. For a system prepared by partially decomposing $CaCO_3$ into an evacuated space, the number of degrees of freedom is
    (a) 2
    (b) 0
    (c) 1
    (d) 3.

165. The number of degrees of freedom for a system prepared by partially decomposing NH_4Cl into an evacuated space is
    (a) 0
    (b) 1
    (c) 2
    (d) 3.

166. For a system of two miscible non-reacting species which exists as an azeotrope in vapour–liquid equilibrium, the number of degrees of freedom is
    (a) 2
    (b) 0
    (c) 3
    (d) 1.

167. The number of degrees of freedom for a system consisting of the gases CO, CO_2, H_2, H_2O and CH_4 in chemical equilibrium is
    (a) 0
    (b) 1
    (c) 3
    (d) 4.

168. For the reaction $2SO_2 + O_2 \Leftrightarrow 2SO_3$, $\Delta H = -42$ kcal, the forward reaction will be favoured by
    (a) Low temperature
    (b) High pressure
    (c) Both (a) and (b)
    (d) Neither (a) nor (b).

169. The efficiency of a Carnot engine working between 500 K and 200 K is
    (a) 80%
    (b) 75%
    (c) 60%
    (d) 70%.

170. Pressure, volume, temperature and entropy are
    (a) Energy properties
    (b) Derived properties
    (c) Reference properties
    (d) None of these.

171. Work functions and free energy functions are
    (a) Derived properties
    (b) Energy properties
    (c) Reference properties
    (d) None of these.

172. Constant boiling mixtures are called
    (a) Distillates
    (b) Refluxes
    (c) Ternary solutions
    (d) Azeotropes.

173. The partial pressure of a component of an ideal solution is directly proportional to its mole fraction in the solution. This is known as
    (a) Henry's law
    (b) Raoult's law
    (c) Amagot's law
    (d) Dalton's law.

174. In a dilute solution, the solute obeys
    (a) Henry's law
    (b) Raoult's law
    (c) Dalton's law
    (d) None of these.

175. The solvent in a dilute solution follows
- (a) Henry's law
- (b) Dalton's law
- (c) Raoult's law
- (d) None of these.

176. The ratio of fugacity to fugacity at standard state is called the
- (a) Activity
- (b) Activity co-efficient
- (c) Fugacity coefficient
- (d) Chemical potential.

177. Entropy is a
- (a) State function
- (b) Path function
- (c) Both (a) and (b)
- (d) Neither (a) nor (b).

178. Entropy can be mathematically expressed as
- (a) $dS = \dfrac{dH_{rev}}{T}$
- (b) $dS = \dfrac{dQ_{irrev}}{T}$
- (c) $dS = \dfrac{dU_{rev}}{T}$
- (d) None of these.

179. Choose the correct one:
- (a) $S_s > S_l > S_g$
- (b) $S_s < S_l < S_g$
- (c) $S_g > S_l > S_s$
- (d) $S_s > S_l < S_g$.

180. When solid iodine is sublimed to gaseous iodine, the entropy of the system
- (a) Decreases
- (b) Increases
- (c) Remains constant
- (d) None of these.

181. For a system
- (a) Gauge pressure = Atmospheric pressure + Absolute pressure
- (b) Gauge pressure = Absolute pressure + Atmospheric pressure
- (c) Gauge pressure = Atmospheric pressure – Absolute pressure
- (d) Gauge pressure = Absolute pressure – Atmospheric pressure.

182. The state of a system is identified by its
- (a) Shape
- (b) Size
- (c) Properties
- (d) Surroundings.

183. Suppose a refrigerator with its door open is operated continuously in a kitchen and the kitchen is isolated from the rest of the house. The temperature of the kitchen will
- (a) Decrease
- (b) Increase
- (c) Not change
- (d) Increase first and then decrease.

184. For a closed system, the first law of thermodynamics tells us that
- (a) $\oint dQ = \oint dW$
- (b) $dQ = dE + dW$
- (c) $dQ - dW$ is an exact differential
- (d) All of these.

185. If an ideal gas undergoes a reversible adiabatic expansion, the work done by the gas is given by
- (a) $\dfrac{P_2 V_2 - P_1 V_1}{\gamma - 1}$
- (b) $\dfrac{P_1 V_1 - P_2 V_2}{\gamma - 1}$

    (c) $\dfrac{R(T_2 - T_1)}{\gamma - 1}$ (d)   None of these.

186. The area enclosed by a thermodynamic cycle on a *T–S* diagram represents the

    (a)   Net work done   (b)   Net heat interaction

    (c)   Both (a) and (b)   (d)   Neither (a) nor (b).

187. When a system undergoes an irreversible (spontaneous) process

    (a) $\Delta S_{\text{System}} + \Delta S_{\text{Surroundings}} > 0$ (b) $\Delta S_{\text{System}} + \Delta S_{\text{Surroundings}} < 0$

    (c) $\Delta S_{\text{System}} + \Delta S_{\text{Surroundings}} \leq 0$ (d) $\Delta S_{\text{System}} + \Delta S_{\text{Surroundings}} \geq 0$.

188. For a pure substance

    (a)   The entropy of saturated vapour decreases with increase in pressure

    (b)   The enthalpy of saturated vapour decreases with increase in pressure

    (c)   The enthalpy of vaporization decreases with increase in pressure

    (d)   None of these.

189. At critical state the compressibility factor for a van der Waals gas is

    (a)   0.75   (b)   0.5

    (c)   0.375   (d)   1.0.

190. Consider two identical closed rigid vessels A and B. Vessel A contains 32 kg oxygen at 373 K while B contains 28 kg nitrogen at 373 K. If p_A and p_B are the corresponding partial pressures in the two containers, then

    (a) $p_A = p_B$ (b) $p_A < p_B$

    (c) $p_A > p_B$ (d)   None of these.

191. For a constant temperature process in an ideal gas

    (a) $\Delta Q = -\text{ve}$ (b) $\Delta U = 0$

    (c) $\Delta W = 0$ (d) $\Delta U = -\text{ve}$.

192. A fixed mass of gas emits 250 J of heat energy and contracts at a constant pressure of 1×10^5 Pa from 2.5×10^{-3} m^3 to 1.0×10^{-3} m^3. What is the change in the internal energy?

    (a)   −250 J   (b)   100 J

    (c)   −100 J   (d)   250 J.

193. The basic purpose of a heat engine is to convert

    (a)   Mechanical energy into thermal energy

    (b)   Thermal energy into electrical energy

    (c)   Thermal energy into mechanical energy

    (d)   None of these.

194. When a system does work on its surroundings, its internal energy will

    (a)   Decrease   (b)   Increase

    (c)   Stay constant   (d)   Become more concentrated.

195. The work output of a 50% efficient heat engine giving off 60 J of energy as waste heat is

    (a)   50 J   (b)   60 J

    (c)   45 J   (d)   70 J.

196. How much work will a heat engine do if it takes in 80 J of heat energy from a high-temperature source and expels 25 J of waste heat into its surroundings?

    (a) 60 J   (b) 50 J
    (c) 55 J   (d) 70 J.

197. How much heat must be transferred into a balloon in order to increase its internal energy by 18 J if it does 12 J of work on the surrounding air as it expands?

    (a) 40 J   (b) 65 J
    (c) 55 J   (d) 30 J.

198. What will be the change in temperature of a 0.2 kg iron bar if 400 J of energy is transferred into it as heat? (C_P for iron = 450 J/kg°C)

    (a) 4.4°C   (b) 7.2°C
    (c) 6.5°C   (d) 3.9°C.

199. What will be the final equilibrium temperature when 100 g of water at 30°C and 200 g of water at 80°C are mixed in an insulated container? (C_P for water = 4185 J/kg°C)

    (a) 45.4°C   (b) 72.2°C
    (c) 63.3°C   (d) 39.9°C.

200. The three commonly used temperature scales are

    (a) Fahrenheit, Celsius and Rankine   (b) Fahrenheit, Celsius and Kelvin
    (c) Fahrenheit, Rankine and Kelvin   (d) Celsius, Rankine and Kelvin.

201. When two bodies are in thermal equilibrium

    (a) They have the same pressure
    (b) They have the same temperature
    (c) They are in a quasi-static state of equilibrium
    (d) They have the same chemical composition.

202. Water starts to vaporize at

    (a) 100°C   (b) Saturated temperature
    (c) Superheated temperature   (d) Both (a) and (b).

203. The driving force of a system to be in chemical equilibrium is

    (a) Temperature   (b) Pressure
    (c) Free energy   (d) Chemical potential.

204. The relation between the Rankine scale and the Fahrenheit scale is

    (a) $T(R) = T(°F) + 459.67$ (b) $T(°F) = T(R) + 459.67$
    (c) $T(R) + T(°F) = 459.67$ (d) None of these.

205. The relation between the Celsius scale and the Fahrenheit scale is

    (a) $T(°F) - 1.8T(°C) = 32$ (b) $T(°F) = 1.8T(°C) - 32$
    (c) $T(°F) = 1.8T(°C) + 32$ (d) None of these.

206. The total energy of a system comprises

    (a) Kinetic energy, potential energy and vibrational energy
    (b) Kinetic energy, potential energy and rotational energy
    (c) Kinetic energy, potential energy and chemical energy
    (d) Kinetic energy, potential energy and internal energy.

207. A rigid tank contains a hot fluid that is cooled while being stirred by a paddle wheel. Initially, the internal energy of the fluid is 800 kJ. During the cooling process the fluid loses 500 kJ of heat and the paddle wheel does 100 kJ of work on the fluid. What will be the final internal energy of the fluid?

    (a) 300 kJ   (b) 400 kJ
    (c) 800 kJ   (d) 500 kJ.

208. A reversible process is a process

    (a) Which proceeds with no driving force
    (b) Which takes place spontaneously
    (c) Which is quasi-static
    (d) Which is frictional process.

209. The amount of heat required to raise the temperature by 1° of 1 g of a substance is known as its

    (a) Heat capacity   (b) Thermal conductivity
    (c) Specific heat   (d) None of these.

210. The substance which has a fixed chemical composition throughout is called a/an

    (a) Colloid   (b) Inert substance
    (c) Pure substance   (d) Ideal substance.

211. The substance whose change in volume is quite significant is known as

    (a) Incompressible   (b) Pure
    (c) Compressible   (d) None of these.

212. A liquid in equilibrium with its own vapour at a specified temperature or pressure is called a

    (a) Superheated liquid   (b) Saturated liquid
    (c) Sub-cooled liquid   (d) None of these.

213. The vapour that is about to condense is called the

    (a) Superheated vapour   (b) Compressed vapour
    (c) Saturated vapour   (d) None of these.

214. The temperature at which a liquid is in equilibrium with its own vapour at a specified pressure is called the

    (a) Adiabatic temperature   (b) Saturated temperature
    (c) Equilibrium temperature   (d) None of these.

215. Water boils at 100°C. Here, 100°C is the

    (a) Saturation temperature   (b) Boiling point
    (c) Neither (a) nor (b)   (d) Both (a) and (b).

216. The pressure at which a pure substance changes its phase is known as the

    (a) Saturated pressure   (b) Gauge pressure
    (c) Absolute pressure   (d) None of these.

217. The latent heat of vaporization is the

    (a) The amount of heat absorbed during melting
    (b) The amount of heat absorbed during vaporization

(c) The amount of heat absorbed during sublimation

(d) None of these.

218. The factor(s) considered in formulating the van der Waals equation is/are the

(a) Intermolecular force of attraction   (b) Excluded volume

(c) Departure volume   (d) Both (a) and (b).

219. The internal energy of an ideal gas depends on the

(a) Pressure and temperature   (b) Volume and temperature

(c) Volume and pressure   (d) Temperature only.

220. The van der Waals equation of state is given by

(a) $\left(P + \dfrac{a}{V}\right)(V - b) = RT$ (b) $\left(P + \dfrac{a}{V^2}\right)(V - b) = RT$

(c) $\left(P + \dfrac{a}{V^2}\right)(V + b) = RT$ (d) None of these.

221. The van der Waals equation of state is applicable for the

(a) Solid phase only   (b) Liquid phase only

(c) Liquid and gas phases only   (d) Solid, liquid and gas phases.

222. At a given pressure and temperature, the van der Waals equation of state gives

(a) One real and two imaginary values   (b) Three real values

(c) One imaginary and two real values   (d) None of these.

223. The equation $P = \dfrac{RT}{V - b} - \dfrac{a}{T^{0.5} V(V + b)}$ is known as the

(a) Peng–Robinson equation of state

(b) Redlich–Kwong equation of state

(c) Redlich–Kwong–Soave equation of state

(d) Benedict–Webb–Rubin equation of state.

224. The second virial coefficient B in the virial equation of state $\dfrac{PV}{RT} = 1 + \dfrac{B}{V} + \dfrac{C}{V^2} + \dfrac{D}{V^3}$ + ... is a

(a) Function of volume only   (b) Function of temperature only

(c) Function of pressure only   (d) None of these.

225. For a van der Waals gas, the value of the critical coefficient $\dfrac{RT_C}{P_C V_C}$ is

(a) 3/8   (b) 8/3

(c) 5/3   (d) 7/3.

226. A liquid at a temperature below its saturation temperature is called a

(a) Sub-cooled liquid   (b) Compressed liquid

(c) Superheated liquid   (d) Both (a) and (b).

227. The vapour at a temperature above its saturation temperature is called the

(a) Sub-cooled vapour   (b) Compressed vapour

(c) Superheated vapour   (d) Saturated vapour.

228. For an ideal gas, the compressibility factor at all temperatures and pressures is
    (a) 1
    (b) 0
    (c) −1
    (d) ∞.
229. The compressibility factor is defined as
    (a) The ratio of the volume of the real gas to the volume occupied by the compressible gas
    (b) The volume occupied by the ideal gas to the ratio of the volume of the real gas
    (c) The ratio of the volume of the real gas to the volume occupied by the ideal gas
    (d) None of these.
230. The residual volume is
    (a) A parameter to express the non-ideality of gases
    (b) The difference between the volume of real gas and the volume predicted by ideal gas
    (c) The difference between the volume predicted by ideal gas and the volume of real gas
    (d) Both (a) and (b).
231. "The compressibility factor Z for all gases is approximately the same at the same reduced temperature and reduced pressure." This is known as the
    (a) Law of compressibility factor
    (b) Law of corresponding state
    (c) Amagot's law
    (d) Charles' law.
232. Law of corresponding state was proposed and formulated by
    (a) R.S. Pitzer
    (b) Gibbs–Duhem
    (c) van der Waals
    (d) Charles Dickens.
233. The acentric factor can be mathematically expressed as
    (a) $\omega = -1 - \log_{10} \left. P_r^{sat} \right|_{T_r = 0.7}$
    (b) $\omega = 1 - \log_{10} \left. P_r^{sat} \right|_{T_r = 0.7}$
    (c) $\omega = -1 + \log_{10} \left. P_r^{sat} \right|_{T_r = 0.7}$
    (d) None of these.
234. The heat capacity ratio is defined as
    (a) $\gamma = \dfrac{C_P}{(C_V - 1)}$
    (b) $\gamma = \dfrac{C_P}{C_V}$
    (c) $\gamma = \dfrac{C_V}{C_P}$
    (d) None of these.
235. In a heat engine the heat always flows from
    (a) The higher-temperature region to the lower-temperature one and produces no work
    (b) The lower-temperature region to the higher-temperature one and produces a net work
    (c) The higher-temperature region to the lower-temperature one and produces a net work
    (d) None of these.

236. Thermal efficiency of a heat engine is the ratio of
    (a) The total heat input to the net work output
    (b) The net work output to the total heat input
    (c) The net work input to the total heat output
    (d) None of these.

237. For a heat engine, the thermal efficiency
    (a) Is always less than unity
    (b) Can never be greater than 60 to 70% in actual practice
    (c) Can be greater than unity
    (d) Both (a) and (b).

238. In the case of a refrigerator, heat flows
    (a) In the direction of increasing temperature
    (b) From a higher-temperature region to a lower-temperature one
    (c) From a lower-temperature region to a higher-temperature one
    (d) From source to sink at the same temperature level.

239. It is not possible to construct a device that operates on a cycle and produces no effect other than the transfer of heat from a low-temperature body to high-temperature body. This is known as the
    (a) Kelvin–Planck statement   (b) Joule's statement
    (c) Clausius statement   (d) Carnot statement.

240. For a heat engine operating between the same two reservoirs
    (a) $\eta_{irrev} < \eta_{rev}$ (b) $\eta_{irrev} > \eta_{rev}$
    (c) $\eta_{irrev} = \eta_{rev}$ (d) $\eta_{irrev} \geq \eta_{rev}$.

241. Entropy is
    (a) A direct measure of the randomness of a system
    (b) An index of the tendency of a system towards spontaneous change
    (c) A measure of energy dispersal at a specific temperature
    (d) A measure of the unavailability of a system's energy to do work
    (e) All of these.

242. For the reversible isothermal change of an ideal gas undergoing a process, the change in entropy is given by

    (a) $(\Delta S)_T = R \ln \dfrac{V_1}{V_2}$ (b) $(\Delta S)_T = R \ln \dfrac{P_1}{P_2}$

    (c) $(\Delta S)_T = R \ln \dfrac{P_2}{P_1}$ (d) None of these.

243. For the reversible isobaric change of an ideal gas undergoing a process in a piston–cylinder assembly, the change in entropy will be

    (a) $(\Delta S)_P = C_P \ln \dfrac{T_2}{T_1}$ (b) $(\Delta S)_P = C_P \ln \dfrac{V_2}{V_1}$

    (c) $(\Delta S)_P = C_P \ln \dfrac{T_1}{T_2}$ (d) None of these.

244. When two non-identical ideal gases are mixed irreversibly, the change in entropy due to mixing will be

    (a) $\Delta S_{mix} = -N \sum x_i \ln x_i$

    (b) $\Delta S_{mix} = NR \sum x_i \ln x_i$

    (c) $\Delta S_{mix} = -NR \sum x_i \ln x_i$

    (d) $\Delta S_{mix} = N \sum x_i \ln x_i$.

245. The principle of increase in entropy is concerned with _____.

    (a) Reversible processes

    (b) Irreversible processes

    (c) Quasi-static process

    (d) None of these.

246. The cyclic relationship between P, V and T for a pure substance is given by

    (a) $\left(\dfrac{\partial P}{\partial V}\right)_T \left(\dfrac{\partial P}{\partial T}\right)_V \left(\dfrac{\partial V}{\partial T}\right)_P = -1$

    (b) $\left(\dfrac{\partial P}{\partial V}\right)_T \left(\dfrac{\partial V}{\partial T}\right)_P \left(\dfrac{\partial T}{\partial P}\right)_V = 1$

    (c) $\left(\dfrac{\partial P}{\partial V}\right)_T \left(\dfrac{\partial V}{\partial T}\right)_P \left(\dfrac{\partial T}{\partial P}\right)_V = -1$

    (d) $\left(\dfrac{\partial P}{\partial V}\right)_T \left(\dfrac{\partial V}{\partial T}\right)_P \left(\dfrac{\partial P}{\partial T}\right)_V = -1$.

247. The variation of free energy with pressure at constant temperature is

    (a) $G_1 - G_2 = nRT \ln \dfrac{P_2}{P_1}$

    (b) $G_2 - G_1 = nRT \ln \dfrac{P_2}{P_1}$

    (c) $G_2 - G_1 = nRT \ln \dfrac{P_1}{P_2}$

    (d) None of these.

248. For an isothermal reversible change of the system, the work function is

    (a) $\left(\dfrac{\partial A}{\partial V}\right)_T = P$

    (b) $\left(\dfrac{\partial A}{\partial U}\right)_T = -P$

    (c) $\left(\dfrac{\partial A}{\partial V}\right)_T = -P$

    (d) $\left(\dfrac{\partial A}{\partial V}\right)_T = -P - 1$.

249. If a system undergoes an isothermal reversible change, the mechanical work involved in the system during transformation can be represented by

    (a) $\Delta A_T = W_{max}$

    (b) $-\Delta A_T = W_{max}$

    (c) $\Delta A_T = -W_{max}$

    (d) $-\Delta A_T = W_{min}$.

250. Which of the following is an extensive property?

    (a) Free energy

    (b) Refractive index

    (c) Specific heat

    (d) Surface tension.

251. "When a compressed gas expands adiabatically and slowly through a porous plug, it undergoes a temperature change which varies in magnitude and sign." This effect is known as the

    (a) Peltier effect

    (b) Joule–Thomson effect

    (c) Claude effect

    (d) Linde–Hampson effect.

252. In Joule–Thomson effect, there will be a cooling effect when

(a) $\mu = -ve$

(b) $\mu = +ve$

(c) $\mu = \alpha$

(d) Indeterminate.

253. The Maxwell's relation $\left(\dfrac{\partial T}{\partial V}\right)_S = -\left(\dfrac{\partial P}{\partial S}\right)_V$ is derived from the fundamental thermodynamic relation

(a) $dH = TdS + VdP$

(b) $dA = -PdV - SdT$

(c) $dG = VdP - SdT$

(d) $dU = VdS - PdV.$

254. The TdS equation $TdS = C_P dT - T\left(\dfrac{\partial V}{\partial T}\right)_P dP$ is valid where entropy is a function of

(a) T and P

(b) P and V

(c) T and V

(d) None of these.

255. The inversion temperature of hydrogen is

(a) 315 K

(b) 202 K

(c) 275 K

(d) 345 K.

256. The inversion temperature of helium is

(a) 460 K

(b) 40 K

(c) 620 K

(d) 823 K.

257. The inversion point of a gas can be mathematically expressed as

(a) $T_i = \dfrac{2ab}{R}$

(b) $T_i = \dfrac{2b}{Ra}$

(c) $T_i = \dfrac{2a}{Rb}$

(d) None of these.

258. Residual free energy is defined as

(a) $G^R = G - G^{ig}$

(b) $G^R = G^{ig} - G$

(c) $G^R = G + G^{ig}$

(d) None of these.

259. Departure functions are useful to calculate the thermodynamic property of real fluids

(a) When the $P\text{-}V\text{-}T$ data of the substance is unavailable

(b) When the $P\text{-}V\text{-}T$ data of the substance is sufficient

(c) When the $P\text{-}V\text{-}T$ data of the substance is available and insufficient

(d) None of these.

260. For a non-ideal gas, fugacity

(a) Is equal to pressure

(b) Is not equal to pressure

(c) Is not concerned with pressure at all

(d) None of these.

261. The effect of temperature on fugacity can be represented by

(a) $\left(\dfrac{\partial \ln f}{\partial P}\right)_T = \dfrac{VT}{R}$

(b) $\left(\dfrac{\partial \ln f}{\partial P}\right)_T = \dfrac{VR}{T}$

(c) $\left(\dfrac{\partial \ln f}{\partial P}\right)_T = \dfrac{V}{RT}$

(d) $\left(\dfrac{\partial \ln P}{\partial f}\right)_T = \dfrac{V}{RT}.$

262. A process is said to be *uniform* if there is
    (a) No change with time
    (b) No change with location over a particular region
    (c) Both (a) and (b)
    (d) Neither (a) nor (b).

263. In a throttling device the
    (a) Isentropic process takes place
    (b) Gas undergoes compression process slowly and adiabatically
    (c) Cooling effect is always obtained
    (d) None of these.

264. The operation of a throttling device follows the
    (a) Zeroth law of thermodynamics   (b) First law of thermodynamics
    (c) Second law of thermodynamics   (d) Third law of thermodynamics.

265. Throttling is an
    (a) Isentropic process   (b) Isochoric process
    (c) Isobaric process   (d) Isenthalpic process.

266. Isothermal efficiency is defined as the ratio of the
    (a) Isothermal work to the actual work, i.e., $\eta_{iso} = \dfrac{W_{iso}}{W_{actual}}$

    (b) Actual work to the isothermal work, i.e., $\eta_{iso} = \dfrac{W_{actual}}{W_{iso}}$

    (c) Actual work to the ideal work, i.e., $\eta_{iso} = \dfrac{W_{actual}}{W_{ideal}}$

    (d) None of these.

267. A nozzle is a device that
    (a) Increases the velocity of a flowing fluid
    (b) Converts mechanical energy to kinetic energy
    (c) Converts kinetic energy to mechanical energy
    (d) Converts mechanical energy to potential energy.

268. A diffuser is a contrivance that
    (a) Increases the pressure of a fluid by decreasing its velocity
    (b) Converts pressure energy to kinetic energy
    (c) Decreases the pressure of a fluid by increasing its velocity
    (d) None of these.

269. The duties of a nozzle and a diffuser are
    (a) Opposite to each other   (b) Identical to each other
    (c) Not comparable at all   (d) None of these.

270. Nozzles and diffusers are widely used in
    (a) Heat exchangers   (b) Refrigeration systems
    (c) Rockets and other space vehicles   (d) None of these.

271. Pick out the correct relation between the COP of a refrigerator (COP_R) and that of the heat pump (COP_{HP})

(a) $COP_{HP} = 1 - COP_R$ (b) $COP_R = COP_{HP} + 1$

(c) $COP_{HP} = COP_R - 1$ (d) $COP_{HP} = COP_R + 1$.

272. 1 ton of refrigeration is equivalent to

(a) 3.517 kW   (b) 4.202 kW

(c) 250 kcal/min   (d) 50000 kcal/min.

273. The relative co-efficient of performance of a heat engine is the ratio of

(a) The theoretical COP to the actual COP

(b) The actual COP to the theoretical COP

(c) The theoretical COP to the ideal COP

(d) None of these.

274. The advantage of the vapour-compression refrigeration system over the absorption refrigeration system is that

(a) The charging of the refrigerant is quite simple

(b) The space requirement for installation of the system is relatively low

(c) There is no chance of leakage of the refrigerant from the system

(d) Thermal energy is supplied as input.

275. NH_3 is not always preferred as a refrigerant due to its

(a) Flammability   (b) Viscosity

(c) Toxicity   (d) Ozone depletion potential.

276. Predict the correct one out of the following:

(a) The boiling point of a refrigerant should be appreciably lower than the temperature levels at which the refrigerator works

(b) The freezing point of the refrigerant should be appreciably lower than the operating temperature of the system

(c) The operating pressure in the condenser should not be high

(d) All of these.

277. The critical temperature and pressure of a refrigerant should be

(a) Below the operating temperature and pressure of the system

(b) Well above the operating temperature and pressure of the system

(c) Equal to the operating temperature and pressure of the system

(d) None of these.

278. Global Warming Potential (GWP) is an index that indicates the ability of a gas

(a) To absorb the ultra-violet rays   (b) To absorb the infra-red rays

(c) To absorb the photo-incidental rays   (d) None of these.

279. Ozone Depletion Potential (ODP) is an index that indicates the ability of a gas

(a) To deplete the tropospheric ozone layer

(b) To deplete the stratospheric ozone layer

(c) To absorb the atmospheric ozone

(d) To increase the amount of ozone in the stratosphere.

280. For a good refrigerant
   (a) Ozone Depletion Potential should be high and Global Warming Potential should be less
   (b) Ozone Depletion Potential should be less and Global Warming Potential should be high
   (c) Both should be high
   (d) Both should be negligible.

281. To estimate the Ozone Depletion Potential of a refrigerant
   (a) CFC-11 is used as a reference gas   (b) CFC-12 is used as a reference gas
   (c) CO_2 is used as a reference gas   (d) SO_2 is used as a reference gas.

282. To compute the Global Warming Potential of a refrigerant
   (a) HFC-134a is used as a reference gas
   (b) Hydrocarbon is used as a reference gas
   (c) CO_2 is used as a reference gas
   (d) CH_4 is used as a reference gas.

283. Liquefaction can be achieved through
   (a) Expansion of gas through a work-producing device (isentropic expansion)
   (b) Joule–Thomson expansion (isenthalpic expansion)
   (c) Exchange of heat at constant pressure
   (d) All of these.

284. The partial molar property of a component can be measured using
   (a) Analytical method only   (b) Graphical method only
   (c) Analytical and graphical methods   (d) Experimental method.

285. At the same temperature and pressure, the chemical potentials of a component in two phases under equilibrium conditions
   (a) Are equal   (b) Are different
   (c) Can not be predicted   (d) None of these.

286. The influence of pressure on chemical potential can be expressed as

   (a) $\left(\dfrac{\partial \overline{V}_i}{\partial P}\right)_{T,\,n_i} = \mu_i$ (b) $\left(\dfrac{\partial \mu_i}{\partial T}\right)_{P,\,n_i} = \overline{V}_i$

   (c) $\left(\dfrac{\partial \mu_i}{\partial P}\right)_{T,\,n_i} = \overline{V}_i$ (d) $\left(\dfrac{\partial \mu_i}{\partial n_i}\right)_{T,\,P} = \overline{V}_i.$

287. The activity of a component
   (a) Is a measure of *effective concentration* of a component
   (b) Results from the interaction between the molecules in a non-ideal gas or solution
   (c) Is dimensionless
   (d) All of these.

288. The activity of a component depends on
   (a) Temperature only   (b) Temperature and pressure only
   (c) Temperature, pressure and composition   (d) None of these.

289. The activity of the component is defined as the ratio of
    (a) The fugacity of the component in the solution to the fugacity of the component in the ideal state
    (b) The fugacity of the component in the standard state to the fugacity of the component in the solution
    (c) The fugacity of the component in the solution to the fugacity of the component in the standard state
    (d) None of these.

290. The activity coefficient is a measure of
    (a) The ideal behaviour of chemical substances in a mixture.
    (b) The deviation from ideal behaviour of chemical substances in a mixture
    (c) The *effective concentration* of a component
    (d) None of these.

291. For an ideal solution, the activity coefficient is defined as
    (a) $\gamma_i = \dfrac{f_i x_i}{\bar{f}_i}$
    (b) $\gamma_i = \dfrac{x_i}{a_i}$
    (c) $\gamma_i = \dfrac{a_i}{x_i}$
    (d) None of these.

292. The temperature dependence of the activity coefficient can be expressed as
    (a) $\left(\dfrac{\partial \ln \gamma_i}{\partial T}\right)_P = \dfrac{\bar{H}_i - H_i}{RT^2}$
    (b) $\left(\dfrac{\partial \ln \gamma_i}{\partial T}\right)_P = \dfrac{H_i - \bar{H}_i}{T^2}$
    (c) $\left(\dfrac{\partial \ln \gamma_i}{\partial T}\right)_P = \dfrac{H_i - \bar{H}_i}{RT}$
    (d) $\left(\dfrac{\partial \ln \gamma_i}{\partial T}\right)_P = \dfrac{H_i - \bar{H}_i}{RT^2}$.

293. The change in partial molar properties with composition at constant temperature and pressure can be better explained with the
    (a) Gibbs–Helmholtz equation
    (b) Gibbs–Duhem equation
    (c) Gibbs equation
    (d) Clausius–Clapeyron equation.

294. Excess property of a component can be schematically represented by
    (a) $H^E = H - H^{id}$
    (b) $H^E = H^{id} - H$
    (c) $H^E - H^{id} = H$
    (d) None of these.

295. "The fugacity of each component in an ideal solution is proportional to the mole fraction of that pure component in the solution." This is known as
    (a) Henry's law
    (b) Raoult's law
    (c) Lewis–Randall rule
    (d) Gibbs law.

296. The Lewis–Randall rule is applicable
    (a) At high pressures
    (b) At low pressures when the gas mixture behaves as an ideal solution
    (c) At low temperature
    (d) At low temperature and high pressure.

297. Raoult's law can be expressed as

(a) $\overline{P}_i = y_i P_i^{sat}$

(b) $\overline{P}_i = y_i^{sat} P_i$

(c) $P_i^{sat} = y_i \overline{P}_i$

(d) None of these.

298. Raoult's law is found to be quite satisfactory for a

(a) Dilute solution

(b) Ideal solution

(c) Non-ideal solution

(d) Perfect gas.

299. For an ideal solution, the enthalpy change of mixing (ΔH_{mix}) is always given by

(a) $\Delta H_{mix} = 1$

(b) $\Delta H_{mix} = 0$

(c) $\Delta H_{mix} = -1$

(d) $\Delta H_{mix} = \infty$.

300. Predict the correct form of the Gibbs–Duhem equation in terms of the activity coefficient:

(a) $x_1 \dfrac{d \ln \gamma_1}{dx_1} = (1 - x_2) \dfrac{d \ln \gamma_2}{dx_2}$

(b) $(1 - x_1) \dfrac{d \ln \gamma_1}{dx_1} = x_2 \dfrac{d \ln \gamma_2}{dx_2}$

(c) $x_1 \dfrac{d \ln \gamma_1}{dx_1} = x_2 \dfrac{d \ln \gamma_2}{dx_2}$

(d) $x_1 \dfrac{d \ln \gamma_1}{dx_1} = (1 + x_2) \dfrac{d \ln \gamma_2}{dx_2}$.

301. For a binary ideal solution, if x_1 and x_2 are the mole fractions of components 1 and 2 respectively, then

(a) $x_1 + x_2 = 0$

(b) $x_1 + x_2 = -1$

(c) $x_1 + x_2 = 1$

(d) $x_1 + x_2 = \infty$.

302. In a binary ideal solution consisting of components A and B

(a) Component A obeys Raoult's law and component B obeys Henry's law

(b) Component A obeys Henry's law and component B obeys Raoult's law

(c) Both the components obey Raoult's law

(d) None of these.

303. Pick up the correct sentence out of the following:

(a) Minimum boiling azeotropes may be formed if a solution exhibits very small positive deviation from ideal behaviour

(b) Maximum boiling azeotropes may be formed if the solution exhibits very large positive deviation from ideal behaviour

(c) Minimum boiling azeotropes may be formed if the solution exhibits very large positive deviation from ideal behaviour

(d) None of these.

304. At constant temperature and pressure, the free energy for a chemically reacting system at equilibrium is

(a) Minimum

(b) Maximum

(c) Can not be predicted

(d) None of these.

305. The criterion of phase equilibrium of a component is

(a) $dG_{T,P,V} = 0$

(b) $dG_{T,P} = 1$

(c) $dG_{T,P} = 0$

(d) $dA_{T,P} = 0$.

306. The vapour–liquid equilibrium of a binary system can be better represented by

(a) Temperature–composition (T–X–Y) diagram
(b) Pressure–composition (P–X–Y) diagram
(c) Pressure–temperature (P–T) diagram
(d) All of these.

307. The *boiling point diagram* of a binary mixture can be represented with the help of the

(a) Temperature versus volume plot   (b) Pressure versus volume plot
(c) Pressure versus composition plot   (d) Temperature versus composition plot.

308. The *boiling point diagram* of a binary solution is used to know how the equilibrium changes with

(a) Pressure   (b) Free energy
(c) Temperature   (d) Entropy.

309. The upper curve in the boiling point diagram is called

(a) The saturated vapour curve   (b) The dew point curve
(c) The saturated liquid curve   (d) Both (a) and (b).

310. The bubble point and dew point curves meet where the mixture turns into

(a) Purely two components   (b) Purely one component
(c) Purely three components   (d) None of these.

311. The term *azeotrope* means

(a) Condensing without changing   (b) Boiling without changing
(c) Both (a) and (b)   (d) Neither (a) nor (b).

312. Excess Gibbs free energy models are used to determine the value of the

(a) Isothermal compressibility   (b) Activity coefficient
(c) Coefficient of volume expansion (d) Gibbs' free energy.

313. For the determination of the activity coefficient for the system comprising relatively simple and preferably non-polar liquids, we generally use the

(a) Wohl's equation   (b) Margules equation
(c) Van Laar equation   (d) Wilson equation.

314. The disadvantage(s) of the Wilson equation over the van Laar equation is/are that

(a) It can not be applied for the systems in which γ shows maxima or minima
(b) It is not suitable for the systems exhibiting limited miscibility
(c) Neither (a) nor (b)
(d) Both (a) and (b).

315. The advantage(s) of the UNIQUAC equation is/are that

(a) It is suitable for solutions having small as well as large molecules
(b) The equation is simple as it consists of only two adjustable parameters
(c) It is also applicable to a good number of non-polar solutions such as aldehydes and ketones
(d) All of these.

316. Bubble point is defined as the point at which

  (a) The first bubble of vapour is formed upon heating a liquid consisting of two or more components at a given pressure

  (b) The first drop of liquid (condensate) is formed upon cooling the vapour at a given pressure

  (c) A saturation point is reached

  (d) A critical point is reached.

317. For checking the thermodynamic consistency of VLE data

  (a) A test is usually performed on the basis of the Gibbs–Duhem equation in terms of the activity coefficients

  (b) The data can be obtained from the pressure–temperature–mole fractions (P–T–x–y) diagram

  (c) A test is usually performed on the basis of the Margules equation

  (d) Both (a) and (b).

318. For a ternary system having a homogeneous single phase, the degree(s) of freedom will be

  (a) 3   (b) 4

  (c) 2   (d) 1.

319. The depression of freezing point is defined as the difference between

  (a) The freezing points of the pure solvent and the solution containing the volatile solute

  (b) The melting points of the pure solvent and the solution containing the non-volatile solute

  (c) The freezing points of the pure solvent and the solution containing the non-volatile solute

  (d) The freezing points of the pure solvent and the solution containing the non-volatile solvent.

320. The magnitude of the osmotic pressure depends on the

  (a) Temperature

  (b) Gibbs' free energy

  (c) Nature of the semi-permeable membrane

  (d) Entropy.

321. The reaction coordinate

  (a) Is a one-dimensional coordinate

  (b) Indicates the progress of a chemical reaction along a pathway

  (c) Is a geometric parameter that changes during the conversion of one or more molecular entities

  (d) All of these.

322. A chemical reaction must proceed in the direction of

  (a) Increasing Gibbs free energy   (b) Decreasing Gibbs free energy

  (c) Constant Gibbs free energy   (d) None of these.

323. The equilibrium criterion of a chemical reaction can be expressed as

  (a) $(dG)_{T,P} = 1$ (b) $(dG)_{T,P} = -1$

  (c) $(dG)_{T,P} = 0$ (d) $(dG)_{T,P} = \infty$.

324. For a chemically reacting system, the relation between Gibbs' free energy change and equilibrium constant is given by

(a) $K = e^{-\frac{RT}{\Delta \mathring{G}}}$

(b) $K = e^{-\frac{\Delta \mathring{G}}{RT}}$

(c) $K = e^{\frac{\Delta \mathring{G}}{RT}}$

(d) $K = e^{\frac{\Delta \mathring{G} R}{T}}$.

325. Heterogeneous reaction is defined as the reaction which requires

(a) At least one phase to proceed

(b) At least two phases to proceed

(c) One initiator to take place

(d) One negative catalyst to take place.

326. The phase rule for a chemically reacting system was formulated by

(a) Gibbs–Duhem

(b) Gibbs

(c) Arrhenius

(d) Van't Hoff.

327. The phase rule for a chemically reacting system differs from the same for a non-reacting system in terms of

(a) The number of phases

(b) The number of independent reactions

(c) The number of components

(d) None of these.

328. A fuel cell is an electrochemical energy conversion device in which

(a) The chemical energy of a fuel is converted directly into electrical energy

(b) The electrical energy of a fuel is converted directly into chemical energy

(c) The chemical energy of a fuel is converted directly into mechanical energy

(d) None of these.

329. The fuel cell

(a) Produces power at low cost at considerable periods

(b) Requires quite extensive maintenance

(c) Has harmful by-products

(d) None of these.

330. The criterion for equilibrium for a chemically reacting system

(a) Is independent of the laws of thermodynamics

(b) Is obtained from the first and second laws of thermodynamics

(c) Can not be obtained unless the fundamental relation is known

(d) None of these.

331. A process is said to be feasible if

(a) $\Delta G > 0$

(b) $\Delta G < 0$

(c) $\Delta G < 1$

(d) $\Delta G > 1$.

332. For a chemically reacting system at constant temperature and pressure, the criterion for equilibrium can be expressed as

(a) $\Sigma \mu_i dn_i = 0$

(b) $\Sigma n_i d\mu_i < 0$

(c) $\Sigma n_i d\mu_i = 0$

(d) $\Sigma n_i d\mu_i > 1$.

333. For an ideal gas the change in Gibbs' free energy at constant temperature is given by

(a) $Rd \ln T$

(b) $RTd \ln V$

(c) $RTd \ln P$

(d) None of these.

334. The activity of component i can be written as

(a) $a_i = \dfrac{f_i^0}{f_i}$

(b) $a_i = \dfrac{f_i}{f_i^0}$

(c) $a_i = \ln\left(\dfrac{f_i}{f_i^0}\right)$

(d) $a_i = \ln(f_i - f_i^0)$.

335. The influence of temperature on the equilibrium constant of a chemical reaction can be obtained from

(a) $\left(\dfrac{\partial \ln K}{\partial T}\right) = -\dfrac{\Delta \overset{\circ}{H}}{T^2}$

(b) $\ln \dfrac{K_2}{K_1} = \dfrac{\Delta \overset{\circ}{H}}{R}\left(\dfrac{1}{T_2} - \dfrac{1}{T_1}\right)$

(c) $\left(\dfrac{\partial \ln K}{\partial T}\right) = -\dfrac{T^2}{\Delta \overset{\circ}{H}}$

(d) $\ln \dfrac{K_2}{K_1} = \dfrac{\Delta \overset{\circ}{H}}{R}\left(\dfrac{1}{T_1} - \dfrac{1}{T_2}\right)$.

336. In the steady flow energy equation, which of the following remains constant?

(a) Entropy

(b) Pressure

(c) Enthalpy

(d) Total energy.

337. For the reaction $CO\ (g) + 2H_2\ (g) \rightarrow CH_3OH\ (g)$, with increase in pressure the degree of conversion

(a) Decreases

(b) Increases

(c) Remains constant

(d) None of these.

338. Methanol can be produced by the following reaction

$$CO\ (g) + 2H_2\ (g) \rightarrow CH_3OH\ (g)$$

If some inert gas is added to the reaction mixture, then the degree of conversion

(a) Decreases

(b) Does not change

(c) Increases

(d) None of the above.

339. Hydrogen gas is produced according to the reaction

$$CH_4\ (g) + H_2O\ (g) \rightarrow CO\ (g) + 3H_2\ (g)$$

With increase in pressure the degree of conversion

(a) Decreases

(b) Increases

(c) Remains constant

(d) None of these.

340. For the reaction $N_2\ (g) + 3H_2\ (g) \rightarrow 2NH_3\ (g)$, maximum conversion can be achieved if the reaction takes place at

(a) Low pressure and high temperature

(b) Low temperature and high pressure

(c) Low pressure and low temperature

(d) High pressure and high temperature.

341. Given that the standard free energy of formation at 298 K for NH_3 is -16.750 kJ/mol, the standard Gibbs' free energy change for the ammonia synthesis reaction $N_2\ (g) + 3H_2$ $(g) \rightarrow 2NH_3\ (g)$ is

(a) -35.7 kJ

(b) -33.5 kJ

(c) -38.2 kJ

(d) -53.4 kJ.

342. The chemical potential of component i in a solution mixture can be expressed as

(a) $\mu_i = RT \ln G_i^0 + a_i$ (b) $\mu_i = RT \ln a_i + G_i^0$

(c) $\mu_i = RT \ln G_i^0 - a_i$ (d) $\mu_i + RT \ln G_i^0 = a_i$.

343. The chemical potential of a pure substance is equal to the

(a) Specific Gibbs' free energy   (b) Molar entropy

(c) The Gibbs' free energy   (d) Molar Gibbs' free energy.

344. In an ideal gas mixture consisting of components A and B, the partial pressure of component A is equal to

(a) $p_A = \dfrac{n_A - n_B}{N}$ (b) $p_A = \dfrac{N}{n_A + n_B}$

(c) $p_A = \dfrac{n_A + n_B}{N}$ (d) $p_A = \dfrac{N}{n_A - n_B}$.

345. In an ideal gas mixture consisting of components A and B, the mole fraction of component A is equal to

(a) $n_A = \dfrac{p_A + p_B}{P}$ (b) $n_A = \dfrac{p_A - p_B}{P}$

(c) $n_A = \dfrac{P}{p_A + p_B}$ (d) $n_A = \dfrac{P}{p_A - p_B}$.

346. In an absorption refrigeration system, the heat energy Q_1 is supplied at temperature T_1 while the system absorbs heat energy Q_3 from a cold space at temperature T_3. If the ambient temperature is T_2, then the COP of the refrigerator (ignoring the negligible amount of heat supplied to the pump) is

(a) $\dfrac{T_3(T_2 - T_1)}{T_1(T_1 - T_2)}$ (b) $\dfrac{T_1(T_2 - T_3)}{T_3(T_1 - T_2)}$

(c) $\dfrac{T_3(T_1 - T_2)}{T_1(T_2 - T_3)}$ (d) $\dfrac{(T_1 - T_2)}{(T_2 - T_3)}$.

347. For any pure substance, the difference between C_P and C_V can be expressed in terms of the isothermal compressibility α and volume expansivity β as

(a) $C_P - C_V = \dfrac{TV\beta}{\alpha^2}$ (b) $C_P - C_V = \dfrac{TV\beta^2}{\alpha}$

(c) $C_P - C_V = \dfrac{TV\alpha}{\beta^2}$ (d) $C_P - C_V = \dfrac{TV\alpha^2}{\beta}$.

348. The availability of a system

(a) Depends upon the conditions of the system only

(b) Is independent of the conditions of the surroundings

(c) Does not depend upon the conditions of the system

(d) Depends upon the conditions of the system and the surroundings.

349. The sudden bursting of a cycle tyre is an
(a) Isothermal process
(b) Isochoric process
(c) Isobaric process
(d) Adiabatic process.

350. Changes of state such as melting, vaporization and freezing are
(a) Adiabatic processes
(b) Isothermal processes
(c) Isochoric processes
(d) Isobaric processes.

351. A process during which a system returns to its initial state through a number of different processes is called a/an
(a) Adiabatic process
(b) Isothermal process
(c) Cyclic process
(d) Iso-volumetric process.

352. When the heat capacity ratio is 1.67, then the ideal gas is
(a) Diatomic
(b) Monatomic
(c) Triatomic
(d) None of these.

353. According to the phase rule, the triple point of a pure substance is
(a) Invariant
(b) Univariant
(c) Divariant
(d) None of these.

354. The internal energy of an element at 1 atm and 25°C is
(a) 0 kcal/kmol
(b) 25 kcal/kmol
(c) 273 kcal/kmol
(d) None of these.

355. Which of the following changes during phase transition?
(a) Pressure
(b) Volume
(c) Temperature
(d) All of these.

356. The acentric factor for all materials is always
(a) 0
(b) 1
(c) <1
(d) >1.

357. For a binary system comprising two miscible non-reacting components in vapour–liquid equilibrium forming an azeotrope, the number of degrees of freedom is
(a) 0
(b) 1
(c) 2
(d) 3.

358. Melting of ice is an example of
(a) Adiabatic process
(b) Isothermal process
(c) Isochoric process
(d) Isobaric process.

359. The equation of state for a certain gas is given by $P(V - (b)) = RT$, where b is a positive constant. The Joule–Thomson coefficient of this gas would be
(a) + ve
(b) – ve
(c) Zero
(d) None of these.

360. As time passes, the entropy of the universe
(a) Decreases
(b) Increases
(c) Remains constant
(d) Can not be predicted.

361. A 4 m³ rigid tank contains nitrogen gas at 400 kPa and 350 K. Now heat is transferred to the nitrogen in the tank and the pressure of nitrogen rises to 700 kPa. The work done during the process is

(a) 900 kJ   (b) 500 kJ

(c) 700 kJ   (d) 0 kJ

362. Air-conditioning machine operates on the principle of

(a) Second law of thermodynamics   (b) Zeroth law of thermodynamics

(c) First law of thermodynamics   (d) Third law of thermodynamics

363. The phase rule is coined by

(a) Gibbs   (b) Helmholtz

(c) Joule   (d) van der Waals

364. A 0.8 m³ rigid tank contains nitrogen gas at 600 kPa and 300 K. Now the gas is compressed isothermally to a volume of 0.1 m³. The work done on the gas during this compression process is

(a) 746 kJ   (b) 998 kJ

(c) 1890 kJ   (d) 420 kJ

365. The temperature and pressure of air are 30°C and 1 atm respectively. The mass of air behaving ideally contained in a room of size 5 m × 5 m × 5 m is

(a) 234.2 kg   (b) 144.9 kg

(c) 204.6 kg   (d) None of these

366. For an ideal gas, the slope of the pressure-volume curve at a given point will be

(a) Steeper for an adiabatic process than for an isothermal process

(b) Steeper for an isothermal process than for an adiabatic process

(c) Identical for both the processes

(d) None of the above

367. The distribution coefficient (k) depends upon

(a) Temperature only

(b) Pressure only

(c) Temperature and pressure only

(d) Temperature, pressure and concentration

368. For an ideal gas mixture undergoing a reversible gaseous phase chemical reaction, the equilibrium constant

(a) Increases with pressure   (b) Decreases with pressure

(c) Is not dependent of pressure   (d) None of these

369. The ratio of the intake volume to the displacement volume in a single-stage compressor is called

(a) Isentropic efficiency   (b) Theoretical volumetric efficiency

(c) Actual volumetric efficiency   (d) Swept volumetric efficiency

370. A car tyre of volume 0.057 m³ is inflated to 300 kPa and 300 K. After driving the car for 10 hours, the pressure in the tyre increases to 330 kPa. Assume air is an ideal gas and C_V for air is 21 J/mol-K. The change in internal energy of air in tyre is

(a) 630   (b) 760

(c) 880   (d) 380

371. The shaft work done by the fluid in a reversible flow process in which there are no changes in the kinetic, potential and surface energies, is given by

(a) $W_{\text{shaft}} = \int_1^2 PdV$ (b) $W_{\text{shaft}} = \int_2^1 PdV$

(c) $W_{\text{shaft}} = \int_2^1 VdP$ (d) $W_{\text{shaft}} = \int_1^2 VdP$

372. The change in entropy for an ideal gas undergoing a thermodynamic process is

(a) $S_2 - S_1 = -R \ln \dfrac{V_1}{V_2}$ (b) $S_2 - S_1 = -R \ln \dfrac{P_1}{P_2}$

(c) $S_2 - S_1 = R \ln \dfrac{V_1}{V_2}$ (d) $S_2 - S_1 = R \ln \dfrac{P_1}{P_2}$

373. For a multi-component system, the chemical potential is equivalent to

(a) Molar free energy   (b) Molar concentration difference

(c) Molar free energy change   (d) Partial molar free energy

374. A domestic refrigerator consists of

(a) A condenser, a compressor and an evaporator

(b) A condenser, a throttling valve, a compressor and an evaporator

(c) A condenser, a throttling valve and an evaporator

(d) A condenser, a generator, a compressor and an evaporator

375. A vertical cylinder containing helium gas is filled with a piston of 50 kg mass and cross-sectional area of 0.025 m^2. If the atmospheric pressure outside the cylinder is 95 kPa and the local gravitational acceleration is 9.79 m/s^2, then the helium pressure is

(a) 114580 Pa   (b) 3479 Pa

(c) 25432 Pa   (d) 12008 Pa

376. A flat balloon is inflated by filling it with helium from a tank of compressed helium. The final volume of the balloon is 6 m^3. The barometer reads 100 kPa. Consider the tank, the balloon, and the connecting pipe as a system. The work done against the atmosphere is

(a) 900 kJ   (b) 600 kJ

(c) 300 kJ   (d) None of these

377. The degree of freedom of a system comprising a gaseous mixture of H$_2$ and NH$_3$ will be

(a) 1   (b) 2

(c) 3   (d) 0

378. The Joule–Thomson coefficient for any gas at inversion point is

(a) 1   (b) 0

(c) 2   (d) 3

379. For an ideal solution, isotherm on an enthalpy-concentration diagram will be
- (a) Parabola
- (b) Hyperbola
- (c) Sine curve
- (d) Straight line

380. An inventor claims to have developed an engine that takes in 100 MJ of heat at 400 K, rejects 40 MJ of heat at 200 K, and delivers 15 kWh of mechanical work. The claim of the inventor is
- (a) True
- (b) False
- (c) Invalid
- (d) None of these

381. When the temperature of an ideal gas is increased from 27 °C to 927 °C, the kinetic energy will be
- (a) Same
- (b) Twice
- (c) Eight times
- (d) Four times

382. The value of $\Delta W = \int_1^2 P dV$ of an ideal gas in a reversible isothermal process is
- (a) 0
- (b) $\dfrac{P_1 V_1 - P_2 V_2}{\gamma - 1}$
- (c) $P_1 V_1 \ln \dfrac{V_2}{V_1}$
- (d) $P_1 (V_2 - V_1)$

383. Compressibility factor for a given vapour or gas can be represented by
- (a) $Z = 1 + B'P + C'P^2 + D'P^3 + \cdots$
- (b) $Z = 1 + \dfrac{B}{V} + \dfrac{C}{V^2} + \dfrac{D}{V^3} + \cdots$
- (c) Either (a) or (b)
- (d) None of these

384. n_1 moles of an ideal monoatomic gas at temperature T_1 and pressure P are in one compartment of an, insulated container. In an adjoining compartment, separated by an insulating partition three are n_2 moles of another ideal monoatomic gas at temperature T_2 and pressure P. When the partition is removed, the final temperature of the mixture is
- (a) $\dfrac{n_1 (T_1 + T_2)}{(n_1 + n_2)}$
- (b) $\dfrac{(T_1 + T_2)}{2}$
- (c) $\dfrac{(n_1 T_1 + n_2 T_2)}{(n_1 + n_2)}$
- (d) $\dfrac{(n_1 T_1 + n_1 T_2)}{(n_1 + n_2)}$

385. For a reversible process, change in entropy of the system
- (a) Approaches to zero
- (b) Increases
- (c) Decreases
- (d) Remains constant

386. The work done in reversible polytropic steady flow processes is given by
- (a) $\dfrac{P_1 V_1 - P_2 V_2}{n - 1}$
- (b) $\dfrac{n(P_1 V_1 - P_2 V_2)}{n - 1}$
- (c) $n(P_1 V_1 - P_2 V_2)$
- (d) $\dfrac{n(P_1 V_1 - P_2 V_2)}{n + 1}$

387. For the slope of an isothermal and adiabatic curves through a point on P-V diagram of an ideal gas, the relation is
    (a) Slope of an isothermal curve = slope of an adiabatic curve
    (b) Slope of an isothermal curve = γ (slope of an adiabatic curve)
    (c) Slope of an adiabatic curve = γ (slope of an isothermal curve)
    (d) Slope of an adiabatic curve = $\left(\dfrac{\gamma}{\gamma-1}\right)$ (slope of an isothermal curve)

388. On a Mollier chart, the slope of the curve representing a reversible isobaric process is equal to

    (a) $T - \beta$

    (b) $T + \dfrac{1}{\beta}$

    (c) T

    (d) $\beta T - 1$

389. On a Mollier diagram, the slope of the curve representing a reversible isothermal process is equal to

    (a) $T - \beta$

    (b) $T - \dfrac{1}{\beta}$

    (c) T

    (d) $T + \dfrac{1}{\beta}$

390. Critical pressure ratio for the flow of gases through a converging nozzle is given by

    (a) $\left(\dfrac{2}{\gamma+1}\right)^{\gamma/\gamma-1}$

    (b) $\left(\dfrac{1}{\gamma+1}\right)^{\gamma/\gamma-1}$

    (c) $\left(\dfrac{2\gamma}{\gamma+1}\right)^{\gamma/\gamma-1}$

    (d) None of these

ANSWERS

1. (d)	2. (d)	3. (d)	4. (b)	5. (a)	6. (c)	7. (b)
8. (b)	9. (a)	10. (b)	11. (c)	12. (c)	13. (d)	14. (c)
15. (b)	16. (d)	17. (c)	18. (c)	19. (c)	20. (b)	21. (c)
22. (a)	23. (a)	24. (a)	25. (d)	26. (c)	27. (a)	28. (d)
29. (b)	30. (c)	31. (a)	32. (a)	33. (d)	34. (a)	35. (c)
36. (d)	37. (b)	38. (d)	39. (c)	40. (a)	41. (a)	42. (a)
43. (b)	44. (c)	45. (d)	46. (c)	47. (d)	48. (a)	49. (a)
50. (c)	51. (d)	52. (b)	53. (c)	54. (a)	55. (b)	56. (b)
57. (b)	58. (a)	59. (a)	60. (b)	61. (c)	62. (d)	63. (b)

64. (b)	65. (c)	66. (b)	67. (c)	68. (d)	69. (b)	70. (c)
71. (a)	72. (b)	73. (b)	74. (a)	75. (c)	76. (d)	77. (a)
78. (c)	79. (a)	80. (d)	81. (b)	82. (d)	83. (d)	84. (b)
85. (b)	86. (a)	87. (b)	88. (d)	89. (d)	90. (c)	91. (d)
92. (d)	93. (b)	94. (c)	95. (b)	96. (d)	97. (b)	98. (c)
99. (a)	100. (c)	101. (c)	102. (c)	103. (d)	104. (a)	105. (b)
106. (c)	107. (a)	108. (d)	109. (d)	110. (a)	111. (a)	112. (b)
113. (a)	114. (b)	115. (b)	116. (b)	117. (a)	118. (c)	119. (b)
120. (c)	121. (d)	122. (d)	123. (b)	124. (c)	125. (b)	126. (d)
127. (c)	128. (b)	129. (d)	130. (a)	131. (b)	132. (a)	133. (c)
134. (a)	135. (d)	136. (b)	137. (c)	138. (d)	139. (b)	140. (a)
141. (b)	142. (b)	143. (a)	144. (d)	145. (c)	146. (b)	147. (a)
148. (b)	149. (a)	150. (b)	151. (a)	152. (c)	153. (c)	154. (b)
155. (c)	156. (b)	157. (b)	158. (d)	159. (c)	160. (a)	161. (a)
162. (a)	163. (b)	164. (c)	165. (b)	166. (a)	167. (d)	168. (c)
169. (c)	170. (c)	171. (b)	172. (d)	173. (b)	174. (a)	175. (c)
176. (c)	177. (d)	178. (b)	179. (c)	180. (b)	181. (d)	182. (c)
183. (b)	184. (d)	185. (b)	186. (c)	187. (a)	188. (c)	189. (c)
190. (c)	191. (b)	192. (b)	193. (c)	194. (a)	195. (b)	196. (c)
197. (d)	198. (a)	199. (c)	200. (b)	201. (b)	202. (d)	203. (d)
204. (a)	205. (c)	206. (d)	207. (b)	208. (a)	209. (c)	210. (c)
211. (c)	212. (b)	213. (c)	214. (b)	215. (d)	216. (a)	217. (b)
218. (d)	219. (d)	220. (b)	221. (c)	222. (b)	223. (b)	224. (b)
225. (b)	226. (d)	227. (c)	228. (a)	229. (c)	230. (d)	231. (b)
232. (c)	233. (c)	234. (b)	235. (c)	236. (b)	237. (d)	238. (c)
239. (c)	240. (a)	241. (e)	242. (b)	243. (a)	244. (c)	245. (a)
246. (c)	247. (b)	248. (c)	249. (b)	250. (a)	251. (b)	252. (b)
253. (d)	254. (a)	255. (b)	256. (b)	257. (c)	258. (a)	259. (a)
260. (b)	261. (c)	262. (b)	263. (b)	264. (b)	265. (d)	266. (a)
267. (b)	268. (a)	269. (a)	270. (c)	271. (d)	272. (a)	273. (b)
274. (a)	275. (c)	276. (d)	277. (b)	278. (b)	279. (b)	280. (d)
281. (a)	282. (c)	283. (d)	284. (c)	285. (a)	286. (c)	287. (d)
288. (c)	289. (c)	290. (b)	291. (c)	292. (d)	293. (b)	294. (a)

295. (c)	296. (b)	297. (a)	298. (a)	299. (b)	300. (c)	301. (c)
302. (c)	303. (c)	304. (a)	305. (c)	306. (d)	307. (d)	308. (c)
309. (d)	310. (b)	311. (b)	312. (b)	313. (c)	314. (d)	315. (d)
316. (a)	317. (d)	318. (b)	319. (c)	320. (a)	321. (d)	322. (b)
323. (c)	324. (b)	325. (b)	326. (b)	327. (b)	328. (a)	329. (a)
330. (b)	331. (b)	332. (a)	333. (c)	334. (b)	335. (d)	336. (d)
337. (b)	338. (a)	339. (a)	340. (b)	341. (b)	342. (b)	343. (d)
344. (c)	345. (a)	346. (c)	347. (b)	348. (d)	349. (d)	350. (b)
351. (c)	352. (b)	353. (a)	354. (a)	355. (b)	356. (c)	357. (b)
358. (b)	359. (b)	360. (b)	361. (d)	362. (a)	363. (a)	364. (c)
365. (b)	366. (a)	367. (d)	368. (c)	369. (b)	370. (a)	371. (c)
372. (b)	373. (d)	374. (b)	375. (a)	376. (b)	377. (c)	378. (b)
379. (d)	380. (b)	381. (d)	382. (c)	383. (c)	384. (c)	385. (a)
386. (b)	387. (b)	388. (c)	389. (b)	390. (a)		

Bibliography

Advanced Centre of Cryogenic Research, Jadavpur University, Kolkata (in collaboration with Mechanical Engineering Department, Jadavpur University), Course material on refrigeration and air-conditioning prepared by Prof. P. Sengupta.

Arora, C.P., *Thermodynamics*, Tata McGraw-Hill, New Delhi, 1998.

Arora, S.C. and S. Domkundwar, *A Course in Refrigeration and Air-conditioning*, 5th ed., Dhanpat Rai and Sons, Delhi, 1989.

Balmer, R.T., *Thermodynamics*, Jaico Publishing House, Delhi, 1999.

Barron, Randall F., *Cryogenic Systems*, 2nd ed., McGraw-Hill, 1985, pp. 291–337.

Cengel, Y.A. and M.A. Boles, *Thermodynamics—An Engineering Approach*, 3rd ed., WCB/McGraw-Hill, 1998.

Denbigh, K., *The Principles of Chemical Equilibrium*, 4th ed., Cambridge University Press, Cambridge, U.K., 1980.

Epstein, P.S., *Textbook of Thermodynamics*, Wiley, New York, 1937.

Esbach, F., *Handbook of Engineering Thermodynamics*, 3rd ed., Wiley Handbook Series, 1975.

Flynn, T., *Cryogenic Engineering*, Marcel Dekker, New York, 1996.

Granet, I. and M. Bluestein, *Thermodynamics and Heat Power*, Pearson Education (Asia), 2001.

Haselden, G.G., *Cryogenic Fundamentals*, Academic Press, London, 1971.

Hougen, O.A., K.M. Watson, and R.A. Ragatz, *Chemical Process Principles*, Part II, 2nd ed., John Wiley, New York, 1960.

Kutepov, A.M., T.I. Bondareva, and M.G. Berengarten, *Basic Chemical Engineering with Practical Applications*, Mir Publishers, Moscow, 1988.

Kyle, B.G., *Chemical and Process Thermodynamics*, 2nd ed., Prentice-Hall of India, New Delhi, 1994.

Lewis, G.N., M., Randall, K.S. Pitzer, and L. Brewer, *Thermodynamics*, McGraw-Hill, New York, 1981.

Moran, M.J. and H.N. Shapiro, *Fundamentals of Engineering Thermodynamics*, 4th ed., John Wiley, 2000.

Nag, P.K., *Engineering Thermodynamics*, 3rd ed., Tata McGraw-Hill, New Delhi, 2005.

Narayanan, K.V., *A Textbook of Chemical Engineering Thermodynamics*, PHI Learning, 2008.

Obert, E.F., *Concept of Thermodynamics*, McGraw-Hill, New York, 1960.

Perry, J.H. and C.H. Chilton, *Chemical Engineers Handbook*, 5th ed., McGraw-Hill, Tokyo, 1973.

Pitzer, K.S., *J. Chem. Phys.*, Vol. 7, 1939, p. 583.

Rakshit, P.C., *Physical Chemistry*, Sarat Book Distributors, Kolkata, 2002.

Rao, Y.V.C., *Chemical Engineering Thermodynamics*, Universities Press (India), 2005.

Rao, Y.V.C., *Engineering Thermodynamics through Problems*, Universities Press (India), 2003.

Rastogi, R.P. and R.R. Mishra, *An Introduction to Chemical Thermodynamics,* Vikas Publishing House, 2006.

Redlich, O. and J.N.S. Kwong, *Chem. Rev.*, Vol. 44, 1949.

Reid, R.C., J.M. Prausnitz, R.N. Lichtenhaler and E.G. Azevedo, *Molecular Thermodynamics of Fluid Phase Equilibria*, 2nd ed., Prentice Hall, New Jersey, 1986.

Report on refrigerant history, *Indian Journal of Cryogenics*, Vol. 19, 1994.

Sarao, A.S. and K.S. Rai, *Applied Thermodynamics*, 7th ed., Satya Prakashan, New Delhi, 1988.

Smith, J.M., H.C. van Ness and M.M. Abott, *Introduction to Chemical Engineering Thermodynamics*, 7th ed., Chemical Engineering Series, McGraw-Hill, 1996.

Treybal, R.E., *Mass Transfer Operations*, 3rd ed., McGraw-Hill, New York, 1981.

Zemansky, M.W., *Heat and Thermodynamics*, 4th ed., McGraw-Hill, New York, 1957.

Index